Temperature, 15, 18, 19, 47, 48, 149, 234, 266, 433

Traffic safety, 513

Travel, 78, 107, 154, 272, 582

Uniform motion problems, 96, 97, 105, 106, 141, 143, 197-203, 216, 218, 219, 320, 321, 324, 325, 336, 337, 338, 339, 383, 436, 483, 484, 487, 488, 489, 498, 499, 500, 578, 581, 582, 592, 594, 595

U.S. Presidents, 233

Utilities, 153, 207

Value mixture problems, 189, 190, 193, 194, 195, 196, 197, 216, 217, 219, 221, 595

Wages, 78, 154, 326

Water-current problems, 321, 324, 325, 336, 337, 338, 339

Work problems, 481, 482, 485, 486, 487, 489, 498, 499, 500, 502, 546, 579, 581, 592

Woodwork, 86

Beginning Algebra
with Applications

Sixth Edition

Richard N. Aufmann
Palomar College, California

Vernon C. Barker
Palomar College, California

Joanne S. Lockwood
Plymouth State College, New Hampshire

Houghton Mifflin Company
Boston   New York

Publisher: Jack Shira

Senior Sponsoring Editor: Lynn Cox

Senior Development Editor: Dawn Nuttall

Editorial Associate: Melissa Parkin

Senior Project Editor: Tamela Ambush

Senior Production/Design Coordinator: Carol Merrigan

Manufacturing Manager: Florence Cadran

Senior Marketing Manager: Ben Rivera

Marketing Associate: Alexandra Shaw

Printed in the U.S.A.

Library of Congress Catalog Card Number: 2002115904

ISBNs
Student Text: 0–618–306056
Instructor's Annotated Edition: 0–618–306064

56789–DOW– 06 05 04

Contents

Preface   *xi*

AIM for Success   *xxv*

1 | Real Numbers   1

Prep Test   *2*

Section 1.1   Introduction to Integers   *2*
Objective 1 Order relations   *2*
Objective 2 Opposites and absolute value   *4*

Section 1.2   Operations with Integers   *9*
Objective 1 Add integers   *9*
Objective 2 Subtract integers   *10*
Objective 3 Multiply integers   *11*
Objective 4 Divide integers   *13*
Objective 5 Application problems   *14*

Section 1.3   Rational Numbers   *21*
Objective 1 Write rational numbers as decimals   *21*
Objective 2 Add and subtract rational numbers   *22*
Objective 3 Multiply and divide rational numbers   *24*
Objective 4 Convert among percents, fractions, and decimals   *26*

Section 1.4   Exponents and the Order of Operations Agreement   *33*
Objective 1 Exponential expressions   *33*
Objective 2 The Order of Operations Agreement   *35*

Focus on Problem Solving: Inductive Reasoning   *39* | **Projects and Group Activities:**
Calculators   *40* | Using Patterns in Mathematics   *41* | www.fedstats.gov   *41* |
Moving Averages   *41* | **Chapter Summary**   *42* | **Chapter Review Exercises**   *44* |
Chapter Test   *47*

2 | Variable Expressions   51

Prep Test   *52*

Section 2.1   Evaluating Variable Expressions   *52*
Objective 1 Evaluate variable expressions   *52*

Section 2.2   Simplifying Variable Expressions   *58*
Objective 1 The Properties of the Real Numbers   *58*
Objective 2 Simplify variable expressions using the Properties of Addition   *61*
Objective 3 Simplify variable expressions using the Properties of Multiplication   *62*
Objective 4 Simplify variable expressions using the Distributive Property   *63*
Objective 5 Simplify general variable expressions   *64*

Section 2.3   Translating Verbal Expressions into Variable Expressions   *68*

Objective 1   Translate a verbal expression into a variable expression   *68*
Objective 2   Translate a verbal expression into a variable expression and then simplify the resulting expression   *70*
Objective 3   Translate application problems   *71*

Focus on Problem Solving: From Concrete to Abstract   *79* | **Projects and Group Activities:** Investigation into Operations with Even and Odd Integers   *80* | Applications of Patterns in Mathematics   *81* | **Chapter Summary**   *82* | **Chapter Review Exercises**   *83* |
Chapter Test   *85* | **Cumulative Review Exercises**   *86*

3 | Solving Equations and Inequalities   89

Prep Test   *90*

Section 3.1   Introduction to Equations   *90*

Objective 1   Determine whether a given number is a solution of an equation   *90*
Objective 2   Solve equations of the form $x + a = b$ *91*
Objective 3   Solve equations of the form $ax = b$ *93*
Objective 4   Application problems: uniform motion   *95*
Objective 5   Applications of percent   *98*

Section 3.2   General Equations   *112*

Objective 1   Solve equations of the form $ax + b = c$ *112*
Objective 2   Solve equations of the form $ax + b = cx + d$ *114*
Objective 3   Solve equations containing parentheses   *115*
Objective 4   Application problems   *117*

Section 3.3   Inequalities   *125*

Objective 1   Solve inequalities using the Addition Property of Inequalities   *125*
Objective 2   Solve inequalities using the Multiplication Property of Inequalities   *128*
Objective 3   Solve general inequalities   *129*

Focus on Problem Solving: Analyzing Data   *135* | **Projects and Group Activities:** Health and Nutrition   *136* | The Consumer Price Index   *137* | **Chapter Summary**   *138* |
Chapter Review Exercises   *140* | **Chapter Test**   *142* | **Cumulative Review Exercises**   *143*

4 | Solving Equations and Inequalities: Applications   147

Prep Test   *148*

Section 4.1   Translating Sentences into Equations   *148*

Objective 1   Translate a sentence into an equation and solve   *148*
Objective 2   Application problems   *149*

Section 4.2   Integer, Coin, and Stamp Problems   *155*
                Objective 1 Consecutive integer problems   *155*
                Objective 2 Coin and stamp problems   *156*

Section 4.3   Geometry Problems   *162*
                Objective 1 Perimeter problems   *162*
                Objective 2 Problems involving angles formed by intersecting lines   *164*
                Objective 3 Problems involving the angles of a triangle   *168*

Section 4.4   Markup and Discount Problems   *176*
                Objective 1 Markup problems   *176*
                Objective 2 Discount problems   *178*

Section 4.5   Investment Problems   *183*
                Objective 1 Investment problems   *183*

Section 4.6   Mixture Problems   *189*
                Objective 1 Value mixture problems   *189*
                Objective 2 Percent mixture problems   *190*

Section 4.7   Uniform Motion Problems   *197*
                Objective 1 Uniform motion problems   *197*

Section 4.8   Inequalities   *203*
                Objective 1 Applications of inequalities   *203*

                Focus on Problem Solving: Trial-and-Error Approach to Problem Solving   *209* |
                Diagramming Problems   *210* | **Projects and Group Activities:** Nielsen Ratings   *211* |
                Changes in Percent Concentration   *212* | **Chapter Summary**   *213* | **Chapter Review**
                Exercises   *215* | **Chapter Test**   *218* | **Cumulative Review Exercises** *220*

---

5   **Linear Equations and Inequalities**   **223**

Prep Test   *224*

Section 5.1   The Rectangular Coordinate System   *224*
                Objective 1 Graph points in a rectangular coordinate system   *224*
                Objective 2 Scatter diagrams   *226*
                Objective 3 Average rate of change   *228*

Section 5.2   Graphs of Straight Lines   *237*
                Objective 1 Determine solutions of linear equations in two variables   *237*
                Objective 2 Graph equations of the form $y = mx + b$ *238*
                Objective 3 Graph equations of the form $Ax + By = C$ *240*

Section 5.3   Slopes of Straight Lines   *250*
                Objective 1 Find the slope of a straight line   *250*
                Objective 2 Graph a line using the slope and y-intercept   *255*

Section 5.4   Equations of Straight Lines   *261*
Objective 1 Find the equation of a line using the equation $y = mx + b$ *261*
Objective 2 Find the equation of a line using the point-slope formula   *262*

Section 5.5   Functions   *267*
Objective 1 Introduction to functions   *267*
Objective 2 Graphs of linear functions   *269*

Section 5.6   Graphing Linear Inequalities   *278*
Objective 1 Graph inequalities in two variables   *278*

Focus on Problem Solving: Counterexamples   *283* | **Projects and Group Activities:** Graphing
Linear Equations with a Graphing Utility   *284* | Population Projections   *285* |
Chapter Summary   *286* | **Chapter Review Exercises**   *289* | **Chapter Test**   *292* |
Cumulative Review Exercises   *295*

6 | Systems of Linear Equations   299

Prep Test   *300*

Section 6.1   Solving Systems of Linear Equations by Graphing   *300*
Objective 1 Solve systems of linear equations by graphing   *300*

Section 6.2   Solving Systems of Linear Equations by the Substitution Method   *309*
Objective 1 Solve systems of linear equations by the substitution method   *309*

Section 6.3   Solving Systems of Linear Equations by the Addition Method   *314*
Objective 1 Solve systems of linear equations by the addition method   *314*

Section 6.4   Application Problems in Two Variables   *319*
Objective 1 Rate-of-wind and water-current problems   *319*
Objective 2 Application problems   *321*

Focus on Problem Solving: Using a Table and Searching for a Pattern   *328* | **Projects and
Group Activities:** Find a Pattern   *329* | Break-even Analysis   *330* | Solving a System
of Equations with a Graphing Calculator   *330* | **Chapter Summary**   *333* | **Chapter
Review Exercises**   *334* | **Chapter Test**   *337* | **Cumulative Review Exercises**   *338*

7 | Polynomials   341

Prep Test   *342*

Section 7.1   Addition and Subtraction of Polynomials   *342*
Objective 1 Add polynomials   *342*
Objective 2 Subtract polynomials   *343*

Section 7.2   Multiplication of Monomials   *347*
 Objective 1 Multiply monomials   *347*
 Objective 2 Simplify powers of monomials   *348*

Section 7.3   Multiplication of Polynomials   *351*
 Objective 1 Multiply a polynomial by a monomial   *351*
 Objective 2 Multiply two polynomials   *352*
 Objective 3 Multiply two binomials   *352*
 Objective 4 Multiply binomials that have special products   *353*
 Objective 5 Application problems   *355*

Section 7.4   Integer Exponents and Scientific Notation   *360*
 Objective 1 Integer exponents   *360*
 Objective 2 Scientific notation   *363*

Section 7.5   Division of Polynomials   *370*
 Objective 1 Divide a polynomial by a monomial   *370*
 Objective 2 Divide polynomials   *371*

 Focus on Problem Solving: Dimensional Analysis   *374* | **Projects and Group Activities:** Pascal's
 Triangle   *376* | Population and Land Allocation   *377* | Properties of
 Polynomials   *377* | **Chapter Summary**   *378* | **Chapter Review Exercises**   *379* |
 Chapter Test   *381* | **Cumulative Review Exercises**   *382*

8 | Factoring   **385**

Prep Test   *386*

Section 8.1   Common Factors   *386*
 Objective 1 Factor a monomial from a polynomial   *386*
 Objective 2 Factor by grouping   *387*

Section 8.2   Factoring Polynomials of the Form $x^2 + bx + c$ *392*
 Objective 1 Factor trinomials of the form $x^2 + bx + c$ *392*
 Objective 2 Factor completely   *394*

Section 8.3   Factoring Polynomials of the Form $ax^2 + bx + c$ *398*
 Objective 1 Factor trinomials of the form $ax^2 + bx + c$ using trial factors   *398*
 Objective 2 Factor trinomials of the form $ax^2 + bx + c$ by grouping   *402*

Section 8.4   Special Factoring   *407*
 Objective 1 Factor the difference of two squares and perfect-square trinomials   *407*
 Objective 2 Factor completely   *409*

Section 8.5   Solving Equations   *414*
 Objective 1 Solve equations by factoring   *414*
 Objective 2 Application problems   *415*

Focus on Problem Solving: Making a Table   *421* | The Trial-and-Error Method   *422* |
Projects and Group Activities: Prime and Composite Numbers   *423* | Search the World
Wide Web   *424* | Problems Involving Falling Objects   *425* | **Chapter Summary**   *428* |
Chapter Review Exercises   *429* | **Chapter Test**   *431* | **Cumulative Review Exercises** *431*

9   Rational Expressions   435

Prep Test   *436*

Section 9.1   Multiplication and Division of Rational Expressions   *436*
Objective 1   Simplify rational expressions   *436*
Objective 2   Multiply rational expressions   *438*
Objective 3   Divide rational expressions   *439*

Section 9.2   Expressing Fractions in Terms of the Least Common Multiple of Their Denominators   *445*
Objective 1   Find the least common multiple (LCM) of two or more polynomials   *445*
Objective 2   Express two fractions in terms of the LCM of their denominators   *446*

Section 9.3   Addition and Subtraction of Rational Expressions   *450*
Objective 1   Add and subtract rational expressions with the same denominator   *450*
Objective 2   Add and subtract rational expressions with different denominators   *451*

Section 9.4   Complex Fractions   *458*
Objective 1   Simplify complex fractions   *458*

Section 9.5   Equations Containing Fractions   *462*
Objective 1   Solve equations containing fractions   *462*
Objective 2   Solve proportions   *464*
Objective 3   Applications of proportions   *466*
Objective 4   Problems involving similar triangles   *467*

Section 9.6   Literal Equations   *477*
Objective 1   Solve a literal equation for one of the variables   *477*

Section 9.7   Application Problems   *480*
Objective 1   Work problems   *480*
Objective 2   Uniform motion problems   *482*

Focus on Problem Solving: Negations and *If . . . then . . .* Sentences   *489* |
Calculators   *491* | **Projects and Group Activities:** Intensity of Illumination   *491* |
Scale Model of the Solar System   *493* | **Chapter Summary**   *494* | **Chapter Review
Exercises**   *496* | **Chapter Test**   *499* | **Cumulative Review Exercises** *500*

10 ┐ Radical Expressions   505

Prep Test   *506*

Section 10.1   Introduction to Radical Expressions   *506*
Objective 1 Simplify numerical radical expressions   *506*
Objective 2 Simplify variable radical expressions   *509*

Section 10.2   Addition and Subtraction of Radical Expressions   *514*
Objective 1 Add and subtract radical expressions   *514*

Section 10.3   Multiplication and Division of Radical Expressions   *517*
Objective 1 Multiply radical expressions   *517*
Objective 2 Divide radical expressions   *519*

Section 10.4   Solving Equations Containing Radical Expressions   *524*
Objective 1 Solve equations containing one or more radical expressions   *524*
Objective 2 Application problems   *527*

Focus on Problem Solving: Deductive Reasoning   *534* | **Projects and Group Activities:** Mean and Standard Deviation   *535* | Distance to the Horizon   *537* | Expressway On-Ramp Curves   *538* | **Chapter Summary**   *539* | **Chapter Review Exercises**   *541* | **Chapter Test**   *543* | **Cumulative Review Exercises**   *544*

11 ┐ Quadratic Equations   549

Prep Test   *550*

Section 11.1   Solving Quadratic Equations by Factoring or by Taking Square Roots   *550*
Objective 1 Solve quadratic equations by factoring   *550*
Objective 2 Solve quadratic equations by taking square roots   *552*

Section 11.2   Solving Quadratic Equations by Completing the Square   *558*
Objective 1 Solve quadratic equations by completing the square   *558*

Section 11.3   Solving Quadratic Equations by Using the Quadratic Formula   *564*
Objective 1 Solve quadratic equations by using the quadratic formula   *564*

Section 11.4   Graphing Quadratic Equations in Two Variables   *570*
Objective 1 Graph a quadratic equation of the form $y = ax^2 + bx + c$ *570*

Section 11.5   Application Problems   *578*
Objective 1 Application problems   *578*

Focus on Problem Solving: Applying Algebraic Manipulation and Graphing Techniques to Business   *583* | **Projects and Group Activities:** Graphical Solutions of Quadratic Equations   *585* | Geometric Construction of Completing the Square   *587* | **Chapter Summary**   *588* | **Chapter Review Exercises**   *590* | **Chapter Test**   *593* | **Cumulative Review Exercises**   *594*

Final Exam   *596*

Appendix   *601*

    Table of Properties   *601*

    Calculator Guide for the TI-83 and TI-83 Plus   *603*

Solutions to Chapter Problems   *S1*

Answers to Selected Exercises   *A1*

Glossary   *G1*

Index   *I1*

Preface

The sixth edition of *Beginning Algebra with Applications* provides mathematically sound and comprehensive coverage of the topics considered essential in an introductory algebra course. The text has been designed not only to meet the needs of the traditional college student, but also to serve the needs of returning students whose mathematical proficiency may have declined during years away from formal education.

In this new edition of *Beginning Algebra with Applications,* we have continued to integrate the approaches suggested by AMATYC. Each chapter opens by illustrating and referencing a mathematical application within the chapter. At the end of each section there are "Applying Concepts" exercises that include writing, synthesis, critical thinking, and challenge problems. At the end of each chapter there is a "Focus on Problem Solving" that introduces students to various problem-solving strategies. This is followed by "Projects and Group Activities" that can be used for cooperative learning activities.

NEW! Changes to this Edition

The material in Section 3 of Chapter 1, *Real Numbers,* has been reorganized to provide for a better flow of the concepts presented. It now begins with writing rational numbers as decimals, then presents operations on rational numbers, and concludes with converting percents, fractions and decimals.

In Section 3 of Chapter 2, *Variable Expressions,* Objectives 1 and 2 were combined in order to present a smoother presentation of translating verbal expressions into variable expressions.

New material has been added to Chapter 3, *Solving Equations and Inequalities.* In Section 1, the following equations are presented as applications of equations of the form $ax = b$.

Uniform motion equation: $d = rt$
Simple interest equation: $I = Prt$
Percent mixture equation: $A = Qr$

This serves not only to reinforce solving linear equations, but it provides exposure to the concepts inherent in uniform motion, interest, and mixture problems, which will serve the student well when these equations are used in more difficult applications later in the text.

In response to user requests, Chapter 5, *Linear Equations and Inequalities,* now includes a discussion of parallel lines. Also new to this chapter is a discussion of average rate of change in Section 1. Students work with data for which the rate of change is not constant. This lays the groundwork for a better understanding of the slope of a linear equation being a constant.

In Chapter 6, *Systems of Equations,* the discussion of solutions of systems of equations has been expanded and improved. This development should lead to greater student understanding of dependent, inconsistent, and independent systems of equations.

The material in the last section in Chapter 7, *Polynomials,* was separated into two sections. Section 7.4 now covers integer exponents and scientific notation. Section 7.5 presents division of polynomials. Students will benefit from the separation of these topics, which are often difficult ones for first-year algebra students.

Previous users of *Beginning Algebra with Applications* will note that all references to "algebraic fractions" in Chapter 9 have been changed to "rational expressions." This change will make the transition to higher-level math courses easier for students.

In Chapter 10, *Radical Expressions*, square roots are simplified by using perfect squares rather than by using prime factorization. Therefore, rather than writing a radicand of 20 as 2^2 times 5, 20 is written as the product of 4 and 5, an approach that better mirrors what the student is thinking mentally when simplifying radical expressions.

Chapter 11, *Quadratic Equations*, now includes a discussion of finding the x- and y-intercepts of a parabola. Students are asked to find the intercepts algebraically and are instructed in finding these points graphically.

Many of the exercise sets now include developmental exercises, the intent of which is to reinforce the concepts underlying the skills presented in the lesson. For example, in Chapter 10, *Radical Expressions*, the student is asked, "Which of the numbers 2, 9, 20, 25, 50, 81, and 100 are *not* perfect squares?"

Throughout the text, data problems have been updated to reflect current data and trends. These application problems will demonstrate to students the variety of problems in real life that require mathematical analysis. Instructors will find that many of these problems may lead to interesting class discussions. Each chapter opener provides a list of the applications in that chapter.

A large number of photos have been added to this text. This addition will help students to make the connection between mathematics and the real world.

Another new feature of this edition is *AIM for Success*, which explains what is required of a student to be successful and how this text has been designed to foster student success. *AIM for Success* can be used as a lesson on the first day of class or as a project for students to complete to strengthen their study skills. There are suggestions for teaching this lesson in the *Instructor's Resource Manual*.

Related to *AIM for Success* are *Prep Tests* which occur at the beginning of each chapter. These tests focus on the particular *prerequisite* skills that will be used in the upcoming chapter. The answers to these questions can be found in the Answer Appendix along with a reference (except for Chapter 1) to the objective from which the question was taken. Students who miss a question are encouraged to review the objective from which the question was taken.

The *Go Figure* problem accompanying the *Prep Test* is a puzzle problem designed to engage students in problem solving.

Chapter Opening Features

NEW! Chapter Opener

New, motivating chapter opener photos and captions have been added, illustrating and referencing a specific application from the chapter. A list of the applications in the chapter is also provided.

The web icon at the bottom of the second page lets students know of additional online resources at **math.college.hmco.com/students.**

Objective-Specific Approach

Each chapter begins with a list of learning objectives which form the framework for a complete learning system. The objectives are woven throughout the text (i.e. Exercises, Prep Tests, Chapter Review Exercises, Chapter Tests, Cumulative Review Exercises) as well as through the print and multimedia ancillaries. This results in a seamless learning system delivered in one consistent voice.

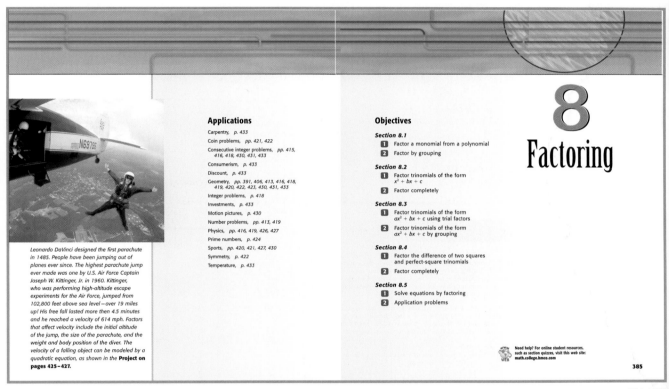

pages 384–385

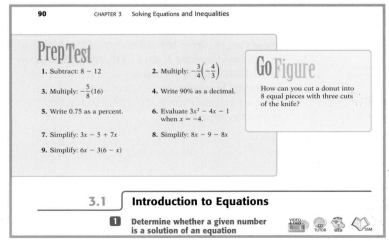

page 90

NEW! Prep Test and Go Figure

Prep Tests occur at the beginning of each chapter and test students on *previously covered* concepts that are required in the coming chapter. Answers are provided in the Answer Appendix. Objective references are also provided if a student needs to review specific concepts.

The **Go Figure** problem that follows the *Prep Test* is a playful puzzle problem designed to engage students in problem solving.

Aufmann Interactive Method (AIM)

An Interactive Approach

Beginning Algebra with Applications uses an interactive style that provides a student with an opportunity to try a skill as it is presented. Each section is divided into objectives, and every objective contains one or more sets of matched-pair examples. The first example in each set is worked out; the second example, called "Problem," is for the student to work. By solving this problem, the student actively practices concepts as they are presented in the text.

There are *complete, worked-out* solutions to these problems in an appendix. By comparing their solution to the solution in the appendix, students obtain immediate feedback on, and reinforcement of, the concept.

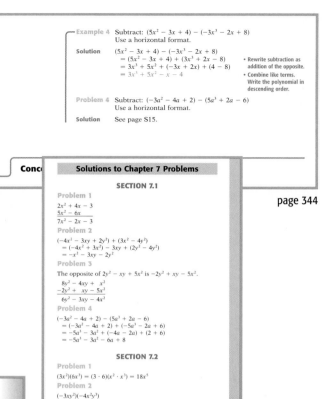

Example 4 Subtract: $(5x^2 - 3x + 4) - (-3x^3 - 2x + 8)$
Use a horizontal format.

Solution $(5x^2 - 3x + 4) - (-3x^3 - 2x + 8)$
$= (5x^2 - 3x + 4) + (3x^3 + 2x - 8)$ • Rewrite subtraction as addition of the opposite.
$= 3x^3 + 5x^2 + (-3x + 2x) + (4 - 8)$ • Combine like terms. Write the polynomial in descending order.
$= 3x^3 + 5x^2 - x - 4$

Problem 4 Subtract: $(-3a^2 - 4a + 2) - (5a^3 + 2a - 6)$
Use a horizontal format.

Solution See page S15.

7.1 Conc

page 344

Solutions to Chapter 7 Problems

SECTION 7.1

Problem 1
$2x^2 + 4x - 3$
$5x^2 - 6x$
─────────
$7x^2 - 2x - 3$

Problem 2
$(-4x^2 - 3xy + 2y^2) + (3x^2 - 4y^2)$
$= (-4x^2 + 3x^2) - 3xy + (2y^2 - 4y^2)$
$= -x^2 - 3xy - 2y^2$

Problem 3
The opposite of $2y^2 - xy + 5x^2$ is $-2y^2 + xy - 5x^2$.
$8y^2 - 4xy + x^2$
$-2y^2 + xy - 5x^2$
─────────
$6y^2 - 3xy - 4x^2$

Problem 4
$(-3a^2 - 4a + 2) - (5a^3 + 2a - 6)$
$= (-3a^2 - 4a + 2) + (-5a^3 - 2a + 6)$
$= -5a^3 - 3a^2 + (-4a - 2a) + (2 + 6)$
$= -5a^3 - 3a^2 - 6a + 8$

SECTION 7.2

Problem 1
$(3x^2)(6x^3) = (3 \cdot 6)(x^2 \cdot x^3) = 18x^5$

Problem 2
$(-3xy^2)(-4x^2y^3)$

page S15

AIM for Success

Welcome to *Beginning Algebra with Applications*. As you begin this course, we know two important facts: (1) We want you to succeed. (2) You want to succeed. To do that requires an effort from each of us. For the next few pages, we are going to show you what is required of you to achieve that success and how you can use the features of this text to be successful.

Motivation

One of the most important keys to success is motivation. We can try to motivate you by offering interesting or important ways mathematics can benefit you. But, in the end, the motivation must come from you. On the first day of class, it is easy to be motivated. Eight weeks into the term, it is harder to keep that motivation.

To stay motivated, there must be outcomes from this course that are worth your time, money, and energy. List some reasons you are taking this course. Do not make a mental list—actually write them out.

Take Note
Motivation alone will not lead to success. For instance, suppose a person who cannot swim is placed in a boat, taken out to the middle of a lake, and then thrown overboard. That person has a lot of motivation but there is a high likelihood the person will drown without some help. Motivation gives us the desire to learn but is not the same as learning.

Although we hope that one of the reasons you listed was an interest in mathematics, we know that many of you are taking this course because it is required to graduate, it is a prerequisite for a course you must take, or because it is required for your major. Although you may not agree that this course is necessary, it is! If you are motivated to graduate or complete the requirements for your major, then use that motivation to succeed in this course. Do not become distracted from your goal to complete your education!

Commitment

To be successful, you must make a commitment to succeed. This means devoting time to math so that you achieve a better understanding of the subject.

List some activities (sports, hobbies, talents such as dance, art, or music) that you enjoy and at which you would like to become better.

ACTIVITY	TIME SPENT	TIME WISHED SPENT

Thinking about these activities, put the number of hours that you spend each week practicing these activities next to the activity. Next to that number, indicate the number of hours per week you would like to spend on these activities.

Whether you listed surfing or sailing, aerobics or restoring cars, or any other activity you enjoy, note how many hours a week you spend doing it. To succeed in math, you must be willing to commit the same amount of time. Success requires some sacrifice.

The "I Can't Do Math" Syndrome

There may be things you cannot do, such as lift a two-ton boulder. You can, however, do math. It is much easier than lifting the two-ton boulder. When

page xxv

NEW! AIM for Success Student Preface

This new student 'how to use this book' preface explains what is required of a student to be successful and how this text has been designed to foster student success, including the Aufmann Interactive Method (AIM). *AIM for Success* can be used as a lesson on the first day of class or as a project for students to complete to strengthen their study skills. There are suggestions for teaching this lesson in the *Instructor's Resource Manual*.

Problem Solving

Focus on Problem Solving

At the end of each chapter is a Focus on Problem Solving feature which introduces the student to various successful problem-solving strategies. Strategies such as drawing a diagram, applying solutions to other problems, working backwards, inductive reasoning, and trial and error are some of the techniques that are demonstrated.

Use the roster method to list the set of integers that are common to the solution sets of the two inequalities.

137. $5x - 12 \le x + 8$
  $3x - 4 \ge 2 + x$

138. $6x - 5 > 9x - 2$
  $5x - 6 < 8x + 9$

139. $4(x - 2) \le 3x + 5$
  $7(x - 3) \ge 5x - 1$

140. $3(x + 2) < 2(x + 4)$
  $4(x + 5) > 3(x + 6)$

Graph.

141. $|x| < 3$

142. $|x| < 4$

143. $|x| > 2$

144. $|x| > 1$

Focus on Problem Solving

Analyzing Data   **Demography** is the statistical study of human populations. Many groups are interested in the size of certain segments of the population and projections of population growth. For example, public school administrators want estimates of the number of school-age children who will be living in their district 10 years from now.

The U.S. government provides estimates of the future population of the United States. You can find these projections at the Census Bureau website at **www.census.gov.** The table below contains data from this website.

	Age	Under 5	5-17	18-24	25-34	35-44	45-54	55-64	65 and older
2010	Male	10,272	26,639	15,388	19,286	19,449	21,706	16,973	16,966
	Female	9,827	25,363	14,774	19,565	19,993	22,455	18,457	22,749
2050	Male	13,748	35,227	18,734	25,034	24,550	22,340	21,127	36,289
	Female	13,165	33,608	18,069	25,425	25,039	23,106	22,517	45,710

Population Projections of the United States, by Age and Sex (in thousands)

For the following exercises, round all percents to the nearest tenth of a percent.

1. Which of the age groups listed are of interest to public school officials?

2. Which age group is of interest to retirement home administrators?

3. Which age groups are of concern to accountants determining future benefits to be paid out by the Social Security Administration?

4. Which age group is of interest to manufacturers of disposable diapers?

5. Which age groups are of primary concern to college and university admissions officers?

page 135

---

466 CHAPTER 9 Rational Expressions

B. $\dfrac{6}{x + 4} = \dfrac{12}{5x - 13}$

$(5x - 13)(x + 4)\dfrac{6}{x + 4} = (5x - 13)(x +$

$(5x - 13)6 = (x + 4)12$

$30x - 78 = 12x + 48$

$18x - 78 = 48$

$18x = 126$

$x = 7$

The solution is 7.

Problem 3 Solve. **A.** $\dfrac{2}{x + 3} = \dfrac{6}{5x + 5}$ **B.** $\dfrac{5}{2x - 3} = \dfrac{10}{x + 3}$

Solution See pages S21–S22.

3 **Applications of proportions**

Example 4 The monthly loan payment for a car is $29.50 for each $1000 borrowed. At this rate, find the monthly payment for a $9000 car loan.

Strategy To find the monthly payment, write and solve a proportion using P to represent the monthly car payment.

Solution $\dfrac{29.50}{1000} = \dfrac{P}{9000}$ • The monthly payments are in the numerators. The loan amounts are in the denominators.

$9000\left(\dfrac{29.50}{1000}\right) = 9000\left(\dfrac{P}{9000}\right)$

$265.50 = P$

The monthly payment is $265.50.

> **Take Note**
> It is also correct to write the proportion with the loan amounts in the numerators and the monthly payments in the denominators. The solution will be the same.

Problem 4 Nine ceramic tiles are required to tile a 4 ft² area. At this rate, how many square feet can be tiled with 270 ceramic tiles?

Solution See page S22.

Example 5 An investment of $1200 earns $96 each year. At the same rate, how much additional money must be invested to earn $128 each year?

Strategy To find the additional amount of money that must be invested, write and solve a proportion using x to represent

page 466

Problem-Solving Strategies

The text features a carefully developed approach to problem solving that emphasizes the importance of *strategy* when solving problems. Students are encouraged to develop their own strategies—to draw diagrams, to write out the solution steps in words—as part of their solution to a problem. In each case, model strategies are presented as guides for students to follow as they attempt the parallel Problem. Having students provide strategies is a natural way to incorporate writing into the math curriculum.

Real Data and Applications

Applications

One way to motivate an interest in mathematics is through applications. Wherever appropriate, the last objective of a section presents applications that require the student to use problem-solving strategies, along with the skills covered in that section, to solve practical problems. This carefully integrated applied approach generates student awareness of the value of algebra as a real-life tool.

Applications are taken from many disciplines including agriculture, business, carpentry, chemistry, construction, Earth science, education, manufacturing, nutrition, real estate, and sports.

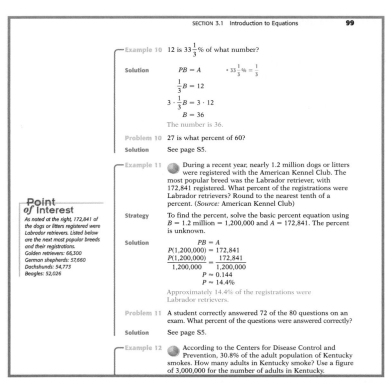

Example 10   12 is $33\frac{1}{3}\%$ of what number?

Solution $PB = A$ • $33\frac{1}{3}\% = \frac{1}{3}$

$\frac{1}{3}B = 12$

$3 \cdot \frac{1}{3}B = 3 \cdot 12$

$B = 36$

The number is 36.

Problem 10   27 is what percent of 60?

Solution   See page S5.

Example 11   During a recent year, nearly 1.2 million dogs or litters were registered with the American Kennel Club. The most popular breed was the Labrador retriever, with 172,841 registered. What percent of the registrations were Labrador retrievers? Round to the nearest tenth of a percent. (*Source:* American Kennel Club)

Point of Interest
As noted at the right, 172,841 of the dogs or litters registered were Labrador retrievers. Listed below are the next most popular breeds and their registrations.
Golden retrievers: 66,300
German shepherds: 57,660
Dachshunds: 54,773
Beagles: 52,026

Strategy   To find the percent, solve the basic percent equation using $B = 1.2$ million $= 1,200,000$ and $A = 172,841$. The percent is unknown.

Solution

$PB = A$

$P(1,200,000) = 172,841$

$\dfrac{P(1,200,000)}{1,200,000} = \dfrac{172,841}{1,200,000}$

$P \approx 0.144$

$P \approx 14.4\%$

Approximately 14.4% of the registrations were Labrador retrievers.

Problem 11   A student correctly answered 72 of the 80 questions on an exam. What percent of the questions were answered correctly?

Solution   See page S5.

Example 12   According to the Centers for Disease Control and Prevention, 30.8% of the adult population of Kentucky smokes. How many adults in Kentucky smoke? Use a figure of 3,000,000 for the number of adults in Kentucky.

page 99

24.   **Postage**   The postage rate for first-class mail is $.37 for the first ounce and $.23 for each additional ounce or fraction of an ounce. Find the largest whole number of ounces a package can weigh if it is to cost you less than $3 to mail the package.

25.   **Fund Raising**   Your class decides to publish a calendar to raise money. The initial cost, regardless of the number of calendars printed, is $900. After the initial cost, each calendar costs $1.50 to produce. What is the minimum number of calendars your class must sell at $6 per calendar to make a profit of at least $1200?

26.   **Debt Carried by Americans**   During a recent year, the mean (average) household income of Americans under age 35 was $48,297. The mean household income of Americans aged 35 to 50 was $60,427. The graph below shows average amounts of debt carried during the same year by those under 35 and by those aged 35 to 50.
a.   In which categories is the average debt carried by Americans under age 35 greater than the average debt carried by Americans aged 35 to 50?
b.   In which categories is the average debt carried by Americans under age 35 greater than $10,000?
c.   For the categories shown in the graph, what was the total average debt of Americans under age 35? How much greater is this debt than that of Americans aged 35 to 50?
d.   Given the answers to parts a, b, and c, what is the significance of the average incomes of these two age groups?
e.   Write a paragraph describing any other observations you can make from the data.

Real Data

Real data examples and exercises, identified by an icon, ask students to analyze and solve problems taken from actual situations. Students are often required to work with tables, graphs, and charts drawn from a variety of disciplines.

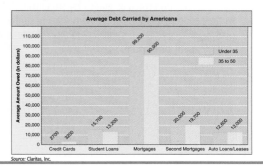

Average Debt Carried by Americans

Source: Claritas, Inc.

page 208

Student Pedagogy

Icons

The icons at each objective head remind students of the many and varied additional resources available for each objective.

Key Terms and Concepts

Key terms, in bold, emphasize important terms. The key terms are also provided in a **Glossary** at the back of the text.

Key Concepts are presented in green boxes in order to highlight these important concepts and to provide for easy reference.

Point of Interest

These margin notes contain interesting sidelights about mathematics, its history, or its application.

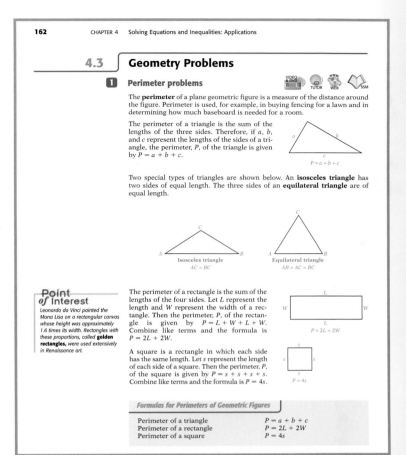

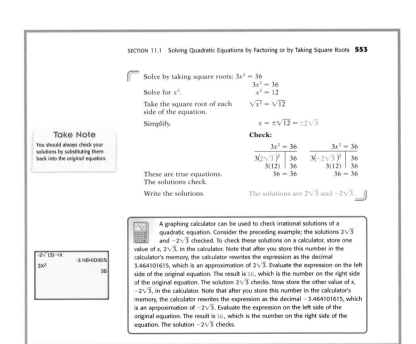

Take Note

These margin notes alert students to a point requiring special attention or are used to amplify the concept under discussion.

Instructional Examples

Colored brackets indicate an instructional example that usually has explanatory remarks to the left of the solution. These examples are used to provide additional examples of important concepts as well as examine important techniques or principles that are often used in the solution of other types of problems.

Graphing Calculator Boxes

Graphing calculators are incorporated as an optional feature at appropriate places throughout the text. This material is designated by the graphing calculator icon.

Exercises and Projects

Exercises

The exercise sets of *Beginning Algebra with Applications* emphasize skill building, skill maintenance, and applications. Icons identify appropriate writing, data analysis, and calculator exercises.

Before each exercise set are **Concept Review** exercises designed to test a student's understanding of the section material. Used as oral exercises, these can lead to interesting class discussions.

Included in each exercise set are **Applying Concepts** that present extensions of topics, require analysis, or offer challenge problems. The writing exercises ask students to explain answers, write about a topic in the section, or research and report on a related topic.

5.5   Applying Concepts

60. a. A function f consists of the ordered pairs $\{(-4, -6), (-2, -2), (0, 2), (2, 6), (4, 10)\}$. Find $f(2)$.
  b. A function f consists of the ordered pairs $\{(0, 9), (1, 8), (2, 7), (3, 6), (4, 5)\}$. Find $f(1)$.

61. 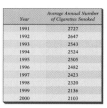 Health The average annual number of cigarettes smoked by an American adult for the years 1991 through 2000 is shown in the table at the right. (*Source:* USDA Economic Research Service Agricultural Outlook)
  a. Do these data represent a function? Why or why not?
  b. If the data represent a function, do they represent a linear function? Why or why not?
  c. If the data represent a function, is the year a function of the number of cigarettes smoked, or is the number of cigarettes smoked a function of the year?
  d. If you had to predict the average annual number of cigarettes smoked in 2001 on the basis of the information in the table, would you predict that it would be more or less than 2103? Explain your answer.

Year	Average Annual Number of Cigarettes Smoked
1991	2727
1992	2647
1993	2543
1994	2524
1995	2505
1996	2482
1997	2423
1998	2320
1999	2136
2000	2103

62. Height-Time Graphs A child's height is a function of the child's age. The graph

page 277

71.  The Postal Service The graph below shows the cost of a first-class postage stamp from 1950 to 2002. A quadratic function that approximately models the data is $y = 0.0087x^2 - 0.6125x + 9.9096$, where $x \geq 50$ and $x = 50$ for the year 1950, and y is the cost in cents of a first-class stamp. What does the model equation predict will be the cost of a first-class stamp in 2020? Round to the nearest cent.

Cost of a First-Class Postage Stamp

(graph: Cost in cents vs. Year, showing step values 3¢, 4¢, 5¢, 6¢, 8¢, 10¢, 13¢, 15¢, 18¢, 20¢, 22¢, 25¢, 29¢, 32¢, 33¢, 34¢, 37¢)

72. The point whose coordinates are (x_1, y_1) lies in quadrant I and is a point on the graph of the equation $y = 2x^2 - 2x + 1$. Given $y_1 = 13$, find x_1.

73. The point whose coordinates are (x_1, y_1) lies in quadrant II and is a point on the graph of the equation $y = 2x^2 - 3x - 2$. Given $y_1 = 12$, find x_1.

page 577

$1 \cdot 1 = 1$	$2 \cdot 1 = 2*$	$3 \cdot 1 = 3*$	4
$1 \cdot 2 = 2$	$2 \cdot 2 = 4*$	$3 \cdot 2 = 6*$	4
$1 \cdot 3 = 3$	$2 \cdot 3 = 6*$	$3 \cdot 3 = 9$	4
$1 \cdot 4 = 8$	$2 \cdot 4 = 8$	$3 \cdot 4 = 12*$	4
$1 \cdot 5 = 5$	$2 \cdot 5 = 10$	$3 \cdot 5 = 15$	$4 \cdot 5 = 20$	$5 \cdot 5 = 25$	$6 \cdot 2 = 30*$
$1 \cdot 6 = 6$	$2 \cdot 6 = 12$	$3 \cdot 6 = 18$	$4 \cdot 6 = 24$	$5 \cdot 6 = 30$	$6 \cdot 6 = 36$

By counting the products that are not repeats, we can see that there are 18 different products.

In this chapter, you may have noticed that a drawing accompanies the strategy step of many application problems. We encourage you to draw a diagram when solving the application problems in these sections, as well as in any other situation where it can prove helpful. Doing so will undoubtedly improve your problem-solving skills.

Projects & Group Activities

Nielsen Ratings

Point of Interest
The top-ranked programs in network prime time for the week of October 1 to October 7, 2001, ranked by Nielsen Media Research, were:
Friends
ER
The West Wing
Law & Order
Everybody Loves Raymond
Inside Schwartz
CSI: Crime Scene Investigation
Will & Grace
Becker
Judging Amy
NFL Monday Night Football
Jag

Nielsen Media Research surveys television viewers to determine the numbers of people watching particular shows. There are an estimated 102.2 million U.S. households with televisions. Each **rating point** represents 1% of that number, or 1,022,000 households. For instance, if *60 Minutes* received a rating of 9.6, then 9.6%, or (0.096)(102,200,000) = 9,811,200, households watched that program.

A rating point does not mean that 1,022,000 people are watching a program. A rating point refers to the number of TV sets tuned to that program; there may be more than one person watching a television set in the household.

Nielsen Media Research also describes a program's share of the market. **Share** is the percent of television sets in use that are tuned to a program. Suppose that the same week *60 Minutes* received 9.6 rating points, the show received a share of 19%. This means that 19% of all households with a television *turned on* were tuned to *60 Minutes*, whereas only 9.6% of all households *with a television* were tuned to the program.

1. If *Law & Order* received a Nielsen rating of 10.8 and a share of 19, how many TV households watched the program that week? How many TV households were watching television during that hour? Round to the nearest hundred thousand.

2. Suppose *Primetime Thursday* received a rating of 16.2 and a share of 28. How many TV households watched the program that week? How many TV

Projects and Group Activities

The Projects and Group Activities feature at the end of each chapter can be used as extra credit or for cooperative learning activities. The projects cover various aspects of mathematics, including the use of calculators, collecting data from the internet, data analysis, and extended applications.

page 211

End of Chapter

Chapter Summary

At the end of each chapter there is a Chapter Summary that includes *Key Words* and *Essential Rules and Procedures*, with accompanying examples, that were covered in the chapter. These chapter summaries provide a single point of reference as the student prepares for a test. Each concept references the objective and the page number from the lesson where the concept is introduced.

Chapter Summary

Key Words	Examples
An **equation** expresses the equality of two mathematical expressions. [3.1.1, p. 90]	$5(3x - 2) = 4x + 7$ is an equation.
A **solution of an equation** is a number that, when substituted for the variable, results in a true equation. [3.1.1, p. 90]	1 is the solution of the equation $6x - 4 = 2$ because $6(1) - 4 = 2$ is a true equation.
To **solve** an equation means to find a solution of the equation. The goal is to rewrite the equation in the form **variable = constant**, because the constant is the solution. [3.1.2, p. 91]	The equation $x = -9$ is in the form *variable = constant*. The constant, -9, is the solution of the equation.
An **inequality** is an expression that contains the symbol $<$, $>$, $\leq$, or $\geq$. [3.3.1, p. 125]	$8x - 1 \geq 5x + 23$ is an inequality.
The **solution set of an inequality** is a set of numbers, each element of which, when substituted for the variable, results in a true inequality. [3.3.1, p. 125]	The solution set of $8x - 1 \geq 5x + 23$ is $x \geq 8$ because every number greater than or equal to 8, when substituted for the variable, results in a true inequality.

Essential Rules and Procedures

The Addition Property of Equations The same number or variable term can be added to each side of an equation without changing the solution of the equation. [3.1.2, p. 91]	$x + 12 = -19$ $x + 12 - 12 = -19 - 12$ $x = -31$

page 138

Chapter Review Exercises

Review exercises are found at the end of each chapter. These exercises are selected to help the student integrate all of the topics presented in the chapter.

140 CHAPTER 3 Solving Equations and Inequalities

Chapter Review Exercises

1. Is 3 a solution of $5x - 2 = 4x + 5$?

2. Solve: $x - 4 = 16$

3. Solve: $8x = -56$

4. Solve: $5x - 6 = 29$

5. Solve: $5x + 3 = 10x - 17$

6. Solve: $3(5x + 2) + 2 = 10x + 5[x - (3x - 1)]$

7. What is 81% of 500?

8. 18 is 72% of what number?

page 140

Chapter Test

The Chapter Test exercises are designed to simulate a possible test of the material in the chapter.

142 CHAPTER 3 Solving Equations and Inequalities

Chapter Test

1. Solve: $\frac{3}{4}x = -9$

2. Solve: $6 - 5x = 5x + 11$

3. Solve: $3x - 5 = -14$

4. Is -2 a solution of $x^2 - 3x = 2x - 6$?

5. Solve: $x + \frac{1}{2} = \frac{5}{8}$

6. Solve: $5x - 2(4x - 3) = 6x + 9$

page 142

Cumulative Review Exercises

Cumulative Review Exercises, which appear at the end of each chapter (beginning with Chapter 2), help students maintain skills learned in previous chapters.

The answers to all Chapter Review Exercises, all Chapter Test exercises, and all Cumulative Review Exercises are given in the Answer Section. Along with the answer, there is a reference to the objective that pertains to each exercise.

Cumulative Review Exercises

1. Subtract: $-6 - (-20) - 8$

2. Multiply: $(-2)(-6)(-4)$

3. Subtract: $-\frac{5}{6} - \left(-\frac{7}{16}\right)$

4. Divide: $-2\frac{1}{3} \div 1\frac{1}{6}$

5. Simplify: $-4^2 \cdot \left(-\frac{3}{2}\right)^3$

6. Simplify: $25 - 3 \cdot \frac{(5 - 2)^2}{2^3 + 1} - (-2)$

7. Evaluate $3(a - c) - 2ab$ when $a = 2$, $b = 3$, and $c = -4$.

8. Simplify: $3x - 8x + (-12x)$

page 143

Resources

INSTRUCTOR RESOURCES

Beginning Algebra with Applications has a complete set of teaching aids for the instructor.

Instructor's Annotated Edition   This edition contains a replica of the student text and additional items just for the instructor. These include: *Instructor Notes, Transparency Master icons, In-Class Examples, Concept Checks, Discuss the Concepts, New Vocabulary/Symbols, etc., Vocabulary/Symbols, etc. to Review, Optional Student Activities, Quick Quizzes, Answers to Writing Exercises*, and *Suggested Assignments*. Answers to all exercises are also provided.

Instructor's Resource Manual with Solutions Manual   The *Instructor's Resource Manual* includes a lesson plan for the *AIM for Success* as well as suggested course sequences. The *Instructor's Solutions Manual* contains worked-out solutions for all exercises in the text.

Printed Test Bank with Chapter Tests   The *Printed Test Bank* provides a print-out of one example of each of the algorithmic items in *HM Testing*. This resource also contains ready-to-use printed Chapter Tests as well as cumulative tests and final exams.

HM³ Tutorial (Instructor version)   This tutorial CD software package for Microsoft Windows was written and developed by the authors specifically for use with this text. For each objective, exercises and quizzes are algorithmically generated, solution steps are animated, lessons and exercises are presented in a colorful, lively manner, and an integrated classroom management system tracks student performance.

NEW! WebCT ePacks   *WebCT ePacks* provide instructors with a flexible, Internet-based education platform providing multiple ways to present learning materials. The *WebCT ePacks* come with a full array of features to enrich the online learning experience.

NEW! BlackBoard Cartridges   The *Houghton Mifflin Blackboard cartridge* allows flexible, efficient, and creative ways to present learning materials and opportunities. In addition to course management benefits, instructors may make use of an electronic grade book, receive papers from students enrolled in the course via the Internet, and track student use of the communication and collaboration functions.

NEW! HMClassPrep w/ HM Testing CD-ROM   *HMClass Prep CD-Rom* contains a multitude of text-specific resources for instructors to use to enhance the classroom experience. These resources can be easily accessed by chapter or resource type and can also link you to the text's web site. *HMTesting* is our computerized test generator and contains a database of algorithmic test items as well as providing online testing and gradebook functions.

NEW! Instructor Text-Specific Web Site Resources on the Class Prep CD are also available on the instructor web site at *math.college.hmco.com/instructors*. Appropriate items are password protected. Instructors also have access to the student part of the text's web site.

STUDENT RESOURCES

Student Solutions Manual The *Student Solutions Manual* contains complete solutions to all odd-numbered exercises in the text.

Math Study Skills Workbook *by Paul D. Nolting* This workbook is designed to reinforce skills and minimize frustration for students in any math class, lab, or study skills course. It offers a wealth of study tips and sound advice on note taking, time mangement, and reducing math anxiety. In addition, numerous opportunities for self assessment enable students to track their own progress.

NEW! HM eduSpace *eduSpace* is a new content delivery system, combining an algorithmic tutorial program, online delivery of course materials, and classroom management functions. The interactive online content correlates directly to this text and can be accessed 24 hours a day.

HM³ Tutorial (Student version) This tutorial CD software package for Microsoft Windows was written and developed by the authors specifically for use with this text. HM³ is an interactive tutorial containing lessons, exercises, and quizzes for every section of the text. Lessons are presented in a colorful, lively manner; solution steps are animated; and exercises and quizzes are algorithmically generated. Next to every objective head, the CD TUTOR serves as a reminder that there is an HM³ tutorial lesson corresponding to that objective.

NEW! SMARTHINKING™ Live, OnLine Tutoring Houghton Mifflin has partnered with SMARTHINKING to provide an easy-to-use and effective online tutorial service. *Whiteboard Simulations* and *Practice Area* promote real-time visual interaction.

Three levels of service are offered.

- **Text-specific Tutoring** provides real-time, one-on-one instruction with a specially qualified 'e-structor.'

- **Questions Any Time** allows students to submit questions to the tutor outside the scheduled hours and receive a reply within 24 hours.

- **Independent Study Resources** connect students with around-the-clock access to additional educational services, including interactive web sites, diagnostic tests and Frequently Asked Questions posed to SMARTHINKING e-structors.

NEW! Videos and DVDs This edition offers brand new text-specific videos and DVDs, hosted by Dana Mosely, covering all sections of the text and providing a valuable resource for further instruction and review. Next to every objective head, the VIDEO & DVD serves as a reminder that the objective is covered in a video/DVD lesson.

NEW! Student Text-Specific Web Site Student resources, such as section quizzes, can be found at this text's web site at *math.college.hmco.com/students*.

Acknowledgments

The authors would like to thank the people who have reviewed this manuscript and provided many valuable suggestions.

Joe Archuleta, *Imperial Valley College, CA*
Don K. Brown, *Georgia Southern University, GA*
Will Clarke, *La Sierra University, CA*
Andres Delgado, *Orange County Community College, NY*
Sudhir Goel, *Valdosta State University, GA*
Yash Manchanda, *Saddleback College, CA*
Rogers Martin, *Louisiana State University in Shreveport, LA*
John Searcy
Gail Ulager, *Robert Morris College, PA*
Tom Wheeler, *Georgia Southwestern State University, GA*
Jean Wilberg, *Morrisville College, NY*
Michael D. Young, *Pitt Community College, NC*

Special thanks to Christi Verity for her diligent preparation of the solutions manuals and for her contribution to the accuracy of the textbooks.

AIM for Success

Welcome to *Beginning Algebra with Applications*. As you begin this course, we know two important facts: (1) We want you to succeed. (2) You want to succeed. To do that requires an effort from each of us. For the next few pages, we are going to show you what is required of you to achieve that success and how you can use the features of this text to be successful.

Motivation

One of the most important keys to success is motivation. We can try to motivate you by offering interesting or important ways mathematics can benefit you. But, in the end, the motivation must come from you. On the first day of class, it is easy to be motivated. Eight weeks into the term, it is harder to keep that motivation.

To stay motivated, there must be outcomes from this course that are worth your time, money, and energy. List some reasons you are taking this course. Do not make a mental list—actually write them out.

_____

_____

_____

Although we hope that one of the reasons you listed was an interest in mathematics, we know that many of you are taking this course because it is required to graduate, it is a prerequisite for a course you must take, or because it is required for your major. Although you may not agree that this course is necessary, it is! If you are motivated to graduate or complete the requirements for your major, then use that motivation to succeed in this course. Do not become distracted from your goal to complete your education!

Commitment

To be successful, you must make a committment to succeed. This means devoting time to math so that you achieve a better understanding of the subject.

List some activities (sports, hobbies, talents such as dance, art, or music) that you enjoy and at which you would like to become better.

ACTIVITY	*TIME SPENT*	*TIME WISHED SPENT*

Thinking about these activities, put the number of hours that you spend each week practicing these activities next to the activity. Next to that number, indicate the number of hours per week you would like to spend on these activities.

Whether you listed surfing or sailing, aerobics or restoring cars, or any other activity you enjoy, note how many hours a week you spend doing it. To succeed in math, you must be willing to commit the same amount of time. Success requires some sacrifice.

The "I Can't Do Math" Syndrome

There may be things you cannot do, such as lift a two-ton boulder. You can, however, do math. It is much easier than lifting the two-ton boulder. When you first learned the activities you listed above, you probably could not do them well. With practice, you got better. With practice, you will be better at math. Stay focused, motivated, and committed to success.

It is difficult for us to emphasize how important it is to overcome the "I Can't Do Math Syndrome." If you listen to interviews of very successful athletes after a particularly bad performance, you will note that they focus on the positive aspect of what they did, not the negative. Sports psychologists encourage athletes to always be positive—to have a "Can Do" attitude. Develop this attitude toward math.

Strategies for Success

Textbook Reconnaissance Right now, do a 15-minute "textbook reconnaissance" of this book. Here's how:

First, read the table of contents. Do it in three minutes or less. Next, look through the entire book, page by page. Move quickly. Scan titles, look at pictures, notice diagrams.

A textbook reconnaissance shows you where a course is going. It gives you the big picture. That's useful because brains work best when going from the general to the specific. Getting the big picture before you start makes details easier to recall and understand later on.

Your textbook reconnaissance will work even better if, as you scan, you look for ideas or topics that are interesting to you. List three facts, topics, or problems that you found interesting during your textbook reconnaissance.

_____

_____

_____

The idea behind this technique is simple: It's easier to work at learning material if you know it's going to be useful to you.

Not all the topics in this book will be "interesting" to you. But that is true of any subject. Surfers find that on some days the waves are better than others, musicians find some music more appealing than other music, computer gamers find some computer games more interesting than others, car enthusiasts find some cars more exciting than others. Some car enthusiasts would rather have a completely restored 1957 Chevrolet than a new Ferrari.

Know the Course Requirements To do your best in this course, you must know exactly what your instructor requires. Course requirements may be stated in a *syllabus*, which is a printed outline of the main topics of the course, or they may be presented orally. When they are listed in a syllabus or on other printed pages, keep them in a safe place. When they are presented orally, make sure to take complete notes. In either case, it is important that you understand them completely and follow them exactly. Be sure you know the answer to the following questions.

1. What is your instructor's name?
2. Where is your instructor's office?
3. At what times does your instructor hold office hours?
4. Besides the textbook, what other materials does your instructor require?
5. What is your instructor's attendance policy?
6. If you must be absent from a class meeting, what should you do before returning to class? What should you do when you return to class?
7. What is the instructor's policy regarding collection or grading of homework assignments?

8. What options are available if you are having difficulty with an assignment? Is there a math tutoring center?
9. Is there a math lab at your school? Where is it located? What hours is it open?
10. What is the instructor's policy if you miss a quiz?
11. What is the instructor's policy if you miss an exam?
12. Where can you get help when studying for an exam?

Remember: Your instructor wants to see you succeed. If you need help, ask! Do not fall behind. If you are running a race and fall behind by 100 yards, you may be able to catch up but it will require more effort than had you not fallen behind.

Time Management We know that there are demands on your time. Family, work, friends, and entertainment all compete for your time. We do not want to see you receive poor job evaluations because you are studying math. However, it is also true that we do not want to see you receive poor math test scores because you devoted too much time to work. When several competing and important tasks require your time and energy, the only way to manage the stress of being successful at both is to manage your time efficiently.

Instructors often advise students to spend twice the amount of time outside of class studying as they spend in the classroom. Time management is important if you are to accomplish this goal and succeed in school. The following activity is intended to help you structure your time more efficiently.

List the name of each course you are taking this term, the number of class hours each course meets, and the number of hours you should spend studying each subject outside of class. Then fill in a weekly schedule like the one below. Begin by writing in the hours spent in your classes, the hours spent at work (if you have a job), and any other commitments that are not flexible with respect to the time that you do them. Then begin to write down commitments that are more flexible, including hours spent studying. Remember to reserve time for activities such as meals and exercise. You should also schedule free time.

> ### Take Note
>
> Besides time management, there must be realistic ideas of how much time is available. There are very few people who can *successfully* work full-time and go to school full-time. If you work 40 hours a week, take 15 units, spend the recommended study time given at the right, and sleep 8 hours a day, you will use over 80% of the available hours in a week. That leaves less than 20% of the hours in a week for family, friends, eating, recreation, and other activities.

	Monday	Tuesday	Wednesday	Thursday	Friday	Saturday	Sunday
7–8 a.m.							
8–9 a.m.							
9–10 a.m.							
10–11 a.m.							
11–12 p.m.							
12–1 p.m.							
1–2 p.m.							
2–3 p.m.							
3–4 p.m.							
4–5 p.m.							
5–6 p.m.							
6–7 p.m.							
7–8 p.m.							
8–9 p.m.							
9–10 p.m.							
10–11 p.m.							
11–12 a.m.							

We know that many of you must work. If that is the case, realize that working 10 hours a week at a part-time job is equivalent to taking a three-unit class. If you must work, consider letting your education progress at a slower rate to allow you to be successful at both work and school. There is no rule that says you must finish school in a certain time frame.

Schedule Study Time As we encouraged you to do by filling out the time management form above, schedule a certain time to study. You should think of this time the way you would the time for work or class—that is, reasons for missing study time should be as compelling as reasons for missing work or class. "I just didn't feel like it" is not a good reason to miss your scheduled study time.

Although this may seem like an obvious exercise, list a few reasons you might want to study.

_____

_____

_____

Of course we have no way of knowing the reasons you listed, but from our experience one reason given quite frequently is "To pass the course." There is nothing wrong with that reason. If that is the most important reason for you to study, then use it to stay focused.

One method of keeping to a study schedule is to form a ***study group***. Look for people who are committed to learning, who pay attention in class, and who are punctual. Ask them to join your group. Choose people with similar educational goals but different methods of learning. You can gain insight from seeing the material from a new perspective. Limit groups to four or five people; larger groups are unwieldy.

There are many ways to conduct a study group. Begin with the following suggestions and see what works best for your group.

1. Test each other by asking questions. Each group member might bring two or three sample test questions to each meeting.
2. Practice teaching each other. Many of us who are teachers learned a lot about our subject when we had to explain it to someone else.
3. Compare class notes. You might ask other students about material in your notes that is difficult for you to understand.
4. Brainstorm test questions.
5. Set an agenda for each meeting. Set approximate time limits for each agenda item and determine a quitting time.

And finally, probably the most important aspect of studying is that it should be done in relatively small chunks. If you can only study three hours a week for this course (probably not enough for most people), do it in blocks of one hour on three separate days, preferably after class. Three hours of studying on a Sunday is not as productive as three hours of paced study.

Text Features That Promote Success

There are 11 chapters in this text. Each chapter is divided into sections, and each section is subdivided into learning objectives. Each learning objective is labeled with a number from 1 to 5.

Preparing for a Chapter Before you begin a new chapter, you should take some time to review previously learned skills. There are two ways to do this. The first is to complete the **Cumulative Review**, which occurs after every chapter (except Chapter 1). For instance, turn to page 143. The questions in this review are taken from the previous chapters. The answers for all these exercises can be found on page A9. Turn to that page now and locate the answers for the Chapter 3 Cumulative Review. After the answer to the first exercise, which is 6, you will see the objective reference [1.2.2]. This means that this question was taken from Chapter 1, Section 2, Objective 2. If you missed this question, you should return to that objective and restudy the material.

A second way of preparing for a new chapter is to complete the ***Prep Test***. This test focuses on the particular skills that will be required for the new chapter. Turn to page 90 to see a Prep Test. The answers for the Prep Test are the first set of answers in the answer section for a chapter. Turn to page A6 to see the answers for the Chapter 3 Prep Test. Note that an objective reference is given for each question. If you answer a question incorrectly, restudy the objective from which the question was taken.

Before the class meeting in which your professor begins a new section, you should read each objective statement for that section. Next, browse through the objective material, being sure to note each word in bold type. These words indicate important concepts that you must know in order to learn the material. Do not worry about trying to understand all the material. Your professor is there to assist you with that endeavor. The purpose of browsing through the material is so that your brain will be prepared to accept and organize the new information when it is presented to you.

Turn to page 2. Write down the title of the first objective in Section 1.1. Under the title of the objective, write down the words in the objective that are in bold print. It is not necessary for you to understand the meaning of these words. You are in this class to learn their meaning.

_____ _____ _____ _____

_____ _____ _____ _____

_____ _____ _____ _____

_____ _____ _____ _____

_____ _____ _____ _____

Math is Not a Spectator Sport To learn mathematics you must be an active participant. Listening and watching your professor do mathematics is not enough. Mathematics requires that you interact with the lesson you are studying. If you filled in the blanks above, you were being interactive. There are other ways this textbook has been designed to help you be an active learner.

Instructional Examples Colored brackets indicate an instructional example that usually has explanatory remarks to the left of the solution. These examples are used for two purposes. The first is to provide additional examples of important concepts. Secondly, these examples examine important techniques or principles that are often used in the solution of other types of problems.

$$y + \frac{1}{2} = \frac{5}{4}$$

$$y + \frac{1}{2} - \frac{1}{2} = \frac{5}{4} - \frac{1}{2}$$

$$y + 0 = \frac{5}{4} - \frac{2}{4}$$

$$y = \frac{3}{4}$$

The solution is $\frac{3}{4}$.

Solve: $y + \frac{1}{2} = \frac{5}{4}$

The goal is to rewrite the equation in the form *variable = constant*.

Add the opposite of the constant term $\frac{1}{2}$ to each side of the equation. This is equivalent to subtracting $\frac{1}{2}$ from each side of the equation.

The solution $\frac{3}{4}$ checks.

$$y + \frac{1}{2} = \frac{5}{4}$$

$$y + \frac{1}{2} - \frac{1}{2} = \frac{5}{4} - \frac{1}{2}$$

$$y + 0 = \frac{5}{4} - \frac{2}{4}$$

$$y = \frac{3}{4}$$

The solution is $\frac{3}{4}$.

page 92

When you complete the example, get a clean sheet of paper. Write down the problem and then try to complete the solution without referring to your notes or the book. When you can do that, move on to the next part of the objective.

Leaf through the book now and write down the page numbers of two other occurrences of an instructional example.

Example/Problem Pairs One of the key instructional features of this text is Example/Problem pairs. Note that each Example is completely worked out and the "Problem" following the example is not. Study the worked-out example carefully by working through each step. Then work the Problem. If you get stuck, refer to the page number following the Problem which directs you to the page on which the Problem is solved—a complete worked-out solution is provided. Try to use the given solution to get a hint for the step you are stuck on. Then try to complete your solution.

Example 4 Solve: $\frac{1}{2} = x + \frac{2}{3}$

Solution

$$\frac{1}{2} = x + \frac{2}{3}$$

$$\frac{1}{2} - \frac{2}{3} = x + \frac{2}{3} - \frac{2}{3}$$ • Subtract $\frac{2}{3}$ from each side of the equation.

$$-\frac{1}{6} = x$$ • Simplify each side of the equation.

The solution is $-\frac{1}{6}$.

Problem 4:

$$-8 = 5 + x$$

$$-8 - 5 = 5 - 5 + x$$

$$-13 = x$$

The solution is -13.

Problem 4 Solve: $-8 = 5 + x$

Solution   See page S4.

page 93

When you have completed your solution, check your work against the solution we provided. (Turn to page S4 to see the solution of Problem 4.) Be aware that frequently there is more than one way to solve a problem. Your answer, however, should be the same as the given answer. If you have any question as to whether your method will "always work," check with your instructor or with someone in the math center.

Browse through the textbook and write down the page numbers where two other Example/Problem pairs occur.

Remember: Be an active participant in your learning process. When you are sitting in class watching and listening to an explanation, you may think that you understand. However, until you actually try to do it, you will have no confirmation of the new knowledge or skill. Most of us have had the experience of sitting in class thinking we knew how to do something only to get home and realize that we didn't.

Word Problems Word problems are difficult because we must read the problem, determine the quantity we must find, think of a method to do that, and then actually solve the problem. In short, we must formulate a *strategy* to solve the problem and then devise a *solution*.

Note in the Example/Problem pair below that part of every word problem is a strategy and part is a solution. The strategy is a written description of how we will solve the problem. In the corresponding Problem, you are asked to formulate a strategy. Do not skip this step, and be sure to write it out.

> **Take Note**
>
> There is a strong connection between reading and being a successful student in math or in any other subject. If you have difficulty reading, consider taking a reading course. Reading is much like other skills. There are certain things you can learn that will make you a better reader.

Problem 11:

Strategy:

To find the percent of questions answers correctly, solve the basic percent equation using $B = 80$ and $A = 72$.

Solution:
$$PB = A$$
$$P(80) = 72$$
$$80P = 72$$
$$\frac{80P}{80} = \frac{72}{80}$$
$$P = 0.9$$

90% of the questions were answered correctly.

Example 11   During a recent year, nearly 1.2 million dogs or litters were registered with the American Kennel Club. The most popular breed was the Labrador retriever, with 172,841 registered. What percent of the registrations were Labrador retrievers? Round to the nearest tenth of a percent. (*Source:* American Kennel Club)

Strategy   To find the percent, solve the basic percent equation using $B = 1.2$ million $= 1,200,000$ and $A = 172,841$. The percent is unknown.

Solution
$$PB = A$$
$$P(1,200,000) = 172,841$$
$$\frac{P(1,200,000)}{1,200,000} = \frac{172,841}{1,200,000}$$
$$P \approx 0.144$$
$$P \approx 14.4\%$$

Approximately 14.4% of the registrations were Labrador retrievers.

Problem 11   A student correctly answered 72 of the 80 questions on an exam. What percent of the questions were answered correctly?

Solution   See page S5.

page 99

Rule Boxes Pay special attention to rules placed in boxes. These rules give you the reasons certain types of problems are solved the way they are. When you see a rule, try to rewrite the rule in your own words.

> **Take Note**
>
> If a rule has more than one part, be sure to make a notation to that effect.

Multiplying each side of an equation by the same number (<u>NOT</u> 0) will not change the solution.

> ### *Multiplication Property of Equations*
>
> Each side of an equation can be multiplied by the same nonzero number without changing the solution of the equation.

page 93

Find and write down two page numbers on which there are examples of rule boxes.

Chapter Exercises When you have completed studying an objective, do the exercises in the exercise set that correspond with that objective. The exercises are labeled with the same number as the objective. Math is a subject that needs to be learned in small sections and practiced continually in order to be mastered. Doing all of the exercises in each exercise set will help you master the problem-solving techniques necessary for success. As you work through the exercises for an objective, check your answers to the odd-numbered exercises with those in the back of the book.

Preparing for a Test There are important features of this text that can be used to prepare for a test.

- Chapter Summary
- Chapter Review Exercises
- Chapter Test

After completing a chapter, read the Chapter Summary. (See page 138 for the Chapter 3 Summary.) This summary highlights the important topics covered in the chapter. The page number following each topic refers you to the page in the text on which you can find more information about the concept.

Following the Chapter Summary are Chapter Review Exercises (see page 140) and a Chapter Test (see page 142). Doing the review exercises is an important way of testing your understanding of the chapter. The answer to each review exercise is given at the back of the book, along with its objective reference. After checking your answers, restudy any objective from which a question you missed was taken. It may be helpful to retry some of the exercises for that objective to reinforce your problem-solving techniques.

The Chapter Test should be used to prepare for an exam. We suggest that you try the Chapter Test a few days before your actual exam. Take the test in a quiet place and try to complete the test in the same amount of time you will be allowed for your exam. When taking the Chapter Test, practice the strategies of successful test takers: (1) scan the entire test to get a feel for the questions; (2) read the directions carefully; (3) work the problems that are easiest for you first; and perhaps most importantly, (4) try to stay calm.

When you have completed the Chapter Test, check your answers. If you missed a question, review the material in that objective and rework some of the exercises from that objective. This will strengthen your ability to perform the skills in that objective.

Your career goal here. →

Is it difficult to be successful? YES! Successful music groups, artists, professional athletes, chefs, and _____ have to work very hard to achieve their goals. They focus on their goals and ignore distractions. The things we ask you to do to achieve success take time and commitment. We are confident that if you follow our suggestions, you will succeed.

Beginning Algebra
with Applications

Applications

Aviation, *p. 19*

Chemistry, *p. 18*

Education, *p. 19*

Employment, *p. 31*

The Federal budget, *p. 41*

Geography, *p. 18, 19*

Government, *p. 32*

Inductive reasoning, *p. 39*

Meteorology, *p. 8*

Patterns in mathematics, *p. 41*

The Postal Service, *p. 32*

The Stock Market, *p. 14, 19, 20, 41*

Temperature, *p. 15, 18, 19, 47, 48*

Signed numbers are used to express the amount of change in the price of a share of stock. Positive numbers indicate an increase in the price, and negative numbers indicate a decrease. Stock market analysts calculate the moving average to a stock, which is the average of the changes in a stock's price over several days. The moving average is used to discover trends in the movement of the price and to help advise investors. The **Project on pages 41–42** *involves calculating moving averages for different stocks.*

1

Real Numbers

Objectives

Section 1.1
1. Order relations
2. Opposites and absolute value

Section 1.2
1. Add integers
2. Subtract integers
3. Multiply integers
4. Divide integers
5. Application problems

Section 1.3
1. Write rational numbers as decimals
2. Add and subtract rational numbers
3. Multiply and divide rational numbers
4. Convert among percents, fractions, and decimals

Section 1.4
1. Exponential expressions
2. The Order of Operations Agreement

 WEB **Need help? For online student resources, such as section quizzes, visit this web site: math.college.hmco.com**

PrepTest

1. What is 127.1649 rounded to the nearest hundredth?

2. Add: 49,147 + 596

3. Subtract: 5004 − 487

4. Multiply: 407 × 28

5. Divide: 456 ÷ 19

6. What is the smallest number into which both 8 and 12 divide evenly?

7. What is the greatest number that divides evenly into both 16 and 20?

8. Without using 1, write 21 as a product of two whole numbers.

9. Represent the shaded portion of the figure as a fraction.

Go Figure

If you multiply the first 20 natural numbers
(1 · 2 · 3 · 4 · 5 · · · · · 17 · 18 · 19 · 20), how many zeros will be at the end of the number?

1.1 Introduction to Integers

1 Order relations

It seems to be a human characteristic to group similar items. For instance, a botanist places plants with similar characteristics in groups called species. Nutritionists classify foods according to food groups; for example, pasta, crackers, and rice are among the foods in the bread group.

Mathematicians place objects with similar properties in groups called sets. A **set** is a collection of objects. The objects in a set are called **elements** of the set.

The **roster method** of writing sets encloses a list of the elements in braces. The set of sections within an orchestra is written {brass, percussion, strings, woodwinds}. When the elements of a set are listed, each element is listed only once. For instance, if the list of numbers 1, 2, 3, 2, 3 were placed in a set, the set would be {1, 2, 3}.

The numbers that we use to count objects, such as the number of students in a classroom or the number of people living in an apartment house, are the natural numbers.

Natural numbers = {1, 2, 3, 4, 5, 6, 7, 8, 9, 10, . . .}

The three dots mean that the list of natural numbers continues on and on and that there is no largest natural number.

Point of Interest

The Alexandrian astronomer Ptolemy began using omicron, o, the first letter of the Greek word that means "nothing," as the symbol for zero in A.D. 150. It was not until the 13th century, however, that Fibonacci introduced 0 to the Western world as a placeholder so that we could distinguish, for example, 45 from 405.

The natural numbers do not include zero. The number zero is used for describing data such as the number of people who have run a two-minute mile and the number of college students at Providence College who are under the age of 10. The set of whole numbers includes the natural numbers and zero.

$$\textbf{Whole numbers} = \{0, 1, 2, 3, 4, 5, 6, 7, \ldots\}$$

The whole numbers do not provide all the numbers that are useful in applications. For instance, a meteorologist also needs numbers below zero.

$$\textbf{Integers} = \{\ldots, -5, -4, -3, -2, -1, 0, 1, 2, 3, 4, 5, \ldots\}$$

Each integer can be shown on a number line. The integers to the left of zero on the number line are called **negative integers.** The integers to the right of zero are called **positive integers,** or natural numbers. Zero is neither a positive nor a negative integer.

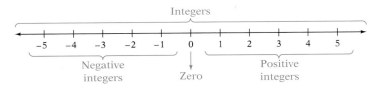

The **graph of an integer** is shown by placing a heavy dot on the number line directly above the number. The graphs of -3 and 4 are shown on the number line below.

Consider the following sentences.

The quarterback threw the football, and the receiver caught *it*.

A student purchased a computer and used *it* to write English and history papers.

In the first sentence, *it* is used to mean the football; in the second sentence, *it* means the computer. In language, the word *it* can stand for many different objects. Similarly, in mathematics, a letter of the alphabet can be used to stand for a number. Such a letter is called a **variable.** Variables are used in the following definition of inequality symbols.

Definition of Inequality Symbols

If a and b are two numbers and a is to the left of b on the number line, then a **is less than** b. This is written $a < b$.

If a and b are two numbers and a is to the right of b on the number line, then a **is greater than** b. This is written $a > b$.

Point of Interest

The symbols for "is less than" and "is greater than" were introduced by Thomas Harriot around 1630. Before that, ⊏ and ⊐ were used for < and >, respectively.

Negative 4 is less than negative 1.

$-4 < -1$

5 is greater than 0.

$5 > 0$

There are also inequality symbols for **is less than or equal to** (≤) and **is greater than or equal to** (≥).

$7 \leq 15$ 7 is less than or equal to 15. $6 \leq 6$ 6 is less than or equal to 6.
This is true because $7 < 15$.   This is true because $6 = 6$.

Example 1 Use the roster method to write the set of negative integers greater than or equal to -6.

Solution $A = \{-6, -5, -4, -3, -2, -1\}$ • A set is designated by a capital letter.

Problem 1 Use the roster method to write the set of positive integers less than 5.

Solution See page S1.

Example 2 Given $A = \{-6, -2, 0\}$, which elements of set A are less than or equal to -2?

Solution $-6 < -2$ • Find the order relation between each element of set A and -2.

$-2 = -2$

$0 > -2$

The elements -6 and -2 are less than or equal to -2.

Problem 2 Given $B = \{-5, -1, 5\}$, which elements of set B are greater than -1?

Solution See page S1.

2 Opposites and absolute value

Two numbers that are the same distance from zero on the number line but are on opposite sides of zero are **opposite numbers,** or **opposites.** The opposite of a number is also called its **additive inverse.**

The opposite of 5 is -5.

The opposite of -5 is 5.

The negative sign can be read "the opposite of."

$-(2) = -2$ The opposite of 2 is -2.

$-(-2) = 2$ The opposite of -2 is 2.

Example 3 Find the opposite number.   **A.** 6   **B.** -51

Solution **A.** -6 **B.** 51

> **Problem 3** Find the opposite number.   **A.** -9 **B.** 62
>
> **Solution**   See page S1.

The **absolute value** of a number is its distance from zero on the number line. Therefore, the absolute value of a number is a positive number or zero. The symbol for absolute value is two vertical bars, $|\;|$.

The distance from 0 to 3 is 3. Therefore, the absolute value of 3 is 3.

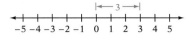

$$|3| = 3$$

The distance from 0 to -3 is 3. Therefore, the absolute value of -3 is 3.

$$|-3| = 3$$

> **Absolute Value**
>
> The absolute value of a positive number is the number itself. The absolute value of zero is zero. The absolute value of a negative number is the opposite of the negative number.

Example 4 Evaluate.   **A.** $|-4|$ **B.** $-|-10|$

Solution   **A.** $|-4| = 4$

B. $-|-10| = -10$ • The absolute value sign does not affect the negative sign in front of the absolute value sign. You can read $-|-10|$ as "the opposite of the absolute value of negative 10."

Problem 4 Evaluate.   **A.** $|-5|$ **B.** $-|-9|$

Solution   See page S1.

1.1   Concept Review

Determine whether the statement is always true, sometimes true, or never true.

1. The absolute value of a number is positive.

2. The absolute value of a number is negative.

3. If x is an integer, then $|x| < -3$.

4. If x is an integer, then $|x| > -2$.

5. The opposite of a number is a positive number.

6. The set of positive integers is $\{0, 1, 2, 3, 4, \ldots\}$.

7. If a is an integer, then $a \le a$.

8. If a and b are integers and $a > b$, then $a \ge b$.

9. If x is a negative integer, then $|x| = -x$.

10. Suppose a and b are integers and $a < b$. Then $|a| < |b|$.

1.1   Exercises

1. How do the whole numbers differ from the natural numbers?

2. Explain the difference between the symbols $<$ and $\le$.

Place the correct symbol, $<$ or $>$, between the two numbers.

3. -2 -5

4. -6 -1

5. -16 1

6. -2 13

7. 3 -7

8. 5 -6

9. 0 -3

10. 8 0

11. -42 27

12. -36 49

13. 21 -34

14. 53 -46

15. -27 -39

16. -51 -20

17. -131 101

18. 127 -150

Use the roster method to write the set.

19. the natural numbers less than 9

20. the natural numbers less than or equal to 6

21. the positive integers less than or equal to 8

22. the positive integers less than 4

23. the negative integers greater than-7

24. the negative integers greater than or equal to -5

Solve.

25. Given $A = \{-7, 0, 2, 5\}$, which elements of set A are greater than 2?

26. Given $B = \{-8, 0, 7, 15\}$, which elements of set B are greater than 7?

27. Given $D = \{-23, -18, -8, 0\}$, which elements of set D are less than -8?

28. Given $C = \{-33, -24, -10, 0\}$, which elements of set C are less than -10?

29. Given $E = \{-35, -13, 21, 37\}$, which elements of set E are greater than -10?

30. Given $F = \{-27, -14, 14, 27\}$, which elements of set F are greater than -15?

31. Given $B = \{-52, -46, 0, 39, 58\}$, which elements of set B are less than or equal to 0?

32. Given $A = \{-12, -9, 0, 12, 34\}$, which elements of set A are greater than or equal to 0?

33. Given $C = \{-23, -17, 0, 4, 29\}$, which elements of set C are greater than or equal to -17?

34. Given $D = \{-31, -12, 0, 11, 45\}$, which elements of set D are less than or equal to -12?

35. Given that set A is the positive integers less than 10, which elements of set A are greater than or equal to 5?

36. Given that set B is the positive integers less than or equal to 12, which elements of set B are greater than 6?

37. Given that set D is the negative integers greater than or equal to -10, which elements of set D are less than -4?

38. Given that set C is the negative integers greater than -8, which elements of set C are less than or equal to -3?

2 **39.** What is the additive inverse of a number?

40. What is the absolute value of a number?

Find the opposite number.

41. 22   **42.** 45   **43.** -31 **44.** -88

45. -168 **46.** -97 **47.** 630   **48.** 450

Evaluate.

49. $-(-18)$ **50.** $-(-30)$ **51.** $-(49)$ **52.** $-(67)$

53. $|16|$ **54.** $|19|$ **55.** $|-12|$ **56.** $|-22|$

57. $-|29|$ **58.** $-|20|$ **59.** $-|-14|$ **60.** $-|-18|$

61. $-|0|$ **62.** $|-30|$ **63.** $-|34|$ **64.** $-|-45|$

Solve.

65. Given $A = \{-8, -5, -2, 1, 3\}$, find
   a. the opposite of each element of set A.
   b. the absolute value of each element of set A.

66. Given $B = \{-11, -7, -3, 1, 5\}$, find
   a. the opposite of each element of set B.
   b. the absolute value of each element of set B.

Place the correct symbol, $<$ or $>$, between the two numbers.

67. $|-83|$ $|58|$ **68.** $|22|$ $|-19|$ **69.** $|43|$ $|-52|$ **70.** $|-71|$ $|-92|$

71. $|-68|$ $|-42|$ **72.** $|12|$ $|-31|$ **73.** $|-45|$ $|-61|$ **74.** $|-28|$ $|43|$

1.1   **Applying Concepts**

Write the given numbers in order from least to greatest.

75. $|-5|, 6, -|-8|, -19$

76. $-4, |-15|, -|-7|, 0$

77. $-(-3), -22, |-25|, |-14|$

78. $-|-26|, -(-8), |-17|, -(5)$

A meteorologist may report a wind-chill temperature. This is the equivalent temperature, including the effects of wind and temperature, that a person would feel in calm air conditions. The table that follows gives the wind-chill temperature for various wind speeds and temperatures. For instance, when the temperature is 5°F and the wind is blowing at 15 mph, the wind-chill temperature is −25°F. Use this table for Exercises 79 and 80. (*Source: Information Please Almanac*)

Wind-Chill Factors																	
Wind Speed (in mph)	Thermometer Reading (in degrees Fahrenheit)																
	35	30	25	20	15	10	5	0	−5	−10	−15	−20	−25	−30	−35	−40	−45
5	33	27	21	19	12	7	0	−5	−10	−15	−21	−26	−31	−36	−42	−47	−52
10	22	16	10	3	−3	−9	−15	−22	−27	−34	−40	−46	−52	−58	−64	−71	−77
15	16	9	2	−5	−11	−18	−25	−31	−38	−45	−51	−58	−65	−72	−78	−85	−92
20	12	4	−3	−10	−17	−24	−31	−39	−46	−53	−60	−67	−74	−81	−88	−95	−103
25	8	1	−7	−15	−22	−29	−36	−44	−51	−59	−66	−74	−81	−88	−96	−103	−110
30	6	−2	−10	−18	−25	−33	−41	−49	−56	−64	−71	−79	−86	−93	−101	−109	−116
35	4	−4	−12	−20	−27	−35	−43	−52	−58	−67	−74	−82	−89	−97	−105	−113	−120
40	3	−5	−13	−21	−29	−37	−45	−53	−60	−69	−76	−84	−92	−100	−107	−115	−123
45	2	−6	−14	−22	−30	−38	−46	−54	−62	−70	−78	−85	−93	−102	−109	−117	−125

79. **Meteorology** Which of the following weather conditions feels colder?

a. a temperature of 5°F with a 20 mph wind or a temperature of 10°F with a 15 mph wind

b. a temperature of −25°F with a 10 mph wind or a temperature of −15°F with a 20 mph wind

80. **Meteorology** Which of the following weather conditions feels warmer?

a. a temperature of 25°F with a 25 mph wind or a temperature of 10°F with a 10 mph wind

b. a temperature of −5°F with a 10 mph wind or a temperature of −15°F with a 5 mph wind

Complete.

81. On the number line, the two points that are four units from 0 are _____ and _____ .

82. On the number line, the two points that are six units from 0 are _____ and _____ .

83. On the number line, the two points that are seven units from 4 are _____ and _____ .

84. On the number line, the two points that are five units from −3 are _____ and _____ .

85. If a is a positive number, then $-a$ is a _____ number.

86. If *a* is a negative number, then −*a* is a _____ number.

87. An integer that is its own opposite is _____ .

88. True or False. If $a \geq 0$, then $|a| = a$.

89. True or False. If $a \leq 0$, then $|a| = -a$.

90. Give examples of some games that use negative integers in the scoring.

91. Give some examples of English words that are used as variables.

92. 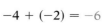 Write an essay describing your interpretation of the data presented in the table at the right.

The Twenty-First Century: An Easier Ladder to Climb Number of 30- to 34-year-olds competing for jobs			
	1992	*1999*	*2005*
Men	10.3 million	9.2 million	8.5 million
Women	8.3 million	7.9 million	7.8 million
Total	18.6 million	17.1 million	16.3 million

1.2 Operations with Integers

1 Add integers

A number can be represented anywhere along the number line by an arrow. A positive number is represented by an arrow pointing to the right, and a negative number is represented by an arrow pointing to the left. The size of the number is represented by the length of the arrow.

Addition is the process of finding the total of two numbers. The numbers being added are called **addends.** The total is called the **sum.** Addition of integers can be shown on the number line. To add integers, find the point on the number line corresponding to the first addend. At that point, draw an arrow representing the second addend. The sum is the number directly below the tip of the arrow.

$4 + 2 = 6$

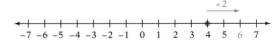

$-4 + (-2) = -6$

$-4 + 2 = -2$

$4 + (-2) = 2$

The pattern for addition shown on the preceding number lines is summarized in the following rules for adding integers.

Addition of Integers

Integers with the same sign
To add two numbers with the same sign, add the absolute values of the numbers. Then attach the sign of the addends.

$2 + 8 = 10$

$-2 + (-8) = -10$

Integers with different signs
To add two numbers with different signs, find the absolute value of each number. Then subtract the smaller of these absolute values from the larger one. Attach the sign of the number with the larger absolute value.

$-2 + 8 = 6$

$2 + (-8) = -6$

Example 1   Add.   **A.** $162 + (-247)$ **B.** $-14 + (-47)$

C. $-4 + (-6) + (-8) + 9$

Solution   **A.** $162 + (-247) = -85$

- The signs are different. Subtract the absolute values of the numbers $(247 - 162)$. Attach the sign of the number with the larger absolute value.

B. $-14 + (-47) = -61$

- The signs are the same. Add the absolute values of the numbers $(14 + 47)$. Attach the sign of the addends.

C. $-4 + (-6) + (-8) + 9$
$= -10 + (-8) + 9$
$= -18 + 9$
$= -9$

- To add more than two numbers, add the first two numbers. Then add the sum to the third number. Continue until all the numbers have been added.

Problem 1   Add.

A. $-162 + 98$ **B.** $-154 + (-37)$ **C.** $-36 + 17 + (-21)$

Solution   See page S1.

---

2   **Subtract integers**

Subtraction is the process of finding the difference between two numbers. Subtraction of an integer is defined as addition of the opposite integer.

Subtract $8 - 3$ by using addition of the opposite.

Subtraction → Addition of the Opposite

$$8 \;-\; (+3) \;=\; 8 \;+\; (-3) \;=\; 5$$

Opposites

To subtract one integer from another, add the opposite of the second integer to the first integer.

$$\boxed{\begin{array}{c}\text{first}\\\text{number}\end{array}} - \boxed{\begin{array}{c}\text{second}\\\text{number}\end{array}} = \boxed{\begin{array}{c}\text{first}\\\text{number}\end{array}} + \boxed{\begin{array}{c}\text{the opposite of}\\\text{the second number}\end{array}}$$

$$
\begin{array}{rcccccr}
40 & - & 60 & = & 40 & + & (-60) & = & -20\\
-40 & - & 60 & = & -40 & + & (-60) & = & -100\\
-40 & - & (-60) & = & -40 & + & 60 & = & 20\\
40 & - & (-60) & = & 40 & + & 60 & = & 100
\end{array}
$$

Example 2 Subtract: $-12 - 8$

Solution $-12 - 8 = -12 + (-8)$ • Rewrite subtraction as addition of the opposite.

$\qquad\qquad = -20$ • Add.

Problem 2 Subtract: $-8 - 14$

Solution   See page S1.

When subtraction occurs several times in a problem, rewrite each subtraction as addition of the opposite. Then add.

Example 3 Subtract: $-8 - 30 - (-12) - 7 - (-14)$

Solution $-8 - 30 - (-12) - 7 - (-14)$
$\quad = -8 + (-30) + 12 + (-7) + 14$ • Rewrite each subtraction as addition of the opposite.

$\quad = -38 + 12 + (-7) + 14$ • Add the first two numbers. Then add the sum to the third number. Continue until all the numbers have been added.
$\quad = -26 + (-7) + 14$
$\quad = -33 + 14$
$\quad = -19$

Problem 3 Subtract: $4 - (-3) - 12 - (-7) - 20$

Solution   See page S1.

3 ## Multiply integers

Point of Interest

The cross $\times$ *was first used as a symbol for multiplication in 1631 in a book titled* The Key to Mathematics. *In that same year, another book,* Practice of the Analytical Art, *advocated the use of a dot to indicate multiplication.*

Multiplication is the process of finding the product of two numbers.

Several different symbols are used to indicate multiplication. The numbers being multiplied are called **factors;** for instance, 3 and 2 are factors in each of the examples at the right. The result is called the **product.** Note that when parentheses are used and there is no arithmetic operation symbol, the operation is multiplication.

$$3 \times 2 = 6$$
$$3 \cdot 2 = 6$$
$$(3)(2) = 6$$
$$3(2) = 6$$
$$(3)2 = 6$$

When 5 is multiplied by a sequence of decreasing integers, each product decreases by 5.

$$(5)(3) = 15$$
$$(5)(2) = 10$$
$$(5)(1) = 5$$
$$(5)(0) = 0$$

The pattern developed can be continued so that 5 is multiplied by a sequence of negative numbers. The resulting products must be negative in order to maintain the pattern of decreasing by 5.

$$(5)(-1) = -5$$
$$(5)(-2) = -10$$
$$(5)(-3) = -15$$
$$(5)(-4) = -20$$

This illustrates that the product of a positive number and a negative number is negative.

When -5 is multiplied by a sequence of decreasing integers, each product increases by 5.

$$(-5)(3) = -15$$
$$(-5)(2) = -10$$
$$(-5)(1) = -5$$
$$(-5)(0) = 0$$

The pattern developed can be continued so that -5 is multiplied by a sequence of negative numbers. The resulting products must be positive in order to maintain the pattern of increasing by 5.

$$(-5)(-1) = 5$$
$$(-5)(-2) = 10$$
$$(-5)(-3) = 15$$
$$(-5)(-4) = 20$$

This illustrates that the product of two negative numbers is positive.

This pattern for multiplication is summarized in the following rules for multiplying integers.

Multiplication of Integers

Integers with the same sign
To multiply two numbers with the same sign, multiply the absolute values of the numbers. The product is positive.

$$4 \cdot 8 = 32$$
$$(-4)(-8) = 32$$

Integers with different signs
To multiply two numbers with different signs, multiply the absolute values of the numbers. The product is negative.

$$-4 \cdot 8 = -32$$
$$(4)(-8) = -32$$

Example 4   Multiply.   **A.** $-42 \cdot 62$ **B.** $2(-3)(-5)(-7)$

Solution   **A.** $-42 \cdot 62$
$\qquad = -2604$

• The signs are different. The product is negative.

B. $2(-3)(-5)(-7)$
$\qquad = -6(-5)(-7)$
$\qquad = 30(-7)$
$\qquad = -210$

• To multiply more than two numbers, multiply the first two numbers. Then multiply the product by the third number. Continue until all the numbers have been multiplied.

Problem 4   Multiply.   **A.** $-38 \cdot 51$ **B.** $-7(-8)(9)(-2)$

Solution   See page S1.

4 **Divide integers**

For every division problem there is a related multiplication problem.

$$\text{Division: } \frac{8}{2} = 4 \qquad \text{Related multiplication: } 4 \cdot 2 = 8$$

This fact can be used to illustrate the rules for dividing signed numbers.

The quotient of two numbers with the same sign is positive.	$\frac{12}{3} = 4$ because $4 \cdot 3 = 12$.
	$\frac{-12}{-3} = 4$ because $4(-3) = -12$.
The quotient of two numbers with different signs is negative.	$\frac{12}{-3} = -4$ because $-4(-3) = 12$.
	$\frac{-12}{3} = -4$ because $-4 \cdot 3 = -12$.

Division of Integers

Integers with the same sign
To divide two numbers with the same sign, divide the absolute values of the numbers. The quotient is positive.

Integers with different signs
To divide two numbers with different signs, divide the absolute values of the numbers. The quotient is negative.

Note that $\frac{-12}{3} = -4$, $\frac{12}{-3} = -4$, and $-\frac{12}{3} = -4$. This suggests the following rule.

If a and b are two integers, and $b \neq 0$, then $\dfrac{a}{-b} = \dfrac{-a}{b} = -\dfrac{a}{b}$.

Read $b \neq 0$ as "b is not equal to 0." The reason why the denominator must not be equal to 0 is explained in the following discussion of zero and one in division.

Zero and One in Division

Zero divided by any number other than zero is zero.	$\frac{0}{a} = 0, a \neq 0$	because $0 \cdot a = 0$.
Division by zero is not defined.	$\frac{4}{0} = ?$	$? \times 0 = 4$ There is no number whose product with zero is 4.
Any number other than zero divided by itself is 1.	$\frac{a}{a} = 1, a \neq 0$	because $1 \cdot a = a$.
Any number divided by 1 is the number.	$\frac{a}{1} = a$	because $a \cdot 1 = a$.

Example 5   Divide.   **A.** $(-120) \div (-8)$ **B.** $\dfrac{95}{-5}$ **C.** $-\dfrac{-81}{3}$

Solution   **A.** $(-120) \div (-8) = 15$

   B. $\dfrac{95}{-5} = -19$

   C. $-\dfrac{-81}{3} = -(-27) = 27$

Problem 5   Divide.   **A.** $(-135) \div (-9)$ **B.** $\dfrac{84}{-6}$ **C.** $-\dfrac{36}{-12}$

Solution   See page S1.

5   Application problems

In many courses, your course grade depends on the *average* of all your test scores. You compute the average by calculating the sum of all your test scores and then dividing that result by the number of tests. Statisticians call this average an **arithmetic mean.** Besides its application to finding the average of your test scores, the arithmetic mean is used in many other situations.

Stock market analysts calculate the **moving average** of a stock. This is the arithmetic mean of the changes in the value of a stock for a given number of days. To illustrate the procedure, we will calculate the 5-day moving average of a stock. In actual practice, a stock market analyst may use 15 days, 30 days, or some other number.

The table below shows the amount of increase or decrease, in cents, in the closing price of a stock for a 10-day period.

Day 1	Day 2	Day 3	Day 4	Day 5	Day 6	Day 7	Day 8	Day 9	Day 10
+50	−175	+225	0	−275	−75	−50	+50	−475	−50

To calculate the 5-day moving average of this stock, determine the average of the stock for days 1 through 5, days 2 through 6, days 3 through 7, and so on.

Days 1–5	Days 2–6	Days 3–7	Days 4–8	Days 5–9	Days 6–10
+50	−175	+225	0	−275	−75
−175	+225	0	−275	−75	−50
+225	0	−275	−75	−50	+50
0	−275	−75	−50	+50	−475
−275	−75	−50	+50	−475	−50
Sum = −175	Sum = −300	Sum = −175	Sum = −350	Sum = −825	Sum = −600
Av = $\dfrac{-175}{5} = -35$	Av = $\dfrac{-300}{5} = -60$	Av = $\dfrac{-175}{5} = -35$	Av = $\dfrac{-350}{5} = -70$	Av = $\dfrac{-825}{5} = -165$	Av = $\dfrac{-600}{5} = -120$

The 5-day moving average is the list of means: $-35, -60, -35, -70, -165,$ and -120. If the numbers in the list tend to increase, the price of the stock is showing an upward trend; if they decrease, the price of the stock is showing a downward trend. For the stock shown here, the price of the stock is showing a downward trend. These trends help analysts determine whether to recommend a stock.

Example 6 The daily low temperatures, in degrees Celsius, during one week were recorded as follows: $-8°, 2°, 0°, -7°, 1°, 6°, -1°$. Find the average daily low temperature for the week.

Strategy To find the average daily low temperature:

▶ Add the seven temperature readings.

▶ Divide the sum by 7.

Solution
$$-8 + 2 + 0 + (-7) + 1 + 6 + (-1)$$
$$= -6 + 0 + (-7) + 1 + 6 + (-1)$$
$$= -6 + (-7) + 1 + 6 + (-1)$$
$$= -13 + 1 + 6 + (-1)$$
$$= -12 + 6 + (-1)$$
$$= -6 + (-1)$$
$$= -7$$

$$-7 \div 7 = -1$$

The average daily low temperature was $-1°C$.

Problem 6 The daily high temperatures, in degrees Celsius, during one week were recorded as follows: $-5°, -6°, 3°, 0°, -4°, -7°, -2°$. Find the average daily high temperature for the week.

Solution See page S1.

1.2 Concept Review

Determine whether the statement is always true, sometimes true, or never true.

1. The sum of two integers is larger than either of the integers being added.

2. The sum of two nonzero integers with the same sign is positive.

3. The quotient of two integers with different signs is negative.

4. To find the opposite of a number, multiply the number by -1.

5. To find the opposite of an integer, divide the integer by -1.

6. To perform the subtraction $a - b$, add the opposite of a to b.

7. $4(-8) = -4$

8. If x and y are two integers and $y = 0$, then $\frac{x}{y} = 0$.

9. If x and y are two different integers, then $x - y = y - x$.

10. If x is an integer and $4x = 0$, then $x = 0$.

1.2 | Exercises

1 **1.** ✎ Explain how to add two integers with the same sign.

2. ✎ Explain how to add two integers with different signs.

Add.

3. $-3 + (-8)$ **4.** $-12 + (-1)$ **5.** $-4 + (-5)$ **6.** $-12 + (-12)$

7. $6 + (-9)$ **8.** $4 + (-9)$ **9.** $-6 + 7$ **10.** $-12 + 6$

11. $2 + (-3) + (-4)$ **12.** $7 + (-2) + (-8)$ **13.** $-3 + (-12) + (-15)$

14. $9 + (-6) + (-16)$ **15.** $-17 + (-3) + 29$ **16.** $13 + 62 + (-38)$

17. $-3 + (-8) + 12$ **18.** $-27 + (-42) + (-18)$

19. $13 + (-22) + 4 + (-5)$ **20.** $-14 + (-3) + 7 + (-6)$

21. $-22 + 10 + 2 + (-18)$ **22.** $-6 + (-8) + 13 + (-4)$

23. $-126 + (-247) + (-358) + 339$ **24.** $-651 + (-239) + 524 + 487$

2 **25.** ✎ Explain the meaning of the words *minus* and *negative*.

26. ✎ Explain how to rewrite $6 - (-9)$ as addition of the opposite.

Subtract.

27. $16 - 8$ **28.** $12 - 3$ **29.** $7 - 14$

30. $7 - (-2)$ **31.** $3 - (-4)$ **32.** $-6 - (-3)$

33. $-4 - (-2)$ **34.** $6 - (-12)$ **35.** $-12 - 16$

36. $-4 - 3 - 2$ **37.** $4 - 5 - 12$ **38.** $12 - (-7) - 8$

39. $-12 - (-3) - (-15)$ **40.** $4 - 12 - (-8)$ **41.** $13 - 7 - 15$

42. $-6 + 19 - (-31)$ **43.** $-30 - (-65) - 29 - 4$ **44.** $42 - (-82) - 65 - 7$

45. $-16 - 47 - 63 - 12$

46. $42 - (-30) - 65 - (-11)$

47. $-47 - (-67) - 13 - 15$

48. $-18 - 49 - (-84) - 27$

3 **49.** ✎ Explain how to multiply two integers with the same sign.

50. ✎ Explain how to multiply two integers with different signs.

51. ✎ Name the operation in each expression. Justify your answer.
a. $8(-7)$ **b.** $8 - 7$ **c.** $8 - (-7)$ **d.** $-xy$ **e.** $x(-y)$ **f.** $-x - y$

52. ✎ Name the operation in each expression. Justify your answer.
a. $(4)(-6)$ **b.** $4 - (6)$ **c.** $4 - (-6)$ **d.** $-ab$ **e.** $a(-b)$ **f.** $-a - b$

Multiply.

53. $14 \cdot 3$

54. $62 \cdot 9$

55. $5(-4)$

56. $4(-7)$

57. $-8(2)$

58. $-9(3)$

59. $(-5)(-5)$

60. $(-3)(-6)$

61. $(-7)(0)$

62. $-32 \cdot 4$

63. $-24 \cdot 3$

64. $19(-7)$

65. $6(-17)$

66. $-8(-26)$

67. $-4(-35)$

68. $-5(23)$

69. $5 \cdot 7(-2)$

70. $8(-6)(-1)$

71. $(-9)(-9)(2)$

72. $-8(-7)(-4)$

73. $-5(8)(-3)$

74. $(-6)(5)(7)$

75. $-1(4)(-9)$

76. $6(-3)(-2)$

77. $4(-4) \cdot 6(-2)$

78. $-5 \cdot 9(-7) \cdot 3$

79. $-9(4) \cdot 3(1)$

80. $8(8)(-5)(-4)$

81. $-6(-5)(12)(0)$

82. $7(9) \cdot 10(-1)$

4 Write the related multiplication problem.

83. $\dfrac{-36}{-12} = 3$

84. $\dfrac{28}{-7} = -4$

85. $\dfrac{-55}{11} = -5$

86. $\dfrac{-20}{-10} = 2$

Divide.

87. $12 \div (-6)$

88. $18 \div (-3)$

89. $(-72) \div (-9)$

90. $(-64) \div (-8)$

91. $0 \div (-6)$

92. $-49 \div 0$

93. $45 \div (-5)$

94. $-24 \div 4$

95. $-36 \div 4$

96. $-56 \div 7$

97. $-81 \div (-9)$

98. $-40 \div (-5)$

99. $72 \div (-3)$

100. $44 \div (-4)$

101. $-60 \div 5$

102. $144 \div 9$ **103.** $78 \div (-6)$ **104.** $84 \div (-7)$

105. $-72 \div 4$ **106.** $-80 \div 5$ **107.** $-114 \div (-6)$

108. $-128 \div 4$ **109.** $-130 \div (-5)$ **110.** $(-280) \div 8$

111. $-132 \div (-12)$ **112.** $-156 \div (-13)$ **113.** $-182 \div 14$

114. $-144 \div 12$ **115.** $143 \div 11$ **116.** $168 \div 14$

5 **117.** Temperature Find the temperature after a rise of 9°C from −6°C.

118. Temperature Find the temperature after a rise of 7°C from −18°C.

119. Temperature The high temperature for the day was 10°C. The low temperature was −4°C. Find the difference between the high and low temperatures for the day.

120. Temperature The low temperature for the day was −2°C. The high temperature was 11°C. Find the difference between the high and low temperatures for the day.

121. Chemistry The temperature at which mercury boils is 360°C. Mercury freezes at −39°C. Find the difference between the temperature at which mercury boils and the temperature at which it freezes.

122. Chemistry The temperature at which radon boils is −62°C. Radon freezes at −71°C. Find the difference between the temperature at which radon boils and the temperature at which it freezes.

The elevation, or height, of places on Earth is measured in relation to sea level, or the average level of the ocean's surface. The following table shows height above sea level as a positive number and depth below sea level as a negative number. (*Source: Information Please Almanac*)

Continent	Highest Elevation (in meters)		Lowest Elevation (in meters)	
Africa	Mt. Kilimanjaro	5895	Qattara Depression	−133
Asia	Mt. Everest	8848	Dead Sea	−400
Europe	Mt. Elbrus	5634	Caspian Sea	−28
North America	Mt. McKinley	6194	Death Valley	−86
South America	Mt. Aconcagua	6960	Salinas Grandes	−40

123. Geography Use the table to find the difference in elevation between Mt. Elbrus and the Caspian Sea.

124. Geography Use the table to find the difference in elevation between Mt. Aconcagua and Salinas Grandes.

125. Geography Use the table to find the difference in elevation between Mt. Kilimanjaro and the Qattara Depression.

126. Geography Use the table to find the difference in elevation between Mt. McKinley and Death Valley.

127. Geography Use the table to find the difference in elevation between Mt. Everest and the Dead Sea.

128. Temperature The daily low temperatures, in degrees Celsius, during one week were recorded as follows: 4°, −5°, 8°, 0°, −9°, −11°, −8°. Find the average daily low temperature for the week.

129. Temperature The daily high temperatures, in degrees Celsius, during one week were recorded as follows: −8°, −9°, 6°, 7°, −2°, −14°, −1°. Find the average daily high temperature for the week.

130. Temperature On January 22, 1943, the temperature at Spearfish, South Dakota, rose from −4°F to 45°F in two minutes. How many degrees did the temperature rise during those two minutes?

131. Temperature In a 24-hour period in January of 1916, the temperature in Browning, Montana, dropped from 44°F to −56°F. How many degrees did the temperature drop during that time?

The table at the right shows the average temperatures at different cruising altitudes for airplanes. Use the table for Exercises 132 and 133.

Cruising Altitude	Average Temperature
12,000 ft	16°F
20,000 ft	−12°F
30,000 ft	−48°F
40,000 ft	−70°F
50,000 ft	−70°F

132. Aviation What is the difference between the average temperatures at 12,000 ft and at 40,000 ft?

133. Aviation How much colder is the average temperature at 30,000 ft than at 20,000 ft?

134. Education To discourage random guessing on a multiple-choice exam, a professor assigns 5 points for a correct answer, −2 points for an incorrect answer, and 0 points for leaving the question blank. What is the score for a student who had 20 correct answers, had 13 incorrect answers, and left 7 questions blank?

135. Education To discourage random guessing on a multiple-choice exam, a professor assigns 7 points for a correct answer, −3 points for an incorrect answer, and −1 point for leaving the question blank. What is the score for a student who had 17 correct answers, had 8 incorrect answers, and left 2 questions blank?

136. The Stock Market The value of a share of Wal-Mart stock on August 10, 2001, was $53.60. The table below shows the amount of increase or decrease, to the nearest 10 cents, from the August 10 closing price of the stock for a 10-day period. Calculate the 5-day moving average for this stock.

Day 1	Day 2	Day 3	Day 4	Day 5	Day 6	Day 7	Day 8	Day 9	Day 10
−140	+20	−40	−30	−90	+80	−170	+30	−50	−120

137.   **The Stock Market** The value of a share of Disney stock on August 10, 2001, was \$27.40. The table below shows the amount of increase or decrease, to the nearest 10 cents, from the August 10 closing price of the stock for a 10-day period. Calculate the 5-day moving average for this stock.

Day 1	Day 2	Day 3	Day 4	Day 5	Day 6	Day 7	Day 8	Day 9	Day 10
−20	−20	−50	−20	−10	+30	−10	+30	0	+50

1.2   Applying Concepts

Simplify.

138. $|-7 + 12|$

139. $|13 - (-4)|$

140. $|-13 - (-2)|$

141. $|18 - 21|$

Assuming the pattern is continued, find the next three numbers in the pattern.

142. $-7, -11, -15, -19, \ldots$

143. $16, 11, 6, 1, \ldots$

144. $7, -14, 28, -56, \ldots$

145. $1024, -256, 64, \ldots$

Solve.

146. 32,844 is divisible by 3. By rearranging the digits, find the largest possible number that is still divisible by 3.

147. 4563 is not divisible by 4. By rearranging the digits, find the largest possible number that is divisible by 4.

148. How many three-digit numbers of the form 8__4 are divisible by 3?

In each exercise, determine which statement is false.

149. a. $|3 + 4| = |3| + |4|$ **b.** $|3 - 4| = |3| - |4|$ **c.** $|4 + 3| = |4| + |3|$ **d.** $|4 - 3| = |4| - |3|$

150. a. $|5 + 2| = |5| + |2|$ **b.** $|5 - 2| = |5| - |2|$ **c.** $|2 + 5| = |2| + |5|$ **d.** $|2 - 5| = |2| - |5|$

Determine which statement is true for all real numbers.

151. a. $|x + y| \leq |x| + |y|$ **b.** $|x + y| = |x| + |y|$ **c.** $|x + y| \geq |x| + |y|$

152. a. $||x| - |y|| \leq |x| - |y|$ **b.** $||x| - |y|| = |x| - |y|$ **c.** $||x| - |y|| \geq |x| - |y|$

153. Is the difference between two integers always smaller than either one of the numbers in the difference? If not, give an example for which the difference between two integers is greater than either integer.

154. If $-4x$ equals a positive integer, is x a positive or a negative integer? Explain your answer.

155. Explain why division by zero is not allowed.

1.3 Rational Numbers

1 **Write rational numbers as decimals**

A **rational number** is the quotient of two integers. Therefore, a rational number is a number that can be written in the form $\frac{a}{b}$, where a and b are integers, and b is not zero. A rational number written in this way is commonly called a **fraction.**

$\dfrac{a}{b}$ ← an integer
← a nonzero integer

$\left.\dfrac{2}{3}, \dfrac{-4}{9}, \dfrac{18}{-5}, \dfrac{4}{1}\right\}$ Rational numbers

Because an integer can be written as the quotient of the integer and 1, every integer is a rational number.

$5 = \dfrac{5}{1} \qquad -3 = \dfrac{-3}{1}$

A number written in **decimal notation** is also a rational number.

three-tenths $0.3 = \dfrac{3}{10}$

thirty-five hundredths $0.35 = \dfrac{35}{100}$

negative four-tenths $-0.4 = \dfrac{-4}{10}$

A rational number written as a fraction can be written in decimal notation.

Example 1 Write $\dfrac{5}{8}$ as a decimal.

Solution

$$
\begin{array}{r}
0.625 \\
8\overline{)5.000} \\
-4\ 8 \\
\hline
20 \\
-16 \\
\hline
40 \\
-40 \\
\hline
0
\end{array}
$$

← This is called a **terminating decimal.**

← The remainder is zero.

$\dfrac{5}{8} = 0.625$

Problem 1 Write $\dfrac{4}{25}$ as a decimal.

Solution   See page S1.

Example 2   Write $\frac{4}{11}$ as a decimal.

Solution
$$
\begin{array}{r}
0.3636\ldots \quad \leftarrow \text{This is called a \textbf{repeating decimal.}} \\
11\overline{)4.0000} \\
-3\,3 \quad\quad \\
\hline
70 \quad\quad \\
-66 \quad\quad \\
\hline
40 \quad\quad \\
-33 \quad\quad \\
\hline
70 \quad\quad \\
-66 \quad\quad \\
\hline
4 \quad\quad
\end{array}
$$
← The remainder is never zero.

$\frac{4}{11} = 0.\overline{36}$ ← The bar over the digits 3 and 6 is used to show that these digits repeat.

Problem 2   Write $\frac{4}{9}$ as a decimal. Place a bar over the repeating digits of the decimal.

Solution   See page S1.

Rational numbers can be written as fractions, such as $-\frac{6}{7}$ or $\frac{8}{3}$, where the numerator and denominator are integers. But every rational number also can be written as a repeating decimal (such as 0.25767676...) or a terminating decimal (such as 1.73). This was illustrated in Examples 1 and 2.

Numbers that cannot be written either as a repeating decimal or as a terminating decimal are called **irrational numbers.** For example, 2.45445444544445... is an irrational number. Two other examples are $\sqrt{2}$ and π.

$$\sqrt{2} = 1.414213562\ldots \quad \pi = 3.141592654\ldots$$

The three dots mean that the digits continue on and on without ever repeating or terminating. Although we cannot write a decimal that is exactly equal to $\sqrt{2}$ or to π, we can give approximations of these numbers. The symbol $\approx$ is read "is approximately equal to." Shown below are $\sqrt{2}$ rounded to the nearest thousandth and π rounded to the nearest hundredth.

$$\sqrt{2} \approx 1.414 \quad \pi \approx 3.14$$

The rational numbers and the irrational numbers taken together are called the **real numbers.**

2   **Add and subtract rational numbers**

To add or subtract fractions, first rewrite each fraction as an equivalent fraction with a common denominator, using the least common multiple (LCM) of the denominators as the common denominator. Then add the numerators, and place the sum over the common denominator. Write the answer in simplest form.

Take Note

No matter how far we carry out the division, the remainder is never zero. The decimal $0.\overline{36}$ is a repeating decimal.

The sign rules for adding integers apply to addition of rational numbers.

Example 3 Add: $-\dfrac{5}{6} + \dfrac{3}{10}$

Solution Prime factorization of 6 and 10:

$$6 = 2 \cdot 3 \qquad 10 = 2 \cdot 5$$

$$\text{LCM} = 2 \cdot 3 \cdot 5 = 30$$

- Find the LCM of the denominators 6 and 10. The LCM of denominators is sometimes called the **least common denominator** (LCD).

$$-\frac{5}{6} + \frac{3}{10} = -\frac{25}{30} + \frac{9}{30}$$

- Rewrite the fractions as equivalent fractions, using the LCM of the denominators as the common denominator.

$$= \frac{-25 + 9}{30}$$

$$= \frac{-16}{30}$$

- Add the numerators, and place the sum over the common denominator.

$$= -\frac{8}{15}$$

- Write the answer in simplest form.

Problem 3 Subtract: $\dfrac{5}{9} - \dfrac{11}{12}$

Solution See page S1.

Example 4 Simplify: $-\dfrac{3}{4} + \dfrac{1}{6} - \dfrac{5}{8}$

Solution $-\dfrac{3}{4} + \dfrac{1}{6} - \dfrac{5}{8} = -\dfrac{18}{24} + \dfrac{4}{24} - \dfrac{15}{24}$

- The LCM of 4, 6, and 8 is 24.

$$= \frac{-18}{24} + \frac{4}{24} + \frac{-15}{24}$$

$$= \frac{-18 + 4 + (-15)}{24}$$

$$= \frac{-29}{24}$$

$$= -\frac{29}{24}$$

Problem 4 Simplify: $-\dfrac{7}{8} - \dfrac{5}{6} + \dfrac{1}{2}$

Solution See page S2.

To add or subtract decimals, write the numbers so that the decimal points are in a vertical line. Then proceed as in the addition or subtraction of integers. Write the decimal point in the answer directly below the decimal points in the problem.

Example 5   Add: $14.02 + 137.6 + 9.852$

Solution

$$\begin{array}{r} 14.02 \\ 137.6 \\ +\ \ 9.852 \\ \hline 161.472 \end{array}$$

- Write the decimals so that the decimal points are in a vertical line.

- Write the decimal point in the sum directly below the decimal points in the problem.

Problem 5   Add: $3.097 + 4.9 + 3.09$

Solution   See page S2.

Example 6   Add: $-114.039 + 84.76$

Solution

$$\begin{array}{r} 114.039 \\ -\ \ 84.76 \\ \hline 29.279 \end{array}$$

$$-114.039 + 84.76$$
$$= -29.279$$

- The signs are different. Subtract the absolute value of the number with the smaller absolute value from the absolute value of the number with the larger absolute value.

- Attach the sign of the number with the larger absolute value.

Problem 6   Subtract: $16.127 - 67.91$

Solution   See page S2.

3   **Multiply and divide rational numbers**

The sign rules for multiplying and dividing integers apply to multiplication and division of rational numbers. The product of two fractions is the product of the numerators divided by the product of the denominators.

Example 7   Multiply: $\dfrac{3}{8} \cdot \dfrac{12}{17}$

Solution

$$\dfrac{3}{8} \cdot \dfrac{12}{17} = \dfrac{3 \cdot 12}{8 \cdot 17}$$

- Multiply the numerators. Multiply the denominators.

$$= \dfrac{3 \cdot \overset{1}{\cancel{2}} \cdot \overset{1}{\cancel{2}} \cdot 3}{2 \cdot \underset{1}{\cancel{2}} \cdot \underset{1}{\cancel{2}} \cdot 17}$$

- Write the prime factorization of each factor. Divide by the common factors.

$$= \dfrac{9}{34}$$

- Multiply the numbers remaining in the numerator. Multiply the numbers remaining in the denominator.

Problem 7   Multiply: $-\dfrac{7}{12} \cdot \dfrac{9}{14}$

Solution   See page S2.

The **reciprocal** of a fraction is the fraction with the numerator and denominator interchanged. For instance, the reciprocal of $\dfrac{2}{3}$ is $\dfrac{3}{2}$, and the reciprocal of $-\dfrac{5}{4}$ is $-\dfrac{4}{5}$. To divide fractions, multiply the dividend by the reciprocal of the divisor.

Example 8   Divide: $\dfrac{3}{10} \div \left(-\dfrac{18}{25}\right)$

Solution $\dfrac{3}{10} \div \left(-\dfrac{18}{25}\right) = -\left(\dfrac{3}{10} \div \dfrac{18}{25}\right)$ • The signs are different. The quotient is negative.

$= -\left(\dfrac{3}{10} \cdot \dfrac{25}{18}\right)$ • Change division to multiplication. Invert the divisor.

$= -\left(\dfrac{3 \cdot 25}{10 \cdot 18}\right)$ • Multiply the numerators. Multiply the denominators.

$= -\left(\dfrac{\overset{1}{\cancel{3}} \cdot \overset{1}{\cancel{5}} \cdot 5}{2 \cdot \underset{1}{\cancel{5}} \cdot 2 \cdot \underset{1}{\cancel{3}} \cdot 3}\right)$

$= -\dfrac{5}{12}$

Problem 8   Divide: $-\dfrac{3}{8} \div \left(-\dfrac{5}{12}\right)$

Solution   See page S2.

To multiply decimals, multiply as in the multiplication of whole numbers. Write the decimal point in the product so that the number of decimal places in the product equals the sum of the decimal places in the factors.

Example 9   Multiply: $(-6.89)(0.00035)$

Solution

$$
\begin{array}{rl}
6.89 & \text{2 decimal places} \\
\times\,0.00035 & \text{5 decimal places} \\
\hline
3445 & \\
2067 & \\
\hline
0.0024115 & \text{7 decimal places}
\end{array}
$$

• Multiply the absolute values.

$(-6.89)(0.00035) = -0.0024115$ • The signs are different. The product is negative.

Problem 9   Multiply: $(-5.44)(3.8)$

Solution   See page S2.

To divide decimals, move the decimal point in the divisor to make it a whole number. Move the decimal point in the dividend the same number of places to the right. Place the decimal point in the quotient directly over the decimal point in the dividend. Then divide as in the division of whole numbers.

Example 10   Divide: $1.32 \div 0.27$. Round to the nearest tenth.

Solution $0.27\overline{)1.32.}$ • Move the decimal point two places to the right in the divisor and in the dividend. Place the decimal point in the quotient.

$$\begin{array}{r} 4.88 \approx 4.9 \\ 27.\overline{)132.00} \\ -108 \\ \hline 24\ 0 \\ -21\ 6 \\ \hline 2\ 40 \\ -2\ 16 \\ \hline 24 \end{array}$$

• The symbol $\approx$ is used to indicate that the quotient is an approximate value that has been rounded off.

Problem 10   Divide $-0.394 \div 1.7$. Round to the nearest hundredth.

Solution   See page S2.

4   ## Convert among percents, fractions, and decimals

VIDEO & DVD   CD TUTOR   WWW WEB   SSM

"A population growth rate of 3%," "a manufacturer's discount of 25%," and "an 8% increase in pay" are typical examples of the many ways in which percent is used in applied problems. **Percent** means "parts of 100." Thus 27% means 27 parts of 100.

In applied problems involving a percent, it is usually necessary either to rewrite the percent as a fraction or a decimal, or to rewrite a fraction or a decimal as a percent.

To write 27% as a fraction, remove the percent sign and multiply by $\frac{1}{100}$.

$$27\% = 27\left(\frac{1}{100}\right) = \frac{27}{100}$$

To write a percent as a decimal, remove the percent sign and multiply by 0.01.

To write 33% as a decimal, remove the percent sign and multiply by 0.01.

$$33\% = 33(0.01) = 0.33$$

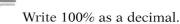

Move the decimal point two places to the left. Then remove the percent sign.

Write 100% as a decimal.

$$100\% = 100(0.01) = 1$$

Example 11   Write 130% as a fraction and as a decimal.

Solution $$130\% = 130\left(\frac{1}{100}\right) = \frac{130}{100} = 1\frac{3}{10}$$

• To write a percent as a fraction, remove the percent sign and multiply by $\frac{1}{100}$.

$$130\% = 130(0.01) = 1.30$$

• To write a percent as a decimal, remove the percent sign and multiply by 0.01.

Problem 11 Write 125% as a fraction and as a decimal.

Solution See page S2.

Example 12 Write $33\frac{1}{3}\%$ as a fraction.

Solution $33\frac{1}{3}\% = 33\frac{1}{3}\left(\frac{1}{100}\right) = \frac{100}{3}\left(\frac{1}{100}\right)$ • Write the mixed number

$$= \frac{1}{3}$$ $33\frac{1}{3}$ as the improper

fraction $\frac{100}{3}$.

Problem 12 Write $16\frac{2}{3}\%$ as a fraction.

Solution See page S2.

Example 13 Write 0.25% as a decimal.

Solution $0.25\% = 0.25(0.01) = 0.0025$ • Remove the percent sign and
multiply by 0.01.

Problem 13 Write 6.08% as a decimal.

Solution See page S2.

A fraction or decimal can be written as a percent by multiplying by 100%. Recall that 100% = 1, and multiplying a number by 1 does not change the number.

To write $\frac{5}{8}$ as a percent, mul- $\frac{5}{8} = \frac{5}{8}(100\%) = \frac{500}{8}\% = 62.5\%$ or $62\frac{1}{2}\%$

tiply by 100%.

To write 0.82 as a percent, $0.82 = 0.82(100\%) = 82\%$
multiply by 100%.

Move the decimal point two places to the right. Then write the percent sign.

Example 14 Write as a percent.   **A.** 0.027   **B.** 1.34

Solution **A.** $0.027 = 0.027(100\%) = 2.7\%$ • To write a decimal as a
percent, multiply by 100%.

B. $1.34 = 1.34(100\%) = 134\%$

Problem 14 Write as a percent.   **A.** 0.043   **B.** 2.57

Solution See page S2.

Example 15 Write $\frac{5}{6}$ as a percent. Round to the nearest tenth of a percent.

Solution $\frac{5}{6} = \frac{5}{6}(100\%) = \frac{500}{6}\% \approx 83.3\%$ • To write a fraction as a
percent, multiply by 100%.

Problem 15 Write $\frac{5}{9}$ as a percent. Round to the nearest tenth of a percent.

Solution See page S2.

Example 16   Write $\frac{7}{16}$ as a percent. Write the remainder in fractional form.

Solution $\dfrac{7}{16} = \dfrac{7}{16}(100\%) = \dfrac{700}{16}\% = 43\dfrac{3}{4}\%$ • **Multiply the fraction by 100%.**

Problem 16   Write $\frac{9}{16}$ as a percent. Write the remainder in fractional form.

Solution   See page S2.

1.3   Concept Review

Determine whether the statement is always true, sometimes true, or never true.

1. To multiply two fractions, you must first rewrite the fractions as equivalent fractions with a common denominator.

2. The number $\frac{\pi}{3}$ is an example of a rational number.

3. A rational number can be written as a terminating decimal.

4. The number $2.585585558\ldots$ is an example of a repeating decimal.

5. An irrational number is a real number.

6. 37%, 0.37, and $\frac{37}{100}$ are three numbers that have the same value.

7. To write a decimal as a percent, multiply the decimal by $\frac{1}{100}$.

8. If a, b, c, and d are natural numbers, then $\frac{a}{b} + \frac{c}{d} = \frac{a+c}{b+d}$.

9. -12 is an example of a number that is both an integer and a rational number.

10. $\frac{7}{9}$ is an example of a number that is both a rational number and an irrational number.

1.3   Exercises

1   Write as a decimal. Place a bar over the repeating digits of a repeating decimal.

1. $\dfrac{1}{3}$ **2.** $\dfrac{2}{3}$ **3.** $\dfrac{1}{4}$ **4.** $\dfrac{3}{4}$ **5.** $\dfrac{2}{5}$

6. $\dfrac{4}{5}$ **7.** $\dfrac{1}{6}$ **8.** $\dfrac{5}{6}$ **9.** $\dfrac{1}{8}$ **10.** $\dfrac{7}{8}$

11. $\dfrac{2}{9}$ **12.** $\dfrac{8}{9}$ **13.** $\dfrac{5}{11}$ **14.** $\dfrac{10}{11}$ **15.** $\dfrac{7}{12}$

16. $\dfrac{11}{12}$ **17.** $\dfrac{4}{15}$ **18.** $\dfrac{8}{15}$ **19.** $\dfrac{7}{16}$ **20.** $\dfrac{15}{16}$

21. $\dfrac{6}{25}$ **22.** $\dfrac{14}{25}$ **23.** $\dfrac{9}{40}$ **24.** $\dfrac{21}{40}$ **25.** $\dfrac{15}{22}$

26. $\dfrac{19}{22}$ **27.** $\dfrac{11}{24}$ **28.** $\dfrac{19}{24}$ **29.** $\dfrac{5}{33}$ **30.** $\dfrac{25}{33}$

31. $\dfrac{3}{37}$ **32.** $\dfrac{14}{37}$

2 Simplify.

33. $\dfrac{2}{3} + \dfrac{5}{12}$ **34.** $\dfrac{1}{2} + \dfrac{3}{8}$ **35.** $\dfrac{5}{8} - \dfrac{5}{6}$ **36.** $\dfrac{1}{9} - \dfrac{5}{27}$

37. $-\dfrac{5}{12} - \dfrac{3}{8}$ **38.** $-\dfrac{5}{6} - \dfrac{5}{9}$ **39.** $-\dfrac{6}{13} + \dfrac{17}{26}$ **40.** $-\dfrac{7}{12} + \dfrac{5}{8}$

41. $-\dfrac{5}{8} - \left(-\dfrac{11}{12}\right)$ **42.** $\dfrac{1}{3} + \dfrac{5}{6} - \dfrac{2}{9}$ **43.** $\dfrac{1}{2} - \dfrac{2}{3} + \dfrac{1}{6}$ **44.** $-\dfrac{3}{8} - \dfrac{5}{12} - \dfrac{3}{16}$

45. $-\dfrac{5}{16} + \dfrac{3}{4} - \dfrac{7}{8}$ **46.** $\dfrac{1}{2} - \dfrac{3}{8} - \left(-\dfrac{1}{4}\right)$ **47.** $\dfrac{3}{4} - \left(-\dfrac{7}{12}\right) - \dfrac{7}{8}$ **48.** $\dfrac{1}{3} - \dfrac{1}{4} - \dfrac{1}{5}$

49. $\dfrac{2}{3} - \dfrac{1}{2} + \dfrac{5}{6}$ **50.** $\dfrac{5}{16} + \dfrac{1}{8} - \dfrac{1}{2}$ **51.** $\dfrac{5}{8} - \left(-\dfrac{5}{12}\right) + \dfrac{1}{3}$ **52.** $\dfrac{1}{8} - \dfrac{11}{12} + \dfrac{1}{2}$

53. $-\dfrac{7}{9} + \dfrac{14}{15} + \dfrac{8}{21}$ **54.** $1.09 + 6.2$ **55.** $-32.1 - 6.7$ **56.** $5.13 - 8.179$

57. $-13.092 + 6.9$ **58.** $2.54 - 3.6$ **59.** $5.43 + 7.925$ **60.** $-16.92 - 6.925$

61. $-3.87 + 8.546$ **62.** $6.9027 - 17.692$

63. $2.09 - 6.72 - 5.4$ **64.** $16.4 + 3.09 - 7.93$

65. $-18.39 + 4.9 - 23.7$ **66.** $19 - (-3.72) - 82.75$

67. $-3.07 - (-2.97) - 17.4$ **68.** $-3.09 - 4.6 - 27.3$

69. $317.09 - 46.902 + 583.0714$ **70.** $71.0235 - 86.0974 + 254.309$

3 Simplify.

71. $\dfrac{1}{2}\left(-\dfrac{3}{4}\right)$ **72.** $-\dfrac{2}{9}\left(-\dfrac{3}{14}\right)$ **73.** $\left(-\dfrac{3}{8}\right)\left(-\dfrac{4}{15}\right)$

74. $\frac{5}{8}\left(-\frac{7}{12}\right)\frac{16}{25}$

75. $\left(\frac{1}{2}\right)\left(-\frac{3}{4}\right)\left(-\frac{5}{8}\right)$

76. $\left(\frac{5}{12}\right)\left(-\frac{8}{15}\right)\left(-\frac{1}{3}\right)$

77. $\frac{3}{8} \div \frac{1}{4}$

78. $\frac{5}{6} \div \left(-\frac{3}{4}\right)$

79. $-\frac{5}{12} \div \frac{15}{32}$

80. $\frac{1}{8} \div \left(-\frac{5}{12}\right)$

81. $-\frac{4}{9} \div \left(-\frac{2}{3}\right)$

82. $-\frac{6}{11} \div \frac{4}{9}$

83. $(1.2)(3.47)$

84. $(-0.8)(6.2)$

85. $(-1.89)(-2.3)$

86. $(6.9)(-4.2)$

87. $(1.06)(-3.8)$

88. $(-2.7)(-3.5)$

89. $(1.2)(-0.5)(3.7)$

90. $(-2.4)(6.1)(0.9)$

91. $(-0.8)(3.006)(-5.1)$

Simplify. Round to the nearest hundredth.

92. $-24.7 \div 0.09$

93. $-1.27 \div (-1.7)$

94. $9.07 \div (-3.5)$

95. $-354.2086 \div 0.1719$

96. $-2658.3109 \div (-0.0473)$

97. $(-3.92) \div (-45.008)$

4 **98.**   **a.** Explain how to convert a fraction to a percent.
    b. Explain how to convert a percent to a fraction.

99.   **a.** Explain how to convert a decimal to a percent.
    b. Explain how to convert a percent to a decimal.

100.   Explain why multiplying a number by 100% does not change the
    value of the number.

Write as a fraction and as a decimal.

101. 75%

102. 40%

103. 50%

104. 10%

105. 64%

106. 88%

107. 175%

108. 160%

109. 19%

110. 87%

111. 5%

112. 2%

113. 450%

114. 380%

115. 8%

116. 4%

Write as a fraction.

117. $11\frac{1}{9}\%$

118. $37\frac{1}{2}\%$

119. $31\frac{1}{4}\%$

120. $66\frac{2}{3}\%$

121. $\frac{1}{2}\%$

122. $5\frac{3}{4}\%$

123. $6\frac{1}{4}\%$

124. $83\frac{1}{3}\%$

Write as a decimal.

125. 7.3%

126. 9.1%

127. 15.8%

128. 0.3%

129. 9.15%

130. 121.2%

131. 18.23%

132. 0.15%

Write as a percent.

133. 0.15

134. 0.37

135. 0.05

136. 0.02

137. 0.175

138. 0.125

139. 1.15

140. 2.142

141. 0.008

142. 0.004

143. 0.065

144. 0.083

Write as a percent. Round to the nearest tenth of a percent.

145. $\dfrac{27}{50}$

146. $\dfrac{83}{100}$

147. $\dfrac{1}{3}$

148. $\dfrac{3}{8}$

149. $\dfrac{4}{9}$

150. $\dfrac{9}{20}$

151. $2\dfrac{1}{2}$

152. $1\dfrac{2}{7}$

Write as a percent. Write the remainder in fractional form.

153. $\dfrac{3}{8}$

154. $\dfrac{7}{16}$

155. $\dfrac{5}{14}$

156. $\dfrac{4}{7}$

157. $1\dfrac{1}{4}$

158. $2\dfrac{5}{8}$

159. $1\dfrac{5}{9}$

160. $1\dfrac{13}{16}$

Employment The graph at the right shows the responses to a survey that asked respondents, "How did you find your most recent job?" Use the graph for Exercises 161 to 163.

161. What fraction of the respondents found their most recent jobs on the Internet?

162. What fraction of the respondents found their most recent jobs through a referral?

163. Did more or fewer than one-quarter of the respondents find their most recent jobs through a newspaper ad?

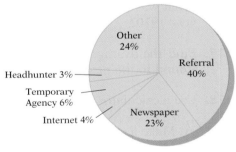

Other 24%

Referral 40%

Headhunter 3%

Temporary Agency 6%

Internet 4%

Newspaper 23%

How Did You Find Your Most Recent Job?

1.3 Applying Concepts

Classify each of the following numbers as a natural number, an integer, a positive integer, a negative integer, a rational number, an irrational number, or a real number. List all that apply.

164. −1

165. 28

166. $-\dfrac{9}{34}$

167. −7.707

168. $5.2\overline{6}$

169. 0.171771777...

Solve.

170. Find the average of $\frac{5}{8}$ and $\frac{3}{4}$.

171. 🔘 **The Postal Service** Postage for first-class mail is $.37 for the first ounce or fraction of an ounce and $.23 for each additional ounce or fraction of an ounce. Find the cost of mailing, by first-class mail, a letter that weighs $4\frac{1}{2}$ oz.

🔘 **Government** The table at the right shows the surplus or deficit, in billions of dollars, for the federal budget for selected years from 1945 through 2000. A negative sign (−) indicates a deficit. Use this table for Exercises 172 to 177. (*Source:* U.S. Office of Management and Budget)

Year	Federal Budget Surplus or Deficit (in billions of dollars)
1945	−47.553
1950	−3.119
1955	−2.993
1960	0.301
1965	−1.411
1970	−2.842
1975	−53.242
1980	−73.835
1985	−212.334
1990	−221.194
1995	−163.899
1996	−107.450
1997	−21.940
1998	69.246
1999	79.263
2000	117.305

172. In which year listed was the deficit lowest?

173. Find the difference between the deficits in the years 1980 and 1985.

174. Calculate the difference between the surplus in 1960 and the deficit in 1955.

175. How many times greater was the deficit in 1985 than in 1975? Round to the nearest whole number.

176. What was the average deficit per quarter, in millions of dollars, for the year 1970?

177. Find the average surplus or deficit for the years 1995 through 2000. Round to the nearest million.

178. Use a calculator to determine the decimal representations of $\frac{17}{99}, \frac{45}{99}$, and $\frac{73}{99}$. Make a conjecture as to the decimal representation of $\frac{83}{99}$. Does your conjecture work for $\frac{33}{99}$? What about $\frac{1}{99}$?

179. If the same positive integer is added to both the numerator and the denominator of $\frac{2}{5}$, is the new fraction less than, equal to, or greater than $\frac{2}{5}$?

180. A magic square is one in which the numbers in every row, column, and diagonal sum to the same number. Complete the magic square at the right.

$\frac{2}{3}$		
	$\frac{1}{6}$	$\frac{5}{6}$
		$-\frac{1}{3}$

181. Let x represent the price of a car. If the sales tax is 6% of the price, express the total of the price of the car and the sales tax in terms of x.

182. Let x represent the price of a suit. If the suit is on sale at a discount rate of 30%, express the price of the suit after the discount in terms of x.

183. The price of a pen was 60¢. During a storewide sale, the price was reduced to a different whole number of cents, and the entire stock was sold for $54.59. Find the price of the pen during the sale.

184. Find three natural numbers a, b, and c such that $\frac{1}{a} + \frac{1}{b} + \frac{1}{c}$ is a natural number.

185. If a and b are rational numbers and $a < b$, is it always possible to find a rational number c such that $a < c < b$? If not, explain why. If so, show how to find one.

186. In your own words, define **a.** a rational number, **b.** an irrational number, and **c.** a real number.

187. Explain why you "invert and multiply" when dividing a fraction by a fraction.

188. Explain why you need a common denominator when adding two fractions and why you don't need a common denominator when multiplying two fractions.

| 1.4 | # Exponents and the Order of Operations Agreement |

1 Exponential expressions

Repeated multiplication of the same factor can be written using an exponent.

$$2 \cdot 2 \cdot 2 \cdot 2 \cdot 2 = 2^5 \leftarrow \text{exponent} \qquad a \cdot a \cdot a \cdot a = a^4 \leftarrow \text{exponent}$$

base   base

The **exponent** indicates how many times the factor, called the **base,** occurs in the multiplication. The multiplication $2 \cdot 2 \cdot 2 \cdot 2 \cdot 2$ is in **factored form.** The exponential expression 2^5 is in **exponential form.**

2^1 is read "the first power of two" or just "two." ⟶ Usually the exponent 1 is not written.

2^2 is read "the second power of two" or "two squared."

2^3 is read "the third power of two" or "two cubed."

2^4 is read "the fourth power of two."

2^5 is read "the fifth power of two."

a^5 is read "the fifth power of a."

There is a geometric interpretation of the first three natural-number powers.

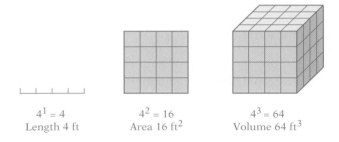

$4^1 = 4$
Length 4 ft

$4^2 = 16$
Area 16 ft^2

$4^3 = 64$
Volume 64 ft^3

To evaluate an exponential expression, write each factor as many times as indicated by the exponent. Then multiply.

$$3^5 = 3 \cdot 3 \cdot 3 \cdot 3 \cdot 3 = 243$$

$$2^3 \cdot 3^2 = (2 \cdot 2 \cdot 2) \cdot (3 \cdot 3) = 8 \cdot 9 = 72$$

Take Note

The -4 is squared only when the negative sign is *inside* the parentheses. In $(-4)^2$, we are squaring -4; in -4^2, we are finding the opposite of 4^2.

Example 1 Evaluate $(-4)^2$ and -4^2.

Solution $(-4)^2 = (-4)(-4) = 16$

$-4^2 = -(4 \cdot 4) = -16$

Problem 1 Evaluate $(-5)^3$ and -5^3.

Solution   See page S2.

Take Note

The product of an even number of negative factors is positive. The product of an odd number of negative factors is negative.

Example 2 Evaluate $(-2)^4$ and $(-2)^5$.

Solution
$$\begin{aligned} (-2)^4 &= (-2)(-2)(-2)(-2) \\ &= 4(-2)(-2) \\ &= -8(-2) \\ &= 16 \end{aligned}$$
$$\begin{aligned} (-2)^5 &= (-2)(-2)(-2)(-2)(-2) \\ &= 4(-2)(-2)(-2) \\ &= -8(-2)(-2) \\ &= 16(-2) \\ &= -32 \end{aligned}$$

Problem 2 Evaluate $(-3)^3$ and $(-3)^4$.

Solution   See page S2.

Example 3 Evaluate $(-3)^2 \cdot 2^3$ and $\left(-\dfrac{2}{3}\right)^3$.

Solution $(-3)^2 \cdot 2^3 = (-3)(-3) \cdot (2)(2)(2) = 9 \cdot 8 = 72$

$$\left(-\frac{2}{3}\right)^3 = \left(-\frac{2}{3}\right)\left(-\frac{2}{3}\right)\left(-\frac{2}{3}\right) = -\frac{2 \cdot 2 \cdot 2}{3 \cdot 3 \cdot 3} = -\frac{8}{27}$$

Problem 3 Evaluate $(3^3)(-2)^3$ and $\left(-\dfrac{2}{5}\right)^2$.

Solution   See page S2.

2 The Order of Operations Agreement

Evaluate $2 + 3 \cdot 5$.

There are two arithmetic operations, addition and multiplication, in this problem. The operations could be performed in different orders.

Add first. $\underbrace{2 + 3} \cdot 5$ Multiply first. $2 + \underbrace{3 \cdot 5}$

Then multiply. $\underbrace{5 \quad \cdot 5}$ Then add. $\underbrace{2 + \quad 15}$

                               25 17

In order to prevent there being more than one answer to the same problem, an Order of Operations Agreement has been established.

The Order of Operations Agreement

Step 1 Perform operations inside grouping symbols. Grouping symbols include parentheses ( ), brackets [ ], absolute value symbols | |, and the fraction bar.

Step 2 Simplify exponential expressions.

Step 3 Do multiplication and division as they occur from left to right.

Step 4 Do addition and subtraction as they occur from left to right.

Example 4 Simplify: $12 - 24(8 - 5) \div 2^2$

Solution $12 - 24(8 - 5) \div 2^2$
  $= 12 - 24(3) \div 2^2$ • Perform operations inside grouping symbols.
  $= 12 - 24(3) \div 4$ • Simplify exponential expressions.
  $= 12 - 72 \div 4$ • Do multiplication and division as they occur
                                                from left to right.
  $= 12 - 18$ • Do addition and subtraction as they occur
                                                from left to right.
  $= -6$

Problem 4 Simplify: $36 \div (8 - 5)^2 - (-3)^2 \cdot 2$

Solution   See page S2.

One or more of the steps shown in Example 4 may not be needed to simplify an expression. In that case, proceed to the next step in the Order of Operations Agreement.

Example 5   Simplify: $\dfrac{4 + 8}{2 + 1} - |3 - 1| + 2$

Solution $\dfrac{4 + 8}{2 + 1} - |3 - 1| + 2$

$= \dfrac{12}{3} - |2| + 2$ • Perform operations inside grouping symbols (above and below the fraction bar and inside the absolute value symbols).

$= \dfrac{12}{3} - 2 + 2$ • Find the absolute value of 2.

$= 4 - 2 + 2$ • Do multiplication and division as they occur from left to right.

$= 2 + 2$ • Do addition and subtraction as they occur from left to right.

$= 4$

Problem 5   Simplify: $27 \div 3^2 + (-3)^2 \cdot 4$

Solution   See page S2.

When an expression has grouping symbols inside grouping symbols, first perform the operations inside the *inner* grouping symbols by following Steps 2, 3, and 4 of the Order of Operations Agreement. Then perform the operations inside the *outer* grouping symbols by following Steps 2, 3, and 4 in sequence.

Example 6   Simplify: $6 \div [4 - (6 - 8)] + 2^2$

Solution $6 \div [4 - (6 - 8)] + 2^2$
$= 6 \div [4 - (-2)] + 2^2$ • Perform operations inside inner grouping symbols.

$= 6 \div 6 + 2^2$ • Perform operations inside outer grouping symbols.

$= 6 \div 6 + 4$ • Simplify exponential expressions.
$= 1 + 4$ • Do multiplication and division.
$= 5$ • Do addition and subtraction.

Problem 6   Simplify: $4 - 3[4 - 2(6 - 3)] \div 2$

Solution   See page S2.

1.4   Concept Review

Determine whether the statement is always true, sometimes true, or never true.

1. $(-5)^2$, -5^2, and $-(5)^2$ all represent the same number.

2. By the Order of Operations Agreement, addition is performed before division.

3. In the expression 3^8, 8 is the base.

4. The expression 9^4 is in exponential form.

5. To evaluate the expression $6 + 7 \cdot 10$ means to determine what one number it is equal to.

6. When using the Order of Operations Agreement, we consider the absolute value symbol a grouping symbol.

7. Using the Order of Operations Agreement, it is possible to get more than one correct answer to a problem.

8. The Order of Operations Agreement is used for natural numbers, integers, rational numbers, and real numbers.

9. The expression $(2^3)^2$ equals $2^{(3^2)}$.

10. Suppose a and b are positive integers. Then $a^b = b^a$.

1.4   **Exercises**

1 Evaluate.

1. 6^2

2. 7^4

3. -7^2

4. -4^3

5. $(-3)^2$

6. $(-2)^3$

7. $(-3)^4$

8. $(-5)^3$

9. $\left(\dfrac{1}{2}\right)^2$

10. $\left(-\dfrac{3}{4}\right)^3$

11. $(0.3)^2$

12. $(1.5)^3$

13. $\left(\dfrac{2}{3}\right)^2 \cdot 3^3$

14. $\left(-\dfrac{1}{2}\right)^3 \cdot 8$

15. $(0.3)^3 \cdot 2^3$

16. $(0.5)^2 \cdot 3^3$

17. $(-3) \cdot 2^2$

18. $(-5) \cdot 3^4$

19. $(-2) \cdot (-2)^3$

20. $(-2) \cdot (-2)^2$

21. $2^3 \cdot 3^3 \cdot (-4)$

22. $(-3)^3 \cdot 5^2 \cdot 10$

23. $(-7) \cdot 4^2 \cdot 3^2$

24. $(-2) \cdot 2^3 \cdot (-3)^2$

25. $\left(\dfrac{2}{3}\right)^2 \cdot \dfrac{1}{4} \cdot 3^3$

26. $\left(\dfrac{3}{4}\right)^2 \cdot (-4) \cdot 2^3$

27. $8^2 \cdot (-3)^5 \cdot 5$

2 **28.** Why do we need an Order of Operations Agreement?

29. Describe each step in the Order of Operations Agreement.

Simplify by using the Order of Operations Agreement.

30. $4 - 8 \div 2$

31. $2^2 \cdot 3 - 3$

32. $2(3 - 4) - (-3)^2$

33. $16 - 32 \div 2^3$

34. $24 - 18 \div 3 + 2$

35. $8 - (-3)^2 - (-2)$

36. $16 + 15 \div (-5) - 2$

37. $14 - 2^2 - |4 - 7|$

38. $3 - 2[8 - (3 - 2)]$

39. $-2^2 + 4[16 \div (3 - 5)]$

40. $6 + \dfrac{16 - 4}{2^2 + 2} - 2$

41. $24 \div \dfrac{3^2}{8 - 5} - (-5)$

42. $96 \div 2[12 + (6 - 2)] - 3^3$

43. $4 \cdot [16 - (7 - 1)] \div 10$

44. $16 \div 2 - 4^2 - (-3)^2$

45. $18 \div |9 - 2^3| + (-3)$

46. $16 - 3(8 - 3)^2 \div 5$

47. $4(-8) \div [2(7 - 3)^2]$

48. $\dfrac{(-10) + (-2)}{6^2 - 30} \div |2 - 4|$

49. $16 - 4 \cdot \dfrac{3^3 - 7}{2^3 + 2} - (-2)^2$

50. $(0.2)^2 \cdot (-0.5) + 1.72$

51. $0.3(1.7 - 4.8) + (1.2)^2$

52. $(1.8)^2 - 2.52 \div 1.8$

53. $(1.65 - 1.05)^2 \div 0.4 + 0.8$

54. $\dfrac{3}{8} \div \left| \dfrac{5}{6} + \dfrac{2}{3} \right|$

55. $\left(\dfrac{3}{4} \right)^2 - \left(\dfrac{1}{2} \right)^3 \div \dfrac{3}{5}$

1.4   Applying Concepts

56. Using the Order of Operations Agreement, describe how to simplify Exercise 46.

Place the correct symbol, $<$ or $>$, between the two numbers.

57. $(0.9)^3$ 1^5

58. $(-3)^3$ $(-2)^5$

59. $(-1.1)^2$ $(0.9)^2$

Simplify.

60. $1^2 + 2^2 + 3^2 + 4^2$

61. $1^3 + 2^3 + 3^3 + 4^3$

62. $(-1)^3 + (-2)^3 + (-3)^3 + (-4)^3$

63. $(-2)^2 + (-4)^2 + (-6)^2 + (-8)^2$

Solve.

64. Find a rational number, r, that satisfies the condition.
   a. $r^2 < r$ **b.** $r^2 = r$ **c.** $r^2 > r$

65. Computers A computer can do 6,000,000 additions in 1 s. To the nearest second, how many seconds will it take the computer to do 10^8 additions?

66. The sum of two natural numbers is 41. Each of the two numbers is the square of a natural number. Find the two numbers.

Determine the ones digit when the expression is evaluated.

67. 34^{202}

68. 23^{502}

69. 27^{622}

70. Prepare a report on the Kelvin scale. Include in your report an explanation of how to convert between the Kelvin scale and the Celsius scale.

Focus on Problem Solving

Inductive Reasoning   Suppose you take 9 credit hours each semester. The total number of credit hours you have taken at the end of each semester can be described in a list of numbers.

$$9, 18, 27, 36, 45, 54, 63, \ldots$$

The list of numbers that indicates the total credit hours is an ordered list of numbers, called a **sequence.** Each number in a sequence is called a **term** of the sequence. The list is ordered because the position of a number in the list indicates the semester in which that number of credit hours has been taken. For example, the 7th term of the sequence is 63, and a total of 63 credit hours have been taken after the 7th semester.

Assuming the pattern is continued, find the next three numbers in the pattern

$$-6, -10, -14, -18, \ldots$$

This list of numbers is a sequence. The first step in solving this problem is to observe the pattern in the list of numbers. In this case, each number in the list is 4 less than the previous number. The next three numbers are $-22, -26, -30$.

This process of discovering the pattern in a list of numbers is inductive reasoning. **Inductive reasoning** involves making generalizations from specific examples; in other words, we reach a conclusion by making observations about particular facts or cases.

Try the following exercises. Each exercise requires inductive reasoning.

Name the next two terms in the sequence.

1. 1, 3, 5, 7, 1, 3, 5, 7, 1, . . .   **2.** 1, 4, 2, 5, 3, 6, 4, . . .

3. 1, 2, 4, 7, 11, 16, . . .   **4.** A, B, C, G, H, I, M, . . .

Draw the next shape in the sequence.

5.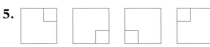

6. |• ||• |||•• • |||•• • |||•• • •

Solve.

7. Convert $\frac{1}{11}, \frac{2}{11}, \frac{3}{11}, \frac{4}{11}$, and $\frac{5}{11}$ to decimals. Then use the pattern you observe to convert $\frac{6}{11}, \frac{7}{11}$, and $\frac{9}{11}$ to decimals.

8. Convert $\frac{1}{33}, \frac{2}{33}, \frac{4}{33}, \frac{5}{33}$, and $\frac{7}{33}$ to decimals. Then use the pattern you observe to convert $\frac{8}{33}, \frac{13}{33}$, and $\frac{19}{33}$ to decimals.

Projects & Group Activities

Calculators

Does your calculator use the Order of Operations Agreement? To find out, try this problem:

$$2 + 4 \cdot 7$$

If your answer is 30, then the calculator uses the Order of Operations Agreement. If your answer is 42, it does not use that agreement.

Even if your calculator does not use the Order of Operations Agreement, you can still correctly evaluate numerical expressions. The parentheses keys, ⎡(⎤ and ⎡)⎤, are used for this purpose.

Remember that $2 + 4 \cdot 7$ means $2 + (4 \cdot 7)$ because the multiplication must be completed before the addition. To evaluate this expression, enter the following:

Enter: 2 ⎡+⎤ ⎡(⎤ 4 ⎡×⎤ 7 ⎡)⎤ ⎡=⎤

Display: 2   2   ⎡(⎤   4   4   7   28   30

When using your calculator to evaluate numerical expressions, insert parentheses around multiplications or divisions. This has the effect of forcing the calculator to do the operations in the order you want rather than in the order the calculator wants.

Evaluate.

1. $3 \cdot (15 - 2 \cdot 3) - 36 \div 3$ **2.** $4 \cdot 2^2 - (12 + 24 \div 6) - 5$

3. $16 \div 4 \cdot 3 + (3 \cdot 4 - 5) + 2$ **4.** $15 \cdot 3 \div 9 + (2 \cdot 6 - 3) + 4$

Using your calculator to simplify numerical expressions sometimes requires use of the ⎡+/−⎤ key or, on some calculators, the negative key, which is frequently shown as ⎡(−)⎤. These keys change the sign of the number currently in the display. To enter −4:

• For those calculators with ⎡+/−⎤, press 4 and then ⎡+/−⎤.

• For those calculators with ⎡(−)⎤, press ⎡(−)⎤ and then 4.

Here are the keystrokes for evaluating the expression $3(-4) - (-5)$.

Calculators with ⎡+/−⎤ key: 3 ⎡×⎤ 4 ⎡+/−⎤ ⎡−⎤ 5 ⎡+/−⎤ ⎡=⎤

Calculators with ⎡(−)⎤ key: 3 ⎡×⎤ ⎡(−)⎤ 4 ⎡−⎤ ⎡(−)⎤ 5 ⎡=⎤

This example illustrates that calculators make a distinction between negative and minus. To perform the operation $3 - (-3)$, you cannot enter 3 ⎡−⎤ ⎡−⎤ 3. This would result in 0, which is not the correct answer. You must enter

3 ⎡−⎤ 3 ⎡+/−⎤ ⎡=⎤   or   3 ⎡−⎤ ⎡(−)⎤ 3 ⎡=⎤

Use a calculator to evaluate each of the following exercises.

5. $-16 \div 2$ **6.** $3(-8)$ **7.** $47 - (-9)$

8. $-50 - (-14)$ **9.** $4 - (-3)^2$ **10.** $-8 + (-6)^2 - 7$

**Using Patterns
in Mathematics**

Suppose you are asked to cut a pie into the greatest number of pieces with only five straight cuts of a knife. An illustration showing how five cuts can produce 13 pieces is given at the left. The correct answer, however, is more than 13 pieces.

A reasonable question to ask is, "How do I know when I have the maximum number of pieces?" To determine the answer, we suggest that you start with one cut, then two cuts, then three cuts, and so on. Try to discover a pattern for the greatest number of pieces that each successive cut can produce.

www.fedstats.gov

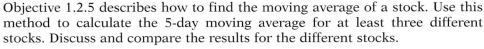

WEB

Information regarding the history of the federal budget can be found on the World Wide Web. Go to the web site **www.fedstats.gov.** Click on "Agencies listed alphabetically," find the "Office of Management and Budget," and click on "Key Statistics." Click on "Budget of the U.S. Government," then "Spreadsheet Files," and then "Historical Tables." Look at Table 1.1. When the federal budget table appears on the screen, look for the column that lists each year's surplus or deficit. You will see that a negative sign (−) is used to show a deficit. Note that it states near the top of the screen that the figures in the table are in millions of dollars.

1. During which years shown in the table was there a surplus?

2. During which year was the deficit the greatest?

3. Find the difference between the surplus or deficit this year and the surplus or deficit 5 years ago.

4. What is the difference between the surplus or deficit this year and the surplus or deficit a decade ago?

5. Determine what two numbers in the table are being subtracted in each row in order to arrive at the number in the surplus or deficit column.

6. Describe the trend of the federal deficit over the last 10 years.

Moving Averages

WEB

Objective 1.2.5 describes how to find the moving average of a stock. Use this method to calculate the 5-day moving average for at least three different stocks. Discuss and compare the results for the different stocks.

For this project, you will need to use stock tables, which are printed in the business section of major newspapers. Your college library should have copies of these publications. In a stock table, the column headed "Chg." provides the change in the price of a share of the stock; that is, it gives the difference between the closing price for the day shown and the closing price for the previous day. The symbol + indicates the change was an increase in price; the symbol − indicates the change was a decrease in price.

You can also look up stock prices on the Internet. A printout from Yahoo! Finance is shown on the following page. It provides data on shares of IBM.

YAHOO! *FINANCE*

Historical Prices - IBM (International Business Machines Corp) As of 26-Mar-02

More Info: Quote | Chart | News | Profile | Research | SEC | Msgs | Insider

	Month	Day	Year		○ Daily
Start:	Mar ▾	12	02		○ Weekly
End:	Mar ▾	26	02		○ Monthly
					○ Dividends

Ticker Symbol: ibm   [Get Data]

Date	Open	High	Low	Close	Volume	Adj. Close
26-Mar-02	103.57	105.70	102.30	**102.90**	8,144,000	102.90
25-Mar-02	105.80	106.66	103.50	**103.56**	6,277,900	103.56
22-Mar-02	106.50	106.70	105.07	**105.60**	5,507,900	105.60
21-Mar-02	105.70	106.78	104.70	**106.78**	5,113,100	106.78
20-Mar-02	106.90	106.90	105.49	**105.50**	4,844,100	105.50
19-Mar-02	106.85	108.05	106.49	**107.49**	4,614,800	107.49
18-Mar-02	107.10	108.64	106.23	**106.35**	5,301,200	106.35
15-Mar-02	106.55	107.45	105.59	**106.79**	10,864,100	106.79
14-Mar-02	107.02	107.95	106.59	**106.60**	5,335,500	106.60
13-Mar-02	108.30	108.65	106.70	**107.18**	6,409,200	107.18
12-Mar-02	105.52	108.85	105.51	**108.50**	11,059,300	108.50

Chapter Summary

Key Words

A **set** is a collection of objects. The objects in the set are called the **elements** of the set. The **roster method** of writing sets encloses a list of the elements in braces.   [1.1.1, p. 2]*

The set of **natural numbers** is {1, 2, 3, 4, 5, 6, 7,…}. The set of **integers** is {…, −4, −3, −2, −1, 0, 1, 2, 3, 4,…}.   [1.1.1, pp. 2–3]

A number a **is less than** another number b, written $a < b$, if a is to the left of b on the number line. A number a **is greater than** another number b, written $a > b$, if a is to the right of b on the number line. The symbol ≤ means **is less than or equal to**. The symbol ≥ means **is greater than or equal to**. [1.1.1, pp. 3–4]

Examples

The set of even natural numbers less than 10 is written $A = \{2, 4, 6, 8\}$. The elements in this set are the numbers 2, 4, 6, and 8.

$-9 < 7$
$4 > -5$

*The numbers in brackets refer to the objective where the Key Word or Essential Rule or Procedure is first introduced. For example, the reference [1.1.1] stands for Chapter 1, Section 1, Objective 1. This notation will be used in all Chapter Summaries throughout the text.

Two numbers that are the same distance from zero on the number line but on opposite sides of zero are **opposite numbers**, or **opposites**. The opposite of a number is also called its **additive inverse**. [1.1.2, p. 4]

5 and -5 are opposites.

The **absolute value** of a number is its distance from zero on the number line. The absolute value of a number is positive or zero. [1.1.2, p. 5]

$|-12| = 12$
$-|7| = -7$

A **rational number** is a number that can be written in the form $\frac{a}{b}$, where a and b are integers and $b \neq 0$. A rational number written in this form is commonly called a **fraction**. [1.3.1, p. 21]

$\frac{3}{8}, \frac{-9}{11}$, and $\frac{4}{1}$ are rational numbers.

An **irrational number** is a number that has a decimal representation that never terminates or repeats. [1.3.1, p. 22]

π, $\sqrt{2}$, and $2.17117111711117\ldots$ are irrational numbers.

The rational numbers and the irrational numbers taken together are called the **real numbers**. [1.3.1, p. 22]

$\frac{3}{8}, \frac{-9}{11}, \frac{4}{1}, \pi, \sqrt{2}$, and $2.17117111711117\ldots$ are real numbers.

Percent means "parts of 100." [1.3.4, p. 26]

63% means 63 of 100 equal parts.

An expression of the form a^n is in **exponential form**, where a is the **base** and n is the **exponent**. [1.4.1, p. 33]

6^3 is an exponential expression in which 6 is the base and 3 is the exponent.

Essential Rules and Procedures

To add two integers with the same sign, add the absolute values of the numbers. Then attach the sign of the addends. [1.2.1, p. 10]

$9 + 3 = 12$
$-9 + (-3) = -12$

To add two integers with different signs, find the absolute value of each number. Then subtract the smaller of these absolute values from the larger one. Attach the sign of the number with the greater absolute value. [1.2.1, p. 11]

$9 + (-3) = 6$
$-9 + 3 = -6$

To subtract one integer from another, add the opposite of the second integer to the first integer. [1.2.2, p. 10]

$6 - 11 = 6 + (-11) = -5$
$-2 - (-8) = -2 + 8 = 6$

To multiply two integers with the same sign, multiply the absolute values of the numbers. The product is positive. [1.2.3, p. 12]

$3(5) = 15$
$(-3)(-5) = 15$

To multiply two integers with different signs, multiply the absolute values of the numbers. The product is negative. [1.2.3, p. 12]

$-3(5) = -15$
$3(-5) = -15$

To divide two integers with the same sign, divide the absolute values of the numbers. The quotient is positive. [1.2.4, p. 13]

$14 \div 2 = 7$
$-14 \div (-2) = 7$

To divide two integers with different signs, divide the absolute values of the numbers. The quotient is negative. [1.2.4, p. 13]	$14 \div (-2) = -7$ $-14 \div 2 = -7$

Zero and One in Division   [1.2.4, p. 13]

$\dfrac{0}{a} = 0, a \neq 0$ $\qquad\qquad\qquad\qquad\qquad$ $\dfrac{0}{3} = 0$

Division by zero is undefined. $\qquad\qquad\qquad$ $\dfrac{9}{0}$ is undefined.

$\dfrac{a}{a} = 1, a \neq 0$ $\qquad\qquad\qquad\qquad\qquad$ $\dfrac{-2}{-2} = 1$

$\dfrac{a}{1} = a$ $\qquad\qquad\qquad\qquad\qquad\qquad$ $\dfrac{-6}{1} = -6$

To convert a percent to a decimal, remove the percent sign and multiply by 0.01. [1.3.4, p. 26]	$85\% = 85(0.01) = 0.85$
To convert a percent to a fraction, remove the percent sign and multiply by $\dfrac{1}{100}$. [1.3.4, p. 26]	$25\% = 25 \cdot \dfrac{1}{100} = \dfrac{25}{100} = \dfrac{1}{4}$
To convert a decimal or a fraction to a percent, multiply by 100%. [1.3.4, p. 26]	$0.5 = 0.5(100\%) = 50\%$ $\dfrac{3}{8} = \dfrac{3}{8} \cdot 100\% = \dfrac{300}{8}\% = 37.5\%$

Order of Operations Agreement   [1.4.2, p. 35]

Step 1 Perform operations inside grouping symbols.
Step 2 Simplify exponential expressions.
Step 3 Do multiplication and division as they occur from left to right.
Step 4 Do addition and subtraction as they occur from left to right.

$16 \div (-2)^3 + (7 - 12)$

$\quad = 16 \div (-2)^3 + (-5)$
$\quad = 16 \div (-8) + (-5)$
$\quad = -2 + (-5)$
$\quad = -7$

Chapter Review Exercises

1. Use the roster method to write the set of natural numbers less than 7.

2. Write $\dfrac{5}{8}$ as a percent.

3. Evaluate: $-|-4|$

4. Subtract: $16 - (-30) - 42$

5. Divide: $-561 \div (-33)$

6. Write $\dfrac{7}{9}$ as a decimal. Place a bar over the repeating digits of the decimal.

7. Simplify: $(6.02)(-0.89)$

8. Simplify: $\dfrac{-10 + 2}{2 + (-4)} \div 2 + 6$

9. Find the opposite of -4 .

10. Subtract: $16 - 30$

11. Write 0.672 as a percent.

12. Write $79\frac{1}{2}\%$ as a fraction.

13. Divide: $-72 \div 8$

14. Write $\frac{17}{20}$ as a decimal.

15. Divide: $\frac{5}{12} \div \left(-\frac{5}{6}\right)$

16. Simplify: $3^2 - 4 + 20 \div 5$

17. Place the correct symbol, $<$ or $>$, between the two numbers.

$$-1 \quad 0$$

18. Add: $-22 + 14 + (-8)$

19. Multiply: $(-5)(-6)(3)$

20. Subtract: $6.039 - 12.92$

21. Given $A = \{-5, -3, 0\}$, which elements of set A are less than or equal to -3?

22. Write 7% as a decimal.

23. Evaluate: $\frac{3}{4} \cdot (4)^2$

24. Place the correct symbol, $<$ or $>$, between the two numbers.

$$-2 \quad -40$$

25. Add: $13 + (-16)$

26. Multiply: $(-4)(12)$

27. Add: $-\frac{2}{5} + \frac{7}{15}$

28. Evaluate: $(-3^3) \cdot 2^2$

29. Write $2\frac{7}{9}$ as a percent. Round to the nearest tenth of a percent.

30. Write 240% as a decimal.

31. Find the opposite of -2 .

32. Subtract: $7 - 21$

33. Divide: $96 \div (-12)$

34. Write $\frac{7}{20}$ as a decimal.

35. Divide: $-\frac{7}{16} \div \frac{3}{8}$

36. Simplify: $2^3 \div 4 - 2(2 - 7)$

37. Evaluate: $|-3|$

38. Subtract: $12 - (-10) - 4$

39. Write $2\frac{8}{9}$ as a percent. Write the remainder in fractional form.

40. Given $C = \{-12, -8, -1, 7\}$, find

  a. the opposite of each element of set C.
  b. the absolute value of each element of set C.

41. Divide: $(-204) \div (-17)$

42. Write $\dfrac{7}{11}$ as a decimal. Place a bar over the repeating digits of the decimal.

43. Divide: $0.2654 \div (-0.023)$
Round to the nearest tenth.

44. Simplify: $(7 - 2)^2 - 5 - 3 \cdot 4$

45. Place the correct symbol, $<$ or $>$, between the two numbers.

$$8 \quad -10$$

46. Add: $-12 + 8 + (-4)$

47. Multiply: $2(-3)(-12)$

48. Add: $-\dfrac{5}{8} + \dfrac{1}{6}$

49. Given $D = \{-24, -17, -9, 0, 4\}$, which elements of set D are greater than -19?

50. Write 0.002 as a percent.

51. Evaluate: $-4^2 \cdot \left(\dfrac{1}{2}\right)^2$

52. Add: $-1.329 + 4.89$

53. Evaluate: $-|17|$

54. Subtract: $-5 - 22 - (-13) - 19 - (-6)$

55. Multiply: $\left(\dfrac{1}{3}\right)\left(-\dfrac{4}{5}\right)\left(\dfrac{3}{8}\right)$

56. Place the correct symbol, $<$ or $>$, between the two numbers.

$$-43 \quad -34$$

57. Write $\dfrac{18}{25}$ as a decimal.

58. Evaluate $(-2)^3 \cdot 4^2$.

59. Write 0.075 as a percent.

60. Write $\dfrac{19}{35}$ as a percent. Write the remainder in fractional form.

61. Add: $14 + (-18) + 6 + (-20)$

62. Multiply: $-4(-8)(12)(0)$

63. Simplify: $2^3 - 7 + 16 \div (-3 + 5)$

64. Simplify: $\dfrac{3}{4} + \dfrac{1}{2} - \dfrac{3}{8}$

65. Divide: $-128 \div (-8)$

66. Place the correct symbol, $<$ or $>$, between the two numbers.

$$-57 \quad 28$$

67. Evaluate: $\left(-\dfrac{1}{3}\right)^3 \cdot 9^2$

68. Add: $-7 + (-3) + (-12) + 16$

69. Multiply: $5(-2)(10)(-3)$

70. Use the roster method to write the set of negative integers greater than -4.

71. Find the temperature after a rise of 14°C from −6°C.

72. The daily low temperatures, in degrees Celsius, for a three-day period were recorded as follows: −8°, 7°, −5°. Find the average low temperature for the three-day period.

73.  Use the table to find the difference between the record high temperature and the record low temperature for January in Bismarck, North Dakota.

Bismarck, North Dakota

	Jan	Feb	Mar	Apr	May	Jun	Jul	Aug	Sep	Oct	Nov	Dec
Record High	62°F (1981)	69°F (1992)	80°F (1967)	93°F (1992)	96°F (1980)	107°F (1988)	109°F (1973)	107°F (1973)	105°F (1959)	95°F (1963)	75°F (1978)	65°F (1979)
Record Low	−44°F (1950)	−43°F (1994)	−28°F (1995)	−12°F (1975)	15°F (1967)	30°F (1969)	35°F (1971)	33°F (1988)	11°F (1974)	−10°F (1991)	−30°F (1985)	−40°F (1983)

Source: **www.weather.com**

74. Find the temperature after a rise of 7°C from −13°C.

75.  The temperature on the surface of the planet Venus is 480°C. The temperature on the surface of the planet Pluto is −234°C. Find the difference between the surface temperatures on Venus and Pluto.

Chapter Test

1. Write 55% as a fraction.

2. Given $B = \{-8, -6, -4, -2\}$, which elements of set B are less than −5?

3. Subtract: $-9 - (-6)$

4. Write $\dfrac{3}{20}$ as a decimal.

5. Multiply: $\dfrac{3}{4}\left(-\dfrac{2}{21}\right)$

6. Divide: $-75 \div 5$

7. Evaluate: $\left(-\dfrac{2}{3}\right)^3 \cdot 3^2$

8. Add: $-7 + (-3) + 12$

9. Use the roster method to write the set of positive integers less than or equal to 6.

10. Write 1.59 as a percent.

11. Evaluate: $|-29|$

12. Place the correct symbol, $<$ or $>$, between the two numbers.

$$-47 \qquad -68$$

13. Subtract: $-\dfrac{4}{9} - \dfrac{5}{6}$

14. Multiply: $-6(-43)$

15. Simplify: $8 + \dfrac{12 - 4}{3^2 - 1} - 6$

16. Divide: $-\dfrac{5}{8} \div \left(-\dfrac{3}{4}\right)$

17. Write $\dfrac{3}{13}$ as a percent. Round to the nearest tenth of a percent.

18. Write 6.2% as a decimal.

19. Subtract: $13 - (-5) - 4$

20. Write $\dfrac{13}{30}$ as a decimal. Place a bar over the repeating digits of the decimal.

21. Multiply: $(-0.9)(2.7)$

22. Divide: $-180 \div (-12)$

23. Evaluate: $2^2 \cdot (-4)^2 \cdot 10$

24. Add: $15 + (-8) + (-19)$

25. Evaluate: $-|-34|$

26. Place the correct symbol, $<$ or $>$, between the two numbers.

$$53 \quad -92$$

27. Given $A = \{-17, -6, 5, 9\}$, find

    a. the opposite of each element of set A.
    b. the absolute value of each element of set A.

28. Write $\dfrac{16}{23}$ as a percent. Write the remainder in fractional form.

29. Add: $-18.354 + 6.97$

30. Multiply: $-4(8)(-5)$

31. Simplify: $9(-4) \div [2(8 - 5)^2]$

32. Find the temperature after a rise of 12°C from -8°C.

33. Use the table to find the average of the record low temperatures in Fairbanks, Alaska, for the first four months of the year.

Fairbanks, Alaska

	Jan	Feb	Mar	Apr	May	Jun	Jul	Aug	Sep	Oct	Nov	Dec
Record High	50°F (1981)	47°F (1987)	56°F (1994)	74°F (1960)	89°F (1960)	96°F (1969)	94°F (1975)	93°F (1994)	84°F (1957)	65°F (1969)	49°F (1997)	44°F (1985)
Record Low	−61°F (1969)	−58°F (1993)	−49°F (1956)	−24°F (1986)	−1°F (1964)	30°F (1950)	35°F (1959)	27°F (1987)	3°F (1992)	−27°F (1975)	−46°F (1990)	−62°F (1961)

Source: **www.weather.com**

At the beginning of a Little League season, a schedule for all the teams in the league is created. Every team must play every other team at least once, and every team should play the same number of games. A pattern can be used to set up a schedule that meets both of these requirements. In the **Project on page 81,** a pattern is used to determine the number of games to be scheduled in a softball league.

Applications

Abstract problems,   *pp. 79, 80*

Agriculture,   *p. 77*

Astronomy,   *pp. 77, 86*

Billing,   *p. 78*

Biology,   *p. 76*

Chemistry,   *p. 78*

Coins,   *p. 78*

Computers,   *pp. 72, 87*

Consumerism,   *p. 78*

Cost of living,   *p. 77*

Demography,   *p. 77*

Earth science,   *p. 76*

Genetics,   *p. 76*

Geometry,   *pp. 54, 55, 57, 77, 85*

Investments,   *p. 72*

Metalwork,   *p. 78*

Mixture problems,   *p. 85*

Music,   *p. 72*

Patterns in mathematics,   *p. 81*

Pulleys,   *p. 78*

Space exploration,   *p. 76*

Sports,   *pp. 72, 76, 77*

Taxes,   *p. 77*

Telecommunications,   *p. 76*

Travel,   *p. 78*

Wages,   *p. 78*

Woodwork,   *p. 86*

2

Variable Expressions

Objectives

Section 2.1

1 Evaluate variable expressions

Section 2.2

1 The Properties of the Real Numbers

2 Simplify variable expressions using the Properties of Addition

3 Simplify variable expressions using the Properties of Multiplication

4 Simplify variable expressions using the Distributive Property

5 Simplify general variable expressions

Section 2.3

1 Translate a verbal expression into a variable expression

2 Translate a verbal expression into a variable expression and then simplify the resulting expression

3 Translate application problems

Need help? For online student resources, such as section quizzes, visit this web site: **math.college.hmco.com**

PrepTest

1. Subtract: $-12 - (-15)$ **2.** Divide: $-36 \div (-9)$

3. Add: $-\dfrac{3}{4} + \dfrac{5}{6}$ **4.** What is the reciprocal of $-\dfrac{9}{4}$?

5. Divide: $\left(-\dfrac{3}{4}\right) \div \left(-\dfrac{5}{2}\right)$ **6.** Evaluate: -2^4

7. Evaluate: $\left(\dfrac{2}{3}\right)^3$ **8.** Evaluate: $3 \cdot 4^2$

9. Evaluate: $7 - 2 \cdot 3$ **10.** Evaluate: $5 - 7(3 - 2^2)$

Go Figure

Two fractions are inserted between $\frac{1}{4}$ and $\frac{1}{2}$ so that the difference between any two successive fractions is the same. Find the sum of the four fractions.

2.1 | Evaluating Variable Expressions

1 **Evaluate variable expressions**

VIDEO & DVD   CD TUTOR   WEB   SSM

Point of Interest

Historical manuscripts indicate that mathematics is at least 4000 years old. Yet it was only 400 years ago that mathematicians started using variables to stand for numbers. The idea that a letter can stand for some number was a critical turning point in mathematics.

Often we discuss a quantity without knowing its exact value, such as the price of gold next month, the cost of a new automobile next year, or the tuition for next semester. In algebra, a letter of the alphabet is used to stand for a quantity that is unknown or one that can change, or *vary*. The letter is called a **variable.** An expression that contains one or more variables is called a **variable expression.**

A variable expression is shown at the right. The expression can be rewritten by writing subtraction as the addition of the opposite.

$$3x^2 - 5y + 2xy - x - 7$$

$$3x^2 + (-5y) + 2xy + (-x) + (-7)$$

Note that the expression has five addends. The **terms** of a variable expression are the addends of the expression. The expression has five terms.

$$\underbrace{\underbrace{3x^2 \quad - 5y \quad + 2xy \quad - x}_{\text{Variable terms}} \quad \overbrace{- 7}^{\text{5 terms}}}$$

Variable terms   Constant term

The terms $3x^2$, $-5y$, $2xy$, and $-x$ are **variable terms.**

The term -7 is a **constant term,** or simply a **constant.**

Each variable term is composed of a **numerical coefficient** and a **variable part** (the variable or variables and their exponents).

When the numerical coefficient is 1 or -1, the 1 is usually not written ($x = 1x$ and $-x = -1x$).

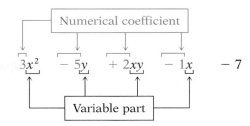

Numerical coefficient

$$3x^2 \quad -5y \quad +2xy \quad -1x \quad -7$$

Variable part

Example 1   Name the variable terms of the expression $2a^2 - 5a + 7$.

Solution $2a^2, -5a$

Problem 1   Name the constant term of the expression $6n^2 + 3n - 4$.

Solution   See page S3.

Variable expressions occur naturally in science. In a physics lab, a student may discover that a weight of 1 pound will stretch a spring $\frac{1}{2}$ inch. A weight of 2 pounds will stretch the spring 1 inch. By experimenting, the student can discover that the distance the spring will stretch is found by multiplying the weight, in pounds, by $\frac{1}{2}$. By letting W represent the number of pounds attached to the spring, the distance, in inches, the spring stretches can be represented by the variable expression $\frac{1}{2}W$.

With a weight of W pounds, the spring will stretch $\frac{1}{2} \cdot W = \frac{1}{2}W$ inches.

With a weight of 10 pounds, the spring will stretch $\frac{1}{2} \cdot 10 = 5$ inches.

With a weight of 3 pounds, the spring will stretch $\frac{1}{2} \cdot 3 = 1\frac{1}{2}$ inches.

Replacing the variable or variables in a variable expression with numbers and then simplifying the resulting numerical expression is called **evaluating the variable expression.**

Example 2   Evaluate $ab - b^2$ when $a = 2$ and $b = -3$.

Solution $ab - b^2$

$2(-3) - (-3)^2$ • Replace each variable in the expression with the number it represents.

$= 2(-3) - 9$ • Use the Order of Operations Agreement to simplify the resulting numerical expression.

$= -6 - 9$

$= -15$

Problem 2   Evaluate $2xy + y^2$ when $x = -4$ and $y = 2$.

Solution   See page S3.

Example 3   Evaluate $\dfrac{a^2 - b^2}{a - b}$ when $a = 3$ and $b = -4$.

Solution $\dfrac{a^2 - b^2}{a - b}$

$\dfrac{(3)^2 - (-4)^2}{3 - (-4)}$ • Replace each variable in the expression with the number it represents.

$= \dfrac{9 - 16}{3 - (-4)}$ • Use the Order of Operations Agreement to simplify the resulting numerical expression.

$= \dfrac{-7}{7}$

$= -1$

Problem 3   Evaluate $\dfrac{a^2 + b^2}{a + b}$ when $a = 5$ and $b = -3$.

Solution   See page S3.

Example 4   Evaluate $x^2 - 3(x - y) - z^2$ when $x = 2$, $y = -1$, and $z = 3$.

Solution $x^2 - 3(x - y) - z^2$
$(2)^2 - 3[2 - (-1)] - (3)^2$ • Replace each variable in the expression with the number it represents.

$= (2)^2 - 3(3) - (3)^2$ • Use the Order of Operations Agreement to simplify the resulting numerical expression.
$= 4 - 3(3) - 9$
$= 4 - 9 - 9$
$= -5 - 9$
$= -14$

Problem 4   Evaluate $x^3 - 2(x + y) + z^2$ when $x = 2$, $y = -4$, and $z = -3$.

Solution   See page S3.

Example 5   The diameter of the base of a right circular cylinder is 5 cm. The height of the cylinder is 8.5 cm. Find the volume of the cylinder. Round to the nearest tenth.

8.5 cm

5 cm

Solution $V = \pi r^2 h$ • Use the formula for the volume of a right circular cylinder.

$V = \pi(2.5)^2(8.5)$ • $r = \dfrac{1}{2}d = \dfrac{1}{2}(5) = 2.5$

$V = \pi(6.25)(8.5)$ • Use the π key on your calculator to enter the value of π.
$V \approx 166.9$

The volume is approximately 166.9 cm³.

Problem 5   The diameter of the base of a right
circular cone is 9 cm. The height of
the cone is 9.5 cm. Find the volume of
the cone. Round to the nearest tenth.

Solution   See page S3.

9.5 cm

9 cm

 A graphing calculator can be used to evaluate variable expressions. When
the value of each variable is stored in the calculator's memory and a
variable expression is then entered on the calculator, the calculator
evaluates that variable expression for the values of the variables stored in its
memory. See the Appendix for a description of keystroking procedures.

2.1   Concept Review

Determine whether the statement is always true, sometimes true, or never
true.

1. The expression $3x^2$ is a variable expression.

2. In the expression $8y^3 - 4y$, the terms are $8y^3$ and $4y$.

3. For the expression x^5, the value of x is 1.

4. For the expression $6a + 7b$, 7 is a constant term.

5. The Order of Operations Agreement is used in evaluating a variable
expression.

6. The result of evaluating a variable expression is a single number.

2.1   Exercises

1 Name the terms of the variable expression. Then underline the constant
term.

1. $2x^2 + 5x - 8$ **2.** $-3n^2 - 4n + 7$ **3.** $6 - a^4$

Name the variable terms of the expression. Then underline the variable part of each term.

4. $9b^2 - 4ab + a^2$

5. $7x^2y + 6xy^2 + 10$

6. $5 - 8n - 3n^2$

Name the coefficients of the variable terms.

7. $x^2 - 9x + 2$

8. $12a^2 - 8ab - b^2$

9. $n^3 - 4n^2 - n + 9$

10. What is the meaning of the phrase "evaluate a variable expression"?

11. What is the difference between the meaning of "the value of the variable" and the meaning of "the value of the variable expression"?

Evaluate the variable expression when $a = 2$, $b = 3$, and $c = -4$.

12. $3a + 2b$

13. $a - 2c$

14. $-a^2$

15. $2c^2$

16. $-3a + 4b$

17. $3b - 3c$

18. $b^2 - 3$

19. $-3c + 4$

20. $16 \div (2c)$

21. $6b \div (-a)$

22. $bc \div (2a)$

23. $-2ab \div c$

24. $a^2 - b^2$

25. $b^2 - c^2$

26. $(a + b)^2$

27. $a^2 + b^2$

28. $2a - (c + a)^2$

29. $(b - a)^2 + 4c$

30. $b^2 - \dfrac{ac}{8}$

31. $\dfrac{5ab}{6} - 3cb$

32. $(b - 2a)^2 + bc$

Evaluate the variable expression when $a = -2$, $b = 4$, $c = -1$, and $d = 3$.

33. $\dfrac{b + c}{d}$

34. $\dfrac{d - b}{c}$

35. $\dfrac{2d + b}{-a}$

36. $\dfrac{b + 2d}{b}$

37. $\dfrac{b - d}{c - a}$

38. $\dfrac{2c - d}{-ad}$

39. $(b + d)^2 - 4a$

40. $(d - a)^2 - 3c$

41. $(d - a)^2 \div 5$

42. $(b - c)^2 \div 5$

43. $b^2 - 2b + 4$

44. $a^2 - 5a - 6$

45. $\dfrac{bd}{a} \div c$

46. $\dfrac{2ac}{b} \div (-c)$

47. $2(b + c) - 2a$

48. $3(b - a) - bc$

49. $\dfrac{b - 2a}{bc^2 - d}$

50. $\dfrac{b^2 - a}{ad + 3c}$

51. $\dfrac{1}{3}d^2 - \dfrac{3}{8}b^2$

52. $\dfrac{5}{8}a^4 - c^2$

53. $\dfrac{-4bc}{2a - b}$

54. $\dfrac{abc}{b - d}$

55. $a^3 - 3a^2 + a$

56. $d^3 - 3d - 9$

57. $-\dfrac{3}{4}b + \dfrac{1}{2}(ac + bd)$

58. $-\dfrac{2}{3}d - \dfrac{1}{5}(bd - ac)$

59. $(b - a)^2 - (d - c)^2$

60. $(b + c)^2 + (a + d)^2$

61. $4ac + (2a)^2$

62. $3dc - (4c)^2$

Evaluate the variable expression when $a = 2.7$, $b = -1.6$, and $c = -0.8$.

63. $c^2 - ab$

64. $(a + b)^2 - c$

65. $\dfrac{b^3}{c} - 4a$

Solve. Round to the nearest tenth.

66. Geometry Find the volume of a sphere that has a radius of 8.5 cm.

67. Geometry Find the volume of a right circular cylinder that has a radius of 1.25 in. and a height of 5.25 in.

68. Geometry The radius of the base of a right circular cylinder is 3.75 ft. The height of the cylinder is 9.5 ft. Find the surface area of the cylinder.

69. Geometry The length of one base of a trapezoid is 17.5 cm, and the length of the other base is 10.25 cm. The height is 6.75 cm. What is the area of the trapezoid?

70. Geometry A right circular cone has a height of 2.75 in. The diameter of the base is 1 in. Find the volume of the cone.

71. Geometry A right circular cylinder has a height of 12.6 m. The diameter of the base is 7 m. Find the volume of the cylinder.

72. The value of z is the value of $a^2 - 2a$ when $a = -3$. Find the value of z^2.

73. The value of a is the value of $3x^2 - 4x - 5$ when $x = -2$. Find the value of $3a - 4$.

74. The value of c is the value of $a^2 + b^2$ when $a = 2$ and $b = -2$. Find the value of $c^2 - 4$.

75. The value of d is the value of $3w^2 - 2v$ when $w = -1$ and $v = 3$. Find the value of $d^2 - 4d$.

8.5 cm

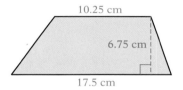

10.25 cm

6.75 cm

17.5 cm

12.6 m

7 m

2.1 **Applying Concepts**

Evaluate the variable expression when $a = -2$ and $b = -3$.

76. $|2a + 3b|$

77. $|-4ab|$

78. $|5a - b|$

Evaluate the variable expression when $a = \dfrac{2}{3}$ and $b = -\dfrac{3}{2}$.

79. $\dfrac{1}{3}a^5b^6$

80. $\dfrac{(2ab)^3}{2a^3b^3}$

81. $|5ab - 8a^2b^2|$

Evaluate the following expressions for $x = 2$, $y = 3$, and $z = -2$.

82. $3^x - x^3$

83. $2^y - y^2$

84. z^y

85. z^x

86. $x^x - y^y$

87. $y^{(x^2)}$

88. For each of the following, determine the first natural number x, greater than 2, for which the second expression is larger than the first. On the basis of your answers, make a conjecture that appears to be true about the expressions x^n and n^x, where $n = 3, 4, 5, 6, 7, \ldots$ and x is a natural number greater than 2.

    a. $x^3, 3^x$ **b.** $x^4, 4^x$

    c. $x^5, 5^x$ **d.** $x^6, 6^x$

2.2   Simplifying Variable Expressions

1   **The Properties of the Real Numbers**

The Properties of the Real Numbers describe the way operations on numbers can be performed. Here are some of the Properties of the Real Numbers and an example of each.

> **The Commutative Property of Addition**
>
> If a and b are real numbers, then $a + b = b + a$.

Two terms can be added in either order; the sum is the same.

$$4 + 3 = 7 \quad \text{and} \quad 3 + 4 = 7$$

> **The Commutative Property of Multiplication**
>
> If a and b are real numbers, then $a \cdot b = b \cdot a$.

Two factors can be multiplied in either order; the product is the same.

$$(5)(-2) = -10 \quad \text{and} \quad (-2)(5) = -10$$

> ### The Associative Property of Addition
>
> If a, b, and c are real numbers, then $(a + b) + c = a + (b + c)$.

When three or more terms are added, the terms can be grouped (with parentheses, for example) in any order; the sum is the same.

$$2 + (3 + 4) = 2 + 7 = 9 \quad \text{and} \quad (2 + 3) + 4 = 5 + 4 = 9$$

> ### The Associative Property of Multiplication
>
> If a, b, and c are real numbers, then $(a \cdot b) \cdot c = a \cdot (b \cdot c)$.

When three or more factors are multiplied, the factors can be grouped in any order; the product is the same.

$$(2 \cdot 3) \cdot 4 = 6 \cdot 4 = 24 \quad \text{and} \quad 2 \cdot (3 \cdot 4) = 2 \cdot 12 = 24$$

> ### The Addition Property of Zero
>
> If a is a real number, then $a + 0 = 0 + a = a$.

The sum of a term and zero is the term.

$$4 + 0 = 4 \qquad 0 + 4 = 4$$

> ### The Multiplication Property of Zero
>
> If a is a real number, then $a \cdot 0 = 0 \cdot a = 0$.

The product of a term and zero is zero.

$$(5)(0) = 0 \qquad (0)(5) = 0$$

> ### The Multiplication Property of One
>
> If a is a real number, then $a \cdot 1 = 1 \cdot a = a$.

The product of a term and 1 is the term.

$$6 \cdot 1 = 6 \qquad 1 \cdot 6 = 6$$

> ### The Inverse Property of Addition
>
> If a is a real number, then $a + (-a) = (-a) + a = 0$.

The sum of a number and its opposite is zero.
The opposite of a number is called its **additive inverse.**

$$8 + (-8) = 0 \qquad (-8) + 8 = 0$$

The Inverse Property of Multiplication

If a is a real number, and $a \neq 0$, then $a \cdot \dfrac{1}{a} = \dfrac{1}{a} \cdot a = 1$.

The product of a number and its reciprocal is 1.

$\dfrac{1}{a}$ is the **reciprocal** of a. $\dfrac{1}{a}$ is also called the **multiplicative inverse** of a.

$$7 \cdot \dfrac{1}{7} = 1 \qquad \dfrac{1}{7} \cdot 7 = 1$$

The Distributive Property

If a, b, and c are real numbers, then $a(b + c) = ab + ac$ or $(b + c)a = ba + ca$.

By the Distributive Property, the term outside the parentheses is multiplied by each term inside the parentheses.

$$2(3 + 4) = 2 \cdot 3 + 2 \cdot 4 \qquad (4 + 5)2 = 4 \cdot 2 + 5 \cdot 2$$
$$2 \cdot 7 = 6 + 8 \qquad\qquad 9 \cdot 2 = 8 + 10$$
$$14 = 14 \qquad\qquad\quad 18 = 18$$

Example 1 Complete the statement by using the Commutative Property of Multiplication.

$(6)(5) = (?)(6)$

Solution $(6)(5) = (5)(6)$ •The Commutative Property of Multiplication states that $a \cdot b = b \cdot a$.

Problem 1 Complete the statement by using the Inverse Property of Addition.

$7 + ? = 0$

Solution See page S3.

Example 2 Identify the property that justifies the statement.

$2(8 + 5) = 16 + 10$

Solution The Distributive Property •The Distributive Property states that $a(b + c) = ab + ac$.

> **Problem 2** Identify the property that justifies the statement.
>
> $$5 + (13 + 7) = (5 + 13) + 7$$
>
> **Solution**   See page S3.

2   Simplify variable expressions using the Properties of Addition

VIDEO & DVD   CD TUTOR   WWW WEB   SSM

Like terms of a variable expression are terms with the same variable part. (Because $x^2 = x \cdot x$, x^2 and x are not like terms.)

Constant terms are like terms. 4 and 9 are like terms.

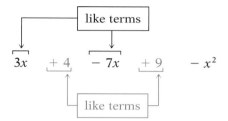

like terms

$3x$ $+ 4$ $- 7x$ $+ 9$ $- x^2$

like terms

To **combine** like terms, use the Distributive Property $ba + ca = (b + c)a$ to add the coefficients.

$$2x + 3x = (2 + 3)x$$
$$= 5x$$

> ┌─ **Example 3** Simplify.   **A.** $-2y + 3y$ **B.** $5x - 11x$
> │
> │ **Solution**   **A.** $-2y + 3y$
> │ $= (-2 + 3)y$ • Use the Distributive Property
> │ $ba + ca = (b + c)a$.
> │ $= 1y$ • Add the coefficients.
> │ $= y$ • Use the Multiplication Property of One.
> │
> │   **B.** $5x - 11x$
> │ $= [5 + (-11)]x$ • Use the Distributive Property
> │ $ba + ca = (b + c)a$.
> │ $= -6x$ • Add the coefficients.
> │
> │ **Problem 3** Simplify.   **A.** $9x + 6x$ **B.** $-4y - 7y$
> │
> │ **Solution**   See page S3.

In simplifying more complicated expressions, use the Properties of Addition.

The Commutative Property of Addition can be used when adding two like terms. The terms can be added in either order. The sum is the same.

$$2x + (-4x) = -4x + 2x$$
$$[2 + (-4)]x = (-4 + 2)x$$
$$-2x = -2x$$

The Associative Property of Addition is used when adding three or more terms. The terms can be grouped in any order. The sum is the same.

$$3x + 5x + 9x = (3x + 5x) + 9x = 3x + (5x + 9x)$$
$$8x + 9x = 3x + 14x$$
$$17x = 17x$$

By the Addition Property of Zero, the sum of a term and zero is the term.

$$5x + 0 = 0 + 5x = 5x$$

By the Inverse Property of Addition, the sum of a term and its additive inverse is zero.

$$7x + (-7x) = -7x + 7x = 0$$

Example 4 Simplify.   **A.** $8x + 3y - 8x$ **B.** $4x^2 + 5x - 6x^2 - 2x$

Solution   **A.** $8x + 3y - 8x$
$$= 3y + 8x - 8x$$
- Use the Commutative Property of Addition to rearrange the terms.
$$= 3y + (8x - 8x)$$
- Use the Associative Property of Addition to group like terms.
$$= 3y + 0$$
- Use the Inverse Property of Addition.
$$= 3y$$
- Use the Addition Property of Zero.

B. $4x^2 + 5x - 6x^2 - 2x$
$$= 4x^2 - 6x^2 + 5x - 2x$$
- Use the Commutative Property of Addition to rearrange the terms.
$$= (4x^2 - 6x^2) + (5x - 2x)$$
- Use the Associative Property of Addition to group like terms.
$$= -2x^2 + 3x$$
- Combine like terms.

Problem 4 Simplify.   **A.** $3a - 2b + 5a$ **B.** $x^2 - 7 + 9x^2 - 14$

Solution   See page S3.

---

3   ## Simplify variable expressions using the Properties of Multiplication

The Properties of Multiplication are used in simplifying variable expressions.

The Associative Property is used when multiplying three or more factors.

$$2(3x) = (2 \cdot 3)x = 6x$$

The Commutative Property can be used to change the order in which factors are multiplied.

$$(3x) \cdot 2 = 2 \cdot (3x) = 6x$$

By the Multiplication Property of One, the product of a term and 1 is the term.

$$(8x)(1) = (1)(8x) = 8x$$

By the Inverse Property of Multiplication, the product of a term and its reciprocal is 1.

$$5x \cdot \frac{1}{5x} = \frac{1}{5x} \cdot 5x = 1, x \neq 0$$

Example 5 Simplify.   **A.** $2(-x)$ **B.** $\frac{3}{2}\left(\frac{2x}{3}\right)$ **C.** $(16x)2$

Solution   **A.** $2(-x) = 2(-1 \cdot x)$
- $-x = -1x = -1 \cdot x$
$$= [2 \cdot (-1)]x$$
- Use the Associative Property of Multiplication to group factors.
$$= -2x$$
- Multiply.

B. $\dfrac{3}{2}\left(\dfrac{2x}{3}\right) = \dfrac{3}{2}\left(\dfrac{2}{3}x\right)$ • Note that $\dfrac{2x}{3} = \dfrac{2}{3} \cdot \dfrac{x}{1} = \dfrac{2}{3}x.$

$\phantom{\dfrac{3}{2}\left(\dfrac{2x}{3}\right)} = \left(\dfrac{3}{2} \cdot \dfrac{2}{3}\right)x$ • Use the Associative Property of Multiplication to group factors.

$\phantom{\dfrac{3}{2}\left(\dfrac{2x}{3}\right)} = 1x$ • Use the Inverse Property of Multiplication.
$\phantom{\dfrac{3}{2}\left(\dfrac{2x}{3}\right)} = x$ • Use the Multiplication Property of One.

C. $(16x)2 = 2(16x)$ • Use the Commutative Property of Multiplication to rearrange factors.

$ = (2 \cdot 16)x$ • Use the Associative Property of Multiplication to group factors.

$ = 32x$ • Multiply.

Problem 5   Simplify.   **A.** $-7(-2a)$ **B.** $-\dfrac{5}{6}(-30y^2)$ **C.** $(-5x)(-2)$

Solution   See page S3.

---

4 ## Simplify variable expressions using the Distributive Property

VIDEO & DVD   CD TUTOR   WEB   SSM

The Distributive Property is used to remove parentheses from a variable expression.

$3(2x - 5)$
$= 3(2x) - 3(5)$
$= 6x - 15$

An extension of the Distributive Property is used when an expression contains more than two terms.

$4(x^2 + 6x - 1)$
$= 4(x^2) + 4(6x) - 4(1)$
$= 4x^2 + 24x - 4$

Example 6   Simplify.   **A.** $-3(5 + x)$ **B.** $-(2x - 4)$
 **C.** $(2y - 6)2$ **D.** $5(3x + 7y - z)$

Solution   **A.** $-3(5 + x)$
$= -3(5) + (-3)x$ • Use the Distributive Property.
$= -15 - 3x$ • Multiply.

B. $-(2x - 4)$
$= -1(2x - 4)$ • Just as $-x = -1x,$
 $-(2x - 4) = -1(2x - 4).$

$= -1(2x) - (-1)(4)$ • Use the Distributive Property.
$= -2x + 4$ • Note: When a negative sign immediately precedes the parentheses, remove the parentheses and change the sign of *each* term inside the parentheses.

C. $(2y - 6)2$
$= (2y)(2) - (6)(2)$ • Use the Distributive Property
$= 4y - 12$ $(b + c)a = ba + ca.$

D. $5(3x + 7y - z)$
$= 5(3x) + 5(7y) - 5(z)$ • Use the Distributive Property.
$= 15x + 35y - 5z$

Problem 6   Simplify.   **A.** $7(4 + 2y)$ **B.** $-(5x - 12)$
C. $(3a - 1)5$ **D.** $-3(6a^2 - 8a + 9)$

Solution   See page S3.

---

5   ## Simplify general variable expressions

When simplifying variable expressions, use the Distributive Property to remove parentheses and brackets used as grouping symbols.

Example 7   Simplify: $4(x - y) - 2(-3x + 6y)$

Solution $4(x - y) - 2(-3x + 6y)$
$= 4x - 4y + 6x - 12y$ • Use the Distributive Property to remove parentheses.

$= 10x - 16y$ • Combine like terms.

Problem 7   Simplify: $7(x - 2y) - 3(-x - 2y)$

Solution   See page S3.

Example 8   Simplify: $2x - 3[2x - 3(x + 7)]$

Solution $2x - 3[2x - 3(x + 7)]$
$= 2x - 3[2x - 3x - 21]$ • Use the Distributive Property to remove the inner grouping symbols.

$= 2x - 3[-x - 21]$ • Combine like terms inside the grouping symbols.

$= 2x + 3x + 63$ • Use the Distributive Property to remove the brackets.

$= 5x + 63$ • Combine like terms.

Problem 8   Simplify: $3y - 2[x - 4(2 - 3y)]$

Solution   See page S3.

2.2   Concept Review

Determine whether the statement is always true, sometimes true, or never true.

1. The Associative Property of Addition states that two terms can be added in either order and the sum will be the same.

2. When three numbers are multiplied, the numbers can be grouped in any order and the product will be the same.

3. The Multiplication Property of One states that multiplying a number by one does not change the number.

4. The sum of a number and its additive inverse is zero.

5. The product of a number and its multiplicative inverse is one.

6. The terms x and x^2 are like terms because both have a coefficient of 1.

7. $3(x + 4) = 3x + 4$ is an example of the correct application of the Distributive Property.

8. To add like terms, add the coefficients; the variable part remains unchanged.

2.2 **Exercises**

1 Use the given property to complete the statement.

1. The Commutative Property of Multiplication
$2 \cdot 5 = 5 \cdot ?$

2. The Commutative Property of Addition
$9 + 17 = ? + 9$

3. The Associative Property of Multiplication
$(4 \cdot 5) \cdot 6 = 4 \cdot (? \cdot 6)$

4. The Associative Property of Addition
$(4 + 5) + 6 = ? + (5 + 6)$

5. The Distributive Property
$2(4 + 3) = 8 + ?$

6. The Addition Property of Zero
$? + 0 = -7$

7. The Inverse Property of Addition
$8 + ? = 0$

8. The Inverse Property of Multiplication
$\dfrac{1}{-5}(-5) = ?$

9. The Multiplication Property of One
$? \cdot 1 = -4$

10. The Multiplication Property of Zero
$12 \cdot ? = 0$

Identify the property that justifies the statement.

11. $-7 + 7 = 0$

12. $(-8)\left(-\dfrac{1}{8}\right) = 1$

13. $23 + 19 = 19 + 23$

14. $-21 + 0 = -21$

15. $2 + (6 + 14) = (2 + 6) + 14$

16. $(-3 + 9)8 = -24 + 72$

17. $3 \cdot 5 = 5 \cdot 3$

18. $-32(0) = 0$

19. $(4 \cdot 3) \cdot 5 = 4 \cdot (3 \cdot 5)$

20. $\dfrac{1}{4}(1) = \dfrac{1}{4}$

21. What are *like terms*? Give an example of two like terms. Give an example of two terms that are not like terms.

22. Explain the meaning of the phrase "simplify a variable expression."

23. Which of the following are like terms?
$3x$, $3x^2$, $5x$, $5xy$

24. Which of the following are like terms?
$-7a$, $-7b$, $-4a$, $-7a^2$

2 Simplify.

25. $6x + 8x$

26. $12x + 13x$

27. $9a - 4a$

28. $12a - 3a$

29. $4y - 10y$

30. $8y - 6y$

31. $-3b - 7$

32. $-12y - 3$

33. $-12a + 17a$

34. $-3a + 12a$

35. $5ab - 7ab$

36. $9ab - 3ab$

37. $-12xy + 17xy$

38. $-15xy + 3xy$

39. $-3ab + 3ab$

40. $-7ab + 7ab$

41. $-\dfrac{1}{2}x - \dfrac{1}{3}x$

42. $-\dfrac{2}{5}y + \dfrac{3}{10}y$

43. $\dfrac{3}{8}x^2 - \dfrac{5}{12}x^2$

44. $\dfrac{2}{3}y^2 - \dfrac{4}{9}y^2$

45. $3x + 5x + 3x$

46. $8x + 5x + 7x$

47. $5a - 3a + 5a$

48. $10a - 17a + 3a$

49. $-5x^2 - 12x^2 + 3x^2$

50. $-y^2 - 8y^2 + 7y^2$

51. $7x - 8x + 3y$

52. $8y - 10x + 8x$

53. $7x - 3y + 10x$

54. $8y + 8x - 8y$

55. $3a - 7b - 5a + b$

56. $-5b + 7a - 7b + 12a$

57. $3x - 8y - 10x + 4x$

58. $3y - 12x - 7y + 2y$

59. $x^2 - 7x - 5x^2 + 5x$

60. $3x^2 + 5x - 10x^2 - 10x$

3 Simplify.

61. $4(3x)$

62. $12(5x)$

63. $-3(7a)$

64. $-2(5a)$

65. $-2(-3y)$

66. $-5(-6y)$

67. $(4x)2$

68. $(6x)12$

69. $(3a)(-2)$

70. $(7a)(-4)$

71. $(-3b)(-4)$

72. $(-12b)(-9)$

73. $-5(3x^2)$

74. $-8(7x^2)$

75. $\dfrac{1}{3}(3x^2)$

76. $\dfrac{1}{5}(5a)$

77. $\dfrac{1}{8}(8x)$

78. $-\dfrac{1}{4}(-4a)$

79. $-\dfrac{1}{7}(-7n)$

80. $\left(\dfrac{3x}{4}\right)\left(\dfrac{4}{3}\right)$

81. $\dfrac{12x}{5}\left(\dfrac{5}{12}\right)$

82. $(-6y)\left(-\dfrac{1}{6}\right)$

83. $(-10n)\left(-\dfrac{1}{10}\right)$

84. $\dfrac{1}{3}(9x)$

85. $\dfrac{1}{7}(14x)$

86. $-\dfrac{1}{5}(10x)$

87. $-\dfrac{1}{8}(16x)$

88. $-\dfrac{2}{3}(12a^2)$

89. $-\dfrac{5}{8}(24a^2)$

90. $-\dfrac{1}{2}(-16y)$

91. $-\dfrac{3}{4}(-8y)$

92. $(16y)\left(\dfrac{1}{4}\right)$

93. $(33y)\left(\dfrac{1}{11}\right)$

94. $(-6x)\left(\dfrac{1}{3}\right)$

95. $(-10x)\left(\dfrac{1}{5}\right)$

96. $(-8a)\left(-\dfrac{3}{4}\right)$

4 Simplify.

97. $-(x + 2)$ **98.** $-(x + 7)$ **99.** $2(4x - 3)$ **100.** $5(2x - 7)$

101. $-2(a + 7)$ **102.** $-5(a + 16)$ **103.** $-3(2y - 8)$ **104.** $-5(3y - 7)$

105. $(5 - 3b)7$ **106.** $(10 - 7b)2$ **107.** $-3(3 - 5x)$ **108.** $-5(7 - 10x)$

109. $3(5x^2 + 2x)$ **110.** $6(3x^2 + 2x)$ **111.** $-2(-y + 9)$ **112.** $-5(-2x + 7)$

113. $(-3x - 6)5$ **114.** $(-2x + 7)7$ **115.** $2(-3x^2 - 14)$ **116.** $5(-6x^2 - 3)$

117. $-3(2y^2 - 7)$ **118.** $-8(3y^2 - 12)$ **119.** $-(6a^2 - 7b^2)$

120. $3(x^2 + 2x - 6)$ **121.** $4(x^2 - 3x + 5)$ **122.** $-2(y^2 - 2y + 4)$

123. $-3(y^2 - 3y - 7)$ **124.** $2(-a^2 - 2a + 3)$ **125.** $4(-3a^2 - 5a + 7)$

126. $-5(-2x^2 - 3x + 7)$ **127.** $-3(-4x^2 + 3x - 4)$ **128.** $3(2x^2 + xy - 3y^2)$

129. $5(2x^2 - 4xy - y^2)$ **130.** $-(3a^2 + 5a - 4)$ **131.** $-(8b^2 - 6b + 9)$

5 Simplify.

132. $4x - 2(3x + 8)$ **133.** $6a - (5a + 7)$ **134.** $9 - 3(4y + 6)$

135. $10 - (11x - 3)$ **136.** $5n - (7 - 2n)$ **137.** $8 - (12 + 4y)$

138. $3(x + 2) - 5(x - 7)$ **139.** $2(x - 4) - 4(x + 2)$ **140.** $12(y - 2) + 3(7 - 3y)$

141. $6(2y - 7) - 3(3 - 2y)$ **142.** $3(a - b) - 4(a + b)$ **143.** $2(a + 2b) - (a - 3b)$

144. $4[x - 2(x - 3)]$ **145.** $2[x + 2(x + 7)]$ **146.** $-2[3x + 2(4 - x)]$

147. $-5[2x + 3(5 - x)]$ **148.** $-3[2x - (x + 7)]$ **149.** $-2[3x - (5x - 2)]$

150. $2x - 3[x - 2(4 - x)]$ **151.** $-7x + 3[x - 7(3 - 2x)]$ **152.** $-5x - 2[2x - 4(x + 7)] - 6$

153. $4a - 2[2b - (b - 2a)] + 3b$ **154.** $2x + 3(x - 2y) + 5(3x - 7y)$ **155.** $5y - 2(y - 3x) + 2(7x - y)$

2.2   **Applying Concepts**

Simplify.

156. $C - 0.7C$ **157.** $\dfrac{1}{3}(3x + y) - \dfrac{2}{3}(6x - y)$ **158.** $-\dfrac{1}{4}[2x + 2(y - 6y)]$

Complete.

159. A number that has no reciprocal is _____ .

160. A number that is its own reciprocal is _____ .

161. The additive inverse of $a - b$ is _____ .

162. Determine whether the statement is true or false. If the statement is false, give an example that illustrates that it is false.
  a. Division is a commutative operation.
  b. Division is an associative operation.
  c. Subtraction is an associative operation.
  d. Subtraction is a commutative operation.
  e. Addition is a commutative operation.

163. Define an operation $\otimes$ as $a \otimes b = (a \cdot b) - (a + b)$.
  For example, $7 \otimes 5 = (7 \cdot 5) - (7 + 5) = 35 - 12 = 23$.
  a. Is $\otimes$ a commutative operation? Support your answer.
  b. Is $\otimes$ an associative operation? Support your answer.

164. Give examples of two operations that occur in everyday experience that are not commutative (for example, putting on socks and then shoes).

165. In your own words, explain the Distributive Property.

2.3 Translating Verbal Expressions into Variable Expressions

1 **Translate a verbal expression into a variable expression**

VIDEO & DVD   CD TUTOR   WEB   SSM

One of the major skills required in applied mathematics is the ability to translate a verbal expression into a variable expression. This requires recognizing the verbal phrases that translate into mathematical operations. Here is a partial list of the phrases used to indicate the different mathematical operations.

Addition	added to	6 added to y	$y + 6$
	more than	8 more than x	$x + 8$
	the sum of	the sum of x and z	$x + z$
	increased by	t increased by 9	$t + 9$
	the total of	the total of 5 and y	$5 + y$
	plus	b plus 17	$b + 17$
Subtraction	minus	x minus 2	$x - 2$
	less than	7 less than t	$t - 7$
	less	7 less t	$7 - t$
	subtracted from	5 subtracted from d	$d - 5$
	decreased by	m decreased by 3	$m - 3$
	the difference between	the difference between y and 4	$y - 4$

Multiplication	times	10 times t	$10t$
	of	one half of x	$\frac{1}{2}x$
	the product of	the product of y and z	yz
	multiplied by	y multiplied by 11	$11y$
	twice	twice n	$2n$
Division	divided by	x divided by 12	$\frac{x}{12}$
	the quotient of	the quotient of y and z	$\frac{y}{z}$
	the ratio of	the ratio of t to 9	$\frac{t}{9}$
Power	the square of	the square of x	x^2
	the cube of	the cube of a	a^3

Translating a phrase that contains the word *sum, difference, product,* or *quotient* can sometimes cause a problem. In the examples at the right, note where the operation symbol is placed.

the *sum* of x and y $x + y$

the *difference* between x and y $x - y$

the *product* of x and y $x \cdot y$

the *quotient* of x and y $\frac{x}{y}$

Example 1   Translate into a variable expression.

A. the total of five times b and c

B. the quotient of eight less than n and fourteen

C. thirteen more than the sum of seven and the square of x

Solution   **A.** the <u>total</u> of five <u>times</u> b and c

• Identify words that indicate mathematical operations.

$5b + c$

• Use the operations to write the variable expression.

B. the <u>quotient</u> of eight <u>less than</u> n and fourteen

• Identify words that indicate mathematical operations.

$\dfrac{n - 8}{14}$

• Use the operations to write the variable expression.

C. thirteen <u>more than</u> the <u>sum</u> of seven and the <u>square</u> of x

$(7 + x^2) + 13$

Problem 1   Translate into a variable expression.

A. eighteen less than the cube of x

B. y decreased by the sum of z and nine

C. the difference between the square of q and the sum of r and t

Solution   See page S3.

In most applications that involve translating phrases into variable expressions, the variable to be used is not given. To translate these phrases, a variable must be assigned to an unknown quantity before the variable expression can be written.

Take Note

The expression $n(6 + n^3)$ must have parentheses. If we write $n \cdot 6 + n^3$, then by the Order of Operations Agreement, only the 6 is multiplied by n, but we want n to be multiplied by the *total* of 6 and n^3.

Example 2   Translate "a number multiplied by the total of six and the cube of the number" into a variable expression.

Solution   the unknown number: n

• Assign a variable to one of the unknown quantities.

the cube of the number: n^3
the total of six and the cube of the number: $6 + n^3$

• Use the assigned variable to write an expression for any other unknown quantity.

$n(6 + n^3)$

• Use the assigned variable to write the variable expression.

Problem 2   Translate "a number added to the product of five and the square of the number" into a variable expression.

Solution   See page S3.

Example 3   Translate "the quotient of twice a number and the difference between the number and twenty" into a variable expression.

Solution   the unknown number: n
twice the number: $2n$
the difference between the number and twenty: $n - 20$

$$\frac{2n}{n - 20}$$

Problem 3   Translate "the product of three and the sum of seven and twice a number" into a variable expression.

Solution   See page S4.

---

2   **Translate a verbal expression into a variable expression and then simplify the resulting expression**

After translating a verbal expression into a variable expression, simplify the variable expression by using the Properties of the Real Numbers.

Example 4   Translate and simplify "the total of four times an unknown number and twice the difference between the number and eight."

Solution   the unknown number: n

• Assign a variable to one of the unknown quantities.

four times the unknown number: $4n$
twice the difference between the number and eight: $2(n - 8)$

• Use the assigned variable to write an expression for any other unknown quantity.

$$4n + 2(n - 8)$$

• Use the assigned variable to write the variable expression.

$$= 4n + 2n - 16$$

• Simplify the variable expression.

$$= 6n - 16$$

Problem 4   Translate and simplify "a number minus the difference between twice the number and seventeen."

Solution   See page S4.

Example 5   Translate and simplify "the difference between five-eighths of a number and two-thirds of the same number."

Solution   the unknown number: n

five-eighths of the number: $\dfrac{5}{8}n$

• Assign a variable to one of the unknown quantities.

two-thirds of the number: $\dfrac{2}{3}n$

• Use the assigned variable to write an expression for any other unknown quantity.

$$\dfrac{5}{8}n - \dfrac{2}{3}n$$

$$= \dfrac{15}{24}n - \dfrac{16}{24}n$$

• Use the assigned variable to write the variable expression.

$$= -\dfrac{1}{24}n$$

• Simplify the variable expression.

Problem 5   Translate and simplify "the sum of three-fourths of a number and one-fifth of the same number."

Solution   See page S4.

3   **Translate application problems**

VIDEO & DVD   CD TUTOR   WEB   SSM

Many of the applications of mathematics require that you identify an unknown quantity, assign a variable to that quantity, and then attempt to express another unknown quantity in terms of that variable.

Suppose we know that the sum of two numbers is 10 and that one of the two numbers is 4. We can find the other number by subtracting 4 from 10.

one number: 4
other number: $10 - 4 = 6$
The two numbers are 4 and 6.

Now suppose we know that the sum of two numbers is 10, we don't know either number, and we want to express *both* numbers in terms of the *same* variable.

Let one number be x. Again, we can find the other number by subtracting x from 10.

one number: x
other number: $10 - x$
The two numbers are x and $10 - x$.

Note that the sum of x and $10 - x$ is 10.

$$x + (10 - x) = x + 10 - x = 10$$

Example 6 The length of a swimming pool is 20 ft longer than the width. Express the length of the pool in terms of the width.

Solution   the width of the pool: W

• Assign a variable to the width of the pool.

the length is 20 more than the width: $W + 20$

• Express the length of the pool in terms of W.

> **Take Note**
>
> Any variable can be used. For example, if the width is y, then the length is $y + 20$.

Problem 6 An older computer takes twice as long to process a set of data as does a newer model. Express the amount of time it takes the older computer to process the data in terms of the amount of time it takes the newer model.

Solution   See page S4.

Example 7 An investor divided $5000 between two accounts, one a mutual fund and the other a money market fund. Use one variable to express the amounts invested in each account.

Solution   the amount invested in the mutual fund: x

• Assign a variable to the amount invested in one account.

the amount invested in the money market fund: $5000 - x$

• Express the amount invested in the other account in terms of x.

> **Take Note**
>
> It is also correct to assign the variable to the amount in the money market fund. Then the amount in the mutual fund is $5000 - x$.

Problem 7 A guitar string 6 ft long was cut into two pieces. Use one variable to express the lengths of the two pieces.

Solution   See page S4.

2.3   Concept Review

Determine whether the statement is always true, sometimes true, or never true.

1. "Five less than n" can be translated as "$5 - n$."

2. A variable expression contains an equals sign.

3. If the sum of two numbers is 12 and one of the two numbers is x, then the other number can be expressed as $x - 12$.

4. The words *total* and *times* both indicate multiplication.

5. The words *quotient* and *ratio* both indicate division.

6. The expressions $7y - 8$ and $(7y) - 8$ are equivalent.

7. The phrase "five times the sum of x and y" and the phrase "the sum of five times x and y" yield the same variable expression.

2.3 Exercises

1 Translate into a variable expression.

1. d less than nineteen

2. the sum of six and c

3. r decreased by twelve

4. w increased by fifty-five

5. a multiplied by twenty-eight

6. y added to sixteen

7. five times the difference between n and seven

8. thirty less than the square of b

9. y less the product of three and y

10. the sum of four-fifths of m and eighteen

11. the product of negative six and b

12. nine increased by the quotient of t and five

13. four divided by the difference between p and six

14. the product of seven and the total of r and eight

15. the quotient of nine less than x and twice x

16. the product of a and the sum of a and thirteen

17. twenty-one less than the product of s and negative four

18. fourteen more than one-half of the square of z

19. the ratio of eight more than d to d

20. the total of nine times the cube of m and the square of m

21. three-eighths of the sum of t and fifteen

22. s decreased by the quotient of s and two

23. w increased by the quotient of seven and w

24. the difference between the square of c and the total of c and fourteen

25. d increased by the difference between sixteen times d and three

26. the product of eight and the total of b and five

27. a number divided by nineteen

28. thirteen less a number

29. forty more than a number

30. three-sevenths of a number

31. the square of the difference between a number and ninety

32. the quotient of twice a number and five

33. the sum of four-ninths of a number and twenty

34. eight subtracted from the product of fifteen and a number

35. the product of a number and ten more than the number

36. six less than the total of a number and the cube of the number

37. fourteen added to the product of seven and a number

38. the quotient of three and the total of four and a number

39. the quotient of twelve and the sum of a number and two

40. eleven plus one-half of a number

41. the ratio of two and the sum of a number and one

42. a number multiplied by the difference between twice the number and nine

43. the difference between sixty and the quotient of a number and fifty

44. the product of nine less than a number and the number

45. the sum of the square of a number and three times the number

46. the quotient of seven more than twice a number and the number

47. the sum of three more than a number and the cube of the number

48. a number decreased by the difference between the cube of the number and ten

49. the square of a number decreased by one-fourth of the number

50. four less than seven times the square of a number

51. twice a number decreased by the quotient of seven and the number

52. eighty decreased by the product of thirteen and a number

53. the cube of a number decreased by the product of twelve and the number

54. the quotient of five and the sum of a number and nineteen

2 Translate into a variable expression. Then simplify the expression.

55. a number increased by the total of the number and ten

56. a number added to the product of five and the number

57. a number decreased by the difference between nine and the number

58. eight more than the sum of a number and eleven

59. the difference between one-fifth of a number and three-eighths of the number

60. a number minus the sum of the number and fourteen

61. four more than the total of a number and nine

62. the sum of one-eighth of a number and one-twelfth of the number

63. twice the sum of three times a number and forty

64. the sum of a number divided by two and the number

65. seven times the product of five and a number

66. sixteen multiplied by one-fourth of a number

67. the total of seventeen times a number and twice the number

68. the difference between nine times a number and twice the number

69. a number plus the product of the number and twelve

70. nineteen more than the difference between a number and five

71. three times the sum of the square of a number and four

72. a number subtracted from the product of the number and seven

73. three-fourths of the sum of sixteen times a number and four

74. the difference between fourteen times a number and the product of the number and seven

75. sixteen decreased by the sum of a number and nine

76. eleven subtracted from the difference between eight and a number

77. five more than the quotient of four times a number and two

78. twenty minus the sum of four-ninths of a number and three

79. six times the total of a number and eight

80. four times the sum of a number and twenty

81. seven minus the sum of a number and two

82. three less than the sum of a number and ten

83. one-third of the sum of a number and six times the number

84. twice the quotient of four times a number and eight

85. the total of eight increased by the cube of a number and twice the cube of the number

86. the sum of five more than the square of a number and twice the square of the number

87. twelve more than a number added to the difference between the number and six

88. a number plus four added to the difference between three and twice the number

89. the sum of a number and nine added to the difference between the number and twenty

90. seven increased by a number added to twice the difference between the number and two

91. fourteen plus the product of three less than a number and ten

92. a number plus the product of the number minus five and seven

3 Write a variable expression.

93. The sum of two numbers is 18. Express the two numbers in terms of the same variable.

94. The sum of two numbers is 20. Express the two numbers in terms of the same variable.

95. Genetics The human genome contains 11,000 more genes than the roundworm genome. Express the number of genes in the human genome in terms of the number of genes in the roundworm genome. (*Source:* Celera, USA TODAY research)

96. Telecommunications In 1947, phone companies began using area codes. According to information found at **555-1212.com**, there were 151 more area codes at the beginning of 2001 than there were in 1947. Express the number of area codes in 2001 in terms of the number of area codes in 1947.

97. Space Exploration A survey in *USA Today* reported that almost three-fourths of Americans think that money should be spent on exploration of Mars. Express the number of Americans who think that money should be spent on exploration of Mars in terms of the total number of Americans.

98. Biology According to the American Podiatric Medical Association, the bones in your foot account for one-fourth of all bones in your body. Express the number of bones in your foot in terms of the number of bones in your body.

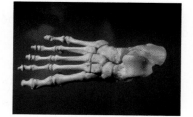

99. Sports In football, the number of points awarded for a touchdown is three times the number of points awarded for a safety. Express the number of points awarded for a touchdown in terms of the number of points awarded for a safety.

100. 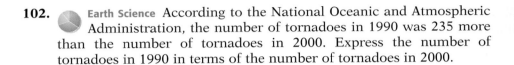 **Cost of Living** A cost-of-living calculator at **HOMECARE.com** shows that a person living in Manhattan would need twice the salary of a person living in San Diego. Express the salary needed to live in Manhattan in terms of the salary needed to live in San Diego.

101. **Astronomy** The planet Saturn has 9 more moons than Jupiter. Express the number of moons of Saturn in terms of the number of moons of Jupiter. (*Source:* NASA)

102. **Earth Science** According to the National Oceanic and Atmospheric Administration, the number of tornadoes in 1990 was 235 more than the number of tornadoes in 2000. Express the number of tornadoes in 1990 in terms of the number of tornadoes in 2000.

103. **Agriculture** In a recent year, Alabama produced one-half the number of pounds of pecans that Texas produced in the same year. Express the amount of pecans produced in Alabama in terms of the amount produced in Texas. (*Source:* National Agricultural Statistics Service)

104. **Taxes** According to the Internal Revenue Service, it takes about one-fifth the time to fill out Schedule B (interest and dividends) than it does to complete Schedule A (itemized deductions). Express the amount of time it takes to fill out Schedule B in terms of the time it takes to fill out Schedule A.

105. **Demography** According to the U.S. Bureau of the Census, the world population in the year 2050 is expected to be twice the world population in 1980. Express the expected world population in 2050 in terms of the world population in 1980.

106. **Geometry** The length of a rectangle is 5 m more than twice the width. Express the length of the rectangle in terms of the width.

107. **Geometry** In a triangle, the measure of the smallest angle is 10 degrees less than one-half the measure of the largest angle. Express the measure of the smallest angle in terms of the measure of the largest angle.

108. **Sports** A halyard 12 ft long is cut into two pieces. Use the same variable to express the lengths of the two pieces.

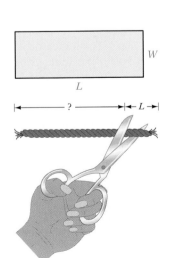

109. **Coins** A coin bank contains thirty-five coins in nickels and dimes. Use the same variable to express the number of nickels and the number of dimes in the coin bank.

110. **Travel** Two cars are traveling in opposite directions and at different rates. Two hours later the cars are 200 mi apart. Express the distance traveled by the faster car in terms of the distance traveled by the slower car.

111. **Wages** An employee is paid $720 per week plus $27 for each hour of overtime worked. Express the employee's weekly pay in terms of the number of hours of overtime worked.

112. **Billing** An auto repair bill is $138 for parts and $45 for each hour of labor. Express the amount of the repair bill in terms of the number of hours of labor.

113. **Consumerism** The cost of a cellular phone service is $19.95 per month plus $.08 per minute for phone calls. Express the monthly cost of the phone service in terms of the number of minutes of phone calls.

2.3 Applying Concepts

114. **Chemistry** The chemical formula for glucose (sugar) is $C_6H_{12}O_6$. This formula means that there are twelve hydrogen atoms, six carbon atoms, and six oxygen atoms in each molecule of glucose. If x represents the number of atoms of oxygen in a pound of sugar, express the number of hydrogen atoms in the pound of sugar in terms of x.

115. **Metalwork** A wire whose length is given as x inches is bent into a square. Express the length of a side of the square in terms of x.

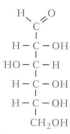

116. **Pulleys** A block-and-tackle system is designed so that pulling one end of a rope five feet will move a weight on the other end a distance of three feet. If x represents the distance the rope is pulled, express the distance the weight moves in terms of x.

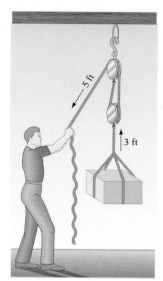

117. Translate the expressions $5x + 8$ and $5(x + 8)$ into phrases.

118. In your own words, explain how variables are used.

119. Explain the similarities and differences between the expressions "the difference between x and 5" and "5 less than x."

Focus on Problem Solving

From Concrete to Abstract

In your study of algebra, you will find that the problems are less concrete than those you studied in arithmetic. Problems that are concrete provide information pertaining to a specific instance. Algebra is more abstract. Abstract problems are theoretical; they are stated without reference to a specific instance. Let's look at an example of an abstract problem.

How many minutes are there in h hours?

A strategy that can be used to solve this problem is to begin by substituting a number for the variable.

How many minutes are there in 5 hours?

You know that there are 60 minutes in 1 hour. To find the number of minutes in 5 hours, multiply 5 by 60.

$$60 \cdot 5 = 300 \qquad \text{There are 300 minutes in 5 hours.}$$

Use the same procedure to find the number of minutes in h hours: multiply h by 60.

$$60 \cdot h = 60h \qquad \text{There are } 60h \text{ minutes in } h \text{ hours.}$$

This problem might be taken a step further:

If you walk one mile in x minutes, how far can you walk in h hours?

Consider the same problem using numbers in place of the variables.

If you walk one mile in 20 minutes, how far can you walk in 3 hours?

To solve this problem, you need to calculate the number of minutes in 3 hours (multiply 3 by 60), and divide the result by the number of minutes it takes to walk one mile (20 minutes).

$$\frac{60 \cdot 3}{20} = \frac{180}{20} = 9 \qquad \text{If you walk one mile in 20 minutes, you can walk 9 miles in 3 hours.}$$

Use the same procedure to solve the related abstract problem. Calculate the number of minutes in h hours (multiply h by 60), and divide the result by the number of minutes it takes to walk one mile (x minutes).

$$\frac{60 \cdot h}{x} = \frac{60h}{x} \qquad \text{If you walk one mile in } x \text{ minutes, you can walk } \frac{60h}{x} \text{ miles in } h \text{ hours.}$$

At the heart of the study of algebra is the use of variables. It is the variables in the problems above that make them abstract. But it is variables that allow us to generalize situations and state rules about mathematics.

Try each of the following problems.

1. How many hours are there in d days?

2. You earn d dollars an hour. What are your wages for working h hours?

3. If p is the price of one share of stock, how many shares can you purchase with d dollars?

4. A company pays a television station d dollars to air a commercial lasting s seconds. What is the cost per second?

5. After every v videotape rentals, you are entitled to one free rental. You have rented t tapes, where $t < v$. How many more do you need to rent before you are entitled to a free rental?

6. Your car gets g miles per gallon. How many gallons of gasoline does your car consume traveling t miles?

7. If you drink j ounces of juice each day, how many days will q quarts of the juice last?

8. A TV station has m minutes of commercials each hour. How many ads lasting s seconds each can be sold for each hour of programming?

9. A factory worker can assemble p products in m minutes. How many products can the factory worker assemble in h hours?

10. If one candy bar costs n nickels, how many candy bars can be purchased with q quarters?

Projects & Group Activities

Investigation into Operations with Even and Odd Integers

An **even integer** is an integer that is divisible by 2. For example, 6, −12, and 480 are even integers. An **odd integer** is an integer that is not evenly divisible by 2. For example, 3, −25, and 971 are odd integers.

Complete each statement with the word *even* or *odd*.

1. If k is an odd integer, then $k + 1$ is an _____ integer.

2. If k is an odd integer, then $k - 2$ is an _____ integer.

3. If n is an integer, then $2n$ is an _____ integer.

4. If m and n are even integers, then $m - n$ is an _____ integer.

5. If m and n are even integers, then mn is an _____ integer.

6. If m and n are odd integers, then $m + n$ is an _____ integer.

7. If m and n are odd integers, then $m - n$ is an _____ integer.

8. If m and n are odd integers, then mn is an _____ integer.

9. If m is an even integer and n is an odd integer, then $m - n$ is an _____ integer.

10. If m is an even integer and n is an odd integer, then $m + n$ is an _____ integer.

Applications of Patterns in Mathematics

For the circle shown at the left, use a straight line to connect each dot on the circle with every other dot on the circle. How many different straight lines are there?

Follow the same procedure for each of the circles shown below. How many different straight lines are there for each?

1. Find a pattern to describe the number of dots on a circle and the corresponding number of different lines drawn.

2. Use the pattern from Exercise 1 to determine the number of different lines that would be drawn in a circle with 7 dots and in a circle with 8 dots.

3. You are arranging a tennis tournament with nine players. How many singles matches will be played among the nine players if each player plays each of the other players once?

Now consider **triangular numbers,** which were studied by ancient Greek mathematicians. The numbers 1, 3, 6, 10, 15, 21 are the first six triangular numbers. Note in the following diagram that a triangle can be formed using the number of dots that correspond to a triangular number.

Observe that the number of dots in a row is one more than the number in the row above. The total number of dots can be found by addition.

$$1 = 1 \qquad\qquad 1 + 2 = 3$$
$$1 + 2 + 3 = 6 \qquad\qquad 1 + 2 + 3 + 4 = 10$$
$$1 + 2 + 3 + 4 + 5 = 15 \qquad 1 + 2 + 3 + 4 + 5 + 6 = 21$$

4. Use this pattern to find the seventh triangular number and the eighth triangular number. Check your answers by drawing the corresponding triangles of dots.

5. Discuss the relationship between triangular numbers and the pattern describing the correspondence between the number of dots on a circle and the number of different lines drawn (see Exercise 1).

6. Suppose you are in charge of scheduling softball games for a league. There are 10 teams in the league. Use the pattern of triangular numbers to determine the number of games that must be scheduled.

Chapter Summary

Key Words

Key Words	Examples

A **variable** is a letter that is used to stand for a quantity that is unknown or that can change. A **variable expression** is an expression that contains one or more variables.   [2.1.1, p. 52]

$5x - 4y + 7z$ is a variable expression. It contains the variables x, y, and z.

The **terms of a variable expression** are the addends of the expression. Each term is a **variable term** or a **constant term**. [2.1.1, p. 52]

The expression $4a^2 - 6b^3 + 7$ has three terms: $4a^2$, $-6b^3$, and 7. $4a^2$ and $-6b^3$ are variable terms. 7 is a constant term.

A variable term is composed of a **numerical coefficient** and a **variable part**.   [2.1.1, p. 53]

For the expression $8p^4r$, 8 is the coefficient and p^4r is the variable part.

Replacing the variables in a variable expression with numbers and then simplifying the numerical expression is called **evaluating the variable expression**.   [2.1.1, p. 53]

To evaluate $a^2 - 2b$ when $a = -3$ and $b = -5$, simplify the expression $(-3)^2 - 2(-5)$.

Like terms of a variable expression are the terms with the same variable part.   [2.2.2, p. 61]

For the expression $2st - 3t + 9s - 11st$, the terms $2st$ and $-11st$ are like terms.

The **additive inverse** of a number is the opposite of the number [2.2.1, p. 60]

The additive inverse of 8 is -8. The additive inverse of -15 is 15.

The **multiplicative inverse** of a number is the reciprocal of the number.   [2.2.1, p. 60]

The multiplicative inverse of $\frac{3}{8}$ is $\frac{8}{3}$.

Essential Rules and Procedures

The Commutative Property of Addition
If a and b are real numbers, then $a + b = b + a$. [2.2.1, p. 58]

$5 + 2 = 7$ and $2 + 5 = 7$

The Commutative Property of Multiplication
If a and b are real numbers, then $ab = ba$. [2.2.1, p. 58]

$6(-3) = -18$ and $-3(6) = -18$

The Associative Property of Addition
If a, b, and c are real numbers, then $(a + b) + c = a + (b + c)$. [2.2.1, p. 59]

$-1 + (4 + 7) = -1 + 11 = 10$
$(-1 + 4) + 7 = 3 + 7 = 10$

The Associative Property of Multiplication
If a, b, and c are real numbers, then $(ab)c = a(bc)$. [2.2.1, p. 59]

$(-2 \cdot 5) \cdot 3 = -10 \cdot 3 = -30$
$-2 \cdot (5 \cdot 3) = -2 \cdot 15 = -30$

The Addition Property of Zero
If a is a real number, then $a + 0 = 0 + a = a$. [2.2.1, p. 59]

$9 + 0 = 9 \qquad 0 + 9 = 9$

The Multiplication Property of Zero
If a is a real number, then $a \cdot 0 = 0 \cdot a = 0$. [2.2.1, p. 59]

$-8(0) = 0 \qquad 0(-8) = 0$

The Multiplication Property of One
If a is a real number, then $1 \cdot a = a \cdot 1 = a$. [2.2.1, p. 59]

$7 \cdot 1 = 7 \qquad 1 \cdot 7 = 7$

The Inverse Property of Addition
If a is a real number, then $a + (-a) = (-a) + a = 0$.
[2.2.1, p. 59]

$4 + (-4) = 0 \qquad -4 + 4 = 0$

The Inverse Property of Multiplication
If a is a real number and $a \neq 0$, then $a \cdot \dfrac{1}{a} = \dfrac{1}{a} \cdot a = 1$.

[2.2.1, p. 60]

$6 \cdot \dfrac{1}{6} = 1 \qquad \dfrac{1}{6} \cdot 6 = 1$

The Distributive Property
If a, b, and c are real numbers, then $a(b + c) = ab + ac$.
[2.2.1, p. 60]

$$5(2 + 3) = 5 \cdot 2 + 5 \cdot 3$$
$$= 10 + 15$$
$$= 25$$

Chapter Review Exercises

1. Simplify: $-7y^2 + 6y^2 - (-2y^2)$

2. Simplify: $(12x)\left(\dfrac{1}{4}\right)$

3. Simplify: $\dfrac{2}{3}(-15a)$

4. Simplify: $-2(2x - 4)$

5. Simplify: $5(2x + 4) - 3(x - 6)$

6. Evaluate $a^2 - 3b$ when $a = 2$ and $b = -4$.

7. Complete the statement by using the Inverse Property of Addition.
$-9 + ? = 0$

8. Simplify: $-4(-9y)$

9. Simplify: $-2(-3y + 9)$

10. Simplify: $3[2x - 3(x - 2y)] + 3y$

11. Simplify: $-4(2x^2 - 3y^2)$

12. Simplify: $3x - 5x + 7x$

13. Evaluate $b^2 - 3ab$ when $a = 3$ and $b = -2$.

14. Simplify: $\dfrac{1}{5}(10x)$

15. Simplify: $5(3 - 7b)$

16. Simplify: $2x + 3[4 - (3x - 7)]$

17. Identify the property that justifies the statement.
$-4(3) = 3(-4)$

18. Simplify: $3(8 - 2x)$

19. Simplify: $-2x^2 - (-3x^2) + 4x^2$

20. Simplify: $-3x - 2(2x - 7)$

21. Simplify: $-3(3y^2 - 3y - 7)$

22. Simplify: $-2[x - 2(x - y)] + 5y$

23. Evaluate $\dfrac{-2ab}{2b - a}$ when $a = -4$ and $b = 6$.

24. Simplify: $(-3)(-12y)$

25. Simplify: $4(3x - 2) - 7(x + 5)$

26. Simplify: $(16x)\left(\dfrac{1}{8}\right)$

27. Simplify: $-3(2x^2 - 7y^2)$

28. Evaluate $3(a - c) - 2ab$ when $a = 2$, $b = 3$, and $c = -4$.

29. Simplify: $2x - 3(x - 2)$

30. Simplify: $2a - (-3b) - 7a - 5b$

31. Simplify: $-5(2x^2 - 3x + 6)$

32. Simplify: $3x - 7y - 12x$

33. Simplify: $\dfrac{1}{2}(12a)$

34. Simplify: $2x + 3[x - 2(4 - 2x)]$

35. Simplify: $3x + (-12y) - 5x - (-7y)$

36. Simplify: $\left(-\dfrac{5}{6}\right)(-36b)$

37. Complete the statement by using the Distributive Property.
$(6 + 3)7 = 42 + ?$

38. Simplify: $4x^2 + 9x - 6x^2 - 5x$

39. Simplify: $-\dfrac{3}{8}(16x^2)$

40. Simplify: $-3[2x - (7x - 9)]$

41. Simplify: $-(8a^2 - 3b^2)$

42. Identify the property that justifies the statement.
$-32(0) = 0$

43. Translate "b decreased by the product of seven and b" into a variable expression.

44. Translate "the sum of a number and twice the square of the number" into a variable expression.

45. Translate "three less than the quotient of six and a number" into a variable expression.

46. Translate "ten divided by the difference between y and two" into a variable expression.

47. Translate and simplify "eight times the quotient of twice a number and sixteen."

48. Translate and simplify "the product of four and the sum of two and five times a number."

49. The length of the base of a triangle is 15 in. more than the height of the triangle. Express the length of the base of the triangle in terms of the height of the triangle.

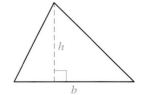

50. A coffee merchant made 20 lb of a blend of coffee using only mocha java beans and espresso beans. Use the same variable to express the amounts of mocha java beans and espresso beans in the coffee blend.

Chapter Test

1. Simplify: $(9y)4$

2. Simplify: $7x + 5y - 3x - 8y$

3. Simplify: $8n - (6 - 2n)$

4. Evaluate $3ab - (2a)^2$ when $a = -2$ and $b = -3$.

5. Identify the property that justifies the statement.
$$\frac{3}{8}(1) = \frac{3}{8}$$

6. Simplify: $-4(-x + 10)$

7. Simplify: $\frac{2}{3}x^2 - \frac{7}{12}x^2$

8. Simplify: $(-10x)\left(-\frac{2}{5}\right)$

9. Simplify: $(-4y^2 + 8)6$

10. Complete the statement by using the Inverse Property of Addition.
$$-19 + ? = 0$$

11. Evaluate $\dfrac{-3ab}{2a + b}$ when $a = -1$ and $b = 4$.

12. Simplify: $5(x + y) - 8(x - y)$

13. Simplify: $6b - 9b + 4b$

14. Simplify: $13(6a)$

15. Simplify: $3(x^2 - 5x + 4)$

16. Evaluate $4(b - a) + bc$ when $a = 2$, $b = -3$, and $c = 4$.

17. Simplify: $6x - 3(y - 7x) + 2(5x - y)$

18. Translate "the quotient of eight more than n and seventeen" into a variable expression.

19. Translate "the difference between the sum of a and b and the square of b" into a variable expression.

20. Translate "the sum of the square of a number and the product of the number and eleven" into a variable expression.

21. Translate and simplify "twenty times the sum of a number and nine."

22. Translate and simplify "two more than a number added to the difference between the number and three."

23. Translate and simplify "a number minus the product of one-fourth and twice the number."

24. The distance from Neptune to the sun is thirty times the distance from Earth to the sun. Express the distance from Neptune to the sun in terms of the distance from Earth to the sun.

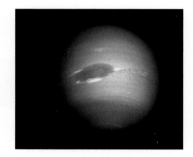

25. A nine-foot board is cut into two pieces of different lengths. Use the same variable to express the lengths of the two pieces.

Cumulative Review Exercises

1. Add: $-4 + 7 + (-10)$

2. Subtract: $-16 - (-25) - 4$

3. Multiply: $(-2)(3)(-4)$

4. Divide: $(-60) \div 12$

5. Write $1\dfrac{1}{4}$ as a decimal.

6. Write 60% as a fraction and as a decimal.

7. Use the roster method to write the set of negative integers greater than or equal to -4.

8. Write $\dfrac{2}{25}$ as a percent.

9. Subtract: $\dfrac{7}{12} - \dfrac{11}{16} - \left(-\dfrac{1}{3}\right)$

10. Divide: $\dfrac{5}{12} \div \left(\dfrac{3}{2}\right)$

11. Multiply: $\left(-\dfrac{9}{16}\right)\left(\dfrac{8}{27}\right)\left(-\dfrac{3}{2}\right)$

12. Simplify: $-3^2 \cdot \left(-\dfrac{2}{3}\right)^3$

13. Simplify: $-2^5 \div (3 - 5)^2 - (-3)$

14. Simplify: $\left(-\dfrac{3}{4}\right)^2 - \left(\dfrac{3}{8} - \dfrac{11}{12}\right)$

15. Evaluate $a - 3b^2$ when $a = 4$ and $b = -2$.

16. Simplify: $-2x^2 - (-3x^2) + 4x^2$

17. Simplify: $8a - 12b - 9a$

18. Simplify: $\dfrac{1}{3}(9a)$

19. Simplify: $\left(-\dfrac{5}{8}\right)(-32b)$

20. Simplify: $5(4 - 2x)$

21. Simplify: $-3(-2y + 7)$

22. Simplify: $-2(3x^2 - 4y^2)$

23. Simplify: $-4(2y^2 - 5y - 8)$

24. Simplify: $-4x - 3(2x - 5)$

25. Simplify: $3(4x - 1) - 7(x + 2)$

26. Simplify: $3x + 2[x - 4(2 - x)]$

27. Simplify: $3[4x - 2(x - 4y)] + 5y$

28. Translate "the difference between six and the product of a number and twelve" into a variable expression.

29. Translate and simplify "the total of five and the difference between a number and seven."

30. The speed of a Digital Subscriber Line (DSL) Internet connection is ten times faster than that of a dial-up connection. Express the speed of the DSL connection in terms of the speed of the dial-up connection.

In what year will these children enter high school? How many children in the United States are in this age group? When will these children be eligible to collect social security? Demographers ask questions such as these when studying population groups. By analyzing these groups today, projections are made for the future. These projections are used by many organizations, including college administrators and managers of retirement communities. The **Focus on Problem Solving on pages 135–136** *provides an example of how demography is used to plan as far as 50 years in to the future.*

Applications

Advertising, *p. 108*

Anthropology, *pp. 122, 123*

Business, *pp. 107, 110, 123, 141, 143*

Chemistry, *pp. 143, 144*

Conservation, *p. 141*

Construction, *p. 107*

Consumerism, *pp. 108, 110*

Consumer Price Index, *pp. 137, 138*

Demography, *pp. 135, 136*

Depreciation, *p. 117*

e-commerce, *p. 107*

Education, *p. 99*

Employment, *p. 123*

Energy, *p. 108*

EPA, *p. 107*

The Federal Government, *p. 108*

Geometry, *pp. 110, 140, 141*

Health, *pp. 99, 108, 136, 137*

Health Care, *p. 108*

The Internet, *p. 107*

Investment, *pp. 100, 101, 108, 109, 117, 141, 143*

Lever systems, *pp. 118, 119, 124, 141, 144*

Mixture problems, *pp. 101, 109, 110, 141, 143*

Monthly mortgage payments, *p. 111*

Natural resources, *p. 107*

Nutrition, *p. 136*

Pets, *p. 99*

Physics, *pp. 122, 141*

Safety, *p. 108*

Space exploration, *p. 142*

Sports, *p. 144*

Taxes, *pp. 100, 107, 144*

Travel, *p. 107*

Uniform motion problems, *pp. 96, 97, 105, 106, 141, 143*

3

Solving Equations and Inequalities

Objectives

Section 3.1

1 Determine whether a given number is a solution of an equation

2 Solve equations of the form $x + a = b$

3 Solve equations of the form $ax = b$

4 Application problems: uniform motion

5 Applications of percent

Section 3.2

1 Solve equations of the form $ax + b = c$

2 Solve equations of the form $ax + b = cx + d$

3 Solve equations containing parentheses

4 Application problems

Section 3.3

1 Solve inequalities using the Addition Property of Inequalities

2 Solve inequalities using the Multiplication Property of Inequalities

3 Solve general inequalities

Need help? For online student resources, such as section quizzes, visit this web site: **math.college.hmco.com**

PrepTest

1. Subtract: $8 - 12$

2. Multiply: $-\dfrac{3}{4}\left(-\dfrac{4}{3}\right)$

3. Multiply: $-\dfrac{5}{8}(16)$

4. Write 90% as a decimal.

5. Write 0.75 as a percent.

6. Evaluate $3x^2 - 4x - 1$ when $x = -4$.

7. Simplify: $3x - 5 + 7x$

8. Simplify: $8x - 9 - 8x$

9. Simplify: $6x - 3(6 - x)$

Go Figure

How can you cut a donut into 8 equal pieces with three cuts of the knife?

3.1   Introduction to Equations

1 **Determine whether a given number is a solution of an equation**

VIDEO & DVD   CD TUTOR   WEB   SSM

Point of Interest

One of the most famous equations ever stated is $E = mc^2$. This equation, stated by Albert Einstein, shows that there is a relationship between mass m and energy E.

An **equation** expresses the equality of two mathematical expressions. The expressions can be either numerical or variable expressions.

$$\left.\begin{array}{l} 9 + 3 = 12 \\ 3x - 2 = 10 \\ y^2 + 4 = 2y - 1 \\ z = 2 \end{array}\right\} \text{Equations}$$

The equation at the right is true if the variable is replaced by 5.

$x + 8 = 13$
$5 + 8 = 13$ A true equation

The equation $x + 8 = 13$ is false if the variable is replaced by 7.

$7 + 8 = 13$ A false equation

A **solution** of an equation is a number that, when substituted for the variable, results in a true equation. 5 is a solution of the equation $x + 8 = 13$. 7 is not a solution of the equation $x + 8 = 13$.

Example 1   Is -3 a solution of the equation $4x + 16 = x^2 - 5$?

Solution

$$4x + 16 = x^2 - 5$$

$4(-3) + 16$	$(-3)^2 - 5$
$-12 + 16$	$9 - 5$
$4 =$	4

• Replace the variable by the given number, -3.

• Evaluate the numerical expressions using the Order of Operations Agreement.

• Compare the results. If the results are equal, the given number is a solution. If the results are not equal, the given number is not a solution.

Yes, -3 is a solution of the equation $4x + 16 = x^2 - 5$.

Problem 1   Is $\frac{1}{4}$ a solution of $5 - 4x = 8x + 2$?

Solution   See page S4.

Example 2   Is -4 a solution of $4 + 5x = x^2 - 2x$?

Solution

$$4 + 5x = x^2 - 2x$$

$4 + 5(-4)$	$(-4)^2 - 2(-4)$
$4 + (-20)$	$16 - (-8)$

$$-16 \neq 24$$

No, -4 is not a solution of the equation $4 + 5x = x^2 - 2x$.

- Replace the variable by the given number, -4.
- Evaluate the numerical expressions using the Order of Operations Agreement.
- Compare the results. If the results are equal, the given number is a solution. If the results are not equal, the given number is not a solution.

Problem 2   Is 5 a solution of $10x - x^2 = 3x - 10$?

Solution   See page S4.

2   **Solve equations of the form $x + a = b$**

To **solve an equation** means to find a solution of the equation. The simplest equation to solve is an equation of the form **variable = constant**, because the constant is the solution.

If $x = 5$, then 5 is the solution of the equation because $5 = 5$ is a true equation.

The solution of the equation shown at the right is 7.

$$x + 2 = 9 \qquad 7 + 2 = 9$$

Note that if 4 is added to each side of the equation, the solution is still 7.

$$x + 2 + 4 = 9 + 4$$
$$x + 6 = 13 \qquad 7 + 6 = 13$$

If -5 is added to each side of the equation, the solution is still 7.

$$x + 2 + (-5) = 9 + (-5)$$
$$x - 3 = 4 \qquad 7 - 3 = 4$$

This illustrates the Addition Property of Equations.

> **Take Note**
>
> Think of an equation as a balance scale. If the weights added to each side of the equation are not the same, the pans no longer balance.

> **Addition Property of Equations**
>
> The same number or variable term can be added to each side of an equation without changing the solution of the equation.

This property is used in solving equations. Note the effect of adding, to each side of the equation $x + 2 = 9$, the opposite of the constant term 2. After each side of the equation is simplified, the equation is in the form variable = constant. The solution is the constant.

$$x + 2 = 9$$
$$x + 2 + (-2) = 9 + (-2)$$
$$x + 0 = 7$$
$$x = 7$$

$$\boxed{\text{variable}} = \boxed{\text{constant}}$$

The solution is 7.

In solving an equation, the goal is to rewrite the given equation in the form variable = constant. The Addition Property of Equations can be used to rewrite an equation in this form. The Addition Property of Equations is used to **remove a term** from one side of an equation **by adding the opposite of that term** to each side of the equation.

Example 3   Solve and check: $y - 6 = 9$

Solution

$y - 6 = 9$	• The goal is to rewrite the equation in the form *variable* = *constant*.
$y - 6 + 6 = 9 + 6$	• Add the opposite of the constant term -6 to each side of the equation (the Addition Property of Equations).
$y + 0 = 15$	• Simplify using the Inverse Property of Addition.
$y = 15$	• Simplify using the Addition Property of Zero. Now the equation is in the form *variable* = *constant*.

Check
$$y - 6 = 9$$
$$\overline{15 - 6 \;|\; 9}$$
$$9 = 9$$ • This is a true equation. The solution checks.

The solution is 15.   • Write the solution.

Problem 3   Solve and check: $x - \dfrac{1}{3} = -\dfrac{3}{4}$

Solution   See page S4.

Because subtraction is defined in terms of addition, the Addition Property of Equations makes it possible to subtract the same number from each side of an equation without changing the solution of the equation.

Solve: $y + \dfrac{1}{2} = \dfrac{5}{4}$

The goal is to rewrite the equation in the form *variable* = *constant*.

$$y + \frac{1}{2} = \frac{5}{4}$$

Add the opposite of the constant term $\dfrac{1}{2}$ to each side of the equation. This is equivalent to subtracting $\dfrac{1}{2}$ from each side of the equation.

$$y + \frac{1}{2} - \frac{1}{2} = \frac{5}{4} - \frac{1}{2}$$
$$y + 0 = \frac{5}{4} - \frac{2}{4}$$
$$y = \frac{3}{4}$$

The solution $\dfrac{3}{4}$ checks.

The solution is $\dfrac{3}{4}$.

Example 4 Solve: $\dfrac{1}{2} = x + \dfrac{2}{3}$

Solution
$$\dfrac{1}{2} = x + \dfrac{2}{3}$$

$$\dfrac{1}{2} - \dfrac{2}{3} = x + \dfrac{2}{3} - \dfrac{2}{3}$$ • Subtract $\dfrac{2}{3}$ from each side of the equation.

$$-\dfrac{1}{6} = x$$ • Simplify each side of the equation.

The solution is $-\dfrac{1}{6}$.

Problem 4 Solve: $-8 = 5 + x$

Solution   See page S4.

Note from the solution to Example 4 that an equation can be rewritten in the form *constant = variable*. Whether the equation is written in the form *variable = constant* or in the form *constant = variable*, the solution is the constant.

3 **Solve equations of the form $ax = b$**

The solution of the equation shown at the right is 3.

$$2x = 6 \qquad\qquad 2 \cdot 3 = 6$$

Note that if each side of the equation is multiplied by 5, the solution is still 3.

$$5 \cdot 2x = 5 \cdot 6$$
$$10x = 30 \qquad\qquad 10 \cdot 3 = 30$$

If each side is multiplied by -4, the solution is still 3.

$$(-4) \cdot 2x = (-4) \cdot 6$$
$$-8x = -24 \qquad\qquad -8 \cdot 3 = -24$$

This illustrates the Multiplication Property of Equations.

> **Multiplication Property of Equations**
>
> Each side of an equation can be multiplied by the same nonzero number without changing the solution of the equation.

This property is used in solving equations. Note the effect of multiplying each side of the equation $2x = 6$ by the reciprocal of the coefficient 2. After each side of the equation is simplified, the equation is in the form variable = constant. The solution is the constant.

$$2x = 6$$
$$\dfrac{1}{2} \cdot 2x = \dfrac{1}{2} \cdot 6$$
$$1x = 3$$
$$x = 3$$

| variable | = | constant |

The solution is 3.

In solving an equation, the goal is to rewrite the given equation in the form variable = constant. The Multiplication Property of Equations can be used to rewrite an equation in this form. The Multiplication Property of Equations is used to **remove a coefficient** from a variable term in an equation **by multiplying each side of the equation by the reciprocal of the coefficient.**

Example 5 Solve: $\dfrac{3x}{4} = -9$

Solution

$$\dfrac{3x}{4} = -9$$

$\bullet \ \dfrac{3x}{4} = \dfrac{3}{4}x$

$$\dfrac{4}{3} \cdot \dfrac{3}{4}x = \dfrac{4}{3}(-9)$$

- Multiply each side of the equation by the reciprocal of the coefficient $\dfrac{3}{4}$ (the Multiplication Property of Equations).

$$1x = -12$$

- Simplify using the Inverse Property of Multiplication.

$$x = -12$$

- Simplify using the Multiplication Property of One. Now the equation is in the form *variable = constant.*

The solution is -12.

- Write the solution.

Problem 5 Solve: $-\dfrac{2x}{5} = 6$

Solution See page S4.

Because division is defined in terms of multiplication, the Multiplication Property of Equations makes it possible to divide each side of an equation by the same number without changing the solution of the equation.

Solve: $8x = 16$

The goal is to rewrite the equation in the form *variable = constant.*

$$8x = 16$$

Multiply each side of the equation by the reciprocal of 8. This is equivalent to dividing each side by 8.

$$\dfrac{8x}{8} = \dfrac{16}{8}$$
$$x = 2$$

The solution 2 checks.

The solution is 2.

When using the Multiplication Property of Equations to solve an equation, multiply each side of the equation by the reciprocal of the coefficient when the coefficient is a fraction. Divide each side of the equation by the coefficient when the coefficient is an integer or a decimal.

Example 6 Solve and check: $4x = 6$

Solution

$$4x = 6$$

$$\dfrac{4x}{4} = \dfrac{6}{4}$$

- Divide each side of the equation by 4, the coefficient of x.

$$x = \dfrac{3}{2}$$

- Simplify each side of the equation.

Check

$$4x = 6$$

$$4\left(\frac{3}{2}\right) \bigg| 6$$

$$6 = 6 \qquad \text{• This is a true equation. The solution checks.}$$

The solution is $\frac{3}{2}$.

Problem 6 Solve and check: $6x = 10$

Solution See page S4.

Before using one of the Properties of Equations, check to see whether one or both sides of the equation can be simplified. In Example 7, like terms appear on the left side of the equation. The first step in solving this equation is to combine the like terms so that there is only one variable term on the left side of the equation.

Example 7 Solve: $5x - 9x = 12$

Solution
$$5x - 9x = 12$$
$$-4x = 12 \qquad \text{• Combine like terms.}$$
$$\frac{-4x}{-4} = \frac{12}{-4} \qquad \text{• Divide each side of the equation by } -4.$$
$$x = -3$$

The solution is -3.

Problem 7 Solve: $4x - 8x = 16$

Solution See page S4.

---

4 ## Application problems: uniform motion

> **Take Note**
>
> A car traveling in a *circle* at a constant speed of 45 mph is *not* in uniform motion because the direction of the car is always changing.

Any object that travels at a constant speed in a straight line is said to be in *uniform motion*. **Uniform motion** means that the speed and direction of an object do not change. For instance, a car traveling at a constant speed of 45 mph on a straight road is in uniform motion.

The solution of a uniform motion problem is based on the equation $d = rt$, where d is the distance traveled, r is the rate of travel, and t is the time spent traveling. Suppose a car travels at 50 mph for 3 h. Because the rate (50 mph) and the time (3 h) are known, we can find the distance traveled by solving the equation $d = rt$ for d.

$$d = rt$$
$$d = 50(3) \qquad \text{• } r = 50, t = 3$$
$$d = 150$$

The car travels a distance of 150 mi.

A jogger runs 3 mi in 45 min. What is the rate of the jogger in miles per hour?

Because the answer must be in miles per *hour* and the time is given in *minutes*, convert 45 min to hours.

$$45 \text{ min} = \frac{3}{4}\text{h}$$

To find the rate of the jogger, solve the equation $d = rt$ for r using $d = 3$ and $t = \frac{3}{4}$.

$$d = rt$$
$$3 = r\left(\frac{3}{4}\right)$$
$$3 = \frac{3}{4}r$$
$$\left(\frac{4}{3}\right)3 = \left(\frac{4}{3}\right)\frac{3}{4}r$$
$$4 = r$$

The rate of the jogger is 4 mph.

If two objects are moving in opposite directions, then the rate at which the distance between them is increasing is the sum of the speeds of the two objects. For instance, in the diagram at the right, two cars start from the same point and travel in opposite directions. The distance between them is changing at the rate of 70 mph.

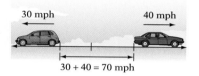

Similarly, if two objects are moving toward each other, the distance between them is decreasing at a rate that is equal to the sum of the speeds. The rate at which the two planes at the right are approaching one another is 800 mph.

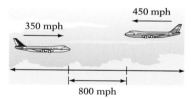

─ **Example 8**   Two cars start from the same point and move in opposite directions. The car moving west is traveling 45 mph, and the car moving east is traveling 60 mph. In how many hours will the cars be 210 mi apart?

Strategy   The distance is 210 mi. Therefore, $d = 210$. The cars are moving in opposite directions, so the rate at which the distance between them is changing is the sum of the rates of the two cars. The rate is 45 mph + 60 mph = 105 mph. Therefore, $r = 105$. To find the time, solve the equation $d = rt$ for t.

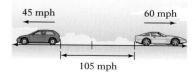

Solution
$$d = rt$$
$$210 = 105t \qquad \bullet\, d = 210, r = 105$$
$$\frac{210}{105} = \frac{105t}{105}$$
$$2 = t$$

In 2 h, the cars will be 210 mi apart.

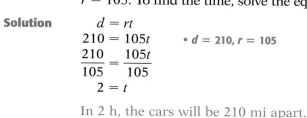

Problem 8   Two cyclists start at the same time at opposite ends of an 80-mile course. One cyclist is traveling 18 mph, and the second cyclist is traveling 14 mph. How long after they begin will they meet?

Solution   See page S5.

If a motorboat is on a river that is flowing at a rate of 4 mph, the boat will float down the river at a speed of 4 mph even though the motor is not on. Now suppose the motor is turned on and the power adjusted so that the boat will travel at 10 mph without the aid of the current. Then, if the boat is moving *with* the current, its effective speed is the speed of the boat using power plus the speed of the current: 10 mph + 4 mph = 14 mph. However, if the boat is moving in the opposite direction of the current, the current slows the boat down, and the effective speed of the boat is the speed of the boat using power minus the speed of the current: 10 mph − 4 mph = 6 mph.

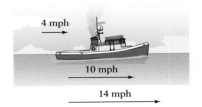

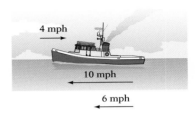

There are other situations in which the above concepts may be applied.

Example 9   An airline passenger is walking between two airline terminals and decides to get on a moving sidewalk that is 150 ft long. If the passenger walks at a rate of 7 ft/s and the moving sidewalk moves at a rate of 9 ft/s, how long, in seconds, will it take the passenger to walk from one end of the moving sidewalk to the other?

Strategy   The distance is 150 ft. Therefore, $d = 150$. The passenger is traveling at 7 ft/s and the moving sidewalk is traveling at 9 ft/s. The rate of the passenger is the sum of the two rates, or 16 ft/s. Therefore, $r = 16$. To find the time, solve the equation $d = rt$ for t.

Solution
$$d = rt$$
$$150 = 16t \qquad \bullet \ d = 150, r = 16$$
$$\frac{150}{16} = \frac{16t}{16}$$
$$9.375 = t$$

It will take the passenger 9.375 s to travel the length of the moving sidewalk.

Problem 9   A plane that can normally travel at 250 mph in calm air is flying into a headwind of 25 mph. How far can the plane fly in 3 h?

Solution   See page S5.

5   Applications of percent

The solution of a problem that involves a percent requires solving the basic percent equation.

> ### The Basic Percent Equation
>
> $$\text{Percent} \cdot \text{Base} = \text{Amount}$$
> $$P \quad \cdot \quad B \quad = \quad A$$

To translate a problem involving a percent into an equation, remember that the word *of* translates into "multiply" and the word *is* translates into "=". The base usually follows the word *of*.

20% of what number is 30?

Given: $P = 20\% = 0.20$
$ A = 30$
Unknown: Base

$$PB = A$$
$$(0.20)B = 30$$
$$\frac{0.20B}{0.20} = \frac{30}{0.20}$$
$$B = 150$$

The number is 150.

What percent of 40 is 30?

Given: $B = 40$
$ A = 30$
Unknown: Percent

$$PB = A$$
$$P(40) = 30$$
$$40P = 30$$
$$\frac{40P}{40} = \frac{30}{40}$$

We must write the fraction as a percent in order to answer the question.

$$P = \frac{3}{4}$$
$$P = 75\%$$

30 is 75% of 40.

Find 25% of 200.

Given: $P = 25\% = 0.25$
$ B = 200$
Unknown: Amount

$$PB = A$$
$$0.25(200) = A$$
$$50 = A$$

25% of 200 is 50.

In most cases, we write a percent as a decimal before solving the basic percent equation. However, some percents are more easily written as a fraction. For example,

$$33\frac{1}{3}\% = \frac{1}{3} \qquad 66\frac{2}{3}\% = \frac{2}{3} \qquad 16\frac{2}{3}\% = \frac{1}{6} \qquad 83\frac{1}{3}\% = \frac{5}{6}$$

Example 10   12 is $33\frac{1}{3}\%$ of what number?

Solution

$$PB = A \qquad \bullet\ 33\frac{1}{3}\% = \frac{1}{3}$$

$$\frac{1}{3}B = 12$$

$$3 \cdot \frac{1}{3}B = 3 \cdot 12$$

$$B = 36$$

The number is 36.

Problem 10   27 is what percent of 60?

Solution   See page S5.

Example 11   During a recent year, nearly 1.2 million dogs or litters were registered with the American Kennel Club. The most popular breed was the Labrador retriever, with 172,841 registered. What percent of the registrations were Labrador retrievers? Round to the nearest tenth of a percent. (*Source:* American Kennel Club)

Strategy   To find the percent, solve the basic percent equation using $B = 1.2$ million $= 1,200,000$ and $A = 172,841$. The percent is unknown.

Solution

$$PB = A$$
$$P(1,200,000) = 172,841$$
$$\frac{P(1,200,000)}{1,200,000} = \frac{172,841}{1,200,000}$$
$$P \approx 0.144$$
$$P \approx 14.4\%$$

Approximately 14.4% of the registrations were Labrador retrievers.

Point
of **Interest**

As noted at the right, 172,841 of the dogs or litters registered were Labrador retrievers. Listed below are the next most popular breeds and their registrations.
Golden retrievers: 66,300
German shepherds: 57,660
Dachshunds: 54,773
Beagles: 52,026

Problem 11   A student correctly answered 72 of the 80 questions on an exam. What percent of the questions were answered correctly?

Solution   See page S5.

Example 12   According to the Centers for Disease Control and Prevention, 30.8% of the adult population of Kentucky smokes. How many adults in Kentucky smoke? Use a figure of 3,000,000 for the number of adults in Kentucky.

Strategy   To find the number of adults who smoke, solve the basic percent equation using $B = 3,000,000$ and $P = 30.8\% = 0.308$. The amount is unknown.

Solution

$$PB = A$$
$$0.308(3,000,000) = A$$
$$924,000 = A$$

Approximately 924,000 adults in Kentucky smoke.

Problem 12 The price of a digital camcorder is $895. A 6% sales tax is added to the price. How much is the sales tax?

Solution   See page S5.

The simple interest that an investment earns is given by the equation $I = Prt$, where I is the simple interest, P is the principal, or amount invested, r is the simple interest rate, and t is the time.

Take Note

$18 \text{ months} = \dfrac{18}{12} \text{ years}$
$= 1.5 \text{ years}$

A $1500 investment has an annual simple interest rate of 7%. Find the simple interest earned on the investment after 18 months.

The time is given in months, but the interest rate is an annual rate. Therefore, we must convert 18 months to years.

$18 \text{ months} = 1.5 \text{ years}$

To find the interest earned, solve the equation $I = Prt$ for I using $P = 1500$, $r = 0.07$, and $t = 1.5$.

$I = Prt$
$I = 1500(0.07)(1.5)$
$I = 157.5$

The interest earned is $157.50.

Example 13 In April, Marshall Wardell was charged $8.72 in interest on an unpaid credit card balance of $545. Find the annual interest rate for this credit card.

Strategy The interest is $8.72. Therefore, $I = 8.72$. The unpaid balance is $545. This is the principal on which interest is calculated. Therefore, $P = 545$. The time is 1 month. Because the *annual* interest rate must be found and the time is given as 1 month, write 1 month as $\dfrac{1}{12}$ year. Therefore, $t = \dfrac{1}{12}$. To find the interest rate, solve the equation $I = Prt$ for r.

Solution

$$I = Prt$$
$$8.72 = 545r\left(\frac{1}{12}\right) \qquad • I = 8.72, P = 545, t = \frac{1}{12}$$
$$8.72 = \frac{545}{12}r$$
$$\frac{12}{545}(8.72) = \frac{12}{545}\left(\frac{545}{12}r\right)$$
$$0.192 = r$$

The annual interest rate is 19.2%.

Take Note

Municipal bonds are frequently sold by government agencies to pay for roads, aqueducts, and other essential public needs. An investor who purchases one of these bonds is paid interest on the cost of the bond for a certain number of years. When that time period expires, the cost of the bond is returned to the investor.

Problem 13   Clarissa Adams purchased a $1000 municipal bond that earns an annual simple interest rate of 6.4%. How much must she deposit into a bank account that earns 8% annual simple interest so that the interest earned from each account after one year is the same?

Solution   See page S5.

The amount of a substance in a solution can be given as a percent of the total solution. For instance, if a certain fruit juice drink is advertised as containing 27% cranberry juice, then 27% of the contents of the bottle must be cranberry juice.

Solving problems involving mixtures is based on the percent mixture equation $Q = Ar$, where Q is the quantity of a substance in the solution, A is the amount of the solution, and r is the percent concentration of the substance.

Point of Interest

In the jewelry industry, the amount of gold in a piece of jewelry is measured in karats. *Pure gold is 24 karats. A necklace that is 18 karats is*

$\dfrac{18}{24} = 0.75 = 75\%$ *gold.*

The formula for a perfume requires that the concentration of jasmine be 1.2% of the total amount of perfume. How many ounces of jasmine are in a 2-ounce bottle of this perfume?

The amount of perfume is 2 oz. Therefore, $A = 2$. The percent concentration is 1.2%, so $r = 0.012$. To find the number of ounces of jasmine, solve the equation $Q = Ar$ for Q.

$Q = Ar$
$Q = 2(0.012)$
$Q = 0.024$

The bottle contains 0.024 oz of jasmine.

Example 14   To make a certain color of blue, 4 oz of cyan must be contained in 1 gal of paint. What is the percent concentration of cyan in the paint?

Strategy   The amount of cyan is given in ounces and the amount of paint is given in gallons; we must convert ounces to gallons or gallons to ounces. For this problem, we will convert gallons to ounces: 1 gal = 128 oz.

Solution   To find the percent concentration of cyan, solve the equation $Q = Ar$ for r using $Q = 4$ and $A = 128$.

$Q = Ar$
$4 = 128r$
$\dfrac{4}{128} = \dfrac{128r}{128}$

$0.03125 = r$

The percent concentration of cyan is 3.125%.

Problem 14   The concentration of sugar in Choco-Pops cereal is 25%. If a bowl of this cereal contains 2 oz of sugar, how many ounces of cereal are in the bowl?

Solution   See page S5.

3.1   Concept Review

Determine whether the statement is always true, sometimes true, or never true.

1. Both sides of an equation can be multiplied by the same number without changing the solution of the equation.

2. Both sides of an equation can be divided by the same number without changing the solution of the equation.

3. For an equation of the form $ax = b, a \neq 0$, multiplying both sides of the equation by the reciprocal of a will result in an equation of the form $x = constant$.

4. Use the Multiplication Property of Equations to remove a term from one side of an equation.

5. Adding a negative 3 to each side of an equation yields the same result as subtracting 3 from each side of the equation.

6. In the basic percent equation, the amount follows the word *of*.

7. An equation contains an equals sign.

3.1   Exercises

1. What is the difference between an equation and an expression?

2. How can you determine whether a number is a solution of an equation?

3. Is 4 a solution of $2x = 8$?

4. Is 3 a solution of $y + 4 = 7$?

5. Is -1 a solution of $2b - 1 = 3$?

6. Is -2 a solution of $3a - 4 = 10$?

7. Is 1 a solution of $4 - 2m = 3$?

8. Is 2 a solution of $7 - 3n = 2$?

9. Is 5 a solution of $2x + 5 = 3x$?

10. Is 4 a solution of $3y - 4 = 2y$?

11. Is 0 a solution of $4a + 5 = 3a + 5$?

12. Is 0 a solution of $4 - 3b = 4 - 5b$?

13. Is -2 a solution of $4 - 2n = n + 10$?

14. Is -3 a solution of $5 - m = 2 - 2m$?

15. Is 3 a solution of $z^2 + 1 = 4 + 3z$?

16. Is 2 a solution of $2x^2 - 1 = 4x - 1$?

17. Is -1 a solution of $y^2 - 1 = 4y + 3$?

18. Is -2 a solution of
$m^2 - 4 = m + 3$?

19. Is 4 a solution of
$x(x + 1) = x^2 + 5$?

20. Is 3 a solution of
$2a(a - 1) = 3a + 3$?

21. Is $-\frac{1}{4}$ a solution of
$8t + 1 = -1$?

22. Is $\frac{1}{2}$ a solution of
$4y + 1 = 3$?

23. Is $\frac{2}{5}$ a solution of
$5m + 1 = 10m - 3$?

24. Is $\frac{3}{4}$ a solution of
$8x - 1 = 12x + 3$?

25. Is 2.1 a solution of
$x^2 - 4x = x + 1.89$?

26. Is 1.5 a solution of
$c^2 - 3c = 4c - 8.25$?

2 **27.** How is the Addition Property of Equations used to solve an equation?

28. Explain why the goal in solving an equation is to write the equation in the form *variable = constant*.

29. Without solving the equation $x + \frac{13}{15} = -\frac{21}{43}$, determine whether x is less than or greater than $-\frac{21}{43}$. Explain your answer.

30. Without solving the equation $x - \frac{11}{16} = \frac{19}{24}$, determine whether x is less than or greater than $\frac{19}{24}$. Explain your answer.

Solve and check.

31. $x + 5 = 7$

32. $y + 3 = 9$

33. $b - 4 = 11$

34. $z - 6 = 10$

35. $2 + a = 8$

36. $5 + x = 12$

37. $m + 9 = 3$

38. $t + 12 = 10$

39. $n - 5 = -2$

40. $x - 6 = -5$

41. $b + 7 = 7$

42. $y - 5 = -5$

43. $a - 3 = -5$

44. $x - 6 = -3$

45. $z + 9 = 2$

46. $n + 11 = 1$

47. $10 + m = 3$

48. $8 + x = 5$

49. $9 + x = -3$

50. $10 + y = -4$

51. $b - 5 = -3$

52. $t - 6 = -4$

53. $2 = x + 7$

54. $-8 = n + 1$

55. $4 = m - 11$

56. $-6 = y - 5$

57. $12 = 3 + w$

58. $-9 = 5 + x$

59. $4 = -10 + b$

60. $-7 = -2 + x$

61. $m + \frac{2}{3} = -\frac{1}{3}$

62. $c + \frac{3}{4} = -\frac{1}{4}$

63. $x - \frac{1}{2} = \frac{1}{2}$

64. $x - \frac{2}{5} = \frac{3}{5}$

65. $\frac{5}{8} + y = \frac{1}{8}$

66. $\frac{4}{9} + a = -\frac{2}{9}$

67. $m + \frac{1}{2} = -\frac{1}{4}$

68. $b + \frac{1}{6} = -\frac{1}{3}$

69. $x + \frac{2}{3} = \frac{3}{4}$

70. $n + \frac{2}{5} = \frac{2}{3}$

71. $-\dfrac{5}{6} = x - \dfrac{1}{4}$

72. $-\dfrac{1}{4} = c - \dfrac{2}{3}$

73. $-\dfrac{1}{21} = m + \dfrac{2}{3}$

74. $\dfrac{5}{9} = b - \dfrac{1}{3}$

75. $\dfrac{5}{12} = n + \dfrac{3}{4}$

76. $d + 1.3619 = 2.0148$

77. $w + 2.932 = 4.801$

78. $-0.813 + x = -1.096$

79. $-1.926 + t = -1.042$

3 **80.** 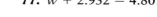 How is the Multiplication Property of Equations used to solve an equation?

81. 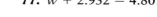 Why, when the Multiplication Property of Equations is used, must the number that multiplies each side of the equation not be zero?

82. 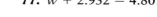 Without solving the equation $-\dfrac{15}{41}x = -\dfrac{23}{25}$, determine whether x is less than or greater than zero. Explain your answer.

83. 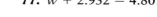 Without solving the equation $\dfrac{5}{28}x = -\dfrac{3}{44}$, determine whether x is less than or greater than zero. Explain your answer.

Solve and check.

84. $5x = 15$

85. $4y = 28$

86. $3b = -12$

87. $2a = -14$

88. $-3x = 6$

89. $-5m = 20$

90. $-3x = -27$

91. $-6n = -30$

92. $20 = 4c$

93. $18 = 2t$

94. $-32 = 8w$

95. $-56 = 7x$

96. $8d = 0$

97. $-5x = 0$

98. $36 = 9z$

99. $35 = -5x$

100. $-64 = 8a$

101. $-32 = -4y$

102. $-42 = 6t$

103. $-12m = -144$

104. $\dfrac{x}{3} = 2$

105. $\dfrac{x}{4} = 3$

106. $-\dfrac{y}{2} = 5$

107. $-\dfrac{b}{3} = 6$

108. $\dfrac{n}{7} = -4$

109. $\dfrac{t}{6} = -3$

110. $\dfrac{2}{5}x = 12$

111. $-\dfrac{4}{3}c = -8$

112. $\dfrac{5}{6}y = -20$

113. $-\dfrac{2}{3}d = 8$

114. $-\dfrac{3}{5}m = 12$

115. $\dfrac{2n}{3} = 2$

116. $\dfrac{5x}{6} = -10$

117. $\dfrac{-3z}{8} = 9$

118. $\dfrac{-4x}{5} = -12$

119. $-6 = -\dfrac{2}{3}y$

120. $-15 = -\dfrac{3}{5}x$

121. $\dfrac{2}{9} = \dfrac{2}{3}y$

122. $-\dfrac{6}{7} = -\dfrac{3}{4}b$

123. $-\dfrac{2}{5}m = -\dfrac{6}{7}$

124. $5x + 2x = 14$

125. $3n + 2n = 20$

126. $7d - 4d = 9$

127. $10y - 3y = 21$

128. $2x - 5x = 9$ **129.** $\dfrac{x}{1.4} = 3.2$ **130.** $\dfrac{z}{2.9} = -7.8$ **131.** $3.4a = 7.004$

132. $2.3m = 2.415$ **133.** $-3.7x = 7.77$ **134.** $-1.6m = 5.44$

4 Uniform Motion Problems

135. A train travels at 45 mph for 3 h and then increases its speed to 55 mph for 2 more hours. How far does the train travel in the 5-hour period?

136. As part of a training program for the Boston Marathon, a runner wants to build endurance by running at a rate of 9 mph for 20 min. How far will the runner travel in that time period?

137. It takes a hospital dietician 40 min to drive from home to the hospital, a distance of 20 mi. What is the dietician's average rate of speed?

138. Marcella leaves home at 9:00 A.M. and drives to school, arriving at 9:45 A.M. If the distance between home and school is 27 mi, what is Marcella's average rate of speed?

139. The Ride for Health Bicycle Club has chosen a 36-mile course for this Saturday's ride. If the riders plan on averaging 12 mph while they are riding and they have a 1-hour lunch break planned, how long will it take them to complete the trip?

140. Palmer's average running speed is 3 kilometers per hour faster than his walking speed. If Palmer can run around a 30-kilometer course in 2 h, how many hours would it take for Palmer to walk the same course?

141. A shopping mall has a moving sidewalk that takes shoppers from the shopping area to the parking garage, a distance of 250 ft. If your normal walking rate is 5 ft/s and the moving sidewalk is traveling at 3 ft/s, how many seconds would it take you to walk on the moving sidewalk from the parking garage to the shopping area?

142. K&B River Tours offers a river trip that takes passengers from the K&B dock to a small island that is 24 mi away. The passengers spend 1 h at the island and then return to the K&B dock. If the speed of the boat is 10 mph in calm water and the rate of the current is 2 mph, how long does the trip last?

143. Two joggers start at the same time from opposite ends of an 8-mile jogging trail and begin running toward each other. One jogger is running at a rate of 5 mph, and the other jogger is running at a rate of 7 mph. How long, in minutes, after they start will the two joggers meet?

144. Two cyclists start from the same point at the same time and move in opposite directions. One cyclist is traveling at 8 mph, and the other cyclist is traveling at 9 mph. After 30 min, how far apart are the two cyclists?

145. Petra, who can paddle her canoe at a rate of 10 mph in calm water, is paddling her canoe on a river whose current is 2 mph. How long will it take her to travel 4 mi upstream against the current?

146. At 8:00 A.M., a train leaves a station and travels at a rate of 45 mph. At 9:00 A.M., a second train leaves the same station on the same track and travels in the direction of the first train at a speed of 60 mph. At 10:00 A.M., how far apart are the two trains?

5 **147.** Employee A had an annual salary of $22,000, Employee B had an annual salary of $28,000, and Employee C had an annual salary of $26,000 before each employee was given a 5% raise. Which of the three employees now has the highest annual salary? Explain how you arrived at your answer.

148. Each of three employees earned an annual salary of $25,000 before Employee A was given a 3% raise, Employee B was given a 6% raise, and Employee C was given a 4.5% raise. Which of the three employees now has the highest annual salary? Explain how you arrived at your answer.

Solve.

149. 12 is what percent of 50?

150. What percent of 125 is 50?

151. Find 18% of 40.

152. What is 25% of 60?

153. 12% of what is 48?

154. 45% of what is 9?

155. What is $33\frac{1}{3}\%$ of 27?

156. Find $16\frac{2}{3}\%$ of 30.

157. What percent of 12 is 3?

158. 10 is what percent of 15?

159. 60% of what is 3?

160. 75% of what is 6?

161. 12 is what percent of 6?

162. 20 is what percent of 16?

163. $5\frac{1}{4}\%$ of what is 21?

164. $37\frac{1}{2}\%$ of what is 15?

165. Find 15.4% of 50.

166. What is 18.5% of 46?

167. 1 is 0.5% of what?

168. 3 is 1.5% of what?

169. $\frac{3}{4}\%$ of what is 3?

170. $\frac{1}{2}\%$ of what is 3?

171. Find 125% of 16.

172. What is 250% of 12?

173. 16.4 is what percent of 20.4? Round to the nearest percent.

174. Find 18.3% of 625. Round to the nearest tenth.

175. Without solving an equation, determine whether 40% of 80 is less than, equal to, or greater than 80% of 40.

176. Without solving an equation, determine whether $\frac{1}{4}\%$ of 80 is less than, equal to, or greater than 25% of 80.

177.  **Construction** The Arthur Ashe Tennis Stadium in New York has a seating capacity of 22,500 people. Of these seats, 1.11% are wheelchair-accessible seats. How many seats in the stadium are reserved for wheelchair accessibility? Round to the nearest whole number.

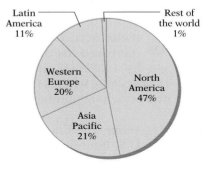

Projected Regional Shares of Global E-Commerce, 2004

178. **e-Commerce** The graph at the right, based on data from Forrester Research, shows the percent share of the global e-commerce market for various regions of the world. The total projected revenue from e-commerce business in 2004 is $7500 billion. How many billion dollars more is Western Europe's projected share than Latin America's projected share?

179. **Natural Resources** On average, a person uses 13.2 gal of water per day for showering. This is 17.8% of the total amount of water used per person per day in the average single-family home. Find the total amount of water used per person per day in the average single-family home. Round to the nearest whole number. (*Source:* American Water Works Association)

180. **Business** During a recent year, 276 billion product coupons were issued by manufacturers. Shoppers redeemed 4.8 billion of these coupons. What percent of the coupons issued were redeemed by customers? Round to the nearest tenth of a percent. (*Source:* NCH NuWorld Consumer Behavior Study, America Coupon Council)

181. **Travel** According to the Travel Industry Association of America, a typical traveler's total vacation bill is $1442. Of that amount, $735 is paid in cash. To the nearest percent, what percent of a typical traveler's vacation bill is paid in cash?

182. **Taxes** In a recent year, 238 U.S. airports collected $1.1 billion in passenger taxes. Of this amount, $88 million was spent on noise reduction. What percent of the passenger taxes collected was spent on noise reduction?

183. **The Internet** The circle graph at the right shows the responses to a question about downloading music from Napster. How many people were surveyed? (*Source:* Taylor Nelson Sofres Intersearch)

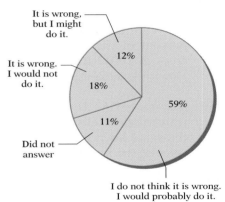

184. **EPA** In January 2001, the Environmental Protection Agency (EPA) required all 54,000 community water systems in the United States to reduce the amount of arsenic in drinking water to 40 parts per billion from 50 parts per billion. What percent decrease does this represent?

185. **Safety** Recently the National Safety Council collected data on the leading causes of accidental death. The findings revealed that for people age 20, 30 died from a fall, 47 from fire, 200 from drowning, and 1950 from motor vehicle accidents. What percent of the accidental deaths were not attributed to motor vehicle accidents? Round to the nearest percent.

186. **Health** According to *Health* magazine, the average American has increased his or her daily consumption of calories from 18 years ago by 11.6%. If the average daily consumption was 1970 calories 18 years ago, what is the average daily consumption today? Round to the nearest whole number.

187. **Energy** The Energy Information Administration reports that if every U.S. household switched 4 h of lighting per day from incandescent bulbs to compact fluorescent bulbs, we would save 31.7 billion kilowatt-hours of electricity a year, or 33% of the total electricity used for home lighting. What is the total electricity used for home lighting in this country? Round to the nearest tenth of a billion.

188. **The Federal Government** To override a presidential veto, at least $66\frac{2}{3}\%$ of the Senate must vote to override the veto. There are 100 senators in the Senate. What is the minimum number of votes needed to override a veto?

189. **Advertising** Suppose that 125 million people saw a particular 30-second commercial for a soft drink during Super Bowl XXXV. The cost for the commercial was approximately $2.2 million.
 a. If the soft drink manufacturer makes a profit of $.08 on each can sold, what percent of the people watching the commercial would have to buy one can of the soft drink for the company to recover the cost of the commercial?
 b. What percent of the people watching the commercial would have to purchase one six-pack of the soft drink for the company to recover the cost of the commercial? Round to the nearest tenth of a percent.

190. **Health Care** The graph at the right shows the average health care costs paid by an employee in the United States. What percent of the total average health care cost was out-of-pocket costs in 2001? Round to the nearest tenth of a percent. (*Source:* Hewitt Associates)

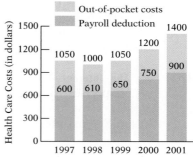

191. **Consumerism** A 16-horsepower lawn tractor with automatic transmission sells for $1579.99. This is 11% more than the price of the same lawn tractor with a standard transmission. What is the price of the less expensive model? Round to the nearest cent.

192. **Investment** If Miranda Perry invests $2500 in an account that earns an 8% annual simple interest rate, how much interest will Miranda have earned after 9 months?

193. Investment Kachina Caron invests $1200 in a simple interest account and earns $72 in 8 months. What is the annual simple interest rate?

194. Investment How much money must Andrea invest for two years in an account that earns an annual simple interest rate of 8% if she wants to earn $300 from the investment?

195. Investment Sal Boxer divided a gift of $3000 into two different accounts. He placed $1000 in one account that earned an annual simple interest rate of 7.5%. The remaining money was placed in an account that earned an annual simple interest rate of 8.25%. How much interest did Sal earn from the two accounts after one year?

196. Investment If Americo invests $2500 at an 8% annual simple interest rate and Octavia invests $3000 at a 7% annual simple interest rate, which of the two will earn the greater amount of interest in one year?

197. Investment Makana invested $900 in an account that earned an annual simple interest rate that was 1% higher than the rate her friend Marlys earned on her investment. If Marlys earned $51 after one year from an investment of $850, how much did Makana earn after one year?

198. Investment A $2000 investment at an annual simple interest rate of 6% earned as much interest after one year as another investment in an account that earned 8% annual simple interest. How much was invested at 8%?

199. Investment An investor places $1000 in an account that earns 9% annual simple interest and $1000 in an account that earns 6% annual simple interest. If each investment is left in the account for the same period of time, is the interest rate on the combined investment less than 6%, between 6% and 9%, or greater than 9%?

200. Mixture Problem The concentration of platinum in a necklace is 15%. The necklace weighs 12 g. Find the amount of platinum in the necklace.

201. Mixture Problem A 250-milliliter solution of a fabric dye contains 5 ml of hydrogen peroxide. What is the percent concentration of the hydrogen peroxide?

202. Mixture Problem A carpet is made of a blend of wool and other fibers. If the concentration of wool in the carpet is 75% and the carpet weighs 175 lb, how much wool is in the carpet?

203. Mixture Problem Apple Dan's 32-ounce apple-flavored fruit drink contains 8 oz of apple juice. Forty ounces of a generic brand of apple-flavored fruit drink contains 9 oz of apple juice. Which of the two brands has the greater concentration of apple juice?

204. Mixture Problem Bakers use simple syrup in many of their recipes. Simple syrup is made by combining 500 g of sugar with 500 g of water and mixing it well until the sugar dissolves. What is the percent concentration of sugar in simple syrup?

205. Mixture Problem A pharmacist has 50 g of a topical cream that contains 75% glycerin. How many grams of the cream are not glycerin?

206. Mixture Problem A chemist has 100 ml of a solution that is 9% acetic acid. If the chemist adds 50 ml of pure water to this solution, what is the percent concentration of the resulting mixture?

207. Mixture Problem A 500-gram salt and water solution contains 50 g of salt. This mixture is left in the open air and 100 g of water evaporates from the solution. What is the percent concentration of salt in the remaining solution?

3.1   Applying Concepts

Solve.

208. $\dfrac{2m + m}{5} = -9$

209. $\dfrac{3y - 8y}{7} = 15$

210. $\dfrac{1}{\frac{1}{x}} = 5$

211. $\dfrac{1}{\frac{1}{x}} + 8 = -19$

212. $\dfrac{4}{\frac{3}{b}} = 8$

213. $\dfrac{5}{\frac{7}{a}} - \dfrac{3}{\frac{7}{a}} = 6$

214. Solve for x: $x \div 28 = 1481$ remainder 25

215. Geometry Find the value of x in the diagram shown at the right.

216. Consumerism Your bill for dinner, including a 7.25% sales tax, was $92.74. You want to leave a 15% tip on the cost of the dinner before the sales tax. Find the amount of the tip to the nearest dollar.

217. Consumerism The total cost for a dinner was $97.52. This included a 15% tip calculated on the cost of the dinner after a 6% sales tax. Find the cost of dinner before the tip and tax.

218. Business A retailer decides to increase the original price of each item in the store by 10%. After the price increase, the retailer notices a significant drop in sales and so decides to reduce the current price of each item in the store by 10%. Are the prices back to the original prices? If not, are the prices lower or higher than the original prices?

219. If a quantity increases by 100%, how many times its original value is the new value?

220. Make up an equation of the form $x + a = b$ that has 2 as a solution.

221. Make up an equation of the form $ax = b$ that has -2 as a solution.

222. **Monthly Mortgage Payments** Suppose you have a $100,000 30-year mortgage. What is the difference between the monthly payment if the interest rate on your loan is 7.75% and the monthly payment if the interest rate is 7.25%? (*Hint:* You will need to discuss the question with an agent at a bank that provides mortgage services, discuss the question with a real estate agent, or find and use the formula for determining the monthly payment on a mortgage.)

223. The following problem does not contain enough information: "How many hours does it take to fly from Los Angeles to New York?" What additional information do we need in order to answer the question?

224. Write an explanation of how to use the tax rate schedule shown below. Calculate the income tax on a single taxpayer's taxable income of $75,000.

2001 Tax Rate Schedule			
If taxable income is more than	*but not more than*	*Tax is*	*of the amount over*
For single taxpayers			
0	$27,050	15%	0
$27,050	$65,550	$4,057.50 plus 27.5%	$27,050
$65,550	$136,750	$14,645.00 plus 30.5%	$65,550
$136,750	$297,350	$36,361.00 plus 35.5%	$136,750
$297,350	—	$93,374.00 plus 39.1%	$297,350
For married taxpayers, filing jointly			
0	$45,200	15%	0
$45,200	$109,250	$6,780.00 plus 27.5%	$45,200
$109,250	$166,500	$24,393.75 plus 30.5%	$109,250
$166,500	$297,350	$41,855.00 plus 35.5%	$166,500
$297,350	—	$88,306.75 plus 39.1%	$297,350

3.2 | General Equations

1 **Solve equations of the form $ax + b = c$**

In solving an equation of the form $ax + b = c$, the goal is to rewrite the equation in the form *variable = constant*. This requires applying both the Addition and the Multiplication Properties of Equations.

Solve: $\dfrac{2}{5}x - 3 = -7$

$$\frac{2}{5}x - 3 = -7$$

Add 3 to each side of the equation.

$$\frac{2}{5}x - 3 + 3 = -7 + 3$$

Simplify.

$$\frac{2}{5}x = -4$$

Multiply each side of the equation by the reciprocal of the coefficient $\frac{2}{5}$.

$$\frac{5}{2} \cdot \frac{2}{5}x = \frac{5}{2}(-4)$$

Simplify. Now the equation is in the form *variable = constant*.

$$x = -10$$

Check $\dfrac{2}{5}x - 3 = -7$

$$\begin{array}{c|c} \frac{2}{5}(-10) - 3 & -7 \\ -4 - 3 & -7 \\ -7 = -7 \end{array}$$

Write the solution.   The solution is -10.

Example 1 Solve: $3x - 7 = -5$

Solution
$$3x - 7 = -5$$
$$3x - 7 + 7 = -5 + 7 \qquad \bullet \text{ Add 7 to each side of the equation.}$$
$$3x = 2 \qquad \bullet \text{ Simplify.}$$
$$\frac{3x}{3} = \frac{2}{3} \qquad \bullet \text{ Divide each side of the equation by 3.}$$
$$x = \frac{2}{3} \qquad \bullet \text{ Simplify. Now the equation is in the form } variable = constant.$$

The solution is $\frac{2}{3}$. $\bullet$ Write the solution.

Problem 1 Solve: $5x + 7 = 10$

Solution   See page S6.

Example 2   Solve: $5 = 9 - 2x$

Solution

$$5 = 9 - 2x$$

$$5 - 9 = 9 - 9 - 2x$$ • Subtract 9 from each side of the equation.

$$-4 = -2x$$ • Simplify.

$$\frac{-4}{-2} = \frac{-2x}{-2}$$ • Divide each side of the equation by −2.

$$2 = x$$ • Simplify.

The solution is 2.   • Write the solution.

Problem 2   Solve: $11 = 11 + 3x$

Solution   See page S6.

Note in Example 4 in the Introduction to Equations section that the original equation, $\frac{1}{2} = x + \frac{2}{3}$, contained fractions with denominators of 2 and 3. The least common multiple of 2 and 3 is 6. The least common multiple has the property that both 2 and 3 will divide evenly into it. Therefore, if both sides of the equation are multiplied by 6, both of the denominators will divide evenly into 6. The result is an equation that does not contain any fractions. Multiplying an equation that contains fractions by the LCM of the denominators is called **clearing denominators.** It is an alternative method of solving an equation that contains fractions.

Solve: $\frac{1}{2} = x + \frac{2}{3}$ $\frac{1}{2} = x + \frac{2}{3}$

Multiply both sides of the equation by 6, the LCM of the denominators. $6\left(\frac{1}{2}\right) = 6\left(x + \frac{2}{3}\right)$

Simplify each side of the equation. Use the Distributive Property on the right side of the equation. $3 = 6(x) + 6\left(\frac{2}{3}\right)$

Note that multiplying both sides of the equation by the LCM of the denominators eliminated the fractions. $3 = 6x + 4$

Solve the resulting equation.

$$3 - 4 = 6x + 4 - 4$$

$$-1 = 6x$$

$$\frac{-1}{6} = \frac{6x}{6}$$

$$-\frac{1}{6} = x$$

The solution is $-\frac{1}{6}$.

Example 3   Solve: $\dfrac{2}{3} + \dfrac{1}{4}x = -\dfrac{1}{3}$

Solution

$$\dfrac{2}{3} + \dfrac{1}{4}x = -\dfrac{1}{3}$$

$$12\left(\dfrac{2}{3} + \dfrac{1}{4}x\right) = 12\left(-\dfrac{1}{3}\right)$$

$$12\left(\dfrac{2}{3}\right) + 12\left(\dfrac{1}{4}x\right) = -4$$

$$8 + 3x = -4$$

$$8 - 8 + 3x = -4 - 8$$

$$3x = -12$$

$$\dfrac{3x}{3} = \dfrac{-12}{3}$$

$$x = -4$$

- The equation contains fractions. Find the LCM of the denominators.
- The LCM of 3 and 4 is 12. Multiply each side of the equation by 12.
- Use the Distributive Property to multiply the left side of the equation by 12.
- The equation now contains no fractions.

The solution is -4.

Problem 3   Solve: $\dfrac{5}{2} - \dfrac{2}{3}x = \dfrac{1}{2}$

Solution   See page S6.

---

2 **Solve equations of the form $ax + b = cx + d$**

In solving an equation of the form $ax + b = cx + d$, the goal is to rewrite the equation in the form *variable* = *constant*. Begin by rewriting the equation so that there is only one variable term in the equation. Then rewrite the equation so that there is only one constant term.

Solve: $4x - 5 = 6x + 11$

$$4x - 5 = 6x + 11$$

Subtract $6x$ from each side of the equation.
$$4x - 6x - 5 = 6x - 6x + 11$$

Simplify. Now there is only one variable term in the equation.
$$-2x - 5 = 11$$

Add 5 to each side of the equation.
$$-2x - 5 + 5 = 11 + 5$$

Simplify. Now there is only one constant term in the equation.
$$-2x = 16$$

Divide each side of the equation by -2.
$$\dfrac{-2x}{-2} = \dfrac{16}{-2}$$

Simplify. Now the equation is in the form *variable* = *constant*.
$$x = -8$$

Check
$$4x - 5 = 6x + 11$$

$$
\begin{array}{c|c}
4(-8) - 5 & 6(-8) + 11 \\
-32 - 5 & -48 + 11 \\
\end{array}
$$
$$-37 = -37$$

Write the solution.   The solution is -8.

Example 4   Solve: $4x - 3 = 8x - 7$

Solution $4x - 3 = 8x - 7$

$4x - 8x - 3 = 8x - 8x - 7$ • Subtract $8x$ from each side of the equation.

$-4x - 3 = -7$ • Simplify. Now there is only one variable term in the equation.

$-4x - 3 + 3 = -7 + 3$ • Add 3 to each side of the equation.
$-4x = -4$ • Simplify. Now there is only one constant term in the equation.

$$\dfrac{-4x}{-4} = \dfrac{-4}{-4}$$ • Divide each side of the equation by -4.

$x = 1$ • Simplify. Now the equation is in the form *variable = constant*.

The solution is 1.   • Write the solution.

Problem 4   Solve: $5x + 4 = 6 + 10x$

Solution   See page S6.

---

3   **Solve equations containing parentheses**

When an equation contains parentheses, one of the steps in solving the equation requires the use of the Distributive Property. The Distributive Property is used to remove parentheses from a variable expression. $a(b + c) = ab + ac$

Solve: $4 + 5(2x - 3) = 3(4x - 1)$

$$4 + 5(2x - 3) = 3(4x - 1)$$

Use the Distributive Property to remove parentheses. $4 + 10x - 15 = 12x - 3$

Simplify. $10x - 11 = 12x - 3$

Subtract $12x$ from each side of the equation. $10x - 12x - 11 = 12x - 12x - 3$

Simplify. Now there is only one variable term in the equation. $-2x - 11 = -3$

Add 11 to each side of the equation. $-2x - 11 + 11 = -3 + 11$

Simplify. Now there is only one constant term in the equation.	$-2x = 8$
Divide each side of the equation by -2.	$\dfrac{-2x}{-2} = \dfrac{8}{-2}$
Simplify. Now the equation is in the form *variable = constant*.	$x = -4$

Check $\dfrac{4 + 5(2x - 3) = 3(4x - 1)}{}$

$$
\begin{array}{c|c}
4 + 5[2(-4) - 3] & 3[4(-4) - 1] \\
4 + 5(-8 - 3) & 3(-16 - 1) \\
4 + 5(-11) & 3(-17) \\
4 - 55 & -51 \\
-51 = -51
\end{array}
$$

Write the solution.   The solution is -4.

In Chapter 2, we discussed the use of a graphing calculator to evaluate variable expressions. The same procedure can be used to check the solution of an equation. Consider the example above. After we divide both sides of the equation by -2, the solution appears to be -4. To check this solution, store the value of x, -4, in the calculator. Evaluate the expression on the left side of the original equation: $4 + 5(2x - 3)$. The result is -51. Now evaluate the expression on the right side of the original equation: $3(4x - 1)$. The result is -51. Because the results are equal, the solution -4 checks. See the Appendix for a description of keystroking procedures for different models of graphing calculators.

Example 5   Solve: $3x - 4(2 - x) = 3(x - 2) - 4$

Solution
$$3x - 4(2 - x) = 3(x - 2) - 4$$
$$3x - 8 + 4x = 3x - 6 - 4$$ • Use the Distributive Property to remove parentheses.

$$7x - 8 = 3x - 10$$ • Simplify.
$$7x - 3x - 8 = 3x - 3x - 10$$ • Subtract $3x$ from each side of the equation.
$$4x - 8 = -10$$
$$4x - 8 + 8 = -10 + 8$$ • Add 8 to each side of the equation.
$$4x = -2$$
$$\dfrac{4x}{4} = \dfrac{-2}{4}$$ • Divide each side of the equation by 4.

$$x = -\dfrac{1}{2}$$ • The equation is in the form *variable = constant*.

The solution is $-\dfrac{1}{2}$.

Problem 5   Solve: $5x - 4(3 - 2x) = 2(3x - 2) + 6$

Solution   See page S6.

Example 6   Solve: $3[2 - 4(2x - 1)] = 4x - 10$

Solution
$$3[2 - 4(2x - 1)] = 4x - 10$$
$$3[2 - 8x + 4] = 4x - 10$$
• Use the Distributive Property to remove the parentheses.

$$3[6 - 8x] = 4x - 10$$
• Simplify inside the brackets.
$$18 - 24x = 4x - 10$$
• Use the Distributive Property to remove the brackets.

$$18 - 24x - 4x = 4x - 4x - 10$$
$$18 - 28x = -10$$
• Subtract $4x$ from each side of the equation.

$$18 - 18 - 28x = -10 - 18$$
• Subtract 18 from each side of the equation.
$$-28x = -28$$

$$\frac{-28x}{-28} = \frac{-28}{-28}$$
• Divide each side of the equation by -28.
$$x = 1$$

The solution is 1.

Problem 6   Solve: $-2[3x - 5(2x - 3)] = 3x - 8$

Solution   See page S6.

4 Application problems

Example 7   A company uses the equation $V = C - 6000t$ to determine the depreciated value V, after t years, of a milling machine that originally cost C dollars. If a milling machine originally cost $50,000, in how many years will the depreciated value of the machine be $38,000?

Strategy   To find the number of years, replace C with 50,000 and V with 38,000 in the given equation, and solve for t.

Solution
$$V = C - 6000t$$
$$38,000 = 50,000 - 6000t$$
$$38,000 - 50,000 = 50,000 - 50,000 - 6000t$$
$$-12,000 = -6000t$$
$$\frac{-12,000}{-6000} = \frac{-6000t}{-6000}$$
$$2 = t$$

The depreciated value of the machine will be $38,000 in 2 years.

Problem 7   The value V of an investment of $7500 at an annual simple interest rate of 6% is given by the equation $V = 450t + 7500$, where t is the amount of time, in years, that the money is invested. In how many years will the value of a $7500 investment be $10,200?

Solution   See page S6.

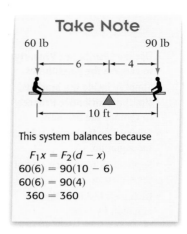

Take Note

This system balances because

$$F_1x = F_2(d - x)$$
$$60(6) = 90(10 - 6)$$
$$60(6) = 90(4)$$
$$360 = 360$$

A lever system is shown below. It consists of a lever, or bar; a fulcrum; and two forces, F_1 and F_2. The distance d represents the length of the lever, x represents the distance from F_1 to the fulcrum, and $d - x$ represents the distance from F_2 to the fulcrum.

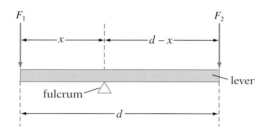

A principle of physics states that when the lever system balances,

$$F_1x = F_2(d - x)$$

Example 8 A lever is 10 ft long. A force of 100 lb is applied to one end of the lever, and a force of 400 lb is applied to the other end. When the system balances, how far is the fulcrum from the 100-pound force?

Strategy Draw a diagram of the situation.

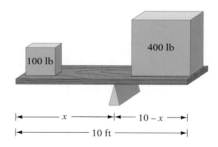

The lever is 10 ft long, so $d = 10$. One force is 100 lb, so $F_1 = 100$. The other force is 400 lb, so $F_2 = 400$. To find the distance of the fulcrum from the 100-pound force, replace the variables F_1, F_2, and d in the lever-system equation with the given values, and solve for x.

Solution
$$F_1x = F_2(d - x)$$
$$100x = 400(10 - x)$$
$$100x = 4000 - 400x$$
$$100x + 400x = 4000 - 400x + 400x$$
$$500x = 4000$$
$$\frac{500x}{500} = \frac{4000}{500}$$
$$x = 8$$

The fulcrum is 8 ft from the 100-pound force.

Problem 8   A lever is 14 ft long. At a distance of 6 ft from the fulcrum, a force of 40 lb is applied. How large a force must be applied to the other end of the lever so that the system will balance?

Solution   See page S6.

3.2 | Concept Review

Determine whether the statement is always true, sometimes true, or never true.

1. The same variable term can be added to both sides of an equation without changing the solution of the equation.

2. The same variable term can be subtracted from both sides of an equation without changing the solution of the equation.

3. An equation of the form $ax + b = c$ cannot be solved if a is a negative number.

4. The solution of the equation $\frac{x}{3} = 0$ is 0.

5. The solution of the equation $\frac{x}{0} = 3$ is 0.

6. In solving an equation of the form $ax + b = cx + d$, the goal is to rewrite the equation in the form $variable = constant$.

7. In solving an equation of the form $ax + b = cx + d$, subtracting cx from each side of the equation results in an equation with only one variable term in it.

3.2 | Exercises

1 Solve and check.

1. $3x + 1 = 10$

2. $4y + 3 = 11$

3. $2a - 5 = 7$

4. $5m - 6 = 9$

5. $5 = 4x + 9$

6. $2 = 5b + 12$

7. $13 = 9 + 4z$

8. $7 - c = 9$

9. $2 - x = 11$

10. $4 - 3w = -2$

11. $5 - 6x = -13$

12. $8 - 3t = 2$

13. $-5d + 3 = -12$

14. $-8x - 3 = -19$

15. $-7n - 4 = -25$

16. $-12x + 30 = -6$

17. $-13 = -11y + 9$

18. $2 = 7 - 5a$

19. $3 = 11 - 4n$

20. $-35 = -6b + 1$

21. $-8x + 3 = -29$

22. $-3m - 21 = 0$

23. $7x - 3 = 3$

24. $8y + 3 = 7$

25. $6a + 5 = 9$

26. $3m + 4 = 11$

27. $11 = 15 + 4n$

28. $4 = 2 - 3c$

29. $9 - 4x = 6$

30. $7 - 8z = 0$

31. $1 - 3x = 0$

32. $6y - 5 = -7$

33. $8b - 3 = -9$

34. $5 - 6m = 2$

35. $7 - 9a = 4$

36. $9 = -12c + 5$

37. $10 = -18x + 7$

38. $2y + \dfrac{1}{3} = \dfrac{7}{3}$

39. $3x - \dfrac{5}{6} = \dfrac{13}{6}$

40. $5y + \dfrac{3}{7} = \dfrac{3}{7}$

41. $9x + \dfrac{4}{5} = \dfrac{4}{5}$

42. $4 = 7 - 2w$

43. $7 = 9 - 5a$

44. $-6y + 5 = 13$

45. $-4x + 3 = 9$

46. $\dfrac{1}{2}a - 3 = 1$

47. $\dfrac{1}{3}m - 1 = 5$

48. $\dfrac{2}{5}y + 4 = 6$

49. $\dfrac{3}{4}n + 7 = 13$

50. $-\dfrac{2}{3}x + 1 = 7$

51. $-\dfrac{3}{8}b + 4 = 10$

52. $\dfrac{x}{4} - 6 = 1$

53. $\dfrac{y}{5} - 2 = 3$

54. $\dfrac{2x}{3} - 1 = 5$

55. $\dfrac{3c}{7} - 1 = 8$

56. $4 - \dfrac{3}{4}z = -2$

57. $3 - \dfrac{4}{5}w = -9$

58. $5 + \dfrac{2}{3}y = 3$

59. $17 + \dfrac{5}{8}x = 7$

60. $17 = 7 - \dfrac{5}{6}t$

61. $\dfrac{2}{3} = y - \dfrac{1}{2}$

62. $\dfrac{3}{4} = \dfrac{1}{12}x + 2$

63. $\dfrac{3}{8} = \dfrac{5}{12} - \dfrac{1}{3}b$

64. $\dfrac{2}{3} = \dfrac{3}{4} - \dfrac{1}{2}y$

65. $7 = \dfrac{2x}{5} + 4$

66. $5 - \dfrac{4c}{7} = 8$

67. $7 - \dfrac{5}{9}y = 9$

68. $6a + 3 + 2a = 11$

69. $5y + 9 + 2y = 23$

70. $7x - 4 - 2x = 6$

71. $11z - 3 - 7z = 9$

72. $2x - 6x + 1 = 9$

73. $b - 8b + 1 = -6$

74. $-1 = 5m + 7 - m$

75. $8 = 4n - 6 + 3n$

76. $0.15y + 0.025 = -0.074$

77. $1.2x - 3.44 = 1.3$

78. $3.5 = 3.5 + 0.076x$

79. $-6.5 = 4.3y - 3.06$

Solve.

80. If $2x - 3 = 7$, evaluate $3x + 4$.

81. If $3x + 5 = -4$, evaluate $2x - 5$.

82. If $4 - 5x = -1$, evaluate $x^2 - 3x + 1$.

83. If $2 - 3x = 11$, evaluate $x^2 + 2x - 3$.

2 Solve and check.

84. $8x + 5 = 4x + 13$

85. $6y + 2 = y + 17$

86. $7m + 4 = 6m + 7$

87. $11n + 3 = 10n + 11$

88. $5x - 4 = 2x + 5$

89. $9a - 10 = 3a + 2$

90. $12y - 4 = 9y - 7$

91. $13b - 1 = 4b - 19$

92. $15x - 2 = 4x - 13$

93. $7a - 5 = 2a - 20$

94. $3x + 1 = 11 - 2x$

95. $n - 2 = 6 - 3n$

96. $2x - 3 = -11 - 2x$

97. $4y - 2 = -16 - 3y$

98. $2b + 3 = 5b + 12$

99. $m + 4 = 3m + 8$

100. $4x - 7 = 5x + 1$

101. $6d - 2 = 7d + 5$

102. $4y - 8 = y - 8$

103. $5a + 7 = 2a + 7$

104. $6 - 5x = 8 - 3x$

105. $10 - 4n = 16 - n$

106. $2x - 4 = 6x$

107. $2b - 10 = 7b$

108. $8m = 3m + 20$

109. $9y = 5y + 16$

110. $-x - 4 = -3x - 16$

111. $8 - 4x = 18 - 5x$

112. $6 - 10a = 8 - 9a$

113. $5 - 7m = 2 - 6m$

114. $8b + 5 = 5b + 7$

115. $6y - 1 = 2y + 2$

116. $7x - 8 = x - 3$

117. $10x - 3 = 3x - 1$

118. $5n + 3 = 2n + 1$

119. $8a - 2 = 4a - 5$

120. $\frac{1}{5}d = \frac{1}{2}d + 3$

121. $\frac{3}{4}x = \frac{1}{12}x + 2$

122. $\frac{2}{3}a = \frac{1}{5}a + 7$

123. $\frac{4}{5}c - 7 = \frac{1}{10}c$

124. $\frac{1}{2}y = 10 - \frac{1}{3}y$

125. $\frac{2}{3}b = 2 - \frac{5}{6}b$

126. $8.7y = 3.9y + 9.6$

127. $4.5x - 5.4 = 2.7x$

128. $5.6x = 7.2x - 6.4$

Solve.

129. If $5x = 3x - 8$, evaluate $4x + 2$.

130. If $7x + 3 = 5x - 7$, evaluate $3x - 2$.

131. If $2 - 6a = 5 - 3a$, evaluate $4a^2 - 2a + 1$.

132. If $1 - 5c = 4 - 4c$, evaluate $3c^2 - 4c + 2$.

3 Solve and check.

133. $5x + 2(x + 1) = 23$

134. $6y + 2(2y + 3) = 16$

135. $9n - 3(2n - 1) = 15$

136. $12x - 2(4x - 6) = 28$

137. $7a - (3a - 4) = 12$

138. $9m - 4(2m - 3) = 11$

139. $5(3 - 2y) + 4y = 3$

140. $4(1 - 3x) + 7x = 9$

141. $10x + 1 = 2(3x + 5) - 1$

142. $5y - 3 = 7 + 4(y - 2)$

143. $4 - 3a = 7 - 2(2a + 5)$

144. $9 - 5x = 12 - (6x + 7)$

145. $3y - 7 = 5(2y - 3) + 4$

146. $2a - 5 = 4(3a + 1) - 2$

147. $5 - (9 - 6x) = 2x - 2$

148. $7 - (5 - 8x) = 4x + 3$

149. $3[2 - 4(y - 1)] = 3(2y + 8)$

150. $5[2 - (2x - 4)] = 2(5 - 3x)$

151. $3a + 2[2 + 3(a - 1)] = 2(3a + 4)$

152. $5 + 3[1 + 2(2x - 3)] = 6(x + 5)$

153. $-2[4 - (3b + 2)] = 5 - 2(3b + 6)$

154. $-4[x - 2(2x - 3)] + 1 = 2x - 3$

155. $0.3x - 2(1.6x) - 8 = 3(1.9x - 4.1)$

156. $0.56 - 0.4(2.1y + 3) = 0.2(2y + 6.1)$

Solve.

157. If $4 - 3a = 7 - 2(2a + 5)$, evaluate $a^2 + 7a$.

158. If $9 - 5x = 12 - (6x + 7)$, evaluate $x^2 - 3x - 2$.

159. If $2z - 5 = 3(4z + 5)$, evaluate $\dfrac{z^2}{z - 2}$.

160. If $3n - 7 = 5(2n + 7)$, evaluate $\dfrac{n^2}{2n - 6}$.

4 Black ice is an ice covering on roads that is especially difficult to see and therefore extremely dangerous for motorists. The distance that a car traveling 30 mph will slide after its brakes are applied is related to the outside temperature by the formula $C = \frac{1}{4}D - 45$, where C is the Celsius temperature and D is the distance in feet that the car will slide.

161. **Physics** Determine the distance a car will slide on black ice when the outside temperature is $-3°C$.

162. **Physics** Determine the distance a car will slide on black ice when the outside temperature is $-11°C$.

The pressure at a certain depth in the ocean can be approximated by the equation $P = \frac{1}{2}D + 15$, where P is the pressure in pounds per square inch, and D is the depth in feet.

163. **Physics** Find the depth of a diver when the pressure on the diver is 35 lb/in².

164. **Physics** Find the depth of a diver when the pressure on the diver is 45 lb/in².

The distance s, in feet, that an object will fall in t seconds is given by $s = 16t^2 + vt$, where v is the initial velocity of the object in feet per second.

165. **Physics** Find the initial velocity of an object that falls 80 ft in 2 s.

166. **Physics** Find the initial velocity of an object that falls 144 ft in 3 s.

Anthropologists can approximate the height of a primate from the size of its humerus (the bone extending from the shoulder to the elbow) by using the equation $H = 1.2L + 27.8$, where L is the length of the humerus and H is the height of the primate, both in inches.

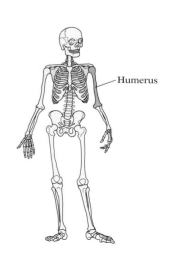

Humerus

167. **Anthropology** An anthropologist estimates the height of a primate to be 66 in. What is the approximate length of the humerus of this primate? Round to the nearest tenth.

168. Anthropology The height of a primate is estimated to be 62 in. Find the approximate length of the humerus of this primate.

The fare, F, to be charged a customer by a taxi company is calculated using the formula $F = 1.50 + 0.95(m - 1)$, where m is the number of miles traveled.

169. Business A customer is charged \$6.25. How many miles was the passenger driven?

170. Business A passenger is charged \$9.10. Find the number of miles the customer was driven.

A person's accurate typing speed can be approximated by the equation $S = \frac{W - 5e}{10}$, where S is the accurate typing speed in words per minute, W is the number of words typed in ten minutes, and e is the number of errors made.

171. Employment After taking a 10-minute typing test, a job candidate was told that he had an accurate speed of 35 words per minute. He had typed a total of 390 words. How many errors did he make?

172. Employment A job applicant took a 10-minute typing test and was told that she had an accurate speed of 37 words per minute. If she had typed a total of 400 words, how many errors did she make?

A telephone company estimates that the number N of phone calls made per day between two cities of populations P_1 and P_2 that are d miles apart is given by the equation $N = \frac{2.51P_1P_2}{d^2}$.

173. Business Estimate the population P_2, given that P_1 is 48,000, the number of phone calls is 1,100,000, and the distance between the cities is 75 mi. Round to the nearest thousand.

174. Business Estimate the population P_1, given that P_2 is 125,000, the number of phone calls is 2,500,000, and the distance between the cities is 50 mi. Round to the nearest thousand.

To determine the break-even point, or the number of units that must be sold so that no profit or loss occurs, an economist uses the equation $Px = Cx + F$, where P is the selling price per unit, x is the number of units sold, C is the cost to make each unit, and F is the fixed cost.

175. Business An economist has determined that the selling price per unit for a television is \$250. The cost to make one television is \$165, and the fixed cost is \$29,750. Find the break-even point.

176. Business A manufacturing engineer determines that the cost per unit for a compact disc is \$3.35 and that the fixed cost is \$6180. The selling price for the compact disc is \$8.50. Find the break-even point.

Use the lever-system equation $F_1 x = F_2(d - x)$.

177. In the lever-system equation $F_1 x = F_2(d - x)$, what does x represent? What does d represent? What does $d - x$ represent?

178. Suppose an 80-pound child is sitting at one end of a 7-foot seesaw, and a 60-pound child is sitting at the other end. The 80-pound child is 3 feet from the fulcrum. Use the lever-system equation to explain why the seesaw balances.

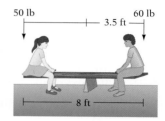

179. Lever Systems Two children are sitting 8 ft apart on a seesaw. One child weighs 60 lb and the second child weighs 50 lb. The fulcrum is 3.5 ft from the child weighing 60 lb. Is the seesaw balanced?

180. Lever Systems An adult and a child are on a seesaw that is 14 ft long. The adult weighs 175 lb and the child weighs 70 lb. How many feet from the child must the fulcrum be placed so that the seesaw balances?

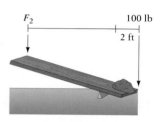

181. Lever Systems A lever 10 ft long is used to balance a 100-pound rock. The fulcrum is placed 2 ft from the rock. What force must be applied to the other end of the lever to balance the rock?

182. Lever Systems A 50-pound weight is applied to the left end of a seesaw that is 10 ft long. The fulcrum is 4 ft from the 50-pound weight. A weight of 30 lb is applied to the right end of the seesaw. Is the 30-pound weight adequate to balance the seesaw?

183. Lever Systems In preparation for a stunt, two acrobats are standing on a plank that is 18 ft long. One acrobat weighs 128 lb and the second acrobat weighs 160 lb. How far from the 128-pound acrobat must the fulcrum be placed so that the acrobats are balanced on the plank?

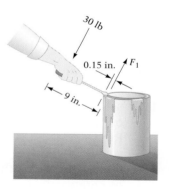

184. Lever Systems A screwdriver 9 in. long is used as a lever to open a can of paint. The tip of the screwdriver is placed under the lip of the lid with the fulcrum 0.15 in. from the lip. A force of 30 lb is applied to the other end of the screwdriver. Find the force on the lip of the lid.

3.2 Applying Concepts

Solve. If the equation has no solution, write "No solution."

185. $3(2x - 1) - (6x - 4) = -9$

186. $\dfrac{1}{5}(25 - 10b) + 4 = \dfrac{1}{3}(9b - 15) - 6$

187. $3[4(w + 2) - (w + 1)] = 5(2 + w)$

188. $\dfrac{2(5x - 6) - 3(x - 4)}{7} = x + 2$

189. Solve for x: $32{,}166 \div x = 518$ remainder 50

190. One-half of a certain number equals two-thirds of the same number. Find the number.

The Addition Property of Inequalities is used when, in order to rewrite an inequality in the form *variable* > *constant* or *variable* < *constant*, a term must be removed from one side of the inequality. Add the opposite of the term to each side of the inequality.

Because subtraction is defined in terms of addition, the Addition Property of Inequalities makes it possible to subtract the same number from each side of an inequality without changing the solution set of the inequality.

Solve: $x + 4 < 5$

$$x + 4 < 5$$

Subtract 4 from each side of the inequality.

$$x + 4 - 4 < 5 - 4$$

Simplify.

$$x < 1$$

The graph of the solution set of $x + 4 < 5$ is shown at the right.

Solve: $5x - 6 \le 4x - 4$

$$5x - 6 \le 4x - 4$$

Subtract $4x$ from each side of the inequality.

$$5x - 4x - 6 \le 4x - 4x - 4$$

Simplify.

$$x - 6 \le -4$$

Add 6 to each side of the inequality.

$$x - 6 + 6 \le -4 + 6$$

Simplify.

$$x \le 2$$

Example 2 Solve and graph the solution set of $x + 5 > 3$.

Solution
$$x + 5 > 3$$
$$x + 5 - 5 > 3 - 5 \qquad \bullet \text{ Subtract 5 from each side of the inequality.}$$
$$x > -2$$

Problem 2 Solve and graph the solution set of $x + 2 < -2$.

Solution   See page S7.

Example 3 Solve: $7x - 14 \le 6x - 16$

Solution
$$7x - 14 \le 6x - 16$$
$$7x - 6x - 14 \le 6x - 6x - 16 \qquad \bullet \text{ Subtract } 6x \text{ from each side of the}$$
$$x - 14 \le -16 \qquad\qquad\qquad \text{inequality.}$$
$$x - 14 + 14 \le -16 + 14 \qquad \bullet \text{ Add 14 to each side of the}$$
$$x \le -2 \qquad\qquad\qquad\qquad \text{inequality.}$$

Problem 3 Solve: $5x + 3 > 4x + 5$

Solution   See page S7.

2 Solve inequalities using the Multiplication Property of Inequalities

In solving an inequality, the goal is to rewrite the given inequality in the form *variable* > *constant* or *variable* < *constant*. The Multiplication Property of Inequalities is used when, in order to rewrite an inequality in this form, a coefficient must be removed from one side of the inequality.

> ### Multiplication Property of Inequalities—Rule 1
>
> Each side of an inequality can be multiplied by the same positive number without changing the solution set of the inequality.
>
> If $a > b$ and $c > 0$, then $ac > bc$.   If $a < b$ and $c > 0$, then $ac < bc$.

Take Note

$c > 0$ means c is a positive number.

The inequality symbols are not changed.

$$5 > 4 \qquad \text{• A true inequality}$$
$$5(2) > 4(2) \qquad \text{• Multiply by } positive \text{ 2.}$$
$$10 > 8 \qquad \text{• A true inequality}$$

> ### Multiplication Property of Inequalities—Rule 2
>
> If each side of an inequality is multiplied by the same negative number and the inequality symbol is reversed, then the solution set of the inequality is not changed.
>
> If $a > b$ and $c < 0$, then $ac < bc$.   If $a < b$ and $c < 0$, then $ac > bc$.

Take Note

$c < 0$ means c is a negative number.

The inequality symbol is reversed in each case.

$$6 < 9 \qquad \text{• A true inequality}$$
$$6(-3) > 9(-3) \qquad \text{• Multiply by } negative \text{ 3 and reverse the inequality symbol.}$$
$$-18 > -27 \qquad \text{• A true inequality}$$

The Multiplication Property of Inequalities also holds true for an inequality containing the symbol $\geq$ or $\leq$.

Solve: $-\dfrac{3}{2}x \leq 6$

Multiply each side of the inequality by the reciprocal of the coefficient $-\dfrac{3}{2}$. Because $-\dfrac{2}{3}$ is a negative number, the inequality symbol must be reversed.

$$-\frac{3}{2}x \leq 6$$

$$-\frac{2}{3}\left(-\frac{3}{2}x\right) \geq -\frac{2}{3}(6)$$

Simplify.

$$x \geq -4$$

The graph of the solution set of $-\dfrac{3}{2}x \leq 6$ is shown at the right.

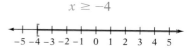

Any time an inequality is multiplied or divided by a negative number, the inequality symbol must be reversed. Compare these two examples:

$2x < -4$	Divide each side by
$\dfrac{2x}{2} < \dfrac{-4}{2}$	*positive* 2. The inequality symbol
$x < -2$	is *not* reversed.

$-2x < 4$	Divide each side by
$\dfrac{-2x}{-2} > \dfrac{4}{-2}$	*negative* 2. The inequality symbol
$x > -2$	*is* reversed.

Recall that division is defined in terms of multiplication. Therefore, the Multiplication Property of Inequalities allows each side of an inequality to be divided by the same number. When each side of an inequality is divided by a positive number, the inequality symbol remains the same. When each side of an inequality is divided by a negative number, the inequality symbol must be reversed.

Solve: $-5x > 8$

Divide each side of the inequality by -5. Because -5 is a negative number, the inequality symbol must be reversed.

Simplify.

$$-5x > 8$$
$$\frac{-5x}{-5} < \frac{8}{-5}$$
$$x < -\frac{8}{5}$$

Example 4   Solve and graph the solution set of $7x < -14$.

Solution
$$7x < -14$$
$$\frac{7x}{7} < \frac{-14}{7}$$ • Divide each side of the inequality by 7. Because 7 is a *positive* number, do not change the inequality symbol.
$$x < -2$$

$$-5 \; -4 \; -3 \; -2 \; -1 \;\; 0 \;\; 1 \;\; 2 \;\; 3 \;\; 4 \;\; 5$$

Problem 4   Solve and graph the solution set of $3x < 9$.

Solution   See page S7.

Example 5   Solve: $-\dfrac{5}{8}x \leq \dfrac{5}{12}$

Solution
$$-\frac{5}{8}x \leq \frac{5}{12}$$
$$-\frac{8}{5}\left(-\frac{5}{8}x\right) \geq -\frac{8}{5}\left(\frac{5}{12}\right)$$ • Multiply each side of the inequality by the reciprocal of $-\dfrac{5}{8}$. $-\dfrac{8}{5}$ is a negative number. Reverse the inequality symbol.
$$x \geq -\frac{2}{3}$$

Problem 5   Solve: $-\dfrac{3}{4}x \geq 18$

Solution   See page S7.

3   **Solve general inequalities**

In solving an inequality, it is often necessary to apply both the Addition and the Multiplication Properties of Inequalities.

Take Note

Solving these inequalities is similar to solving the equations in Section 3.2 *except* that when you multiply or divide the inequality by a negative number, you must reverse the inequality symbol.

Solve: $3x - 2 < 5x + 4$

Subtract $5x$ from each side of the inequality. Simplify.

$$3x - 2 < 5x + 4$$
$$3x - 5x - 2 < 5x - 5x + 4$$
$$-2x - 2 < 4$$

Add 2 to each side of the inequality. Simplify.

$$-2x - 2 + 2 < 4 + 2$$
$$-2x < 6$$

Divide each side of the inequality by -2. Because -2 is a negative number, the inequality symbol must be reversed. Simplify.

$$\frac{-2x}{-2} > \frac{6}{-2}$$

$$x > -3$$

Example 6 Solve: $7x - 3 \le 3x + 17$

Solution
$$7x - 3 \le 3x + 17$$
$$7x - 3x - 3 \le 3x - 3x + 17$$
$$4x - 3 \le 17$$
$$4x - 3 + 3 \le 17 + 3$$
$$4x \le 20$$
$$\frac{4x}{4} \le \frac{20}{4}$$
$$x \le 5$$

• Subtract $3x$ from each side of the inequality.

• Add 3 to each side of the inequality.

• Divide each side of the inequality by 4.

Problem 6 Solve: $5 - 4x > 9 - 8x$

Solution   See page S7.

When an inequality contains parentheses, one of the steps in solving the inequality requires the use of the Distributive Property.

Solve: $-2(x - 7) > 3 - 4(2x - 3)$

Use the Distributive Property to remove parentheses. Simplify.

$$-2(x - 7) > 3 - 4(2x - 3)$$
$$-2x + 14 > 3 - 8x + 12$$
$$-2x + 14 > 15 - 8x$$

Add $8x$ to each side of the inequality. Simplify.

$$-2x + 8x + 14 > 15 - 8x + 8x$$
$$6x + 14 > 15$$

Subtract 14 from each side of the inequality. Simplify.

$$6x + 14 - 14 > 15 - 14$$
$$6x > 1$$

Divide each side of the inequality by 6.

$$\frac{6x}{6} > \frac{1}{6}$$

Simplify.

$$x > \frac{1}{6}$$

Example 7   Solve: $3(3 - 2x) \geq -5x - 2(3 - x)$

Solution

$$3(3 - 2x) \geq -5x - 2(3 - x)$$
$$9 - 6x \geq -5x - 6 + 2x$$ • Use the Distributive Property.
$$9 - 6x \geq -3x - 6$$
$$9 - 6x + 3x \geq -3x + 3x - 6$$ • Add 3x to each side of the inequality.
$$9 - 3x \geq -6$$
$$9 - 9 - 3x \geq -6 - 9$$ • Subtract 9 from each side of the inequality.
$$-3x \geq -15$$
$$\frac{-3x}{-3} \leq \frac{-15}{-3}$$ • Divide each side of the inequality by −3. Reverse the inequality symbol.
$$x \leq 5$$

Problem 7   Solve: $8 - 4(3x + 5) \leq 6(x - 8)$

Solution   See page S7.

3.3   Concept Review

Determine whether the statement is always true, sometimes true, or never true.

1. The same variable term can be added to both sides of an inequality without changing the solution set of the inequality.

2. The same variable term can be subtracted from both sides of an inequality without changing the solution set of the inequality.

3. Both sides of an inequality can be multiplied by the same number without changing the solution set of the inequality.

4. Both sides of an inequality can be divided by the same number without changing the solution set of the inequality.

5. Suppose $a > 0$ and $b < 0$. Then $ab > 0$.

6. Suppose $a < 0$. Then $a^2 > 0$.

7. Suppose $a > 0$ and $b < 0$. Then $a^2 > b$.

8. Suppose $a > b$. Then $-a > -b$.

9. Suppose $a < b$. Then $ac < bc$.

10. Suppose $a \neq 0$, $b \neq 0$, and $a > b$. Then $\frac{1}{a} > \frac{1}{b}$.

3.3 | **Exercises**

1 Graph.

1. $x > 2$

2. $x \geq -1$

3. $x \leq 0$

4. $x < 4$

5. Which numbers are solutions of the inequality $x + 7 \leq -3$?
   a. -17 **b.** 8 **c.** -10 **d.** 0

6. Which numbers are solutions of the inequality $x - 5 < -6$?
   a. 1 **b.** -1 **c.** 12 **d.** -5

Solve and graph the solution set.

7. $x + 1 < 3$

8. $y + 2 < 2$

9. $x - 5 > -2$

10. $x - 3 > -2$

11. $n + 4 \geq 7$

12. $x + 5 \geq 3$

13. $x - 6 \leq -10$

14. $y - 8 \leq -11$

15. $5 + x \geq 4$

16. $-2 + n \geq 0$

Solve.

17. $y - 3 \geq -12$

18. $x + 8 \geq -14$

19. $3x - 5 < 2x + 7$

20. $5x + 4 < 4x - 10$

21. $8x - 7 \geq 7x - 2$

22. $3n - 9 \geq 2n - 8$

23. $2x + 4 < x - 7$

24. $9x + 7 < 8x - 7$

25. $4x - 8 \leq 2 + 3x$

26. $5b - 9 < 3 + 4b$

27. $6x + 4 \geq 5x - 2$

28. $7x - 3 \geq 6x - 2$

29. $2x - 12 > x - 10$

30. $3x + 9 > 2x + 7$

31. $d + \dfrac{1}{2} < \dfrac{1}{3}$

32. $x - \dfrac{3}{8} < \dfrac{5}{6}$

33. $x + \dfrac{5}{8} \geq -\dfrac{2}{3}$

34. $y + \dfrac{5}{12} \geq -\dfrac{3}{4}$

35. $2x - \dfrac{1}{2} < x + \dfrac{3}{4}$

36. $6x - \dfrac{1}{3} \leq 5x - \dfrac{1}{2}$

37. $3x + \dfrac{5}{8} > 2x + \dfrac{5}{6}$

38. $4b - \dfrac{7}{12} \geq 3b - \dfrac{9}{16}$

39. $x + 5.8 \leq 4.6$

40. $n - 3.82 \leq 3.95$

41. $x - 0.23 \leq 0.47$

42. $3.8x < 2.8x - 3.8$

43. $1.2x < 0.2x - 7.3$

2 **44.** Which numbers are solutions of the inequality $5x > 15$?
   a. 6 **b.** -4 **c.** 3 **d.** 5

45. Which numbers are solutions of the inequality $-4x \leq 12$?
   a. 0 **b.** 3 **c.** -3 **d.** -4

Solve and graph the solution set.

46. $3x < 12$

47. $8x \le -24$

48. $5y \ge 15$

49. $24x > -48$

50. $16x \le 16$

51. $3x > 0$

52. $-8x > 8$

53. $-2n \le -8$

54. $-6b > 24$

55. $-4x < 8$

Solve.

56. $-5y \ge 20$

57. $3x < 5$

58. $7x > 2$

59. $-8x \le -40$

60. $-6x \le -40$

61. $10x > -25$

62. $-3x \ge \dfrac{6}{7}$

63. $-5x \ge \dfrac{10}{3}$

64. $\dfrac{5}{6}n < 15$

65. $\dfrac{2}{3}x < -12$

66. $\dfrac{5}{6}x < -20$

67. $-\dfrac{3}{8}x < 6$

68. $\dfrac{3}{4}x < 12$

69. $\dfrac{2}{3}y \ge 4$

70. $\dfrac{5}{8}x \ge 10$

71. $-\dfrac{2}{3}x \le 4$

72. $-\dfrac{3}{7}x \le 6$

73. $-\dfrac{2}{11}b \ge -6$

74. $-\dfrac{4}{7}x \ge -12$

75. $\dfrac{2}{3}n < \dfrac{1}{2}$

76. $\dfrac{3}{5}x > \dfrac{7}{10}$

77. $-\dfrac{2}{3}x \ge \dfrac{4}{7}$

78. $-\dfrac{3}{8}x \ge \dfrac{9}{14}$

79. $-\dfrac{3}{4}y \ge -\dfrac{5}{8}$

80. $-\dfrac{8}{9}x \ge -\dfrac{16}{27}$

81. $\dfrac{2}{3}x \le \dfrac{9}{14}$

82. $\dfrac{7}{12}x \le \dfrac{9}{14}$

83. $-\dfrac{3}{5}y < \dfrac{9}{10}$

84. $-\dfrac{5}{12}y < \dfrac{1}{6}$

85. $-0.27x < 0.135$

86. $-0.63x < 4.41$

87. $8.4y \ge -6.72$

88. $3.7y \ge -1.48$

89. $1.5x \le 6.30$

90. $2.3x \le 5.29$

91. $-3.9x \ge -19.5$

92. $0.035x < -0.0735$

93. $0.07x < -0.378$

94. $-11.7x \le 4.68$

3 **95.**   In your own words, state the Addition Property of Inequalities and the Multiplication Property of Inequalities.

96. What differentiates solving linear equations from solving linear inequalities?

Solve.

97. $4x - 8 < 2x$ **98.** $7x - 4 < 3x$ **99.** $2x - 8 > 4x$

100. $3y + 2 > 7y$ **101.** $8 - 3x \le 5x$ **102.** $10 - 3x \le 7x$

103. $3x + 2 \ge 5x - 8$ **104.** $2n - 9 \ge 5n + 4$ **105.** $5x - 2 < 3x - 2$

106. $8x - 9 > 3x - 9$ **107.** $0.1(180 + x) > x$ **108.** $x > 0.2(50 + x)$

109. $0.15x + 55 > 0.10x + 80$ **110.** $-3.6b + 16 < 2.8b + 25.6$ **111.** $2(3x - 1) > 3x + 4$

112. $5(2x + 7) > -4x - 7$ **113.** $3(2x - 5) \ge 8x - 5$ **114.** $5x - 8 \ge 7x - 9$

115. $2(2y - 5) \le 3(5 - 2y)$ **116.** $2(5x - 8) \le 7(x - 3)$

117. $5(2 - x) > 3(2x - 5)$ **118.** $4(3d - 1) > 3(2 - 5d)$

119. $5(x - 2) > 9x - 3(2x - 4)$ **120.** $3x - 2(3x - 5) > 4(2x - 1)$

121. $4 - 3(3 - n) \le 3(2 - 5n)$ **122.** $15 - 5(3 - 2x) \le 4(x - 3)$

123. $2x - 3(x - 4) \ge 4 - 2(x - 7)$ **124.** $4 + 2(3 - 2y) \le 4(3y - 5) - 6y$

125. $\dfrac{1}{2}(9x - 10) \le -\dfrac{1}{3}(12 - 6x)$ **126.** $\dfrac{1}{4}(8 - 12d) < \dfrac{2}{5}(10d + 15)$

127. $\dfrac{2}{3}(9t - 15) + 4 < 6 + \dfrac{3}{4}(4 - 12t)$ **128.** $\dfrac{3}{8}(16 - 8c) - 9 \ge \dfrac{3}{5}(10c - 15) + 7$

129. $3[4(n - 2) - (1 - n)] > 5(n - 4)$ **130.** $2(m + 7) \le 4[3(m - 2) - 5(1 + m)]$

131. What number is a solution of $3x - 4 \ge 5$ but not a solution of $3x - 4 > 5$?

132. What number is a solution of $8 - 2(x + 6) \le 4$ but not a solution of $8 - 2(x + 6) < 4$?

3.3 Applying Concepts

Use the roster method to list the set of positive integers that are solutions of the inequality.

133. $7 - 2b \le 15 - 5b$ **134.** $13 - 8a \ge 2 - 6a$

135. $2(2c - 3) < 5(6 - c)$ **136.** $-6(2 - d) \ge 4(4d - 9)$

Use the roster method to list the set of integers that are common to the solution sets of the two inequalities.

137. $5x - 12 \leq x + 8$
$3x - 4 \geq 2 + x$

138. $6x - 5 > 9x - 2$
$5x - 6 < 8x + 9$

139. $4(x - 2) \leq 3x + 5$
$7(x - 3) \geq 5x - 1$

140. $3(x + 2) < 2(x + 4)$
$4(x + 5) > 3(x + 6)$

Graph.

141. $|x| < 3$

142. $|x| < 4$

143. $|x| > 2$

144. $|x| > 1$

Focus on Problem Solving

Analyzing Data

WEB

Demography is the statistical study of human populations. Many groups are interested in the size of certain segments of the population and projections of population growth. For example, public school administrators want estimates of the number of school-age children who will be living in their district 10 years from now.

The U.S. government provides estimates of the future population of the United States. You can find these projections at the Census Bureau website at **www.census.gov**. The table below contains data from this website.

	Age	Under 5	5-17	18-24	25-34	35-44	45-54	55-64	65 and older
2010	Male	10,272	26,639	15,388	19,286	19,449	21,706	16,973	16,966
	Female	9,827	25,363	14,774	19,565	19,993	22,455	18,457	22,749
2050	Male	13,748	35,227	18,734	25,034	24,550	22,340	21,127	36,289
	Female	13,165	33,608	18,069	25,425	25,039	23,106	22,517	45,710

Population Projections of the United States, by Age and Sex (in thousands)

For the following exercises, round all percents to the nearest tenth of a percent.

1. Which of the age groups listed are of interest to public school officials?

2. Which age group is of interest to retirement home administrators?

3. Which age groups are of concern to accountants determining future benefits to be paid out by the Social Security Administration?

4. Which age group is of interest to manufacturers of disposable diapers?

5. Which age groups are of primary concern to college and university admissions officers?

6. In which age groups do males outnumber females? In which do females outnumber males?

7. What percent of the projected population aged 65 and older in the year 2010 is female? Does this percent decrease for the projected population in 2050? If so, by how much?

8. Find the difference between the percent of the population that will be under 18 in 2010 and the percent that will be under 18 in 2050.

9. Assume that the work force consists of people aged 25 to 64. Find the decrease in the percent of the population represented by that age group from 2010 to 2050.

10. Find the increase in the percent of the population represented by those aged 65 and older from 2010 to 2050.

11. Why are the sizes of the current work force and the population aged 65 and older of concern to the Social Security Administration?

12. Describe any patterns you see in the table.

13. Calculate a statistic based on the data in the table and explain why it would be of interest to an institution (such as a school system) or a manufacturer of consumer goods (such as a baby food manufacturer).

Projects & Group Activities

Health and Nutrition

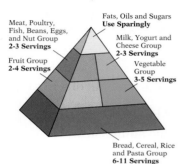

Food Pyramid
A Guide to Daily Food Choices

Fats, Oils and Sugars
Use Sparingly

Meat, Poultry, Fish, Beans, Eggs, and Nut Group
2-3 Servings

Milk, Yogurt and Cheese Group
2-3 Servings

Fruit Group
2-4 Servings

Vegetable Group
3-5 Servings

Bread, Cereal, Rice and Pasta Group
6-11 Servings

A formula used to determine recommended daily protein intake is

$$P = \frac{W}{2.2}$$

where P is the amount of protein in grams and W is the individual's weight in pounds. However, if you are sedentary, the result should be multiplied by 80%. If you exercise regularly (at least 30 min five times per week), the result should be multiplied by 120%.

1. Explain why the result is multiplied by 80% for a sedentary person and by 120% for an active person.

2. Calculate the amount of protein you should consume each day.

The American College of Sports Medicine (ACSM) recommends that you know how to determine your target heart rate in order to get the full benefit of exercise. Your **target heartbeat rate** is the heart rate you should achieve during any aerobic exercise such as running, cycling, fast walking, or an aerobics class. According to the ACSM, you should reach your target rate and maintain it for 20 min or more to achieve cardiovascular fitness. The intensity level varies for different individuals. A sedentary person might begin at the 60% level and gradually work up to 70%, whereas athletes and very fit individuals might work at the 85% level. The ACSM suggests you calculate both

50% and 85% of your maximum heart rate. This will give you the low and high ends of the range that your heart rate should stay within.

To calculate your target heart rate:

	Example
Subtract your age from 220. This is your maximum heart rate.	$220 - 20 = 200$
Multiply your maximum heart rate by 50%. This is the low end of your range.	$200(0.50) = 100$
Divide the low end rate by 6. This is your low 10-second heart rate.	$100 \div 6 \approx 17$
Multiply your maximum heart rate by 85%. This is the high end of your range.	$200(0.85) = 170$
Divide the high end rate by 6. This is your high 10-second heart rate.	$170 \div 6 \approx 28$

3. Why are the low-end and high-end rates divided by 6 in order to determine the 10-second heart rate?

4. Calculate your target heart rate, including the low and high end of your range.

5. What is the increase in your 10-second heart rate from the low end to the high end?

The Consumer Price Index

WEB

The Consumer Price Index (CPI) is a percent that is written without the percent sign. For instance, a CPI of 160.1 means 160.1%. This number means that an item that cost $100 between 1982 and 1984 (the base years) would cost $160.10 today. Determining the cost is an application of the basic percent equation.

$$PB = A$$

(CPI)(cost in base year) = cost today

$$1.601(100) = 160.1 \qquad \bullet \ 160.1\% = 1.601$$

The table below gives the CPI for various products in July 2001. If you have Internet access, you can obtain current data for the items below plus other items not on this list by visiting the website of the Bureau of Labor Statistics at **www.bls.gov.**

Product	CPI
All items	177.5
Food and beverages	174.0
Housing	177.6
Clothes	122.6
Transportation	154.4
Medical Care	273.1
Energy	132.4
Education: Tuition	337.2

1. Of the items listed, are there any items that cost more than twice as much in 2001 as they cost during the base years? If so, which ones?

2. Of the items listed, are there any items that cost more than one-and-one-half times as much in 2001 as they cost during the base years, but less than twice as much as they cost during the base years? If so, which ones?

3. If the average cost for tuition for one semester at a public college was $4000 in the base years, what was the average cost for tuition for one semester at a public college in 2001?

4. If a new car cost $20,000 in 2001, what would a comparable new car have cost during the base years? Use the transportation category.

5. If rent for an urban apartment in 2001 was $3000, what was the rent for the same apartment during the base years?

6. The CPI for all items in the summer of 1970 was 39.0. What salary in 2001 would have had the same purchasing power as a salary of $10,000 in 1970?

Chapter Summary

Key Words

An **equation** expresses the equality of two mathematical expressions.   [3.1.1, p. 90]

Examples

$5(3x - 2) = 4x + 7$ is an equation.

A **solution of an equation** is a number that, when substituted for the variable, results in a true equation.   [3.1.1, p. 90]

1 is the solution of the equation $6x - 4 = 2$ because $6(1) - 4 = 2$ is a true equation.

To **solve** an equation means to find a solution of the equation. The goal is to rewrite the equation in the form **variable = constant**, because the constant is the solution.   [3.1.2, p. 91]

The equation $x = -9$ is in the form *variable = constant*. The constant, -9, is the solution of the equation.

An **inequality** is an expression that contains the symbol $<$, $>$, $\leq$, or $\geq$.   [3.3.1, p. 125]

$8x - 1 \geq 5x + 23$ is an inequality.

The **solution set of an inequality** is a set of numbers, each element of which, when substituted for the variable, results in a true inequality.   [3.3.1, p. 125]

The solution set of $8x - 1 \geq 5x + 23$ is $x \geq 8$ because every number greater than or equal to 8, when substituted for the variable, results in a true inequality.

Essential Rules and Procedures

The Addition Property of Equations
The same number or variable term can be added to each side of an equation without changing the solution of the equation. [3.1.2, p. 91]

$$x + 12 = -19$$
$$x + 12 - 12 = -19 - 12$$
$$x = -31$$

The Multiplication Property of Equations
Each side of an equation can be multiplied by the same non-zero number without changing the solution of the equation. [3.1.3, p. 93]

$$-6x = 24$$
$$\frac{-6x}{-6} = \frac{24}{-6}$$
$$x = -4$$

To solve an equation of the form *ax* + *b* = *c* or an equation of the form *ax* + *b* = *cx* + *d*, use both the Addition and the Multiplication Properties of Equations.   [3.2.1 and 3.2.2, pp. 112, 114]

$$7x - 4 = 3x + 16$$
$$7x - 3x - 4 = 3x - 3x + 16$$
$$4x - 4 = 16$$
$$4x - 4 + 4 = 16 + 4$$
$$4x = 20$$
$$\frac{4x}{4} = \frac{20}{4}$$
$$x = 5$$

To solve an equation containing parentheses, use the Distributive Property to remove parentheses.   [3.2.3, p. 115]

$$3(2x - 1) = 15$$
$$6x - 3 = 15$$
$$6x - 3 + 3 = 15 + 3$$
$$6x = 18$$
$$\frac{6x}{6} = \frac{18}{6}$$
$$x = 3$$

The Addition Property of Inequalities
The same number or variable term can be added to each side of an inequality without changing the solution set of the inequality.

If $a > b$, then $a + c > b + c$.
If $a < b$, then $a + c < b + c$.   [3.3.1, p. 126]

$$x - 9 \le 14$$
$$x - 9 + 9 \le 14 + 9$$
$$x \le 23$$

The Multiplication Property of Inequalities   [3.3.2, p. 128]

Rule 1 Each side of an inequality can be multiplied by the same **positive number** without changing the solution set of the inequality.
If $a > b$ and $c > 0$, then $ac > bc$.
If $a < b$ and $c > 0$, then $ac < bc$.

$$7x < -21$$
$$\frac{7x}{7} < \frac{-21}{7}$$
$$x < -3$$

Rule 2 If each side of an inequality is multiplied by the same **negative number** and the inequality symbol is reversed, then the solution of the inequality is not changed.
If $a > b$ and $c < 0$, then $ac < bc$.
If $a < b$ and $c < 0$, then $ac > bc$.

$$-7x \ge 21$$
$$\frac{-7x}{-7} \le \frac{21}{-7}$$
$$x \le -3$$

The Basic Percent Equation
Percent · Base = Amount
$$P \cdot B = A$$
[3.1.5, p. 98]

40% of what number is 16?
$$PB = A$$
$$0.40B = 16$$
$$\frac{0.40B}{0.40} = \frac{16}{0.40}$$
$$B = 40$$

The number is 40.

Chapter Review Exercises

1. Is 3 a solution of $5x - 2 = 4x + 5$?

2. Solve: $x - 4 = 16$

3. Solve: $8x = -56$

4. Solve: $5x - 6 = 29$

5. Solve: $5x + 3 = 10x - 17$

6. Solve: $3(5x + 2) + 2 = 10x + 5[x - (3x - 1)]$

7. What is 81% of 500?

8. 18 is 72% of what number?

9. 27 is what percent of 40?

10. Graph: $x \le -2$

11. Solve and graph the solution set of $x - 3 > -1$.

12. Solve and graph the solution set of $3x > -12$.

13. Solve: $3x + 4 \ge -8$

14. Solve: $7x - 2(x + 3) \ge x + 10$

15. Is 2 a solution of $x^2 + 4x + 1 = 3x + 7$?

16. Solve: $4.6 = 2.1 + x$

17. Solve: $\dfrac{x}{7} = -7$

18. Solve: $14 + 6x = 17$

19. Solve: $12y - 1 = 3y + 2$

20. Solve: $x + 5(3x - 20) = 10(x - 4)$

21. What is $66\dfrac{2}{3}\%$ of 24?

22. 60 is 48% of what number?

23. 0.5 is what percent of 3?

24. Solve and graph the solution set of $2 + x < -2$.

25. Solve and graph the solution set of $5x \le -10$.

26. Solve: $6x + 3(2x - 1) = -27$

27. Solve: $a - \dfrac{1}{6} = \dfrac{2}{3}$

28. Solve: $\dfrac{3}{5}a = 12$

29. Solve: $32 = 9x - 4 - 3x$

30. Solve: $-4[x + 3(x - 5)] = 3(8x + 20)$

31. Solve: $4x - 12 < x + 24$

32. What is $\dfrac{1}{2}\%$ of 3000?

33. Solve: $3x + 7 + 4x = 42$

34. Solve: $5x - 6 > 19$

35. 8 is what percent of 200?

36. Solve: $6x - 9 < 4x + 3(x + 3)$

37. Solve: $5 - 4(x + 9) > 11(12x - 9)$

38. Find the measure of the third angle of a triangle if the first angle is 20° and the second angle is 50°. Use the equation $A + B + C = 180°$, where A, B, and C are the measures of the angles of a triangle.

39. A lever is 12 ft long. At a distance of 2 ft from the fulcrum, a force of 120 lb is applied. How large a force must be applied to the other end of the lever so that the system will balance? Use the lever-system equation $F_1x = F_2(d - x)$.

40. Use the equation $P = 2L + 2W$, where P is the perimeter of a rectangle, L is its length, and W is its width, to find the width of a rectangle that has a perimeter of 49 ft and a length of 18.5 ft.

41. Find the discount on a 26-inch, 21-speed dual-index grip-shifting bicycle if the sale price is $128 and the regular price is $159.99. Use the equation $S = R - D$, where S is the sale price, R is the regular price, and D is the discount.

42. In Central America and Mexico, 1184 plants and animals are at risk of extinction. This represents approximately 10.7% of all the species at risk of extinction on Earth. Approximately how many plants and animals are at risk of extinction on Earth? (*Source:* World Conservation Union)

43. The pressure at a certain depth in the ocean can be approximated by the equation $P = 15 + \frac{1}{2}D$, where P is the pressure in pounds per square inch and D is the depth in feet. Use this equation to find the depth when the pressure is 55 lb/in².

44. A lever is 8 ft long. A force of 25 lb is applied to one end of the lever, and a force of 15 lb is applied to the other end. Find the location of the fulcrum when the system balances. Use the lever-system equation $F_1x = F_2(d - x)$.

45. Find the length of a rectangle when the perimeter is 84 ft and the width is 18 ft. Use the equation $P = 2L + 2W$, where P is the perimeter of a rectangle, L is the length, and W is the width.

46. A motorboat that can travel 15 mph in calm water is traveling with the current of a river that is moving at 5 mph. How long will it take the motorboat to travel 30 mi?

47. Cathy Serano would like to earn $125 per year from two investments. She has found one investment that will pay her $75 per year. How much must she invest in an account that earns an annual simple interest rate of 8% to reach her goal?

48. A 150-milliliter acid solution contains 9 ml of hydrochloric acid. What is the percent concentration of hydrochloric acid?

Chapter Test

1. Solve: $\frac{3}{4}x = -9$

2. Solve: $6 - 5x = 5x + 11$

3. Solve: $3x - 5 = -14$

4. Is -2 a solution of $x^2 - 3x = 2x - 6$?

5. Solve: $x + \frac{1}{2} = \frac{5}{8}$

6. Solve: $5x - 2(4x - 3) = 6x + 9$

7. Solve: $7 - 4x = -13$

8. Solve: $11 - 4x = 2x + 8$

9. Solve: $x - 3 = -8$

10. Solve: $3x - 2 = 5x + 8$

11. Solve: $-\frac{3}{8}x = 5$

12. Solve: $6x - 3(2 - 3x) = 4(2x - 7)$

13. Solve: $6 - 2(5x - 8) = 3x - 4$

14. Solve: $9 - 3(2x - 5) = 12 + 5x$

15. Solve: $3(2x - 5) = 8x - 9$

16. 20 is what percent of 16?

17. 30% of what is 12?

18. Graph: $x > -2$

19. Solve and graph the solution set of $-2 + x \le -3$.

20. Solve and graph the solution set of $\frac{3}{8}x > -\frac{3}{4}$.

21. Solve: $x + \frac{1}{3} > \frac{5}{6}$

22. Solve: $3(x - 7) \ge 5x - 12$

23. Solve: $-\frac{3}{8}x \le 6$

24. Solve: $4x - 2(3 - 5x) \le 6x + 10$

25. Solve: $3(2x - 5) \ge 8x - 9$

26. Solve: $15 - 3(5x - 7) < 2(7 - 2x)$

27. Solve: $-6x + 16 = -2x$

28. 20 is $83\frac{1}{3}\%$ of what number?

29. Solve and graph the solution set of $\frac{2}{3}x \ge 2$.

30. Is 5 a solution of $x^2 + 2x + 1 = (x + 1)^2$?

31. A person's weight on the moon is $16\frac{2}{3}\%$ of the person's weight on Earth. If an astronaut weighs 180 lb on Earth, how much would the astronaut weigh on the moon?

32. A chemist mixes 100 g of water at 80°C with 50 g of water at 20°C. Use the equation $m_1 \cdot (T_1 - T) = m_2 \cdot (T - T_2)$ to find the final temperature of the water after mixing. In this equation, m_1 is the quantity of water at the hotter temperature, T_1 is the temperature of the hotter water, m_2 is the quantity of water at the cooler temperature, T_2 is the temperature of the cooler water, and T is the final temperature of the water after mixing.

33. A financial manager has determined that the cost per unit for a calculator is $15 and that the fixed costs per month are $2000. Find the number of calculators produced during a month in which the total cost was $5000. Use the equation $T = U \cdot N + F$, where T is the total cost, U is the cost per unit, N is the number of units produced, and F is the fixed cost.

34. Two hikers start at the same time from opposite ends of a 15-mile course. One hiker is walking at a rate of 3.5 mph, and the second hiker is walking at a rate of 4 mph. How long after they begin will they meet?

35. Victor Jameson has invested $750 in a simple interest account that has an annual interest rate of 6.2%. How much must he invest in a second account that earns 5% simple interest so that the interest earned on the second account is the same as that earned on the first account?

36. An 8-ounce glass of chocolate milk contains 2 oz of chocolate syrup. Find the percent concentration of chocolate syrup in the chocolate milk.

Cumulative Review Exercises

1. Subtract: $-6 - (-20) - 8$

2. Multiply: $(-2)(-6)(-4)$

3. Subtract: $-\dfrac{5}{6} - \left(-\dfrac{7}{16}\right)$

4. Divide: $-2\dfrac{1}{3} \div 1\dfrac{1}{6}$

5. Simplify: $-4^2 \cdot \left(-\dfrac{3}{2}\right)^3$

6. Simplify: $25 - 3 \cdot \dfrac{(5 - 2)^2}{2^3 + 1} - (-2)$

7. Evaluate $3(a - c) - 2ab$ when $a = 2$, $b = 3$, and $c = -4$.

8. Simplify: $3x - 8x + (-12x)$

9. Simplify: $2a - (-3b) - 7a - 5b$

10. Simplify: $(16x)\left(\dfrac{1}{8}\right)$

11. Simplify: $-4(-9y)$

12. Simplify: $-2(-x^2 - 3x + 2)$

13. Simplify: $-2(x - 3) + 2(4 - x)$

14. Simplify: $-3[2x - 4(x - 3)] + 2$

15. Use the roster method to write the set of negative integers greater than −8.

16. Write $\dfrac{7}{8}$ as a percent. Write the remainder in fractional form.

17. Write 342% as a decimal.

18. Write $62\dfrac{1}{2}\%$ as a fraction.

19. Is −3 a solution of $x^2 + 6x + 9 = x + 3$?

20. Solve: $x - 4 = -9$

21. Solve: $\dfrac{3}{5}x = -15$

22. Solve: $13 - 9x = -14$

23. Solve: $5x - 8 = 12x + 13$

24. Solve: $8x - 3(4x - 5) = -2x - 11$

25. Solve: $-\dfrac{3}{4}x > \dfrac{2}{3}$

26. Solve: $5x - 4 \geq 4x + 8$

27. Solve: $3x + 17 < 5x - 1$

28. Translate "the difference between eight and the quotient of a number and twelve" into a variable expression.

29. Translate and simplify "the sum of a number and two more than the number."

30. A club treasurer has some five-dollar bills and some ten-dollar bills. The treasurer has a total of 35 bills. Use one variable to express the number of each denomination of bill.

31. A fishing wire 3 ft long is cut into two pieces, one shorter than the other. Express the length of the shorter piece in terms of the length of the longer piece.

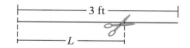

32. A computer programmer receives a weekly wage of $650, and $110.50 is deducted for income tax. Find the percent of the computer programmer's salary deducted for income tax.

33. The world record time for a 1-mile race can be approximated by the equation $t = 17.08 - 0.0067y$, where t is the time in minutes and y is the year of the race. Use this equation to predict the year in which the first "4-minute mile" was run. (The actual year was 1954.) Round to the nearest whole number.

34. A chemist mixes 300 g of water at 75°C with 100 g of water at 15°C. Use the equation $m_1 \cdot (T_1 - T) = m_2 \cdot (T - T_2)$ to find the final temperature of the water. In this equation, m_1 is the quantity of water at the hotter temperature, T_1 is the temperature of the hotter water, m_2 is the quantity of water at the cooler temperature, T_2 is the temperature of the cooler water, and T is the final temperature of the water after mixing.

35. A lever is 25 ft long. At a distance of 12 ft from the fulcrum, a force of 26 lb is applied. How large a force must be applied to the other end of the lever so that the system will balance?

By surveying television viewers, Nielsen Media Research can determine the number of people watching particular TV shows. The larger the audience, the greater the number of rating points the show receives. Programs with a high number of rating points can demand and receive higher fees from sponsors who pay to have their advertisements aired during the show. The **Project on pages 211–212** describes the meaning of Nielsen rating points and their relationship to market share.

Applications

Agriculture, *p. 152*
Auto repair, *p. 220*
Aviation, *p. 207*
Business, *pp. 154, 155, 204, 206*
Carepentry, *pp. 150, 153, 154, 217, 218*
Chemistry, *p. 149*
Coin and stamp problems, *pp. 156, 157, 158, 159, 160, 161, 217, 219, 221*
Coins, *p. 154*
Commissions, *p. 152*
Compensation, *pp. 206, 217*
Consecutive integer problems, *pp. 155, 156, 158, 159, 161, 217, 218, 219, 221*
Construction, *pp. 154, 220*
Consumerism, *pp. 153, 204, 206, 207, 216, 219*
Debt carried by Americans, *p. 208*
Depreciation, *p. 152*
Discount, *pp. 178, 179, 181, 182, 216, 218, 219*
Education, *pp. 153, 203, 206, 217*
Fund raising, *p. 208*
Geography, *p. 153*
Geometry, *pp. 153, 163–167, 204, 207, 216, 217, 218, 219, 221*
Health, *p. 205*
History, *p. 152*
Income taxes, *p. 205*
Integer problems, *pp. 149, 151, 152, 205, 206, 207, 215, 217, 218, 220*
Investment problems, *pp. 183–188, 216, 217, 219, 221*
Labor unions, *p. 154*
Libraries, *p. 221*
Manufacturing, *p. 150*
Markup, *pp. 177, 178, 179, 180, 182, 216, 217, 218, 221*
Metalwork, *p. 154*
Music, *p. 215*
Nielsen ratings, *pp. 211–212*
Nutrition, *pp. 152, 207*
Percent mixture problems, *pp. 191, 192, 195, 196, 197, 212, 216, 217, 218, 222*
Postage, *pp. 154, 208*
Recycling, *p. 206*
Safety, *p. 153*
Sports, *p. 205*
Structures, *pp. 152, 217*
Temperature, *p. 149*
Travel, *p. 154*
Uniform motion problems, *pp. 197–203, 216, 218, 219*
Utilities, *pp. 153, 207*
Value mixture problems, *pp. 189, 190, 193, 194, 195, 196, 197, 216, 217, 219, 221*
Wages, *p. 154*

4

Solving Equations and Inequalities: Applications

Objectives

Section 4.1
1. Translate a sentence into an equation and solve
2. Application problems

Section 4.2
1. Consecutive integer problems
2. Coin and stamp problems

Section 4.3
1. Perimeter problems
2. Problems involving angles formed by intersecting lines
3. Problems involving the angles of a triangle

Section 4.4
1. Markup problems
2. Discount problems

Section 4.5
1. Investment problems

Section 4.6
1. Value mixture problems
2. Percent mixture problems

Section 4.7
1. Uniform motion problems

Section 4.8
1. Applications of inequalities

Need help? For online student resources, such as section quizzes, visit this web site: **math.college.hmco.com**

PrepTest

1. Simplify: $R - 0.35R$

2. Simplify: $0.08x + 0.05(400 - x)$

3. Simplify: $n + (n + 2) + (n + 4)$

4. Translate into a variable expression: "The difference between 5 and twice a number."

5. Write 0.4 as a percent.

6. Solve: $25x + 10(9 - x) = 120$

7. Solve: $36 = 48 - 48r$

8. Solve: $4(2x - 5) < 12$

9. Twenty ounces of a snack mixture contain nuts and pretzels. Let n represent the number of ounces of nuts in the mixture. Express the number of ounces of pretzels in the mixture in terms of n.

Go Figure

If $x + y$, xy, and $\dfrac{x}{y}$ all equal the same number, find the values of x and y.

4.1 | **Translating Sentences into Equations**

1 **Translate a sentence into an equation and solve**

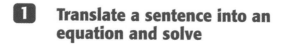

An equation states that two mathematical expressions are equal. Therefore, to translate a sentence into an equation requires recognizing the words or phrases that mean "equals." Besides "equals," some of these phrases are "is," "is equal to," "amounts to," and "represents."

Once the sentence is translated into an equation, the equation can be solved by rewriting the equation in the form *variable = constant*.

Example 1   Translate "two times the sum of a number and eight equals the sum of four times the number and six" into an equation and solve.

Solution   the unknown number: n

- Assign a variable to the unknown quantity.

| two times the sum of a number and eight | equals | the sum of four times the number and six |

- Find two verbal expressions for the same value.

$$2(n + 8) = 4n + 6$$

- Write a mathematical expression for each verbal expression. Write the equals sign.

$$2n + 16 = 4n + 6$$
$$2n - 4n + 16 = 4n - 4n + 6$$
$$-2n + 16 = 6$$
$$-2n + 16 - 16 = 6 - 16$$
$$-2n = -10$$
$$\frac{-2n}{-2} = \frac{-10}{-2}$$
$$n = 5$$

- Solve the equation.

The number is 5.

Problem 1   Translate "nine less than twice a number is five times the sum of the number and twelve" into an equation and solve.

Solution   See page S7.

2   Application problems

Example 2   The temperature of the sun on the Kelvin scale is 6500 K. This is 4740° less than the temperature on the Fahrenheit scale. Find the Fahrenheit temperature.

Strategy   To find the Fahrenheit temperature, write and solve an equation using F to represent the Fahrenheit temperature.

Solution

| 6500 | is | 4740° less than the Fahrenheit temperature |

$$6500 = F - 4740$$
$$6500 + 4740 = F - 4740 + 4740$$
$$11{,}240 = F$$

The temperature of the sun is 11,240°F.

Problem 2   A molecule of octane gas has eight carbon atoms. This represents twice the number of carbon atoms in a butane gas molecule. Find the number of carbon atoms in a butane molecule.

Solution   See page S7.

Take Note

We could also let x represent the length of the longer piece and $10 - x$ represent the length of the shorter piece. Then our equation would be

$$3(10 - x) = 2x$$

Show that this equation results in the same solutions.

Example 3   A board 10 ft long is cut into two pieces. Three times the length of the shorter piece is twice the length of the longer piece. Find the length of each piece.

Strategy   Draw a diagram. To find the length of each piece, write and solve an equation using x to represent the length of the shorter piece and $10 - x$ to represent the length of the longer piece.

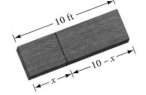

Solution

three times the length of the shorter piece	is	twice the length of the longer piece

$$3x = 2(10 - x)$$
$$3x = 20 - 2x$$
$$3x + 2x = 20 - 2x + 2x$$
$$5x = 20$$
$$\frac{5x}{5} = \frac{20}{5}$$
$$x = 4$$

• The length of the shorter piece is 4 ft.

$$10 - x = 10 - 4 = 6$$

• Substitute the value of x into the variable expression for the longer piece and evaluate.

The length of the shorter piece is 4 ft. The length of the longer piece is 6 ft.

Problem 3   A company manufactures 160 bicycles per day. Four times the number of 3-speed bicycles made each day equals 30 less than the number of 10-speed bicycles made each day. Find the number of 10-speed bicycles manufactured each day.

Solution   See page S7.

4.1   **Concept Review**

Determine whether the statement is always true, sometimes true, or never true.

1. When two expressions represent the same value, we say that the expressions are equal to each other.

2. Both "four less than a number" and "the difference between four and a number" are translated as "$4 - n$."

3. When translating a sentence into an equation, we can use any variable to represent an unknown number.

4. When translating a sentence into an equation, translate the word "is" as "$=$."

5. If a rope 19 in. long is cut into two pieces and one of the pieces has length x inches, then the length of the other piece can be expressed as $(x - 19)$ inches.

6. In addition to a number, the answer to an application problem must have a unit, such as meters, dollars, minutes, or miles per hour.

4.1   Exercises

1 Translate into an equation and solve.

1. The difference between a number and fifteen is seven. Find the number.

2. The sum of five and a number is three. Find the number.

3. The product of seven and a number is negative twenty-one. Find the number.

4. The quotient of a number and four is two. Find the number.

5. Four less than three times a number is five. Find the number.

6. The difference between five and twice a number is one. Find the number.

7. Four times the sum of twice a number and three is twelve. Find the number.

8. Twenty-one is three times the difference between four times a number and five. Find the number.

9. Twelve is six times the difference between a number and three. Find the number.

10. The difference between six times a number and four times the number is negative fourteen. Find the number.

11. Twenty-two is two less than six times a number. Find the number.

12. Negative fifteen is three more than twice a number. Find the number.

13. Seven more than four times a number is three more than two times the number. Find the number.

14. The difference between three times a number and four is five times the number. Find the number.

15. Eight less than five times a number is four more than eight times the number. Find the number.

16. The sum of a number and six is four less than six times the number. Find the number.

17. Twice the difference between a number and twenty-five is three times the number. Find the number.

18. Four times a number is three times the difference between thirty-five and the number. Find the number.

19. The sum of two numbers is twenty. Three times the smaller is equal to two times the larger. Find the two numbers.

20. The sum of two numbers is fifteen. One less than three times the smaller is equal to the larger. Find the two numbers.

21. The sum of two numbers is eighteen. The total of three times the smaller and twice the larger is forty-four. Find the two numbers.

22. The sum of two numbers is two. The difference between eight and twice the smaller number is two less than four times the larger. Find the two numbers.

2 Write an equation and solve.

23. **Depreciation** As a result of depreciation, the value of a car is now $9600. This is three-fifths of its original value. Find the original value of the car.

24. **Structures** The length of the Royal Gorge Bridge in Colorado is 320 m. This is one-fourth the length of the Golden Gate Bridge. Find the length of the Golden Gate Bridge.

25. **Nutrition** One slice of cheese pizza contains 290 calories. A medium-size orange has one-fifth that number of calories. How many calories are in a medium-size orange?

26. **History** John D. Rockefeller died in 1937. At the time of his death, Rockefeller had accumulated a wealth of $1,400 million, which was equal to one-sixty-fifth of the gross national product of the United States at that time. What was the U.S. gross national product in 1937? (*Source: The Wealthy 100: A Ranking of the Richest Americans, Past and Present*)

27. **Agriculture** A soil supplement that weighs 24 lb contains iron, potassium, and a mulch. There is five times as much mulch as iron and twice as much potassium as iron. Find the amount of mulch in the soil supplement.

28. **Commissions** A real estate agent sold two homes and received commissions totaling $6000. The agent's commission on one home was one and one-half times the commission on the second home. Find the agent's commission on each home.

29. Safety Loudness, or the intensity of sound, is measured in decibels. The sound level of a television is about 70 decibels, which is considered a safe hearing level. A food blender runs at 20 decibels higher than a TV, and a jet engine's decibel reading is 40 less than twice that of a blender. At this level, exposure can cause hearing loss. Find the intensity of the sound of a jet engine.

Point of Interest

The low-frequency pulses or whistles made by blue whales have been measured at up to 188 decibels, making them the loudest sounds produced by a living organism.

30. Education A university employs a total of 600 teaching assistants and research assistants. There are three times as many teaching assistants as research assistants. Find the number of research assistants employed by the university.

31. Consumerism The purchase price of a new big-screen TV, including finance charges, was $3276. A down payment of $450 was made. The remainder was paid in 24 equal monthly installments. Find the monthly payment.

32. Consumerism The purchase price of a new computer system, including finance charges, was $6350. A down payment of $350 was made. The remainder was paid in 24 equal monthly installments. Find the monthly payment.

33. Geometry Greek architects considered a rectangle whose length was approximately 1.6 times its width to be the most visually appealing. Find the length and width of a rectangle constructed in this manner if the sum of the length and width is 130 ft.

Greenland

Iceland

34. 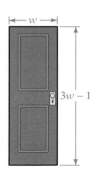 Geography Greenland, the largest island in the world, is 21 times larger than Iceland. The combined area of Greenland and Iceland is 880,000 mi². Find the area of Greenland.

35. Consumerism The cost to replace a water pump in a sports car was $600. This included $375 for the water pump and $45 per hour for labor. How many hours of labor were required to replace the water pump?

36. Utilities The cost of electricity in a certain city is $.08 for each of the first 300 kWh (kilowatt-hours) and $.13 for each kWh over 300 kWh. Find the number of kilowatt-hours used by a family that receives a $51.95 electric bill.

37. Consumerism The fee charged by a ticketing agency for a concert is $5.50 plus $37.50 for each ticket purchased. If your total charge for tickets is $343, how many tickets are you purchasing?

38. Carpentry A carpenter is building a wood door frame. The height of the frame is 1 ft less than three times the width. What is the width of the largest door frame that can be constructed from a board 19 ft long? (*Hint:* A door frame consists of only three sides; there is no frame below a door.)

39. Construction The length of a rectangular cement patio is 3 ft less than three times the width. The sum of the length and width of the patio is 21 ft. Find the length of the patio.

40. Carpentry A 20-foot board is cut into two pieces. Twice the length of the shorter piece is 4 ft more than the length of the longer piece. Find the length of the shorter piece.

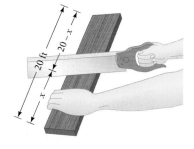

41. Labor Unions A union charges monthly dues of $4.00 plus $.15 for each hour worked during the month. A union member's dues for March were $29.20. How many hours did the union member work during the month of March?

42. Business The cellular phone service for a business executive is $35 per month plus $.40 per minute of phone use. In a month when the executive's cellular phone bill was $99.80, how many minutes did the executive use the phone?

43. Business A technical information hotline charges a customer $9.00 plus $.50 per minute to answer questions about software. How many minutes did a customer who received a bill for $14.50 use this service?

44. Metalwork A wire 12 ft long is cut into two pieces. Each piece is bent into the shape of a square. The perimeter of the larger square is twice the perimeter of the smaller square. Find the perimeter of the larger square.

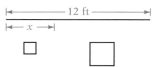

4.1 Applying Concepts

45. The amount of liquid in a container triples every minute. The container becomes completely filled at 3:40 P.M. What fractional part of the container is filled at 3:39 P.M.?

46. Travel A cyclist traveling at a constant speed completes $\frac{3}{5}$ of a trip in $1\frac{1}{2}$ h. In how many additional hours will the cyclist complete the entire trip?

47. ● Postage The rate for first-class mail is $.37 for the first ounce and $.23 for each additional ounce or fraction of an ounce. If the cost to mail a package first class is $3.13, how much does the package weigh?

48. Wages Four employees are paid at four consecutive levels on a wage scale. The difference between any two consecutive levels is $320 per month. The average of the four employees' monthly wages is $3880. What is the monthly wage of the highest-paid employee?

49. Coins A coin bank contains nickels, dimes, and quarters. There are 14 nickels in the bank, $\frac{1}{6}$ of the coins are dimes, and $\frac{3}{5}$ of the coins are quarters. How many coins are in the coin bank?

50. Business During one day at an office, one-half of the amount of money in the petty cash drawer was used in the morning, and one-third of the remaining money was used in the afternoon, leaving $5 in the petty cash drawer at the end of the day. How much money was in the petty cash drawer at the start of the day?

51.  A formula is an equation that relates variables in a known way. Find two examples of formulas that are used in your college major. Explain what each of the variables represents.

4.2   Integer, Coin, and Stamp Problems

1   **Consecutive integer problems**

Recall that the integers are the numbers ..., −4, −3, −2, −1, 0, 1, 2, 3, 4, ...

An **even integer** is an integer that is divisible by 2. Examples of even integers are −8, 0, and 22.

An **odd integer** is an integer that is not divisible by 2. Examples of odd integers are −17, 1, and 39.

Consecutive integers are integers that follow one another in order. Examples of consecutive integers are shown at the right. (Assume the variable n represents an integer.)	11, 12, 13 −8, −7, −6 $n, n + 1, n + 2$
Examples of **consecutive even integers** are shown at the right. (Assume the variable n represents an even integer.)	24, 26, 28 −10, −8, −6 $n, n + 2, n + 4$
Examples of **consecutive odd integers** are shown at the right. (Assume the variable n represents an odd integer.)	19, 21, 23, −1, 1, 3 $n, n + 2, n + 4$

Solve: The sum of three consecutive odd integers is 51. Find the integers.

STRATEGY   *for solving a consecutive integer problem*

▶ Let a variable represent one of the integers. Express each of the other integers in terms of that variable. Remember that consecutive integers differ by 1. Consecutive even or consecutive odd integers differ by 2.

First odd integer: n
Second odd integer: $n + 2$
Third odd integer: $n + 4$

▶ Determine the relationship among the integers.

The sum of the three odd integers is 51.

$$n + (n + 2) + (n + 4) = 51$$
$$3n + 6 = 51$$
$$3n = 45$$
$$n = 15$$

$n + 2 = 15 + 2 = 17$ • Substitute the value of *n* into the variable expressions for the
$n + 4 = 15 + 4 = 19$ second and third integers.

The three consecutive odd integers are 15, 17, and 19.

Example 1   Find three consecutive even integers such that three times
the second integer is six more than the sum of the first and
third integers.

Strategy   ▶ First even integer: n
Second even integer: $n + 2$
Third even integer: $n + 4$
▶ Three times the second integer equals six more than the
sum of the first and third integers.

Solution $3(n + 2) = n + (n + 4) + 6$
$3n + 6 = 2n + 10$
$n + 6 = 10$
$n = 4$ • The first even integer is 4.

$n + 2 = 4 + 2 = 6$ • Substitute the value of *n* into the
$n + 4 = 4 + 4 = 8$ variable expressions for the second
and third integers.

The three consecutive even integers are 4, 6, and 8.

Problem 1   Find three consecutive integers whose sum is -12.

Solution   See page S8.

---

2   **Coin and stamp problems**

3 • 25 = 75 cents

d • 10 = 10d cents

In solving problems that deal with coins or stamps of different values, it is necessary to represent the values of the coins or stamps in the same unit of money. Frequently, the unit of money is cents.

The value of 3 quarters in cents is $3 \cdot 25$ or 75 cents.
The value of 4 nickels in cents is $4 \cdot 5$ or 20 cents.
The value of d dimes in cents is $d \cdot 10$ or $10d$ cents.

Solve: A coin bank contains $1.20 in dimes and quarters. In all, there are nine coins in the bank. Find the number of quarters in the bank.

STRATEGY   *for solving a coin problem*

▶ For each denomination of coin, write a numerical or variable expression
for the number of coins, the value of the coin in cents, and the total
value of the coins in cents. The results can be recorded in a table.

The total number of coins is 9.

Number of quarters: x
Number of dimes: $9 - x$

Coin	Number of coins	·	Value of coin in cents	=	Total value in cents
Quarter	x	·	25	=	$25x$
Dime	$9 - x$	·	10	=	$10(9 - x)$

▶ Determine the relationship between the total values of the different coins. Use the fact that the sum of the total values of the different denominations of coins is equal to the total value of all the coins.

The sum of the total values of the different denominations of coins is equal to the total value of all the coins (120 cents).

$$25x + 10(9 - x) = 120$$
$$25x + 90 - 10x = 120$$
$$15x + 90 = 120$$
$$15x = 30$$
$$x = 2$$

• The total value of the quarters plus the total value of the dimes equals the amount of money in the bank.

There are 2 quarters in the bank.

Example 2   A collection of stamps consists of 3¢ stamps and 8¢ stamps. The number of 8¢ stamps is five more than three times the number of 3¢ stamps. The total value of all the stamps is $1.75. Find the number of each type of stamp in the collection.

Strategy   ▶ Number of 3¢ stamps: x
Number of 8¢ stamps: $3x + 5$

Stamp	Number	Value	Total value
3¢	x	3	$3x$
8¢	$3x + 5$	8	$8(3x + 5)$

▶ The sum of the total values of the different types of stamps equals the total value of all the stamps (175 cents).

Solution $3x + 8(3x + 5) = 175$
$3x + 24x + 40 = 175$
$27x + 40 = 175$
$27x = 135$
$x = 5$

• The total value of the 3¢ stamps plus the total value of the 8¢ stamps equals the total value of all the stamps.

$3x + 5 = 3(5) + 5$
$= 15 + 5 = 20$

• Find the number of 8¢ stamps.

There are five 3¢ stamps and twenty 8¢ stamps in the collection.

Problem 2　A coin bank contains nickels, dimes, and quarters. There are five times as many nickels as dimes and six more quarters than dimes. The total value of all the coins is $6.30. Find the number of each kind of coin in the bank.

Solution　See page S8.

4.2　Concept Review

Determine whether the statement is always true, sometimes true, or never true.

1. An even integer is a multiple of 2.

2. When $10d$ is used to represent the value of d dimes, 10 is written as the coefficient of d because it is the number of cents in one dime.

3. Given the consecutive odd integers -5 and -3, the next consecutive odd integer is -1.

4. If the first of three consecutive odd integers is n, then the second and third consecutive odd integers are represented as $n + 1$ and $n + 3$.

5. You have a total of 20 coins in nickels and dimes. If n represents the number of nickels you have, then $n - 20$ represents the number of dimes you have.

6. You have a total of seven coins in dimes and quarters. The total value of the coins is $1.15. If d represents the number of dimes you have, then $10d + 25(7 - d) = 1.15$.

4.2　Exercises

1 **Consecutive Integer Problems**

1. Explain how to represent three consecutive integers using only one variable.

2. Explain why both consecutive even integers and consecutive odd integers are represented algebraically as $n, n + 2, n + 4, \ldots$.

3. The sum of three consecutive integers is 54. Find the integers.

4. The sum of three consecutive integers is 75. Find the integers.

5. The sum of three consecutive even integers is 84. Find the integers.

6. The sum of three consecutive even integers is 48. Find the integers.

7. The sum of three consecutive odd integers is 57. Find the integers.

8. The sum of three consecutive odd integers is 81. Find the integers.

9. Find two consecutive even integers such that five times the first integer is equal to four times the second integer.

10. Find two consecutive even integers such that six times the first integer equals three times the second integer.

11. Nine times the first of two consecutive odd integers equals seven times the second. Find the integers.

12. Five times the first of two consecutive odd integers is three times the second. Find the integers.

13. Find three consecutive integers whose sum is negative twenty-four.

14. Find three consecutive even integers whose sum is negative twelve.

15. Three times the smallest of three consecutive even integers is two more than twice the largest. Find the integers.

16. Twice the smallest of three consecutive odd integers is five more than the largest. Find the integers.

17. Find three consecutive odd integers such that three times the middle integer is six more than the sum of the first and third integers.

18. Find three consecutive odd integers such that four times the middle integer is equal to two less than the sum of the first and third integers.

2 Coin and Stamp Problems

19. Explain how to represent the total value of x nickels using the equation

Number of coins · Value of the coin in cents = Total value in cents

20. Suppose a coin purse contains only dimes and quarters. In the context of this situation, explain the meaning of the statement "The sum of the total values of the different denominations of coins is equal to the total value of all the coins."

21. A bank contains 27 coins in dimes and quarters. The coins have a total value of $4.95. Find the number of dimes and quarters in the bank.

22. A coin purse contains 18 coins in nickels and dimes. The coins have a total value of $1.15. Find the number of nickels and dimes in the coin purse.

23. A business executive bought 40 stamps for $13.68. The purchase included 37¢ stamps and 23¢ stamps. How many of each type of stamp were bought?

24. A postal clerk sold some 34¢ stamps and some 23¢ stamps. Altogether, 15 stamps were sold for a total cost of $4.44. How many of each type of stamp were sold?

25. A drawer contains 29¢ stamps and 3¢ stamps. The number of 29¢ stamps is four less than three times the number of 3¢ stamps. The total value of all the stamps is $1.54. How many 29¢ stamps are in the drawer?

26. The total value of the dimes and quarters in a bank is $6.05. There are six more quarters than dimes. Find the number of each type of coin in the bank.

27. A child's piggy bank contains 44 coins in quarters and dimes. The coins have a total value of $8.60. Find the number of quarters in the bank.

28. A coin bank contains nickels and dimes. The number of dimes is 10 less than twice the number of nickels. The total value of all the coins is $2.75. Find the number of each type of coin in the bank.

29. A total of 26 bills are in a cash box. Some of the bills are one-dollar bills, and the rest are five-dollar bills. The total amount of cash in the box is $50. Find the number of each type of bill in the cash box.

30. A bank teller cashed a check for $200 using twenty-dollar bills and ten-dollar bills. In all, 12 bills were handed to the customer. Find the number of twenty-dollar bills and the number of ten-dollar bills.

31. A coin bank contains pennies, nickels, and dimes. There are six times as many nickels as pennies and four times as many dimes as pennies. The total amount of money in the bank is $7.81. Find the number of pennies in the bank.

32. A coin bank contains pennies, nickels, and quarters. There are seven times as many nickels as pennies and three times as many quarters as pennies. The total amount of money in the bank is $5.55. Find the number of pennies in the bank.

33. A collection of stamps consists of 22¢ stamps and 40¢ stamps. The number of 22¢ stamps is three more than four times the number of 40¢ stamps. The total value of the stamps is $8.34. Find the number of 22¢ stamps in the collection.

34. A collection of stamps consists of 2¢ stamps, 8¢ stamps, and 14¢ stamps. The number of 2¢ stamps is five more than twice the number of 8¢ stamps. The number of 14¢ stamps is three times the number of 8¢ stamps. The total value of the stamps is $2.26. Find the number of each type of stamp in the collection.

35. A collection of stamps consists of 3¢ stamps, 7¢ stamps, and 12¢ stamps. The number of 3¢ stamps is five less than the number of 7¢ stamps. The number of 12¢ stamps is one-half the number of 7¢ stamps. The total value of all the stamps is $2.73. Find the number of each type of stamp in the collection.

36. A collection of stamps consists of 2¢ stamps, 5¢ stamps, and 7¢ stamps. There are nine more 2¢ stamps than 5¢ stamps and twice as many 7¢ stamps as 5¢ stamps. The total value of the stamps is $1.44. Find the number of each type of stamp in the collection.

37. A collection of stamps consists of 6¢ stamps, 8¢ stamps, and 15¢ stamps. The number of 6¢ stamps is three times the number of 8¢ stamps. There are six more 15¢ stamps than there are 6¢ stamps. The total value of all the stamps is $5.16. Find the number of each type of stamp.

38. A child's piggy bank contains nickels, dimes, and quarters. There are twice as many nickels as dimes and four more quarters than nickels. The total value of all the coins is $9.40. Find the number of each type of coin.

4.2   Applying Concepts

39. Find four consecutive even integers whose sum is -36.

40. Find four consecutive odd integers whose sum is -48.

41. A coin bank contains only dimes and quarters. The number of quarters in the bank is two less than twice the number of dimes. There are 34 coins in the bank. How much money is in the bank?

42. A postal clerk sold twenty stamps to a customer. The number of 37¢ stamps purchased was two more than twice the number of 23¢ stamps purchased. If the customer bought only 37¢ stamps and 23¢ stamps, how much money did the clerk collect from the customer?

43. Find three consecutive odd integers such that the sum of the first and third integers is twice the second integer.

44. Find four consecutive integers such that the sum of the first and fourth integers equals the sum of the second and third integers.

4.3 Geometry Problems

1 Perimeter problems

The **perimeter** of a plane geometric figure is a measure of the distance around the figure. Perimeter is used, for example, in buying fencing for a lawn and in determining how much baseboard is needed for a room.

The perimeter of a triangle is the sum of the lengths of the three sides. Therefore, if a, b, and c represent the lengths of the sides of a triangle, the perimeter, P, of the triangle is given by $P = a + b + c$.

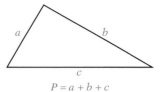

$$P = a + b + c$$

Two special types of triangles are shown below. An **isosceles triangle** has two sides of equal length. The three sides of an **equilateral triangle** are of equal length.

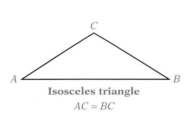

Isosceles triangle
$AC = BC$

Equilateral triangle
$AB = AC = BC$

The perimeter of a rectangle is the sum of the lengths of the four sides. Let L represent the length and W represent the width of a rectangle. Then the perimeter, P, of the rectangle is given by $P = L + W + L + W$. Combine like terms and the formula is $P = 2L + 2W$.

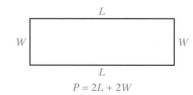

$$P = 2L + 2W$$

A square is a rectangle in which each side has the same length. Let s represent the length of each side of a square. Then the perimeter, P, of the square is given by $P = s + s + s + s$. Combine like terms and the formula is $P = 4s$.

$$P = 4s$$

Formulas for Perimeters of Geometric Figures

Perimeter of a triangle	$P = a + b + c$
Perimeter of a rectangle	$P = 2L + 2W$
Perimeter of a square	$P = 4s$

The perimeter of a rectangle is 26 ft. The length of the rectangle is 1 ft more than twice the width. Find the width and length of the rectangle.

Strategy

Let a variable represent the width. Represent the length in terms of that variable. Use the formula for the perimeter of a rectangle.

Width: W
Length: $2W + 1$

2W + 1

W

Solution

$$P = 2L + 2W$$

$P = 26$. Substitute $2W + 1$ for L. $\quad 26 = 2(2W + 1) + 2W$

Use the Distributive Property. $\quad 26 = 4W + 2 + 2W$

Combine like terms. $\quad 26 = 6W + 2$

Subtract 2 from each side of the equation. $\quad 24 = 6W$

Divide each side of the equation by 6. $\quad 4 = W$

Find the length of the rectangle by substituting 4 for W in $2W + 1$.

$L = 2W + 1$
$\quad = 2(4) + 1 = 8 + 1 = 9$

The width is 4 ft. The length is 9 ft.

Example 1 The perimeter of an isosceles triangle is 25 ft. The length of the third side is 2 ft less than the length of one of the equal sides. Find the measures of the three sides of the triangle.

Strategy ▶ Each equal side: x
The third side: $x - 2$
▶ Use the equation for the perimeter of a triangle.

x x

$x - 2$

Solution
$$P = a + b + c$$
$$25 = x + x + (x - 2)$$
$$25 = 3x - 2$$
$$27 = 3x$$
$$9 = x$$

$x - 2 = 9 - 2 = 7$ • Substitute the value of x into the variable expression for the length of the third side.

Each of the equal sides measures 9 ft.
The third side measures 7 ft.

Problem 1 A carpenter is designing a square patio with a perimeter of 52 ft. What is the length of each side?

Solution See page S8.

2 Problems involving angles formed by intersecting lines

A unit used to measure angles is the **degree**. The symbol for degree is °. ∠ is the symbol for angle.

One complete revolution is 360°. It is probable that the Babylonians chose 360° for a circle because they knew that there are 365 days in one year, and 360 is the closest number to 365 that is divisible by many numbers.

A 90° angle is called a **right angle.** The symbol ∟ represents a right angle. Angle C ($\angle C$) is a right angle.

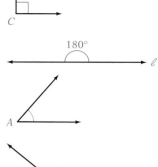

A 180° angle is called a **straight angle.** The angle at the right is a straight angle.

An **acute angle** is an angle whose measure is between 0° and 90°. $\angle A$ at the right is an acute angle.

An **obtuse angle** is an angle whose measure is between 90° and 180°. $\angle B$ at the right is an obtuse angle.

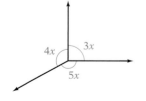

Given the diagram at the left, find x.

Strategy   The sum of the measures of the three angles is 360°. To find x, write an equation and solve for x.

Solution $3x + 4x + 5x = 360°$
$$12x = 360°$$
$$x = 30°$$

The measure of x is 30°.

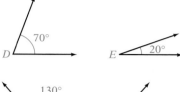

Complementary angles are two angles whose measures have the sum 90°.

$\angle D + \angle E = 70° + 20° = 90°$

$\angle D$ and $\angle E$ are complementary angles.

Supplementary angles are two angles whose measures have the sum 180°.

$\angle F + \angle G = 130° + 50° = 180°$

$\angle F$ and $\angle G$ are supplementary angles.

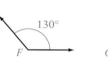

Example 2   Find the complement of a 39° angle.

Strategy   To find the complement, let x represent the complement of a 39° angle. Use the fact that complementary angles are two angles whose sum is 90° to write an equation. Solve for x.

Solution $39° + x = 90°$
$x = 51°$ • Subtract 39 from each side of the equation.

The complement of a 39° angle is a 51° angle.

Problem 2   Find the supplement of a 107° angle.

Solution   See page S8.

Parallel lines never meet. The distance between them is always the same. The symbol ‖ means "is parallel to." In the figure at the right, $\ell_1 \parallel \ell_2$.

Perpendicular lines are intersecting lines that form right angles. The symbol ⊥ means "is perpendicular to." In the figure at the right, $p \perp q$.

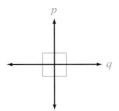

Four angles are formed by the intersection of two lines. If the two lines are perpendicular, then each of the four angles is a right angle. If the two lines are not perpendicular, then two of the angles formed are acute angles and two of the angles formed are obtuse angles. The two acute angles are always opposite each other, and the two obtuse angles are always opposite each other.

In the figure at the right, $\angle w$ and $\angle y$ are acute angles, and $\angle x$ and $\angle z$ are obtuse angles.

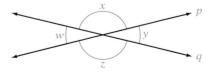

Two angles that are on opposite sides of the intersection of two lines are called **vertical angles.** Vertical angles have the same measure. $\angle w$ and $\angle y$ are vertical angles. $\angle x$ and $\angle z$ are vertical angles.

Vertical angles have the same measure.

$$\angle w = \angle y$$
$$\angle x = \angle z$$

Two angles that share a common side are called **adjacent angles.** For the figure shown above, $\angle x$ and $\angle y$ are adjacent angles, as are $\angle y$ and $\angle z$, $\angle z$ and $\angle w$, and $\angle w$ and $\angle x$. Adjacent angles of intersecting lines are supplementary angles.

Adjacent angles of intersecting lines are supplementary angles.

$$\angle x + \angle y = 180°$$
$$\angle y + \angle z = 180°$$
$$\angle z + \angle w = 180°$$
$$\angle w + \angle x = 180°$$

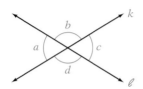

In the diagram at the left, $\angle b = 115°$. Find the measures of angles a, c, and d.

$\angle a$ is supplementary to $\angle b$ because $\angle a$ and
$\angle b$ are adjacent angles of intersecting lines.

$$\angle a + \angle b = 180°$$
$$\angle a + 115° = 180°$$
$$\angle a = 65°$$

$\angle c = \angle a$ because $\angle c$ and $\angle a$ are vertical angles. $\angle c = 65°$
$\angle d = \angle b$ because $\angle d$ and $\angle b$ are vertical angles. $\angle d = 115°$

Example 3   Find x.

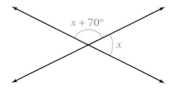

Strategy   The angles labeled are adjacent angles of intersecting lines
and are therefore supplementary angles. To find x, write an
equation and solve for x.

Solution $$x + (x + 70°) = 180°$$
$$2x + 70° = 180°$$
$$2x = 110°$$
$$x = 55°$$

Problem 3   Find x.

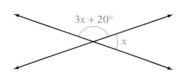

Solution   See page S8.

A line that intersects two other lines at
different points is called a **transversal**.
If the lines cut by a transversal t are parallel lines and the transversal is not perpendicular to the parallel lines, all four
acute angles have the same measure and
all four obtuse angles have the same
measure. In the figure at the right,

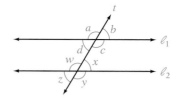

$$\angle b = \angle d = \angle x = \angle z$$
$$\angle a = \angle c = \angle w = \angle y$$

Alternate interior angles are two non-
adjacent angles that are on opposite
sides of the transversal and between the
lines. In the figure above, $\angle c$ and $\angle w$ are
alternate interior angles, and $\angle d$ and $\angle x$
are alternate interior angles. Alternate
interior angles have the same measure.

**Alternate interior angles have
the same measure.**

$$\angle c = \angle w$$
$$\angle d = \angle x$$

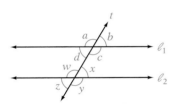

Alternate exterior angles are two non-adjacent angles that are on opposite sides of the transversal and outside the parallel lines. In the figure at the left, $\angle a$ and $\angle y$ are alternate exterior angles, and $\angle b$ and $\angle z$ are alternate exterior angles. Alternate exterior angles have the same measure.

Alternate exterior angles have the same measure.

$$\angle a = \angle y$$
$$\angle b = \angle z$$

Corresponding angles are two angles that are on the same side of the transversal and are both acute angles or are both obtuse angles. In the figure above, the following pairs of angles are corresponding angles: $\angle a$ and $\angle w$, $\angle d$ and $\angle z$, $\angle b$ and $\angle x$, and $\angle c$ and $\angle y$. Corresponding angles have the same measure.

Corresponding angles have the same measure.

$$\angle a = \angle w$$
$$\angle d = \angle z$$
$$\angle b = \angle x$$
$$\angle c = \angle y$$

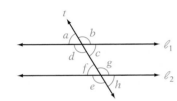

In the diagram at the left, $\ell_1 \| \ell_2$ and $\angle f = 58°$. Find the measures of $\angle a$, $\angle c$, and $\angle d$.

$\angle a$ and $\angle f$ are corresponding angles. $\angle a = \angle f = 58°$

$\angle c$ and $\angle f$ are alternate interior angles. $\angle c = \angle f = 58°$

$\angle d$ is supplementary to $\angle a$. $\angle d + \angle a = 180°$
$$\angle d + 58° = 180°$$
$$\angle d = 122°$$

Example 4   Given $\ell_1 \| \ell_2$, find x.

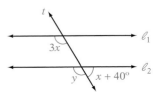

Strategy $3x = y$ because corresponding angles have the same measure. $y + (x + 40°) = 180°$ because adjacent angles of intersecting lines are supplementary angles.

Substitute $3x$ for y and solve for x.

Solution $y + (x + 40°) = 180°$
$$3x + (x + 40°) = 180°$$
$$4x + 40° = 180°$$
$$4x = 140°$$
$$x = 35°$$

Problem 4   Given $\ell_1 \| \ell_2$, find x.

Solution   See page S8.

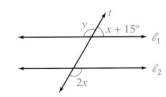

3   Problems involving the angles of a triangle

If the lines cut by a transversal are not parallel lines, the three lines will intersect at three points. In the figure at the right, the transversal t intersects lines p and q. The three lines intersect at points A, B, and C. The geometric figure formed by line segments AB, BC, and AC is a **triangle.**

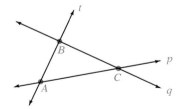

The angles within the region enclosed by the triangle are called **interior angles.** In the figure at the right, angles a, b, and c are interior angles. The sum of the measures of the interior angles is 180°.

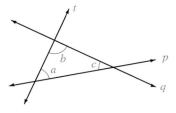

$$\angle a + \angle b + \angle c = 180°$$

The Sum of the Measures of the Interior Angles of a Triangle

The sum of the measures of the interior angles of a triangle is 180°.

An angle adjacent to an interior angle is an **exterior angle.** In the figure at the right, angles m and n are exterior angles for angle a. The sum of the measures of an interior and an exterior angle is 180°.

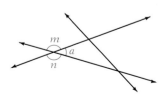

$$\angle a + \angle m = 180°$$
$$\angle a + \angle n = 180°$$

Given that $\angle c = 40°$ and $\angle e = 60°$, find the measure of $\angle d$.

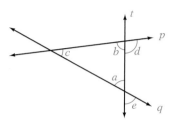

$\angle a$ and $\angle e$ are vertical angles.

The sum of the interior angles is 180°.

$$\angle a = \angle e = 60°$$

$$\angle c + \angle a + \angle b = 180°$$
$$40° + 60° + \angle b = 180°$$
$$100° + \angle b = 180°$$
$$\angle b = 80°$$

∠*b* and ∠*d* are supplementary angles.

$$\angle b + \angle d = 180°$$
$$80° + \angle d = 180°$$
$$\angle d = 100°$$

Example 5   Given that ∠*a* = 45° and
∠*x* = 100°, find the measures
of angles *b*, *c*, and *y*.

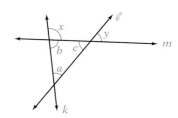

Strategy   ▶ To find the measure of ∠*b*, use the fact that ∠*b* and ∠*x*
are supplementary angles.
▶ To find the measure of ∠*c*, use the fact that the sum of
the measures of the interior angles of a triangle is 180°.
▶ To find the measure of ∠*y*, use the fact that ∠*c* and ∠*y*
are vertical angles.

Solution
$$\angle b + \angle x = 180°$$
$$\angle b + 100° = 180°$$
$$\angle b = 80°$$

$$\angle a + \angle b + \angle c = 180°$$
$$45° + 80° + \angle c = 180°$$
$$125° + \angle c = 180°$$
$$\angle c = 55°$$

$$\angle y = \angle c = 55°$$

Problem 5   Given that ∠*y* = 55°, find the
measures of angles *a*, *b*, and *d*.

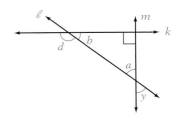

Solution   See page S8.

Example 6   Two angles of a triangle measure 43° and 86°. Find the
measure of the third angle.

Strategy   To find the measure of the third angle, use the fact that the
sum of the measures of the interior angles of a triangle is
180°. Write an equation using *x* to represent the measure of
the third angle. Solve the equation for *x*.

Solution
$$x + 43° + 86° = 180°$$
$$x + 129° = 180°$$
$$x = 51°$$

The measure of the third angle is 51°.

Problem 6   One angle of a triangle is a right angle, and one angle
measures 27°. Find the measure of the third angle.

Solution   See pages S8–S9.

4.3 Concept Review

Determine whether the statement is always true, sometimes true, or never true.

1. The formula for the perimeter of a rectangle is $P = 2L + 2W$, where L represents the length and W represents the width of a rectangle.

2. In the formula for the perimeter of a triangle, $P = a + b + c$, the variables a, b, and c represent the measures of the three angles of a triangle.

3. If two angles of a triangle measure 52° and 73°, then the third angle of the triangle measures 55°.

4. An isosceles triangle has two sides of equal measure and two angles of equal measure.

5. The perimeter of a geometric figure is a measure of the area of the figure.

6. In a right triangle, the sum of the measures of the two acute angles is 90°.

4.3 Exercises

1 Geometry Problems

1. What is the difference between an isosceles triangle and an equilateral triangle?

2. Which of the following units can be used to measure perimeter?
a. feet
b. square miles
c. meters
d. ounces
e. cubic inches
f. centimeters

3. In an isosceles triangle, the third side is 50% of the length of one of the equal sides. Find the length of each side when the perimeter is 125 ft.

4. In an isosceles triangle, the length of one of the equal sides is three times the length of the third side. The perimeter is 21 m. Find the length of each side.

5. The perimeter of a rectangle is 42 m. The length of the rectangle is 3 m less than twice the width. Find the length and width of the rectangle.

$2W - 3$

W

6. The width of a rectangle is 25% of the length. The perimeter is 250 cm. Find the length and width of the rectangle.

7. The perimeter of a rectangle is 120 ft. The length of the rectangle is twice the width. Find the length and width of the rectangle.

8. The perimeter of a rectangle is 50 m. The width of the rectangle is 5 m less than the length. Find the length and width of the rectangle.

9. The perimeter of a triangle is 110 cm. One side is twice the second side. The third side is 30 cm more than the second side. Find the length of each side.

10. The perimeter of a triangle is 33 ft. One side of the triangle is 1 ft longer than the second side. The third side is 2 ft longer than the second side. Find the length of each side.

11. The width of the rectangular foundation of a building is 30% of the length. The perimeter of the foundation is 338 ft. Find the length and width of the foundation.

12. The perimeter of a rectangular playground is 440 ft. If the width is 100 ft, what is the length of the playground?

13. A rectangular vegetable garden has a perimeter of 64 ft. The length of the garden is 20 ft. What is the width of the garden?

14. Each of two sides of a triangular banner measures 18 in. If the perimeter of the banner is 46 in., what is the length of the third side of the banner?

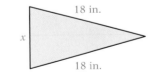

15. The perimeter of a square picture frame is 48 in. Find the length of each side of the frame.

16. A square rug has a perimeter of 32 ft. Find the length of each side of the rug.

2 **Geometry Problems**

17. Name the number of degrees in each of the following.
   a. a right angle
   b. a straight angle
   c. one complete revolution
   d. an acute angle
   e. an obtuse angle
   f. complementary angles
   g. supplementary angles

18. Determine whether enough information is given to solve the problem.
   a. Find the sum of the measures of the two acute angles in a right triangle.
   b. Find the measures of the two acute angles in a right triangle.

19. Find the complement of a 28° angle.

20. Find the complement of a 46° angle.

21. Find the supplement of a 73° angle.

22. Find the supplement of a 119° angle.

Find the measure of $\angle x$.

23.
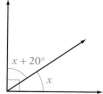

24.
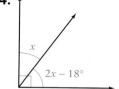

25.
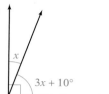

Find the measure of $\angle a$.

26.
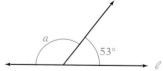

27.
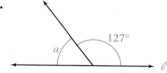

28.
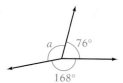

29.
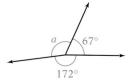

Find x.

30.
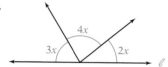

31.
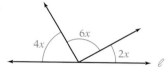

32.
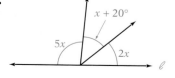

33.
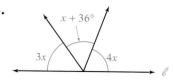

34.

35.

Find the measure of ∠x.

36.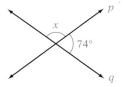

37.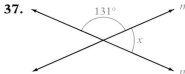

Find x.

38.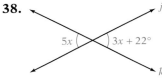

39.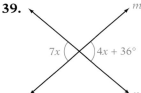

Given that $\ell_1 \parallel \ell_2$, find the measures of angles a and b.

40.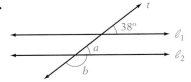

41.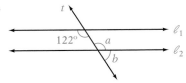

42.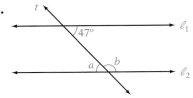

43.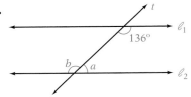

Given that $\ell_1 \parallel \ell_2$, find x.

44.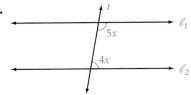

45.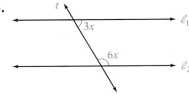

46.

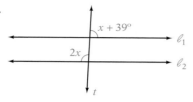

47.

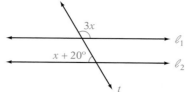

48. Given that $\angle a = 51°$, find the measure of $\angle b$.

49. Given that $\angle a = 38°$, find the measure of $\angle b$.

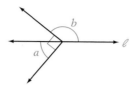

---

3 Geometry Problems

50. Given that $\angle a = 95°$ and $\angle b = 70°$, find the measures of angles x and y.

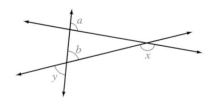

51. Given that $\angle a = 35°$ and $\angle b = 55°$, find the measures of angles x and y.

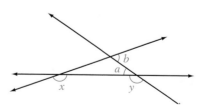

52. Given that $\angle y = 45°$, find the measures of angles a and b.

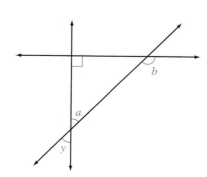

53. Given that $\angle y = 130°$, find the measures of angles a and b.

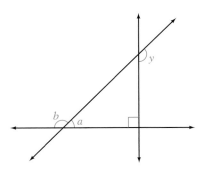

54. Given that $AO \perp OB$, express in terms of x the number of degrees in $\angle BOC$.

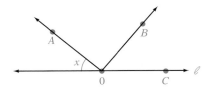

55. Given that $AO \perp OB$, express in terms of x the number of degrees in $\angle AOC$.

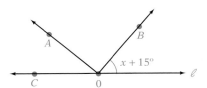

56. One angle in a triangle is a right angle, and one angle measures 30°. What is the measure of the third angle?

57. A triangle has a 45° angle and a right angle. Find the measure of the third angle.

58. Two angles of a triangle measure 42° and 103°. Find the measure of the third angle.

59. Two angles of a triangle measure 62° and 45°. Find the measure of the third angle.

60. A triangle has a 13° angle and a 65° angle. What is the measure of the third angle?

61. A triangle has a 105° angle and a 32° angle. What is the measure of the third angle?

62. In an isosceles triangle, one angle is three times the measure of one of the equal angles. Find the measure of each angle.

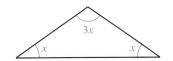

63. In an isosceles triangle, one angle is 10° less than three times the measure of one of the equal angles. Find the measure of each angle.

64. In an isosceles triangle, one angle is 16° more than twice the measure of one of the equal angles. Find the measure of each angle.

65. In a triangle, one angle is twice the measure of the second angle. The third angle is three times the measure of the second angle. Find the measure of each angle.

4.3 Applying Concepts

66. A rectangle and an equilateral triangle have the same perimeter. The length of the rectangle is three times the width. Each side of the triangle is 8 cm. Find the length and width of the rectangle.

67. The length of a rectangle is 1 cm more than twice the width. If the length of the rectangle is decreased by 2 cm and the width is decreased by 1 cm, the perimeter is 20 cm. Find the length and width of the original rectangle.

68. For the figure at the right, find the sum of the measures of angles x, y, and z.

69.  For the figure at the right, explain why $\angle a + \angle b = \angle x$. Write a rule that describes the relationship between an exterior angle of a triangle and the opposite interior angles. Use the rule to write an equation involving angles a, c, and z.

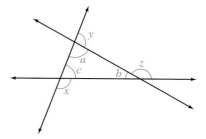

70. The width of a rectangle is $8x$. The perimeter is $48x$. Find the length of the rectangle in terms of the variable x.

71.  Discuss the concepts of accuracy and precision.

72.  Write a report stating the arguments of the proponents of changing to the metric system in the United States or a report stating the arguments of those opposed to switching from the U.S. Customary System of measurement to the metric system.

73.  Prepare a report on the use of geometric form in architecture.

4.4   Markup and Discount Problems

1   **Markup problems**

VIDEO & DVD   CD TUTOR   WEB   SSM

Cost is the price that a business pays for a product. **Selling price** is the price for which a business sells a product to a customer. The difference between selling price and cost is called **markup**. Markup is added to a retailer's cost to cover the expenses of operating a business. Markup is usually expressed as a percent of the retailer's cost. This percent is called the **markup rate**.

The basic markup equations used by a business are

$$\text{Selling price} = \text{Cost} + \text{Markup}$$
$$S = C + M$$

$$\text{Markup} = \text{Markup rate} \cdot \text{Cost}$$
$$M = r \cdot C$$

By substituting $r \cdot C$ for M in the first equation, we can also write selling price as

$$S = C + M$$
$$S = C + (r \cdot C)$$
$$S = C + rC$$

The equation $S = C + rC$ is the equation used to solve the markup problems in this section.

Example 1   The manager of a clothing store buys a suit for $180 and sells the suit for $252. Find the markup rate.

Strategy   Given: $C = \$180$
$S = \$252$
Unknown markup rate: r
Use the equation $S = C + rC$.

Solution

$$S = C + rC$$
$$252 = 180 + 180r$$

• Substitute the values of C and S into the equation.

$$252 - 180 = 180 - 180 + 180r$$
$$72 = 180r$$

• Subtract 180 from each side of the equation.

$$\frac{72}{180} = \frac{180r}{180}$$

• Divide each side of the equation by 180.

$$0.4 = r$$

• The decimal must be changed to a percent.

The markup rate is 40%.

Problem 1   The cost to the manager of a sporting goods store for a tennis racket is $120. The selling price of the racket is $180. Find the markup rate.

Solution   See page S9.

Example 2   The manager of a furniture store uses a markup rate of 45% on all items. The selling price of a chair is $232. Find the cost of the chair.

Strategy   Given: $r = 45\% = 0.45$
$S = \$232$
Unknown cost: C
Use the equation $S = C + rC$.

Solution

$$S = C + rC$$
$$232 = C + 0.45C$$
$$232 = 1.45C$$
$$160 = C$$

• $C + 0.45C = 1C + 0.45C = (1 + 0.45)C$

The cost of the chair is $160.

Problem 2   A hardware store employee uses a markup rate of 40% on all items. The selling price of a lawnmower is $266. Find the cost.

Solution   See page S9.

2 **Discount problems**

Discount is the amount by which a retailer reduces the regular price of a product for a promotional sale. Discount is usually expressed as a percent of the regular price. This percent is called the **discount rate.**

The basic discount equations used by a business are

$$\text{Sale price} = \text{Regular price} - \text{Discount}$$
$$S = R - D$$

$$\text{Discount} = \text{Discount rate} \cdot \text{Regular price}$$
$$D = r \cdot R$$

By substituting $r \cdot R$ for D in the first equation, we can also write sale price as

$$S = R - D$$
$$S = R - (r \cdot R)$$
$$S = R - rR$$

The equation $S = R - rR$ is the equation used to solve the discount problems in this section.

Example 3   In a garden supply store, the regular price of a 100-foot garden hose is $48. During an "after-summer sale," the hose is being sold for $36. Find the discount rate.

Strategy   Given: $R = \$48$
               $S = \$36$
Unknown discount rate: r
Use the equation $S = R - rR$.

Solution
$$S = R - rR$$
$$36 = 48 - 48r$$
  • Substitute the values of R and S into the equation.

$$36 - 48 = 48 - 48 - 48r$$
$$-12 = -48r$$
  • Subtract 48 from each side of the equation.

$$\frac{-12}{-48} = \frac{-48r}{-48}$$
  • Divide each side of the equation by -48.

$$0.25 = r$$
  • The decimal must be changed to a percent.

The discount rate is 25%.

Problem 3   A case of motor oil that regularly sells for $39.80 is on sale for $29.85. What is the discount rate?

Solution   See page S9.

Example 4   The sale price for a chemical sprayer is $27.30. This price is 35% off the regular price. Find the regular price.

Strategy	Given: $S = \$27.30$
	$r = 35\% = 0.35$
	Unknown regular price: R
	Use the equation $S = R - rR$.
Solution	$S = R - rR$

$$27.30 = R - 0.35R$$
$$27.30 = 0.65R \qquad \bullet\ R - 0.35R = 1R - 0.35R = (1 - 0.35)R$$
$$42 = R$$

The regular price is $42.00.

Problem 4   The sale price for a telephone is $43.50. This price is 25% off the regular price. Find the regular price.

Solution   See page S9.

4.4   Concept Review

Determine whether the statement is always true, sometimes true, or never true.

1. In the markup equation $S = C + rC$, the variable r represents the markup rate.

2. Markup is an amount of money, whereas markup rate is a percent.

3. In the discount equation $S = R - rR$, the variable R represents the discount rate.

4. In the markup equation, S represents sale price, and in the discount equation, S represents selling price.

5. $R - 0.45R = 1R - 0.45R = (1 - 0.45)R = 0.55R$ is an example of the application of the Distributive Property.

4.4   Exercises

1 Markup Problems

1. Explain the difference between the cost and the selling price of a product.

2. Explain the difference between markup and markup rate.

3. A computer software retailer uses a markup rate of 40%. Find the selling price of a computer game that costs the retailer $40.

4. A car dealer advertises a 5% markup over cost. Find the selling price of a car that costs the dealer $26,000.

5. An electronics store uses a markup rate of 58%. Find the selling price of a camcorder that costs the store owner $358.

6. The cost to a landscape architect for a 25-gallon tree is $65. Find the selling price of the tree if the markup rate used by the architect is 30%.

7. A set of golf clubs costing $360 is sold for $630. Find the markup rate on the set of golf clubs.

8. A jeweler purchases a diamond ring for $2399.50. The selling price is $4799. Find the markup rate.

9. A freezer costing $360 is sold for $520. Find the markup rate. Round to the nearest tenth of a percent.

10. A grocer purchases a can of fruit juice for $1.96. The selling price of the fruit juice is $2.45. Find the markup rate.

11. A watch costing $98 is sold for $156.80. Find the markup rate on the watch.

12. A sofa costing $320 is sold for $479. Find the markup rate. Round to the nearest tenth of a percent.

13. The selling price of a compact disc player is $168. The markup rate used by the seller is 40%. Find the cost of the compact disc player.

14. A manufacturer of exercise equipment uses a markup rate of 45%. One of the manufacturer's treadmills has a selling price of $580. Find the cost of the treadmill.

15. A markup rate of 40% was used on a basketball with a selling price of $82.60. Find the cost of the basketball.

16. A digitally recorded compact disc has a selling price of $18.90. The markup rate is 40%. Find the cost of the CD.

17. A markup rate of 25% is used on a computer that has a selling price of $2187.50. Find the cost of the computer.

18. The selling price of a portable CD player is $132. The markup rate used by the retailer is 60%. Find the cost to the retailer.

2 Discount Problems

19. ✎ Explain the meaning of sale price and discount.

20. ✎ Explain the difference between discount and discount rate.

21. A tennis racket that regularly sells for $135 is on sale for 25% off the regular price. Find the sale price.

22. A fax machine that regularly sells for $219.99 is on sale for $33\frac{1}{3}$% off the regular price. Find the sale price.

23. A supplier of electrical equipment offers a 5% discount for a purchase that is paid for within 30 days. A transformer regularly sells for $230. Find the discount price of a transformer that is paid for 10 days after the purchase.

24. A clothing wholesaler offers a discount of 10% per shirt when 10 to 20 shirts are purchased and a discount of 15% per shirt when 21 to 50 shirts are purchased. A shirt regularly sells for $27. Find the sale price per shirt when 35 shirts are purchased.

25. A car stereo system that regularly sells for $425 is on sale for $318.75. Find the discount rate.

26. A pair of roller blades that regularly sells for $99.99 is on sale for $79.99. Find the discount rate. Round to the nearest percent.

27. An oak bedroom set with a regular price of $1295 is on sale for $995. Find the discount rate. Round to the nearest tenth of a percent.

28. A stereo set with a regular price of $495 is on sale for $395. Find the markdown rate. Round to the nearest tenth of a percent.

29. A digital camera with a regular price of $325 is on sale for $201.50. Find the markdown rate.

30. A luggage set with a regular price of $178 is on sale for $103.24. Find the discount rate.

31. The sale price of a free-weight home gym is $408, which is 20% off the regular price. Find the regular price.

32. The sale price of a toboggan is $77, which is 30% off the regular price. Find the regular price.

33. A mechanic's tool set is on sale for $180 after a markdown of 40% off the regular price. Find the regular price.

34. The sale price of a modem is $126, which is 30% off the regular price. Find the regular price.

35. A telescope is on sale for $165 after a markdown of 40% off the regular price. Find the regular price.

36. An exercise bike is on sale for $390, having been marked down 25% off the regular price. Find the regular price.

4.4   Applying Concepts

37. A pair of shoes that now sells for $63 has been marked up 40%. Find the markup on the pair of shoes.

38. The sale price of a motorcycle helmet is 25% off the regular price. The discount is $70. Find the sale price.

39. A refrigerator selling for $770 has a markup of $220. Find the markup rate.

40. The sale price of a copy machine is $765 after a discount of $135. Find the discount rate.

41. The manager of a camera store uses a markup rate of 30%. Find the cost of a camera selling for $299.

42. The sale price of a television is $180. Find the regular price if the sale price was computed by taking $\frac{1}{3}$ off the regular price followed by an additional 25% discount on the reduced price.

43. A customer buys four tires, three at the regular price and one for 20% off the regular price. The four tires cost $304. What was the regular price of a tire?

44. A lamp, originally priced at under $100, was on sale for 25% off the regular price. When the regular price, a whole number of dollars, was discounted, the discounted price was also a whole number of dollars. Find the largest possible number of dollars in the regular price of the lamp.

45. A used car is on sale for a discount of 20% off the regular price of $8500. An additional 10% discount on the sale price was offered. Is the result a 30% discount? What is the single discount that would give the same sale price?

46. Write a report on series trade discounts. Explain how to convert a series discount to a single-discount equivalent.

4.5   Investment Problems

1   Investment problems

Recall that the annual simple interest that an investment earns is given by the equation $I = Pr$, where I is the simple interest, P is the principal, or the amount invested, and r is the simple interest rate.

Point of Interest

You may be familiar with the simple interest formula $I = Prt$. *If so, you know that* t *represents time. In the problems in this section, time is always 1 (one year), so the formula* $I = Prt$ *simplifies to*

$$I = Pr(1)$$
$$I = Pr$$

The annual simple interest rate on a $4500 investment is 8%. Find the annual simple interest earned on the investment.

Given: $P = \$4500$ $I = Pr$
$\quad\quad r = 8\% = 0.08$ $I = 4500(0.08)$
Unknown interest: I $I = 360$

The annual simple interest earned is $360.

Solve: An investor has a total of $10,000 to deposit into two simple interest accounts. On one account, the annual simple interest rate is 7%. On the second account, the annual simple interest rate is 8%. How much should be invested in each account so that the total annual interest earned is $785?

STRATEGY | *for solving a problem involving money deposited in two simple interest accounts*

▶ For each amount invested, use the equation $Pr = I$. Write a numerical or variable expression for the principal, the interest rate, and the interest earned. The results can be recorded in a table.

The sum of the amounts invested is $10,000.

Amount invested at 7%: x
Amount invested at 8%: $\$10,000 - x$

Take Note

Use the information given in the problem to fill in the principal and interest rate columns of the table. Fill in the interest earned column by multiplying the two expressions you wrote in each row.

	Principal, P	·	Interest rate, r	=	Interest earned, I
Amount at 7%	x	·	0.07	=	$0.07x$
Amount at 8%	$10,000 - x$	·	0.08	=	$0.08(10,000 - x)$

▶ Determine how the amounts of interest earned on the individual investments are related. For example, the total interest earned by both accounts may be known, or it may be known that the interest earned on one account is equal to the interest earned on the other account.

The sum of the interest earned by the two investments equals the total annual interest earned ($785).

$$0.07x + 0.08(10,000 - x) = 785$$
$$0.07x + 800 - 0.08x = 785$$
$$-0.01x + 800 = 785$$
$$-0.01x = -15$$
$$x = 1500$$

- The interest earned on the 7% account plus the interest earned on the 8% account equals the total annual interest earned.

$$10,000 - x = 10,000 - 1500 = 8500$$

- Substitute the value of x into the variable expression for the amount invested at 8%.

The amount invested at 7% is $1500.
The amount invested at 8% is $8500.

Example 1 An investment counselor invested 75% of a client's money into a 9% annual simple interest money market fund. The remainder was invested in 6% annual simple interest government securities. Find the amount invested in each if the total annual interest earned is $3300.

Strategy ▶ Amount invested: x
Amount invested at 9%: $0.75x$
Amount invested at 6%: $0.25x$

	Principal	·	Rate	=	Interest
Amount at 9%	$0.75x$	·	0.09	=	$0.09(0.75x)$
Amount at 6%	$0.25x$	·	0.06	=	$0.06(0.25x)$

▶ The sum of the interest earned by the two investments equals the total annual interest earned ($3300).

Solution
$$0.09(0.75x) + 0.06(0.25x) = 3300$$
$$0.0675x + 0.015x = 3300$$
$$0.0825x = 3300$$

- The interest earned on the 9% account plus the interest earned on the 6% account equals the total annual interest earned.

$$x = 40,000$$

- The amount invested is $40,000.

$$0.75x = 0.75(40,000) = 30,000$$

- Find the amount invested at 9%.

$$0.25x = 0.25(40,000) = 10,000$$

- Find the amount invested at 6%.

The amount invested at 9% is $30,000.
The amount invested at 6% is $10,000.

Problem 1 An investment of $2500 is made at an annual simple interest rate of 7%. How much additional money must be invested at 10% so that the total interest earned will be 9% of the total investment?

Solution See page S9.

4.5 Concept Review

Determine whether the statement is always true, sometimes true, or never true.

1. In the simple interest formula $I = Pr$, the variable I represents the simple interest earned.

2. For one year, you have $1000 deposited in an account that pays 5% annual simple interest. You will earn exactly $5 in simple interest on this account.

3. For one year, you have x dollars deposited in an account that pays 7% annual simple interest. You will earn $0.07x$ in simple interest on this account.

4. If you have a total of $8000 deposited in two accounts and you represent the amount you have in the first account as x, then the amount in the second account is represented as $8000 - x$.

5. The amount of interest earned on one account is $0.05x$ and the amount of interest earned on a second account is $0.08(9000 - x)$. If the two accounts earn the same amount of interest, then we can write the equation $0.05x + 0.08(9000 - x)$.

6. If the amount of interest earned on one account is $0.06x$ and the amount of interest earned on a second account is $0.09(4000 - x)$, then the total interest earned on the two accounts can be represented as $0.06x + 0.09(4000 - x)$.

4.5 Exercises

1 Investment Problems

1. A dentist invested a portion of $15,000 in a 7% annual simple interest account and the remainder in a 6.5% annual simple interest government bond. The two investments earn $1020 in interest annually. How much was invested in each account?

2. A university alumni association invested part of $20,000 in preferred stock that earns 8% annual simple interest and the remainder in a municipal bond that earns 7% annual simple interest. The amount of interest earned each year is $1520. How much was invested in each account?

3. A professional athlete deposited an amount of money into a high-yield mutual fund that earns 13% annual simple interest. A second deposit, $2500 more than the first, was placed in a certificate of deposit earning 7% annual simple interest. In one year, the total interest earned on both investments was $475. How much money was invested in the mutual fund?

4. Jan Moser made a deposit into a 7% annual simple interest account. She made another deposit, $1500 less than the first, in a certificate of deposit earning 9% annual simple interest. The total interest earned on both investments for one year was $505. How much money was deposited in the certificate of deposit?

5. A team of cancer research specialists received a grant of $300,000 to be used for cancer research. They deposited some of the money in a 10% simple interest account and the remainder in an 8.5% annual simple interest account. How much was deposited in each account if the annual interest is $28,500?

6. Reggie Means invested part of $30,000 in municipal bonds that earn 6.5% annual simple interest and the remainder of the money in 8.5% corporate bonds. How much is invested in each account if the total annual interest earned is $2190?

7. To provide for retirement income, Teresa Puelo purchases a $5000 bond that earns 7.5% annual simple interest. How much money does Teresa have invested in bonds that earn 8% annual simple interest if the total annual interest earned from the two investments is $615?

8. After the sale of some income property, Jeremy Littlefield invested $40,000 in a certificate of deposit that earns 7.25% annual simple interest. How much money does he have invested in certificates that earn an annual simple interest rate of 8.5% if the total annual interest earned from the two investments is $5025?

9. Suki Hiroshi has made an investment of $2500 at an annual simple interest rate of 7%. How much money does she have invested at an annual simple interest rate of 11% if the total interest earned is 9% of the total investment?

10. Mae Jackson has a total of $6000 invested in two simple interest accounts. The annual simple interest rate on one account is 9%. The annual simple interest rate on the second account is 6%. How much is invested in each account if both accounts earn the same amount of interest?

11. A charity deposited a total of $54,000 in two simple interest accounts. The annual simple interest rate on one account is 8%. The annual simple interest rate on the second account is 12%. How much was invested in each account if the total annual interest earned is 9% of the total investment?

12. A sports foundation deposited a total of $24,000 into two simple interest accounts. The annual simple interest rate on one account is 7%. The annual simple interest rate on the second account is 11%. How much is invested in each account if the total annual interest earned is 10% of the total investment?

13. Wayne Miller, an investment banker, invested 55% of the bank's available cash in an account that earns 8.25% annual simple interest. The remainder of the cash was placed in an account that earns 10% annual simple interest. The interest earned in one year was $58,743.75. Find the total amount invested.

14. Mohammad Aran, a financial planner, recommended that 40% of a client's cash account be invested in preferred stock earning 9% annual simple interest. The remainder of the client's cash was placed in Treasury bonds earning 7% annual simple interest. The total annual interest earned from the two investments was $2496. Find the total amount invested.

15. Sarah Ontkean is the manager of a mutual fund. She placed 30% of the fund's available cash in a 6% annual simple interest account, 25% in 8% corporate bonds, and the remainder in a money market fund earning 7.5% annual simple interest. The total annual interest earned from the investments was $35,875. Find the total amount invested.

16. Joseph Abruzzio is the manager of a trust. He invested 30% of a client's cash in government bonds that earn 6.5% annual simple interest, 30% in utility stocks that earn 7% annual simple interest, and the remainder in an account that earns 8% annual simple interest. The total annual interest earned from the investments was $5437.50. Find the total amount invested.

4.5   Applying Concepts

17. A sales representative invests in a stock paying 9% dividends. A research consultant invests $5000 more than the sales representative in bonds paying 8% annual simple interest. The research consultant's income from the investment is equal to the sales representative's. Find the amount of the research consultant's investment.

18. A financial manager invested 20% of a client's money in bonds paying 9% annual simple interest, 35% in an 8% simple interest account, and the remainder in 9.5% corporate bonds. Find the amount invested in each if the total annual interest earned is $5325.

19. A plant manager invested $3000 more in stocks than in bonds. The stocks paid 8% annual simple interest, and the bonds paid 9.5% annual simple interest. Both investments yielded the same income. Find the total annual interest received on both investments.

20. A bank offers a customer a 2-year certificate of deposit (CD) that earns 8% compound annual interest. This means that the interest earned each year is added to the principal before the interest for the next year is calculated. Find the value in 2 years of a nurse's investment of $2500 in this CD.

21. A bank offers a customer a 3-year certificate of deposit (CD) that earns 8.5% compound annual interest. This means that the interest earned each year is added to the principal before the interest for the next year is calculated. Find the value in 3 years of an accountant's investment of $3000 in this CD.

22. Write an essay on the topic of annual percentage rates.

23. Financial advisors may predict how much money we should have saved for retirement by the ages of 35, 45, 55, and 65. One such prediction is included in the table below.

a. According to the estimates in the table, how much should a couple who have earnings of $75,000 have saved for retirement by age 55?

b. Write an explanation of how interest and interest rates affect the level of savings required at any given age. What effect do inflation rates have on savings?

Minimum Levels of Savings Required for Married Couples to Be Prepared for Retirement *Savings Accumulation by Age*			
	35	45	55	65
Earnings = $50,000	8,000	23,000	90,000	170,000
Earnings = $75,000	17,000	60,000	170,000	310,000
Earnings = $100,000	34,000	110,000	280,000	480,000
Earnings = $150,000	67,000	210,000	490,000	840,000

4.6 Mixture Problems

1 Value mixture problems

A value mixture problem involves combining two ingredients that have different prices into a single blend. For example, a coffee merchant may blend two types of coffee into a single blend, or a candy manufacturer may combine two types of candy to sell as a "variety pack."

The solution of a value mixture problem is based on the equation $V = AC$, where V is the value of an ingredient, A is the amount of the ingredient, and C is the cost per unit of the ingredient.

Find the value of 5 oz of a gold alloy that costs $185 per ounce.

Given: $A = 5$ oz $\qquad$ $V = AC$
$\qquad\quad$ $C = \$185$ $\qquad$ $V = 5(185)$
Unknown value: V $\qquad$ $V = 925$

The value of the 5 oz of gold alloy is $925.

Solve: A coffee merchant wants to make 9 lb of a blend of coffee costing $6 per pound. The blend is made using a $7 grade and a $4 grade of coffee. How many pounds of each of these grades should be used?

STRATEGY | for solving a mixture problem

▶ For each ingredient in the mixture, write a numerical or variable expression for the amount of the ingredient used, the unit cost of the ingredient, and the value of the amount used. For the blend, write a numerical or variable expression for the amount, the unit cost of the blend, and the value of the amount. The results can be recorded in a table.

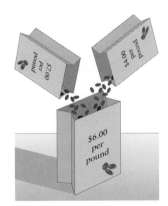

The sum of the amounts is 9 lb.

Amount of $7 coffee: x

Amount of $4 coffee: $9 - x$

Take Note

Use the information given in the problem to fill in the amount and unit cost columns of the table. Fill in the value column by multiplying the two expressions you wrote in each row. Use the expressions in the last column to write the equation.

	Amount, A	·	Unit cost, C	=	Value, V
$7 grade	x	·	7	=	$7x$
$4 grade	$9 - x$	·	4	=	$4(9 - x)$
$6 blend	9	·	6	=	$6(9)$

▶ Determine how the values of the individual ingredients are related. Use the fact that the sum of the values of these ingredients is equal to the value of the blend.

The sum of the values of the $7 grade and the $4 grade is equal to the value of the $6 blend.

$$7x + 4(9 - x) = 6(9)$$
$$7x + 36 - 4x = 54$$
$$3x + 36 = 54$$
$$3x = 18$$
$$x = 6$$

• The value of the $7 grade plus the value of the $4 grade equals the value of the blend.

$$9 - x = 9 - 6 = 3$$

• Substitute the value of x into the variable expression for the amount of $4 grade.

The merchant must use 6 lb of the $7 coffee and 3 lb of the $4 coffee.

Example 1   How many ounces of a silver alloy that costs $6 per ounce must be mixed with 10 oz of a silver alloy that costs $8 per ounce to make a mixture that costs $6.50 per ounce?

Strategy   ▶ Ounces of $6 alloy: x

	Amount	Cost	Value
$6 alloy	x	6	$6x$
$8 alloy	10	8	$8(10)$
$6.50 mixture	$10 + x$	6.50	$6.50(10 + x)$

▶ The sum of the values before mixing equals the value after mixing.

Solution
$$6x + 8(10) = 6.50(10 + x)$$
$$6x + 80 = 65 + 6.5x$$
$$-0.5x + 80 = 65$$
$$-0.5x = -15$$
$$x = 30$$

• The value of the $6 alloy plus the value of the $8 alloy equals the value of the mixture.

30 oz of the $6 silver alloy must be used.

Problem 1   A gardener has 20 lb of a lawn fertilizer that costs $.90 per pound. How many pounds of a fertilizer that costs $.75 per pound should be mixed with this 20 lb of lawn fertilizer to produce a mixture that costs $.85 per pound?

Solution   See page S9.

2 # Percent mixture problems

Recall that the amount of a substance in a solution can be given as a percent of the total solution. For example, in a 5% saltwater solution, 5% of the total solution is salt. The remaining 95% is water.

The solution of a percent mixture problem is based on the equation $Q = Ar$, where Q is the quantity of a substance in the solution, r is the percent of concentration, and A is the amount of solution.

A 500-milliliter bottle contains a 3% solution of hydrogen peroxide. Find the amount of hydrogen peroxide in the solution.

Given: $A = 500$ $Q = Ar$
            $r = 3\% = 0.03$ $Q = 500(0.03)$
Unknown amount: Q $Q = 15$

The bottle contains 15 ml of hydrogen peroxide.

Solve: How many gallons of 15% salt solution must be mixed with 4 gal of a 20% salt solution to make a 17% salt solution?

STRATEGY | *for solving a percent mixture problem*

▶ For each solution, use the equation $Ar = Q$. Write a numerical or variable expression for the amount of solution, the percent of concentration, and the quantity of the substance in the solution. The results can be recorded in a table.

The unknown quantity of 15% solution: x

Take Note

Use the information given in the problem to fill in the amount and percent columns of the table. Fill in the quantity column by multiplying the two expressions you wrote in each row. Use the expressions in the last column to write the equation.

	Amount of solution, A	·	Percent of concentration, r	=	Quantity of substance, Q
15% solution	x	·	0.15	=	$0.15x$
20% solution	4	·	0.20	=	$0.20(4)$
17% solution	$x + 4$	·	0.17	=	$0.17(x + 4)$

▶ Determine how the quantities of the substance in the individual solutions are related. Use the fact that the sum of the quantities of the substances being mixed is equal to the quantity of the substance after mixing.

The sum of the quantities of salt in the 15% solution and in the 20% solution is equal to the quantity of salt in the 17% solution.

$0.15x + 0.20(4) = 0.17(x + 4)$ • The amount of salt in the 15% solution plus the
      $0.15x + 0.8 = 0.17x + 0.68$ amount of salt in the 20% solution equals the
    $-0.02x + 0.8 = 0.68$ amount of salt in the 17% solution.
              $-0.02x = -0.12$
                      $x = 6$

6 gal of the 15% solution are required.

Example 2 A chemist wishes to make 3 L of a 7% acid solution by mixing a 9% acid solution and a 4% acid solution. How many liters of each solution should the chemist use?

Strategy   ▶ Liters of 9% solution: x
Liters of 4% solution: $3 - x$

	Amount	Percent	Quantity
9%	x	0.09	$0.09x$
4%	$3 - x$	0.04	$0.04(3 - x)$
7%	3	0.07	$0.07(3)$

▶ The sum of the quantities before mixing is equal to the quantity after mixing.

Solution
$$0.09x + 0.04(3 - x) = 0.07(3)$$
$$0.09x + 0.12 - 0.04x = 0.21$$
$$0.05x + 0.12 = 0.21$$
$$0.05x = 0.09$$
$$x = 1.8$$

• The amount of acid in the 9% solution plus the amount of acid in the 4% solution equals the amount of acid in the 7% solution.
• 1.8 L of the 9% solution are needed.

$$3 - x = 3 - 1.8 = 1.2$$

• Find the amount of the 4% solution needed.

The chemist needs 1.8 L of the 9% solution and 1.2 L of the 4% solution.

Problem 2   How many quarts of pure orange juice are added to 5 qt of fruit drink that is 10% orange juice to make an orange drink that is 25% orange juice?

Solution   See pages S9–S10.

4.6   Concept Review

Determine whether the statement is always true, sometimes true, or never true.

1. Both value mixture and percent mixture problems involve combining two or more ingredients into a single substance.

2. In the value mixture equation $V = AC$, the variable A represents the quantity of an ingredient.

3. Suppose we are mixing two salt solutions. Then the variable Q in the percent mixture equation $Q = Ar$ represents the amount of salt in a solution.

4. If we combine an alloy that costs $8 an ounce with an alloy that costs $5 an ounce, the cost of the resulting mixture will be greater than $8 an ounce.

5. If we combine a 9% acid solution with a solution that is 4% acid, the resulting solution will be less than 4% acid.

6. If 8 L of a solvent costs $75 per liter, then the value of the 8 L of solvent is $600.

7. If 100 oz of a silver alloy is 25% silver, then the alloy contains 25 oz of silver.

4.6 | **Exercises**

1 Value Mixture Problems

1. Explain the meaning of each variable in the equation $V = AC$. Give an example of how this equation is used.

2. Suppose you are mixing peanuts that cost $4 per pound with raisins that cost $2 per pound. In the context of this situation, explain the meaning of the statement "The sum of the values of these ingredients is equal to the value of the blend."

3. At a veterinary clinic, a special high-protein dog food that costs $6.75 per pound is mixed with a vitamin supplement that costs $3.25 per pound. How many pounds of each should be used to make 5 lb of a mixture that costs $4.65 per pound?

4. A wild bird seed mix is made by combining 100 lb of millet seed costing $.60 per pound with sunflower seeds costing $1.10 per pound. How many pounds of sunflower seeds are needed to make a mixture that costs $.70 per pound?

5. Find the cost per pound of a coffee mixture made from 8 lb of coffee that costs $9.20 per pound and 12 lb of coffee that costs $5.50 per pound.

6. How many pounds of chamomile tea that cost $4.20 per pound must be mixed with 12 lb of orange tea that cost $2.25 per pound to make a mixture that costs $3.40 per pound?

7. A goldsmith combined an alloy that costs $4.30 per ounce with an alloy that costs $1.80 per ounce. How many ounces of each were used to make a mixture of 200 oz costing $2.50 per ounce?

8. An herbalist has 30 oz of herbs costing $2 per ounce. How many ounces of herbs costing $1 per ounce should be mixed with these 30 oz of herbs to produce a mixture costing $1.60 per ounce?

9. A health food store has a section in which you can create your own chocolate-covered fruit mixes. You choose 2 lb of the cherries that cost $4.50 per pound. You also choose blueberries that cost $3.75 per pound. How many pounds of blueberries can you add to create a mixture that costs $4.25 per pound?

10. Find the cost per ounce of a mixture of 200 oz of a cologne that costs $7.50 per ounce and 500 oz of a cologne that costs $4.00 per ounce.

11. How many kilograms of hard candy that cost $7.50 per kilogram must be mixed with 24 kg of jelly beans that cost $3.25 per kilogram to make a mixture that costs $4.50 per kilogram?

12. A grocery store offers a cheese sampler that includes a pepper cheddar cheese that costs $16 per kilogram and Pennsylvania Jack that costs $6 per kilogram. How many kilograms of each were used to make a 5-kilogram mixture that costs $9 per kilogram?

13. A snack food is made by mixing 5 lb of popcorn that costs $.80 per pound with caramel that costs $2.40 per pound. How much caramel is needed to make a mixture that costs $1.40 per pound?

14. A lumber company combined oak wood chips that cost $3.10 per pound with pine wood chips that cost $2.50 per pound. How many pounds of each were used to make an 80-pound mixture costing $2.65 per pound?

15. The manager of a farmer's market has 500 lb of grain that costs $1.20 per pound. How many pounds of meal costing $.80 per pound should be mixed with the 500 lb of grain to produce a mixture that costs $1.05 per pound?

16. A caterer made an ice cream punch by combining fruit juice that costs $4.50 per gallon with ice cream that costs $6.50 per gallon. How many gallons of each were used to make 100 gal of punch costing $5 per gallon?

17. The manager of a specialty food store combined almonds that cost $4.50 per pound with walnuts that cost $2.50 per pound. How many pounds of each were used to make a 100-pound mixture that costs $3.24 per pound?

18. Find the cost per pound of a "house blend" of coffee that is made from 12 lb of Central American coffee that costs $8 per pound and 30 lb of South American coffee that costs $4.50 per pound.

19. Find the cost per pound of a sugar-coated breakfast cereal made from 40 lb of sugar that cost $2.00 per pound and 120 lb of corn flakes that cost $1.20 per pound.

20. Find the cost per ounce of a gold alloy made from 25 oz of pure gold that costs $482 per ounce and 40 oz of an alloy that costs $300 per ounce.

21. Find the cost per ounce of a sunscreen made from 100 oz of a lotion that costs $2.50 per ounce and 50 oz of a lotion that costs $4.00 per ounce.

22. How many liters of a blue dye that costs $1.60 per liter must be mixed with 18 L of anil that costs $2.50 per liter to make a mixture that costs $1.90 per liter?

2 Percent Mixture Problems

23. Explain the meaning of each variable in the equation $Q = Ar$. Give an example of how this equation is used.

24. Suppose you are mixing a 5% acid solution with a 10% acid solution. In the context of this situation, explain the meaning of the statement "The sum of the quantities of the substances being mixed is equal to the quantity of the substance after mixing."

25. Forty ounces of a 30% gold alloy are mixed with 60 oz of a 20% gold alloy. Find the percent concentration of the resulting gold alloy.

26. One hundred ounces of juice that is 50% tomato juice are added to 200 oz of a vegetable juice that is 25% tomato juice. What is the percent concentration of tomato juice in the resulting mixture?

27. How many gallons of a 15% acid solution must be mixed with 5 gal of a 20% acid solution to make a 16% acid solution?

28. How many pounds of a chicken feed that is 50% corn must be mixed with 400 lb of a feed that is 80% corn to make a chicken feed that is 75% corn?

29. A rug is made by weaving 20 lb of yarn that is 50% wool with a yarn that is 25% wool. How many pounds of the yarn that is 25% wool must be used if the finished rug is to be 35% wool?

30. Five gallons of a dark green latex paint that is 20% yellow paint are combined with a lighter green latex paint that is 40% yellow paint. How many gallons of the lighter green paint must be used to create a green paint that is 25% yellow paint?

31. How many gallons of a plant food that is 9% nitrogen must be combined with another plant food that is 25% nitrogen to make 10 gal of a plant food that is 15% nitrogen?

32. A chemist wants to make 50 ml of a 16% acid solution by mixing a 13% acid solution and an 18% acid solution. How many milliliters of each solution should the chemist use?

33. Five grams of sugar are added to a 45-gram serving of a breakfast cereal that is 10% sugar. What is the percent concentration of sugar in the resulting mixture?

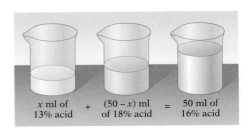

x ml of   + $(50 - x)$ ml   =   50 ml of
13% acid   of 18% acid   16% acid

34. Thirty ounces of pure silver are added to 50 oz of a silver alloy that is 20% silver. What is the percent concentration of the resulting alloy?

35. To make the potpourri mixture sold at a florist shop, 70 oz of a potpourri that is 80% lavender are combined with a potpourri that is 60% lavender. The resulting potpourri is 74% lavender. How much of the potpourri that is 60% lavender is used?

36. The manager of a garden shop mixes grass seed that is 40% rye grass with 40 lb of grass seed that is 60% rye grass to make a mixture that is 56% rye grass. How much of the 40% rye grass is used?

37. A hair dye is made by blending a 7% hydrogen peroxide solution and a 4% hydrogen peroxide solution. How many milliliters of each are used to make a 300-milliliter solution that is 5% hydrogen peroxide?

38. At a cosmetics company, 40 L of pure aloe cream are mixed with 50 L of a moisturizer that is 64% aloe. What is the percent concentration of aloe in the resulting mixture?

39. How many ounces of pure chocolate must be added to 150 oz of chocolate topping that is 50% chocolate to make a topping that is 75% chocolate?

40. How many ounces of pure bran flakes must be added to 50 oz of cereal that is 40% bran flakes to produce a mixture that is 50% bran flakes?

41. A hair stylist combines 12 oz of shampoo that is 20% conditioner with an 8-ounce bottle of pure shampoo. What is the percent concentration of conditioner in the 20-ounce mixture?

42. A clothing manufacturer has some pure silk thread and some thread that is 85% silk. How many kilograms of each must be woven together to make 75 kg of cloth that is 96% silk?

43. How many ounces of dried apricots must be added to 18 oz of a snack mix that contains 20% dried apricots to make a mixture that is 25% dried apricots?

44. A recipe for a rice dish calls for 12 oz of a rice mixture that is 20% wild rice and 8 oz of pure wild rice. What is the percent concentration of wild rice in the 20-ounce mixture?

4.6   Applying Concepts

45. Find the cost per ounce of a mixture of 30 oz of an alloy that costs $4.50 per ounce, 40 oz of an alloy that costs $3.50 per ounce, and 30 oz of an alloy that costs $3.00 per ounce.

46. A grocer combined walnuts that cost $2.60 per pound and cashews that cost $3.50 per pound with 20 lb of peanuts that cost $2.00 per pound. Find the amount of walnuts and the amount of cashews used to make the 50-pound mixture costing $2.72 per pound.

47. How many ounces of water evaporated from 50 oz of a 12% salt solution to produce a 15% salt solution?

48. A chemist mixed pure acid with water to make 10 L of a 30% acid solution. How much pure acid and how much water did the chemist use?

49. How many grams of pure water must be added to 50 g of pure acid to make a solution that is 40% acid?

50. A radiator contains 15 gal of a 20% antifreeze solution. How many gallons must be drained from the radiator and replaced by pure antifreeze so that the radiator will contain 15 gal of a 40% antifreeze solution?

51. Tickets to a performance by the community theater cost $5.50 for adults and $2.75 for children. A total of 120 tickets were sold for $563.75. How many adults and how many children attended the performance?

52. Explain why we look for patterns and relationships in mathematics. Include a discussion of the relationship between value mixture problems and percent mixture problems and how understanding one of these can make it easier to understand the other. Also discuss why understanding how to solve the value mixture problems in this section can be helpful in solving Exercise 51.

4.7   Uniform Motion Problems

1   **Uniform motion problems**

A train that travels constantly in a straight line at 50 mph is in *uniform motion*. Recall that **uniform motion** means the speed of an object does not change.

The solution of a uniform motion problem is based on the equation $d = rt$, where d is the distance traveled, r is the rate of travel, and t is the time spent traveling.

A car travels at 50 mph for 3 h.

$$d = rt$$
$$d = 50(3)$$
$$d = 150$$

The car travels a distance of 150 mi.

Solve: A car leaves a town traveling at 35 mph. Two hours later, a second car leaves the same town, on the same road, traveling at 55 mph. In how many hours after the second car leaves will the second car pass the first car?

STRATEGY | *for solving a uniform motion problem*

▶ For each object, use the equation $rt = d$. Write a numerical or variable expression for the rate, time, and distance. The results can be recorded in a table.

The first car traveled 2 h longer than the second car.

Unknown time for the second car: t
Time for the first car: $t + 2$

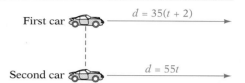

First car $d = 35(t + 2)$

Second car $d = 55t$

	Rate, r	·	Time, t	=	Distance, d
First car	35	·	$t + 2$	=	$35(t + 2)$
Second car	55	·	t	=	$55t$

▶ Determine how the distances traveled by the individual objects are related. For example, the total distance traveled by both objects may be known, or it may be known that the two objects traveled the same distance.

The two cars travel the same distance.

$$35(t + 2) = 55t$$
$$35t + 70 = 55t$$
$$70 = 20t$$
$$3.5 = t$$

• The distance traveled by the first car equals the distance traveled by the second car.

The second car will pass the first car in 3.5 h.

Example 1 Two cars, one traveling 10 mph faster than the other, start at the same time from the same point and travel in opposite directions. In 3 h, they are 288 mi apart. Find the rate of the second car.

Strategy ▶ Rate of second car: r
Rate of first car: $r + 10$

	Rate	Time	Distance
First car	$r + 10$	3	$3(r + 10)$
Second car	r	3	$3r$

▶ The total distance traveled by the two cars is 288 mi.

First car $d = 3(r + 10)$ $d = 3r$ Second car
← 288 mi →

Solution $3(r + 10) + 3r = 288$ • The distance traveled by the first car plus
         $3r + 30 + 3r = 288$ the distance traveled by the second car is
              $6r + 30 = 288$ **288 mi.**
                   $6r = 258$
                     $r = 43$

The second car is traveling 43 mph.

Problem 1   Two trains, one traveling at twice the speed of the other, start at the same time from stations that are 306 mi apart and travel toward each other. In 3 h, the trains pass each other. Find the rate of each train.

Solution   See page S10.

Example 2   A bicycling club rides out into the country at a speed of 16 mph and returns over the same road at 12 mph. How far does the club ride out into the country if it travels a total of 7 h?

Strategy   ▶ Time spent riding out: t
         Time spent riding back: $7 - t$

	Rate	Time	Distance
Out	16	t	$16t$
Back	12	$7 - t$	$12(7 - t)$

▶ The distance out equals the distance back.

$d = 16t$

$d = 12(7 - t)$

Solution $16t = 12(7 - t)$
         $16t = 84 - 12t$
         $28t = 84$
            $t = 3$ • The time is 3 h. Find the distance.

The distance out $= 16t = 16(3) = 48$.

The club rides 48 mi into the country.

Problem 2   On a survey mission, a pilot flew out to a parcel of land and back in 7 h. The rate out was 120 mph. The rate back was 90 mph. How far away was the parcel of land?

Solution   See page S10.

4.7 | Concept Review

Determine whether the statement is always true, sometimes true, or never true.

1. The solution of a uniform motion problem is based on the equation $d = rt$, where r is a percent.

2. If you drive at a constant speed of 60 mph for 4 h, you will travel a total distance of 240 mi.

3. It takes a student a total of 5 h to drive to a friend's house and back. If the variable t is used to represent the time it took to drive to the friend's house, then the expression $t - 5$ is used to represent the time spent on the return trip.

4. If the speed of one train is 20 mph slower than that of a second train, then the speeds of the two trains can be represented as r and $20 - r$.

5. A car and a bus both travel from city A to city B. The car leaves city A 1 h later than the bus, but both vehicles arrive in city B at the same time. If t represents the time the car was traveling, then $t + 1$ represents the time the bus was traveling. If t represents the time the bus was traveling, then $t - 1$ represents the time the car was traveling.

6. Two cars travel the same distance. If the distance traveled by the first car is represented by the expression $35t$, and the distance traveled by the second car is represented by the expression $45(t + 1)$, then we can write the equation $35t = 45(t + 1)$.

4.7 | Exercises

1 Uniform Motion Problems

1. Two planes start from the same point and fly in opposite directions. The first plane is flying 25 mph slower than the second plane. In 2 h, the planes are 470 mi apart. Find the rate of each plane.

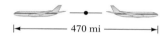
470 mi

2. Two cyclists start from the same point and ride in opposite directions. One cyclist rides twice as fast as the other. In 3 h, they are 81 mi apart. Find the rate of each cyclist.

3. One speed skater starts across a frozen lake at an average speed of 8 meters per second. Ten seconds later, a second speed skater starts from the same point and skates in the same direction at an average speed of 10 meters per second. How many seconds after the second skater starts will the second skater overtake the first skater?

4. A long-distance runner started on a course running at an average speed of 6 mph. Half an hour later, a second runner began the same course at an average speed of 7 mph. How long after the second runner starts will the second runner overtake the first runner?

5. Michael Chan leaves a dock in his motorboat and travels at an average speed of 9 mph toward the Isle of Shoals, a small island off the coast of Massachusetts. Two hours later a tour boat leaves the same dock and travels at an average speed of 18 mph toward the same island. How many hours after the tour boat leaves will Michael's boat be alongside the tour boat?

6. A 555-mile, 5-hour plane trip was flown at two speeds. For the first part of the trip, the average speed was 105 mph. For the remainder of the trip, the average speed was 115 mph. How long did the plane fly at each speed?

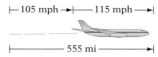

7. An executive drove from home at an average speed of 30 mph to an airport where a helicopter was waiting. The executive boarded the helicopter and flew to the corporate offices at an average speed of 60 mph. The entire distance was 150 mi. The entire trip took 3 h. Find the distance from the airport to the corporate offices.

8. A jogger starts from one end of a 15-mile nature trail at 8:00 A.M. One hour later, a cyclist starts from the other end of the trail and rides toward the jogger. If the rate of the jogger is 6 mph and the rate of the cyclist is 9 mph, at what time will the two meet?

9. After a sailboat had been on the water for 3 h, a change in the wind direction reduced the average speed of the boat by 5 mph. The entire distance sailed was 57 mi. The total time spent sailing was 6 h. How far did the sailboat travel in the first 3 h?

10. A car and a bus set out at 3 P.M. from the same point headed in the same direction. The average speed of the car is twice the average speed of the bus. In 2 h, the car is 68 mi ahead of the bus. Find the rate of the car.

11. A passenger train leaves a train depot 2 h after a freight train leaves the same depot. The freight train is traveling 20 mph slower than the passenger train. Find the rate of each train if the passenger train overtakes the freight train in 3 h.

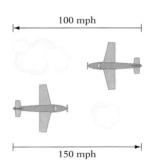

12. A stunt driver was needed at the production site of a Hollywood movie. The average speed of the stunt driver's flight to the site was 150 mph, and the average speed of the return trip was 100 mph. Find the distance of the round trip if the total flying time was 5 h.

13. A ship traveling east at 25 mph is 10 mi from a harbor when another ship leaves the harbor traveling east at 35 mph. How long does it take the second ship to catch up to the first ship?

14. At 10 A.M. a plane leaves Boston, Massachusetts, for Seattle, Washington, a distance of 3000 mi. One hour later a plane leaves Seattle for Boston. Both planes are traveling at a speed of 500 mph. How many hours after the plane leaves Seattle will the planes pass each other?

15. At noon a train leaves Washington, D.C., headed for Charleston, South Carolina, a distance of 500 mi. The train travels at a speed of 60 mph. At 1 P.M. a second train leaves Charleston headed for Washington, D.C., traveling at 50 mph. How long after the train leaves Charleston will the two trains pass each other?

16. A race car driver starts along a 50-mile race course traveling at an average speed of 90 mph. Fifteen minutes later a second driver starts along the same course at an average speed of 120 mph. Will the second car overtake the first car before the drivers reach the end of the course?

17. A bus traveled on a straight road for 2 h at an average speed that was 20 mph faster than its average speed on a winding road. The time spent on the winding road was 3 h. Find the average speed on the winding road if the total trip was 210 mi.

18. A bus traveling at a rate of 60 mph overtakes a car traveling at a rate of 45 mph. If the car had a 1-hour head start, how far from the starting point does the bus overtake the car?

19. A car traveling at 48 mph overtakes a cyclist who, riding at 12 mph, had a 3-hour head start. How far from the starting point does the car overtake the cyclist?

20. A plane left Kennedy Airport Tuesday morning for a 605-mile, 5-hour trip. For the first part of the trip, the average speed was 115 mph. For the remainder of the trip, the average speed was 125 mph. How long did the plane fly at each speed?

4.7   Applying Concepts

21. A car and a cyclist start at 10 A.M. from the same point, headed in the same direction. The average speed of the car is 5 mph more than three times the average speed of the cyclist. In 1.5 h, the car is 46.5 mi ahead of the cyclist. Find the rate of the cyclist.

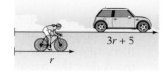

22. A cyclist and a jogger set out at 11 A.M. from the same point, headed in the same direction. The average speed of the cyclist is twice the average speed of the jogger. In 1 h, the cyclist is 7 mi ahead of the jogger. Find the rate of the cyclist.

23. A car and a bus set out at 2 P.M. from the same point, headed in the same direction. The average speed of the car is 30 mph slower than twice the average speed of the bus. In 2 h, the car is 30 mi ahead of the bus. Find the rate of the car.

24. At 10 A.M., two campers left their campsite by canoe and paddled downstream at an average speed of 12 mph. They then turned around and paddled back upstream at an average rate of 4 mph. The total trip took 1 h. At what time did the campers turn around downstream?

25. At 7 A.M., two joggers start from opposite ends of an 8-mile course. One jogger is running at a rate of 4 mph, and the other is running at a rate of 6 mph. At what time will the joggers meet?

26. A truck leaves a depot at 11 A.M. and travels at a speed of 45 mph. At noon, a van leaves the same place and travels the same route at a speed of 65 mph. At what time does the van overtake the truck?

27. A bicyclist rides for 2 h at a speed of 10 mph and then returns at a speed of 20 mph. Find the cyclist's average speed for the trip.

28. A car travels a 1-mile track at an average speed of 30 mph. At what average speed must the car travel the next mile so that the average speed for the 2 mi is 60 mph?

29. Explain why the motion problems in this section are restricted to *uniform* motion.

30. Explain why 60 mph is the same as 88 ft/s.

4.8   Inequalities

1   Applications of inequalities

Solving application problems requires recognition of the verbal phrases that translate into mathematical symbols. Here is a partial list of the phrases used to indicate each of the four inequality symbols.

$<$ is less than	$>$ is greater than
	is more than
	exceeds
$\leq$ is less than or equal to	$\geq$ is greater than or equal to
maximum	minimum
at most	at least
or less	or more

Example 1   A student must have at least 450 points out of 500 points on five tests to receive an A in a course. One student's results on the first four tests were 93, 79, 87, and 94. What scores on the last test will enable this student to receive an A in the course?

Strategy   To find the scores, write and solve an inequality using N to represent the score on the last test.

Solution

| total number of points on the five tests | is greater than or equal to | 450 |

$$93 + 79 + 87 + 94 + N \geq 450$$
$$353 + N \geq 450$$
$$353 - 353 + N \geq 450 - 353$$
$$N \geq 97$$

The student's score on the last test must be equal to or greater than 97.

Problem 1   An appliance dealer will make a profit on the sale of a television set if the cost of the new set is less than 70% of the selling price. What minimum selling price will enable the dealer to make a profit on a television set that costs the dealer $340?

Solution   See page S10.

Example 2   The base of a triangle is 8 in., and the height is $(3x - 5)$ in. Express as an integer the maximum height of the triangle when the area is less than 112 in².

Strategy   To find the maximum height:
▶ Replace the variables in the area formula by the given values and solve for x.
▶ Replace the variable in the expression $3x - 5$ with the value found for x.

Solution

| one-half the base times the height | is less than | 112 in² |

$$\frac{1}{2}(8)(3x - 5) < 112$$
$$4(3x - 5) < 112$$
$$12x - 20 < 112$$
$$12x - 20 + 20 < 112 + 20$$
$$12x < 132$$
$$\frac{12x}{12} < \frac{132}{12}$$
$$x < 11$$

$$3x - 5 = 3(11) - 5 = 28$$

• Substitute the value of x into the variable expression for the height. Note that the height <u>is less than</u> 28 because $x < 11$.

The maximum height of the triangle is 27 in.

Problem 2   Company A rents cars for $9 per day and 10¢ per mile driven. Company B rents cars for $12 per day and 8¢ per mile driven. You want to rent a car for one week. What is the maximum number of miles you can drive a Company A car if it is to cost you less than a Company B car?

Solution   See page S10.

4.8   Concept Review

Determine whether the statement is always true, sometimes true, or never true.

1. Both "is greater than" and "is more than" indicate the inequality symbol $\geq$.

2. A minimum refers to a lower limit, whereas a maximum refers to an upper limit.

3. Given that $x > \frac{32}{6}$, the minimum integer that satisfies the inequality is 6.

4. Given that $x < \frac{25}{4}$, the maximum integer that satisfies the inequality is 7.

5. A rental car costs $15 per day and 20¢ per mile driven. If m represents the number of miles the rental car is driven, then the expression $15 + 0.20m$ represents the cost to rent the car for one week.

6. A patient's physician recommends a maximum cholesterol level of 205 units. The patient's cholesterol level is 199 units and therefore must be lowered 6 units in order to meet the physician's recommended level.

4.8   Exercises

1. **Integer Problems** Three-fifths of a number is greater than two-thirds. Find the smallest integer that satisfies the inequality.

2. **Sports** To be eligible for a basketball tournament, a basketball team must win at least 60% of its remaining games. If the team has 17 games remaining, how many games must the team win to qualify for the tournament?

3. **Income Taxes** To avoid a tax penalty, at least 90% of a self-employed person's total annual income tax liability must be paid by April 15. What amount of income tax must be paid by April 15 by a person with an annual income tax liability of $3500?

4. **Health** A health official recommends a maximum cholesterol level of 220 units. A patient has a cholesterol level of 275. By how many units must this patient's cholesterol level be reduced to satisfy the recommended maximum level?

5. Recycling A service organization will receive a bonus of $200 for collecting more than 1850 lb of aluminum cans during its four collection drives. On the first three drives, the organization collected 505 lb, 493 lb, and 412 lb. How many pounds of cans must the organization collect on the fourth drive to receive the bonus?

6. Education A professor scores all tests with a maximum of 100 points. To earn an A in this course, a student must have an average of at least 92 on four tests. One student's grades on the first three tests were 89, 86, and 90. Can this student earn an A grade?

7. Education A student must have an average of at least 80 points on five tests to receive a B in a course. The student's grades on the first four tests were 75, 83, 86, and 78. What scores on the last test will enable this student to receive a B in the course?

8. Compensation A car sales representative receives a commission that is the greater of $250 or 8% of the selling price of a car. What dollar amounts in the sale price of a car will make the commission offer more attractive than the $250 flat fee?

9. Compensation A sales representative for a stereo store has the option of a monthly salary of $2000 or a 35% commission on the selling price of each item sold by the representative. What dollar amounts in sales will make the commission more attractive than the monthly salary?

10. Integer Problem Four times the sum of a number and five is less than six times the number. Find the smallest integer that satisfies the inequality.

11. Compensation The sales agent for a jewelry company is offered a flat monthly salary of $3200 or a salary of $1000 plus an 11% commission on the selling price of each item sold by the agent. If the agent chooses the $3200 salary, what dollar amount does the agent expect to sell in one month?

12. Compensation A baseball player is offered an annual salary of $200,000 or a base salary of $100,000 plus a bonus of $1000 for each hit over 100 hits. How many hits must the baseball player make to earn more than $200,000?

13. Consumerism A computer bulletin board service charges a flat fee of $10 per month or $4 per month plus $.10 for each minute the service is used. For how many minutes must a person use this service for the cost to exceed $10?

14. Business A site licensing fee for a computer program is $1500. The fee allows a company to use the program at any computer terminal within the company. Alternatively, a company can choose to pay $200 for each computer terminal it has. How many computer terminals must a company have for the site licensing fee to be the more economical plan?

15. Nutrition For a product to be labeled orange juice, a state agency requires that at least 80% of the drink be real orange juice. How many ounces of artificial flavors can be added to 32 oz of real orange juice if the drink is to be labeled orange juice?

16. Nutrition Grade A hamburger cannot contain more than 20% fat. How much fat can a butcher mix with 300 lb of lean meat to meet the 20% requirement?

17. Consumerism A shuttle service taking skiers to a ski area charges $8 per person each way. Four skiers are debating whether to take the shuttle bus or rent a car for $45 plus $.25 per mile. Assuming that the skiers will share the cost of the car and that they want the least expensive method of transportation, how far away is the ski area if they choose to take the shuttle service?

18. Utilities A residential water bill is based on a flat fee of $10 plus a charge of $.75 for each 1000 gal of water used. Find the number of gallons of water a family can use and have a monthly water bill that is less than $55.

19. Consumerism Company A rents cars for $25 per day and 8¢ per mile driven. Company B rents cars for $15 per day and 14¢ per mile driven. You want to rent a car for one day. Find the maximum number of miles you can drive a Company B car if it is to cost you less than a Company A car.

20. Geometry A rectangle is 8 ft wide and $(2x + 7)$ ft long. Express as an integer the maximum length of the rectangle when the area is less than 152 ft^2.

4.8   **Applying Concepts**

21. Integer Problems Find three positive consecutive odd integers such that three times the sum of the first two integers is less than four times the third integer. (*Hint:* There is more than one solution.)

22. Integer Problems Find three positive consecutive even integers such that four times the sum of the first two integers is less than or equal to five times the third integer. (*Hint:* There is more than one solution.)

23. Aviation A maintenance crew requires between 30 min and 45 min to prepare an aircraft for its next flight. How many aircraft can this crew prepare for flight in a 6-hour period of time?

24. ⬤ Postage The postage rate for first-class mail is $.37 for the first ounce and $.23 for each additional ounce or fraction of an ounce. Find the largest whole number of ounces a package can weigh if it is to cost you less than $3 to mail the package.

25. Fund Raising Your class decides to publish a calendar to raise money. The initial cost, regardless of the number of calendars printed, is $900. After the initial cost, each calendar costs $1.50 to produce. What is the minimum number of calendars your class must sell at $6 per calendar to make a profit of at least $1200?

26. ⬤ Debt Carried by Americans During a recent year, the mean (average) household income of Americans under age 35 was $48,297. The mean household income of Americans aged 35 to 50 was $60,427. The graph below shows average amounts of debt carried during the same year by those under 35 and by those aged 35 to 50.
   a. In which categories is the average debt carried by Americans under age 35 greater than the average debt carried by Americans aged 35 to 50?
   b. In which categories is the average debt carried by Americans under age 35 greater than $10,000?
   c. For the categories shown in the graph, what was the total average debt of Americans under age 35? How much greater is this debt than that of Americans aged 35 to 50?
   d. ✎ Given the answers to parts a, b, and c, what is the significance of the average incomes of these two age groups?
   e. ✎ Write a paragraph describing any other observations you can make from the data.

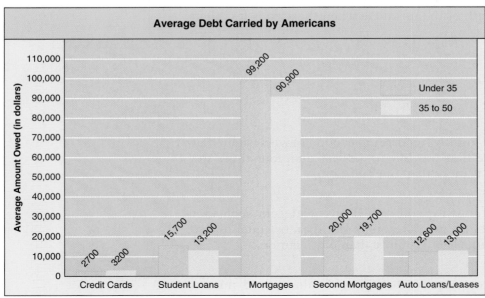

Source: Claritas, Inc.

Focus on Problem Solving

Trial-and-Error Approach to Problem Solving

The questions below require an answer of always true, sometimes true, or never true. These problems are best solved by the trial-and-error method. The trial-and-error method of arriving at a solution to a problem involves repeated tests or experiments.

For example, consider the statement

> Both sides of an equation can be divided by the same number without changing the solution of the equation.

The solution of the equation $6x = 18$ is 3. If we divide both sides of the equation by 2, the result is $3x = 9$ and the solution is still 3. Thus the answer "never true" has been eliminated. We still need to determine whether there is a case for which the statement is not true. Is there a number by which we could divide both sides of the equation and the result would be an equation for which the solution is not 3? If we divide both sides of the equation by 0, the result is $\frac{6x}{0} = \frac{18}{0}$; the solution of this equation is not 3 because the expressions on either side of the equals sign are undefined. Thus the statement is true for some numbers and not true for 0. The statement is sometimes true.

Determine whether the statement is always true, sometimes true, or never true.

1. Both sides of an equation can be multiplied by the same number without changing the solution of the equation.

2. For an equation of the form $ax = b$, $a \neq 0$, multiplying both sides of the equation by the reciprocal of a will result in an equation of the form $x = constant$.

3. The Multiplication Property of Equations is used to remove a term from one side of an equation.

4. Adding -3 to each side of an equation yields the same result as subtracting 3 from each side of the equation.

5. An equation contains an equals sign.

6. The same variable term can be added to both sides of an equation without changing the solution of the equation.

7. An equation of the form $ax + b = c$ cannot be solved if a is a negative number.

8. The solution of the equation $\frac{x}{0} = 0$ is 0.

9. For the diagram at the left, $\angle b = \angle d = \angle e = \angle g$.

10. In solving an equation of the form $ax + b = cx + d$, subtracting cx from each side of the equation results in an equation with only one variable term in it.

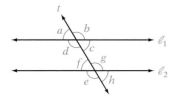

11. If a rope 8 meters long is cut into two pieces and one of the pieces has length x meters, then the length of the other piece can be represented as $(x - 8)$ meters.

12. An even integer is divisible by 2.

13. If the first of two consecutive odd integers is n, then the second consecutive odd integer is represented as $n + 1$.

14. Suppose we are mixing two acid solutions. Then the variable A in the percent mixture equation $Q = Ar$ represents the amount of acid in a solution.

15. If 200 oz of a gold alloy is 75% gold, then the alloy contains 150 oz of gold.

16. If we combine an alloy that costs $6 per ounce with an alloy that costs $3 per ounce, the cost of the resulting mixture will be greater than $3 per ounce.

17. If we combine an 8% salt solution with a solution that is 5% salt, the resulting solution will be less than 8% salt.

18. If the speed of one runner is 2 mph slower than that of a second runner, then the speeds of the two runners can be represented as r and $2 - r$.

Diagramming Problems

How do you best remember something? Do you remember best what you hear? The word *aural* means "pertaining to the ear"; people with a strong aural memory remember best those things that they hear. The word *visual* means "pertaining to the sense of sight"; people with a strong visual memory remember best that which they see written down. Some people claim their memory is in their writing hand—they remember something only if they write it down! The method by which you best remember something is probably also the method by which you can best learn something new.

In problem-solving situations, try to capitalize on your strengths. If you tend to understand material better when you hear it spoken, read application problems aloud or have someone else read them to you. If writing helps you to organize ideas, rewrite application problems in your own words.

No matter what your main strength, visualizing a problem can be a valuable aid in problem solving. A drawing, sketch, diagram, or chart can be a useful tool in problem solving, just as calculators and computers are tools. Making a diagram can help you gain an understanding of the relationships inherent in a problem-solving situation. A sketch will help you to organize the given information and devise or discover a method by which the solution can be determined.

A tour bus drives 5 mi south, then 4 mi west, then 3 mi north, and then 4 mi east. How far is the tour bus from the starting point?

Draw a diagram of the given information.

From the diagram, we can see that the solution can be determined by subtracting 3 from 5: $5 - 3 = 2$.

The bus is 2 mi from the starting point.

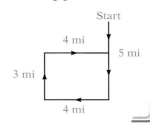

If you roll two ordinary six-sided dice and multiply the two numbers that appear on top, how many different products are there?

Make a chart of the possible products. In the chart below, repeated products are marked with an asterisk.

$1 \cdot 1 = 1$ $2 \cdot 1 = 2*$ $3 \cdot 1 = 3*$ $4 \cdot 1 = 4*$ $5 \cdot 1 = 5*$ $6 \cdot 1 = 6*$
$1 \cdot 2 = 2$ $2 \cdot 2 = 4*$ $3 \cdot 2 = 6*$ $4 \cdot 2 = 8*$ $5 \cdot 2 = 10*$ $6 \cdot 2 = 12*$
$1 \cdot 3 = 3$ $2 \cdot 3 = 6*$ $3 \cdot 3 = 9$ $4 \cdot 3 = 12*$ $5 \cdot 3 = 15*$ $6 \cdot 3 = 18*$
$1 \cdot 4 = 4$ $2 \cdot 4 = 8$ $3 \cdot 4 = 12*$ $4 \cdot 4 = 16$ $5 \cdot 4 = 20*$ $6 \cdot 4 = 24*$
$1 \cdot 5 = 5$ $2 \cdot 5 = 10$ $3 \cdot 5 = 15$ $4 \cdot 5 = 20$ $5 \cdot 5 = 25$ $6 \cdot 2 = 30*$
$1 \cdot 6 = 6$ $2 \cdot 6 = 12$ $3 \cdot 6 = 18$ $4 \cdot 6 = 24$ $5 \cdot 6 = 30$ $6 \cdot 6 = 36$

By counting the products that are not repeats, we can see that there are 18 different products.

In this chapter, you may have noticed that a drawing accompanies the strategy step of many application problems. We encourage you to draw a diagram when solving the application problems in these sections, as well as in any other situation where it can prove helpful. Doing so will undoubtedly improve your problem-solving skills.

Projects & Group Activities

Nielsen Ratings

WEB

Point
of **Interest**

The top-ranked programs in network prime time for the week of October 1 to October 7, 2001, ranked by Nielsen Media Research, were:

Friends
ER
The West Wing
Law & Order
Everybody Loves Raymond
Inside Schwartz
CSI: Crime Scene Investigation
Will & Grace
Becker
Judging Amy
NFL Monday Night Football
Jag

Nielsen Media Research surveys television viewers to determine the numbers of people watching particular shows. There are an estimated 102.2 million U.S. households with televisions. Each **rating point** represents 1% of that number, or 1,022,000 households. For instance, if *60 Minutes* received a rating of 9.6, then 9.6%, or $(0.096)(102,200,000) = 9,811,200$, households watched that program.

A rating point does not mean that 1,022,000 people are watching a program. A rating point refers to the number of TV sets tuned to that program; there may be more than one person watching a television set in the household.

Nielsen Media Research also describes a program's share of the market. **Share** is the percent of television sets in use that are tuned to a program. Suppose that the same week *60 Minutes* received 9.6 rating points, the show received a share of 19%. This means that 19% of all households with a television *turned on* were tuned to *60 Minutes*, whereas only 9.6% of all households *with a television* were tuned to the program.

1. If *Law & Order* received a Nielsen rating of 10.8 and a share of 19, how many TV households watched the program that week? How many TV households were watching television during that hour? Round to the nearest hundred thousand.

2. Suppose *Primetime Thursday* received a rating of 16.2 and a share of 28. How many TV households watched the program that week? How many TV

households were watching television during that hour? Round to the nearest hundred thousand.

3. Suppose *Who Wants to be a Millionaire—Thursday* received a rating of 13.6 during a week in which 34,750,000 people watched the show. Find the average number of people per TV household who watched the program. Round to the nearest tenth.

The cost to advertise during a program is related to its Nielsen rating. The sponsor (the company paying for the advertisement) pays a certain number of dollars for each rating point a show receives.

4. Suppose a television network charges $35,000 per rating point for a 30-second commercial on a daytime talk show. Determine the cost for three 30-second commercials if the Nielsen rating of the show is 11.5.

Nielsen Media Research also tracks the exposure of advertisements. For example, it might be reported that commercials for McDonald's had 500,000,000 household exposures during a week when its advertisement was aired 90 times.

5. Information regarding household exposure of advertisements can be found in *USA Today* each Monday. For a recent week, find the information for the top four advertised products. For each product, calculate the average household exposure for each time the ad was aired.

Nielsen Media Research has a web site on the Internet. You can locate the site by using a search engine. The site does not list rating points or market share, but these statistics can be found on other web sites by using a search engine.

6. Find the top two prime-time television shows for last week. Calculate the number of TV households that watched the programs. Compare the figures with the figures for the top two sports programs for last week.

Changes in Percent Concentration

Suppose you begin with a 30-milliliter solution that is 10% hydrochloric acid.

1. If 10 milliliters of a 30% solution of hydrochloric acid solution are added to this solution, what is the percent concentration of the resulting solution?

2. If x milliliters of a 30% hydrochloric acid solution are added to the original solution, what is the percent concentration of the resulting solution?

3. Let y represent the percent concentration and x the number of milliliters of the 30% hydrochloric acid solution that are being added. Use a graphing calculator to graph the equation. Use a viewing window of Xmin = 0, Xmax = 100, Ymin = 0, and Ymax = 30. Use the graph to determine whether the change in percent concentration from X = 10 to X = 20 is the same as the change in percent concentration from X = 20 to X = 30.

4. If more and more of the 30% hydrochloric acid solution is added to the original mixture, will the percent concentration of the resulting solution ever exceed 30%? Explain your answer.

Chapter Summary

Key Words

Consecutive integers follow one another in order.
[4.2.1, p. 155]

The **perimeter** of a geometric figure is a measure of the
distance around the figure.   [4.3.1, p. 162]

An angle is measured in **degrees**. A 90° angle is a **right angle**. A
180° angle is a **straight angle**. One complete revolution is 360°.
An **acute angle** is an angle whose measure is between 0° and
90°. An **obtuse angle** is an angle whose measure is between 90°
and 180°. An **isosceles triangle** has two sides of equal measure.
The three sides of an **equilateral triangle** are of equal measure.
Complementary angles are two angles whose measures have
the sum 90°. **Supplementary angles** are two angles whose
measures have the sum 180°.   [4.3.1, 4.3.2, pp. 162, 164]

Parallel lines never meet; the distance between them is always
the same. **Perpendicular lines** are intersecting lines that form
right angles. Two angles that are on opposite sides of the
intersection of two lines are **vertical angles;** vertical angles
have the same measure. Two angles that share a common side
are **adjacent angles;** adjacent angles of intersecting lines are
supplementary angles.   [4.3.2, p. 165]

A line that intersects two other lines at two different points is
a **transversal**. If the lines cut by a transversal are parallel lines,
then pairs of equal angles are formed: **alternate interior angles**,
alternate exterior angles, and **corresponding angles**.
[4.3.2, pp. 166–167]

Cost is the price that a business pays for a product. **Selling
price** is the price for which a business sells a product to a
customer. **Markup** is the difference between selling price and
cost. **Markup rate** is the markup expressed as a percent of the
retailer's cost.   [4.4.1, p. 176]

Discount is the amount by which a retailer reduces the regular
price of a product. **Discount rate** is the discount expressed as a
percent of the regular price.   [4.4.2, p. 178]

Examples

11, 12, 13 are consecutive integers.
$-9, -8, -7$ are consecutive integers.

The perimeter of a triangle is the sum
of the lengths of the three sides.
The perimeter of a rectangle is the
sum of the lengths of the four sides.

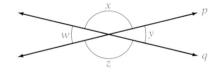

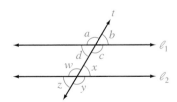

$\angle b = \angle d = \angle x = \angle z$
$\angle a = \angle c = \angle w = \angle y$

If a business pays $100 for a product
and sells that product for $140, then
the cost of the product is $100, the
selling price is $140, and the markup
is $140 − $100 = $40.

The regular price of a product is
$100. The product is now on sale for
$75. The discount on the product is
$100 − $75 = $25.

Essential Rules and Procedures

Consecutive Integers
$n, n + 1, n + 2, \ldots$ [4.2.1, p. 155]

The sum of three consecutive integers is 135.

$$n + (n + 1) + (n + 2) = 135$$

Consecutive Even or Consecutive Odd Integers
$n, n + 2, n + 4, \ldots$ [4.2.1, p. 155]

The sum of three consecutive odd integers is 135.

$$n + (n + 2) + (n + 4) = 135$$

Coin and Stamp Equation

$$\frac{\text{Number}}{\text{of coins}} \cdot \frac{\text{Value of}}{\text{coin in cents}} = \frac{\text{Total value}}{\text{in cents}} \quad \text{[4.2.2, p. 157]}$$

A coin bank contains $1.80 in dimes and quarters. In all, there are twelve coins in the bank. How many dimes are in the bank?

$$10d + 25(12 - d) = 180$$

Formulas for Perimeter
Triangle: $P = a + b + c$
Rectangle: $P = 2L + 2W$
Square: $P = 4s$
[4.3.1, p. 162]

The perimeter of a rectangle is 108 m. The length is 6 m more than three times the width. Find the width of the rectangle.

$$108 = 2(3W + 6) + 2W$$

The perimeter of an isosceles triangle is 34 ft. The length of the third side is 5 ft less than the length of one of the equal sides. Find the length of one of the equal sides.

$$34 = x + x + (x - 5)$$

Sum of the Angles of a Triangle
The sum of the measures of the interior angles of a triangle is 180°.
$\angle A + \angle B + \angle C = 180°$
The sum of an interior and corresponding exterior angle is 180°.
[4.3.3, p. 168]

In a right triangle, the measure of one acute angle is 12° more than the measure of the smallest angle. Find the measure of the smallest angle.

$$x + (x + 12) + 90 = 180$$

Basic Markup Equation
$S = C + rC$ [4.4.1, p. 177]

The manager of a department buys a pasta maker machine for $70 and sells the machine for $98. Find the markup rate.

$$98 = 70 + 70r$$

Basic Discount Equation
$S = R - rR$ [4.4.2, p. 178]

The sale price for a leather bomber jacket is $95. This price is 24% off the regular price. Find the regular price.

$$95 = R - 0.24R$$

Annual Simple Interest Equation
$I = Pr$ [4.5.1, p. 183]

You invest a portion of $10,000 in a 7% annual simple interest account and the remainder in a 6% annual simple interest bond. The two investments earn total annual interest of $680. How much is invested in the 7% account?

$$0.07x + 0.06(10,000 - x) = 680$$

Value Mixture Equation
$V = AC$ [4.6.1, p. 189]

An herbalist has 30 oz of herbs costing $2 per ounce. How many ounces of herbs costing $1 per ounce should be mixed with the 30 oz to produce a mixture costing $1.60 per ounce?

$$30(2) + 1x = 1.60(30 + x)$$

Percent Mixture Equation
$Q = Ar$ [4.6.2, p. 190]

Forty ounces of a 30% gold alloy are mixed with 60 oz of a 20% gold alloy. Find the percent concentration of the resulting gold alloy.

$$0.30(40) + 0.20(60) = x(100)$$

Uniform Motion Equation
$d = rt$ [4.7.1, p. 197]

A boat traveled from a harbor to an island at an average speed of 20 mph. The average speed on the return trip was 15 mph. The total trip took 3.5 h. How long did it take to travel to the island?

$$20t = 15(3.5 - t)$$

Chapter Review Exercises

1. Translate "seven less than twice a number is equal to the number" into an equation and solve.

2. A piano wire is 35 in. long. A note can be produced by dividing this wire into two parts so that three times the length of the shorter piece is twice the length of the longer piece. Find the length of the shorter piece.

3. The sum of two numbers is twenty-one. Three times the smaller number is two less than twice the larger number. Find the two numbers.

4. Translate "the sum of twice a number and six equals four times the difference between the number and two" into an equation and solve.

5. A ceiling fan that regularly sells for $90 is on sale for $60. Find the discount rate.

6. A total of $15,000 is deposited into two simple interest accounts. The annual simple interest rate on one account is 6%. The annual simple interest rate on the second account is 7%. How much should be invested in each account so that the total interest earned is $970?

7. In an isosceles triangle, the measure of one of the two equal angles is 15° more than the measure of the third angle. Find the measure of each angle.

8. A motorcyclist and a bicyclist set out at 8 A.M. from the same point, headed in the same direction. The speed of the motorcyclist is three times the speed of the bicyclist. In 2 h, the motorcyclist is 60 mi ahead of the bicyclist. Find the rate of the motorcyclist.

9. In an isosceles triangle, the measure of one angle is 25° less than half the measure of one of the equal angles. Find the measure of each angle.

10. The manager of a sporting goods store buys packages of cleats for $5.25 each and sells them for $9.45 each. Find the markup rate.

11. A dairy owner mixes 5 gal of cream that is 30% butterfat with 8 gal of milk that is 4% butterfat. Find the percent concentration of butterfat in the resulting mixture.

12. The length of a rectangle is four times the width. The perimeter is 200 ft. Find the length and width of the rectangle.

13. The manager of an accounting firm is investigating contracts whereby large quantities of material can be photocopied at minimum cost. The contract from the Copy Center offers a fee of $50 per week and $.03 per page. The Prints 4 U Company offers a fee of $27 per week and $.05 per page. What is the minimum number of copies per week that could be ordered from the Copy Center if these copies are to cost less than copies from Prints 4 U?

14. The sale price for a video game is $26.56, which is 17% off the regular price. Find the regular price.

15. The owner of a health food store combined cranberry juice that cost $1.79 per quart with apple juice that cost $1.19 per quart. How many quarts of each were used to make 10 qt of a cranapple juice mixture costing $1.61 per quart?

16. A famous mystery writer is contracted to autograph copies of her new book at a bookstore. As a special promotion that day, the book is selling for $33.74, which is only 30% above the cost of the book. Find the cost of the book.

17. The sum of two numbers is thirty-six. The difference between the larger number and eight equals the total of four and three times the smaller number. Find the two numbers.

18. An engineering consultant invested $14,000 in an individual retirement account paying 8.15% annual simple interest. How much money does the consultant have deposited in an account paying 12% annual simple interest if the total interest earned is 9.25% of the total investment?

19. You have removed all the hearts from a standard deck of playing cards and have arranged them in numerical order. Suppose I pick two consecutive cards and tell you their sum is 15. Name the two cards I picked.

20. A pharmacist has 15 L of an 80% alcohol solution. How many liters of pure water should be added to the alcohol solution to make an alcohol solution that is 75% alcohol?

21. A student's grades on five sociology exams were 68, 82, 90, 73, and 95. What is the lowest score this student can receive on the sixth exam and still have earned a total of at least 480 points?

22. Translate "the opposite of seven is equal to one-half a number less ten" into an equation and solve.

23.  The Empire State Building is 1472 ft tall. This is 654 ft less than twice the height of the Eiffel Tower. Find the height of the Eiffel Tower.

24. One angle of a triangle is 15° more than the measure of the second angle. The third angle is 15° less than the measure of the second angle. Find the measure of each angle.

25. The manager of a refreshment stand at a local stadium tallied the receipts for one evening. There was a total of $263 in one-dollar bills and five-dollar bills. In all, there were 155 bills. Find the number of one-dollar bills.

26. A board 10 ft long is cut into two pieces. Four times the length of the shorter piece is 2 ft less than two times the length of the longer piece. Find the length of the longer piece.

27. An optical engineer's consulting fee was $600. This included $80 for supplies and $65 for each hour of consultation. Find the number of hours of consultation.

28. A stamp collector has 12¢ stamps and 17¢ stamps with a total face value of $15.53. The number of 12¢ stamps is three more than twice the number of 17¢ stamps. How many of each type are in the collection?

29. A furniture store uses a markup rate of 60%. The store sells a solid oak curio cabinet for $1074. Find the cost of the curio cabinet.

30. The largest swimming pool in the world is located in Casablanca, Morocco, and is 480 m long. Two swimmers start at the same time from opposite ends of the pool and swim toward each other. One swimmer's rate is 65 m/min. The other swimmer's rate is 55 m/min. In how many minutes after they begin will they meet?

31. The perimeter of a triangle is 35 in. The second side is 4 in. longer than the first side. The third side is 1 in. shorter than twice the first side. Find the measure of each side.

32. The sum of three consecutive odd integers is -45. Find the integers.

33. A rectangle is 15 ft long and $(2x - 4)$ ft wide. Express as an integer the maximum width of the rectangle when the area is less than 180 ft^2.

34. Given that $\ell_1 \| \ell_2$, find the measures of angles a and b.

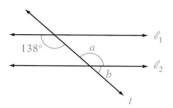

Chapter Test

1. Translate "the sum of six times a number and thirteen is five less than the product of three and the number" into an equation and solve.

2. Translate "the difference between three times a number and fifteen is twenty-seven" into an equation and solve.

3. The sum of two numbers is 18. The difference between four times the smaller number and seven is equal to the sum of two times the larger number and five. Find the two numbers.

4. A board 18 ft long is cut into two pieces. Two feet less than the product of five and the length of the shorter piece is equal to the difference between three times the length of the longer piece and eight. Find the length of each piece.

5. The manager of a sports shop uses a markup rate of 50%. The selling price for a set of golf clubs is $300. Find the cost of the golf clubs.

6. A pair of running shoes that regularly sells for $100 is on sale for $80. Find the discount rate.

7. How many gallons of a 15% acid solution must be mixed with 5 gal of a 20% acid solution to make a 16% acid solution?

8. The perimeter of a rectangle is 38 m. The length of the rectangle is 1 m less than three times the width. Find the length and width of the rectangle.

9. Find three consecutive odd integers such that three times the first integer is one less than the sum of the second and third integers.

10. Florist A charges a $3 delivery fee plus $21 per bouquet delivered. Florist B charges a $15 delivery fee plus $18 per bouquet delivered. An organization wants to send each resident of a small nursing home a bouquet for Valentine's Day. How many residents are in the nursing home if it is more economical for the organization to use Florist B?

11. A total of $7000 is deposited into two simple interest accounts. On one account, the annual simple interest rate is 10%, and on the second account, the annual simple interest rate is 15%. How much should be invested in each account so that the total annual interest earned is $800?

12. A coffee merchant wants to make 12 lb of a blend of coffee costing $6 per pound. The blend is made using a $7 grade and a $4 grade of coffee. How many pounds of each of these grades should be used?

13. Two planes start at the same time from the same point and fly in opposite directions. The first plane is flying 100 mph faster than the second plane. In 3 h, the two planes are 1050 mi apart. Find the rate of each plane.

14. In a triangle, the first angle is 15° more than the second angle. The third angle is three times the second angle. Find the measure of each angle.

15. A coin bank contains 50 coins in nickels and quarters. The total amount of money in the bank is $9.50. Find the number of nickels and the number of quarters in the bank.

16. A club treasurer deposited $2400 in two simple interest accounts. The annual simple interest rate on one account is 6.75%. The annual simple interest rate on the other account is 9.45%. How much should be deposited in each account so that the same interest is earned on each account?

17. The width of a rectangle is 12 ft. The length is $(3x + 5)$ ft. Express as an integer the minimum length of the rectangle if the area is greater than 276 ft^2.

18. A file cabinet that normally sells for $99 is on sale for 20% off. Find the sale price.

Cumulative Review Exercises

1. Given $B = \{-12, -6, -3, -1\}$, which elements of set B are less than -4?

2. Simplify: $-2 + (-8) - (-16)$

3. Simplify: $\left(-\dfrac{2}{3}\right)^3\left(-\dfrac{3}{4}\right)^2$

4. Simplify: $\dfrac{5}{6} - \left(\dfrac{2}{3}\right)^2 \div \left(\dfrac{1}{2} - \dfrac{1}{3}\right)$

5. Evaluate $-|-18|$.

6. Evaluate $b^2 - (a - b)^2$ when $a = 4$ and $b = -1$.

7. Simplify: $5x - 3y - (-4x) + 7y$

8. Simplify: $-4(3 - 2x - 5x^3)$

9. Simplify: $-2[x - 3(x - 1) - 5]$

10. Simplify: $-3x^2 - (-5x^2) + 4x^2$

11. Is 2 a solution of $4 - 2x - x^2 = 2 - 4x$?

12. Solve: $9 - x = 12$

13. Solve: $-\dfrac{4}{5}x = 12$

14. Solve: $8 - 5x = -7$

15. Solve: $-6x - 4(3 - 2x) = 4x + 8$

16. Write 40% as a fraction.

17. Solve and graph the solution set of $4x \geq 16$.

18. Solve: $-15x \leq 45$

19. Solve: $2x - 3 > x + 15$

20. Solve: $12 - 4(x - 1) \leq 5(x - 4)$

21. Write 0.025 as a percent.

22. Write $\dfrac{3}{25}$ as a percent.

23. Find $16\dfrac{2}{3}\%$ of 18.

24. 40% of what is 18?

25. Translate "the sum of eight times a number and twelve is equal to the product of four and the number" into an equation and solve.

26. The area of the cement foundation of a house is 2000 ft². This is 200 ft² more than three times the area of the garage. Find the area of the garage.

27. An auto repair bill was $313. This includes $88 for parts and $45 for each hour of labor. Find the number of hours of labor.

28. A survey of 250 librarians showed that 50 of the libraries had a particular reference book on their shelves. What percent of the libraries had the reference book?

29. A deposit of $4000 is made into an account that earns 11% annual simple interest. How much money is also deposited in an account that pays 14% annual simple interest if the total annual interest earned is 12% of the total investment?

30. The manager of a department store buys a chain necklace for $80 and sells it for $140. Find the markup rate.

31. How many grams of a gold alloy that costs $4 per gram must be mixed with 30 g of a gold alloy that costs $7 per gram to make an alloy costing $5 per gram?

32. How many ounces of pure water must be added to 70 oz of a 10% salt solution to make a 7% salt solution?

33. In an isosceles triangle, the third angle is 8° less than twice the measure of one of the equal angles. Find the measure of one of the equal angles.

34. Three times the second of three consecutive even integers is 14 more than the sum of the first and third integers. Find the middle integer.

35. A coin bank contains dimes and quarters. The number of quarters is five less than four times the number of dimes. The total amount in the bank is $6.45. Find the number of dimes in the bank.

Every year thousands of people gather in New Orleans to celebrate Mardi Gras. Huge crowds take over the streets throughout the French Quarter to be a part of the festivities, including parades, performers, and elaborate costumes. All of the people are just a small portion of our steadily growing population in this country. By comparing the population over the years, we are able to determine the rate of change in the population and then use that rate to predict populations in the future. The **Project on pages 284–286** *shows how to use rate of change to predict population growth.*

Applications

Airline industry,   *p. 291*
Airports,   *p. 234*
Automotive industry,   *pp. 235, 292, 294*
Aviation,   *pp. 227, 254*
Biology,   *p. 274*
Business,   *p. 276*
Child development,   *pp. 230, 231, 235*
Computer science,   *pp. 236, 259*
Construction,   *p. 292*
Consumerism,   *p. 294*
Demography,   *pp. 234, 235*
Depreciation,   *pp. 254, 276*
Discount,   *p. 296*
Education,   *pp. 236, 294*
Fire science,   *p. 294*
Fuel consumption,   *p. 259*
Fuel efficiency,   *p. 233*
Geometry,   *p. 296*
Health,   *pp. 258, 273, 277*
Height-time graphs,   *p. 277*
Housing,   *p. 273*
Integer problems,   *p. 296*
Investments,   *p. 271*
Lever systems,   *p. 296*
Life expectancy,   *p. 291*
Manufacturing,   *p. 294*
Medicine,   *p. 227*
Minimum wage,   *p. 234*
The Olympics,   *pp. 233, 273*
Physics,   *p. 273*
Population,   *pp. 228–230, 285–286*
Postal service,   *p. 258*
Remodeling,   *p. 271*
Salaries,   *p. 292*
Sports,   *pp. 226, 234, 235, 254*
Temperature,   *pp. 234, 266*
Travel,   *p. 272*
U.S. Presidents,   *p. 233*

5

Linear Equations and Inequalities

Objectives

Section 5.1
1. Graph points in a rectangular coordinate system
2. Scatter diagrams
3. Average rate of change

Section 5.2
1. Determine solutions of linear equations in two variables
2. Graph equations of the form $y = mx + b$
3. Graph equations of the form $Ax + By = C$

Section 5.3
1. Find the slope of a straight line
2. Graph a line using the slope and y-intercept

Section 5.4
1. Find the equation of a line using the equation $y = mx + b$
2. Find the equation of a line using the point-slope formula

Section 5.5
1. Introduction to functions
2. Graphs of linear functions

Section 5.6
1. Graph inequalities in two variables

Need help? For online student resources, such as section quizzes, visit this web site:
math.college.hmco.com

WEB

223

PrepTest

1. Simplify: $\dfrac{5 - (-7)}{4 - 8}$

2. Evaluate $\dfrac{a - b}{c - d}$ when $a = 3$, $b = -2$, $c = -3$, and $d = 2$.

3. Simplify: $-3(x - 4)$

4. Solve: $3x + 6 = 0$

5. Solve $4x + 5y = 20$ when $y = 0$.

6. Solve $3x - 7y = 11$ when $x = -1$.

7. Given $y = -4x + 5$, find the value of y when $x = -2$.

8. Simplify: $\dfrac{1}{4}(3x - 16)$

9. Which of the following are solutions of the inequality $-4 < x + 3$?
   a. 0   **b.** -3   **c.** 5   **d.** -7   **d.** -10

Go Figure

Points A, B, C, and D lie on the same line and in that order. The ratio of $\overline{AB}$ to $\overline{AC}$ is 1:4, and the ratio of $\overline{BC}$ to $\overline{CD}$ is 1:2. Find the ratio of $\overline{AB}$ to $\overline{CD}$.

5.1 | The Rectangular Coordinate System

1 **Graph points in a rectangular coordinate system**

Before the 15th century, geometry and algebra were considered separate branches of mathematics. That all changed when René Descartes, a French mathematician who lived from 1596 to 1650, founded **analytic geometry.** In this geometry, a *coordinate system* is used to study the relationships between variables.

A **rectangular coordinate system** is formed by two number lines, one horizontal and one vertical, that intersect at the zero point of each line. The point of intersection is called the **origin.** The two axes are called the **coordinate axes,** or simply the **axes.** Generally, the horizontal axis is labeled the **x-axis,** and the vertical axis is labeled the **y-axis.**

The axes determine a **plane,** which can be thought of as a large, flat sheet of paper. The

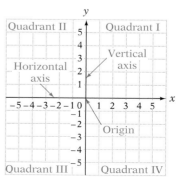

two axes divide the plane into four regions called **quadrants,** which are numbered counterclockwise from I to IV starting at the upper right.

Each point in the plane can be identified by a pair of numbers called an **ordered pair.** The first number of the ordered pair measures a horizontal distance and is called the **abscissa,** or **x-coordinate.** The second number of the pair measures a vertical distance and is called the **ordinate,** or **y-coordinate.** The ordered pair (x, y) associated with a point is also called the **coordinates** of the point.

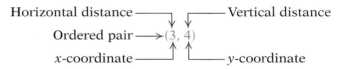

To **graph,** or **plot, a point in the plane,** place a dot at the location given by the ordered pair. For example, to graph the point $(4, 1)$, start at the origin. Move 4 units to the right and then 1 unit up. Draw a dot. To graph $(-3, -2)$, start at the origin. Move 3 units left and then 2 units down. Draw a dot.

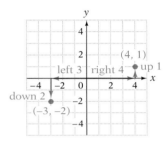

The **graph of an ordered pair** is the dot drawn at the coordinates of the point in the plane. The graphs of the ordered pairs $(4, 1)$ and $(-3, -2)$ are shown at the right.

The graphs of the points whose coordinates are $(2, 3)$ and $(3, 2)$ are shown at the right. Note that they are different points. The order in which the numbers in an ordered pair appear *is* important.

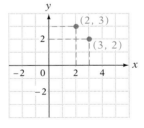

Each point in the plane is associated with an ordered pair, and each ordered pair is associated with a point in the plane. Although only integers are labeled on the coordinate grid, any ordered pair can be graphed by approximating its location. The graphs of the ordered pairs $\left(\frac{3}{2}, -\frac{4}{3}\right)$ and $(-2.4, 3.5)$ are shown at the right.

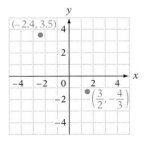

Example 1 Graph the ordered pairs $(-2, -3)$, $(3, -2)$, $(1, 3)$, and $(4, 1)$.

Solution

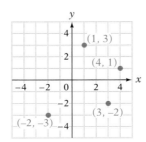

Problem 1   Graph the ordered pairs $(-1, 3)$, $(1, 4)$, $(-4, 0)$, and $(-2, -1)$.

Solution   See page S10.

Example 2   Find the coordinates of each of the points.

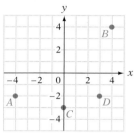

Solution $A(-4, -2)$, $B(4, 4)$, $C(0, -3)$, $D(3, -2)$

Problem 2   Find the coordinates of each of the points.

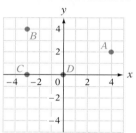

Solution   See page S11.

2   Scatter diagrams

There are many situations in which the relationship between two variables may be of interest. For example, a college admissions director wants to know the relationship between SAT scores (one variable) and success in college (the second variable). An employer is interested in the relationship between scores on a preemployment test and ability to perform a job.

A researcher can investigate the relationship between two variables by means of regression analysis, which is a branch of statistics. The study of the relationship between two variables may begin with a **scatter diagram,** which is a graph of some of the known data.

The following table gives the record times for races of different lengths at a junior high track meet. Record times are rounded to the nearest second. Lengths are given in meters.

Length, x	100	200	400	800	1000	1500
Time, y	20	40	100	200	260	420

The scatter diagram for these data is shown at the right. Each ordered pair represents the length of the race and the record time. For example, the ordered pair (400, 100) indicates that the record for the 400 m race is 100 s.

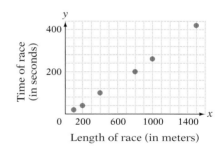

Example 3   To test a heart medicine, a doctor measures the heart rate, in beats per minute, of five patients before and after they take the medication. The results are recorded in the following table. Graph the scatter diagram for these data.

Before medicine, x	85	80	85	75	90
After medicine, y	75	70	70	80	80

Strategy   Graph the ordered pairs on a rectangular coordinate system where the horizontal axis represents the heart rate before taking the medication and the vertical axis represents the heart rate after taking the medication.

Solution

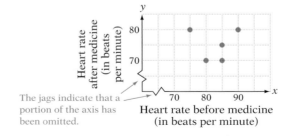

The jags indicate that a portion of the axis has been omitted.

Problem 3   A study by the Federal Aviation Administration showed that narrow, over-the-wing emergency exit rows slow passenger evacuation. The table below shows the space between seats, in inches, and the evacuation time, in seconds, for a group of 35 passengers. The longest evacuation times are recorded in the table. Graph the scatter diagram for these data.

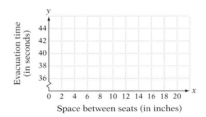

Space between seats, x	6	10	13	15	20
Evacuation time, y	44	43	38	36	37

| Solution   See page S11.

3 Average rate of change

Population of the United States

Year	Population (in millions)
1800	5
1810	7
1820	10
1830	13
1840	17
1850	23
1860	31
1870	40
1880	50
1890	63
1900	76
1910	92
1920	106
1930	123
1940	132
1950	151
1960	179
1970	203
1980	227
1990	249
2000	281

The table at the left shows the population of the United States for each decade from 1800 to 2000. (*Source:* U.S. Bureau of the Census) These data are graphed in the scatter diagram below.

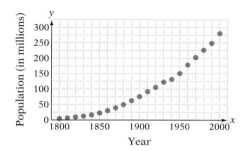

One way to describe how the population of the United States has changed over time is to calculate the change in population for a given time period. For example, we can find the change in population from 1990 to 2000.

Population in 2000 − population in 1990 = 281 million − 249 million

= 32 million

The amount of change in the population, 32 million, does not describe how rapid the growth in population was. To describe the rate of growth, or the *rate of change*, we must consider the number of years over which that growth took place. To determine how rapidly the population grew from 1990 to 2000, divide the change in population by the number of years during which that change took place.

$$\frac{\text{Change in population}}{\text{Change in years}} = \frac{\text{population in 2000} - \text{population in 1990}}{2000 - 1990}$$

$$= \frac{281 - 249}{2000 - 1990}$$

$$= \frac{32}{10} = 3.2$$

The average rate of change was 3.2 million people per year. This means that *on average*, from 1990 to 2000, the population of the United States increased by 3.2 million people per year.

Consider the two points on the graph that correspond to these years:

$$(1990, 249) \text{ and } (2000, 281)$$

Note that **the average rate of change is calculated by dividing the difference between the y values of the two points by the difference between the x values of the two points.**

Average Rate of Change

The average rate of change of y with respect to x is $\dfrac{\text{change in } y}{\text{change in } x}$.

To calculate the average rate of change, think

$$\frac{y \text{ value of second point} - y \text{ value of first point}}{x \text{ value of second point} - x \text{ value of first point}}$$

 Find the average rate of change per year in the U.S. population from 1980 to 1990.

In 1980, the population was 227 million: $(1980, 227)$
In 1990, the population was 249 million: $(1990, 249)$

$$
\begin{aligned}
\text{Average rate of change} &= \frac{\text{change in } y}{\text{change in } x} \\[2mm]
&= \frac{\text{population in 1990} - \text{population in 1980}}{1990 - 1980} \\[2mm]
&= \frac{249 - 227}{1990 - 1980} \\[2mm]
&= \frac{22}{10} = 2.2
\end{aligned}
$$

The average rate of change in the population was 2.2 million people per year.

From 1990 to 2000, the average rate of change was 3.2 million people per year, whereas from 1980 to 1990, the average rate of change was 2.2 million people per year. The rate of change in the U.S. population is not constant; it varies for different intervals of time.

Example 4 Find the average rate of change per year in the U.S. population from 1800 to 1900.

Solution   In 1800, the population was 5 million: (1800, 5)
In 1900, the population was 76 million: (1900, 76)

$$\frac{\text{Average rate}}{\text{of change}} = \frac{\text{change in } y}{\text{change in } x}$$

$$= \frac{\text{population in 1900} - \text{population in 1800}}{1900 - 1800}$$

$$= \frac{76 - 5}{1900 - 1800}$$

$$= \frac{71}{100} = 0.71$$

The average rate of change in the population was 0.71 million people per year, or 710,000 people per year.

Note that if both the ordinates and the abscissas are subtracted in the reverse order, the result is the same.

$$\frac{\text{Average rate}}{\text{of change}} = \frac{\text{population in 1800} - \text{population in 1900}}{1800 - 1900}$$

$$= \frac{5 - 76}{1800 - 1900} = \frac{-71}{-100} = 0.71$$

Problem 4 Find the average rate of change per year in the U.S. population from 1900 to 2000.

Solution   See page S11.

The table below shows weights of fetuses during various stages of prenatal development in humans. Use these data for Example 5 and Problem 5.

Number of weeks after conception	21	24	28	30	38
Weight, in pounds	1	2	3	4	7.5

Example 5   Find the average rate of change per week in the weight of a fetus from 21 weeks after conception to 30 weeks after conception.

Solution   21 weeks after conception, the weight is 1 lb: (21, 1)
30 weeks after conception, the weight is 4 lb: (30, 4)

$$\frac{\text{Average rate}}{\text{of change}} = \frac{\text{change in } y}{\text{change in } x}$$

$$= \frac{\text{weight after 30 weeks} - \text{weight after 21 weeks}}{30 - 21}$$

$$= \frac{4 - 1}{30 - 21} = \frac{3}{9}$$

$$= \frac{1}{3} = 0.\overline{3}$$

The average rate of change in weight is $0.\overline{3}$ lb per week.

Problem 5   Find the average rate of change per week in the weight of a fetus from 28 weeks after conception to 38 weeks after conception.

Solution   See page S11.

5.1   Concept Review

Determine whether the statement is always true, sometimes true, or never true.

1. The point (0, 0) on a rectangular coordinate system is at the origin.

2. When a point is plotted in the rectangular coordinate system, the first number in the ordered pair indicates a movement up or down from the origin. The second number in the ordered pair indicates a movement left or right.

3. In an xy-coordinate system, the first number is the y-coordinate and the second number is the x-coordinate.

4. When a point is plotted in the rectangular coordinate system, a negative x value indicates a movement to the left. A negative y value indicates a movement down.

5. Any point plotted in quadrant III has a positive y value.

6. In the figure at the right, point C is the graph of the point whose coordinates are $(-4, -2)$.

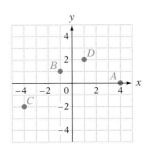

7. Any point on the x-axis has y-coordinate 0.

8. Any point on the y-axis has x-coordinate 0.

5.1 | Exercises

1. Describe the signs of the coordinates of a point in **a.** quadrant I, **b.** quadrant II, **c.** quadrant III, and **d.** quadrant IV.

2. In which quadrant is the point located?
a. $(2, -4)$ **b.** $(-3, 2)$ **c.** $(-1, -6)$

3. Graph the ordered pairs $(-2, 1)$, $(3, -5)$, $(-2, 4)$, and $(0, 3)$.

4. Graph the ordered pairs $(5, -1)$, $(-3, -3)$, $(-1, 0)$, and $(1, -1)$.

5. Graph the ordered pairs $(0, 0)$, $(0, -5)$, $(-3, 0)$, and $(0, 2)$.

6. Graph the ordered pairs $(-4, 5)$, $(-3, 1)$, $(3, -4)$, and $(5, 0)$.

7. Graph the ordered pairs $(-1, 4)$, $(-2, -3)$, $(0, 2)$, and $(4, 0)$.

8. Graph the ordered pairs $(5, 2)$, $(-4, -1)$, $(0, 0)$, and $(0, 3)$.

9. Find the coordinates of each of the points.

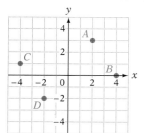

10. Find the coordinates of each of the points.

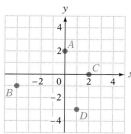

11. Find the coordinates of each of the points.

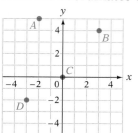

12. Find the coordinates of each of the points.

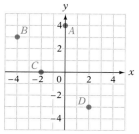

13. a. Name the abscissas of points *A* and *C*.
  b. Name the ordinates of points *B* and *D*.

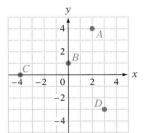

14. a. Name the abscissas of points *A* and *C*.
  b. Name the ordinates of points *B* and *D*.

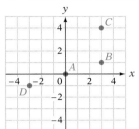

2 15. **Fuel Efficiency** The American Council for an Energy-Efficient Economy releases rankings of environmentally friendly and unfriendly cars and trucks sold in the United States. The following table shows the fuel usage, in miles per gallon of gasoline, in the city and on the highway, for six of the twelve 2001-model vehicles ranked worst for the environment. Graph the scatter diagram for these data.

Fuel use in the city, *x*	11	8	12	10	13	10
Fuel use on the highway, *y*	14	13	17	15	17	13

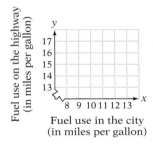

Fuel use on the highway
(in miles per gallon)

Fuel use in the city
(in miles per gallon)

16. **Fuel Efficiency** The American Council for an Energy-Efficient Economy releases rankings of environmentally friendly and unfriendly cars and trucks sold in the United States. The following table shows the fuel usage, in miles per gallon of gasoline, in the city and on the highway, for the six 2001 manual transmission subcompact cars ranked best for the environment. Graph the scatter diagram for these data.

Fuel use in the city, *x*	36	36	30	28	26	24
Fuel use on the highway, *y*	42	44	37	40	33	31

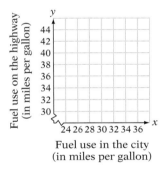

Fuel use on the highway
(in miles per gallon)

Fuel use in the city
(in miles per gallon)

17. **U.S. Presidents** The table below pairs number of children with the number of U.S. presidents with that number of children. Graph the scatter diagram for these data.

Number of children, *x*	0	1	2	3	4	5	6	7	8	10	15
Number of U.S. presidents with *x* children, *y*	6	2	8	7	7	3	5	1	1	1	1

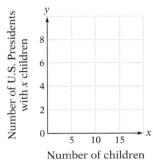

Number of U.S. Presidents
with *x* children

Number of children

18. **The Olympics** The number of gold medals and the number of silver medals won by the European countries Germany, France, Italy, Great Britain, the Netherlands, and Romania in the 2000 Summer Olympics are shown in the table below. Graph the scatter diagram for these data.

Number of gold medals won, *x*	13	13	13	11	12	11
Number of silver medals won, *y*	17	14	8	10	9	6

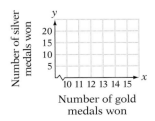

Number of silver
medals won

Number of gold
medals won

19. 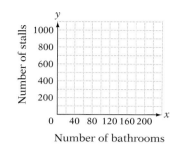 **Airports** The table below records the number of public bathrooms and the number of stalls at each of the following airports: Miami International, Chicago O'Hare, Orlando International, Dallas/Fort Worth, and Seattle/Tacoma. (*Source:* USA TODAY research; Airports Council International) Graph the scatter diagram for these data.

Number of public bathrooms, x	161	125	120	106	68
Number of stalls, y	721	605	600	439	500

20. **Sports** A basketball coach wants to determine whether there is a relationship between the number of shots attempted and the number of shots made. The table below shows the performances of six players on the team for one game. Graph the scatter diagram for these data.

Shots attempted, x	13	11	8	5	4	1
Shots made, y	11	7	5	4	3	1

3 21. **Temperature** On September 10 in midstate New Hampshire, the temperature at 6 A.M. was 45°F. At 2 P.M. on the same day, the temperature was 77°F. Find the average rate of change in temperature per hour.

22. 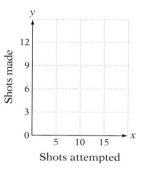 **Demography** The table at the right shows the population of Colonial America for each decade from 1610 to 1780. (*Source:* **infoplease.com**) These figures are estimates, as this period was prior to the establishment of the U.S. Census in 1790.
   a. Find the average annual rate of change in the population of the Colonies from 1650 to 1750.
   b. Was the average annual rate of change in the population from 1650 to 1750 greater than or less than the average annual rate of change in the population from 1700 to 1750?
   c. During which decade was the average annual rate of change the least? What was the average annual rate of change during that decade?

23. **Minimum Wage** The table below shows the minimum hourly wage for the year in which that wage was enacted. (*Source:* U.S. Department of Labor)
   a. What was the average annual rate of change in the minimum wage from 1978 to 1990? Round to the nearest cent.
   b. Find the average annual rate of change in the minimum wage from 1980 to 1990. Is this more or less than the average annual rate of change in the minimum wage from 1978 to 1990?

Population of Colonial America

Year	Population
1610	350
1620	2,300
1630	4,600
1640	26,600
1650	50,400
1660	75,100
1670	111,900
1680	151,500
1690	210,400
1700	250,900
1710	331,700
1720	466,200
1730	629,400
1740	905,600
1750	1,170,800
1760	1,593,600
1770	2,148,100
1780	2,780,400

Year	1978	1979	1980	1981	1990	1991	1996	1997
Minimum wage	2.65	2.90	3.10	3.35	3.80	4.25	4.75	5.15

24. **Child Development** The table below shows average lengths of fetuses for selected stages of prenatal development in humans. (*Source:* **Surebaby.com**)
    a. What is the average rate of change per month in the length of a fetus from the sixth month of pregnancy to the eighth month of pregnancy?
    b. Find the average rate of change per month in the length of a fetus from the second month of pregnancy to the fourth month of pregnancy. Is this greater than or less than the average rate of change per month from the sixth month to the eighth month?

Month of pregnancy	2	4	6	8
Length, in inches	1	7	12	16

25. **Demography** The table below shows the projected population of baby boomers for the years 2000, 2031, and 2046. (*Source:* U.S. Census Bureau)
    a. Find the average rate of change per year in the population of the baby boomer generation from 2031 to 2046 (*Note:* A negative average rate of change denotes a decrease.)
    b. Find the average rate of change per year in the population of baby boomers from 2000 to 2046. Round to the nearest hundred thousand.

Year	2000	2031	2046
Population, in millions	79	51	19

26. **Automotive Industry** The table below shows Oldsmobile sales for selected years from 1985 to 2000. (*Source:* Autodata)
    a. Find the average rate of change per year in Oldsmobile sales from 1985 to 2000. (*Note:* A negative average rate of change denotes a decrease.)
    b. During which three-year period was the absolute value of the average annual rate of change greatest?

Year	1985	1988	1991	1994	1997	2000
Oldsmobile sales	1,066,122	715,270	458,124	448,945	304,759	265,878

27. **Sports** The table at the right shows the prices of 30 seconds of advertising time during selected Super Bowl games. (*Source:* NFL Research)
    a. Find the average rate of change per year in the price of 30 seconds of advertising time from 1967 to 2001. Round to the nearest thousand.
    b. Which was greater, the average annual rate of change from 1971 to 1991 or the average annual rate of change from 1981 to 2001? How much greater?

Super Bowl	Year	Price for 30 seconds of ad time (in thousands of dollars)
I	1967	42
V	1971	72
X	1976	110
XV	1981	275
XX	1986	550
XXV	1991	800
XXX	1996	1,085
XXXV	2001	2,300

28. 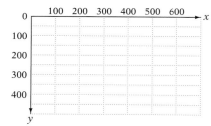 **Education** The table below shows annual per-pupil spending on public elementary and secondary school students for selected years from 1986 to 2001. (*Source:* National Center for Education Statistics)

   a. Find the average rate of change per year in per-pupil spending from 1986 to 1996.

   b. Find the average rate of change per year in per-pupil spending from 1991 to 2001. Is this greater than or less than the average rate of change per year in per-pupil spending from 1986 to 1996?

Year	1986	1991	1996	2001
Per-pupil spending, in dollars	3479	4902	5689	7489

5.1 Applying Concepts

What is the distance from the given point to the horizontal axis?

29. $(-5, 1)$ **30.** $(3, -4)$ **31.** $(-6, 0)$

What is the distance from the given point to the vertical axis?

32. $(-2, 4)$ **33.** $(1, -3)$ **34.** $(5, 0)$

35. Name the coordinates of a point plotted at the origin of the rectangular coordinate system.

36. A computer screen has a coordinate system that is different from the xy-coordinate system we have discussed. In one mode, the origin of the coordinate system is the top left point of the screen, as shown at the right. Plot the points whose coordinates are $(200, 400)$, $(0, 100)$, and $(100, 300)$.

37. Write a paragraph explaining how to plot points in a rectangular coordinate system.

38. Decide on two quantities that may be related, and collect at least ten pairs of values. Here are some examples: height and weight, time studying for a test and the test grade, age of a car and its cost. Draw a scatter diagram for the data. Is there any trend? That is, as the values on the horizontal axis increase, do the values on the vertical axis increase or decrease?

39. There is a coordinate system on Earth that consists of *longitude* and *latitude*. Write a report on how location is determined on the surface of Earth.

5.2 | Graphs of Straight Lines

1 **Determine solutions of linear equations in two variables**

An equation of the form $y = mx + b$, where m is the coefficient of x and b is a constant, is a **linear equation in two variables.** Examples of linear equations follow.

$$y = 3x + 4 \qquad (m = 3, b = 4)$$
$$y = 2x - 3 \qquad (m = 2, b = -3)$$
$$y = -\frac{2}{3}x + 1 \qquad \left(m = -\frac{2}{3}, b = 1\right)$$
$$y = -2x \qquad (m = -2, b = 0)$$
$$y = x + 2 \qquad (m = 1, b = 2)$$

In a linear equation, the exponent of each variable is 1. The equations $y = 2x^2 - 1$ and $y = \dfrac{1}{x}$ are not linear equations.

A solution of a linear equation in two variables is an ordered pair of numbers (x, y) that makes the equation a true statement.

> **Take Note**
>
> An ordered pair is of the form (x, y). For the ordered pair $(1, -2)$, 1 is the x value and -2 is the y value. Substitute 1 for x and -2 for y.

Is $(1, -2)$ a solution of $y = 3x - 5$?

Replace x with 1. Replace y with -2.

Compare the results. If the results are equal, the given ordered pair is a solution. If the results are not equal, the given ordered pair is not a solution.

$$\begin{array}{c|c} \multicolumn{2}{c}{y = 3x - 5} \\ \hline -2 & 3(1) - 5 \\ & 3 - 5 \end{array}$$

$$-2 = -2$$

Yes, $(1, -2)$ is a solution of the equation $y = 3x - 5$.

Besides the ordered pair $(1, -2)$, there are many other ordered-pair solutions of the equation $y = 3x - 5$. For example, the method used above can be used to show that $(2, 1)$, $(-1, -8)$, $\left(\frac{2}{3}, -3\right)$, and $(0, -5)$ are also solutions.

Example 1 Is $(-3, 2)$ a solution of $y = 2x + 2$?

Solution
$$\begin{array}{c|c} \multicolumn{2}{c}{y = 2x + 2} \\ \hline 2 & 2(-3) + 2 \\ & -6 + 2 \\ & -4 \end{array}$$
    • Replace x with -3 and y with 2.

$$2 \neq -4$$

No, $(-3, 2)$ is not a solution of $y = 2x + 2$.

Problem 1 Is $(2, -4)$ a solution of $y = -\dfrac{1}{2}x - 3$?

Solution See page S11.

In general, a linear equation in two variables has an infinite number of ordered-pair solutions. By choosing any value for x and substituting that value into the linear equation, we can find a corresponding value of y.

Find the ordered-pair solution of $y = 2x - 5$ that corresponds to $x = 1$.

Substitute 1 for x.
Solve for y.

$$y = 2x - 5$$
$$y = 2 \cdot 1 - 5$$
$$y = 2 - 5$$
$$y = -3$$

The ordered-pair solution is $(1, -3)$.

Example 2   Find the ordered-pair solution of $y = \frac{2}{3}x - 1$ that corresponds to $x = 3$.

Solution $y = \frac{2}{3}x - 1$

$y = \frac{2}{3}(3) - 1$ • Substitute 3 for x.

$y = 2 - 1$ • Solve for y.

$y = 1$ • When $x = 3$, $y = 1$.

The ordered-pair solution is $(3, 1)$.

Problem 2   Find the ordered-pair solution of $y = -\frac{1}{4}x + 1$ that corresponds to $x = 4$.

Solution   See page S11.

2   ## Graph equations of the form $y = mx + b$

The **graph of an equation in two variables** is a drawing of the ordered-pair solutions of the equation. For a linear equation in two variables, the graph is a straight line.

To graph a linear equation, find ordered-pair solutions of the equation. Do this by choosing any value of x and finding the corresponding value of y. Repeat this procedure, choosing different values for x, until you have found the number of solutions desired.

Because the graph of a linear equation in two variables is a straight line, and a straight line is determined by two points, it is necessary to find only two solutions. However, finding at least three points will ensure accuracy.

Graph $y = 2x + 1$.

Choose any values of x, and then find the corresponding values of y. The numbers 0, 2, and -1 were chosen arbitrarily for x. It is convenient to record these solutions in a table.

x	$y = 2x + 1$	y
0	$2(0) + 1$	1
2	$2(2) + 1$	5
-1	$2(-1) + 1$	-1

Graph the ordered-pair solutions $(0, 1)$, $(2, 5)$, and $(-1, -1)$. Draw a line through the ordered-pair solutions.

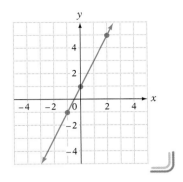

Note that the points whose coordinates are $(-2, -3)$ and $(1, 3)$ are on the graph and that these ordered pairs are solutions of the equation $y = 2x + 1$.

Remember that a graph is a drawing of the ordered-pair solutions of an equation. Therefore, every point on the graph is a solution of the equation, and every solution of the equation is a point on the graph.

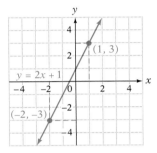

Example 3   Graph: $y = 3x - 2$

Solution

x	y
0	-2
2	4
-1	-5

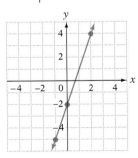

Problem 3   Graph: $y = 3x + 1$

Solution   See page S11.

Graphing utilities create graphs by plotting points and then connecting the points to form a curve. Using a graphing utility, enter the equation $y = 3x - 2$ and verify the graph drawn in Example 3. (Refer to the Appendix for instructions on using a graphing calculator to graph a linear equation.) Trace along the graph and verify that $(0, -2)$, $(2, 4)$, and $(-1, -5)$ are coordinates of points on the graph. Now enter the equation $y = 3x + 1$ given in Problem 3. Verify that the ordered pairs you found for this equation are the coordinates of points on the graph.

When m is a fraction in the equation $y = mx + b$, choose values of x that will simplify the evaluation. For example, to graph $y = \frac{1}{3}x - 1$, we might choose the numbers 0, 3, and -3 for x. Note that these numbers are multiples of the denominator of $\frac{1}{3}$, the coefficient of x.

x	y
0	-1
3	0
-3	-2

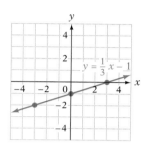

Example 4 Graph: $y = \frac{1}{2}x - 1$

Solution

x	y
0	-1
2	0
-2	-2

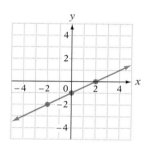

Problem 4 Graph: $y = \frac{1}{3}x - 3$

Solution See page S11.

Using a graphing utility, enter the equation $y = \frac{1}{2}x - 1$ and verify the graph drawn in Example 4. Trace along the graph and verify that $(0, -1)$, $(2, 0)$, and $(-2, -2)$ are the coordinates of points on the graph. Follow the same procedure for Problem 4. (See the Appendix for instructions on entering a fractional coefficient of x.)

---

3 **Graph equations of the form**
$Ax + By = C$

VIDEO & DVD   CD TUTOR   WEB   SSM

An equation in the form $Ax + By = C$, where A and B are coefficients and C is a constant, is also a linear equation. Examples of these equations are shown below.

$2x + 3y = 6$	$(A = 2, B = 3, C = 6)$
$x - 2y = -4$	$(A = 1, B = -2, C = -4)$
$2x + y = 0$	$(A = 2, B = 1, C = 0)$
$4x - 5y = 2$	$(A = 4, B = -5, C = 2)$

One method of graphing an equation of the form $Ax + By = C$ involves first solving the equation for y and then following the same procedure used for graphing an equation of the form $y = mx + b$. To solve the equation for y means to rewrite the equation so that y is alone on one side of the equation and the term containing x and the constant are on the other side of the equation. The Addition and Multiplication Properties of Equations are used to rewrite an equation of the form $Ax + By = C$ in the form $y = mx + b$.

Solve the equation $3x + 2y = 4$ for y.

The equation is in the form $Ax + By = C$.

$$3x + 2y = 4$$

Use the Addition Property of Equations to subtract the term $3x$ from each side of the equation.

$$3x - 3x + 2y = -3x + 4$$

Simplify. Note that on the right side of the equation, the term containing x is first, followed by the constant.

$$2y = -3x + 4$$

Use the Multiplication Property of Equations to multiply each side of the equation by the reciprocal of the coefficient of y. (The coefficient of y is 2; the reciprocal of 2 is $\frac{1}{2}$.)

$$\frac{1}{2} \cdot 2y = \frac{1}{2}(-3x + 4)$$

Simplify. Use the Distributive Property on the right side of the equation.

$$y = \frac{1}{2}(-3x) + \frac{1}{2}(4)$$

The equation is now in the form $y = mx + b$, with $m = -\frac{3}{2}$ and $b = 2$.

$$y = -\frac{3}{2}x + 2$$

In solving the equation $3x + 2y = 4$ for y, where we multiplied both sides of the equation by $\frac{1}{2}$, we could have divided both sides of the equation by 2, as shown at the right. In simplifying the right side after dividing both sides by 2, be sure to divide *each term* by 2.

$$2y = -3x + 4$$

$$\frac{2y}{2} = \frac{-3x + 4}{2}$$

$$y = \frac{-3x}{2} + \frac{4}{2}$$

$$y = -\frac{3}{2}x + 2$$

Being able to solve an equation of the form $Ax + By = C$ for y is important because most graphing utilities require that an equation be in the form $y = mx + b$ when the equation of the line is entered for graphing.

Example 5 Write $3x - 4y = 12$ in the form $y = mx + b$.

Solution

$$3x - 4y = 12$$

$$3x - 3x - 4y = -3x + 12$$ • Subtract 3x from each side of the equation.

$$-4y = -3x + 12$$ • Simplify.

$$\frac{-4y}{-4} = \frac{-3x + 12}{-4}$$ • Divide each side of the equation by 24.

$$y = \frac{-3x}{-4} + \frac{12}{-4}$$

$$y = \frac{3}{4}x - 3$$

Problem 5 Write $5x - 2y = 10$ in the form $y = mx + b$.

Solution See page S11.

To graph an equation of the form $Ax + By = C$, we can first solve the equation for y and then follow the same procedure used for graphing an equation of the form $y = mx + b$. An example follows.

Graph: $3x + 4y = 12$

Write the equation in the form $y = mx + b$. $3x + 4y = 12$

$$4y = -3x + 12$$

$$y = -\frac{3}{4}x + 3$$

Find at least three solutions.
Display the ordered pairs in a table.

x	y
0	3
4	0
−4	6

Graph the ordered pairs on a rectangular coordinate system, and draw a straight line through the points.

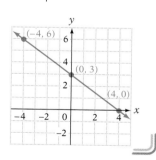

Example 6 Graph.   **A.** $2x - 5y = 10$ **B.** $x + 2y = 6$

Solution   **A.** $2x - 5y = 10$ **B.** $x + 2y = 6$

          $-5y = -2x + 10$ $2y = -x + 6$

            $y = \frac{2}{5}x - 2$ $y = -\frac{1}{2}x + 3$

x	y
0	-2
5	0
-5	-4

x	y
0	3
-2	4
4	1

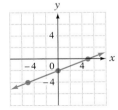

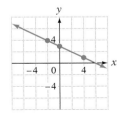

Problem 6   Graph.   **A.** $5x - 2y = 10$ **B.** $x - 3y = 9$

Solution   See page S12.

The graph of the equation $2x + 3y = 6$ is shown at the right. The graph crosses the x-axis at $(3, 0)$. This point is called the **x-intercept.** The graph crosses the y-axis at $(0, 2)$. This point is called the **y-intercept.**

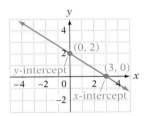

Find the x-intercept and the y-intercept of the graph of the equation $2x + 3y = 6$ algebraically.

To find the x-intercept, let $y = 0$.
(Any point on the x-axis has y-coordinate 0.)

$$2x + 3y = 6$$
$$2x + 3(0) = 6$$
$$2x = 6$$
$$x = 3$$

The x-intercept is $(3, 0)$.

To find the y-intercept, let $x = 0$.
(Any point on the y-axis has x-coordinate 0.)

$$2x + 3y = 6$$
$$2(0) + 3y = 6$$
$$3y = 6$$
$$y = 2$$

The y-intercept is $(0, 2)$.

Take Note

To find the x-intercept, let $y = 0$.
To find the y-intercept, let $x = 0$.

Another method of graphing some equations of the form $Ax + By = C$ is to find the x- and y-intercepts, plot both intercepts, and then draw a line through the two points.

Example 7 Find the x- and y-intercepts for $x - 2y = 4$. Graph the line.

Solution

x-intercept: $x - 2y = 4$

$x - 2(0) = 4$ • To find the x-intercept, let $y = 0$.

$x = 4$

The x-intercept is $(4, 0)$.

y-intercept: $x - 2y = 4$

$0 - 2y = 4$ • To find the y-intercept, let $x = 0$.

$-2y = 4$

$y = -2$

The y-intercept is $(0, -2)$.

• Graph the ordered pairs $(4, 0)$ and $(0, -2)$. Draw a straight line through the points.

Problem 7 Find the x- and y-intercepts for $4x - y = 4$. Graph the line.

Solution See page S12.

The graph of an equation in which one of the variables is missing is either a horizontal line or a vertical line.

The equation $y = 2$ could be written $0 \cdot x + y = 2$. No matter what value of x is chosen, y is always 2. Some solutions to the equation are $(3, 2)$, $(-1, 2)$, $(0, 2)$, and $(-4, 2)$. The graph is shown at the right.

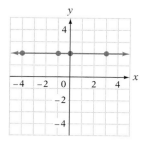

The **graph of $y = b$** is a horizontal line passing through the point whose coordinates are $(0, b)$.

Note that $(0, b)$ is the y-intercept of the graph of $y = b$. An equation of the form $y = b$ does not have an x-intercept.

The equation $x = -2$ could be written $x + 0 \cdot y = -2$. No matter what value of y is chosen, x is always -2. Some solutions to the equation are $(-2, 3)$, $(-2, -2)$, $(-2, 0)$, and $(-2, 2)$. The graph is shown at the right.

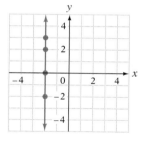

The **graph of $x = a$** is a vertical line passing through the point whose coordinates are $(a, 0)$.

Note that $(a, 0)$ is the x-intercept of the graph of $x = a$. An equation of the form $x = a$ does not have a y-intercept.

Example 8 Graph.   **A.** $y = -2$ **B.** $x = 3$

Solution   **A.** The graph of an equation of the form $y = b$ is a horizontal line with y-intercept $(0, b)$.

B. The graph of an equation of the form $x = a$ is a vertical line with x-intercept $(a, 0)$.

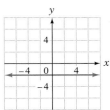

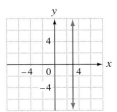

Problem 8 Graph.   **A.** $y = 3$ **B.** $x = -4$

Solution   See page S12.

| 5.2 | **Concept Review** |

Determine whether the statement is always true, sometimes true, or never true.

1. The equation $y = 3x^2 - 6$ is an example of a linear equation in two variables.

2. The value of m in the equation $y = 4x + 9$ is 9.

3. $(3, 5)$ is a solution of the equation $y = x + 2$.

4. The graph of a linear equation in two variables is a straight line.

5. To find the x-intercept of the graph of a linear equation, let $x = 0$.

6. The graph of an equation of the form $y = b$ is a horizontal line.

7. The graph of the equation $x = -4$ has an x-intercept of $(0, -4)$.

8. The graph shown at the right has a y-intercept of $(0, 3)$.

9. The graph of the equation $x = 2$ is a line that goes through every point that has an x-coordinate of 2.

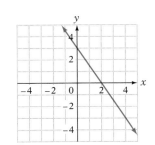

5.2 Exercises

1 For the given linear equation, find the value of m and the value of b.

1. $y = 4x + 1$

2. $y = -3x - 2$

3. $y = \dfrac{5}{6}x + \dfrac{1}{6}$

4. $y = -x$

5. Is $(3, 4)$ a solution of $y = -x + 7$?

6. Is $(2, -3)$ a solution of $y = x + 5$?

7. Is $(-1, 2)$ a solution of $y = \dfrac{1}{2}x - 1$?

8. Is $(1, -3)$ a solution of $y = -2x - 1$?

9. Is $(4, 1)$ a solution of $y = \dfrac{1}{4}x + 1$?

10. Is $(-5, 3)$ a solution of $y = -\dfrac{2}{5}x + 1$?

11. Is $(0, 4)$ a solution of $y = \dfrac{3}{4}x + 4$?

12. Is $(-2, 0)$ a solution of $y = -\dfrac{1}{2}x - 1$?

13. Is $(0, 0)$ a solution of $y = 3x + 2$?

14. Is $(0, 0)$ a solution of $y = -\dfrac{3}{4}x$?

15. Find the ordered-pair solution of $y = 3x - 2$ that corresponds to $x = 3$.

16. Find the ordered-pair solution of $y = 4x + 1$ that corresponds to $x = -1$.

17. Find the ordered-pair solution of $y = \dfrac{2}{3}x - 1$ that corresponds to $x = 6$.

18. Find the ordered-pair solution of $y = \dfrac{3}{4}x - 2$ that corresponds to $x = 4$.

19. Find the ordered-pair solution of $y = -3x + 1$ that corresponds to $x = 0$.

20. Find the ordered-pair solution of $y = \dfrac{2}{5}x - 5$ that corresponds to $x = 0$.

21. Find the ordered-pair solution of $y = \dfrac{2}{5}x + 2$ that corresponds to $x = -5$.

22. Find the ordered-pair solution of $y = -\dfrac{1}{6}x - 2$ that corresponds to $x = 12$.

2 Graph.

23. $y = 2x - 3$

24. $y = -2x + 2$

25. $y = \dfrac{1}{3}x$

26. $y = -3x$

27. $y = \frac{2}{3}x - 1$

28. $y = \frac{3}{4}x + 2$

29. $y = -\frac{1}{4}x + 2$

30. $y = -\frac{1}{3}x + 1$

31. $y = -\frac{2}{5}x + 1$

32. $y = -\frac{1}{2}x + 3$

33. $y = 2x - 4$

34. $y = 3x - 4$

35. $y = -x + 2$

36. $y = -x - 1$

37. $y = -\frac{2}{3}x + 1$

38. $y = 5x - 4$

39. $y = -3x + 2$

40. $y = -x + 3$

 Graph using a graphing utility.

41. $y = 3x - 4$

42. $y = -2x + 3$

43. $y = 2x - 3$

44. $y = -2x - 3$

45. $y = -\frac{2}{3}x$

46. $y = \frac{3}{2}x$

47. $y = \frac{3}{4}x + 2$

48. $y = -\frac{1}{3}x + 2$

49. $y = -\frac{3}{2}x - 3$

50. $y = -\frac{2}{5}x + 2$

51. $y = \frac{1}{4}x - 2$

52. $y = \frac{2}{3}x - 4$

3 Solve for y. Write the equation in the form $y = mx + b$.

53. $3x + y = 10$ **54.** $2x + y = 5$ **55.** $4x - y = 3$ **56.** $5x - y = 7$

57. $3x + 2y = 6$ **58.** $2x + 3y = 9$ **59.** $2x - 5y = 10$ **60.** $5x - 2y = 4$

61. $2x + 7y = 14$ **62.** $6x - 5y = 10$ **63.** $x + 3y = 6$ **64.** $x - 4y = 12$

Graph.

65. $3x + y = 3$ **66.** $2x + y = 4$ **67.** $2x + 3y = 6$ **68.** $3x + 2y = 4$

69. $x - 2y = 4$ **70.** $x - 3y = 6$ **71.** $2x - 3y = 6$ **72.** $3x - 2y = 8$

73. $y = 4$ **74.** $y = -4$ **75.** $x = -2$ **76.** $x = 3$

Find the x- and y-intercepts.

77. $x - y = 3$ **78.** $3x + 4y = 12$ **79.** $y = 2x - 6$ **80.** $y = 2x + 10$

81. $x - 5y = 10$ **82.** $3x + 2y = 12$ **83.** $y = 3x + 12$ **84.** $y = 5x + 10$

85. $2x - 3y = 0$ **86.** $3x + 4y = 0$ **87.** $y = \frac{1}{2}x + 3$ **88.** $y = \frac{2}{3}x - 4$

Graph by using x- and y-intercepts.

89. $5x + 2y = 10$ **90.** $2x - y = 4$ **91.** $3x - 4y = 12$

92. $2x - 3y = -6$ **93.** $2x + 3y = 6$ **94.** $x + 2y = 4$

95. $2x + 5y = 10$ **96.** $3x + 4y = 12$ **97.** $x - 3y = 6$

98. $3x - y = 6$ **99.** $x - 4y = 8$ **100.** $4x + 3y = 12$

101. $2x + y = 3$ **102.** $3x + y = -5$ **103.** $4x - 3y = 6$

Graph using a graphing utility. Verify that the graph has the correct x- and y-intercepts.

104. $x - 2y = -4$ **105.** $3x + 4y = -12$ **106.** $2x - 3y = -6$

107. $2x - y = 4$ **108.** $3x - 4y = 4$ **109.** $2x - 3y = 9$

5.2 Applying Concepts

110. a. Show that the equation $y + 3 = 2(x + 4)$ is a linear equation by writing it in the form $y = mx + b$.
  b. Find the ordered-pair solution that corresponds to $x = -4$.

111. a. Show that the equation $y + 4 = -\frac{1}{2}(x + 2)$ is a linear equation by writing it in the form $y = mx + b$.
  b. Find the ordered-pair solution that corresponds to $x = -2$.

112. For the linear equation $y = 2x - 3$, what is the increase in y that results when x is increased by 1?

113. For the linear equation $y = -x - 4$, what is the decrease in y that results when x is increased by 1?

114. Write the equation of a line that has $(0, 0)$ as both the x-intercept and the y-intercept.

115. Explain **a.** why the y-coordinate of any point on the x-axis is 0 and **b.** why the x-coordinate of any point on the y-axis is 0.

<table><tr><td>**5.3**</td><td>## Slopes of Straight Lines</td></tr></table>

1 **Find the slope of a straight line**

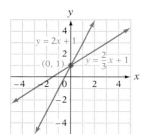

The graphs of $y = \frac{2}{3}x + 1$ and $y = 2x + 1$ are shown at the left. Each graph crosses the y-axis at $(0, 1)$, but the graphs have different slants. The **slope** of a line is a measure of the slant of the line. The symbol for slope is m.

The slope of a line is the ratio of the change in the y-coordinates between any two points on the line to the change in the x-coordinates.

The line containing the points whose coordinates are $(-2, -3)$ and $(6, 1)$ is graphed at the right. The change in y is the difference between the two y-coordinates.

Change in $y = 1 - (-3) = 4$

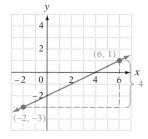

Take Note

The change in the y values can be thought of as the *rise* of the line, and the change in the x values can be thought of as the *run*. Then

slope $= m = \dfrac{\text{rise}}{\text{run}}$.

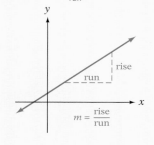

The change in x is the difference between the two x-coordinates.

Change in $x = 6 - (-2) = 8$

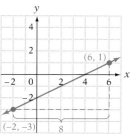

For the line containing the points whose coordinates are $(-2, -3)$ and $(6, 1)$,

$$\text{Slope} = m = \frac{\text{change in } y}{\text{change in } x} = \frac{4}{8} = \frac{1}{2}$$

The slope of the line can also be described as the ratio of the vertical change (4 units) to the horizontal change (8 units) from the point whose coordinates are $(-2, -3)$ to the point whose coordinates are $(6, 1)$.

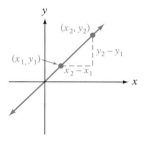

> **Slope Formula**
>
> The slope of a line containing two points P_1 and P_2, whose coordinates are (x_1, y_1) and (x_2, y_2), is given by
>
> $$\text{Slope} = m = \frac{y_2 - y_1}{x_2 - x_1}, \; x_1 \neq x_2$$

In the slope formula, the points P_1 and P_2 are any two points on the line. The slope of a line is constant; therefore, the slope calculated using any two points on the line will be the same.

In Section 1, Objective 3 of this chapter, we calculated the average rate of change in the U.S. population for different intervals of time. In each case, the average rate of change was different. Note that the graph of the population data did not lie on a straight line.

The graph of an equation of the form $y = mx + b$ is a straight line, and the average rate of change is constant. We can calculate the average rate of change between any two points on the line and it will always be the same. For a linear equation, the rate of change is referred to as slope.

 Find the slope of the line containing the points whose coordinates are $(-1, 1)$ and $(2, 3)$.

Let $P_1 = (-1, 1)$ and $P_2 = (2, 3)$. Then $x_1 = -1$, $y_1 = 1$, $x_2 = 2$, and $y_2 = 3$.

$$m = \frac{y_2 - y_1}{x_2 - x_1} = \frac{3 - 1}{2 - (-1)} = \frac{2}{3}$$

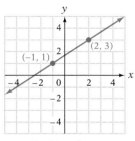

Positive slope

It does not matter which point is named P_1 and which P_2; the slope of the line will be the same. If the points are reversed, then $P_1 = (2, 3)$ and $P_2 = (-1, 1)$.

$$m = \frac{y_2 - y_1}{x_2 - x_1} = \frac{1 - 3}{-1 - 2} = \frac{-2}{-3} = \frac{2}{3}$$

This is the same result. Here the slope is a positive number. A line that slants upward to the right has a **positive slope.**

Take Note

Positive slope means that the value of y increases as the value of x increases.

Find the slope of the line containing the points whose coordinates are $(-3, 4)$ and $(2, -2)$.

Let $P_1 = (-3, 4)$ and $P_2 = (2, -2)$.

$$m = \frac{y_2 - y_1}{x_2 - x_1} = \frac{-2 - 4}{2 - (-3)} = \frac{-6}{5} = -\frac{6}{5}$$

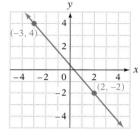

Here the slope is a negative number. A line that slants downward to the right has a **negative slope.**

Negative slope

Find the slope of the line containing the points whose coordinates are $(-1, 3)$ and $(4, 3)$.

Let $P_1 = (-1, 3)$ and $P_2 = (4, 3)$.

$$m = \frac{y_2 - y_1}{x_2 - x_1} = \frac{3 - 3}{4 - (-1)} = \frac{0}{5} = 0$$

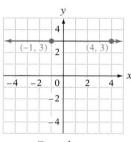

When $y_1 = y_2$, the graph is a horizontal line.

A horizontal line has **zero slope.**

Zero slope

Find the slope of the line containing the points whose coordinates are $(2, -2)$ and $(2, 4)$.

Let $P_1 = (2, -2)$ and $P_2 = (2, 4)$.

$$m = \frac{y_2 - y_1}{x_2 - x_1} = \frac{4 - (-2)}{2 - 2} = \frac{6}{0} \leftarrow \begin{array}{l}\text{Not a real}\\\text{number}\end{array}$$

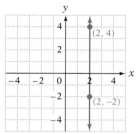

When $x_1 = x_2$, the denominator of $\frac{y_2 - y_1}{x_2 - x_1}$ is 0 and the graph is a vertical line. Because division by zero is not defined, the slope of the line is not defined.

The slope of a vertical line is **undefined.**

Undefined slope

Example 1 Find the slope of the line containing the points P_1 and P_2.

   A. $P_1(-2, -1)$, $P_2(3, 4)$ **B.** $P_1(-3, 1)$, $P_2(2, -2)$
   C. $P_1(-1, 4)$, $P_2(-1, 0)$ **D.** $P_1(-1, 2)$, $P_2(4, 2)$

Solution   **A.** $m = \dfrac{y_2 - y_1}{x_2 - x_1} = \dfrac{4 - (-1)}{3 - (-2)} = \dfrac{5}{5} = 1$

   The slope is 1.

   B. $m = \dfrac{y_2 - y_1}{x_2 - x_1} = \dfrac{-2 - 1}{2 - (-3)} = \dfrac{-3}{5}$

   The slope is $-\dfrac{3}{5}$.

C. $m = \dfrac{y_2 - y_1}{x_2 - x_1} = \dfrac{0 - 4}{-1 - (-1)} = \dfrac{-4}{0}$

The slope is undefined.

D. $m = \dfrac{y_2 - y_1}{x_2 - x_1} = \dfrac{2 - 2}{4 - (-1)} = \dfrac{0}{5} = 0$

The slope is 0.

Problem 1   Find the slope of the line containing the points P_1 and P_2.

A. $P_1(-1, 2), P_2(1, 3)$ **B.** $P_1(1, 2), P_2(4, -5)$

C. $P_1(2, 3), P_2(2, 7)$ **D.** $P_1(1, -3), P_2(-5, -3)$

Solution   See page S12.

Two lines in the rectangular coordinate system that never intersect are called **parallel lines.** The lines l_1 and l_2 graphed at the right are parallel lines. Calculating the slope of each line, we have

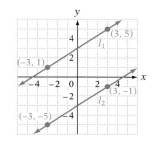

$$\text{Slope of } l_1 = \frac{y_2 - y_1}{x_2 - x_1} = \frac{5 - 1}{3 - (-3)} = \frac{4}{6} = \frac{2}{3}$$

$$\text{Slope of } l_2 = \frac{y_2 - y_1}{x_2 - x_1} = \frac{-1 - (-5)}{3 - (-3)} = \frac{4}{6} = \frac{2}{3}$$

Note that these parallel lines have the same slope. This is true of all parallel lines.

Take Note

We must distinguish between vertical and nonvertical lines in the description of parallel lines at the right because vertical lines in the rectangular coordinate system are parallel, but their slopes are undefined.

Parallel Lines

Two nonvertical lines in the rectangular coordinate system are parallel if and only if they have the same slope. Any two vertical lines in the rectangular coordinate system are parallel lines.

Example 2   Is the line containing the points $(-2, 1)$ and $(-5, -1)$ parallel to the line that contains the points $(1, 0)$ and $(4, 2)$?

Solution   Find the slope of the line through $(-2, 1)$ and $(-5, -1)$.

$$\frac{y_2 - y_1}{x_2 - x_1} = \frac{-1 - 1}{-5 - (-2)} = \frac{-2}{-3} = \frac{2}{3}$$

Find the slope of the line through $(1, 0)$ and $(4, 2)$.

$$\frac{y_2 - y_1}{x_2 - x_1} = \frac{2 - 0}{4 - 1} = \frac{2}{3}$$

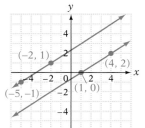

The slopes are equal.
The lines are parallel.

Problem 2   Is the line containing the points $(-2, -3)$ and $(7, 1)$ parallel to the line that contains the points $(1, 4)$ and $(-5, 6)$?

Solution   See page S12.

There are many applications of the concept of slope. Here is one example.

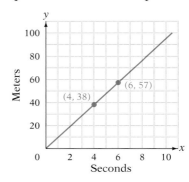

When Florence Griffith-Joyner set the world record for the 100-meter dash, her rate of speed was approximately 9.5 m/s. The graph at the right shows the distance she ran during her record-setting run. From the graph, note that after 4 s she had traveled 38 m and that after 6 s she had traveled 57 m. The slope of the line between these two points is

$$m = \frac{57 - 38}{6 - 4} = \frac{19}{2} = 9.5$$

Note that the slope of the line is the same as the rate at which she was running, 9.5 m/s. The average speed of an object is related to slope.

Example 3   The graph at the right shows the height of a plane above an airport during its 30-minute descent from cruising altitude to landing. Find the slope of the line. Write a sentence that explains the meaning of the slope.

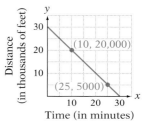

Solution

$$m = \frac{y_2 - y_1}{x_2 - x_1}$$

$$= \frac{5000 - 20{,}000}{25 - 10} = \frac{-15{,}000}{15} = -1000$$

A slope of -1000 means that the height of the plane is *decreasing* at the rate of 1000 ft per minute.

Problem 3   The graph at the right shows the decline in the value of a used car over a five-year period. Find the slope of the line. Write a sentence that states the meaning of the slope.

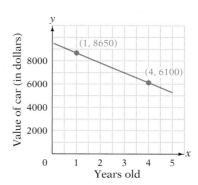

Solution   See page S12.

2 **Graph a line using the slope and y-intercept**

Recall that we can find the y-intercept of a linear equation by letting $x = 0$. To find the y-intercept of $y = 3x + 4$, let $x = 0$.

$$y = 3x + 4$$
$$y = 3(0) + 4$$
$$y = 4$$

The y-intercept is $(0, 4)$.

The constant term of $y = 3x + 4$ is the y-coordinate of the y-intercept.

In general, for any equation of the form $y = mx + b$, the y-intercept is $(0, b)$.

The graph of the equation $y = \frac{2}{3}x + 1$ is shown at the right. The points whose coordinates are $(-3, -1)$ and $(3, 3)$ are on the graph. The slope of the line is

$$m = \frac{3 - (-1)}{3 - (-3)} = \frac{4}{6} = \frac{2}{3}$$

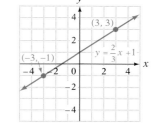

Note that the slope of the line has the same value as the coefficient of x.

Slope-Intercept Form of a Straight Line

For any equation of the form $y = mx + b$, the slope of the line is m, the coefficient of x. The y-intercept is $(0, b)$. The equation

$$y = mx + b$$

is called the **slope-intercept form of a straight line.**

The slope of the graph of $y = -\frac{3}{4}x + 1$ is $-\frac{3}{4}$. The y-intercept is $(0, 1)$.

$$y = \boxed{m}\, x + \boxed{b}$$
$$y = \boxed{-\frac{3}{4}}\, x + \boxed{1}$$

$$\text{Slope} = m = -\frac{3}{4} \qquad\qquad y\text{-intercept} = (0, b) = (0, 1)$$

Find the slope and y-intercept of the graph of $y = \frac{5}{2}x - 6$.

$$\text{Slope} = m = \frac{5}{2} \qquad b = -6$$

The slope is $\frac{5}{2}$.   The y-intercept is $(0, -6)$.

When the equation of a straight line is in the form $y = mx + b$, the graph can be drawn using the slope and y-intercept. First locate the y-intercept. Use the slope to find a second point on the line. Then draw a line through the two points.

Graph $y = 2x - 3$ by using the slope and y-intercept.

y-intercept $= (0, b) = (0, -3)$

$$m = 2 = \frac{2}{1} = \frac{\text{change in } y}{\text{change in } x}$$

Beginning at the y-intercept, move right 1 unit (change in x) and then up 2 units (change in y).

$(1, -1)$ are the coordinates of a second point on the graph.

Draw a line through $(0, -3)$ and $(1, -1)$.

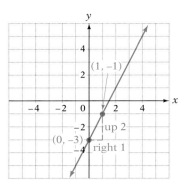

Using a graphing utility, enter the equation $y = 2x - 3$ and verify the graph shown above. Trace along the graph and verify that $(0, -3)$ is the y-intercept and that the point whose coordinates are $(1, -1)$ is on the graph.

Example 4 Graph $y = -\frac{2}{3}x + 1$ by using the slope and y-intercept.

Solution y-intercept $= (0, b) = (0, 1)$

$$m = -\frac{2}{3} = \frac{-2}{3}$$

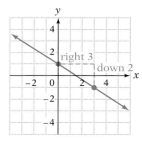

• A slope of $\dfrac{-2}{3}$ means to move right 3 units and then down 2 units.

Problem 4 Graph $y = -\frac{1}{4}x - 1$ by using the slope and y-intercept.

Solution See page S12.

Example 5 Graph $2x - 3y = 6$ by using the slope and y-intercept.

Solution $2x - 3y = 6$

$\qquad -3y = -2x + 6$ • Solve the equation for y.

$\qquad\quad y = \dfrac{2}{3}x - 2$

y-intercept $= (0, -2)$

$m = \dfrac{2}{3}$

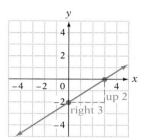

Problem 5   Graph $x - 2y = 4$ by using the slope and y-intercept.

Solution   See page S12.

5.3   Concept Review

Determine whether the statement is always true, sometimes true, or never true.

1. In the equation $y = mx + b$, m is the symbol for slope.

2. The formula for slope is $m = \dfrac{y_2 - y_1}{x_2 - x_1}$.

3. A vertical line has zero slope, and the slope of a horizontal line is undefined.

4. Any two points on a line can be used to find the slope of the line.

5. Shown at the right is the graph of a line with negative slope.

6. The y-intercept of the graph of the equation $y = -2x + 5$ is (5, 0).

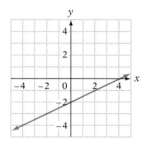

5.3   Exercises

1. Explain how to find the slope of a line when you know two points on the line.

2. What is the difference between a line that has zero slope and a line whose slope is undefined?

Find the slope of the line containing the points P_1 and P_2.

3. $P_1(4, 2)$, $P_2(3, 4)$

4. $P_1(2, 1)$, $P_2(3, 4)$

5. $P_1(-1, 3)$, $P_2(2, 4)$

6. $P_1(-2, 1)$, $P_2(2, 2)$

7. $P_1(2, 4)$, $P_2(4, -1)$

8. $P_1(1, 3)$, $P_2(5, -3)$

9. $P_1(-2, 3)$, $P_2(2, 1)$

10. $P_1(5, -2)$, $P_2(1, 0)$

11. $P_1(8, -3)$, $P_2(4, 1)$

12. $P_1(0, 3)$, $P_2(2, -1)$

13. $P_1(3, -4)$, $P_2(3, 5)$

14. $P_1(-1, 2)$, $P_2(-1, 3)$

15. $P_1(4, -2)$, $P_2(3, -2)$

16. $P_1(5, 1)$, $P_2(-2, 1)$

17. $P_1(0, -1)$, $P_2(3, -2)$

18. $P_1(3, 0)$, $P_2(2, -1)$

19. $P_1(-2, 3)$, $P_2(1, 3)$

20. $P_1(4, -1)$, $P_2(-3, -1)$

21. $P_1(-2, 4)$, $P_2(-1, -1)$

22. $P_1(6, -4)$, $P_2(4, -2)$

23. $P_1(-2, -3)$, $P_2(-2, 1)$

24. $P_1(5, 1)$, $P_2(5, -2)$

25. $P_1(-1, 5)$, $P_2(5, 1)$

26. $P_1(-1, 5)$, $P_2(7, 1)$

27. Is the line containing the points $(-4, 2)$ and $(1, 6)$ parallel to the line that contains the points $(2, -4)$ and $(7, 0)$?

28. Is the line containing the points $(-5, 0)$ and $(0, 2)$ parallel to the line that contains the points $(5, 1)$ and $(0, -1)$?

29. Is the line containing the points $(-2, -3)$ and $(7, 1)$ parallel to the line that contains the points $(6, -5)$ and $(4, 1)$?

30. Is the line containing the points $(4, -3)$ and $(2, 5)$ parallel to the line that contains the points $(-2, -3)$ and $(-4, 1)$?

31. **Postal Service** The graph at the right shows the work accomplished by an electronic mail sorter. Find the slope of the line. Write a sentence that explains the meaning of the slope.

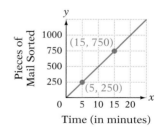

32. **Health** The graph at the right shows the relationship between distance walked and calories burned. Find the slope of the line. Write a sentence that explains the meaning of the slope.

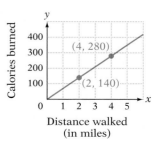

33. Fuel Consumption The graph at the right shows how the amount of gasoline in the tank of a car decreases as the car is driven at a constant speed of 60 mph. Find the slope of the line. Write a sentence that states the meaning of the slope.

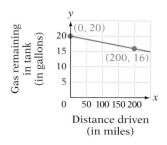

34. Computer Science The graph at the right shows the relationship between time and the number of kilobytes of a file remaining to be downloaded. Find the slope of the line. Write a sentence that states the meaning of the slope.

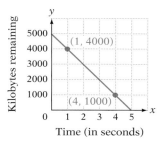

2 **35.** Why is an equation of the form $y = mx + b$ said to be in slope-intercept form?

36. Explain how to graph the equation $y = \frac{2}{3}x + 4$ by using the slope and the y-intercept.

Graph by using the slope and y-intercept.

37. $y = 3x + 1$

38. $y = -2x - 1$

39. $y = \frac{2}{5}x - 2$

40. $y = \frac{3}{4}x + 1$

41. $2x + y = 3$

42. $3x - y = 1$

43. $x - 2y = 4$

44. $x + 3y = 6$

45. $y = \frac{2}{3}x$

46. $y = \frac{1}{2}x$

47. $y = -x + 1$

48. $y = -x - 3$

49. $3x - 4y = 12$ **50.** $5x - 2y = 10$ **51.** $y = -4x + 2$

52. $y = 5x - 2$ **53.** $4x - 5y = 20$ **54.** $x - 3y = 6$

5.3   Applying Concepts

55. What effect does increasing the coefficient of x have on the graph of $y = mx + b, m > 0$?

56. What effect does decreasing the coefficient of x have on the graph of $y = mx + b, m > 0$?

57. What effect does increasing the constant term have on the graph of $y = mx + b$?

58. What effect does decreasing the constant term have on the graph of $y = mx + b$?

59. Match each equation with its graph.

i. $y = -2x + 4$

ii. $y = 2x - 4$

iii. $y = 2$

iv. $2x + 4y = 0$

v. $y = \dfrac{1}{2}x + 4$

vi. $y = -\dfrac{1}{4}x - 2$

A. 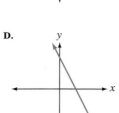   **B.** 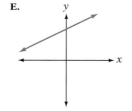   **C.**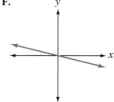

D.   **E.**   **F.**

60. Which of the graphs shows a constant rate of change?

a. 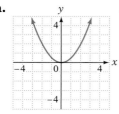   **b.** 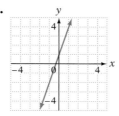   **c.** 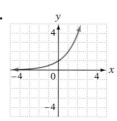   **d.**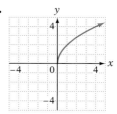

61. Do the graphs of all straight lines have a *y*-intercept? If not, give an example of one that does not.

62. If two lines have the same slope and the same *y*-intercept, must the graphs of the lines be the same? If not, give an example.

63.  What does the highway sign shown at the right have to do with slope?

6% GRADE
7 MILES

5.4   Equations of Straight Lines

1 **Find the equation of a line using the equation $y = mx + b$**

When the slope of a line and a point on the line are known, the equation of the line can be written using the slope-intercept form, $y = mx + b$. In the first example shown below, the known point is the *y*-intercept. In the second example, the known point is a point other than the *y*-intercept.

Find the equation of the line that has slope 3 and *y*-intercept (0, 2).

The given slope, 3, is *m*.	$y = mx + b$
Replace *m* with 3.	$y = 3x + b$
The given point, (0, 2), is the *y*-intercept. Replace *b* with 2.	$y = 3x + 2$

The equation of the line that has slope 3 and *y*-intercept 2 is $y = 3x + 2$.

Find the equation of the line that has slope $\frac{1}{2}$ and contains the point whose coordinates are (−2, 4).

The given slope, $\frac{1}{2}$, is *m*.	$y = mx + b$
Replace *m* with $\frac{1}{2}$.	$y = \frac{1}{2}x + b$
The given point, (−2, 4), is a solution of the equation of the line. Replace *x* and *y* in the equation with the coordinates of the point.	$4 = \frac{1}{2}(-2) + b$

Take Note

Every ordered pair is of the form (*x, y*). For the point (−2, 4), −2 is the *x* value and 4 is the *y* value. Substitute −2 for *x* and 4 for *y*.

Solve for b, the y-intercept.

$$4 = -1 + b$$
$$5 = b$$

Write the equation of the line by replacing m and b in the equation by their values.

$$y = mx + b$$
$$y = \frac{1}{2}x + 5$$

The equation of the line that has slope $\frac{1}{2}$ and contains the point whose coordinates are $(-2, 4)$ is $y = \frac{1}{2}x + 5$.

Example 1 Find the equation of the line that contains the point whose coordinates are $(3, -3)$ and has slope $\frac{2}{3}$.

Solution $y = mx + b$

$y = \frac{2}{3}x + b$ • Replace m with the given slope.

$-3 = \frac{2}{3}(3) + b$ • Replace x and y in the equation with the coordinates of the given point.

$-3 = 2 + b$ • Solve for b.
$-5 = b$

$y = \frac{2}{3}x - 5$ • Write the equation of the line by replacing m and b in $y = mx + b$ by their values.

Problem 1 Find the equation of the line that contains the point whose coordinates are $(4, -2)$ and has slope $\frac{3}{2}$.

Solution See page S13.

2 Find the equation of a line using the point-slope formula

VIDEO & DVD CD TUTOR WEB SSM

An alternative method for finding the equation of a line, given the slope and the coordinates of a point on the line, involves use of the point-slope formula. The point-slope formula is derived from the formula for slope.

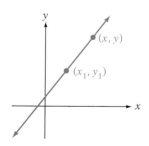

Let (x_1, y_1) be the coordinates of the given point on the line, and let (x, y) be the coordinates of any other point on the line.

Use the formula for slope.

$$\frac{y - y_1}{x - x_1} = m$$

Multiply both sides of the equation by $(x - x_1)$.

$$\frac{y - y_1}{x - x_1}(x - x_1) = m(x - x_1)$$

Simplify.

$$y - y_1 = m(x - x_1)$$

Point-Slope Formula

The equation of the line that has slope m and contains the point whose coordinates are (x_1, y_1) can be found by the point-slope formula:

$$y - y_1 = m(x - x_1)$$

Example 2 Use the point-slope formula to find the equation of a line that passes through the point whose coordinates are $(-2, -1)$ and has slope $\frac{3}{2}$.

Solution $(x_1, y_1) = (-2, -1)$ • Let (x_1, y_1) be the given point.

$$m = \frac{3}{2}$$ • m is the given slope.

$$y - y_1 = m(x - x_1)$$ • This is the point-slope formula.

$$y - (-1) = \frac{3}{2}[x - (-2)]$$ • Substitute -2 for x_1, -1 for y_1, and $\frac{3}{2}$ for m.

$$y + 1 = \frac{3}{2}(x + 2)$$ • Rewrite the equation in the form $y = mx + b$.

$$y + 1 = \frac{3}{2}x + 3$$

$$y = \frac{3}{2}x + 2$$

Problem 2 Use the point-slope formula to find the equation of a line that passes through the point whose coordinates are $(5, 4)$ and has slope $\frac{2}{5}$.

Solution See page S13.

5.4   Concept Review

Determine whether the statement is always true, sometimes true, or never true.

1. The graph of the equation $y = 5x + 7$ has slope 5 and y-intercept $(0, 7)$.

2. If the equation of a line has y-intercept $(0, 4)$, then 4 can be substituted for b in the equation $y = mx + b$.

3. Suppose the equation of a line contains the point $(-3, 1)$. Then when y is -3, x is 1.

4. The point-slope formula is $y - y_1 = mx - x_1$.

5. The point-slope formula can be used to find the equation of a line with zero slope.

6. If it is stated that the y-intercept is 2, then the y-intercept is the point $(2, 0)$.

5.4 | Exercises

1 Use the slope-intercept form.

1. Find the equation of the line that contains the point whose coordinates are (0, 2) and has slope 2.

2. Find the equation of the line that contains the point whose coordinates are (0, −1) and has slope −2.

3. Find the equation of the line that contains the point whose coordinates are (−1, 2) and has slope −3.

4. Find the equation of the line that contains the point whose coordinates are (2, −3) and has slope 3.

5. Find the equation of the line that contains the point whose coordinates are (3, 1) and has slope $\frac{1}{3}$.

6. Find the equation of the line that contains the point whose coordinates are (−2, 3) and has slope $\frac{1}{2}$.

7. Find the equation of the line that contains the point whose coordinates are (4, −2) and has slope $\frac{3}{4}$.

8. Find the equation of the line that contains the point whose coordinates are (2, 3) and has slope $-\frac{1}{2}$.

9. Find the equation of the line that contains the point whose coordinates are (5, −3) and has slope $-\frac{3}{5}$.

10. Find the equation of the line that contains the point whose coordinates are (5, −1) and has slope $\frac{1}{5}$.

11. Find the equation of the line that contains the point whose coordinates are (2, 3) and has slope $\frac{1}{4}$.

12. Find the equation of the line that contains the point whose coordinates are (−1, 2) and has slope $-\frac{1}{2}$.

13. Find the equation of the line that contains the point whose coordinates are (−3, −5) and has slope $-\frac{2}{3}$.

14. Find the equation of the line that contains the point whose coordinates are $(-4, 0)$ and has slope $\frac{5}{2}$.

2 **15.** When is the point-slope formula used?

16. Explain the meaning of each variable in the point-slope formula.

Use the point-slope formula.

17. Find the equation of the line that passes through the point whose coordinates are $(1, -1)$ and has slope 2.

18. Find the equation of the line that passes through the point whose coordinates are $(2, 3)$ and has slope -1.

19. Find the equation of the line that passes through the point whose coordinates are $(-2, 1)$ and has slope -2.

20. Find the equation of the line that passes through the point whose coordinates are $(-1, -3)$ and has slope -3.

21. Find the equation of the line that passes through the point whose coordinates are $(0, 0)$ and has slope $\frac{2}{3}$.

22. Find the equation of the line that passes through the point whose coordinates are $(0, 0)$ and has slope $-\frac{1}{5}$.

23. Find the equation of the line that passes through the point whose coordinates are $(2, 3)$ and has slope $\frac{1}{2}$.

24. Find the equation of the line that passes through the point whose coordinates are $(3, -1)$ and has slope $\frac{2}{3}$.

25. Find the equation of the line that passes through the point whose coordinates are $(-4, 1)$ and has slope $-\frac{3}{4}$.

26. Find the equation of the line that passes through the point whose coordinates are $(-5, 0)$ and has slope $-\frac{1}{5}$.

27. Find the equation of the line that passes through the point whose coordinates are $(-2, 1)$ and has slope $\frac{3}{4}$.

28. Find the equation of the line that passes through the point whose coordinates are $(3, -2)$ and has slope $\frac{1}{6}$.

29. Find the equation of the line that passes through the point whose coordinates are $(-3, -5)$ and has slope $-\frac{4}{3}$.

30. Find the equation of the line that passes through the point whose coordinates are $(3, -1)$ and has slope $\frac{3}{5}$.

5.4   Applying Concepts

Is there a linear equation that contains all the given ordered pairs? If there is, find the equation.

31. $(5, 1), (4, 2), (0, 6)$ **32.** $(-2, -4), (0, -3), (4, -1)$

33. $(-1, -5), (2, 4), (0, 2)$ **34.** $(3, -1), (12, -4), (-6, 2)$

The given ordered pairs are solutions of the same linear equation. Find n.

35. $(0, 1), (4, 9), (3, n)$ **36.** $(2, 2), (-1, 5), (3, n)$

37. $(2, -2), (-2, -4), (4, n)$ **38.** $(1, -2), (-2, 4), (4, n)$

Temperature The relationship between Celsius and Fahrenheit temperature can be given by a linear equation. Water freezes at 0°C or at 32°F. Water boils at 100°C or at 212°F.

39. Write a linear equation expressing Fahrenheit temperature in terms of Celsius temperature.

40. Graph the equation found in Exercise 39.

The formula $y - y_1 = \dfrac{y_2 - y_1}{x_2 - x_1}(x - x_1)$, where $x_1 \neq x_2$, is called the **two-point formula** for a straight line. This formula can be used to find the equation of a line given two points. Use this formula for Exercises 41 and 42.

41. Find the equation of the line that passes through $(-2, 3)$ and $(4, -1)$.

42. Find the equation of the line that passes through the points $(3, -1)$ and $(4, -3)$.

43. Explain why the condition $x_1 \neq x_2$ is placed on the two-point formula given above.

44. Explain how the two-point formula given above can be derived from the point-slope formula.

5.5 | **Functions**

1 | **Introduction to functions**

VIDEO & DVD   CD TUTOR   WEB   SSM

The definition of *set* given in the chapter "Real Numbers" stated that a set is a collection of objects. Recall that the objects in a set are called the elements of the set. The elements in a set can be anything.

The set of planets in our solar system is

{Mercury, Venus, Earth, Mars, Jupiter, Saturn, Uranus, Neptune, Pluto}

The set of colors in a rainbow is

{red, orange, yellow, green, blue, indigo, violet}

The objects in a set can be ordered pairs. When the elements in a set are ordered pairs, the set is called a relation. A **relation** is any set of ordered pairs.

The set {(1, 1), (2, 4), (3, 9), (4, 16), (5, 25)} is a relation. There are five elements in the set. The elements are the ordered pairs (1, 1), (2, 4), (3, 9), (4, 16), and (5, 25).

The following table shows the number of hours that each of eight students spent in the math lab during the week of the math midterm exam and the score that each of these students received on the math midterm.

Hours	2	3	4	4	5	6	6	7
Score	60	70	70	80	85	85	95	90

This information can be written as the relation

{(2, 60), (3, 70), (4, 70), (4, 80), (5, 85), (6, 85), (6, 95), (7, 90)}

where the first coordinate of each ordered pair is the hours spent in the math lab and the second coordinate is the score on the midterm exam.

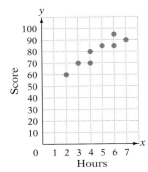

The **domain** of a relation is the set of first coordinates of the ordered pairs. The **range** is the set of second coordinates of the ordered pairs. For the relation above,

Domain = {2, 3, 4, 5, 6, 7}   Range = {60, 70, 80, 85, 90, 95}

The **graph of a relation** is the graph of the ordered pairs that belong to the relation. The graph of the relation given above is shown at the left. The horizontal axis represents the domain (the hours spent in the math lab) and the vertical axis represents the range (the exam score).

A **function** is a special type of relation in which no two ordered pairs have the same first coordinate and different second coordinates. The relation above is not a function because the ordered pairs (4, 70) and (4, 80) have the same first coordinate and different second coordinates. The ordered pairs (6, 85) and (6, 95) also have the same first coordinate and different second coordinates.

The table at the right describes a grading scale that defines a relationship between a test score and a letter grade. Some of the ordered pairs in this relation are (38, F), (73, C), and (94, A).

Score	Letter Grade
90–100	A
80–89	B
70–79	C
60–69	D
0–59	F

This relation defines a function because no two ordered pairs can have the *same* first coordinate and *different* second coordinates. For instance, it is not possible to have an average of 73 paired with any grade other than C. Both (73, C) and (73, A) cannot be ordered pairs belonging to the function, or two students with the same score would receive different grades. Note that (81, B) and (88, B) are ordered pairs of this function. Ordered pairs of a function may have the same *second* coordinate paired with different *first* coordinates.

The domain of this function is $\{0, 1, 2, 3, \ldots, 98, 99, 100\}$.

The range of this function is $\{A, B, C, D, F\}$.

Example 1 Find the domain and range of the relation $\{(-5, 1), (-3, 3), (-1, 5)\}$. Is the relation a function?

Solution The domain is $\{-5, -3, -1\}$.

- The domain of the relation is the set of the first components of the ordered pairs.

The range is $\{1, 3, 5\}$.

- The range of the relation is the set of the second components of the ordered pairs.

No two ordered pairs have the same first coordinate. The relation is a function.

Problem 1 Find the domain and range of the relation $\{(1, 0), (1, 1), (1, 2), (1, 3), (1, 4)\}$. Is the relation a function?

Solution See page S13.

Although a function can be described in terms of ordered pairs or in a table, functions are often described by an equation. The letter f is commonly used to represent a function, but any letter can be used.

The "square" function assigns to each real number its square. The square function is described by the equation

$$f(x) = x^2 \qquad \text{read } f(x) \text{ as "} f \text{ of } x \text{" or "the value of } f \text{ at } x.\text{"}$$

$f(x)$ is the symbol for the number that is paired with x. In terms of ordered pairs, this is written $(x, f(x))$. $f(x)$ is the **value of the function** at x because it is the result of evaluating the variable expression. For example, $f(4)$ means to replace x by 4 and then simplify the resulting numerical expression. This process is called **evaluating the function.**

The notation $f(4)$ is used to indicate the number that is paired with 4. To evaluate $f(x) = x^2$ at 4, replace x with 4 and simplify.

$$f(x) = x^2$$
$$f(4) = 4^2$$
$$f(4) = 16$$

The square function squares a number, and when 4 is squared, the result is 16. For the square function, the number 4 is paired with 16. In other words, when x is 4, $f(x)$ is 16. The ordered pair (4, 16) is an element of the function.

It is important to remember that $f(x)$ does not mean f times x. The letter f stands for the function, and $f(x)$ is the number that is paired with x.

Example 2   Evaluate $f(x) = 2x - 4$ at $x = 3$. Write an ordered pair that is an element of the function.

Solution
$$f(x) = 2x - 4$$
$$f(3) = 2(3) - 4 \qquad \bullet\ \textbf{\textit{f}(3) is the number that is paired with 3.}$$
$$f(3) = 6 - 4 \qquad\qquad \textbf{Replace \textit{x} by 3 and evaluate.}$$
$$f(3) = 2$$

The ordered pair $(3, 2)$ is an element of the function.

Problem 2   Evaluate $f(x) = -5x + 1$ at $x = 2$. Write an ordered pair that is an element of the function.

Solution   See page S13.

When a function is described by an equation and the domain is specified, the range of the function can be found by evaluating the function at each point of the domain.

Example 3   Find the range of the function given by the equation $f(x) = -3x + 2$ if the domain is $\{-4, -2, 0, 2, 4\}$. Write five ordered pairs that belong to the function.

Solution
$$f(x) = -3x + 2$$
$$f(-4) = -3(-4) + 2 = 12 + 2 = 14 \qquad \bullet\ \textbf{Replace \textit{x} by each}$$
$$f(-2) = -3(-2) + 2 = 6 + 2 = 8 \qquad\qquad \textbf{member of the}$$
$$f(0) = -3(0) + 2 = 0 + 2 = 2 \qquad\qquad\ \textbf{domain.}$$
$$f(2) = -3(2) + 2 = -6 + 2 = -4$$
$$f(4) = -3(4) + 2 = -12 + 2 = -10$$

The range is $\{-10, -4, 2, 8, 14\}$.

The ordered pairs $(-4, 14)$, $(-2, 8)$, $(0, 2)$, $(2, -4)$, and $(4, -10)$ belong to the function.

Problem 3   Find the range of the function given by the equation $f(x) = 4x - 3$ if the domain is $\{-5, -3, -1, 1\}$. Write four ordered pairs that belong to the function.

Solution   See page S13.

2 Graphs of linear functions

The solutions of the equation

$$y = 7x - 3$$

are ordered pairs (x, y). For example, the ordered pairs $(-1, -10)$, $(0, -3)$, and $(1, 4)$ are solutions of the equation. Therefore, this equation defines a relation.

It is not possible to substitute one value of x into the equation $y = 7x - 3$ and get two different values of y. For example, the number 1 in the domain cannot be paired with any number other than 4 in the range. (*Remember:* A function cannot have ordered pairs in which the same first coordinate is paired with different second coordinates.) Therefore, the equation defines a function.

The equation $y = 7x - 3$ is an equation of the form $y = mx + b$. In general, any equation of the form $y = mx + b$ is a function.

In the equation $y = 7x - 3$, the variable y is called the **dependent variable** because its value *depends* on the value of x. The variable x is called the **independent variable.** We choose a value for x and substitute that value into the equation to determine the value of y. We say that y is a function of x.

> **Take Note**
>
> When y is a function of x, y and $f(x)$ are interchangeable.

When an equation defines y as a function of x, functional notation is frequently used to emphasize that the relation is a function. In this case, it is common to use the notation $f(x)$. Therefore, we can write the equation

$$y = 7x - 3$$

in functional notation as

$$f(x) = 7x - 3$$

The **graph of a function** is a graph of the ordered pairs (x, y) of the function. Because the graph of the equation $y = mx + b$ is a straight line, a function of the form $f(x) = mx + b$ is a **linear function.**

Graph: $f(x) = \frac{2}{3}x - 1$

Think of the function as the equation $y = \frac{2}{3}x - 1$.

This is the equation of a straight line.

The y-intercept is $(0, -1)$. The slope is $\frac{2}{3}$.

Graph the point $(0, -1)$. From the y-intercept, go right 3 units and then up 2 units. Graph the point $(3, 1)$.

Draw a line through the two points.

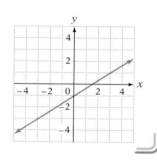

Example 4 Graph: $f(x) = \frac{3}{4}x + 2$

Solution $f(x) = \frac{3}{4}x + 2$

$y = \frac{3}{4}x + 2$

• Think of the function as the equation $y = \frac{3}{4}x + 2$.

• The graph is a straight line with y-intercept $(0, 2)$ and slope $\frac{3}{4}$.

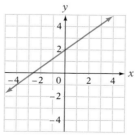

Problem 4 Graph: $f(x) = -\frac{1}{2}x - 3$

Solution   See page S13.

Graphing utilities are used to graph functions. Using a graphing utility,

enter the equation $y = \frac{3}{4}x + 2$ and verify the graph drawn in Example 4.

Trace along the graph and verify that $(-4, -1)$, $(0, 2)$, and $(4, 5)$ are coordinates of points on the graph. Now enter the equation given in Problem 4 and verify the graph you drew.

There are a variety of applications of linear functions. For example, suppose an installer of marble kitchen countertops charges $250 plus $180 per foot of countertop. The equation that describes the total cost, C (in dollars), for having x ft of countertop installed is $C = 180x + 250$. In this situation, C is a function of x; the total cost depends on how many feet of countertop are installed. Therefore, we could rewrite the equation

$$C = 180x + 250$$

as the function

$$f(x) = 180x + 250$$

To graph this equation, we first choose a reasonable domain—for example, values of x between 0 and 25. It would not be reasonable to have $x \leq 0$, because no one would order the installation of 0 ft of countertop, and any amount less than 0 would be a negative amount of countertop. The upper limit of 25 is chosen because most kitchens have less than 25 ft of countertop.

Choosing $x = 5$, 10, and 20 results in the ordered pairs $(5, 1150)$, $(10, 2050)$, and $(20, 3850)$. Graph these points and draw a line through them. The graph is shown at the left.

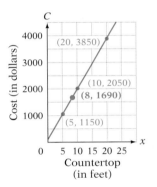

The point whose coordinates are $(8, 1690)$ is on the graph. This ordered pair can be interpreted to mean that 8 ft of countertop costs $1690 to install.

Example 5   The value, V, of an investment of $2500 at an annual simple interest rate of 6% is given by the equation $V = 150t + 2500$, where t is the amount of time, in years, that the money is invested.

A. Write the equation in functional notation.

B. Graph the equation for values of t between 0 and 10.

C. The point whose coordinates are $(5, 3250)$ is on the graph. Write a sentence that explains the meaning of this ordered pair.

Solution   **A.** $V = 150t + 2500$
$f(t) = 150t + 2500$

- The value V of the investment depends on the amount of time t it is invested. The value V is a function of the time t.

B.

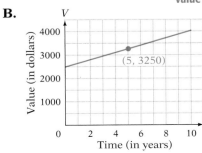

- Some ordered pairs of the function are $(2, 2800)$, $(4, 3100)$, and $(6, 3400)$.

C. The ordered pair (5, 3250) means that in 5 years the value of the investment will be $3250.

Problem 5   A car is traveling at a uniform speed of 40 mph. The distance, d (in miles), the car travels in t hours is given by the equation $d = 40t$.

A. Write the equation in functional notation.

B. Use the coordinate axes at the right to graph this equation for values of t between 0 and 5.

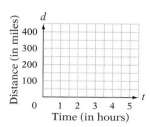

C. The point whose coordinates are (3, 120) is on the graph. Write a sentence that explains the meaning of this ordered pair.

Solution   See page S13.

5.5   Concept Review

Determine whether the statement is always true, sometimes true, or never true.

1. The graph shown at the right is the graph of a function.

2. For the relation graphed at the right, the domain is {−2, −1, 2, 3}.

3. The graph of the function $f(x) = -\frac{2}{3}x + 5$ has a positive slope.

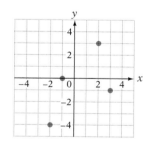

4. The graphs of $y = \frac{1}{4}x - 6$ and $f(x) = \frac{1}{4}x - 6$ are identical.

5. The value of the function {(−3, 3), (−2, 2), (−1, 1), (0, 0)} at −2 is 2.

6. A function can be represented in a table, by an equation, by a graph, or in a chart.

7. The graph of a function of the form $f(x) = mx + b$ is a straight line.

8. The equation $y = 7x - 1$ defines a function.

5.5 Exercises

1. How are relations and functions similar? How are they different?

2. What is the domain of a function? What is the range?

3. **The Olympics** The numbers of gold medals earned by several countries during the 1996 and 2000 Olympic Summer Games are recorded in the table. Write a relation in which the first coordinate is the number of gold medals earned in 1996 and the second coordinate is the number earned in 2000. Is the relation a function?

Country	1996	2000
United States	44	40
Canada	3	3
China	16	28
Italy	13	13
France	15	13
Germany	20	17

4. **Health** The life expectancy in five countries and the prevalence of smoking, as a percent of the inhabitants, are recorded in the table at the right. Write a relation in which the first coordinate is the life expectancy and the second coordinate is the prevalence of smoking, as a percent. Is the relation a function?

Country	Life Expectancy	Percent of Smokers
Iceland	76.6	31
Japan	76.5	59
Costa Rica	75.9	35
Israel	75.9	45
Sweden	75.5	22

5. **Housing** The number of sunny days in a year in five different cities and the average cost of a four-bedroom house (in thousands of dollars) in those cities are given in the table at the right. Write a relation in which the first component is the number of sunny days annually and the second component is the cost of a four-bedroom house. Is the relation a function?

City	Sunny Days Annually	4-Bedroom House Cost (in thousands of dollars)
Nashua, NH	197	125
Portsmouth, NH	205	122
San Jose, CA	257	498
Jacksonville, FL	226	108
Manchester, NH	205	150

6. **Health** Six students decided to go on a diet and fitness program over the summer. Their weights, in pounds, at the beginning and end of the program are given in the table. Write a relation in which the first coordinate is the weight at the beginning of the summer and the second coordinate is the weight at the end of the summer. Is the relation a function?

Weight at Beginning	Weight at End
145	140
140	125
150	130
165	150
140	130
165	160

7. **Physics** The speeds of a car in miles per hour and the distances, in feet, needed to stop at those speeds are recorded in the table at the right. (*Source:* AAA Foundation for Traffic Safety) Write a relation in which the first coordinate is the speed in miles per hour and the second coordinate is the distance needed to stop. Is the relation a function?

Speed (in miles per hour)	Distance Needed to Stop (in feet)
55	273
65	355
75	447
85	549

8. **Biology** The table below shows the length of the humerus (the long bone from the shoulder to the elbow), in centimeters, and the total wingspan, in centimeters, of several pterosaurs, which are extinct flying reptiles of the order Pterosauria. Write a relation in which the first coordinate is the length of the humerus and the second coordinate is the wingspan. Is the relation a function?

Length of humerus, in centimeters	24	32	22	15	4.4	17	15
Wingspan, in centimeters	600	750	430	300	68	370	310

Find the domain and range of the relation. State whether or not the relation is a function.

9. {(0, 0), (2, 0), (4, 0), (6, 0)}

10. {(−2, 2), (0, 2), (1, 2), (2, 2)}

11. {(2, 2), (2, 4), (2, 6), (2, 8)}

12. {(−4, 4), (−2, 2), (0, 0), (−2, −2)}

13. {(0, 0), (1, 1), (2, 2), (3, 3)}

14. {(0, 5), (1, 4), (2, 3), (3, 2), (4, 1), (5, 0)}

15. {(−2, −3), (2, 3), (−1, 2), (1, 2), (−3, 4), (3, 4)}

16. {(−1, 0), (0, −1), (1, 0), (2, 3), (3, 5)}

Evaluate the function at the given value of x. Write an ordered pair that is an element of the function.

17. $f(x) = 4x$; $x = 10$

18. $f(x) = 8x$; $x = 11$

19. $f(x) = x − 5$; $x = −6$

20. $f(x) = x + 7$; $x = −9$

21. $f(x) = 3x^2$; $x = −2$

22. $f(x) = x^2 − 1$; $x = −8$

23. $f(x) = 5x + 1$; $x = \dfrac{1}{2}$

24. $f(x) = 2x − 6$; $x = \dfrac{3}{4}$

25. $f(x) = \dfrac{2}{5}x + 4$; $x = −5$

26. $f(x) = \dfrac{3}{2}x − 5$; $x = 2$

27. $f(x) = 2x^2$; $x = −4$

28. $f(x) = 4x^2 + 2$; $x = −3$

Find the range of the function defined by the given equation. Write five ordered pairs that belong to the function.

29. $f(x) = 3x − 4$; domain = {−5, −3, −1, 1, 3}

30. $f(x) = 2x + 5$; domain = {−10, −5, 0, 5, 10}

31. $f(x) = \dfrac{1}{2}x + 3$; domain = {−4, −2, 0, 2, 4}

32. $f(x) = \dfrac{3}{4}x − 1$; domain = {−8, −4, 0, 4, 8}

33. $f(x) = x^2 + 6$; domain = {−3, −1, 0, 1, 3}

34. $f(x) = 3x^2 + 6$; domain = {−2, −1, 0, 1, 2}

2 Graph.

35. $f(x) = 5x$

36. $f(x) = -4x$

37. $f(x) = x + 2$

38. $f(x) = x - 3$

39. $f(x) = 6x - 1$

40. $f(x) = 3x + 4$

41. $f(x) = -2x + 3$

42. $f(x) = -5x - 2$

43. $f(x) = \dfrac{1}{3}x - 4$

44. $f(x) = \dfrac{3}{5}x + 1$

45. $f(x) = 4$

46. $f(x) = -3$

Graph using a graphing utility.

47. $f(x) = 2x - 1$

48. $f(x) = -3x - 1$

49. $f(x) = -\dfrac{1}{2}x + 1$

50. $f(x) = \dfrac{2}{3}x + 4$

51. $f(x) = \dfrac{3}{4}x + 1$

52. $f(x) = -\dfrac{1}{2}x - 2$

53. $f(x) = -\dfrac{4}{3}x + 5$

54. $f(x) = \dfrac{5}{2}x$

55. $f(x) = -1$

56. Depreciation Depreciation is the declining value of an asset. For instance, a company that purchases a truck for $20,000 has an asset worth $20,000. In 5 years, however, the value of the truck will have declined and it may be worth only $4000. An equation that represents this decline is $V = 20,000 - 3200x$, where V is the value, in dollars, of the truck after x years.

a. Write the equation in functional notation.

b. Use the coordinate axes at the right to graph the equation for values of x between 0 and 5.

c. The point $(4, 7200)$ is on the graph. Write a sentence that explains the meaning of this ordered pair.

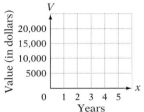

57. Depreciation A company uses the equation $V = 30,000 - 5000x$ to estimate the depreciated value, in dollars, of a computer. (See Exercise 56.)

a. Write the equation in functional notation.

b. Use the coordinate axes at the right to graph the equation for values of x between 0 and 5.

c. The point $(1, 25,000)$ is on the graph. Write a sentence that explains the meaning of this ordered pair.

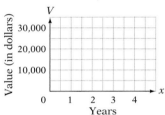

58. Business An architect charges a fee of $500 plus $2.65 per square foot to design a house. The equation that represents the architect's fee is given by $F = 2.65s + 500$, where F is the fee, in dollars, and s is the number of square feet in the house.

a. Write the equation in functional notation.

b. Use the coordinate axes at the right to graph the equation for values of s between 0 and 5000.

c. The point $(3500, 9775)$ is on the graph. Write a sentence that explains the meaning of this ordered pair.

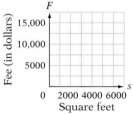

59. Business A rental car company charges a "drop-off" fee of $50 to return a car to a location different than that from which it was rented. In addition, it charges a fee of $.18 per mile the car is driven. An equation that represents the total cost to rent a car from this company is $C = 0.18m + 50$, where C is the total cost, in dollars, and m is the number of miles the car is driven.

a. Write the equation in functional notation.

b. Use the coordinate axes at the right to graph the equation for values of m between 0 and 1000.

c. The point $(500, 140)$ is on the graph. Write a sentence that explains the meaning of this ordered pair.

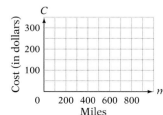

5.5 Applying Concepts

60. a. A function *f* consists of the ordered pairs {(−4, −6), (−2, −2), (0, 2), (2, 6), (4, 10)}. Find *f*(2).
   b. A function *f* consists of the ordered pairs {(0, 9), (1, 8), (2, 7), (3, 6), (4, 5)}. Find *f*(1).

61. Health The average annual number of cigarettes smoked by an American adult for the years 1991 through 2000 is shown in the table at the right. (*Source:* USDA Economic Research Service Agricultural Outlook)
   a. Do these data represent a function? Why or why not?
   b. If the data represent a function, do they represent a linear function? Why or why not?
   c. If the data represent a function, is the year a function of the number of cigarettes smoked, or is the number of cigarettes smoked a function of the year?
   d. If you had to predict the average annual number of cigarettes smoked in 2001 on the basis of the information in the table, would you predict that it would be more or less than 2103? Explain your answer.

Year	Average Annual Number of Cigarettes Smoked
1991	2727
1992	2647
1993	2543
1994	2524
1995	2505
1996	2482
1997	2423
1998	2320
1999	2136
2000	2103

62. Height-Time Graphs A child's height is a function of the child's age. The graph of this function is not linear, because children go through growth spurts as they develop. However, for the graph to be reasonable, the function must be an increasing function (that is, as the age increases, the height increases), because children do not get shorter as they grow older. Match each function described below with a reasonable graph of the function.
   a. The height of a plane above the ground during take-off depends on how long it has been since the plane left the gate.
   b. The height of a football above the ground is related to the number of seconds that have passed since it was punted.
   c. A basketball player is dribbling a basketball. The basketball's distance from the floor is related to the number of seconds that have passed since the player began dribbling the ball.
   d. Two children are seated together on a roller coaster. The height of the children above the ground depends on how long they have been on the ride.

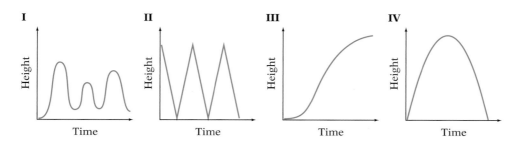

63. Investigating a relationship between two variables is an important task in the application of mathematics. For example, botanists study the relationship between the number of bushels of wheat yielded per acre and the amount of watering per acre. Environmental scientists study the relationship between the incidence of skin cancer and the amount of ozone in the atmosphere. Business analysts study the relationship between the price of a product and the number of products that are sold at that price. Describe a relationship that is important to your major field of study.

64. Functions are a part of our everyday lives. For example, the cost to mail a package via first-class mail is a function of the weight of the package. The tuition paid by a part-time student is a function of the number of credit hours the student registers for. Provide other examples of functions.

5.6   Graphing Linear Inequalities

1   **Graph inequalities in two variables**

The graph of the linear equation $y = x - 2$ separates a plane into three sets:

Point of Interest

Linear inequalities play an important role in applied mathematics. They are used in a branch of mathematics called linear programming, *which was developed during World War II to solve problems in supplying the Air Force with the machine parts necessary to keep planes flying. Today its applications have been broadened to many other disciplines.*

the set of points on the line

the set of points above the line

the set of points below the line

The point whose coordinates are (3, 1) is a solution of $y = x - 2$.

$$y = x - 2$$
$$1 \mid 3 - 2$$
$$1 = 1$$

The point whose coordinates are (3, 3) is a solution of $y > x - 2$.

$$y > x - 2$$
$$3 \mid 3 - 2$$
$$3 > 1$$

Any point above the line is a solution of $y > x - 2$.

The point whose coordinates are (3, −1) is a solution of $y < x - 2$.

$$y < x - 2$$
$$-1 \mid 3 - 2$$
$$-1 < 1$$

Any point below the line is a solution of $y < x - 2$.

The solution set of $y = x - 2$ is all points on the line. The solution set of $y > x - 2$ is all points above the line. The solution set of $y < x - 2$ is all points below the line. The solution set of an inequality in two variables is a **half-plane.**

The following illustrates the procedure for graphing a linear inequality.

Graph the solution set of $2x + 3y \le 6$.

Solve the inequality for y.

$$2x + 3y \le 6$$
$$2x - 2x + 3y \le -2x + 6$$
$$3y \le -2x + 6$$
$$\frac{3y}{3} \le \frac{-2x + 6}{3}$$
$$y \le -\frac{2}{3}x + 2$$

Change the inequality to an equality and graph the line. If the inequality is $\ge$ or $\le$, the line is in the solution set and is shown by a **solid line.** If the inequality is $>$ or $<$, the line is not a part of the solution set and is shown by a **dashed line.**

$$y = -\frac{2}{3}x + 2$$

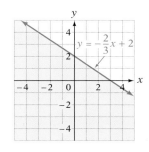

If the inequality is in the form $y > mx + b$ or $y \ge mx + b$, shade the **upper half-plane.** If the inequality is in the form $y < mx + b$ or $y \le mx + b$, shade the **lower half-plane.**

The equation $y \le -\frac{2}{3}x + 2$ is in the form $y \le mx + b$. Draw a solid line and shade the lower half-plane.

The inequality $2x + 3y \le 6$ can also be graphed as shown below.

Graph the solution set of $2x + 3y \le 6$.

Change the inequality to an equality.

$$2x + 3y = 6$$

Find the x- and y-intercepts of the equation. To find the x-intercept, let $y = 0$.

$$2x + 3(0) = 6$$
$$2x = 6$$
$$x = 3$$

The x-intercept is $(3, 0)$.

To find the y-intercept, let $x = 0$.

$$2(0) + 3y = 6$$
$$3y = 6$$
$$y = 2$$

The y-intercept is $(0, 2)$.

Graph the ordered pairs $(3, 0)$ and $(0, 2)$. Draw a solid line through the points because the inequality is $\le$.

Take Note

Any ordered pair is of the form (x, y). For the point $(0, 0)$, substitute 0 for x and 0 for y in the inequality.

The point (0, 0) can be used to determine which region to shade. If (0, 0) is a solution of the inequality, then shade the region that includes the point (0, 0).

$$2x + 3y \leq 6$$
$$2(0) + 3(0) \leq 6$$
$$0 \leq 6 \quad \text{True}$$

If (0, 0) is not a solution of the inequality, then shade the region that does not include the point (0, 0). For this example, (0, 0) is a solution of the inequality. The region that contains the point (0, 0) is shaded.

If the line passes through the point (0, 0), another point must be used to determine which region to shade. For example, use the point (1, 0).

It is important to note that every point in the shaded region is a solution of the inequality and that every solution of the inequality is a point in the shaded region. No point outside of the shaded region is a solution of the inequality.

Example 1   Graph the solution set of $3x + y > -2$.

Solution

$$3x + y > -2$$
$$3x - 3x + y > -3x - 2 \qquad \bullet \text{ Solve the inequality for } y.$$
$$y > -3x - 2$$

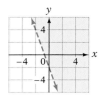

$\bullet$ **Graph $y = -3x - 2$ as a dashed line. Shade the upper half-plane.**

Problem 1   Graph the solution set of $x - 3y < 2$.

Solution   See page S13.

Example 2   Graph the solution set of $y > 3$.

Solution

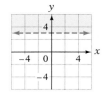

$\bullet$ **The inequality is solved for y. Graph $y = 3$ as a dashed line. Shade the upper half-plane.**

Problem 2   Graph the solution set of $x < 3$.

Solution   See page S13.

5.6 | Concept Review

Determine whether the statement is always true, sometimes true, or never true.

1. $y \geq \frac{1}{2}x - 5$ is an example of a linear equation in two variables.

2. $(-2, 4)$ is a solution of $y > x + 2$.

3. $(3, 0)$ is a solution of $y \leq x - 1$.

4. Every point above the line $y = x - 3$ is a solution of $y < x - 3$.

5. $(0, 0)$ is not a solution of $y > x + 4$. Therefore, the point $(0, 0)$ should not be in the shaded region that indicates the solution set of the inequality.

6. In the graph of an inequality in two variables, any point in the shaded region is a solution of the inequality, and any point not in the shaded region is not a solution of the inequality.

5.6 | Exercises

1 **1.** In the graph of a linear inequality, what does a solid line represent? What does a dashed line represent?

2. In the graph of a linear inequality, what does the shaded portion of the graph represent?

Graph the solution set.

3. $y > 2x + 3$

4. $y > 3x - 9$

5. $y > \frac{3}{2}x - 4$

6. $y > -\frac{5}{4}x + 1$

7. $y \leq -\frac{3}{4}x - 1$

8. $y \leq -\frac{5}{2}x - 4$

9. $y \leq -\frac{6}{5}x - 2$

10. $y < \frac{4}{5}x + 3$

11. $x + y > 4$ **12.** $x - y > -3$ **13.** $2x + y \geq 4$ **14.** $3x + y \geq 6$

15. $y \leq -2$ **16.** $y > 3$ **17.** $2x + 2y \leq -4$ **18.** $-4x + 3y < -12$

5.6 Applying Concepts

Write the inequality given its graph.

19.

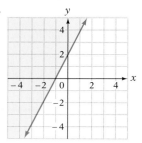

20.

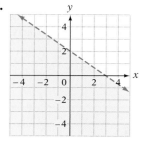

21.

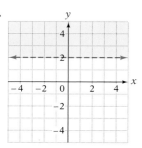

22.

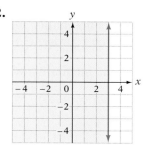

Graph the solution set.

23. $y - 5 < 4(x - 2)$ **24.** $y + 3 < 6(x + 1)$

25. $3x - 2(y + 1) \leq y - (5 - x)$ **26.** $2x - 3(y + 1) \geq y - (4 - x)$

27. Does an inequality in two variables define a relation? Why or why not? Does an inequality in two variables define a function? Why or why not?

Focus on Problem Solving

Counterexamples   Some of the exercises in this text ask you to determine whether a statement is true or false. For instance, the statement "Every real number has a reciprocal" is false because 0 is a real number and 0 does not have a reciprocal.

Finding an example, such as 0 has no reciprocal, to show that a statement is not always true is called "finding a counterexample." A counterexample is an example that shows that a statement is not always true.

Consider the statement "The product of two numbers is greater than either factor." A counterexample to this statement is the factors $\frac{2}{3}$ and $\frac{3}{4}$. The product of these numbers is $\frac{1}{2}$, and $\frac{1}{2}$ is smaller than $\frac{2}{3}$ or $\frac{3}{4}$. There are many other counterexamples to the given statement.

Here are some counterexamples to the statement "The square of a number is always larger than the number."

$$\left(\frac{1}{2}\right)^2 = \frac{1}{4} \quad \text{but} \quad \frac{1}{4} < \frac{1}{2} \qquad 1^2 = 1 \quad \text{but} \quad 1 = 1$$

For each of the next five statements, find at least one counterexample to show that the statement, or conjecture, is false.

1. The product of two integers is always a positive number.

2. The sum of two prime numbers is never a prime number.

3. For all real numbers, $|x + y| = |x| + |y|$.

4. If x and y are nonzero real numbers and $x > y$, then $x^2 > y^2$.

5. The quotient of any two nonzero real numbers is less than either one of the numbers.

When a problem is posed, it may not be known whether the statement is true or false. For instance, Christian Goldbach (1690–1764) stated that every even integer greater than 2 can be written as the sum of two prime numbers. No one has ever been able to find a counterexample to this statement, but neither has anyone been able to prove that it is always true.

In the next five exercises, answer true if the statement is always true. If there is an instance when the statement is false, give a counterexample.

6. The reciprocal of a positive number is always smaller than the number.

7. If $x < 0$, then $|x| = -x$.

8. For any two real numbers x and y, $x + y > x - y$.

9. For any positive integer n, $n^2 + n + 17$ is a prime number.

10. The list of numbers 1, 11, 111, 1111, 11111, ... contains infinitely many composite numbers. (*Hint:* A number is divisible by 3 if the sum of the digits of the number is divisible by 3.)

Projects & Group Activities

Graphing Linear Equations with a Graphing Utility

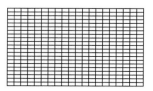

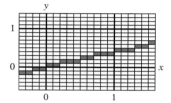

A computer or graphing calculator screen is divided into *pixels*. There are approximately 6000 to 790,000 pixels available on the screen (depending on the computer or calculator). The greater the number of pixels, the smoother a graph will appear. A portion of a screen is shown at the left. Each little rectangle represents one pixel.

The graphing utilities that are used by computers or calculators to graph an equation do basically what we have shown in the text: They choose values of x and, for each, calculate the corresponding value of y. The pixel corresponding to the ordered pair is then turned on. The graph is jagged because pixels are much larger than the dots we draw on paper.

The graph of $y = 0.45x$ is shown at the left as the calculator drew it (jagged).

The x- and y-axes have been chosen so that each pixel represents $\frac{1}{10}$ of a unit.

Consider the region of the graph where $x = 1, 1.1,$ and 1.2.

The corresponding values of y are 0.45, 0.495, and 0.54. Because the y-axis is in tenths, the numbers 0.45, 0.495, and 0.54 are rounded to the nearest tenth before plotting. Rounding 0.45, 0.495, and 0.54 to the nearest tenth results in 0.5 for each number. Thus the ordered pairs (1, 0.45), (1.1, 0.495), and (1.2, 0.54) are graphed as (1, 0.5), (1.1, 0.5), and (1.2, 0.5). These points appear as three illuminated horizontal pixels. The graph of the line appears horizontal. However, if you use the TRACE feature of the calculator (see the Appendix), the actual y-coordinate for each value of x is displayed.

Here are the keystrokes to graph $y = \frac{2}{3}x + 1$. First the equation is entered.

Then the domain (Xmin to Xmax) and the range (Ymin to Ymax) are entered. This is called the **viewing window.**

> **Take Note**
>
> Xmin and Xmax are the smallest and largest values of x that will be shown on the screen. Ymin and Ymax are the smallest and largest values of y that will be shown on the screen.

| Y= | CLEAR | 2 | X, T, Θ, n | ÷ | 3 | + | 1 | WINDOW | (−) | 10 |

| ENTER | 10 | ENTER | 1 | ENTER | (−) | 10 | ENTER | 10 | ENTER | 1 |

| ENTER | GRAPH |

By changing the keystrokes 2 [X, T, Θ, n] [÷] 3 [+] 1, you can graph different equations.

1. $y = 2x + 1$ For $2x$, you may enter $2 \times x$ or just $2x$. The times sign $\times$ is not necessary on many graphing calculators.

2. $y = -\frac{1}{2}x - 2$ Use the [(−)] key to enter a negative sign.

3. $3x + 2y = 6$ Solve for y. Then enter the equation.

4. $4x + 3y = 75$ You must adjust the viewing window.
Suggestion: Xmin = −25, Xmax = 25, Xscl = 5, Ymin = −35, Ymax = 35, Yscl = 5. See the Appendix for assistance.

Population Projections

Shown again in the table at the right is the population of the United States for each decade from 1800 to 2000. The U.S. Bureau of the Census provides not only population counts, but also population projections. One method of projecting future population uses average rate of change.

Suppose we want to predict what the population of the United States will be in 2050. We could assume that the rate of growth in the population from 2000 to 2050 will be the same as it was from 1950 to 2000.

Population of the United States

Year	Population (in millions)
1800	5
1810	7
1820	10
1830	13
1840	17
1850	23
1860	31
1870	40
1880	50
1890	63
1900	76
1910	92
1920	106
1930	123
1940	132
1950	151
1960	179
1970	203
1980	227
1990	249
2000	281

Point
of **Interest**

One definition of project, as given in the American Heritage Dictionary of the English Language, Fourth Edition, is "to calculate, estimate, or predict (something in the future), based on present data or trends."

 Calculate the average annual rate of change in the population from 1950 to 2000.

Average rate of change

$$= \frac{\text{change in } y}{\text{change in } x}$$

$$= \frac{\text{population in 2000} - \text{population in 1950}}{2000 - 1950}$$

$$= \frac{281 - 151}{2000 - 1950}$$

$$= \frac{130}{50} = 2.6$$

The average rate of change in the population from 1950 to 2000 was 2.6 million people per year.

To find the increase in the population from 2000 to 2050, multiply the number of years from 2000 to 2050 (50) by this average rate of change per year (2.6).

2.6(50) = 130

Add this increase to the population in 2000 (281).

281 + 130 = 411

The population in 2050 is projected to be 411 million.

Instead of assuming that the rate of growth in the population from 2000 to 2050 will be the same as it was from 1950 to 2000, we could assume that the rate of growth will be the same as it was from 1990 to 2000.

Calculate the average annual rate of change in the population from 1990 to 2000.

$$\text{Average rate of change} = \frac{281 - 249}{2000 - 1990} = \frac{32}{10} = 3.2$$

Multiply the number of years from 2000 to 2050 (50) by this average rate of change per year (3.2).

3.2(50) = 160

Add this increase to the population in 2000 (281).

281 + 160 = 441

The population in 2050 is projected to be 441 million.

1. The website of the U.S. Bureau of the Census is **www.census.gov.** This site provides low, middle, and high projections for the population of the United States. Which of the projections calculated above is closer to the Census Bureau's middle projection for the population in 2050?
2. Suppose that in 1950 you wanted to predict what the population of the United States would be in the year 2000.
   a. Predict the population assuming that the rate of growth in the population from 1950 to 2000 will be the same as the rate of growth from 1900 to 1950.
   b. Predict the population assuming that the rate of growth in the population from 1950 to 2000 will be the same as the rate of growth from 1940 to 1950.
   c. Predict the population assuming that the rate of growth in the population from 1950 to 2000 will be the same as the rate of growth from 1949 to 1950. According to the U.S. Bureau of the Census, the population in 1949 was 149 million.
   d. Which of the three predictions is closest to the actual population in 2000? Why?
3. The table at the left shows world population milestones. (*Source:* United Nations Population Division)
   a. Write a relation in which the first coordinate is the population of the world, in billions, and the second coordinate is the year.
   b. Is the relation a function? If it is a function, is it a linear function? Why or why not?
   c. If the rate of growth after 1999 continues at the same rate as the rate of growth from 1987 to 1999, when will the world population reach 7 billion?
   d. How might you predict when the world population will reach 8 billion? Use your method to predict when the population will reach 8 billion.

World Population Milestones	
Billions of People	Year
1	1804
2	1927
3	1960
4	1974
5	1987
6	1999

Chapter Summary

Key Words

A **rectangular coordinate system** is formed by two number lines, one horizontal and one vertical, that intersect at the zero point of each line. The number lines that make up a rectangular coordinate system are called the **coordinate axes**, or simply the **axes**. The **origin** is the point of intersection of the two coordinate axes. Generally, the horizontal axis is labeled the **x-axis**, and the vertical axis is labeled the **y-axis**. A rectangular coordinate system divides the plane determined by the axes into four regions called **quadrants**.   [5.1.1, pp. 224–225]

Examples

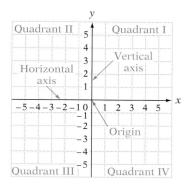

The coordinate axes determine a **plane**. Every point in the plane can be identified by an **ordered pair** (x, y). The first number in an ordered pair is called the **x-coordinate** or the **abscissa**. The second number is called the **y-coordinate** or the **ordinate**. The **coordinates** of a point are the numbers in the ordered pair associated with the point.   [5.1.1, pp. 224–225]

The ordered pair $(4, -9)$ has x-coordinate 4 and y-coordinate -9.

An equation of the form $y = mx + b$, where m is the coefficient of x and b is a constant, is a **linear equation in two variables**. A **solution of a linear equation in two variables** is an ordered pair (x, y) that makes the equation a true statement. [5.2.1, p. 237]

$y = 2x + 3$ is an example of a linear equation in two variables. The ordered pair $(1, 5)$ is a solution of this equation because when 1 is substituted for x and 5 is substituted for y, the result is a true equation.

The point at which a graph crosses the x-axis is called the **x-intercept**. At the x-intercept, the y-coordinate is 0. The point at which a graph crosses the y-axis is called the **y-intercept**. At the y-intercept, the x-coordinate is 0.   [5.2.3, p. 243]

An equation of the form $Ax + By = C$ is also a **linear equation in two variables**.   [5.2.3, p. 240]

$5x - 2y = 10$ is an example of an equation in $Ax + By = C$ form.

The **graph of $y = b$** is a horizontal line with y-intercept $(0, b)$. The **graph of $x = a$** is a vertical line with x-intercept $(a, 0)$. [5.2.3, p. 244]

The graph of $y = -3$ is a horizontal line with y-intercept $(0, -3)$. The graph of $x = 2$ is a vertical line with x-intercept $(2, 0)$.

The **slope** of a line is a measure of the slant of the line. The symbol for slope is m. A line that slants upward to the right has a **positive slope**. A line that slants downward to the right has a **negative slope**. A horizontal line has **zero slope**. The slope of a vertical line is **undefined**.   [5.3.1, pp. 250–252]

For the graph of $y = 4x - 3$, $m = 4$; the slope is positive.

For the graph of $y = -6x + 1$, $m = -6$; the slope is negative.

For the graph of $y = -7$, $m = 0$; the slope is 0.

For the graph of $x = 8$, the slope is undefined.

Two lines in the rectangular coordinate system that never intersect are **parallel lines**. Parallel lines have the same slope. [5.3.1, p. 253]

The lines $y = -2x + 1$ and $y = -2x - 3$ have the same slope and different y-intercepts. The lines are parallel.

A **relation** is any set of ordered pairs. The **domain** of a relation is the set of first coordinates of the ordered pairs. The **range** is the set of second coordinates of the ordered pairs.   [5.5.1, p. 267]

For the relation $\{(-2, -5), (0, -3), (4, -1)\}$, the domain is $\{-2, 0, 4\}$ and the range is $\{-5, -3, -1\}$.

A **function** is a relation in which no two ordered pairs have the same first coordinate and different second coordinates. [5.5.1, p. 267]

$\{(-2, -5), (0, -3), (4, -1)\}$ is a function. $\{(-2, -5), (-2, -3), (4, -1)\}$ is not a function because two ordered pairs have x-coordinate, -2, and different y-coordinates.

A function of the form $f(x) = mx + b$ is a **linear function**. Its graph is a straight line. [5.5.2, p. 270]	$f(x) = -3x + 8$ is an example of a linear function.
The solution set of an inequality in two variables is a **half-plane**. [5.6.1, p. 278]	The solution set of $y > x - 2$ is all points above the line $y = x - 2$. The solution set of $y < x - 2$ is all points below the line $y = x - 2$.

Essential Rules and Procedures

Average Rate of Change

The average rate of change of y with respect to x is $\dfrac{\text{change in } y}{\text{change in } x}$. [5.1.3, p. 229]	The average U.S. household income in 1980 was \$41,910, whereas in 2000 it was \$57,045. (*Source:* U.S. Census Bureau; adjusted for inflation and given in 2000 dollars) $$\frac{57{,}045 - 41{,}910}{2000 - 1980} = 756.75$$ The average annual rate of change in household income from 1980 to 2000 was \$756.75.

To find the x-intercept, let $y = 0$. **To find the y-intercept,** let $x = 0$. [5.2.3, p. 243]	To find the x-intercept of $4x - 3y = 12$, let $y = 0$. To find the y-intercept, let $x = 0$.

$$
\begin{array}{ll}
4x - 3y = 12 & 4x - 3y = 12 \\
4x - 3(0) = 12 & 4(0) - 3y = 12 \\
4x - 0 = 12 & 0 - 3y = 12 \\
4x = 12 & -3y = 12 \\
x = 3 & y = -4
\end{array}
$$

The x-intercept is $(3, 0)$.   The y-intercept is $(0, -4)$.

Slope of a Linear Equation

$\text{Slope} = m = \dfrac{y_2 - y_1}{x_2 - x_1},\ x_2 \neq x_1$ [5.3.1, p. 251]	To find the slope of the line between the points $(2, -3)$ and $(-1, 6)$, let $P_1 = (2, -3)$ and $P_2 = (-1, 6)$. Then $(x_1, y_1) = (2, -3)$ and $(x_2, y_2) = (-1, 6)$.

$$m = \frac{y_2 - y_1}{x_2 - x_1} = \frac{6 - (-3)}{-1 - 2} = \frac{9}{-3} = -3$$

Slope-Intercept Form of a Linear Equation

$y = mx + b$ [5.3.2, p. 255]	For the equation $y = -4x + 7$, the slope $= m = -4$ and the y-intercept $= (0, b) = (0, 7)$.

Point-Slope Formula

$y - y_1 = m(x - x_1)$ [5.4.2, p. 262]

To find the equation of the line that contains the point $(6, -1)$ and has slope 2, let $(x_1, y_1) = (6, -1)$ and $m = 2$.

$$y - y_1 = m(x - x_1)$$
$$y - (-1) = 2(x - 6)$$
$$y + 1 = 2x - 12$$
$$y = 2x - 13$$

Chapter Review Exercises

1. Find the ordered-pair solution of $y = -\frac{2}{3}x + 2$ that corresponds to $x = 3$.

2. Find the equation of the line that contains the point whose coordinates are $(0, -1)$ and has slope 3.

3. Graph: $x = -3$

4. Graph the ordered pairs $(-3, 1)$ and $(0, 2)$.

5. Evaluate $f(x) = 3x^2 + 4$ at $x = -5$.

6. Find the slope of the line that contains the points whose coordinates are $(3, -4)$ and $(1, -4)$.

7. Graph: $y = 3x + 1$

8. Graph: $3x - 2y = 6$

9. Find the equation of the line that contains the point whose coordinates are $(-1, 2)$ and has slope $-\frac{2}{3}$.

10. Find the x- and y-intercepts for $6x - 4y = 12$.

11. Graph the line that has slope $\frac{1}{2}$ and y-intercept $(0, -1)$.

12. Graph: $f(x) = -\frac{2}{3}x + 4$

13. Find the equation of the line that contains the point whose coordinates are $(-3, 1)$ and has slope $\frac{2}{3}$.

14. Find the slope of the line that contains the points whose coordinates are $(2, -3)$ and $(4, 1)$.

15. Evaluate $f(x) = \frac{3}{5}x + 2$ at $x = -10$.

16. Find the domain and range of the relation $\{(-20, -10), (-10, -5), (0, 0), (10, 5)\}$. Is the relation a function?

17. Graph: $y = -\frac{3}{4}x + 3$

18. Graph the line that has slope 2 and y-intercept -2.

19. Graph the ordered pairs $(-2, -3)$ and $(2, 4)$.

20. Graph: $f(x) = 5x + 1$

21. Find the x- and y-intercepts for $2x - 3y = 12$.

22. Find the equation of the line that contains the point whose coordinates are $(-1, 0)$ and has slope 2.

23. Graph: $y = 3$

24. Graph the solution set of $3x + 2y \leq 12$.

25. Is the line containing the points $(-2, 1)$ and $(3, 5)$ parallel to the line that contains the points $(4, -5)$ and $(9, -1)$?

26. Find the equation of the line that contains the point whose coordinates are $(2, 3)$ and has slope $\frac{1}{2}$.

27. Graph the line that has slope -1 and y-intercept $(0, 2)$.

28. Graph: $2x - 3y = 6$

29. Find the ordered-pair solution of $y = 2x - 1$ that corresponds to $x = -2$.

30. Find the equation of the line that contains the point $(0, 2)$ and has slope -3.

31. Graph: $f(x) = 2x$

32. Evaluate $f(x) = 3x - 5$ at $x = \frac{5}{3}$.

33. Find the slope of the line that contains the points whose coordinates are $(3, -2)$ and $(3, 5)$.

34. Find the equation of the line that contains the point whose coordinates are $(2, -1)$ and has slope $\frac{1}{2}$.

35. Is $(-10, 0)$ a solution of $y = \frac{1}{5}x + 2$?

36. Find the ordered-pair solution of $y = 4x - 9$ that corresponds to $x = 2$.

37. Find the x- and y-intercepts for $4x - 3y = 0$.

38. Find the equation of the line that contains the point whose coordinates are $(-2, 3)$ and has zero slope.

39. Graph the solution set of $6x - y > 6$.

40. Graph the line that has slope -3 and y-intercept $(0, 1)$.

41. Find the equation of the line that contains the point whose coordinates are $(0, -4)$ and has slope 3.

42. Graph: $x + 2y = -4$

43. Find the domain and range of the relation $\{(-10, -5), (-5, 0), (5, 0), (-10, 0)\}$. Is the relation a function?

44. Find the range of the function given by the equation $f(x) = 3x + 7$ if the domain is $\{-20, -10, 0, 10, 20\}$.

45. Find the range of the function given by the equation $f(x) = \frac{1}{3}x + 4$ if the domain is $\{-6, -3, 0, 3, 6\}$.

46. The table below shows the average number of years women and men were predicted to live beyond age 65 in 1950, 1960, 1970, 1980, 1990, and 2000. (*Source:* 2000 Social Security Trustees Report) Figures are rounded to the nearest whole number. Graph the scatter diagram for these data.

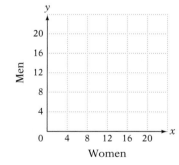

Women, x	15	16	17	18	19	19
Men, y	13	13	13	14	15	16

47. The major airlines routinely overbook flights, but generally passengers voluntarily give up their seats when offered free airline tickets as compensation. The table below gives the number of passengers who voluntarily gave up their seats and the number who involuntarily gave up their seats during a five-year period. Numbers are rounded to the nearest ten thousand. (*Source:* USA TODAY analysis of Department of Transportation data) Graph the scatter diagram for these data.

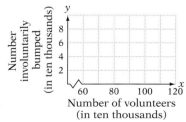

Number of volunteers, in ten thousands, x	60	77	79	90	102
Number involuntarily bumped, in ten thousands, y	4	5	5	6	5

48. The table below shows the tread depth, in millimeters, of a tire and the number of miles, in thousands, that have been driven on that tire.

Miles driven	25	35	40	20	45
Tread depth	4.8	3.5	2.1	5.5	1.0

Write a relation in which the first coordinate is the number of miles driven and the second coordinate is the tread depth. Is the relation a function?

49. A contractor uses the equation $C = 70s + 40,000$ to estimate the cost of building a new house. In this equation, C is the total cost, in dollars, and s is the number of square feet in the home.
  a. Write the equation in functional notation.
  b. Use the coordinate axes at the right to graph the equation for values of s between 0 and 5000.
  c. The point (1500, 145,000) is on the graph. Write a sentence that explains the meaning of this ordered pair.

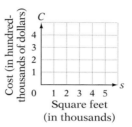

50. The average annual salary of public school classroom teachers was $23,600 in 1985 and $44,047 in 2001. (*Source:* National Center for Education Statistics) Find the average annual rate of change in salary from 1985 to 2001. Round to the nearest cent.

Chapter Test

1. Find the equation of the line that contains the point whose coordinates are (9, −3) and has slope $-\frac{1}{3}$.

2. Find the slope of the line that contains the points whose coordinates are (9, 8) and (−2, 1).

3. Find the x- and y-intercepts for $3x - 2y = 24$.

4. Find the ordered-pair solution of $y = -\frac{4}{3}x - 1$ that corresponds to $x = 9$.

5. Graph: $5x + 3y = 15$

6. Graph: $y = \frac{1}{4}x + 3$

7. Evaluate $f(x) = 4x + 7$ at $x = \frac{3}{4}$.

8. Is (6, 3) a solution of $y = \frac{2}{3}x + 1$?

9. Graph the line that has slope $-\frac{2}{3}$ and y-intercept (0, 4).

10. Graph the solution set of $2x - y \geq 2$.

11. Graph the ordered pairs (3, −2) and (0, 4).

12. Graph the line that has slope 2 and y-intercept (0, −4).

13. Find the slope of the line that contains the points whose coordinates are (4, −3) and (−2, −3).

14. Find the equation of the line that contains the point whose coordinates are (0, 7) and has slope $-\frac{2}{5}$.

15. Find the equation of the line that contains the point whose coordinates are (2, 1) and has slope 4.

16. Evaluate $f(x) = 4x^2 - 3$ at $x = -2$.

17. Graph: $y = -2x - 1$

18. Graph the solution set of $y > 2$.

19. Graph: $x = -3$

20. Graph the line that has slope $\frac{1}{2}$ and y-intercept (0, −3).

21. Graph: $f(x) = -\frac{2}{3}x + 2$

22. Graph: $f(x) = -5x$

23. Evaluate $f(x) = -3x + 7$ at $x = -6$.

24. Find the range of the function given by the equation $f(x) = -3x + 5$ if the domain is {−7, −2, 0, 4, 9}.

25. Is the line containing the points (−3, −2) and (6, 0) parallel to the line that contains the points (6, −4) and (3, 2)?

26. Drivers over age 70 number about 18 million today, up from about 13 million ten years ago. (*Source:* National Highway Traffic Safety Administration) Find the average annual rate of change in the number of drivers over age 70 for the past decade.

27. The distance, in miles, a house is from a fire station and the amount, in thousands of dollars, of fire damage that house sustained in a fire are given in the following table. Graph the scatter diagram from these data.

Distance (in miles), x	3.5	4.0	5.5	6.0
Damage (in thousands of dollars), y	25	30	40	35

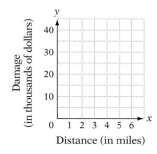

28. A company that manufactures toasters has fixed costs of $1000 each month. The manufacturing cost per toaster is $8. An equation that represents the total cost to manufacture the toasters is $C = 8t + 1000$, where C is the total cost, in dollars, and t is the number of toasters manufactured each month.
   a. Write the equation in functional notation.
   b. Use the coordinate axes at the right to graph the equation for values of t between 0 and 500.

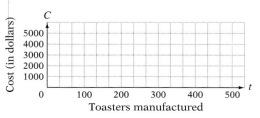

   c. The point (340, 3720) is on the graph. Write a sentence that explains the meaning of this ordered pair.

29. The data in the following table show a reading test grade and the final exam grade in a history class.

Reading test grade	8.5	9.4	10.1	11.4	12.0
History exam	64	68	76	87	92

Write a relation in which the first coordinate is the reading test grade and the second coordinate is the score on the history final exam. Is the relation a function?

30. The graph at the right shows the cost, in dollars, per 1000 board feet of lumber over a six-month period. Find the slope of the line. Write a sentence that states the meaning of the slope.

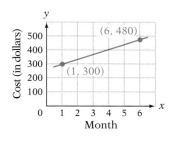

Cumulative Review Exercises

1. Simplify: $12 - 18 \div 3 \cdot (-2)^2$

2. Evaluate $\dfrac{a - b}{a^2 - c}$ when $a = -2$, $b = 3$, and $c = -4$.

3. Simplify: $4(2 - 3x) - 5(x - 4)$

4. Solve: $2x - \dfrac{2}{3} = \dfrac{7}{3}$

5. Solve: $3x - 2[x - 3(2 - 3x)] = x - 6$

6. Write $6\dfrac{2}{3}\%$ as a fraction.

7. Use the roster method to write the set of natural numbers less than 9.

8. Given $D = \{-23, -18, -4, 0, 5\}$, which elements of set D are greater than -16?

9. Solve: $8a - 3 \geq 5a - 6$

10. Solve $4x - 5y = 15$ for y.

11. Find the ordered-pair solution of $y = 3x - 1$ that corresponds to $x = -2$.

12. Find the slope of the line that contains the points whose coordinates are $(2, 3)$ and $(-2, 3)$.

13. Find the x- and y-intercepts for $5x + 2y = 20$.

14. Find the equation of the line that contains the point whose coordinates are $(3, 2)$ and has slope -1.

15. Graph: $y = \dfrac{1}{2}x + 2$

16. Graph: $3x + y = 2$

17. Graph: $f(x) = -4x - 1$

18. Graph the solution set of $x - y \leq 5$.

19. Find the domain and range of the relation $\{(0, 4), (1, 3), (2, 2), (3, 1), (4, 0)\}$. Is the relation a function?

20. Evaluate $f(x) = -4x + 9$ at $x = 5$.

21. Find the range of the function given by the equation $f(x) = -\frac{5}{3}x + 3$ if the domain is $\{-9, -6, -3, 0, 3, 6\}$.

22. The sum of two numbers is 24. Twice the smaller number is three less than the larger number. Find the two numbers.

23. A lever is 8 ft long. A force of 80 lb is applied to one end of the lever, and a force of 560 lb is applied to the other end. Where is the fulcrum located when the system balances?

24. The perimeter of a triangle is 49 ft. The length of the first side is twice the length of the third side, and the length of the second side is 5 ft more than the length of the third side. Find the length of the first side.

25. A suit that regularly sells for $89 is on sale for 30% off the regular price. Find the sale price.

How many rocking chairs can this company manufacture each month? Can the company sell all the chairs it manufactures? If the company sells all the chairs it manufactures, will it make a profit, simply break even, or end up in debt? In order to answer these questions, the company must know the cost to produce the chairs, the selling price per chair, and how many chairs can be sold at that price. Manufacturers benefit from having the knowledge to analyze profit, loss, and breaking-even points, as shown in the **Project on page 333.**

Applications

Break-even analysis,   *p. 330*

Business,   *pp. 325, 336, 339*

Coin problems,   *pp. 323, 326, 327, 336, 338, 339*

Consumerism,   *pp. 325, 326, 332*

Geometry,   *p. 327*

Investments,   *pp. 327, 331–332, 336, 339*

Mixture problems,   *pp. 321–323, 325, 326, 336*

Motion problems,   *pp. 320, 321, 324, 325, 336, 337, 338, 339*

Rate-of-wind problems,   *pp. 320, 321, 324, 325, 336, 339*

Shipping,   *p. 336*

Sports,   *p. 326*

Wages,   *p. 326*

Water-current problems,   *pp. 321, 324, 325, 336, 337, 338, 339*

6

Systems of Linear Equations

Objectives

Section 6.1
1 Solve systems of linear equations by graphing

Section 6.2
1 Solve systems of linear equations by the substitution method

Section 6.3
1 Solve systems of linear equations by the addition method

Section 6.4
1 Rate-of-wind and water-current problems
2 Application problems

Need help? For online student resources, such as section quizzes, visit this web site: math.college.hmco.com

PrepTest

1. Solve $3x - 4y = 24$ for y.

2. Solve: $50 + 0.07x = 0.05(x + 1400)$

3. Simplify: $-3(2x - 7y) + 3(2x + 4y)$

4. Simplify: $4x + 2(3x - 5)$

5. Is $(-4, 2)$ a solution of $3x - 5y = -22$?

6. Find the x- and y-intercepts of $3x - 4y = 12$.

7. Are the graphs of $y = -3x + 6$ and $y = -3x - 4$ parallel?

8. Graph: $y = \dfrac{5}{4}x - 2$

9. One hiker starts along a trail walking at a speed of 3 mph. One-half hour later, another hiker starts on the same trail walking at a speed of 4 mph. How long after the second hiker starts will the two hikers be side-by-side?

Go Figure

Two children are on their way home from school. Carla runs half the time and walks half the time. James runs half the distance and walks half the distance. If Carla and James walk at the same speed and run at the same speed, which child arrives home first?

6.1   Solving Systems of Linear Equations by Graphing

1   **Solve systems of linear equations by graphing**

Equations considered together are called a **system of equations.** A system of equations is shown at the right.

$$2x + y = 3$$
$$x + y = 1$$

A solution of a system of linear equations can be found by graphing the lines of the system on the same coordinate axes. Three examples of **systems of linear equations in two variables** are shown below, along with the graphs of the equations of each system.

System I

$x - 2y = -8$
$2x + 5y = 11$

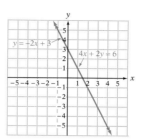

System II

$4x + 2y = 6$
$y = -2x + 3$

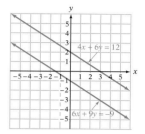

System III

$4x + 6y = 12$
$6x + 9y = -9$

For System I, the two lines intersect at a single point whose coordinates are $(-2, 3)$. Because this point lies on both lines, it is a solution of each equation of the system of equations. We can check this by replacing x by -2 and y by 3 in each equation. The check is shown below.

$$\begin{array}{c|c} x - 2y = -8 \\ \hline -2 - 2(3) & -8 \\ -2 - 6 & -8 \\ & -8 = -8 \ \checkmark \end{array} \qquad \begin{array}{c|c} 2x + 5y = 11 \\ \hline 2(-2) + 5(3) & 11 \\ -4 + 15 & 11 \\ & 11 = 11 \ \checkmark \end{array}$$

• **Replace x by -2 and y by 3.**

A **solution of a system of equations in two variables** is an ordered pair that is a solution of each equation of the system. The ordered pair $(-2, 3)$ is a solution of System I.

Is $(-1, 4)$ a solution of the following system of equations?

$7x + 3y = 5$
$3x - 2y = 12$

$$\begin{array}{c|c} 7x + 3y = 5 \\ \hline 7(-1) + 3(4) & 5 \\ -7 + 12 & 5 \\ & 5 = 5 \ \checkmark \end{array} \qquad \begin{array}{c|c} 3x - 2y = 12 \\ \hline 3(-1) - 2(4) & 12 \\ -3 - 8 & 12 \\ & -11 \neq 12 \quad \text{Does not check.} \end{array}$$

• **Replace x by -1 and y by 4.**

Using the system of equations above and the graph at the right, note that the graph of the ordered pair $(-1, 4)$ lies on the graph of $7x + 3y = 5$ but *not on both lines*. The ordered pair $(-1, 4)$ is *not* a solution of the system of equations. The ordered pair $(2, -3)$, however, does lie on both lines and therefore is a solution of the system of equations.

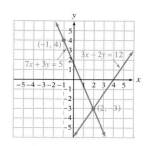

Example 1 Is $(1, -3)$ a solution of the system $3x + 2y = -3$
$$x - 3y = 6?$$

Solution

$3x + 2y = -3$		$x - 3y = 6$	
$3 \cdot 1 + 2(-3)$	-3	$1 - 3(-3)$	6
$3 + (-6)$	-3	$1 - (-9)$	6
$-3 = -3$		$10 \neq 6$	

No, $(1, -3)$ is not a solution of the system of equations.

Problem 1 Is $(-1, -2)$ a solution of the system $2x - 5y = 8$
$$-x + 3y = -5?$$

Solution   See page S14.

Take Note

The fact that there are an infinite number of ordered pairs that are solutions of the system of equations at the right does not mean that *every* ordered pair is a solution. For instance, $(0, 3)$, $(-2, 7)$, and $(2, -1)$ are solutions. However, $(3, 1)$, $(-1, 4)$, and $(1, 6)$ are not solutions. You should verify these statements.

System II from the preceding page and the graph of the equations of that system are shown again at the right. Note that the graph of $y = -2x + 3$ lies directly on top of the graph of $4x + 2y = 6$. Thus the two lines intersect at an infinite number of points. Because the graphs intersect at an infinite number of points, there are an infinite number of solutions of this system of equations. Since each equation represents the same set of points, the solutions of the system of equations can be stated by using the ordered pairs of either one of the equations. Therefore, we can say "The solutions are the ordered pairs that satisfy $4x + 2y = 6$," or we can say "The solutions are the ordered pairs that satisfy $y = -2x + 3$."

$$4x + 2y = 6$$
$$y = -2x + 3$$

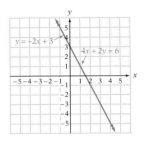

System III from the previous page and the graph of the equations of that system are shown again at the right. Note that in this case the graphs of the lines are parallel and do not intersect. Since the graphs do not intersect, there is no point that is on both lines. Therefore, the system of equations has no solution.

$$4x + 6y = 12$$
$$6x + 9y = -9$$

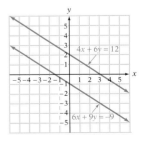

For a system of linear equations in two variables, the graphs can intersect at one point, the graphs can intersect at infinitely many points (the graphs are the same line), or the lines can be parallel and never intersect. Such systems are called **independent, dependent,** and **inconsistent,** respectively.

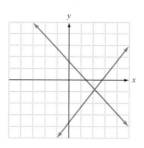

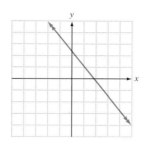

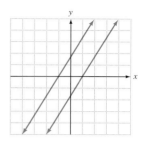

Independent   **Dependent**   **Inconsistent**

One solution   Infinitely many solutions   No solutions

Solving a system of equations means finding the ordered-pair solutions of the system. One way to do this is to draw the graphs of the equations in the system and determine where the graphs intersect.

Solve by graphing: $2x + 3y = 6$
$2x + y = -2$

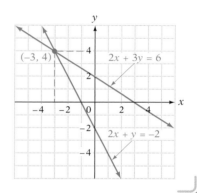

Graph each line.

The graphs of the equations intersect at one point.

The system of equations is independent.

Find the point of intersection.

The solution is $(-3, 4)$.

The INTERSECT feature on a graphing calculator can be used to find the solution of the system of equations given above. However, if this feature is used to estimate the solution, the estimated solution might be $(-3.03, 4.05)$, or some other ordered pair that contains decimals. When these values are rounded to the nearest integer, the ordered pair becomes $(-3, 4)$, which is the solution of the system. This solution can be verified by replacing x by -3 and y by 4 in the system of equations.

Solve by graphing: $2x - y = 1$
$6x - 3y = 12$

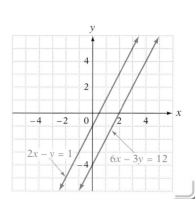

Graph each line.

The lines are parallel and therefore do not intersect.

The system of equations is inconsistent and has no solution.

Solve by graphing: $2x + 3y = 6$
$6x + 9y = 18$

Graph each line.

The two equations represent the same line. The system of equations is dependent and, therefore, has an infinite number of solutions.

The solutions are the ordered pairs that are solutions of the equation $2x + 3y = 6$.

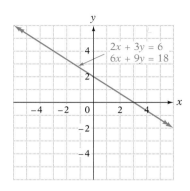

By choosing values for x, substituting these values into the equation $2x + 3y = 6$, and finding the corresponding values for y, we can find some specific ordered-pair solutions. For example, $(3, 0)$, $(0, 2)$, and $(6, -2)$ are solutions of this system of equations.

Example 2   Solve by graphing.   **A.** $x - 2y = 2$ **B.** $4x - 2y = 6$
                                            $x + y = 5$ $y = 2x - 3$

Solution   **A.**

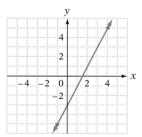

The solution is $(4, 1)$.

B.

The system of equations is dependent. The solutions are the ordered pairs that satisfy the equation $y = 2x - 3$.

Problem 2   Solve by graphing.   **A.** $x + 3y = 3$ **B.** $y = 3x - 1$
                                           $-x + y = 5$ $6x - 2y = -6$

Solution   See page S14.

6.1   Concept Review

Determine whether the statement is always true, sometimes true, or never true.

1. A solution of a system of linear equations in two variables is an ordered pair (x, y).

2. Graphically, the solution of an independent system of linear equations in two variables is the point of intersection of the graphs of the two equations.

3. Suppose an ordered pair is a solution of one equation in a system of linear equations and not of the other equation. Then the system has no solution.

4. Either a system of linear equations has one solution, represented graphically by two lines intersecting at exactly one point, or no solutions, represented graphically by two parallel lines.

5. The system of two linear equations graphed at the right has no solution.

6. An independent system of equations has no solution.

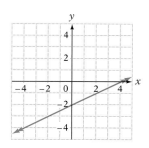

6.1   **Exercises**

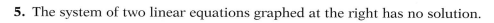

1 1. What is a system of equations?

2. How can you determine whether an ordered pair is a solution of a system of equations in two variables?

3. Is (2, 3) a solution of the system $3x + 4y = 18$
$2x - y = 1$?

4. Is (2, −1) a solution of the system $x - 2y = 4$
$2x + y = 3$?

5. Is (1, −2) a solution of the system
$3x - y = 5$
$2x + 5y = -8$?

6. Is (−1, −1) a solution of the system
$x - 4y = 3$
$3x + y = 2$?

7. Is (4, 3) a solution of the system $5x - 2y = 14$
$x + y = 8$?

8. Is (2, 5) a solution of the system $3x + 2y = 16$
$2x - 3y = 4$?

9. Is (−1, 3) a solution of the system $4x - y = -5$
$2x + 5y = 13$?

10. Is (4, −1) a solution of the system
$x - 4y = 9$
$2x - 3y = 11$?

11. Is (0, 0) a solution of the system $4x + 3y = 0$
$2x - y = 1$?

12. Is (2, 0) a solution of the system $3x - y = 6$
$x + 3y = 2$?

13. Is (2, −3) a solution of the system
$y = 2x - 7$
$3x - y = 9$?

14. Is (−1, −2) a solution of the system
$3x - 4y = 5$
$y = x - 1$?

15. Is (5, 2) a solution of the system $y = 2x - 8$
$y = 3x - 13$?

16. Is (−4, 3) a solution of the system
$y = 2x + 11$
$y = 5x - 19$?

17. Explain how to solve a system of two equations in two variables by graphing.

18. Explain each of the following in terms of a system of linear equations in two variables: an independent system, an inconsistent system, a dependent system.

For Exercises 19–24, state whether the system of equations is independent, inconsistent, or dependent.

19.
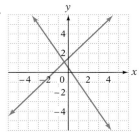

20.
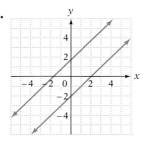

21.
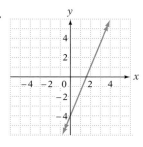

22.
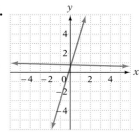

23.
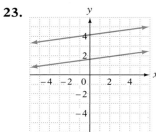

24.
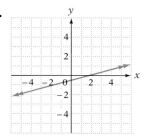

For Exercises 25–30, use the graph of the equations in the system to determine the solution of the system of equations.

25.
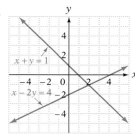

26.
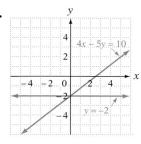

27.
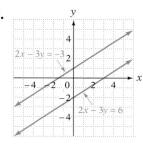

28.
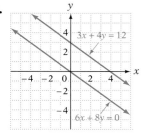

29.
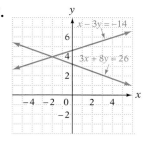

30.
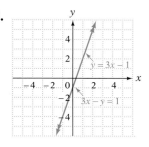

Solve by graphing.

31. $x - y = 3$
$x + y = 5$

32. $2x - y = 4$
$x + y = 5$

33. $x + 2y = 6$
$x - y = 3$

34. $3x - y = 3$
$2x + y = 2$

35. $3x - 2y = 6$
$y = 3$

36. $x = 2$
$3x + 2y = 4$

37. $x = 3$
$y = -2$

38. $x + 1 = 0$
$y - 3 = 0$

39. $y = 2x - 6$
$x + y = 0$

40. $5x - 2y = 11$
$y = 2x - 5$

41. $2x + y = -2$
$6x + 3y = 6$

42. $x + y = 5$
$3x + 3y = 6$

43. $4x - 2y = 4$
$y = 2x - 2$

44. $2x + 6y = 6$
$y = -\dfrac{1}{3}x + 1$

45. $x - y = 5$
$2x - y = 6$

46. $5x - 2y = 10$
$3x + 2y = 6$

47. $3x + 4y = 0$
$2x - 5y = 0$

48. $2x - 3y = 0$
$y = -\dfrac{1}{3}x$

49. $x - 3y = 3$
$2x - 6y = 12$

50. $4x + 6y = 12$
$6x + 9y = 18$

51. $3x + 2y = -4$
$x = 2y + 4$

52. $5x + 2y = -14$
$3x - 4y = 2$

53. $4x - y = 5$
$3x - 2y = 5$

54. $2x - 3y = 9$
$4x + 3y = -9$

Solve by graphing. Then use a graphing utility to verify your solution.

55. $5x - 2y = 10$
$3x + 2y = 6$

56. $x - y = 5$
$2x + y = 4$

57. $2x - 5y = 4$
$x - y = -1$

58. $x - 2y = -5$
$3x + 4y = -15$

59. $2x + 3y = 6$
$y = -\dfrac{2}{3}x + 1$

60. $2x - 5y = 10$
$y = \dfrac{2}{5}x - 2$

6.1   Applying Concepts

Write a system of equations given the graph.

61.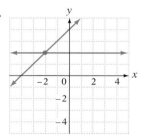

62.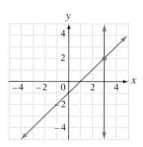

63.

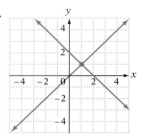

64.

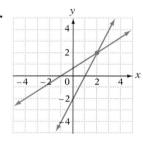

65. Determine whether the statement is always true, sometimes true, or never true.

a. Two parallel lines have the same slope.

b. Two different lines with the same y-intercept are parallel.

c. Two different lines with the same slope are parallel.

66. Write three different systems of equations: **a.** one that has $(-3, 5)$ as its only solution, **b.** one for which there is no solution, and **c.** one that is a dependent system of equations.

67. Explain how you can determine from the graph of a system of two equations in two variables whether it is an independent system of equations. Explain how you can determine whether it is an inconsistent system of equations.

68. The following graph shows the life expectancy at birth for males and females in the United States. Write an essay describing your interpretation of the data presented. Be sure to include in your discussion an interpretation of the point at which the two lines intersect.

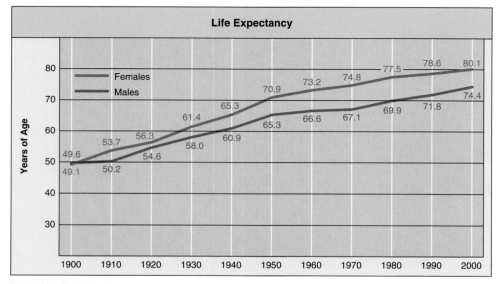

Source: U.S. Census Bureau

6.2 — Solving Systems of Linear Equations by the Substitution Method

1 **Solve systems of linear equations by the substitution method**

A graphical solution of a system of equations is based on approximating the coordinates of a point of intersection. However, the point $\left(\frac{1}{4}, \frac{1}{2}\right)$ would be difficult to read from a graph. An algebraic method called the **substitution method** can be used to find an exact solution of a system of equations. To use the substitution method, we must write one of the equations of the system in terms of x or in terms of y.

Solve by substitution: $2x + 5y = -11$ (1)
$\qquad\qquad\qquad\qquad y = 3x - 9$ (2)

Equation (2) states that $y = 3x - 9$.

Substitute $3x - 9$ for y in equation (1).

$$2x + 5(3x - 9) = -11$$

Solve for x.

$$2x + 15x - 45 = -11$$
$$17x - 45 = -11$$
$$17x = 34$$
$$x = 2$$

Note that $x = 2$. Now we must find y.

Substitute the value of x into equation (2) and solve for y.

$$(2)$$

$$y = 3x - 9$$
$$y = 3 \cdot 2 - 9$$
$$y = 6 - 9$$
$$y = -3$$

The solution is $(2, -3)$.

The graph of the system of equations given above is shown at the right. Note that the lines intersect at the point whose coordinates are $(2, -3)$, which is the algebraic solution we determined by the substitution method.

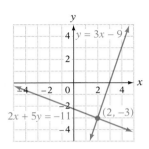

Solve by substitution: $5x + y = 4$ (1)
$\qquad\qquad\qquad\qquad 2x - 3y = 5$ (2)

Equation (1) is the easier equation to solve for one variable in terms of the other.

Solve equation (1) for y.

$$5x + y = 4$$
$$y = -5x + 4$$

Substitute $-5x + 4$ for y in equation (2).
Solve for x.

$$2x - 3(-5x + 4) = 5$$
$$2x + 15x - 12 = 5$$
$$17x - 12 = 5$$
$$17x = 17$$
$$x = 1$$

Substitute the value of x in equation (1) and solve for y.

$$(1) \qquad 5x + y = 4$$
$$5(1) + y = 4$$
$$5 + y = 4$$
$$y = -1$$

The solution is $(1, -1)$.

Solve by substitution: $\quad y = 3x - 1 \quad (1)$
$\qquad\qquad\qquad\qquad\quad y = -2x - 6 \quad (2)$

Substitute $-2x - 6$ for y in equation (1).

Solve for x.

$$y = 3x - 1$$
$$-2x - 6 = 3x - 1$$
$$-5x - 6 = -1$$
$$-5x = 5$$
$$x = -1$$

Substitute the value of x into either equation and solve for y. Equation (1) is used here.

$$y = 3x - 1$$
$$y = 3(-1) - 1$$
$$y = -3 - 1$$
$$y = -4$$

The solution is $(-1, -4)$.

Example 1   Solve by substitution: $\quad 3x + 4y = -2 \quad (1)$
$\qquad\qquad\qquad\qquad\qquad\qquad\qquad -x + 2y = 4 \quad (2)$

Solution

$$-x + 2y = 4$$
$$-x = -2y + 4$$
$$x = 2y - 4$$

• Solve equation (2) for x.

$$3x + 4y = -2$$
$$3(2y - 4) + 4y = -2$$
$$6y - 12 + 4y = -2$$
$$10y - 12 = -2$$
$$10y = 10$$
$$y = 1$$

• Substitute $2y - 4$ for x in equation (1).
• Solve for y.

$$-x + 2y = 4$$
$$-x + 2(1) = 4$$
$$-x + 2 = 4$$
$$-x = 2$$
$$x = -2$$

• Substitute the value of y into equation (2).
• Solve for x.

The solution is $(-2, 1)$.

Problem 1   Solve by substitution: $7x - y = 4$
$3x + 2y = 9$

Solution   See page S14.

Example 2   Solve by substitution: $4x + 2y = 5$ (1)
$y = -2x + 1$ (2)

Take Note	

Solve equation (1) for y. The resulting equation is $y = -2x + \frac{5}{2}$. Thus both lines have the same slope, -2, but different y-intercepts. The lines are parallel.

Solution $\begin{aligned} 4x + 2y &= 5 \\ 4x + 2(-2x + 1) &= 5 \\ 4x - 4x + 2 &= 5 \\ 2 &= 5 \end{aligned}$ • Substitute $-2x + 1$ for y in equation (1).

$2 = 5$ is not a true equation. The system of equations is inconsistent. The system does not have a solution.

Problem 2   Solve by substitution: $3x - y = 4$
$y = 3x + 2$

Solution   See page S14.

Example 3   Solve by substitution: $6x - 2y = 4$ (1)
$y = 3x - 2$ (2)

Take Note

Solve equation (1) for y. The resulting equation is $y = 3x - 2$, which is the same as equation (2).

Solution $\begin{aligned} 6x - 2y &= 4 \\ 6x - 2(3x - 2) &= 4 \\ 6x - 6x + 4 &= 4 \\ 4 &= 4 \end{aligned}$ • Substitute $3x - 2$ for y in equation (1).

$4 = 4$ is a true equation. The system of equations is dependent. The solutions are the ordered pairs that satisfy the equation $y = 3x - 2$.

Problem 3   Solve by substitution: $y = -2x + 1$
$6x + 3y = 3$

Solution   See page S14.

Note from Examples 2 and 3 above: If, when you are solving a system of equations, the variable is eliminated and the result is a false equation, such as $2 = 5$, the system is inconsistent and does not have a solution. If the result is a true equation, such as $4 = 4$, the system is dependent and has an infinite number of solutions.

6.2   **Concept Review**

Determine whether the statement is always true, sometimes true, or never true.

1. If one of the equations in a system of two linear equations is $y = x + 2$, then $x + 2$ can be substituted for y in the other equation in the system.

2. If a system of equations contains the equations $y = 2x + 1$ and $x + y = 5$, then $x + 2x + 1 = 5$.

3. If the equation $x = 4$ results from solving a system of equations by the substitution method, then the solution of the system of equations is 4.

4. If the true equation $6 = 6$ results from solving a system of equations by the substitution method, then the system of equations has an infinite number of solutions.

5. If the false equation $0 = 7$ results from solving a system of equations by the substitution method, then the system of equations is independent.

6. The ordered pair $(0, 0)$ is a solution of a system of linear equations.

6.2 Exercises

1 Solve by substitution.

1. $2x + 3y = 7$
$x = 2$

2. $y = 3$
$3x - 2y = 6$

3. $y = x - 3$
$x + y = 5$

4. $y = x + 2$
$x + y = 6$

5. $x = y - 2$
$x + 3y = 2$

6. $x = y + 1$
$x + 2y = 7$

7. $2x + 3y = 9$
$y = x - 2$

8. $3x + 2y = 11$
$y = x + 3$

9. $3x - y = 2$
$y = 2x - 1$

10. $2x - y = -5$
$y = x + 4$

11. $x = 2y - 3$
$2x - 3y = -5$

12. $x = 3y - 1$
$3x + 4y = 10$

13. $y = 4 - 3x$
$3x + y = 5$

14. $y = 2 - 3x$
$6x + 2y = 7$

15. $x = 3y + 3$
$2x - 6y = 12$

16. $x = 2 - y$
$3x + 3y = 6$

17. $3x + 5y = -6$
$x = 5y + 3$

18. $y = 2x + 3$
$4x - 3y = 1$

19. $x = 4y - 3$
$2x - 3y = 0$

20. $x = 2y$
$-2x + 4y = 6$

21. $y = 2x - 9$
$3x - y = 2$

22. $y = x + 4$
$2x - y = 6$

23. $3x + 4y = 7$
$x = 3 - 2y$

24. $3x - 5y = 1$
$y = 4x - 2$

25. $2x - y = 4$
$3x + 2y = 6$

26. $x + y = 12$
$3x - 2y = 6$

27. $4x - 3y = 5$
$x + 2y = 4$

28. $3x - 5y = 2$
$2x - y = 4$

29. $7x - y = 4$
$5x + 2y = 1$

30. $x - 7y = 4$
$-3x + 2y = 6$

31. $4x - 3y = -1$
$y = 2x - 3$

32. $3x - 7y = 28$
$x = 3 - 4y$

33. $7x + y = 14$
$2x - 5y = -33$

34. $3x + y = 4$
 $4x - 3y = 1$

35. $x - 4y = 9$
 $2x - 3y = 11$

36. $3x - y = 6$
 $x + 3y = 2$

37. $4x - y = -5$
 $2x + 5y = 13$

38. $3x - y = 5$
 $2x + 5y = -8$

39. $3x + 4y = 18$
 $2x - y = 1$

40. $4x + 3y = 0$
 $2x - y = 0$

41. $5x + 2y = 0$
 $x - 3y = 0$

42. $6x - 3y = 6$
 $2x - y = 2$

43. $3x + y = 4$
 $9x + 3y = 12$

44. $x - 5y = 6$
 $2x - 7y = 9$

45. $x + 7y = -5$
 $2x - 3y = 5$

46. $y = 2x + 11$
 $y = 5x - 1$

47. $y = 2x - 8$
 $y = 3x - 13$

48. $y = -4x + 2$
 $y = -3x - 1$

49. $x = 3y + 7$
 $x = 2y - 1$

50. $x = 4y - 2$
 $x = 6y + 8$

51. $x = 3 - 2y$
 $x = 5y - 10$

52. $y = 2x - 7$
 $y = 4x + 5$

53. $3x - y = 11$
 $2x + 5y = -4$

54. $-x + 6y = 8$
 $2x + 5y = 1$

6.2 Applying Concepts

Rewrite each equation so that the coefficients are integers. Then solve the system of equations.

55. $0.1x - 0.6y = -0.4$
 $-0.7x + 0.2y = 0.5$

56. $0.8x - 0.1y = 0.3$
 $0.5x - 0.2y = -0.5$

57. $0.4x + 0.5y = 0.2$
 $0.3x - 0.1y = 1.1$

58. $-0.1x + 0.3y = 1.1$
 $0.4x - 0.1y = -2.2$

59. $1.2x + 0.1y = 1.9$
 $0.1x + 0.3y = 2.2$

60. $1.25x - 0.01y = 1.5$
 $0.24x - 0.02y = -1.52$

For what value of k does the system of equations have no solution?

61. $2x - 3y = 7$
 $kx - 3y = 4$

62. $8x - 4y = 1$
 $2x - ky = 3$

63. $ x = 4y + 4$
 $kx - 8y = 4$

64. The following was offered as a solution to the system of equations shown at the right.

$$2x + 5y = 10 \qquad \text{• Equation (2)}$$
$$2x + 5\left(\tfrac{1}{2}x + 2\right) = 10 \qquad \text{• Substitute } \tfrac{1}{2}x + 2 \text{ for } y.$$
$$2x + \tfrac{5}{2}x + 10 = 10 \qquad \text{• Solve for } x.$$
$$\tfrac{9}{2}x = 0$$
$$x = 0$$

(1) $ y = \dfrac{1}{2}x + 2$

(2) $ 2x + 5y = 10$

At this point the student stated that because $x = 0$, the system of equations has no solution. If this assertion is correct, is the system of equations independent, dependent, or inconsistent? If the assertion is not correct, what is the correct solution?

65. Describe in your own words the process of solving a system of equations by the substitution method.

66. When you solve a system of equations by the substitution method, how do you determine whether the system of equations is dependent? How do you determine whether the system of equations is inconsistent?

6.3 Solving Systems of Linear Equations by the Addition Method

1 **Solve systems of linear equations by the addition method**

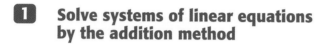

Another algebraic method for solving a system of equations is called the **addition method.** It is based on the Addition Property of Equations.

In the system of equations at the right, note the effect of adding equation (2) to equation (1). Because $2y$ and $-2y$ are opposites, adding the equations results in an equation with only one variable.

$$(1) \quad 3x + 2y = 4$$
$$(2) \quad 4x - 2y = 10$$
$$7x + 0y = 14$$
$$7x = 14$$

The solution of the resulting equation is the first coordinate of the ordered-pair solution of the system.

$$7x = 14$$
$$x = 2$$

The second coordinate is found by substituting the value of x into equation (1) or (2) and then solving for y. Equation (1) is used here.

$$(1) \quad 3x + 2y = 4$$
$$3 \cdot 2 + 2y = 4$$
$$6 + 2y = 4$$
$$2y = -2$$
$$y = -1$$

The solution is $(2, -1)$.

Sometimes adding the two equations does not eliminate one of the variables. In this case, use the Multiplication Property of Equations to rewrite one or both of the equations so that when the equations are added, one of the variables is eliminated.

To do this, first choose which variable to eliminate. The coefficients of that variable must be opposites. Multiply each equation by a constant that will produce coefficients that are opposites.

Solve by the addition method: $3x + 2y = 7$ (1)
$5x - 4y = 19$ (2)

To eliminate y, multiply each side of equation (1) by 2.

$$2(3x + 2y) = 2 \cdot 7$$
$$5x - 4y = 19$$

Now the coefficients of the y terms are opposites.

$$6x + 4y = 14$$
$$5x - 4y = 19$$

Add the equations.
Solve for x.

$$11x + 0y = 33$$
$$11x = 33$$
$$x = 3$$

Substitute the value of x into one of the equations and solve for y. Equation (2) is used here.

$$(2) \quad 5x - 4y = 19$$
$$5 \cdot 3 - 4y = 19$$
$$15 - 4y = 19$$
$$-4y = 4$$
$$y = -1$$

The solution is $(3, -1)$.

Solve by the addition method: $5x + 6y = 3$ (1)
$2x - 5y = 16$ (2)

To eliminate x, multiply each side of equation (1) by 2 and each side of equation (2) by -5. Note how the constants are selected. The negative sign is used so that the coefficients will be opposites.

$$2 \quad (5x + 6y) = 2 \cdot 3$$
$$-5 \quad (2x - 5y) = -5 \cdot 16$$

Now the coefficients of the x terms are opposites.

$$10x + 12y = 6$$
$$-10x + 25y = -80$$

Add the equations.
Solve for y.

$$0x + 37y = -74$$
$$37y = -74$$
$$y = -2$$

Substitute the value of y into one of the equations and solve for x. Equation (1) is used here.

$$(1) \quad 5x + 6y = 3$$
$$5x + 6(-2) = 3$$
$$5x - 12 = 3$$
$$5x = 15$$
$$x = 3$$

The solution is $(3, -2)$.

Solve by the addition method: $5x = 2y - 7$ (1)
$3x + 4y = 1$ (2)

Write equation (1) in the form $Ax + By = C$.

$$5x - 2y = -7$$
$$3x + 4y = 1$$

Eliminate y. Multiply each side of equation (1) by 2.

$$2(5x - 2y) = 2(-7)$$
$$3x + 4y = 1$$

Now the coefficients of the y terms are opposites.

$$10x - 4y = -14$$
$$3x + 4y = 1$$

Add the equations.
Solve for x.

$$13x + 0y = -13$$
$$13x = -13$$
$$x = -1$$

Substitute the value of x into one of the equations and solve for y. Equation (1) is used here.

$$5x = 2y - 7$$
$$5(-1) = 2y - 7$$
$$-5 = 2y - 7$$
$$2 = 2y$$
$$1 = y$$

The solution is $(-1, 1)$.

Solve by the addition method: $2x + y = 2$ (1)
                                      $4x + 2y = -5$ (2)

To eliminate y, multiply each side of equation (1) by -2.

$$-4x - 2y = -4$$
$$4x + 2y = -5$$

Add the equations.
This is not a true equation.

$$0x + 0y = -9$$
$$0 = -9$$

The system of equations is inconsistent. The system does not have a solution.

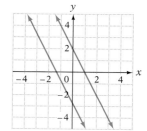

The graphs of the two equations in the preceding system of equations are shown at the left. Note that the graphs are parallel and therefore do not intersect. Thus the system of equations has no solution.

Example 1   Solve by the addition method: $2x + 4y = 7$ (1)
                                                   $5x - 3y = -2$ (2)

Solution

$$5(2x + 4y) = 5 \cdot 7$$
$$-2(5x - 3y) = -2 \cdot (-2)$$

• Eliminate x. Multiply each side of equation (1) by 5 and each side of equation (2) by -2.

$$10x + 20y = 35$$
$$-10x + 6y = 4$$
$$26y = 39$$

• Add the equations.

$$y = \frac{39}{26} = \frac{3}{2}$$

• Solve for y.

$$2x + 4\left(\frac{3}{2}\right) = 7$$

• Substitute the value of y in equation (1).

$$2x + 6 = 7$$
$$2x = 1$$

• Solve for x.

$$x = \frac{1}{2}$$

The solution is $\left(\frac{1}{2}, \frac{3}{2}\right)$.

Problem 1   Solve by the addition method: $x - 2y = 1$
                                                 $2x + 4y = 0$

Solution   See page S14.

Example 2 Solve by the addition method: $6x + 9y = 15$ (1)
$$4x + 6y = 10 \quad (2)$$

Solution

$4(6x + 9y) = 4 \cdot 15$ • **Eliminate** x. **Multiply each side of**
$-6(4x + 6y) = -6 \cdot 10$ **equation (1) by 4 and each side of**
                 equation (2) by -6.

$24x + 36y = 60$
$\underline{-24x - 36y = -60}$
           $0 = 0$ • **Add the equations.**

$0 = 0$ is a true equation. The system of equations is dependent. The solutions are ordered pairs that satisfy the equation $6x + 9y = 15$.

Problem 2 Solve by the addition method: $2x - 3y = 4$
$$-4x + 6y = -8$$

Solution   See page S14.

6.3   Concept Review

Determine whether the statement is always true, sometimes true, or never true.

1. When using the addition method of solving a system of linear equations, if you multiply one side of an equation by a number, then you must multiply the other side of the equation by the same number.

2. When a system of linear equations is being solved by the addition method, in order for one of the variables to be eliminated, the coefficients of one of the variables must be opposites.

3. If the two equations in a system of equations have the same slope and different y-intercepts, then the system of equations has an infinite number of solutions.

4. You are using the addition method to solve the system $3x + 2y = 6$
of equations shown at the right. The first step you will $2x + 3y = -6$
perform is to add the two equations.

5. If the equation $y = -2$ results from solving a system of equations by the addition method, then the solution of the system of equations is -2.

6. If the true equation $-3 = -3$ results from solving a system of equations by the addition method, then the system of equations is dependent.

6.3 Exercises

1 Solve by the addition method.

1. $x + y = 4$
$x - y = 6$

2. $2x + y = 3$
$x - y = 3$

3. $x + y = 4$
$2x + y = 5$

4. $x - 3y = 2$
$x + 2y = -3$

5. $2x - y = 1$
$x + 3y = 4$

6. $x - 2y = 4$
$3x + 4y = 2$

7. $4x - 5y = 22$
$x + 2y = -1$

8. $3x - y = 11$
$2x + 5y = 13$

9. $2x - y = 1$
$4x - 2y = 2$

10. $x + 3y = 2$
$3x + 9y = 6$

11. $4x + 3y = 15$
$2x - 5y = 1$

12. $3x - 7y = 13$
$6x + 5y = 7$

13. $2x - 3y = 5$
$4x - 6y = 3$

14. $2x + 4y = 3$
$3x + 6y = 8$

15. $5x - 2y = -1$
$x + y = 4$

16. $4x - 3y = 1$
$8x + 5y = 13$

17. $5x + 7y = 10$
$3x - 14y = 6$

18. $7x + 10y = 13$
$4x + 5y = 6$

19. $3x - 2y = 0$
$6x + 5y = 0$

20. $5x + 2y = 0$
$3x + 5y = 0$

21. $2x - 3y = 16$
$3x + 4y = 7$

22. $3x + 4y = 10$
$4x + 3y = 11$

23. $x + 3y = 4$
$2x + 5y = 1$

24. $-2x + 7y = 9$
$3x + 2y = -1$

25. $7x - 2y = 13$
$5x + 3y = 27$

26. $3x + 5y = -11$
$2x - 7y = 3$

27. $8x - 3y = 11$
$6x - 5y = 11$

28. $4x - 8y = 36$
$3x - 6y = 27$

29. $5x + 15y = 20$
$2x + 6y = 8$

30. $2x - 3y = 4$
$-x + 4y = 3$

31. $3x = 2y + 7$
$5x - 2y = 13$

32. $2y = 4 - 9x$
$9x - y = 25$

33. $2x + 9y = 5$
$5x = 6 - 3y$

34. $3x - 4 = y + 18$
$4x + 5y = -21$

35. $2x + 3y = 7 - 2x$
$7x + 2y = 9$

36. $5x - 3y = 3y + 4$
$4x + 3y = 11$

37. $3x + y = 1$
$5x + y = 2$

38. $2x - y = 1$
$2x - 5y = -1$

39. $4x + 3y = 3$
$x + 3y = 1$

40. $2x - 5y = 4$
$x + 5y = 1$

41. $3x - 4y = 1$
$4x + 3y = 1$

42. $2x - 7y = -17$
$3x + 5y = 17$

43. $y = 2x - 3$
$4x + 4y = -1$

44. $4x - 2y = 5$
$2x + 3y = 4$

6.3   **Applying Concepts**

Solve.

45. $x - 0.2y = 0.2$
$0.2x + 0.5y = 2.2$

46. $0.5x - 1.2y = 0.3$
$0.2x + y = 1.6$

47. $1.25x - 1.5y = -1.75$
$2.5x - 1.75y = -1$

48. The point of intersection of the graphs of the equations $Ax + 2y = 2$ and $2x + By = 10$ is $(2, -2)$. Find A and B.

49. The point of intersection of the graphs of the equations $Ax - 4y = 9$ and $4x + By = -1$ is $(-1, -3)$. Find A and B.

50. Given that the graphs of the equations $2x - y = 6$, $3x - 4y = 4$, and $Ax - 2y = 0$ all intersect at the same point, find A.

51. Given that the graphs of the equations $3x - 2y = -2$, $2x - y = 0$, and $Ax + y = 8$ all intersect at the same point, find A.

52. For what value of k is the system of equations dependent?

  a. $2x + 3y = 7$
   $4x + 6y = k$

  b. $y = \dfrac{2}{3}x - 3$
   $y = kx - 3$

  c. $x = ky - 1$
   $y = 2x + 2$

53. For what values of k is the system of equations independent?

  a. $x + y = 7$
   $kx + y = 3$

  b. $x + 2y = 4$
   $kx + 3y = 2$

  c. $2x + ky = 1$
   $x + 2y = 2$

54.   Describe in your own words the process of solving a system of equations by the addition method.

6.4   **Application Problems in Two Variables**

1   **Rate-of-wind and water-current problems**

Solving motion problems that involve an object moving with or against a wind or current normally requires two variables. One variable represents the speed of the moving object in calm air or still water, and a second variable represents the rate of the wind or current.

A plane flying with the wind will travel a greater distance per hour than it would travel without the wind. The resulting rate of the plane is represented by the sum of the plane's speed and the rate of the wind.

A plane traveling against the wind, on the other hand, will travel a shorter distance per hour than it would travel without the wind. The resulting rate of the plane is represented by the difference between the plane's speed and the rate of the wind.

The same principle is used to describe the rate of a boat traveling with or against a water current.

Solve: Flying with the wind, a small plane can fly 750 mi in 3 h. Against the wind, the plane can fly the same distance in 5 h. Find the rate of the plane in calm air and the rate of the wind.

STRATEGY: | *For solving rate-of-wind and water-current problems*

▶ Choose one variable to represent the rate of the object in calm conditions and a second variable to represent the rate of the wind or current. Use these variables to express the rate of the object with and against the wind or current. Then use these expressions for the rate, as well as the time traveled with and against the wind or current and the fact that $rt = d$, to write expressions for the distance traveled by the object. The results can be recorded in a table.

Rate of plane in calm air: p
Rate of wind: w

With wind: 750 mi in 3 h

Against wind: 750 mi in 5 h

	Rate	·	Time	=	Distance
With the wind	$p + w$	·	3	=	$3(p + w)$
Against the wind	$p - w$	·	5	=	$5(p - w)$

▶ Determine how the expressions for distance are related.

The distance traveled with the wind is 750 mi. $\quad\quad 3(p + w) = 750$
The distance traveled against the wind is 750 mi. $\quad 5(p - w) = 750$

Solve the system of equations.

$3(p + w) = 750 \quad\quad\quad \dfrac{3(p + w)}{3} = \dfrac{750}{3} \quad\quad\quad p + w = 250$

$5(p - w) = 750 \quad\quad\quad \dfrac{5(p - w)}{5} = \dfrac{750}{5} \quad\quad\quad p - w = 150$

$$2p = 400$$
$$p = 200$$

Substitute the value of p in the equation $p + w = 250$. $\quad\quad p + w = 250$
Solve for w. $\quad\quad\quad\quad\quad\quad\quad\quad\quad\quad\quad\quad\quad\quad 200 + w = 250$
$$w = 50$$

The rate of the plane in calm air is 200 mph.
The rate of the wind is 50 mph.

Example 1 A 600-mile trip from one city to another takes 4 h when a plane is flying with the wind. The return trip against the wind takes 5 h. Find the rate of the plane in still air and the rate of the wind.

Strategy

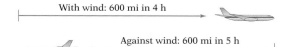

With wind: 600 mi in 4 h

Against wind: 600 mi in 5 h

▶ Rate of the plane in still air: p
Rate of the wind: w

	Rate	Time	Distance
With wind	$p + w$	4	$4(p + w)$
Against wind	$p - w$	5	$5(p - w)$

▶ The distance traveled with the wind is 600 mi.
The distance traveled against the wind is 600 mi.

Solution

$4(p + w) = 600$ (1)
$5(p - w) = 600$ (2)

$\dfrac{4(p + w)}{4} = \dfrac{600}{4}$ • Simplify equation (1) by dividing each side of the equation by 4.

$\dfrac{5(p - w)}{5} = \dfrac{600}{5}$ • Simplify equation (2) by dividing each side of the equation by 5.

$p + w = 150$
$p - w = 120$
$\quad\ \ 2p = 270$ • Add the two equations.
$\quad\ \ \ \ p = 135$ • Solve for p, the rate of the plane in still air.
$p + w = 150$ • Substitute the value of p into one of the
$135 + w = 150$ equations.
$\quad\ \ \ \ w = 15$ • Solve for w, the rate of the wind.

The rate of the plane in still air is 135 mph.
The rate of the wind is 15 mph.

Problem 1 A canoeist paddling with the current can travel 24 mi in 3 h. Against the current, it takes 4 h to travel the same distance. Find the rate of the current and the rate of the canoeist in calm water.

Solution See page S15.

2 Application problems

The application problems in this section are varieties of those problems solved earlier in the text. Each of the strategies for the problems in this section will result in a system of equations.

Solve: A jeweler purchased 5 oz of a gold alloy and 20 oz of a silver alloy for a total cost of $700. The next day, at the same prices per ounce, the jeweler purchased 4 oz of the gold alloy and 30 oz of the silver alloy for a total cost of $630. Find the cost per ounce of the silver alloy.

STRATEGY: | *For solving an application problem in two variables*

▶ Choose one variable to represent one of the unknown quantities and a second variable to represent the other unknown quantity. Write numerical or variable expressions for all the remaining quantities. These results can be recorded in two tables, one for each of the conditions.

Cost per ounce of gold: g Cost per ounce of silver: s

First day

	Amount	·	Unit cost	=	Value
Gold	5	·	g	=	$5g$
Silver	20	·	s	=	$20s$

Second day

	Amount	·	Unit cost	=	Value
Gold	4	·	g	=	$4g$
Silver	30	·	s	=	$30s$

▶ Determine a system of equations. The strategies presented in the chapter "Solving Equations and Inequalities: Applications" can be used to determine the relationships between the expressions in the tables. Each table will give one equation of the system.

The total value of the purchase on the first day was $700. $5g + 20s = 700$
The total value of the purchase on the second day was $4g + 30s = 630$
$630.

Solve the system of equations.

$$5g + 20s = 700 \qquad 4(5g + 20s) = 4 \cdot 700 \qquad 20g + 80s = 2800$$
$$4g + 30s = 630 \qquad -5(4g + 30s) = -5 \cdot 630 \qquad -20g - 150s = -3150$$
$$-70s = -350$$
$$s = 5$$

The cost per ounce of the silver alloy was $5.

Example 2 A store owner purchased 20 incandescent light bulbs and 30 fluorescent bulbs for a total cost of $40. A second purchase, at the same prices, included 30 incandescent bulbs and 10 fluorescent bulbs for a total cost of $25. Find the cost of an incandescent bulb and of a fluorescent bulb.

Strategy   ▶ Cost of an incandescent bulb: I
               Cost of a fluorescent bulb: F

First purchase	Amount	Unit cost	Value
Incandescent	20	I	$20I$
Fluorescent	30	F	$30F$

Second purchase	Amount	Unit cost	Value
Incandescent	30	I	$30I$
Fluorescent	10	F	$10F$

▶ The total of the first purchase was $40.
The total of the second purchase was $25.

Solution

$20I + 30F = 40$ (1)
$30I + 10F = 25$ (2)

$3(20I + 30F) = 3(40)$
$-2(30I + 10F) = -2(25)$

• Eliminate *I*. Multiply each side of equation (1) by 3 and each side of equation (2) by −2.

$60I + 90F = 120$
$-60I - 20F = -50$
$70F = 70$
$F = 1$

• Add the two equations.
• Solve for *F*, the cost of a fluorescent bulb.

$20I + 30F = 40$
$20I + 30(1) = 40$
$20I = 10$

$I = \dfrac{1}{2}$

• Substitute the value of *F* into one of the equations.
• Solve for *I*.

The cost of an incandescent bulb was $.50.
The cost of a fluorescent bulb was $1.00.

Problem 2   Two coin banks contain only dimes and quarters. In the first bank, the total value of the coins is $3.90. In the second bank, there are twice as many dimes as in the first bank and one-half the number of quarters. The total value of the coins in the second bank is $3.30. Find the number of dimes and the number of quarters in the first bank.

Solution   See page S15.

6.4   Concept Review

Determine whether the statement is always true, sometimes true, or never true.

1. A plane flying with the wind is traveling faster than it would be traveling without the wind.

2. The uniform motion equation $r = dt$ is used to solve rate-of-wind and water-current problems.

3. If b represents the rate of a boat in calm water and c represents the rate of the water current, then $b + c$ represents the rate of the boat while it is traveling against the current.

4. Both sides of an equation can be divided by the same number without changing the solution of the equation.

5. If, in a system of equations, p represents the rate of a plane in calm air and w represents the rate of the wind, and the solution of the system is $p = 100$, this means that the rate of the wind is 100.

6. The system of equations at the right represents the following problem:

$$2(p + w) = 600$$
$$3(p - w) = 600$$

A plane flying with the wind flew 600 mi in 2 h. Flying against the wind, the plane could fly the same distance in 3 h. Find the rate of the plane in calm air.

6.4 Exercises

1 **Rate-of-Wind and Water-Current Problems**

1. A whale swimming against an ocean current traveled 60 mi in 2 h. Swimming in the opposite direction, with the current, the whale was able to travel the same distance in 1.5 h. Find the speed of the whale in calm water and the rate of the ocean current.

2. A plane flying with the jet stream flew from Los Angeles to Chicago, a distance of 2250 mi, in 5 h. Flying against the jet stream, the plane could fly only 1750 mi in the same amount of time. Find the rate of the plane in calm air and the rate of the wind.

3. A rowing team rowing with the current traveled 40 km in 2 h. Rowing against the current, the team could travel only 16 km in 2 h. Find the team's rowing rate in calm water and the rate of the current.

4. The bird capable of the fastest flying speed is the swift. A swift flying with the wind to a favorite feeding spot traveled 26 mi in 0.2 h. On the return trip, against the wind, the swift was able to travel only 16 mi in the same amount of time. Find the rate of the swift in calm air and the rate of the wind.

5. A motorboat traveling with the current went 35 mi in 3.5 h. Traveling against the current, the boat went 12 mi in 3 h. Find the rate of the boat in calm water and the rate of the current.

6. A small plane, flying into a headwind, flew 270 mi in 3 h. Flying with the wind, the plane traveled 260 mi in 2 h. Find the rate of the plane in calm air and the rate of the wind.

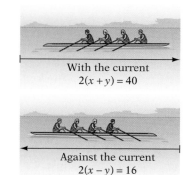

With the current
$2(x + y) = 40$

Against the current
$2(x - y) = 16$

7. A private Learjet 31A transporting passengers was flying with a tailwind and traveled 1120 mi in 2 h. Flying against the wind on the return trip, the jet was able to travel only 980 mi in 2 h. Find the speed of the jet in calm air and the rate of the wind.

8. A rowing team rowing with the current traveled 18 mi in 2 h. Against the current, the team rowed a distance of 8 mi in the same amount of time. Find the rate of the rowing team in calm water and the rate of the current.

9. A seaplane flying with the wind flew from an ocean port to a lake, a distance of 240 mi, in 2 h. Flying against the wind, it made the trip from the lake to the ocean port in 3 h. Find the rate of the plane in calm air and the rate of the wind.

10. Rowing with the current, a canoeist paddled 14 mi in 2 h. Against the current, the canoeist could paddle only 10 mi in the same amount of time. Find the rate of the canoeist in calm water and the rate of the current.

11. A Boeing Apache Longbow military helicopter traveling directly into a strong headwind was able to travel 450 mi in 2.5 h. The return trip, now with a tailwind, took 1 h 40 min. Find the speed of the helicopter in calm air and the rate of the wind.

12. With the wind, a quarterback passes a football 140 ft in 2 s. Against the wind, the same pass would have traveled 80 ft in 2 s. Find the rate of the pass and the rate of the wind.

2 13. **Business** The manager of a computer software store received two shipments of software. The cost of the first shipment, which contained 12 identical word processing programs and 10 identical spreadsheet programs, was $6190. The second shipment, at the same prices, contained 5 copies of the word processing program and 8 copies of the spreadsheet program. The cost of the second shipment was $3825. Find the cost for one copy of the word processing program.

14. **Business** The manager of a discount clothing store received two shipments of fall clothing. The cost of the first shipment, which contained 10 identical sweaters and 20 identical jackets, was $800. The second shipment, at the same prices, contained 5 of the same sweaters and 15 of the same jackets. The cost of the second shipment was $550. Find the cost of one jacket.

15. **Business** A baker purchased 12 lb of wheat flour and 15 lb of rye flour for a total cost of $18.30. A second purchase, at the same prices, included 15 lb of wheat flour and 10 lb of rye flour. The cost of the second purchase was $16.75. Find the cost per pound of the wheat and rye flours.

16. **Consumerism** A computer on-line service charges one hourly rate for regular use but a higher hourly rate for designated "premium" areas. One customer was charged $28 after spending 2 h in premium areas and 9 regular hours. Another customer spent 3 h in premium areas and 6 regular hours and was charged $27. What is the service charge per hour for regular and premium services?

17. **Wages** Shelly Egan works as a stocker at a grocery store during her summer vacation. She gets paid a standard hourly rate for her day hours but a higher hourly rate for any hours she works during the night shift. One week she worked 17 daylight hours and 8 nighttime hours and earned $191. The next week she earned $219 for a total of 12 daytime and 15 nighttime hours. What is the rate she is being paid for daytime hours and what is the rate for nighttime hours?

18. **Consumerism** For using a computerized financial news network for 25 min during prime time and 35 min during non-prime time, a customer was charged $10.75. A second customer was charged $13.35 for using the network for 30 min of prime time and 45 min of non-prime time. Find the cost per minute for using the financial news network during prime time.

19. **Sports** A basketball team scored 87 points in two-point baskets and three-point baskets. If the two-point baskets had been three-point baskets and the three-point baskets had been two-point baskets, the team would have scored 93 points. Find how many two-point baskets and how many three-point baskets the team scored.

20. **Consumerism** The employees of a hardware store ordered lunch from a local delicatessen. The lunch consisted of 4 turkey sandwiches and 7 orders of french fries, for a total cost of $23.30. The next day, the employees ordered 5 turkey sandwiches and 5 orders of french fries totaling $25.75. What does the delicatessen charge for a turkey sandwich? What is the charge for an order of french fries?

21. **Coin Problems** Two coin banks contain only nickels and quarters. The total value of the coins in the first bank is $2.90. In the second bank, there are two more quarters than in the first bank and twice as many nickels. The total value of the coins in the second bank is $3.80. Find the number of nickels and the number of quarters in the first bank.

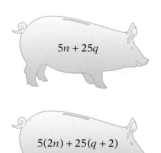

22. **Coin Problems** Two coin banks contain only nickels and dimes. The total value of the coins in the first bank is $3. In the second bank, there are 4 more nickels than in the first bank and one-half as many dimes. The total value of the coins in the second bank is $2. Find the number of nickels and the number of dimes in the first bank.

23. **Coin Problems** The total value of the dimes and quarters in a coin bank is $3.30. If the quarters were dimes and the dimes were quarters, the total value of the coins would be $3. Find the number of dimes and the number of quarters in the bank.

24. **Coin Problems** The total value of the nickels and dimes in a coin bank is $3. If the nickels were dimes and the dimes were nickels, the total value of the coins would be $3.75. Find the number of nickels and the number of dimes in the bank.

6.4 Applying Concepts

25. Geometry Two angles are supplementary. The larger angle is 15° more than twice the measure of the smaller angle. Find the measures of the two angles. (Supplementary angles are two angles whose sum is 180°.)

26. Geometry Two angles are complementary. The larger angle is four times the measure of the smaller angle. Find the measure of the two angles. (Complementary angles are two angles whose sum is 90°.)

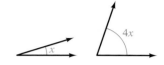

27. Investments An investment club placed a portion of its funds in a 9% annual simple interest account and the remainder in an 8% annual simple interest account. The amount of interest earned for one year was $860. If the amounts placed in each account had been reversed, the interest earned would have been $840. How much was invested in each account?

28. Investments An investor has $5000 to invest in two accounts. The first account earns 8% annual simple interest, and the second account earns 10% annual simple interest. How much money should be invested in each account so that the annual interest earned is $600?

29. Coin Problems The value of the nickels and dimes in a coin bank is $.25. If the number of nickels and the number of dimes were doubled, the value of the coins would be $.50. How many nickels and how many dimes are in the bank?

30. Coin Problems A coin bank contains nickels and dimes, but there are no more than 27 coins. The value of the coins is $2.10. How many different combinations of nickels and/or dimes could be in the bank?

31. The coin and stamp problems in the chapter "Solving Equations and Inequalities: Applications" can be solved using a system of equations. For example, look at the first problem in the section on coin and stamp problems in the chapter. Let q represent the number of quarters in the bank, and let d represent the number of dimes. There is a total of 9 coins, so $q + d = 9$. The total value of the money in the bank is $1.20, so we can write the equation $25q + 10d = 120$. The solution of the system of equations.

$$q + d = 9$$
$$25q + 10d = 120$$

is (2, 7), so there are 2 quarters and 7 dimes in the bank. Explain how to use a system of equations to solve the investment problems in the chapter "Solving Equations and Inequalities: Applications." You might use the first problem in the section on investment problems in that chapter as a basis for your explanation.

Focus on Problem Solving

Using a Table and Searching for a Pattern

Consider the numbers 10, 12, and 28 and the sum of the proper factors (the natural-number factors less than the number) of those numbers.

$$10: 1 + 2 + 5 = 8 \quad 12: 1 + 2 + 3 + 4 + 6 = 16$$
$$28: 1 + 2 + 4 + 7 + 14 = 28$$

10 is called a *deficient number* because the sum of its proper factors is less than the number ($8 < 10$). 12 is called an *abundant number* because the sum of its proper factors is greater than the number ($16 > 12$), and 28 is called a *perfect number* because the sum of its proper divisors equals the number ($28 = 28$).

Our goal for this "Focus on Problem Solving" is to try to find a method that will determine whether a number is deficient, abundant, or perfect without having to first find all the factors and then add them up. We will use a table and search for a pattern.

Before we begin, recall that a prime number is a number greater than 1 whose only factors are itself and 1 and that each natural number greater than 1 has a unique prime factorization. For instance, the prime factorization of 36 is given by $36 = 2^2 \cdot 3^2$. Note that the proper factors of 36 (1, 2, 3, 4, 6, 9, 12, 18) can be represented in terms of the same prime numbers.

$$1 = 2^0, \quad 2 = 2^1, \quad 3 = 3^1, \quad 4 = 2^2, \quad 6 = 2 \cdot 3, \quad 9 = 3^2, \quad 12 = 2^2 \cdot 3,$$
$$18 = 2 \cdot 3^2$$

Now let us consider a trial problem of determining whether 432 is deficient, abundant, or perfect.

> **Take Note**
>
> This table contains the factor $432 = 2^4 \cdot 3^3$, which is not a proper factor.

We write the prime factorization of 432 as $2^4 \cdot 3^3$ and place the factors of 432 in a table as shown at the right. This table contains *all* the factors of 432 represented in terms of the prime-number factors. The sum of each column is shown at the bottom.

	1	$1 \cdot 3$	$1 \cdot 3^2$	$1 \cdot 3^3$
	2	$2 \cdot 3$	$2 \cdot 3^2$	$2 \cdot 3^3$
	2^2	$2^2 \cdot 3$	$2^2 \cdot 3^2$	$2^2 \cdot 3^3$
	2^3	$2^3 \cdot 3$	$2^3 \cdot 3^2$	$2^3 \cdot 3^3$
	2^4	$2^4 \cdot 3$	$2^4 \cdot 3^2$	$2^4 \cdot 3^3$
Sum	31	$31 \cdot 3$	$31 \cdot 3^2$	$31 \cdot 3^3$

Here is the calculation of the sum for the column headed by $1 \cdot 3$.

$$1 \cdot 3 + 2 \cdot 3 + 2^2 \cdot 3 + 2^3 \cdot 3 + 2^4 \cdot 3 = (1 + 2 + 2^2 + 2^3 + 2^4)3 = 31(3)$$

For the column headed by $1 \cdot 3^2$ there is a similar situation.

$$1 \cdot 3^2 + 2 \cdot 3^2 + 2^2 \cdot 3^2 + 2^3 \cdot 3^2 + 2^4 \cdot 3^2 = (1 + 2 + 2^2 + 2^3 + 2^4)3^2 = 31(3^2)$$

The sum of *all* the factors (including 432) is the sum of the last row.

$$\text{Sum of all factors} = 31 + 31 \cdot 3 + 31 \cdot 3^2 + 31 \cdot 3^3 = 31(1 + 3 + 3^2 + 3^3)$$
$$= 31(40) = 1240$$

To find the sum of the *proper* factors, we must subtract 432 from 1240; we get 808 (see "Take Note" on the preceding page). Thus 432 is abundant.

We now look for some pattern for the sum of all the factors. Note that

Sum of left $1 + 2 + 2^2 + 2^3 + 2^4 =$ ⌐――――――⌐ $= 1 + 3 + 3^2 + 3^3$ Sum of
column   top row
$$31(40) = 1240$$

This suggests that the sum of the proper factors can be found by finding the sum of all the prime-power factors for each prime, multiplying those numbers, and then subtracting the original number. Although we have not proved this for all cases, it is a true statement.

For instance, to find the sum of the proper factors of 3240, first find the prime factorization.

$$3240 = 2^3 \cdot 3^4 \cdot 5$$

Now find the following sums:

$$1 + 2 + 2^2 + 2^3 = 15 \qquad 1 + 3 + 3^2 + 3^3 + 3^4 = 121 \qquad 1 + 5 = 6$$

The sum of the proper factors $= (15)(121)6 - 3240 = 7650$. 3240 is abundant.

Determine whether the number is deficient, abundant, or perfect.

1. 200   **2.** 3125   **3.** 8128   **4.** 10,000
5. Is a prime number deficient, abundant, or perfect?

Projects & Group Activities

Find a Pattern

The "Focus on Problem Solving" involved finding the sum of $1 + 3 + 3^2$ and $1 + 2 + 2^2 + 2^3$. For sums that contain a larger number of terms, it may be difficult or time-consuming to try to evaluate the sum. Perhaps there is a pattern for these sums that can be used to calculate them without having to evaluate each exponential expression and then add the results.

Look at the following from the calculation of the sum of the factors of 3240.

$$1 + 2 + 2^2 + 2^3 = 1 + 2 + 4 + 8 = 15$$
$$\frac{2^{3+1} - 1}{2 - 1} = \frac{2^4 - 1}{1} = 16 - 1 = 15$$
$$1 + 3 + 3^2 + 3^3 + 3^4 = 1 + 3 + 9 + 27 + 81 = 121$$
$$\frac{3^{4+1} - 1}{3 - 1} = \frac{3^5 - 1}{2} = \frac{243 - 1}{2} = \frac{242}{2} = 121$$

Consider another sum of this type.

$$1 + 7 + 7^2 + 7^3 + 7^4 + 7^5 = 1 + 7 + 49 + 343 + 2401 + 16{,}807 = 19{,}608$$
$$\frac{7^{5+1} - 1}{7 - 1} = \frac{7^6 - 1}{6} = \frac{117{,}649 - 1}{6} = \frac{117{,}648}{6} = 19{,}608$$

On the basis of the examples shown above, make a conjecture as to the value of $1 + n + n^2 + n^3 + n^4 + \cdots + n^k$, where n and k are natural numbers greater than 1.

Break-even Analysis

Break-even analysis is a method used to determine the sales volume required for a company to break even, or experience neither a profit nor a loss, on the sale of its product. The **break-even point** represents the number of units that must be made and sold in order for income from sales to equal the cost of the product.

The break-even point can be determined by graphing two equations on the same coordinate grid. The first equation is $R = SN$, where R is the revenue earned, S is the selling price per unit, and N is the number of units sold. The second equation is $T = VN + F$, where T is the total cost, F is the fixed costs, V is the variable costs per unit, and N is the number of units sold. The break-even point is the point at which the graphs of the two equations intersect, which represents the point at which revenue is equal to cost.

1. A company manufactures and sells digital watches. The fixed costs are $20,000, the variable costs per unit are $25, and the selling price per watch is $125. Write two equations to represent this information.

2. Graph the two equations on the same co-ordinate grid, using only quadrant I. The horizontal axis is the number of units sold. Use the model shown at the right.

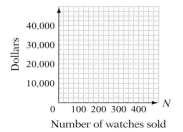

3. What does the point of intersection of the two graphs represent?

4. How many watches must the company sell to break even? What is the company's revenue for this sales figure? What is the company's total cost for this sales figure?

5. Shade the region between the two lines of the graph to the right of the intersection point. What does this shaded region represent? What does the unshaded region between the two lines to the left of the point of intersection represent?

6. You are a marketing manager for a manufacturing company that produces rocking chairs. Sales have been declining during the past year. Management, concerned about breaking even during the coming year, has asked you to prepare some figures. You have spoken to the company's accountant and learned that the company's fixed costs are $48,000 and that the variable costs per chair are $160.
   a. You want to determine the break-even point when the selling price of a chair is $460. What is the break-even point?
   b. During the past year, the company made and sold 130 rocking chairs at a selling price of $500 per chair. Did the company lose money, break even, or earn a profit?

Solving a System of Equations with a Graphing Calculator

A graphing calculator can be used to approximate the solution of a system of equations in two variables. Graph each equation of the system of equations, and then approximate the coordinates of the point of intersection. The process by which you approximate the solution depends on the make and model of calculator you have. In all cases, however, you must first solve each equation in the system of equations for y.

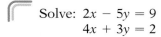

Solve: $2x - 5y = 9$
$\qquad\quad 4x + 3y = 2$

Solve each equation for y.

$2x - 5y = 9$ $\qquad\qquad 4x + 3y = 2$
$\quad\; -5y = -2x + 9$ $\qquad\quad\; 3y = -4x + 2$
$\qquad\quad y = \dfrac{2}{5}x - \dfrac{9}{5}$ $\qquad\qquad y = -\dfrac{4}{3}x + \dfrac{2}{3}$

The following graphing calculator screens are taken from a TI-83 Plus. Similar screens would appear if a different graphing calculator were used.

For the TI-83 and TI-83 Plus, press $\boxed{\text{Y=}}$. Enter one equation as Y1 and the other as Y2. The result should be similar to the screen at the right.

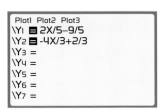

Press $\boxed{\text{GRAPH}}$. The graphs of the two equations should appear on the screen. If the point of intersection is not on the screen, adjust the viewing window by pressing the $\boxed{\text{WINDOW}}$ key.

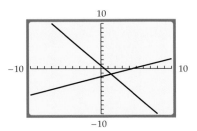

Press $\boxed{\text{2nd}}$ CALC 5 $\boxed{\text{ENTER}}$ $\boxed{\text{ENTER}}$ $\boxed{\text{ENTER}}$. After a few seconds, the point of intersection will appear at the bottom of the screen as X = 1.4230769, Y = −1.230769.

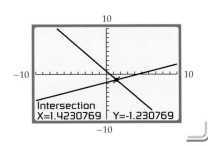

Solve by using a graphing calculator.

1. $4x - 5y = 48$ $\quad$ **2.** $3x + 2y = 11$ $\quad$ **3.** $x = 3y + 2$ $\quad$ **4.** $x = 2y - 5$
$\quad\;\; 5x + 7y = 7$ $\qquad\qquad 7x - 6y = 13$ $\qquad\quad y = 4x - 2$ $\qquad\quad x = 3y + 2$

Here is an example of using a graphing calculator to solve an investment problem.

An account executive deposited $4000 in two simple interest accounts, one with an annual simple interest rate of 3.5% and the other with an annual simple interest rate of 4.5%. How much is deposited in each account if both accounts earn the same annual interest?

Amount invested at 3.5%: x
Amount invested at 4.5%: $4000 - x$

Interest earned on the 3.5% account: $0.035x$
Interest earned on the 4.5% account: $0.045(4000 - x)$

Enter $0.035x$ into Y1. Enter $0.045(4000 - x)$ into Y2. Graph the equations. (We used a window of Xmin = 0, Xmax = 4000, Ymin = 0, Ymax = 250.) Use the INTERSECT feature to find the point of intersection. At the point of intersection, $x = 2250$. This is the amount in the 3.5% account. The amount in the 4.5% account is $4000 - x = 4000 - 2250 = 1750$.

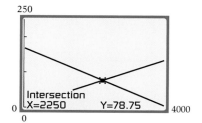

$2250 is invested at 3.5% and $1750 is invested at 4.5%.

Solve each of the following problems using a graphing calculator.

5. Suppose a breadmaker costs $180 and the ingredients and electricity needed to make one loaf of bread cost $.95. If a comparable loaf of bread at a grocery store costs $1.95, how many loaves of bread must you make before the breadmaker pays for itself?

6. Suppose a natural gas clothes washer costs $275 and uses $.25 of gas and water to wash a large load of clothes. If a laundromat charges $1.50 to do a load of laundry, how many loads of clothes must you wash before purchasing the washing machine is more economical?

7. Suppose a natural gas clothes dryer costs $240 and uses $.40 of gas to dry a load of clothes for 1 h. If a laundromat charges $1.75 to use a dryer for 1 h, how many loads of clothes must you dry before purchasing the gas dryer is more economical?

8. When Sara Whitehorse changed jobs, she rolled over the $6000 in her retirement account into two simple interest accounts. On one account, the annual simple interest rate is 9%; on the second account, the annual simple interest rate is 6%. How much must be invested in each account if the accounts are to earn the same amount of annual interest?

Chapter Summary

Key Words

Equations considered together are called a **system of equations**. [6.1.1, p. 300]

A **solution of a system of equations in two variables** is an ordered pair that is a solution of each equation of the system. [6.1.1, p. 301]

An **independent system of equations** has one solution. The graphs of the equations in an independent system of linear equations intersect at one point. [6.1.1, pp. 302–303]

An **inconsistent system of equations** has no solution. The graphs of the equations in an inconsistent system of linear equations are parallel lines. If, when you are solving a system of equations algebraically, the variable is eliminated and the result is a false equation, such as $-3 = 8$, the system is inconsistent. [6.1.1/6.2.1/6.3.1, pp. 302–303, 311, 316]

A **dependent system of equations** has an infinite number of solutions. The graphs of the equations in a dependent system of linear equations represent the same line. If, when you are solving a system of equations algebraically, the variable is eliminated and the result is a true equation, such as $1 = 1$, the system is dependent. [6.1.1/6.2.1/6.3.1, pp. 302–303, 311, 317]

Examples

An example of a system of equations is

$$4x + y = 6$$
$$3x + 2y = 7$$

The solution of the system of equations shown above is the ordered pair $(1, 2)$ because it is a solution of each equation in the system.

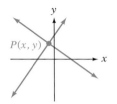

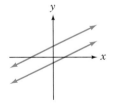

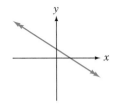

Essential Rules and Procedures

To solve a system of linear equations in two variables by graphing, graph each equation on the same coordinate system. If the lines graphed intersect at one point, the point of intersection is the ordered pair that is a solution of the system. If the lines graphed are parallel, the system is inconsistent. If the lines represent the same line, the system is dependent. [6.1.1, pp. 303–304]

Solve by graphing: $x + 2y = 4$
$2x + y = -1$

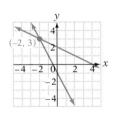

The solution is $(-2, 3)$.

To solve a system of linear equations by the substitution method, one variable must be written in terms of the other variable. [6.2.1, p. 309]

Solve by substitution: $2x - y = 5$ (1)
$3x + y = 5$ (2)

$3x + y = 5$
$\quad\quad\; y = -3x + 5$ • Solve equation (2) for y.

$\quad\quad\quad\quad\quad 2x - y = 5$ • Substitute for y in
$2x - (-3x + 5) = 5$ equation (1).
$\quad\; 2x + 3x - 5 = 5$
$\quad\quad\quad\quad\quad\quad 5x = 10$
$\quad\quad\quad\quad\quad\quad\; x = 2$

$y = -3x + 5$ • Substitute the
$y = -3(2) + 5$ value of x to
$y = -1$ find y.

The solution is $(2, -1)$.

To solve a system of linear equations by the addition method, use the Multiplication Property of Equations to rewrite one or both of the equations so that the coefficients of one variable are opposites. Then add the two equations and solve for the variables. [6.3.1, pp. 314–315]

Solve by the addition method:

$3x + y = 4$ (1)
$\; x + y = 6$ (2)

$\; 3x + y = 4$
$-x - y = -6$ • Multiply both sides of
                 equation (2) by -1.
$\quad\quad 2x = -2$ • Add the two equations.
$\quad\quad\; x = -1$ • Solve for x.

$\quad x + y = 6$ • Substitute the value of x
$-1 + y = 6$ to find y.
$\quad\quad\; y = 7$

The solution is $(-1, 7)$.

Chapter Review Exercises

1. Solve by substitution: $4x + 7y = 3$
$\quad\quad\quad\quad\quad\quad\quad\quad\quad\quad x = y - 2$

2. Solve by graphing: $3x + y = 3$
$\quad\quad\quad\quad\quad\quad\quad\quad\quad x = 2$

3. Solve by the addition method: $3x + 8y = -1$
$\quad\quad\quad\quad\quad\quad\quad\quad\quad\quad\quad\quad\; x - 2y = -5$

4. Solve by substitution: $8x - y = 2$
$\quad\quad\quad\quad\quad\quad\quad\quad\quad\quad y = 5x + 1$

5. Solve by graphing: $x + y = 2$
$\quad\quad\quad\quad\quad\quad\quad\quad x - y = 0$

6. Solve by the addition method: $4x - y = 9$
$\quad\quad\quad\quad\quad\quad\quad\quad\quad\quad\quad 2x + 3y = -13$

7. Solve by substitution: $-2x + y = -4$
$\quad\quad\quad\quad\quad\quad\quad\quad\quad\quad\; x = y + 1$

8. Solve by the addition method:
$8x - y = 25$
$32x - 4y = 100$

9. Solve by the addition method: $5x - 15y = 30$
$2x + 6y = 0$

10. Solve by graphing: $3x - y = 6$
$y = -3$

11. Solve by the addition method: $7x - 2y = 0$
$2x + y = -11$

12. Solve by substitution: $x - 5y = 4$
$y = x - 4$

13. Is $(-1, -3)$ a solution of the system
$5x + 4y = -17$
$2x - y = 1?$

14. Solve by the addition method:
$6x + 4y = -3$
$12x - 10y = -15$

15. Solve by the addition method: $5x + 2y = -9$
$12x - 7y = 2$

16. Solve by graphing: $x - 3y = 12$
$y = x - 6$

17. Solve by the addition method:
$5x + 7y = 21$
$20x + 28y = 63$

18. Solve by substitution: $9x + 12y = -1$
$x - 4y = -1$

19. Solve by graphing: $4x - 2y = 8$
$y = 2x - 4$

20. Solve by the addition method:
$3x + y = -2$
$-9x - 3y = 6$

21. Solve by the addition method: $11x - 2y = 4$
$25x - 4y = 2$

22. Solve by substitution: $4x + 3y = 12$
$y = -\dfrac{4}{3}x + 4$

23. Solve by graphing: $y = -\dfrac{1}{4}x + 3$
$2x - y = 6$

24. Solve by the addition method: $2x - y = 5$
$10x - 5y = 20$

25. Solve by substitution: $6x + 5y = -2$
$y = 2x - 2$

26. Is $(-2, 0)$ a solution of the system
$-x + 9y = 2$
$6x - 4y = 12?$

27. Solve by the addition method: $6x - 18y = 7$
$9x + 24y = 2$

28. Solve by substitution: $12x - 9y = 18$
$y = \dfrac{4}{3}x - 3$

29. Solve by substitution: $9x - y = -3$
$18x - y = 0$

30. Solve by graphing: $x + 2y = 3$
$y = -\dfrac{1}{2}x + 1$

31. Solve by the addition method: $7x - 9y = 9$
$3x - y = 1$

32. Solve by substitution: $7x + 3y = -16$
$x - 2y = 5$

33. Solve by substitution: $5x - 3y = 6$
$x - y = 2$

34. Solve by the addition method: $6x + y = 12$
$9x + 2y = 18$

35. Solve by the addition method: $5x + 12y = 4$
$x + 6y = 8$

36. Solve by substitution: $6x - y = 0$
$7x - y = 1$

37. Flying with the wind, a plane can travel 800 mi in 4 h. Against the wind, the plane requires 5 h to fly the same distance. Find the rate of the plane in calm air and the rate of the wind.

38. Admission to a movie theater is $7 for adults and $5 for children. If the receipts from 200 tickets were $1120, how many adult tickets and how many children's tickets were sold?

39. A canoeist traveling with the current traveled the 30 mi between two river-side campsites in 3 h. The return trip took 5 h. Find the rate of the canoeist in still water and the rate of the current.

40. A small wood carving company mailed 190 advertisements, some requiring 33¢ in postage and others requiring 55¢ in postage. If the total cost for the mailings was $75.90, how many mailings that required 33¢ were sent?

41. A boat traveling with the current went 48 km in 3 h. Against the current, the boat traveled 24 km in 2 h. Find the rate of the boat in calm water and the rate of the current.

42. A music shop sells some compact discs for $15 and some for $10. A customer spent $120 on 10 compact discs. How many at each price did the customer purchase?

43. With a tailwind, a flight crew flew 420 km in 3 h. Flying against the tailwind, the crew flew 440 km in 4 h. Find the rate of the plane in calm air and the rate of the wind.

44. A paddle boat can travel 4 mi downstream in 1 h. Paddling upstream, against the current, the paddle boat traveled 2 mi in 1 h. Find the rate of the boat in calm water and the rate of the current.

45. A silo contains a mixture of lentils and corn. If 50 bushels of lentils were added to this mixture, there would be twice as many bushels of lentils as bushels of corn. If 150 bushels of corn were added to the original mixture, there would be the same amount of corn as lentils. How many bushels of each are in the silo?

46. Flying with the wind, a small plane flew 360 mi in 3 h. Against the wind, the plane took 4 h to fly the same distance. Find the rate of the plane in calm air and the rate of the wind.

47. A coin purse contains 80¢ in nickels and dimes. If there are 12 coins in all, how many dimes are in the coin purse?

48. A pilot flying with the wind flew 2100 mi from one city to another in 6 h. The return trip against the wind took 7 h. Find the rate of the plane in calm air and the rate of the wind.

49. An investor buys 1500 shares of stock, some costing $6 per share and the rest costing $25 per share. If the total cost of the stock is $12,800, how many shares of each did the investor buy?

50. Rowing with the current, a sculling team went 24 mi in 2 h. Rowing against the current, the team went 18 mi in 3 h. Find the rate of the sculling team in calm water and the rate of the current.

Chapter Test

1. Solve by substitution: $4x - y = 11$
$$y = 2x - 5$$

2. Solve by the addition method: $4x + 3y = 11$
$$5x - 3y = 7$$

3. Is $(-2, 3)$ a solution of the system
$2x + 5y = 11$
$x + 3y = 7$?

4. Solve by substitution: $x = 2y + 3$
$$3x - 2y = 5$$

5. Solve by the addition method: $2x - 5y = 6$
$$4x + 3y = -1$$

6. Solve by graphing: $3x + 2y = 6$
$$5x + 2y = 2$$

7. Solve by substitution: $4x + 2y = 3$
$$y = -2x + 1$$

8. Solve by substitution: $3x + 5y = 1$
$$2x - y = 5$$

9. Solve by the addition method: $7x + 3y = 11$
$$2x - 5y = 9$$

10. Solve by substitution: $3x - 5y = 13$
$$x + 3y = 1$$

11. Solve by the addition method: $5x + 6y = -7$
$$3x + 4y = -5$$

12. Is $(2, 1)$ a solution of the system $3x - 2y = 8$
$$4x + 5y = 3?$$

13. Solve by substitution: $3x - y = 5$
$$y = 2x - 3$$

14. Solve by the addition method: $3x + 2y = 2$
$$5x - 2y = 14$$

15. Solve by graphing: $3x + 2y = 6$
$$3x - 2y = 6$$

16. Solve by substitution: $x = 3y + 1$
$$2x + 5y = 13$$

17. Solve by the addition method: $5x + 4y = 7$
$$3x - 2y = 13$$

18. Solve by graphing: $3x + 6y = 2$
$$y = -\frac{1}{2}x + \frac{1}{3}$$

19. Solve by substitution: $4x - 3y = 1$
$$2x + y = 3$$

20. Solve by the addition method: $5x - 3y = 29$
$$4x + 7y = -5$$

21. Solve by substitution: $3x - 5y = -23$
$$x + 2y = -4$$

22. Solve by the addition method: $9x - 2y = 17$
$$5x + 3y = -7$$

23. With the wind, a plane flies 240 mi in 2 h. Against the wind, the plane requires 3 h to fly the same distance. Find the rate of the plane in calm air and the rate of the wind.

24. With the current, a motorboat can travel 48 mi in 3 h. Against the current, the boat requires 4 h to travel the same distance. Find the rate of the boat in calm water and the rate of the current.

25. Two coin banks contain only dimes and nickels. In the first bank, the total value of the coins is $5.50. In the second bank, there are one-half as many dimes as in the first bank and 10 fewer nickels. The total value of the coins in the second bank is $3. Find the number of dimes and the number of nickels in the first bank.

Cumulative Review Exercises

1. Given $A = \{-8, -4, 0\}$, which elements of set A are less than or equal to -4?

2. Use the roster method to write the set of positive integers less than or equal to 10.

3. Simplify: $12 - 2(7 - 5)^2 \div 4$

4. Simplify: $2[5a - 3(2 - 5a) - 8]$

5. Evaluate $\dfrac{a^2 - b^2}{2a}$ when $a = 4$ and $b = -2$.

6. Solve: $-\dfrac{3}{4}x = \dfrac{9}{8}$

7. Solve: $4 - 3(2 - 3x) = 7x - 9$

8. Solve: $3[2 - 4(x + 1)] = 6x - 2$

9. Solve: $-7x - 5 > 4x + 50$

10. Solve: $5 + 2(x + 1) \le 13$

11. What percent of 50 is 12?

12. Find the x- and y-intercepts of $3x - 6y = 12$.

13. Find the slope of the line that contains the points whose coordinates are $(2, -3)$ and $(-3, 4)$.

14. Find the equation of the line that contains the point whose coordinates are $(-2, 3)$ and has slope $-\dfrac{3}{2}$.

15. Graph: $3x - 2y = 6$

16. Graph: $y = -\dfrac{1}{3}x + 3$

17. Graph the solution set of $y > -3x + 4$.

18. Graph: $f(x) = \dfrac{3}{4}x + 2$

19. Find the domain and range of the relation $\{(-5, 5), (0, 5), (1, 5), (5, 5)\}$. Is the relation a function?

20. Evaluate $f(x) = -2x - 5$ at $x = -4$.

21. Is (2, 0) a solution of the system $5x - 3y = 10$
$4x + 7y = 8$?

22. Solve by substitution: $2x - 3y = -7$
$x + 4y = 2$

23. Solve by graphing: $2x + 3y = 6$
$3x + y = 2$

24. Solve by the addition method: $5x - 2y = 8$
$4x + 3y = 11$

25. Find the range of the function given by the equation $f(x) = 5x - 2$ if the domain is $\{-4, -2, 0, 2, 4\}$.

26. A business manager has determined that the cost per unit for a camera is $90 and that the fixed costs per month are $3500. Find the number of cameras produced during a month in which the total cost was $21,500. Use the equation $T = U \cdot N + F$, where T is the total cost, U is the cost per unit, N is the number of units produced, and F is the fixed cost.

27. A total of $8750 is invested in two simple interest accounts. On one account, the annual simple interest rate is 9.6%. On the second account, the annual simple interest rate is 7.2%. How much should be invested in each account so that both accounts earn the same amount of interest?

28. Flying with the wind, a plane can fly 570 mi in 3 h. Flying against the wind, it takes the same amount of time to fly 390 mi. Find the rate of the plane in calm air and the rate of the wind.

29. With the current, a motorboat can travel 24 mi in 2 h. Against the current, the boat requires 3 h to travel the same distance. Find the rate of the boat in calm water.

30. Two coin banks contain only dimes and nickels. In the first bank, the total value of the coins is $5.25. In the second bank, there are one-half as many dimes as in the first bank and 15 fewer nickels. The total value of the coins in the second bank is $2.50. Find the number of dimes in the first bank.

As noted in the chapter 'Linear Equations and Inequalities,' the population of the United States of America rose from 76 million in 1900 to 281 million in 2000. Our growing population has resulted in a drastic decrease in woodlands and farmland and a dramatic increase in land used for housing developments and shopping centers. It has lead demographers (people who study the characteristics of human populations, such as their size, growth, density, and distribution) to ask questions about population and land allocation. The exercises in the **Project on page 377** ask you to determine the answers to questions concerning land allocation.

Applications

Aquariums,   *p. 375*

Astronomy,   *p. 368*

Chemistry,   *p. 368*

Computers,   *p. 368*

Construction,   *p. 376*

Electricity,   *p. 368*

Geology,   *p. 368*

Geometry,   *pp. 351, 355, 358, 374, 381, 382, 383*

Integer problems,   *p. 383*

Interior decorating,   *p. 375*

Light,   *p. 368*

Markup,   *p. 383*

Mixture problems,   *p. 383*

Oceanography,   *p. 376*

Physics,   *p. 368*

Population and land allocation,   *p. 377*

Real estate,   *pp. 375, 376*

Sports,   *pp. 359, 375, 376*

Uniform motion problems,   *p. 383*

7

Polynomials

Objectives

Section 7.1
1. Add polynomials
2. Subtract polynomials

Section 7.2
1. Multiply monomials
2. Simplify powers of monomials

Section 7.3
1. Multiply a polynomial by a monomial
2. Multiply two polynomials
3. Multiply two binomials
4. Multiply binomials that have special products
5. Application problems

Section 7.4
1. Integer exponents
2. Scientific notation

Section 7.5
1. Divide a polynomial by a monomial
2. Divide polynomials

PrepTest

1. Subtract: $-2 - (-3)$

2. Multiply: $-3(6)$

3. Simplify: $-\dfrac{24}{-36}$

4. Evaluate $3n^4$ when $n = -2$.

5. If $\dfrac{a}{b}$ is a fraction in simplest form, what number is not a possible value of b?

6. Are $2x^2$ and $2x$ like terms?

7. Simplify: $3x^2 - 4x + 1 + 2x^2 - 5x - 7$

8. Simplify: $-4y + 4y$

9. Simplify: $-3(2x - 8)$

10. Simplify: $3xy - 4y - 2(5xy - 7y)$

Go Figure

I have two more sisters than brothers. Each of my sisters has two more sisters than brothers. How many more sisters than brothers does my youngest brother have?

7.1 | Addition and Subtraction of Polynomials

1 **Add polynomials**

A **monomial** is a number, a variable, or a product of numbers and variables. For instance,

7	b	$\dfrac{2}{3}a$	$12xy^2$
A number	A variable	A product of a number and a variable	A product of a number and variables

The expression $3\sqrt{x}$ is not a monomial because $\sqrt{x}$ cannot be written as a product of variables. The expression $\dfrac{2x}{y^2}$ is not a monomial because it is a *quotient* of variables.

A **polynomial** is a variable expression in which the terms are monomials.

A polynomial of *one* term is a **monomial.** $-7x^2$ is a monomial.

A polynomial of *two* terms is a **binomial.** $4x + 2$ is a binomial.

A polynomial of *three* terms is a **trinomial.** $7x^2 + 5x - 7$ is a trinomial.

The terms of a polynomial in one variable are usually arranged so that the exponents of the variable decrease from left to right. This is called **descending order.**

$$4x^3 - 3x^2 + 6x - 1$$

$$5y^4 - 2y^3 + y^2 - 7y + 8$$

The **degree** of a polynomial in one variable is the value of the largest exponent on the variable. The degree of $4x^3 - 3x^2 + 6x - 1$ is 3. The degree of $5y^4 - 2y^3 + y^2 - 7y + 8$ is 4. The degree of a nonzero constant is zero. For instance, the degree of 7 is zero. The number zero has no degree.

Polynomials can be added, using either a vertical or a horizontal format, by combining like terms.

Example 1   Add: $(2x^2 + x - 1) + (3x^3 + 4x^2 - 5)$
Use a vertical format.

Solution

$$\begin{array}{r} 2x^2 + x - 1 \\ \underline{3x^3 + 4x^2 \qquad - 5} \\ 3x^3 + 6x^2 + x - 6 \end{array}$$

• Arrange the terms of each polynomial in descending order with like terms in the same column.
• Combine the terms in each column.

Problem 1   Add: $(2x^2 + 4x - 3) + (5x^2 - 6x)$
Use a vertical format.

Solution   See page S15.

Example 2   Add: $(3x^3 - 7x + 2) + (7x^2 + 2x - 7)$
Use a horizontal format.

Solution $(3x^3 - 7x + 2) + (7x^2 + 2x - 7)$
$= 3x^3 + 7x^2 + (-7x + 2x) + (2 - 7)$

$= 3x^3 + 7x^2 - 5x - 5$

• Use the Commutative and Associative Properties of Addition to rearrange and group like terms.
• Combine like terms, and write the polynomial in descending order.

Problem 2   Add: $(-4x^2 - 3xy + 2y^2) + (3x^2 - 4y^2)$
Use a horizontal format.

Solution   See page S15.

2   **Subtract polynomials**

The **opposite** of the polynomial $x^2 - 2x + 3$ is $-(x^2 - 2x + 3)$.

To simplify the opposite of a polynomial, remove the parentheses and change the sign of every term inside the parentheses.

$$-(x^2 - 2x + 3) = -x^2 + 2x - 3$$

Polynomials can be subtracted using either a vertical or a horizontal format. To subtract, add the opposite of the second polynomial to the first.

Example 3   Subtract: $(-3x^2 - 7) - (-8x^2 + 3x - 4)$
Use a vertical format.

Solution   The opposite of $-8x^2 + 3x - 4$ is $8x^2 - 3x + 4$.

$$
\begin{array}{l}
-3x^2 \qquad\ -\ 7 \\
\underline{8x^2 - 3x + 4} \\
5x^2 - 3x - 3
\end{array}
$$

• Write the terms of each polynomial in descending order with like terms in the same column.

• Combine the terms in each column.

Problem 3   Subtract: $(8y^2 - 4xy + x^2) - (2y^2 - xy + 5x^2)$
Use a vertical format.

Solution   See page S15.

Example 4   Subtract: $(5x^2 - 3x + 4) - (-3x^3 - 2x + 8)$
Use a horizontal format.

Solution
$$
\begin{aligned}
&(5x^2 - 3x + 4) - (-3x^3 - 2x + 8) \\
&= (5x^2 - 3x + 4) + (3x^3 + 2x - 8) \\
&= 3x^3 + 5x^2 + (-3x + 2x) + (4 - 8) \\
&= 3x^3 + 5x^2 - x - 4
\end{aligned}
$$

• Rewrite subtraction as addition of the opposite.

• Combine like terms. Write the polynomial in descending order.

Problem 4   Subtract: $(-3a^2 - 4a + 2) - (5a^3 + 2a - 6)$
Use a horizontal format.

Solution   See page S15.

7.1   Concept Review

Determine whether the statement is always true, sometimes true, or never true.

1. The terms of a polynomial are monomials.

2. The polynomial $5x^3 + 4x^6 + 3x + 2$ is written in descending order.

3. The degree of the polynomial $7x^2 - 8x + 1$ is 3.

4. Like terms have the same coefficient and the same variable part.

5. The opposite of the polynomial $ax^3 - bx^2 + cx + d$ is $-ax^3 + bx^2 - cx - d$.

6. Subtraction is addition of the opposite.

7.1   Exercises

1 State whether the polynomial is a monomial, a binomial, or a trinomial.

1. $8x^4 - 6x^2$

2. $4a^2b^2 + 9ab + 10$

3. $7x^3y^4$

State whether or not the expression is a monomial.

4. $3\sqrt{x}$

5. $\dfrac{4}{x}$

6. x^2y^2

State whether or not the expression is a polynomial.

7. $\dfrac{1}{5}x^3 + \dfrac{1}{2}x$

8. $\dfrac{1}{5x^2} + \dfrac{1}{2x}$

9. $x + \sqrt{5}$

10. In your own words, explain the terms *monomial*, *binomial*, *trinomial*, and *polynomial*. Give an example of each.

Add. Use a vertical format.

11. $(x^2 + 7x) + (-3x^2 - 4x)$

12. $(3y^2 - 2y) + (5y^2 + 6y)$

13. $(y^2 + 4y) + (-4y - 8)$

14. $(3x^2 + 9x) + (6x - 24)$

15. $(2x^2 + 6x + 12) + (3x^2 + x + 8)$

16. $(x^2 + x + 5) + (3x^2 - 10x + 4)$

17. $(x^3 - 7x + 4) + (2x^2 + x - 10)$

18. $(3y^3 + y^2 + 1) + (-4y^3 - 6y - 3)$

19. $(2a^3 - 7a + 1) + (-3a^2 - 4a + 1)$

20. $(5r^3 - 6r^2 + 3r) + (r^2 - 2r - 3)$

Add. Use a horizontal format.

21. $(4x^2 + 2x) + (x^2 + 6x)$

22. $(-3y^2 + y) + (4y^2 + 6y)$

23. $(4x^2 - 5xy) + (3x^2 + 6xy - 4y^2)$

24. $(2x^2 - 4y^2) + (6x^2 - 2xy + 4y^2)$

25. $(2a^2 - 7a + 10) + (a^2 + 4a + 7)$

26. $(-6x^2 + 7x + 3) + (3x^2 + x + 3)$

27. $(5x^3 + 7x - 7) + (10x^2 - 8x + 3)$

28. $(3y^3 + 4y + 9) + (2y^2 + 4y - 21)$

29. $(2r^2 - 5r + 7) + (3r^3 - 6r)$

30. $(3y^3 + 4y + 14) + (-4y^2 + 21)$

31. $(3x^2 + 7x + 10) + (-2x^3 + 3x + 1)$

32. $(7x^3 + 4x - 1) + (2x^2 - 6x + 2)$

2 Subtract. Use a vertical format.

33. $(x^2 - 6x) - (x^2 - 10x)$

34. $(y^2 + 4y) - (y^2 + 10y)$

35. $(2y^2 - 4y) - (-y^2 + 2)$

36. $(-3a^2 - 2a) - (4a^2 - 4)$

37. $(x^2 - 2x + 1) - (x^2 + 5x + 8)$

38. $(3x^2 + 2x - 2) - (5x^2 - 5x + 6)$

39. $(4x^3 + 5x + 2) - (-3x^2 + 2x + 1)$ **40.** $(5y^2 - y + 2) - (-2y^3 + 3y - 3)$

41. $(2y^3 + 6y - 2) - (y^3 + y^2 + 4)$ **42.** $(-2x^2 - x + 4) - (-x^3 + 3x - 2)$

Subtract. Use a horizontal format.

43. $(y^2 - 10xy) - (2y^2 + 3xy)$ **44.** $(x^2 - 3xy) - (-2x^2 + xy)$

45. $(3x^2 + x - 3) - (x^2 + 4x - 2)$ **46.** $(5y^2 - 2y + 1) - (-3y^2 - y - 2)$

47. $(-2x^3 + x - 1) - (-x^2 + x - 3)$ **48.** $(2x^2 + 5x - 3) - (3x^3 + 2x - 5)$

49. $(4a^3 - 2a + 1) - (a^3 - 2a + 3)$ **50.** $(b^2 - 8b + 7) - (4b^3 - 7b - 8)$

51. $(4y^3 - y - 1) - (2y^2 - 3y + 3)$ **52.** $(3x^2 - 2x - 3) - (2x^3 - 2x^2 + 4)$

7.1   Applying Concepts

Simplify.

53. $\left(\dfrac{2}{3}a^2 + \dfrac{1}{2}a - \dfrac{3}{4}\right) - \left(\dfrac{5}{3}a^2 + \dfrac{1}{2}a + \dfrac{1}{4}\right)$ **54.** $\left(\dfrac{3}{5}x^2 + \dfrac{1}{6}x - \dfrac{5}{8}\right) + \left(\dfrac{2}{5}x^2 + \dfrac{5}{6}x - \dfrac{3}{8}\right)$

Solve.

55. What polynomial must be added to $3x^2 - 4x - 2$ so that the sum is $-x^2 + 2x + 1$?

56. What polynomial must be added to $-2x^3 + 4x - 7$ so that the sum is $x^2 - x - 1$?

57. What polynomial must be subtracted from $6x^2 - 4x - 2$ so that the difference is $2x^2 + 2x - 5$?

58. What polynomial must be subtracted from $2x^3 - x^2 + 4x - 2$ so that the difference is $x^3 + 2x - 8$?

59. Is it possible to subtract two polynomials, each of degree 3, and have the difference be a polynomial of degree 2? If so, give an example. If not, explain why not.

60. Is it possible to add two polynomials, each of degree 3, and have the sum be a polynomial of degree 2? If so, give an example. If not, explain why not.

7.2   Multiplication of Monomials

1   Multiply monomials

Recall that in the exponential expression x^5, x is the base and 5 is the exponent. The exponent indicates the number of times the base occurs as a factor.

The product of exponential expressions with the *same* base can be simplified by writing each expression in factored form and writing the result with an exponent.

$$x^3 \cdot x^2 = \overbrace{(x \cdot x \cdot x)}^{3 \text{ factors}} \cdot \overbrace{(x \cdot x)}^{2 \text{ factors}}$$
$$\underbrace{}_{5 \text{ factors}}$$
$$= x \cdot x \cdot x \cdot x \cdot x$$
$$= x^5$$

Adding the exponents results in the same product.

$$x^3 \cdot x^2 = x^{3+2} = x^5$$

> **Rule for Multiplying Exponential Expressions**
>
> If m and n are integers, then $x^m \cdot x^n = x^{m+n}$.

For example, in the expression $a^2 \cdot a^6 \cdot a$, the bases are the same. The expression can be simplified by adding the exponents. Recall that $a = a^1$.

$$a^2 \cdot a^6 \cdot a = a^{2+6+1} = a^9$$

Take Note

The Rule for Multiplying Exponential Expressions requires that the bases be the same. The expression $x^3 y^2$ cannot be simplified.

Example 1   Multiply: $(2xy)(3x^2y)$

Solution $(2xy)(3x^2y)$
$= (2 \cdot 3)(x \cdot x^2)(y \cdot y)$ • Use the Commutative and Associative Properties of Multiplication to rearrange and group factors.
$= 6x^{1+2}y^{1+1}$ • Multiply variables with the same base by adding the exponents.
$= 6x^3y^2$

Problem 1   Multiply: $(3x^2)(6x^3)$

Solution   See page S15.

Example 2   Multiply: $(2x^2y)(-5xy^4)$

Solution $(2x^2y)(-5xy^4)$
$= [2(-5)](x^2 \cdot x)(y \cdot y^4)$ • Use the Properties of Multiplication to rearrange and group factors.
$= -10x^3y^5$ • Multiply variables with the same base by adding the exponents.

Problem 2   Multiply: $(-3xy^2)(-4x^2y^3)$

Solution   See page S15.

2 Simplify powers of monomials

A power of a monomial can be simplified by rewriting the expression in factored form and then using the Rule for Multiplying Exponential Expressions.

$$(x^2)^3 = x^2 \cdot x^2 \cdot x^2$$
$$= x^{2+2+2}$$
$$= x^6$$

$$(x^4y^3)^2 = (x^4y^3)(x^4y^3)$$
$$= x^4 \cdot y^3 \cdot x^4 \cdot y^3$$
$$= (x^4 \cdot x^4)(y^3 \cdot y^3)$$
$$= x^{4+4}y^{3+3}$$
$$= x^8y^6$$

Note that multiplying each exponent inside the parentheses by the exponent outside the parentheses gives the same result.

$$(x^2)^3 = x^{2 \cdot 3} = x^6 \qquad (x^4y^3)^2 = x^{4 \cdot 2}y^{3 \cdot 2} = x^8y^6$$

Rule for Simplifying Powers of Exponential Expressions

If m and n are integers, then $(x^m)^n = x^{mn}$.

Simplify: $(x^5)^2$

Multiply the exponents.

$$(x^5)^2 = x^{5 \cdot 2} = x^{10}$$

Rule for Simplifying Powers of Products

If m, n, and p are integers, then $(x^m y^n)^p = x^{mp}y^{np}$.

Simplify: $(c^5d^3)^6$

Multiply each exponent inside the parentheses by the exponent outside the parentheses.

$$(c^5d^3)^6 = c^{5 \cdot 6}d^{3 \cdot 6}$$
$$= c^{30}d^{18}$$

Simplify: $(3a^2b)^3$

Multiply each exponent inside the parentheses by the exponent outside the parentheses.

$$(3a^2b)^3 = 3^{1 \cdot 3}a^{2 \cdot 3}b^{1 \cdot 3}$$
$$= 3^3a^6b^3$$
$$= 27a^6b^3$$

Example 3 Simplify: $(-2x)(-3xy^2)^3$

Solution

$$(-2x)(-3xy^2)^3$$
$$= (-2x)(-3)^3x^3y^6$$
$$= (-2x)(-27)x^3y^6$$
$$= [-2(-27)](x \cdot x^3)y^6$$
$$= 54x^4y^6$$

- Multiply each exponent in $-3xy^2$ by the exponent outside the parentheses.
- Simplify $(-3)^3$.
- Use the Properties of Multiplication to rearrange and group factors.
- Multiply variables with the same base by adding the exponents.

> **Problem 3**   Simplify: $(3x)(2x^2y)^3$
>
> **Solution**   See page S15.

7.2   Concept Review

Determine whether the statement is always true, sometimes true, or never true.

1. In the expression $(6x)^4$, x is the base and 4 is the exponent.

2. To multiply $x^m \cdot x^n$, multiply the exponents.

3. The expression "a power of a monomial" means the monomial is the base of an exponential expression.

4. $x^6 \cdot x \cdot x^8 = x^{14}$

5. $(4x^3y^5)^2 = 4x^6y^{10}$

6. $a^2 \cdot b^7 = ab^9$

7.2   Exercises

1. ✎ Explain how to multiply two exponential expressions with the same base. Provide an example.

2. ✎ Explain how to simplify the power of an exponential expression. Provide an example.

State whether the expression is the product of two exponential expressions or the power of an exponential expression.

3. a. $b^4 \cdot b^8$ **b.** $(b^4)^8$ **4. a.** $(2z)^2$ **b.** $2z \cdot z$

5. a. $(3a^4)^5$ **b.** $(3a^4)(5a)$ **6. a.** $x(-xy^4)$ **b.** $(-xy)^4$

Multiply.

7. $(x)(2x)$ **8.** $(-3y)(y)$ **9.** $(3x)(4x)$

10. $(7y^3)(7y^2)$ **11.** $(-2a^3)(-3a^4)$ **12.** $(5a^6)(-2a^5)$

13. $(x^2y)(xy^4)$ **14.** $(x^2y^4)(xy^7)$ **15.** $(-2x^4)(5x^5y)$

16. $(-3a^3)(2a^2b^4)$ **17.** $(x^2y^4)(x^5y^4)$ **18.** $(a^2b^4)(ab^3)$

19. $(2xy)(-3x^2y^4)$ **20.** $(-3a^2b)(-2ab^3)$ **21.** $(x^2yz)(x^2y^4)$

22. $(-ab^2c)(a^2b^5)$ **23.** $(a^2b^3)(ab^2c^4)$ **24.** $(x^2y^3z)(x^3y^4)$

25. $(-a^2b^2)(a^3b^6)$ **26.** $(xy^4)(-xy^3)$ **27.** $(-6a^3)(a^2b)$

28. $(2a^2b^3)(-4ab^2)$ **29.** $(-5y^4z)(-8y^6z^5)$ **30.** $(3x^2y)(-4xy^2)$

31. $(10ab^2)(-2ab)$ **32.** $(x^2y)(yz)(xyz)$ **33.** $(xy^2z)(x^2y)(z^2y^2)$

34. $(-2x^2y^3)(3xy)(-5x^3y^4)$ **35.** $(4a^2b)(-3a^3b^4)(a^5b^2)$ **36.** $(3ab^2)(-2abc)(4ac^2)$

2 Simplify.

37. $(2^2)^3$ **38.** $(3^2)^2$ **39.** $(-2)^2$ **40.** $(-3)^3$

41. $(-2^2)^3$ **42.** $(-2^3)^3$ **43.** $(x^3)^3$ **44.** $(y^4)^2$

45. $(x^7)^2$ **46.** $(y^5)^3$ **47.** $(-x^2)^2$ **48.** $(-x^2)^3$

49. $(2x)^2$ **50.** $(3y)^3$ **51.** $(-2x^2)^3$ **52.** $(-3y^3)^2$

53. $(x^2y^3)^2$ **54.** $(x^3y^4)^5$ **55.** $(3x^2y)^2$ **56.** $(-2ab^3)^4$

57. $(a^2)(3a^2)^3$ **58.** $(b^2)(2a^3)^4$ **59.** $(-2x)(2x^3)^2$

60. $(2y)(-3y^4)^3$ **61.** $(x^2y)(x^2y)^3$ **62.** $(a^3b)(ab)^3$

63. $(ab^2)^2(ab)^2$ **64.** $(x^2y)^2(x^3y)^3$ **65.** $(-2x)(-2x^3y)^3$

66. $(-3y)(-4x^2y^3)^3$ **67.** $(-2x)(-3xy^2)^2$ **68.** $(-3y)(-2x^2y)^3$

69. $(ab^2)(-2a^2b)^3$ **70.** $(a^2b^2)(-3ab^4)^2$ **71.** $(-2a^3)(3a^2b)^3$

72. $(-3b^2)(2ab^2)^3$ **73.** $(-3ab)^2(-2ab)^3$ **74.** $(-3a^2b)^3(-3ab)^3$

7.2 **Applying Concepts**

Simplify.

75. $(6x)(2x^2) + (4x^2)(5x)$ **76.** $(2a^7)(7a^2) - (6a^3)(5a^6)$

77. $(3a^2b^2)(2ab) - (9ab^2)(a^2b)$ **78.** $(3x^2y^2)^2 - (2xy)^4$

79. $(5xy^3)(3x^4y^2) - (2x^3y)(x^2y^4)$ **80.** $a^2(ab^2)^3 - a^3(ab^3)^2$

81. $4a^2(2ab)^3 - 5b^2(a^5b)$ **82.** $9x^3(3x^2y)^2 - x(x^3y)^2$

83. $-2xy(x^2y)^3 - 3x^5(xy^2)^2$ **84.** $5a^2b(ab^2)^2 + b^3(2a^2b)^2$

85. $a^n \cdot a^n$ **86.** $(a^n)^2$ **87.** $(a^2)^n$ **88.** $a^2 \cdot a^n$

For Exercises 89–92, answer true or false. If the answer is false, correct the right side of the equation.

89. $(-a)^5 = -a^5$ **90.** $(-b)^8 = b^8$

91. $(x^2)^5 = x^{2+5} = x^7$ **92.** $x^3 + x^3 = 2x^{3+3} = 2x^6$

93. Evaluate $(2^3)^2$ and $2^{(3^2)}$. Are the results the same? If not, which expression has the larger value?

94. What is the Order of Operations for the expression x^{m^n}?

95. If n is a positive integer and $x^n = y^n$, when is $x = y$?

96. Geometry The length of a rectangle is $4ab$. The width is $2ab$. Find the perimeter of the rectangle in terms of ab.

4ab

2ab

97. The distance a rock will fall in t seconds is $16t^2$ ft (neglecting air resistance). Find other examples of quantities that can be expressed in terms of an exponential expression, and explain where the expression is used.

7.3 | **Multiplication of Polynomials**

1 **Multiply a polynomial by a monomial**

To multiply a polynomial by a monomial, use the Distributive Property and the Rule for Multiplying Exponential Expressions.

Simplify: $-2x(x^2 - 4x - 3)$

$-2x(x^2 - 4x - 3)$
$= -2x(x^2) - (-2x)(4x) - (-2x)(3)$ • Use the Distributive Property.
$= -2x^3 + 8x^2 + 6x$ • Use the Rule for Multiplying Exponential Expressions.

Example 1 Multiply.   **A.** $(5x + 4)(-2x)$ **B.** $x^3(2x^2 - 3x + 2)$

Solution   **A.** $(5x + 4)(-2x)$
$= 5x(-2x) + 4(-2x)$ • Use the Distributive Property.
$= -10x^2 - 8x$ • Use the Rule for Multiplying Exponential Expressions.

B. $x^3(2x^2 - 3x + 2)$
$= 2x^5 - 3x^4 + 2x^3$

Problem 1 Multiply.   **A.** $(-2y + 3)(-4y)$ **B.** $-a^2(3a^2 + 2a - 7)$

Solution   See page S16.

2 Multiply two polynomials

Multiplication of two polynomials requires the repeated application of the Distributive Property.

$$(y - 2)(y^2 + 3y + 1) = (y - 2)(y^2) + (y - 2)(3y) + (y - 2)(1)$$
$$= y^3 - 2y^2 + 3y^2 - 6y + y - 2$$
$$= y^3 + y^2 - 5y - 2$$

A convenient method of multiplying two polynomials is to use a vertical format similar to that used for multiplication of whole numbers.

$$
\begin{array}{r}
y^2 + 3y + 1 \\
y - 2 \\
\hline
-2y^2 - 6y - 2 \\
y^3 + 3y^2 + \; y \\
\hline
y^3 + \; y^2 - 5y - 2
\end{array}
$$

Multiply each term in the trinomial by -2.
Multiply each term in the trinomial by y.
Like terms must be in the same column.
Add the terms in each column.

Example 2 Multiply: $(2b^3 - b + 1)(2b + 3)$

Solution

$$
\begin{array}{r}
2b^3 - \; b + 1 \\
2b + 3 \\
\hline
6b^3 \qquad\;\; - 3b + 3 \\
4b^4 \qquad - 2b^2 + 2b \\
\hline
4b^4 + 6b^3 - 2b^2 - \; b + 3
\end{array}
$$

- Multiply $2b^3 - b + 1$ by 3.
- Multiply $2b^3 - b + 1$ by $2b$. Arrange the terms in descending order.
- Add the terms in each column.

Problem 2 Multiply: $(2y^3 + 2y^2 - 3)(3y - 1)$

Solution See page S16.

Example 3 Multiply: $(4a^3 - 5a - 2)(3a - 2)$

Solution

$$
\begin{array}{r}
4a^3 - \; 5a - 2 \\
3a - 2 \\
\hline
-8a^3 \qquad\;\; + 10a + 4 \\
12a^4 \qquad - 15a^2 - \; 6a \\
\hline
12a^4 - 8a^3 - 15a^2 + \; 4a + 4
\end{array}
$$

- Multiply $4a^3 - 5a - 2$ by -2.
- Multiply $4a^3 - 5a - 2$ by $3a$.
- Add the terms in each column.

Problem 3 Multiply: $(3x^3 - 2x^2 + x - 3)(2x + 5)$

Solution See page S16.

---

3 Multiply two binomials

It is often necessary to find the product of two binomials. The product can be found using a method called **FOIL**, which is based on the Distributive Property. The letters of FOIL stand for **F**irst, **O**uter, **I**nner, and **L**ast.

Multiply: $(2x + 3)(x + 5)$

Multiply the First terms.	$(2x + 3)(x + 5)$	$2x \cdot x = 2x^2$
Multiply the Outer terms.	$(2x + 3)(x + 5)$	$2x \cdot 5 = 10x$
Multiply the Inner terms.	$(2x + 3)(x + 5)$	$3 \cdot x = 3x$
Multiply the Last terms.	$(2x + 3)(x + 5)$	$3 \cdot 5 = 15$

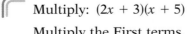

Take Note

FOIL is not really a different way of multiplying. It is based on the Distributive Property.

$(2x + 3)(x + 5)$
$= 2x(x + 5) + 3(x + 5)$
　　F　　O　　I　　L
$= 2x^2 + 10x + 3x + 15$
$= 2x^2 + 13x + 15$

Add the products.
Combine like terms.

$$\begin{array}{ll} & \textbf{F}\quad\;\textbf{O}\quad\textbf{I}\quad\;\textbf{L} \\ (2x + 3)(x + 5) & = 2x^2 + 10x + 3x + 15 \\ & = 2x^2 + 13x + 15 \end{array}$$

Example 4   Multiply: $(4x - 3)(3x - 2)$

Solution $(4x - 3)(3x - 2)$
$\quad = 4x(3x) + 4x(-2) + (-3)(3x) + (-3)(-2)$ • Use the FOIL method.
$\quad = 12x^2 - 8x - 9x + 6$
$\quad = 12x^2 - 17x + 6$ • Combine like terms.

Problem 4   Multiply: $(4y - 5)(3y - 3)$

Solution   See page S16.

Example 5   Multiply: $(3x - 2y)(x + 4y)$

Solution $(3x - 2y)(x + 4y)$
$\quad = 3x(x) + 3x(4y) + (-2y)(x) + (-2y)(4y)$ • Use the FOIL method.
$\quad = 3x^2 + 12xy - 2xy - 8y^2$
$\quad = 3x^2 + 10xy - 8y^2$ • Combine like terms.

Problem 5   Multiply: $(3a + 2b)(3a - 5b)$

Solution   See page S16.

4   ## Multiply binomials that have special products

The expression $(a + b)(a - b)$ is the product of the sum and difference of two terms. The first binomial in the expression is a sum; the second is a difference. The two terms are a and b. The first term in each binomial is a. The second term in each binomial is b.

The expression $(a + b)^2$ is the square of a binomial. The first term in the binomial is a. The second term in the binomial is b.

Using FOIL, it is possible to find a pattern for the product of the sum and difference of two terms and for the square of a binomial.

The Sum and Difference of Two Terms

$$\boldsymbol{(a + b)(a - b)} = a^2 - ab + ab - b^2$$
$$= \boldsymbol{a^2 - b^2}$$

Square of first term ——————————
Square of second term ——————————

The Square of a Binomial

$$(a + b)^2 = (a + b)(a + b) = a^2 + ab + ab + b^2$$
$$= a^2 + 2ab + b^2$$

Square of first term ———————————————

Twice the product of the two terms ———————

Square of last term ———————————————

$$(a - b)^2 = (a - b)(a - b) = a^2 - ab - ab + b^2$$
$$= a^2 - 2ab + b^2$$

Square of first term ———————————————

Twice the product of the two terms ———————

Square of last term ———————————————

Example 6 Multiply: $(2x + 3)(2x - 3)$

Solution $(2x + 3)(2x - 3)$ • $(2x + 3)(2x - 3)$ is the product of the sum and difference of two terms.

$= (2x)^2 - 3^2$ • Square the first term. Square the second term.
$= 4x^2 - 9$ • Simplify.

Problem 6 Multiply: $(2a + 5c)(2a - 5c)$

Solution See page S16.

Example 7 Multiply: $(4c + 5d)^2$

Solution $(4c + 5d)^2$ • $(4c + 5d)^2$ is the square of a binomial.

$= (4c)^2 + 2(4c)(5d) + (5d)^2$ • Square the first term. Find twice the product of the two terms. Square the second term.

$= 16c^2 + 40cd + 25d^2$ • Simplify.

Problem 7 Multiply: $(3x + 2y)^2$

Solution See page S16.

Example 8 Multiply: $(3x - 2)^2$

Solution $(3x - 2)^2$ • $(3x - 2)^2$ is the square of a binomial.

$= (3x)^2 + 2(3x)(-2) + (-2)^2$ • Square the first term. Find twice the product of the two terms. Square the last term.

$= 9x^2 - 12x + 4$ • Simplify.

Problem 8 Multiply: $(6x - y)^2$

Solution See page S16.

5   **Application problems**

Example 9   The radius of a circle is $(x - 4)$ ft. Find the area of the circle in terms of the variable x. Leave the answer in terms of π.

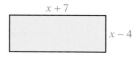

Strategy   To find the area, replace the variable r in the formula $A = \pi r^2$ with the given value, and solve for A.

Solution $A = \pi r^2$
$A = \pi(x - 4)^2$ • **This is the square of a binomial.**
$A = \pi(x^2 - 8x + 16)$
$A = \pi x^2 - 8\pi x + 16\pi$

The area is $(\pi x^2 - 8\pi x + 16\pi)$ ft².

Problem 9   The length of a rectangle is $(x + 7)$ m. The width is $(x - 4)$ m. Find the area of the rectangle in terms of the variable x.

Solution   See page S16.

Example 10   The length of a side of a square is $(3x + 5)$ in. Find the area of the square in terms of the variable x.

Strategy   To find the area of the square, replace the variable s in the formula $A = s^2$ with the given value and simplify.

Solution $A = s^2$
$A = (3x + 5)^2$ • **This is the square of a binomial.**
$A = 9x^2 + 30x + 25$

The area is $(9x^2 + 30x + 25)$ in².

Problem 10   The base of a triangle is $(x + 3)$ cm and the height is $(4x - 6)$ cm. Find the area of the triangle in terms of the variable x.

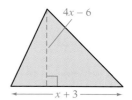

Solution   See page S16.

7.3   **Concept Review**

Determine whether the statement is always true, sometimes true, or never true.

1. To multiply a monomial times a polynomial, use the Distributive Property to multiply each term of the polynomial by the monomial.

2. To multiply two polynomials, multiply each term of one polynomial by the other polynomial.

3. A binomial is a polynomial of degree 2.

4. $(x + 7)(x - 7)$ is the product of the sum and difference of the same two terms.

5. To square a binomial means to multiply it times itself.

6. The square of a binomial is a trinomial.

7. The FOIL method is used to multiply two polynomials.

8. Using the FOIL method, the terms $3x$ and 5 are the "First" terms in $(3x + 5)(2x + 7)$.

7.3 Exercises

1 Multiply.

1. $x(x - 2)$

2. $y(3 - y)$

3. $-x(x + 7)$

4. $-y(7 - y)$

5. $3a^2(a - 2)$

6. $4b^2(b + 8)$

7. $-5x^2(x^2 - x)$

8. $-6y^2(y + 2y^2)$

9. $-x^3(3x^2 - 7)$

10. $-y^4(2y^2 - y^6)$

11. $2x(6x^2 - 3x)$

12. $3y(4y - y^2)$

13. $(2x - 4)3x$

14. $(2x + 1)2x$

15. $-xy(x^2 - y^2)$

16. $-x^2y(2xy - y^2)$

17. $x(2x^3 - 3x + 2)$

18. $y(-3y^2 - 2y + 6)$

19. $-a(-2a^2 - 3a - 2)$

20. $-b(5b^2 + 7b - 35)$

21. $x^2(3x^4 - 3x^2 - 2)$

22. $y^3(-4y^3 - 6y + 7)$

23. $2y^2(-3y^2 - 6y + 7)$

24. $4x^2(3x^2 - 2x + 6)$

25. $(a^2 + 3a - 4)(-2a)$

26. $(b^3 - 2b + 2)(-5b)$

27. $-3y^2(-2y^2 + y - 2)$

28. $-5x^2(3x^2 - 3x - 7)$

29. $xy(x^2 - 3xy + y^2)$

30. $ab(2a^2 - 4ab - 6b^2)$

2 Multiply.

31. $(x^2 + 3x + 2)(x + 1)$

32. $(x^2 - 2x + 7)(x - 2)$

33. $(a - 3)(a^2 - 3a + 4)$

34. $(2x - 3)(x^2 - 3x + 5)$

35. $(-2b^2 - 3b + 4)(b - 5)$

36. $(-a^2 + 3a - 2)(2a - 1)$

37. $(3x - 5)(-2x^2 + 7x - 2)$

38. $(2a - 1)(-a^2 - 2a + 3)$

39. $(x^3 - 3x + 2)(x - 4)$

40. $(y^3 + 4y^2 - 8)(2y - 1)$

41. $(3y - 8)(5y^2 + 8y - 2)$

42. $(4y - 3)(3y^2 + 3y - 5)$

43. $(5a^3 - 15a + 2)(a - 4)$

44. $(3b^3 - 5b^2 + 7)(6b - 1)$

45. $(y + 2)(y^3 + 2y^2 - 3y + 1)$

46. $(2a - 3)(2a^3 - 3a^2 + 2a - 1)$

3 For the product of the binomials, name **a.** the first terms, **b.** the outer terms, **c.** the inner terms, and **d.** the last terms.

47. $(y - 8)(2y + 3)$

48. $(3d + 4)(d - 1)$

Multiply.

49. $(x + 1)(x + 3)$

50. $(y + 2)(y + 5)$

51. $(a - 3)(a + 4)$

52. $(b - 6)(b + 3)$

53. $(y + 3)(y - 8)$

54. $(x + 10)(x - 5)$

55. $(y - 7)(y - 3)$

56. $(a - 8)(a - 9)$

57. $(2x + 1)(x + 7)$

58. $(y + 2)(5y + 1)$

59. $(3x - 1)(x + 4)$

60. $(7x - 2)(x + 4)$

61. $(4x - 3)(x - 7)$

62. $(2x - 3)(4x - 7)$

63. $(3y - 8)(y + 2)$

64. $(5y - 9)(y + 5)$

65. $(3x + 7)(3x + 11)$

66. $(5a + 6)(6a + 5)$

67. $(7a - 16)(3a - 5)$

68. $(5a - 12)(3a - 7)$

69. $(3b + 13)(5b - 6)$

70. $(x + y)(2x + y)$

71. $(2a + b)(a + 3b)$

72. $(3x - 4y)(x - 2y)$

73. $(2a - b)(3a + 2b)$

74. $(5a - 3b)(2a + 4b)$

75. $(2x + y)(x - 2y)$

76. $(3x - 7y)(3x + 5y)$

77. $(2x + 3y)(5x + 7y)$

78. $(5x + 3y)(7x + 2y)$

79. $(3a - 2b)(2a - 7b)$

80. $(5a - b)(7a - b)$

81. $(a - 9b)(2a + 7b)$

82. $(2a + 5b)(7a - 2b)$

83. $(10a - 3b)(10a - 7b)$

84. $(12a - 5b)(3a - 4b)$

85. $(5x + 12y)(3x + 4y)$

86. $(11x + 2y)(3x + 7y)$

87. $(2x - 15y)(7x + 4y)$

88. $(5x + 2y)(2x - 5y)$

89. $(8x - 3y)(7x - 5y)$

90. $(2x - 9y)(8x - 3y)$

4 Multiply.

91. $(y - 5)(y + 5)$

92. $(y + 6)(y - 6)$

93. $(2x + 3)(2x - 3)$

94. $(4x - 7)(4x + 7)$

95. $(x + 1)^2$

96. $(y - 3)^2$

97. $(3a - 5)^2$

98. $(6x - 5)^2$

99. $(3x - 7)(3x + 7)$

100. $(9x - 2)(9x + 2)$ **101.** $(2a + b)^2$ **102.** $(x + 3y)^2$

103. $(x - 2y)^2$ **104.** $(2x - 3y)^2$ **105.** $(4 - 3y)(4 + 3y)$

106. $(4x - 9y)(4x + 9y)$ **107.** $(5x + 2y)^2$ **108.** $(2a - 9b)^2$

5 Geometry Problems

109. The length of a rectangle is $5x$ ft. The width is $(2x - 7)$ ft. Find the area of the rectangle in terms of the variable x.

110. The width of a rectangle is $(x - 6)$ m. The length is $(2x + 3)$ m. Find the area of the rectangle in terms of the variable x.

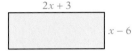

111. The width of a rectangle is $(3x + 1)$ in. The length of the rectangle is twice the width. Find the area of the rectangle in terms of the variable x.

112. The width of a rectangle is $(4x - 3)$ cm. The length of the rectangle is twice the width. Find the area of the rectangle in terms of the variable x.

113. The length of a side of a square is $(2x + 1)$ km. Find the area of the square in terms of the variable x.

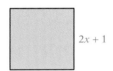

114. The length of a side of a square is $(2x - 3)$ yd. Find the area of the square in terms of the variable x.

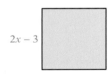

115. The base of a triangle is $4x$ m and the height is $(2x + 5)$ m. Find the area of the triangle in terms of the variable x.

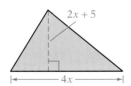

116. The base of a triangle is $(2x + 6)$ in. and the height is $(x - 8)$ in. Find the area of the triangle in terms of the variable x.

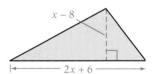

117. The radius of a circle is $(x + 4)$ cm. Find the area of the circle in terms of the variable x. Leave the answer in terms of π.

118. The radius of a circle is $(x - 3)$ ft. Find the area of the circle in terms of the variable x. Leave the answer in terms of π.

119. **Sports** A softball diamond has dimensions 45 ft by 45 ft. A base path border x feet wide lies on both the first-base side and the third-base side of the diamond. Express the total area of the softball diamond and the base path in terms of the variable x.

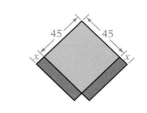

120. **Sports** An athletic field has dimensions 30 yd by 100 yd. An end zone that is w yards wide borders each end of the field. Express the total area of the field and the end zones in terms of the variable w.

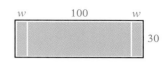

7.3 Applying Concepts

Simplify.

121. $(a + b)^2 - (a - b)^2$

122. $(x + 3y)^2 + (x + 3y)(x - 3y)$

123. $(3a^2 - 4a + 2)^2$

124. $(x + 4)^3$

125. $3x^2(2x^3 + 4x - 1) - 6x^3(x^2 - 2)$

126. $(3b + 2)(b - 6) + (4 + 2b)(3 - b)$

127. $x^n(x^n + 1)$

128. $(x^n + 1)(x^n - 1)$

129. $(x^n + 1)(x^n + 1)$

130. $(x^n - 1)^2$

131. $(x + 1)(x - 1)$

132. $(x + 1)(-x^2 + x - 1)$

133. $(x + 1)(x^3 - x^2 + x - 1)$

134. $(x + 1)(-x^4 + x^3 - x^2 + x - 1)$

Use the pattern of the answers to Exericses 131–134 to write the product.

135. $(x + 1)(x^5 - x^4 + x^3 - x^2 + x - 1)$

136. $(x + 1)(-x^6 + x^5 - x^4 + x^3 - x^2 + x - 1)$

Solve.

137. Find $(4n^3)^2$ if $2n - 3 = 4n - 7$.

138. What polynomial has quotient $x^2 + 2x - 1$ when divided by $x + 3$?

139. What polynomial has quotient $3x - 4$ when divided by $4x + 5$?

140. Subtract $4x^2 - x - 5$ from the product of $x^2 + x + 3$ and $x - 4$.

141. Add $x^2 + 2x - 3$ to the product of $2x - 5$ and $3x + 1$.

142. If a polynomial of degree 3 is multiplied by a polynomial of degree 2, what is the degree of the resulting polynomial?

143. Is it possible to multiply a polynomial of degree 2 by a polynomial of degree 2 and have the product be a polynomial of degree 3? If so, give an example. If not, explain why not.

7.4 | Integer Exponents and Scientific Notation

1 **Integer exponents**

The quotient of two exponential expressions with the *same* base can be simplified by writing each expression in factored form, dividing by the common factors, and then writing the result with an exponent.

$$\frac{x^5}{x^2} = \frac{\overset{1}{\cancel{x}} \cdot \overset{1}{\cancel{x}} \cdot x \cdot x \cdot x}{\underset{1}{\cancel{x}} \cdot \underset{1}{\cancel{x}}} = x^3$$

Note that subtracting the exponents results in the same quotient.

$$\frac{x^5}{x^2} = x^{5-2} = x^3$$

To divide two monomials with the same base, subtract the exponents of the like bases.

Simplify: $\dfrac{a^7}{a^3}$

The bases are the same. Subtract the exponent in the denominator from the exponent in the numerator.

$$\frac{a^7}{a^3} = a^{7-3} = a^4$$

Simplify: $\dfrac{r^8 s^6}{r^7 s}$

Subtract the exponents of the like bases.

$$\frac{r^8 s^6}{r^7 s} = r^{8-7} s^{6-1} = rs^5$$

Recall that for any number a, $a \neq 0$, $\dfrac{a}{a} = 1$. This property is true for exponential expressions as well. For example, for $x \neq 0$, $\dfrac{x^4}{x^4} = 1$.

This expression also can be simplified using the rule for dividing exponential expressions with the same base.

$$\frac{x^4}{x^4} = x^{4-4} = x^0$$

Because $\dfrac{x^4}{x^4} = 1$ and $\dfrac{x^4}{x^4} = x^{4-4} = x^0$, the following definition of zero as an exponent is used.

Zero as an Exponent

If $x \neq 0$, then $x^0 = 1$. The expression 0^0 is not defined.

Simplify: $(12a^3)^0$, $a \neq 0$

Any nonzero expression to the zero power is 1. $(12a^3)^0 = 1$

Simplify: $-(xy^4)^0$, $x \neq 0$, $y \neq 0$

Any nonzero expression to the zero power is 1. Because the negative sign is outside the parentheses, the answer is -1. $-(xy^4)^0 = -(1) = -1$

The meaning of a negative exponent can be developed by examining the quotient $\dfrac{x^4}{x^6}$.

The expression can be simplified by writing the numerator and denominator in factored form, dividing by the common factors, and then writing the result with an exponent.

$$\frac{x^4}{x^6} = \frac{\overset{1}{\cancel{x}} \cdot \overset{1}{\cancel{x}} \cdot \overset{1}{\cancel{x}} \cdot \overset{1}{\cancel{x}}}{\underset{1}{\cancel{x}} \cdot \underset{1}{\cancel{x}} \cdot \underset{1}{\cancel{x}} \cdot \underset{1}{\cancel{x}} \cdot x \cdot x} = \frac{1}{x^2}$$

Now simplify the same expression by subtracting the exponents of the like bases.

$$\frac{x^4}{x^6} = x^{4-6} = x^{-2}$$

Because $\dfrac{x^4}{x^6} = \dfrac{1}{x^2}$ and $\dfrac{x^4}{x^6} = x^{-2}$, the expressions $\dfrac{1}{x^2}$ and x^{-2} must be equal. The following definition of a negative exponent is used.

Definition of Negative Exponents

If n is a positive integer and $x \neq 0$, then $x^{-n} = \dfrac{1}{x^n}$ and $\dfrac{1}{x^{-n}} = x^n$.

Evaluate 2^{-4}.

Write the expression with a positive exponent. Then simplify. $2^{-4} = \dfrac{1}{2^4} = \dfrac{1}{16}$

Now that negative exponents have been defined, the Rule for Dividing Exponential Expressions can be stated.

Rule for Dividing Exponential Expressions

If m and n are integers and $x \neq 0$, then $\dfrac{x^m}{x^n} = x^{m-n}$.

Example 1 Write $\dfrac{3^{-3}}{3^2}$ with a positive exponent. Then evaluate.

Solution $\dfrac{3^{-3}}{3^2} = 3^{-3-2}$ • 3^{-3} and 3^2 have the same base. Subtract the exponents.

$= 3^{-5}$

$= \dfrac{1}{3^5}$ • Use the Definition of Negative Exponents to write the expression with a positive exponent.

$= \dfrac{1}{243}$ • Evaluate.

Problem 1 Write $\dfrac{2^{-2}}{2^3}$ with a positive exponent. Then evaluate.

Solution See page S16.

The rules for simplifying exponential expressions and powers of exponential expressions are true for all integers. These rules are restated here.

Rules for Exponents

If m, n, and p are integers, then

$$x^m \cdot x^n = x^{m+n} \qquad (x^m)^n = x^{mn} \qquad (x^m y^n)^p = x^{mp} y^{np}$$

$$\dfrac{x^m}{x^n} = x^{m-n}, \ x \neq 0 \qquad x^{-n} = \dfrac{1}{x^n}, \ x \neq 0 \qquad x^0 = 1, \ x \neq 0$$

An exponential expression is in simplest form when it is written with only positive exponents.

Simplify: $a^{-7}b^3$

Rewrite a^{-7} with a positive exponent. $\qquad a^{-7}b^3 = \dfrac{b^3}{a^7}$

Example 2 Simplify: $\dfrac{x^{-4}y^6}{xy^2}$

Solution $\dfrac{x^{-4}y^6}{xy^2} = x^{-4-1}y^{6-2}$ • Divide variables with the same base by subtracting the exponents.

$= x^{-5}y^4$

$= \dfrac{y^4}{x^5}$ • Write the expression with only positive exponents.

Problem 2 Simplify: $\dfrac{b^8}{a^{-5}b^6}$

Solution See page S16.

Simplify: $6d^{-4}$, $d \neq 0$

$6d^{-4} = 6 \cdot \dfrac{1}{d^4} = \dfrac{6}{d^4}$ • Use the Definition of Negative Exponents to rewrite the expression with a positive exponent.

Take Note

In the example at the bottom of this page, the exponent on d is -4 (negative 4). The d^{-4} is written in the denominator as d^4. The exponent on 6 is 1 (positive 1). The 6 remains in the numerator.

Note that we indicated $d \neq 0$. This is necessary because division by zero is not defined. In this textbook, we will assume that values of the variables are chosen so that division by zero does not occur.

Example 3 Simplify: **A.** $\dfrac{-35a^6b^{-2}}{25a^{-2}b^5}$ **B.** $(-2x)(3x^{-2})^{-3}$

Solution **A.** $\dfrac{-35a^6b^{-2}}{25a^{-2}b^5} = -\dfrac{35a^6b^{-2}}{25a^{-2}b^5}$ • A negative sign is placed in front of a fraction.

$$= -\dfrac{\overset{1}{\cancel{5}} \cdot 7a^{6-(-2)}b^{-2-5}}{\underset{1}{\cancel{5}} \cdot 5}$$ • Factor the coefficients. Divide by the common factors. Divide variables with the same base by subtracting the exponents.

$$= -\dfrac{7a^8b^{-7}}{5}$$

$$= -\dfrac{7a^8}{5b^7}$$ • Write the expression with only positive exponents.

B. $(-2x)(3x^{-2})^{-3} = (-2x)(3^{-3}x^6)$ • Use the Rule for Simplifying Powers of Products.

$$= \dfrac{-2x \cdot x^6}{3^3}$$ • Write the expression with positive exponents.

$$= -\dfrac{2x^7}{27}$$ • Use the Rule for Multiplying Exponential Expressions, and simplify the numerical exponential expression.

Problem 3 Simplify: **A.** $\dfrac{12x^{-8}y^4}{-16xy^{-3}}$ **B.** $(-3ab)(2a^3b^{-2})^{-3}$

Solution See page S16.

2 Scientific notation

Very large and very small numbers are encountered in the fields of science and engineering. For example, the charge of an electron is 0.00000000000000000000160 coulomb. These numbers can be written more easily in scientific notation. In **scientific notation,** a number is expressed as a product of two factors, one a number between 1 and 10 and the other a power of 10.

To change a number written in decimal notation to scientific notation, write it in the form $a \cdot 10^n$, where a is a number between 1 and 10 and n is an integer.

For numbers greater than 10, move the decimal point to the right of the first digit. The exponent n is positive and equal to the number of places the decimal point has been moved.

$$240{,}000 = 2.4 \cdot 10^5$$

$$93{,}000{,}000 = 9.3 \cdot 10^7$$

For numbers less than 1, move the decimal point to the right of the first nonzero digit. The exponent n is negative. The absolute value of the exponent is equal to the number of places the decimal point has been moved.

$$0.00030 = 3.0 \cdot 10^{-4}$$

$$0.0000832 = 8.32 \cdot 10^{-5}$$

Look at the last example above: $0.0000832 = 8.32 \cdot 10^{-5}$. Using the Definition of Negative Exponents,

$$10^{-5} = \dfrac{1}{10^5} = \dfrac{1}{100{,}000} = 0.00001$$

Take Note

There are two steps to writing a number in scientific notation: (1) determine the number between 1 and 10, and (2) determine the exponent on 10.

Because $10^{-5} = 0.00001$, we can write

$$8.32 \cdot 10^{-5} = 8.32 \cdot 0.00001 = 0.0000832$$

which is the number we started with. We have not changed the value of the number; we have just written it in another form.

Example 4   Write the number in scientific notation.

     A. 824,300,000,000   **B.** 0.000000961

Solution   **A.** $824{,}300{,}000{,}000 = 8.243 \cdot 10^{11}$ • Move the decimal point 11 places to the left. The exponent on 10 is 11.

          B. $0.000000961 = 9.61 \cdot 10^{-7}$ • Move the decimal point 7 places to the right. The exponent on 10 is −7.

Problem 4   Write the number in scientific notation.

     A. 57,000,000,000   **B.** 0.000000017

Solution   See page S16.

Changing a number written in scientific notation to decimal notation also requires moving the decimal point.

When the exponent on 10 is positive, move the decimal point to the right the same number of places as the exponent.

$$3.45 \cdot 10^9 = 3{,}450{,}000{,}000$$

$$2.3 \cdot 10^8 = 230{,}000{,}000$$

When the exponent on 10 is negative, move the decimal point to the left the same number of places as the absolute value of the exponent.

$$8.1 \cdot 10^{-3} = 0.0081$$

$$6.34 \cdot 10^{-6} = 0.00000634$$

Example 5   Write the number in decimal notation.

     A. $7.329 \cdot 10^6$ **B.** $6.8 \cdot 10^{-10}$

Solution   **A.** $7.329 \cdot 10^6 = 7{,}329{,}000$ • The exponent on 10 is positive. Move the decimal point 6 places to the right.

          B. $6.8 \cdot 10^{-10} = 0.00000000068$ • The exponent on 10 is negative. Move the decimal point 10 places to the left.

Problem 5   Write the number in decimal notation.

     A. $5 \cdot 10^{12}$ **B.** $4.0162 \cdot 10^{-9}$

Solution   See page S16.

The rules for multiplying and dividing with numbers in scientific notation are the same as those for calculating with algebraic expressions. The power of 10 corresponds to the variable, and the number between 1 and 10 corresponds to the coefficient of the variable.

	Algebraic Expressions	**Scientific Notation**
Multiplication	$(4x^{-3})(2x^5) = 8x^2$	$(4 \cdot 10^{-3})(2 \cdot 10^5) = 8 \cdot 10^2$
Division	$\dfrac{6x^5}{3x^{-2}} = 2x^{5-(-2)} = 2x^7$	$\dfrac{6 \cdot 10^5}{3 \cdot 10^{-2}} = 2 \cdot 10^{5-(-2)} = 2 \cdot 10^7$

Example 6 Multiply or divide.   **A.** $(3.0 \cdot 10^5)(1.1 \cdot 10^{-8})$

B. $\dfrac{7.2 \cdot 10^{13}}{2.4 \cdot 10^{-3}}$

Solution   **A.** $(3.0 \cdot 10^5)(1.1 \cdot 10^{-8}) = 3.3 \cdot 10^{-3}$ • Multiply 3.0 and 1.1. Add the exponents on 10.

B. $\dfrac{7.2 \cdot 10^{13}}{2.4 \cdot 10^{-3}} = 3 \cdot 10^{16}$ • Divide 7.2 by 2.4. Subtract the exponents on 10.

Problem 6 Multiply or divide.   **A.** $(2.4 \cdot 10^{-9})(1.6 \cdot 10^3)$

B. $\dfrac{5.4 \cdot 10^{-2}}{1.8 \cdot 10^{-4}}$

Solution   See page S16.

7.4 | Concept Review

Determine whether the statement is always true, sometimes true, or never true.

1. The expression $\dfrac{x^5}{y^3}$ can be simplified by subtracting the exponents.

2. The rules of exponents can be applied to expressions that contain an exponent of zero or contain negative exponents.

3. The expression 3^{-2} represents the reciprocal of 3^2.

4. $5x^0 = 0$

5. The expression 4^{-3} represents a negative number.

6. To be in simplest form, an exponential expression cannot contain any negative exponents.

7. $2x^{-5} = \dfrac{1}{2x^5}$

8. $x^4 y^{-6} = \dfrac{1}{x^4 y^6}$

7.4 | Exercises

1 Write with a positive or zero exponent. Then evaluate.

1. 5^{-2}

2. 3^{-3}

3. $\dfrac{1}{8^{-2}}$

4. $\dfrac{1}{12^{-1}}$

5. $\dfrac{3^{-2}}{3}$

6. $\dfrac{5^{-3}}{5}$

7. $\dfrac{2^3}{2^3}$

8. $\dfrac{3^{-2}}{3^{-2}}$

9. Explain how to rewrite a variable that has a negative exponent as an expression with a positive exponent.

10. Explain how to divide two nonzero exponential expressions with the same base. Provide an example.

Simplify.

11. $\dfrac{y^7}{y^3}$

12. $\dfrac{z^9}{z^2}$

13. $\dfrac{a^8}{a^5}$

14. $\dfrac{c^{12}}{c^5}$

15. $\dfrac{p^5}{p}$

16. $\dfrac{w^9}{w}$

17. $\dfrac{4x^8}{2x^5}$

18. $\dfrac{12z^7}{4z^3}$

19. $\dfrac{22k^5}{11k^4}$

20. $\dfrac{14m^{11}}{7m^{10}}$

21. $\dfrac{m^9n^7}{m^4n^5}$

22. $\dfrac{y^5z^6}{yz^3}$

23. $\dfrac{6r^4}{4r^2}$

24. $\dfrac{8x^9}{12x^6}$

25. $\dfrac{-16a^7}{24a^6}$

26. $\dfrac{-18b^5}{27b^4}$

27. x^{-2}

28. y^{-10}

29. $\dfrac{1}{a^{-6}}$

30. $\dfrac{1}{b^{-4}}$

31. $4x^{-7}$

32. $-6y^{-1}$

33. $\dfrac{5}{b^{-8}}$

34. $\dfrac{-3}{v^{-3}}$

35. $\dfrac{1}{3x^{-2}}$

36. $\dfrac{2}{5c^{-6}}$

37. $(ab^5)^0$

38. $(32x^3y^4)^0$

39. $\dfrac{y^3}{y^8}$

40. $\dfrac{z^4}{z^6}$

41. $\dfrac{a^5}{a^{11}}$

42. $\dfrac{m}{m^7}$

43. $\dfrac{4x^2}{12x^5}$

44. $\dfrac{6y^8}{8y^9}$

45. $\dfrac{-12x}{-18x^6}$

46. $\dfrac{-24c^2}{-36c^{11}}$

47. $\dfrac{x^6y^5}{x^8y}$

48. $\dfrac{a^3b^2}{a^2b^3}$

49. $\dfrac{2m^6n^2}{5m^9n^{10}}$

50. $\dfrac{5r^3t^7}{6r^5t^7}$

51. $\dfrac{pq^3}{p^4q^4}$

52. $\dfrac{a^4b^5}{a^5b^6}$

53. $\dfrac{3x^4y^5}{6x^4y^8}$

54. $\dfrac{14a^3b^6}{21a^5b^6}$

55. $\dfrac{14x^4y^6z^2}{16x^3y^9z}$

56. $\dfrac{24a^2b^7c^9}{36a^7b^5c}$

57. $\dfrac{15mn^9p^3}{30m^4n^9p}$

58. $\dfrac{25x^4y^7z^2}{20x^5y^9z^{11}}$

59. $(-2xy^{-2})^3$

60. $(-3x^{-1}y^2)^2$

61. $(3x^{-1}y^{-2})^2$

62. $(5xy^{-3})^{-2}$

63. $(2x^{-1})(x^{-3})$

64. $(-2x^{-5})x^7$

65. $(-5a^2)(a^{-5})^2$

66. $(2a^{-3})(a^7b^{-1})^3$

67. $(-2ab^{-2})(4a^{-2}b)^{-2}$

68. $(3ab^{-2})(2a^{-1}b)^{-3}$

69. $(-5x^{-2}y)(-2x^{-2}y^2)$

70. $\dfrac{a^{-3}b^{-4}}{a^2b^2}$

71. $\dfrac{3x^{-2}y^2}{6xy^2}$

72. $\dfrac{2x^{-2}y}{8xy}$

73. $\dfrac{3x^{-2}y}{xy}$

74. $\dfrac{2x^{-1}y^4}{x^2y^3}$

75. $\dfrac{2x^{-1}y^{-4}}{4xy^2}$

76. $\dfrac{12a^2b^3}{-27a^2b^2}$

77. $\dfrac{-16xy^4}{96x^4y^4}$

78. $\dfrac{-8x^2y^4}{44y^2z^5}$

 79. Why might a number be written in scientific notation instead of decimal notation?

80.   **a.** Explain how to write 0.00000076 in scientific notation.
   b. Explain how to write $4.3 \cdot 10^8$ in decimal notation.

Determine whether the number is written in scientific notation. If not, explain why not.

81. $39.4 \cdot 10^3$

82. $0.8 \cdot 10^{-6}$

83. $7.1 \cdot 10^{2.4}$

84. $5.8 \cdot 10^{-132}$

Write the number in scientific notation.

85. 2,370,000

86. 75,000

87. 0.00045

88. 0.000076

89. 309,000

90. 819,000,000

91. 0.000000601

92. 0.00000000096

Write the number in decimal notation.

93. $7.1 \cdot 10^5$

94. $2.3 \cdot 10^7$

95. $4.3 \cdot 10^{-5}$

96. $9.21 \cdot 10^{-7}$

97. $6.71 \cdot 10^8$

98. $5.75 \cdot 10^9$

99. $7.13 \cdot 10^{-6}$

100. $3.54 \cdot 10^{-8}$

101. **Physics** Light travels approximately 16,000,000,000 mi in one day. Write this number in scientific notation.

102. **Chemistry** Avogadro's number is used in chemistry. Its value is approximately 602,300,000,000,000,000,000,000. Write this number in scientific notation.

103. **Geology** The mass of Earth is 5,980,000,000,000,000,000,000,000 kg. Write this number in scientific notation.

104. **Astronomy** A parsec is a distance measurement that is used by astronomers. One parsec is 3,086,000,000,000,000,000 cm. Write this number in scientific notation.

105. **Electricity** The electric charge on an electron is approximately 0.00000000000000000016 coulomb. Write this number in scientific notation.

106. **Light** The length of an infrared light wave is approximately 0.0000037 m. Write this number in scientific notation.

107. **Computers** One unit used to measure the speed of a computer is the picosecond. One picosecond is 0.000000000001 second. Write this number in scientific notation.

108. **Astronomy** One light-year is the distance traveled by light in one year. One light-year is 5,880,000,000,000 mi. Write this number in scientific notation.

Multiply or divide.

109. $(1.9 \cdot 10^{12})(3.5 \cdot 10^7)$

110. $(4.2 \cdot 10^7)(1.8 \cdot 10^{-5})$

111. $(2.3 \cdot 10^{-8})(1.4 \cdot 10^{-6})$

112. $(3 \cdot 10^{-20})(2.4 \cdot 10^9)$

113. $\dfrac{6.12 \cdot 10^{14}}{1.7 \cdot 10^9}$

114. $\dfrac{6 \cdot 10^{-8}}{2.5 \cdot 10^{-2}}$

115. $\dfrac{5.58 \cdot 10^{-7}}{3.1 \cdot 10^{11}}$

116. $\dfrac{9.03 \cdot 10^6}{4.3 \cdot 10^{-5}}$

7.4   Applying Concepts

Evaluate.

117. $8^{-2} + 2^{-5}$

118. $9^{-2} + 3^{-3}$

119. Evaluate 2^x and 2^{-x} when $x = -2, -1, 0, 1,$ and 2.

120. Evaluate 3^x and 3^{-x} when $x = -2, -1, 0, 1,$ and 2.

Write in decimal notation.

121. 2^{-4}

122. 25^{-2}

Simplify.

123. $\left(\dfrac{9x^2y^4}{3xy^2}\right) - \left(\dfrac{12x^5y^6}{6x^4y^4}\right)$

124. $\left(\dfrac{6x^2 + 9x}{3x}\right) + \left(\dfrac{8xy^2 + 4y^2}{4y^2}\right)$

125. $\left(\dfrac{6x^4yz^3}{2x^2y^3}\right)\left(\dfrac{2x^2z^3}{4y^2z}\right) \div \left(\dfrac{6x^2y^3}{x^4y^2z}\right)$

126. $\left(\dfrac{5x^2yz^3}{3x^4yz^3}\right) \div \left(\dfrac{10x^2y^5z^4}{2y^3z}\right) \div \left(\dfrac{5y^4z^2}{x^2y^6z}\right)$

Complete.

127. If $m = n$ and $a \neq 0$, then $\dfrac{a^m}{a^n} = $ _____.

128. If $m = n + 1$ and $a \neq 0$, then $\dfrac{a^m}{a^n} = $ _____.

Solve.

129. $(-4.8)^x = 1$

130. $-6.3^x = -1$

Determine whether each equation is true or false. If the equation is false, change the right side of the equation to make a true equation.

131. $(2a)^{-3} = \dfrac{2}{a^3}$

132. $((a^{-1})^{-1})^{-1} = \dfrac{1}{a}$

133. $(2 + 3)^{-1} = 2^{-1} + 3^{-1}$

134. If $x \neq \dfrac{1}{3}$, then $(3x - 1)^0 = (1 - 3x)^0$.

135. ✎ Why is the condition $x \neq \dfrac{1}{3}$ given in Exercise 134?

136. ✎ If x is a nonzero real number, is x^{-2} always positive, always negative, or positive or negative depending on whether x is positive or negative? Explain your answer.

137. ✎ If x is a nonzero real number, is x^{-3} always positive, always negative, or positive or negative depending on whether x is positive or negative? Explain your answer.

7.5 Division of Polynomials

1 ### Divide a polynomial by a monomial

Note that $\frac{8+4}{2}$ can be simplified by first adding the terms in the numerator and then dividing the result. It can also be simplified by first dividing each term in the numerator by the denominator and then adding the result.

$$\frac{8+4}{2} = \frac{12}{2} = 6$$

$$\frac{8+4}{2} = \frac{8}{2} + \frac{4}{2} = 4 + 2 = 6$$

To divide a polynomial by a monomial, divide each term in the numerator by the denominator, and write the sum of the quotients.

$$\frac{a+b}{c} = \frac{a}{c} + \frac{b}{c}$$

Divide: $\dfrac{6x^2 + 4x}{2x}$

Divide each term of the polynomial $6x^2 + 4x$ by the monomial $2x$. Simplify.

$$\frac{6x^2 + 4x}{2x} = \frac{6x^2}{2x} + \frac{4x}{2x}$$

$$= 3x + 2$$

Example 1 Divide: $\dfrac{6x^3 - 3x^2 + 9x}{3x}$

Solution

$$\frac{6x^3 - 3x^2 + 9x}{3x} = \frac{6x^3}{3x} - \frac{3x^2}{3x} + \frac{9x}{3x}$$

$$= 2x^2 - x + 3$$

• Divide each term of the polynomial by the monomial $3x$.
• Simplify each expression.

Problem 1 Divide: $\dfrac{4x^3y + 8x^2y^2 - 4xy^3}{2xy}$

Solution See page S16.

Example 2 Divide: $\dfrac{12x^2y - 6xy + 4x^2}{2xy}$

Solution

$$\frac{12x^2y - 6xy + 4x^2}{2xy}$$

$$= \frac{12x^2y}{2xy} - \frac{6xy}{2xy} + \frac{4x^2}{2xy}$$

$$= 6x - 3 + \frac{2x}{y}$$

• Divide each term of the polynomial by the monomial $2xy$.

• Simplify each expression.

Problem 2 Divide: $\dfrac{24x^2y^2 - 18xy + 6y}{6xy}$

Solution See page S16.

2 Divide polynomials

To divide polynomials, use a method similar to that used for division of whole numbers. The same equation used to check division of whole numbers is used to check polynomial division.

$$\textbf{(Quotient} \times \textbf{Divisor)} + \textbf{Remainder} = \textbf{Dividend}$$

For example:

$$
\begin{array}{r}
3 \\
5\overline{)17} \\
-15 \\
\hline
2
\end{array}
$$

(Quotient $\times$ Divisor)	+	Remainder	=	Dividend
(3×5)	+	2	=	17

Divide: $(x^2 - 5x + 8) \div (x - 3)$

Step 1

$$
\begin{array}{r}
x \\
x - 3\overline{)x^2 - 5x + 8} \\
\underline{x^2 - 3x} \\
-2x + 8
\end{array}
$$

Think: $x\overline{)x^2} = \dfrac{x^2}{x} = x$

Multiply: $x(x - 3) = x^2 - 3x$
Subtract: $(x^2 - 5x) - (x^2 - 3x) = -2x$
Bring down $+ 8$.

Step 2

$$
\begin{array}{r}
x - 2 \\
x - 3\overline{)x^2 - 5x + 8} \\
\underline{x^2 - 3x} \\
-2x + 8 \\
\underline{-2x + 6} \\
2
\end{array}
$$

Think: $x\overline{)-2x} = \dfrac{-2x}{x} = -2$

Multiply: $-2(x - 3) = -2x + 6$
Subtract: $(-2x + 8) - (-2x + 6) = 2$
The remainder is 2.

Check:

Quotient $\times$ Divisor + Remainder =   Dividend
$(x - 2)(x - 3) + 2 = x^2 - 3x - 2x + 6 + 2 = x^2 - 5x + 8$

$(x^2 - 5x + 8) \div (x - 3) = x - 2 + \dfrac{2}{x - 3}$

If a term is missing in the dividend, insert the term with zero as its coefficient. This helps keep like terms in the same column. This is illustrated in Example 3.

Example 3   Divide: $(6x + 2x^3 + 26) \div (x + 2)$

Solution

$$
\begin{array}{r}
2x^2 - 4x + 14 \\
x + 2\overline{)2x^3 + 0x^2 + 6x + 26} \\
\underline{2x^3 + 4x^2} \\
-4x^2 + 6x \\
\underline{-4x^2 - 8x} \\
14x + 26 \\
\underline{14x + 28} \\
-2
\end{array}
$$

• Arrange the terms in descending order. There is no term of x^2 in $2x^3 + 6x + 26$. Insert $0x^2$ for the missing term so that like terms will be in columns.

$(6x + 2x^3 + 26) \div (x + 2) = 2x^2 - 4x + 14 - \dfrac{2}{x + 2}$

Problem 3	Divide: $(x^3 - 2x - 4) \div (x - 2)$
Solution	See page S17.

7.5   Concept Review

Determine whether the statement is always true, sometimes true, or never true.

1. For $c \neq 0$, $\dfrac{a - b}{c} = \dfrac{a}{c} - \dfrac{b}{c}$.

2. For $x \neq 0$, $\dfrac{9x^2 + 6x}{3x} = 3x + 6x$.

3. Given $(6x^2 + x - 6) \div (3x + 2) = 2x - 1 - \dfrac{4}{3x + 2}$, then

$(3x + 2)(2x - 1) - \dfrac{4}{3x + 2} = 6x^2 + x - 6$.

7.5   Exercises

1 Divide.

1. $\dfrac{2x + 2}{2}$

2. $\dfrac{5y + 5}{5}$

3. $\dfrac{10a - 25}{5}$

4. $\dfrac{16b - 40}{8}$

5. $\dfrac{3a^2 + 2a}{a}$

6. $\dfrac{6y^2 + 4y}{y}$

7. $\dfrac{4b^3 - 3b}{b}$

8. $\dfrac{12x^2 - 7x}{x}$

9. $\dfrac{3x^2 - 6x}{3x}$

10. $\dfrac{10y^2 - 6y}{2y}$

11. $\dfrac{5x^2 - 10x}{-5x}$

12. $\dfrac{3y^2 - 27y}{-3y}$

13. $\dfrac{x^3 + 3x^2 - 5x}{x}$

14. $\dfrac{a^3 - 5a^2 + 7a}{a}$

15. $\dfrac{x^6 - 3x^4 - x^2}{x^2}$

16. $\dfrac{a^8 - 5a^5 - 3a^3}{a^2}$

17. $\dfrac{5x^2y^2 + 10xy}{5xy}$

18. $\dfrac{8x^2y^2 - 24xy}{8xy}$

19. $\dfrac{9y^6 - 15y^3}{-3y^3}$

20. $\dfrac{4x^4 - 6x^2}{-2x^2}$

21. $\dfrac{3x^2 - 2x + 1}{x}$

22. $\dfrac{8y^2 + 2y - 3}{y}$

23. $\dfrac{-3x^2 + 7x - 6}{x}$

24. $\dfrac{2y^2 - 6y + 9}{y}$

25. $\dfrac{16a^2b - 20ab + 24ab^2}{4ab}$

26. $\dfrac{22a^2b + 11ab - 33ab^2}{11ab}$

27. $\dfrac{9x^2y + 6xy - 3xy^2}{xy}$

2 Divide.

28. $(b^2 - 14b + 49) \div (b - 7)$

29. $(x^2 - x - 6) \div (x - 3)$

30. $(y^2 + 2y - 35) \div (y + 7)$

31. $(2x^2 + 5x + 2) \div (x + 2)$

32. $(2y^2 - 13y + 21) \div (y - 3)$

33. $(4x^2 - 16) \div (2x + 4)$

34. $(2y^2 + 7) \div (y - 3)$

35. $(x^2 + 1) \div (x - 1)$

36. $(x^2 + 4) \div (x + 2)$

37. $(6x^2 - 7x) \div (3x - 2)$

38. $(6y^2 + 2y) \div (2y + 4)$

39. $(5x^2 + 7x) \div (x - 1)$

40. $(6x^2 - 5) \div (x + 2)$

41. $(a^2 + 5a + 10) \div (a + 2)$

42. $(b^2 - 8b - 9) \div (b - 3)$

43. $(2y^2 - 9y + 8) \div (2y + 3)$

44. $(3x^2 + 5x - 4) \div (x - 4)$

45. $(8x + 3 + 4x^2) \div (2x - 1)$

46. $(10 + 21y + 10y^2) \div (2y + 3)$

47. $(15a^2 - 8a - 8) \div (3a + 2)$

48. $(12a^2 - 25a - 7) \div (3a - 7)$

49. $(5 - 23x + 12x^2) \div (4x - 1)$

50. $(24 + 6a^2 + 25a) \div (3a - 1)$

51. $(x^3 + 3x^2 + 5x + 3) \div (x + 1)$

52. $(x^3 - 6x^2 + 7x - 2) \div (x - 1)$

53. $(x^4 - x^2 - 6) \div (x^2 + 2)$

54. $(x^4 + 3x^2 - 10) \div (x^2 - 2)$

7.5   **Applying Concepts**

In Exercises 55 and 56, replace the question marks with expressions to make a true statement.

55. If $\dfrac{x^2 - x - 6}{x - 3} = x + 2$, then $x^2 - x - 6 = (?)(?)$.

56. If $\dfrac{x^2 + 2x - 3}{x - 2} = x + 4 + \dfrac{5}{x - 2}$, then $x^2 + 2x - 3 = (?)(?) + ?$.

Solve.

57. The product of a monomial and $4b$ is $12a^2b$. Find the monomial.

58. The product of a monomial and $6x$ is $24xy^2$. Find the monomial.

59. The quotient of a polynomial and $2x + 1$ is $2x - 4 + \dfrac{7}{2x + 1}$. Find the polynomial.

60. The quotient of a polynomial and $x - 3$ is $x^2 - x + 8 + \dfrac{22}{x - 3}$. Find the polynomial.

Focus on Problem Solving

Dimensional Analysis

In solving application problems, it may be useful to include units in order to organize the problem so that the answer is in the proper units. Using units to organize and check the correctness of an application is called **dimensional analysis**. We use the operations of multiplying units and dividing units in applying dimensional analysis to application problems.

The Rule for Multiplying Exponential Expressions states that we multiply two expressions with the same base by adding the exponents.

$$x^4 \cdot x^6 = x^{4+6} = x^{10}$$

In calculations that involve quantities, the units are operated on algebraically.

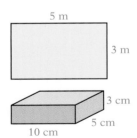

A rectangle measures 3 m by 5 m. Find the area of the rectangle.

$$A = LW = (3 \text{ m})(5 \text{ m}) = (3 \cdot 5)(\text{m} \cdot \text{m}) = 15 \text{ m}^2$$

The area of the rectangle is 15 m² (square meters).

A box measures 10 cm by 5 cm by 3 cm. Find the volume of the box.

$$V = LWH = (10 \text{ cm})(5 \text{ cm})(3 \text{ cm}) = (10 \cdot 5 \cdot 3)(\text{cm} \cdot \text{cm} \cdot \text{cm}) = 150 \text{ cm}^3$$

The volume of the box is 150 cm³ (cubic centimeters).

Find the area of a square whose side measures $(3x + 5)$ in.

$$A = s^2 = [(3x + 5) \text{ in.}]^2 = (3x + 5)^2 \text{ in}^2 = (9x^2 + 30x + 25) \text{ in}^2$$

The area of the square is $(9x^2 + 30x + 25)$ in² (square inches).

Dimensional analysis is used in the conversion of units.

The following example converts the unit miles to feet. The equivalent measures 1 mi and 5280 ft are used to form the following rates, which are called

conversion factors: $\dfrac{1 \text{ mi}}{5280 \text{ ft}}$ and $\dfrac{5280 \text{ ft}}{1 \text{ mi}}$. Because 1 mi = 5280 ft, both of the conversion factors $\dfrac{1 \text{ mi}}{5280 \text{ ft}}$ and $\dfrac{5280 \text{ ft}}{1 \text{ mi}}$ are equal to 1.

To convert 3 mi to feet, multiply 3 mi by the conversion factor $\frac{5280 \text{ ft}}{1 \text{ mi}}$.

3 mi
15,840 ft

$$3 \text{ mi} = 3 \text{ mi} \cdot \boxed{1} = \frac{3 \text{ mi}}{1} \cdot \boxed{\frac{5280 \text{ ft}}{1 \text{ mi}}} = \frac{3 \text{ mi} \cdot 5280 \text{ ft}}{1 \text{ mi}} = 3 \cdot 5280 \text{ ft} = 15{,}840 \text{ ft}$$

There are two important points in the above illustration. First, you can think of dividing the numerator and denominator by the common unit "mile" just as you would divide the numerator and denominator of a fraction by a common factor. Second, the conversion factor $\frac{5280 \text{ ft}}{1 \text{ mi}}$ is equal to 1, and multiplying an expression by 1 does not change the value of the expression.

In the application problem that follows, the units are kept in the problem while the problem is worked.

 In 2000, a horse named Fusaichi Pegasus ran a 1.25-mile race in 2.02 min. Find Fusaichi Pegasus's average speed for the race in miles per hour. Round to the nearest tenth.

Strategy   To find the average speed, use the formula $r = \frac{d}{t}$, where r is the speed, d is the distance, and t is the time. Use the conversion factor $\frac{60 \text{ min}}{1 \text{ h}}$.

Solution $r = \dfrac{d}{t} = \dfrac{1.25 \text{ mi}}{2.02 \text{ min}} = \dfrac{1.25 \text{ mi}}{2.02 \text{ min}} \cdot \dfrac{60 \text{ min}}{1 \text{ h}}$

$= \dfrac{75 \text{ mi}}{2.02 \text{ h}} \approx 37.1 \text{ mph}$

Fusaichi Pegasus's average speed was 37.1 mph.

Try each of the following problems. Round to the nearest tenth or to the nearest cent.

1. Convert 88 ft/s to miles per hour.

2. Convert 8 m/s to kilometers per hour (1 km = 1000 m).

3. A carpet is to be placed in a meeting hall that is 36 ft wide and 80 ft long. At $21.50 per square yard, how much will it cost to carpet the meeting hall?

4. A carpet is to be placed in a room that is 20 ft wide and 30 ft long. At $22.25 per square yard, how much will it cost to carpet the area?

5. Find the number of gallons of water in a fish tank that is 36 in. long and 24 in. wide and is filled to a depth of 16 in. (1 gal = 231 in³).

6. Find the number of gallons of water in a fish tank that is 24 in. long and 18 in. wide and is filled to a depth of 12 in. (1 gal = 231 in³).

7. A $\frac{1}{4}$-acre commercial lot is on sale for $2.15 per square foot. Find the sale price of the commercial lot (1 acre = 43,560 ft²).

8. A 0.75-acre industrial parcel was sold for $98,010. Find the parcel's price per square foot (1 acre = 43,560 ft²).

9. A new driveway will require 800 ft³ of concrete. Concrete is ordered by the cubic yard. How much concrete should be ordered?

10. A piston-engined dragster traveled 1320 ft in 4.477 s at Joliet, Illinois, on June 2, 2001. Find the average speed of the dragster in miles per hour.

11. The Marianas Trench in the Pacific Ocean is the deepest part of the ocean. Its depth is 6.85 mi. The speed of sound under water is 4700 ft/s. Find the time it takes sound to travel from the surface to the bottom of the Marianas Trench and back.

Projects & Group Activities

Pascal's Triangle

Point of Interest

Pascal did not invent the triangle of numbers known as Pascal's Triangle. It was known to mathematicians in China probably as early as A.D. 1050. But Pascal's Traite du triangle arithmetique (Treatise Concerning the Arithmetical Triangle) *brought together all the different aspects of the numbers for the first time.*

Simplifying the power of a binomial is called *expanding the binomial*. The expansion of the first three powers of a binomial is shown below.

$(a + b)^1 = a + b$

$(a + b)^2 = (a + b)(a + b) = a^2 + 2ab + b^2$

$(a + b)^3 = (a + b)^2(a + b) = (a^2 + 2ab + b^2)(a + b) = a^3 + 3a^2b + 3ab^2 + b^3$

Find $(a + b)^4$. [*Hint:* $(a + b)^4 = (a + b)^3(a + b)$]

Find $(a + b)^5$. [*Hint:* $(a + b)^5 = (a + b)^4(a + b)$]

If we continue in this way, the results for $(a + b)^6$ are

$$(a + b)^6 = a^6 + 6a^5b + 15a^4b^2 + 20a^3b^3 + 15a^2b^4 + 6ab^5 + b^6$$

Now expand $(a + b)^8$. Before you begin, see if you can find a pattern that will help you write the expansion of $(a + b)^8$ without having to multiply it out. Here are some hints.

1. Write out the variable terms of each binomial expansion from $(a + b)^1$ through $(a + b)^6$. Observe how the exponents on the variables change.

2. Write out the coefficients of all the terms without the variable parts. It will be helpful to make a triangular arrangement as shown at the left. Note that each row begins and ends with a 1. Also note in the two shaded regions that any number in a row is the sum of the two closest numbers above it. For instance, $1 + 5 = 6$ and $6 + 4 = 10$.

```
      1   1
    1   2   1
  1   3   3   1
 1   4   6   4   1
1   5  10  10   5   1
1  6  15  20  15   6   1
```

The triangle of numbers shown at the left is called Pascal's Triangle. To find the expansion of $(a + b)^8$, you need to find the eighth row of Pascal's Triangle. First find row seven. Then find row eight and use the patterns you have observed to write the expansion of $(a + b)^8$.

WEB

Pascal's Triangle has been the subject of extensive analysis, and many patterns have been found. See if you can find some of them. You might check the Internet, where you will find some web sites with information on Pascal's Triangle.

Population and Land Allocation*

Texas

Rhode Island

One of the Projects and Group Activities in the chapter "Linear Equations and Inequalities" involved population projections. You were asked to make predictions about the population of the United States and of the world. In this project, you are asked to determine hypothetical land allocation for the world's population today. Use the figure 6×10^9 for the current world population and the figure 2.9×10^8 for the current U.S. population. One square mile is approximately 2.8×10^7 ft².

1. If every person in the world moved to Texas and each person were given an equal amount of land, how many square feet of land would each person have? The area of Texas is 2.619×10^5 mi².

2. If every person in the United States moved to Rhode Island and each person were given an equal amount of land, how many square feet of land would each person have? The area of Rhode Island is 1.0×10^3 mi². Round to the nearest whole number.

3. Suppose every person in the world were given a plot of land the size of a two-car garage (22 ft × 22 ft).
   a. How many people would fit in a square mile? Round to the nearest hundred.
   b. How many square miles would be required to accommodate the entire world population? Round to the nearest hundred.

4. If the total land area of Earth were divided equally, how many acres of land would each person be allocated? Use a figure of 5.7×10^7 mi² for the land area of Earth. One acre is 43,560 ft². Round to the nearest tenth.

5. If every person on Earth were given a plot of land the size of a two-car garage, what would be the carrying capacity of Earth? Round to the nearest hundred billion.

*Source: **www.infoplease.com**

Properties of Polynomials

Take Note

Suggestions for graphing equations can be found in the Appendix. For this particular activity, it may be necessary to adjust the viewing window so that you can get a reasonably accurate graph of the equation. Try
Ymax = 50.
Ymin = −50, and
Yscl = 5.

The graph of a fifth-degree polynomial with 4 *turning points* is shown at the right. In this project you will graph various polynomials and try to make a conjecture as to the relationship between the degree of a polynomial and the number of turning points in its graph. For each of the following, graph the equation. Record the degree of the equation and the number of turning points.

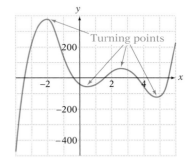

1. Graph: $y = x^3 + 1$

2. Graph: $y = x^3 + 2x^2 - 5x - 6$

3. Graph: $y = x^4 - x^3 - 11x^2 - x - 12$

4. Graph: $y = x^4 - 2x^3 - 13x^2 + 14x + 24$

5. Graph: $y = x^5 - 2$

6. Graph: $y = x^5 - 3x^4 - 11x^3 + 27x^2 + 10x - 24$

7. Graph: $y = x^5 - 2x^4 - 10x^3 + 10x^2 - 11x + 12$

Make a conjecture as to the relationship between the degree of a polynomial and the number of turning points in its graph. Graph a few more polynomials of your choosing and see whether your conjecture is valid for those graphs. If not, refine your conjecture and test it again.

Chapter Summary

Key Words	Examples
A **monomial** is a number, a variable, or a product of numbers and variables. [7.1.1, p. 342]	5 is a number. y is a variable. $8a^2b^3$ is a product of numbers and variables. 5, y, and $8a^2b^3$ are monomials.
A **polynomial** is a variable expression in which the terms are monomials. [7.1.1, p. 342]	As shown above, 5, y, and $8a^2b^3$ are monomials. Therefore, $5 + y + 8a^2b^3$ is a polynomial.
A polynomial of one term is a **monomial**. [7.1.1, p. 342]	5, y, and $8a^2b^3$ are monomials.
A polynomial of two terms is a **binomial**. [7.1.1, p. 342]	$x + 9$, $y^2 - 3$, and $6a + 7b$ are binomials.
A polynomial of three terms is a **trinomial**. [7.1.1, p. 342]	$x^2 + 2x - 1$ is a trinomial.
The **degree of a polynomial in one variable** is the largest exponent on a variable. [7.1.1, p. 343]	The degree of $8x^3 - 5x^2 + 4x - 12$ is 3.

Essential Rules and Procedures

Rule for Multiplying Exponential Expressions
If m and n are integers, then $x^m \cdot x^n = x^{m+n}$.   [7.2.1, p. 347]

$b^5 \cdot b^4 = b^{5+4} = b^9$

Rule for Simplifying Powers of Exponential Expressions
If m and n are integers, then $(x^m)^n = x^{mn}$.   [7.2.2, p. 348]

$(y^3)^7 = y^{3(7)} = y^{21}$

Rule for Simplifying Powers of Products
If m, n, and p are integers, then $(x^m y^n)^p = x^{mp} y^{np}$.   [7.2.2, p. 348]

$(a^6 b^2)^3 = a^{6(3)} b^{2(3)} = a^{18} b^6$

FOIL Method
To find the product of two binomials, add the products of the **F**irst terms, the **O**uter terms, the **I**nner terms, and the **L**ast terms.   [7.3.3, p. 352–353]

$$
\begin{aligned}
&(4x + 3)(2x - 1) \\
&= (4x)(2x) + (4x)(-1) \\
&\quad + (3)(2x) + (3)(-1) \\
&= 8x^2 - 4x + 6x - 3 \\
&= 8x^2 + 2x - 3
\end{aligned}
$$

The Sum and Difference of Two Terms
$(a + b)(a - b) = a^2 - b^2$ [7.3.4, p. 353]

$$(3x + 4)(3x - 4) = (3x)^2 - 4^2$$
$$= 9x^2 - 16$$

The Square of a Binomial
$(a + b)^2 = a^2 + 2ab + b^2$

$$(2x + 5)^2 = (2x)^2 + 2(2x)(5) + 5^2$$
$$= 4x^2 + 20x + 25$$

$(a - b)^2 = a^2 - 2ab + b^2$ [7.3.4, p. 354]

$$(2x - 5)^2 = (2x)^2 - 2(2x)(5) + (-5)^2$$
$$= 4x^2 - 20x + 25$$

Zero as an Exponent
Any nonzero expression to the zero power equals 1.
[7.4.1, p. 360]

$17^0 = 1$ $(5y)^0 = 1, y \neq 0$

Definition of Negative Exponents
If n is a positive integer and $x \neq 0$, then $x^{-n} = \dfrac{1}{x^n}$ and $\dfrac{1}{x^{-n}} = x^n$.

$x^{-6} = \dfrac{1}{x^6}$ and $\dfrac{1}{x^{-6}} = x^6$

[7.4.1, p. 361]

Rule for Dividing Exponential Expressions
If m and n are integers and $x \neq 0$, then $\dfrac{x^m}{x^n} = x^{m-n}$.

$\dfrac{y^8}{y^3} = y^{8-3} = y^5$

[7.4.1, p. 361]

Scientific Notation
To express a number in scientific notation, write it in the form $a \cdot 10^n$, where a is a number between 1 and 10 and n is an integer. If the number is greater than 10, the exponent on 10 will be positive. If the number is less than 1, the exponent on 10 will be negative.

$367{,}000{,}000 = 3.67 \cdot 10^8$

$0.0000059 = 5.9 \cdot 10^{-6}$

To change a number written in scientific notation to decimal notation, move the decimal point to the right if the exponent on 10 is positive and to the left if the exponent on 10 is negative. Move the decimal point the same number of places as the absolute value of the exponent on 10.   [7.4.2, p. 363–364]

$2.418 \cdot 10^7 = 24{,}180{,}000$

$9.06 \cdot 10^{-5} = 0.0000906$

Chapter Review Exercises

1. Add: $(12y^2 + 17y - 4) + (9y^2 - 13y + 3)$

2. Multiply: $(5xy^2)(-4x^2y^3)$

3. Multiply: $-2x(4x^2 + 7x - 9)$

4. Multiply: $(5a - 7)(2a + 9)$

5. Divide: $\dfrac{36x^2 - 42x + 60}{6}$

6. Subtract: $(5x^2 - 2x - 1) - (3x^2 - 5x + 7)$

7. Simplify: $(-3^2)^3$

8. Multiply: $(x^2 - 5x + 2)(x - 1)$

9. Multiply: $(a + 7)(a - 7)$

10. Evaluate: $\dfrac{6^2}{6^{-2}}$

11. Divide: $(x^2 + x - 42) \div (x + 7)$

12. Add: $(2x^3 + 7x^2 + x) + (2x^2 - 4x - 12)$

13. Multiply: $(6a^2b^5)(3a^6b)$

14. Multiply: $x^2y(3x^2 - 2x + 12)$

15. Multiply: $(2b - 3)(4b + 5)$

16. Divide: $\dfrac{16y^2 - 32y}{-4y}$

17. Subtract: $(13y^3 - 7y - 2) - (12y^2 - 2y - 1)$

18. Simplify: $(2^3)^2$

19. Multiply: $(3y^2 + 4y - 7)(2y + 3)$

20. Multiply: $(2b - 9)(2b + 9)$

21. Simplify: $(a^{-2}b^3c)^2$

22. Divide: $(6y^2 - 35y + 36) \div (3y - 4)$

23. Write 0.00000397 in scientific notation.

24. Multiply: $(xy^5z^3)(x^3y^3z)$

25. Multiply: $(6y^2 - 2y + 9)(-2y^3)$

26. Multiply: $(6x - 12)(3x - 2)$

27. Write $6.23 \cdot 10^{-5}$ in decimal notation.

28. Subtract: $(8a^2 - a) - (15a^2 - 4)$

29. Simplify: $(-3x^2y^3)^2$

30. Multiply: $(4a^2 - 3)(3a - 2)$

31. Simplify: $(5y - 7)^2$

32. Simplify: $(-3x^{-2}y^{-3})^{-2}$

33. Divide: $(x^2 + 17x + 64) \div (x + 12)$

34. Write $2.4 \cdot 10^5$ in decimal notation.

35. Multiply: $(a^2b^7c^6)(ab^3c)(a^3bc^2)$

36. Multiply: $2ab^3(4a^2 - 2ab + 3b^2)$

37. Multiply: $(3x + 4y)(2x - 5y)$

38. Divide: $\dfrac{12b^7 + 36b^5 - 3b^3}{3b^3}$

39. Subtract: $(b^2 - 11b + 19) - (5b^2 + 2b - 9)$

40. Simplify: $(5a^7b^6)^2(4ab)$

41. Multiply: $(6b^3 - 2b^2 - 5)(2b^2 - 1)$

42. Multiply: $(6 - 5x)(6 + 5x)$

43. Simplify: $\dfrac{6x^{-2}y^4}{3xy}$

44. Divide: $(a^3 + a^2 + 18) \div (a + 3)$

45. Add: $(4b^3 - 7b^2 + 10) + (2b^2 - 9b - 3)$

46. Multiply: $(2a^{12}b^3)(-9b^2c^6)(3ac)$

47. Multiply: $-9x^2(2x^2 + 3x - 7)$

48. Multiply: $(10y - 3)(3y - 10)$

49. Write 9,176,000,000,000 in scientific notation.

50. Subtract: $(6y^2 + 2y + 7) - (8y^2 + y + 12)$

51. Simplify: $(6x^4y^7z^2)^2(-2x^3y^2z^6)^2$

52. Multiply: $(-3x^3 - 2x^2 + x - 9)(4x + 3)$

53. Simplify: $(8a + 1)^2$

54. Simplify: $\dfrac{4a^{-2}b^{-8}}{2a^{-1}b^{-2}}$

55. Divide: $(b^3 - 2b^2 - 33b - 7) \div (b - 7)$

56. The length of a rectangle is $5x$ m. The width is $(4x - 7)$ m. Find the area of the rectangle in terms of the variable x.

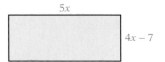

57. The length of a side of a square is $(5x + 4)$ in. Find the area of the square in terms of the variable x.

58. The base of a triangle is $(3x - 2)$ ft and the height is $(6x + 4)$ ft. Find the area of the triangle in terms of the variable x.

59. The radius of a circle is $(x - 6)$ cm. Find the area of the circle in terms of the variable x. Leave the answer in terms of π.

60. The width of a rectangle is $(3x - 8)$ mi. The length is $(5x + 4)$ mi. Find the area of the rectangle in terms of the variable x.

Chapter Test

1. Add: $(3x^3 - 2x^2 - 4) + (8x^2 - 8x + 7)$

2. Multiply: $(-2x^3 + x^2 - 7)(2x - 3)$

3. Multiply: $2x(2x^2 - 3x)$

4. Simplify: $(-2a^2b)^3$

5. Simplify: $\dfrac{12x^2}{-3x^{-4}}$

6. Simplify: $(2ab^{-3})(3a^{-2}b^4)$

7. Subtract: $(3a^2 - 2a - 7) - (5a^3 + 2a - 10)$

8. Multiply: $(a - 2b)(a + 5b)$

9. Divide: $\dfrac{16x^5 - 8x^3 + 20x}{4x}$

10. Divide: $(4x^2 - 7) \div (2x - 3)$

11. Multiply: $(-2xy^2)(3x^2y^4)$

12. Multiply: $-3y^2(-2y^2 + 3y - 6)$

13. Simplify: $\dfrac{27xy^3}{3x^4y^3}$

14. Simplify: $(2x - 5)^2$

15. Multiply: $(2x - 7y)(5x - 4y)$

16. Multiply: $(x - 3)(x^2 - 4x + 5)$

17. Simplify: $(a^2b^{-3})^2$

18. Write 0.000029 in scientific notation.

19. Multiply: $(4y - 3)(4y + 3)$

20. Subtract: $(3y^3 - 5y + 8) - (-2y^2 + 5y + 8)$

21. Simplify: $(-3a^3b^2)^2$

22. Multiply: $(2a - 7)(5a^2 - 2a + 3)$

23. Simplify: $(3b + 2)^2$

24. Simplify: $\dfrac{-2a^2b^3}{8a^4b^8}$

25. Divide: $(8x^2 + 4x - 3) \div (2x - 3)$

26. Multiply: $(a^2b^5)(ab^2)$

27. Multiply: $(a - 3b)(a + 4b)$

28. Write $3.5 \cdot 10^{-8}$ in decimal notation.

29. The length of a side of a square is $(2x + 3)$ m. Find the area of the square in terms of the variable x.

30. The radius of a circle is $(x - 5)$ in. Find the area of the circle in terms of the variable x. Leave the answer in terms of π.

Cumulative Review Exercises

1. Simplify: $\dfrac{3}{16} - \left(-\dfrac{3}{8}\right) - \dfrac{5}{9}$

2. Simplify: $-5^2 \cdot \left(\dfrac{2}{3}\right)^3 \cdot \left(-\dfrac{3}{8}\right)$

3. Simplify: $\left(-\dfrac{1}{2}\right)^2 \div \left(\dfrac{5}{8} - \dfrac{5}{6}\right) + 2$

4. Find the opposite of -87.

5. Write $\dfrac{31}{40}$ as a decimal.

6. Evaluate $\dfrac{b - (a - b)^2}{b^2}$ when $a = 3$ and $b = -2$.

7. Simplify: $-3x - (-xy) + 2x - 5xy$

8. Simplify: $(16x)\left(-\dfrac{3}{4}\right)$

9. Simplify: $-2[3x - 4(3 - 2x) + 2]$

10. Complete the statement by using the Inverse Property of Addition.
$-8 + ? = 0$

11. Solve: $12 = -\dfrac{2}{3}x$

12. Solve: $3x - 7 = 2x + 9$

13. Solve: $3 - 4(2 - x) = 3x + 7$

14. Solve: $-\dfrac{4}{5}x = 16 - x$

15. 38.4 is what percent of 160?

16. Solve: $7x - 8 \geq -29$

17. Find the slope of the line that contains the points whose coordinates are $(3, -4)$ and $(-2, 5)$.

18. Find the equation of the line that contains the point whose coordinates are $(1, -3)$ and has slope $-\dfrac{3}{2}$.

19. Graph: $3x - 2y = -6$

20. Graph the solution set of
$y \leq \dfrac{4}{5}x - 3$.

21. Find the domain and range of the relation $\{(-8, -7), (-6, -5), (-4, -2), (-2, 0)\}$. Is the relation a function?

22. Evaluate $f(x) = -2x + 10$ at $x = 6$.

23. Solve by substitution: $x = 3y + 1$
$$2x + 5y = 13$$

24. Solve by the addition method: $9x - 2y = 17$
$$5x + 3y = -7$$

25. Subtract: $(5b^3 - 4b^2 - 7) - (3b^2 - 8b + 3)$

26. Multiply: $(3x - 4)(5x^2 - 2x + 1)$

27. Multiply: $(4b - 3)(5b - 8)$

28. Simplify: $(5b + 3)^2$

29. Simplify: $\dfrac{-3a^3b^2}{12a^4b^{-2}}$

30. Divide: $\dfrac{-15y^2 + 12y - 3}{-3y}$

31. Divide: $(a^2 - 3a - 28) \div (a + 4)$

32. Simplify: $(-3x^{-4}y)(-3x^{-2}y)$

33. Find the range of the function given by the equation $f(x) = -\dfrac{4}{3}x + 9$ if the domain is $\{-12, -9, -6, 0, 6\}$.

34. Translate and simplify "the product of five and the difference between a number and twelve."

35. Translate "the difference between eight times a number and twice the number is eighteen" into an equation and solve.

36. The width of a rectangle is 40% of the length. The perimeter of the rectangle is 42 m. Find the length and width of the rectangle.

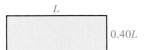

37. A calculator costs a retailer $24. Find the selling price when the markup rate is 80%.

38. Fifty ounces of pure orange juice are added to 200 oz of a fruit punch that is 10% orange juice. What is the percent concentration of orange juice in the resulting mixture?

39. A car traveling at 50 mph overtakes a cyclist who, riding at 10 mph, has had a 2-hour head start. How far from the starting point does the car overtake the cyclist?

40. The length of a side of a square is $(3x + 2)$ ft. Find the area of the square in terms of the variable x.

Leonardo DaVinci designed the first parachute in 1485. People have been jumping out of planes ever since. The highest parachute jump ever made was one by U.S. Air Force Captain Joseph W. Kittinger, Jr. in 1960. Kittinger, who was performing high-altitude escape experiments for the Air Force, jumped from 102,800 feet above sea level — over 19 miles up! His free fall lasted more then 4.5 minutes and he reached a velocity of 614 mph. Factors that affect velocity include the initial altitude of the jump, the size of the parachute, and the weight and body position of the diver. The velocity of a falling object can be modeled by a quadratic equation, as shown in the **Project on pages 425–427.**

Applications

Carpentry, *p. 433*

Coin problems, *pp. 421, 422*

Consecutive integer problems, *pp. 415, 416, 418, 430, 431, 433*

Consumerism, *p. 433*

Discount, *p. 433*

Geometry, *pp. 391, 406, 413, 416, 418, 419, 420, 422, 423, 430, 431, 433*

Integer problems, *p. 418*

Investments, *p. 433*

Motion pictures, *p. 430*

Number problems, *pp. 413, 419*

Physics, *pp. 416, 419, 426, 427*

Prime numbers, *p. 424*

Sports, *pp. 420, 421, 427, 430*

Symmetry, *p. 422*

Temperature, *p. 433*

8
Factoring

Objectives

Section 8.1
1 Factor a monomial from a polynomial
2 Factor by grouping

Section 8.2
1 Factor trinomials of the form $x^2 + bx + c$
2 Factor completely

Section 8.3
1 Factor trinomials of the form $ax^2 + bx + c$ using trial factors
2 Factor trinomials of the form $ax^2 + bx + c$ by grouping

Section 8.4
1 Factor the difference of two squares and perfect-square trinomials
2 Factor completely

Section 8.5
1 Solve equations by factoring
2 Application problems

Need help? For online student resources, such as section quizzes, visit this web site: **math.college.hmco.com**

PrepTest

1. Write 30 as a product of prime numbers.

2. Simplify: $-3(4y - 5)$

3. Simplify: $-(a - b)$

4. Simplify: $2(a - b) - 5(a - b)$

5. Solve: $4x = 0$

6. Solve: $2x + 1 = 0$

7. Multiply: $(x + 4)(x - 6)$

8. Multiply: $(2x - 5)(3x + 2)$

9. Simplify: $\dfrac{x^5}{x^2}$

10. Simplify: $\dfrac{6x^4y^3}{2xy^2}$

Go Figure

Write the expressions in order from greatest to least. Do not use a calculator

$2^{150}, 3^{100}, 5^{50}$

8.1 Common Factors

1 ### Factor a monomial from a polynomial

> **Take Note**
>
> 12 is the GCF of 24 and 60 because 12 is the largest integer that divides evenly into both 24 and 60.

The **greatest common factor (GCF)** of two or more integers is the greatest integer that is a factor of all the integers.

$24 = 2 \cdot 2 \cdot 2 \cdot 3$
$60 = 2 \cdot 2 \cdot 3 \cdot 5$
$\text{GCF} = 2 \cdot 2 \cdot 3 = 12$

The GCF of two or more monomials is the product of the GCF of the coefficients and the common variable factors.

$6x^3y = 2 \cdot 3 \cdot x \cdot x \cdot x \cdot y$
$8x^2y^2 = 2 \cdot 2 \cdot 2 \cdot x \cdot x \cdot y \cdot y$
$\text{GCF} = 2 \cdot x \cdot x \cdot y = 2x^2y$

Note that the exponent of each variable in the GCF is the same as the *smallest* exponent of that variable in either of the monomials.

The GCF of $6x^3y$ and $8x^2y^2$ is $2x^2y$.

Example 1 Find the GCF of $12a^4b$ and $18a^2b^2c$.

Solution
$12a^4b = 2 \cdot 2 \cdot 3 \cdot a^4 \cdot b$ • Factor each monomial.
$18a^2b^2c = 2 \cdot 3 \cdot 3 \cdot a^2 \cdot b^2 \cdot c$

$\text{GCF} = 2 \cdot 3 \cdot a^2 \cdot b = 6a^2b$ • The common variable factors are a^2 and b. c is not a common factor.

Problem 1 Find the GCF of $4x^6y$ and $18x^2y^6$.

Solution See page S17.

The Distributive Property is used to multiply factors of a polynomial. To **factor a polynomial** means to write the polynomial as a product of other polynomials.

$$\begin{array}{ccc} & \text{— Multiply —} & \downarrow \\ \textbf{Factors} & & \textbf{Polynomial} \\ 2x(x+5) & = & 2x^2 + 10x \\ & \text{— Factor —} & \end{array}$$

In the example above, $2x$ is the GCF of the terms $2x^2$ and $10x$. It is a **common monomial factor** of the terms. $x + 5$ is a **binomial factor** of $2x^2 + 10x$.

Example 2 Factor.   **A.** $5x^3 - 35x^2 + 10x$ **B.** $16x^2y + 8x^4y^2 - 12x^4y^5$

Solution   **A.** The GCF is $5x$.

$$\frac{5x^3}{5x} = x^2, \frac{-35x^2}{5x} = -7x, \frac{10x}{5x} = 2$$

$$5x^3 - 35x^2 + 10x$$

$$= 5x(x^2) + 5x(-7x) + 5x(2)$$

$$= 5x(x^2 - 7x + 2)$$

- Find the GCF of the terms of the polynomial.
- Divide each term of the polynomial by the GCF.
- Use the quotients to rewrite the polynomial, expressing each term as a product with the GCF as one of the factors.
- Use the Distributive Property to write the polynomial as a product of factors.

B. $16x^2y = 2 \cdot 2 \cdot 2 \cdot 2 \cdot x^2 \cdot y$
$8x^4y^2 = 2 \cdot 2 \cdot 2 \cdot x^4 \cdot y^2$
$12x^4y^5 = 2 \cdot 2 \cdot 3 \cdot x^4 \cdot y^5$
The GCF is $4x^2y$.

$$16x^2y + 8x^4y^2 - 12x^4y^5$$
$$= 4x^2y(4) + 4x^2y(2x^2y) + 4x^2y(-3x^2y^4)$$
$$= 4x^2y(4 + 2x^2y - 3x^2y^4)$$

Problem 2 Factor.   **A.** $14a^2 - 21a^4b$ **B.** $6x^4y^2 - 9x^3y^2 + 12x^2y^4$

Solution   See page S17.

2 Factor by grouping

In the examples below, the binomials in parentheses are called **binomial factors.**

$$2a(a + b)$$
$$3xy(x - y)$$

The Distributive Property is used to factor a common binomial factor from an expression.

In the expression at the right, the common binomial factor is $y - 3$. The Distributive Property is used to write the expression as a product of factors.

$$x(y - 3) + 4(y - 3)$$
$$= (y - 3)(x + 4)$$

Example 3 Factor: $y(x + 2) + 3(x + 2)$

Solution $y(x + 2) + 3(x + 2)$ • The common binomial factor is $x + 2$.
$$= (x + 2)(y + 3)$$

> **Problem 3**   Factor: $a(b - 7) + b(b - 7)$
>
> **Solution**   See page S17.

Sometimes a binomial factor must be rewritten before a common binomial factor can be found.

Factor: $a(a - b) + 5(b - a)$

$a - b$ and $b - a$ are different binomials.

Note that $(b - a) = (-a + b) = -(a - b)$.

Rewrite $(b - a)$ as $-(a - b)$ so that there is a common factor.

$$a(a - b) + 5(b - a) = a(a - b) + 5[-(a - b)]$$
$$= a(a - b) - 5(a - b)$$
$$= (a - b)(a - 5)$$

Example 4   Factor: $2x(x - 5) + y(5 - x)$

Solution $2x(x - 5) + y(5 - x)$
$$= 2x(x - 5) - y(x - 5)$$ • Rewrite $5 - x$ as $-(x - 5)$ so that there is a common factor.
$$= (x - 5)(2x - y)$$ • Write the expression as a product of factors.

Problem 4   Factor: $3y(5x - 2) - 4(2 - 5x)$

Solution   See page S17.

Some polynomials can be factored by grouping the terms so that a common binomial factor is found.

Factor: $2x^3 - 3x^2 + 4x - 6$

Group the first two terms and the last two terms.

Factor out the GCF from each group.

Write the expression as a product of factors.

$$2x^3 - 3x^2 + 4x - 6$$
$$= (2x^3 - 3x^2) + (4x - 6)$$
$$= x^2(2x - 3) + 2(2x - 3)$$
$$= (2x - 3)(x^2 + 2)$$

Example 5   Factor: $3y^3 - 4y^2 - 6y + 8$

Solution $3y^3 - 4y^2 - 6y + 8$
$$= (3y^3 - 4y^2) - (6y - 8)$$ • Group the first two terms and the last two terms. Note that $-6y + 8 = -(6y - 8)$.
$$= y^2(3y - 4) - 2(3y - 4)$$ • Factor out the GCF from each group.
$$= (3y - 4)(y^2 - 2)$$ • Write the expression as a product of factors.

Problem 5   Factor: $y^5 - 5y^3 + 4y^2 - 20$

Solution   See page S17.

8.1 Concept Review

Determine whether the statement is always true, sometimes true, or never true.

1. To factor a polynomial means to rewrite it as multiplication.

2. The expression $3x(x + 5)$ is a sum, and the expression $3x^2 + 15x$ is a product.

3. The greatest common factor of two numbers is the largest number that divides evenly into both numbers.

4. x is a monomial factor of $x(2x - 1)$. $2x - 1$ is a binomial factor of $x(2x - 1)$.

5. $y(y + 4) + 6$ is in factored form.

6. A common monomial factor is a factor of every term of a polynomial.

7. $a^2 - 8a - ab + 8b = a(a - 8) - b(a + 8)$

8. The binomial $d + 9$ is a common binomial factor of $2c(d + 9) - 7(d + 9)$.

8.1 Exercises

1. Explain the meaning of "a factor of a polynomial" and the meaning of "to factor a polynomial."

2. Explain why the statement is true.

   a. The terms of the binomial $3x - 6$ have a common factor.
   b. The expression $3x^2 + 15$ is not in factored form.
   c. $5y - 7$ is a factor of $y(5y - 7)$.

Find the greatest common factor.

3. x^7, x^3 4. y^6, y^{12} 5. x^2y^4, xy^6 6. a^5b^3, a^3b^8

7. $x^2y^4z^6, xy^8z^2$ 8. ab^2c^3, a^3b^2c 9. $14a^3, 49a^7$ 10. $12y^2, 27y^4$

11. $3x^2y^2, 5ab^2$ 12. $8x^2y^3, 7ab^4$ 13. $9a^2b^4, 24a^4b^2$ 14. $15a^4b^2, 9ab^5$

15. $ab^3, 4a^2b, 12a^2b^3$ 16. $12x^2y, x^4y, 16x$ 17. $2x^2y, 4xy, 8x$

18. $16x^2, 8x^4y^2, 12xy$ 19. $3x^2y^2, 6x, 9x^3y^3$ 20. $4a^2b^3, 8a^3, 12ab^4$

Factor.

21. $5a + 5$

22. $7b - 7$

23. $16 - 8a^2$

24. $12 + 12y^2$

25. $8x + 12$

26. $16a - 24$

27. $30a - 6$

28. $20b + 5$

29. $7x^2 - 3x$

30. $12y^2 - 5y$

31. $3a^2 + 5a^5$

32. $9x - 5x^2$

33. $14y^2 + 11y$

34. $6b^3 - 5b^2$

35. $2x^4 - 4x$

36. $3y^4 - 9y$

37. $10x^4 - 12x^2$

38. $12a^5 - 32a^2$

39. $x^2y - xy^3$

40. $a^2b + a^4b^2$

41. $2a^5b + 3xy^3$

42. $5x^2y - 7ab^3$

43. $6a^2b^3 - 12b^2$

44. $8x^2y^3 - 4x^2$

45. $6a^2bc + 4ab^2c$

46. $10x^2yz^2 + 15xy^3z$

47. $18x^2y^2 - 9a^2b^2$

48. $9a^2x - 27a^3x^3$

49. $6x^3y^3 - 12x^6y^6$

50. $3a^2b^2 - 12a^5b^5$

51. $x^3 - 3x^2 - x$

52. $a^3 + 4a^2 + 8a$

53. $2x^2 + 8x - 12$

54. $a^3 - 3a^2 + 5a$

55. $b^3 - 5b^2 - 7b$

56. $5x^2 - 15x + 35$

57. $8y^2 - 12y + 32$

58. $3x^3 + 6x^2 + 9x$

59. $5y^3 - 20y^2 + 10y$

60. $2x^4 - 4x^3 + 6x^2$

61. $3y^4 - 9y^3 - 6y^2$

62. $2x^3 + 6x^2 - 14x$

63. $3y^3 - 9y^2 + 24y$

64. $2y^5 - 3y^4 + 7y^3$

65. $6a^5 - 3a^3 - 2a^2$

66. $x^3y - 3x^2y^2 + 7xy^3$

67. $8x^2y^2 - 4x^2y + x^2$

68. $x^4y^4 - 3x^3y^3 + 6x^2y^2$

69. $4x^5y^5 - 8x^4y^4 + x^3y^3$

70. $16x^2y - 8x^3y^4 - 48x^2y^2$

2   Factor.

71. $x(a + b) + 2(a + b)$

72. $a(x + y) + 4(x + y)$

73. $x(b + 2) - y(b + 2)$

74. $d(z - 8) + 5(z - 8)$

75. $a(y - 4) - b(y - 4)$

76. $c(x - 6) - 7(x - 6)$

Rewrite the expression so that there is a common binomial factor.

77. $a(x - 2) - b(2 - x)$

78. $a(x - 2) + b(2 - x)$

79. $b(a - 7) + 3(7 - a)$

80. $b(a - 7) - 3(7 - a)$

81. $x(a - 2b) - y(2b - a)$

82. $x(a - 2b) + y(2b - a)$

Factor.

83. $a(x - 2) + 5(2 - x)$

84. $a(x - 7) + b(7 - x)$

85. $b(y - 3) + 3(3 - y)$

86. $c(a - 2) - b(2 - a)$

87. $a(x - y) - 2(y - x)$

88. $3(a - b) - x(b - a)$

89. $z(c + 5) - 8(5 + c)$

90. $y(ab + 4) - 7(4 + ab)$

91. $w(3x - 4) - (4 - 3x)$

92. $d(6y - 1) - (1 - 6y)$

93. $x^3 + 4x^2 + 3x + 12$

94. $x^3 - 4x^2 - 3x + 12$

95. $2y^3 + 4y^2 + 3y + 6$

96. $3y^3 - 12y^2 + y - 4$

97. $ab + 3b - 2a - 6$

98. $yz + 6z - 3y - 18$

99. $x^2a - 2x^2 - 3a + 6$

100. $x^2y + 4x^2 + 3y + 12$

101. $3ax - 3bx - 2ay + 2by$

102. $8 + 2c + 4a^2 + a^2c$

103. $x^2 - 3x + 4ax - 12a$

104. $t^2 + 4t - st - 4s$

105. $xy - 5y - 2x + 10$

106. $2y^2 - 10y + 7xy - 35x$

107. $21x^2 + 6xy - 49x - 14y$

108. $4a^2 + 5ab - 10b - 8a$

109. $2ra + a^2 - 2r - a$

110. $2ab - 3b^2 - 3b + 2a$

111. $4x^2 + 3xy - 12y - 16x$

112. $8s + 12r - 6s^2 - 9rs$

113. $10xy^2 - 15xy + 6y - 9$

114. $10a^2b - 15ab - 4a + 6$

8.1   Applying Concepts

Factor by grouping.

115. a. $2x^2 + 6x + 5x + 15$

b. $2x^2 + 5x + 6x + 15$

116. a. $3x^2 + 3xy - xy - y^2$

b. $3x^2 - xy + 3xy - y^2$

117. a. $2a^2 - 2ab - 3ab + 3b^2$

b. $2a^2 - 3ab - 2ab + 3b^2$

Compare your answers to parts (a) and (b) of Exercises 115–117 in order to answer Exercise 118.

118. Do different groupings of the terms in a polynomial affect the binomial factoring?

A whole number is a perfect number if it is the sum of all of its factors that are less than itself. For example, 6 is a perfect number because all the factors of 6 that are less than 6 are 1, 2, and 3, and $1 + 2 + 3 = 6$.

119. Find a perfect number between 20 and 30.

120. Find a perfect number between 490 and 500.

Solve.

121. Geometry In the equation $P = 2L + 2W$, what is the effect on P when the quantity $L + W$ doubles?

122. Geometry Write an expression in factored form for the shaded portion in the diagram.

a.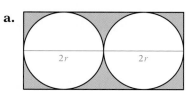

b.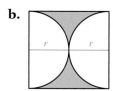

8.2 Factoring Polynomials of the Form $x^2 + bx + c$

1 ### Factor trinomials of the form $x^2 + bx + c$

VIDEO & DVD   CD TUTOR   WEB   SSM

Trinomials of the form $x^2 + bx + c$, where b and c are integers, are shown at the right.

$x^2 + 9x + 14,$ $b = 9,$ $c = 14$
$x^2 - x - 12,$ $b = -1,$ $c = -12$
$x^2 - 2x - 15,$ $b = -2,$ $c = -15$

To factor a trinomial of this form means to express the trinomial as the product of two binomials. Some trinomials expressed as the product of binomials (factored form) are shown at the right.

Trinomial	Factored Form
$x^2 + 9x + 14$	$= (x + 2)(x + 7)$
$x^2 - x - 12$	$= (x + 3)(x - 4)$
$x^2 - 2x - 15$	$= (x + 3)(x - 5)$

The method by which the factors of a trinomial are found is based on FOIL. Consider the following binomial products, noting the relationship between the constant terms of the binomials and the terms of the trinomials.

Signs in the binomials are the same

$$(x + 6)(x + 2) = x^2 + 2x + 6x + (6)(2) = x^2 + 8x + 12$$
Sum of 6 and 2
Product of 6 and 2

$$(x - 3)(x - 4) = x^2 - 4x - 3x + (-3)(-4) = x^2 - 7x + 12$$
Sum of -3 and -4
Product of -3 and -4

Signs in the binomials are opposite

$$(x + 3)(x - 5) = x^2 - 5x + 3x + (3)(-5) = x^2 - 2x - 15$$
Sum of 3 and -5
Product of 3 and -5

$$(x - 4)(x + 6) = x^2 + 6x - 4x + (-4)(6) = x^2 + 2x - 24$$
Sum of -4 and 6
Product of -4 and 6

Points to Remember in Factoring $x^2 + bx + c$

1. In the trinomial, the coefficient of x is the sum of the constant terms of the binomials.

2. In the trinomial, the constant term is the product of the constant terms of the binomials.

3. When the constant term of the trinomial is positive, the constant terms of the binomials have the same sign as the coefficient of x in the trinomial.

4. When the constant term of the trinomial is negative, the constant terms of the binomials have opposite signs.

Success at factoring a trinomial depends on remembering these four points. For example, to factor

$$x^2 - 2x - 24$$

find two numbers whose sum is -2 and whose product is -24 [Points 1 and 2]. Because the constant term of the trinomial is negative (-24), the numbers will have opposite signs [Point 4].

A systematic method of finding these numbers involves listing the factors of the constant term of the trinomial and the sum of those factors.

Factors of -24	Sum of the Factors
1, -24	$1 + (-24) = -23$
-1, 24	$-1 + 24 = 23$
2, -12	$2 + (-12) = -10$
-2, 12	$-2 + 12 = 10$
3, -8	$3 + (-8) = -5$
-3, 8	$-3 + 8 = 5$
4, -6	**$4 + (-6) = -2$**
-4, 6	$-4 + 6 = 2$

Take Note

Always check your proposed factorization to ensure accuracy.

4 and -6 are two numbers whose sum is -2 and whose product is -24. Write the binomial factors of the trinomial.

$$x^2 - 2x - 24 = (x + 4)(x - 6)$$

Check: $(x + 4)(x - 6) = x^2 - 6x + 4x - 24 = x^2 - 2x - 24$

By the Commutative Property of Multiplication, the binomial factors can also be written as

$$x^2 - 2x - 24 = (x - 6)(x + 4)$$

Example 1 Factor: $x^2 + 18x + 32$

Solution

Factors of 32	Sum
1, 32	33
2, 16	**18**
4, 8	12

$x^2 + 18x + 32 = (x + 2)(x + 16)$

• Try only positive factors of 32 [Point 3].

• Once the correct pair is found, the other factors need not be tried.

• Write the factors of the trinomial.

Check $(x + 2)(x + 16) = x^2 + 16x + 2x + 32$
$$= x^2 + 18x + 32$$

Problem 1 Factor: $x^2 - 8x + 15$

Solution See page S17.

Example 2 Factor: $x^2 - 6x - 16$

Solution

Factors of -16	Sum
1, -16	-15
-1, 16	15
2, -8	**-6**
-2, 8	6
4, -4	0

$x^2 - 6x - 16 = (x + 2)(x - 8)$

• The factors must be of opposite signs [Point 4].

• Write the factors of the trinomial.

Check $(x + 2)(x - 8) = x^2 - 8x + 2x - 16$
$$= x^2 - 6x - 16$$

Problem 2 Factor: $x^2 + 3x - 18$

Solution See page S17.

Not all trinomials can be factored when using only integers. Consider the trinomial $x^2 - 6x - 8$.

Factors of -8	Sum
1, -8	-7
-1, 8	7
2, -4	-2
-2, 4	2

Because none of the pairs of factors of -8 have a sum of -6, the trinomial is not factorable using integers. The trinomial is said to be **nonfactorable over the integers.**

2 Factor completely

A polynomial is **factored completely** when it is written as a product of factors that are nonfactorable over the integers.

Take Note

The first step in *any* factoring problem is to determine whether the terms of the polynomial have a *common factor*. If they do, factor it out first.

Example 3 Factor: $3x^3 + 15x^2 + 18x$

Solution The GCF of $3x^3$, $15x^2$, and $18x$ is $3x$.
  • Find the GCF of the terms of the polynomial.

$3x^3 + 15x^2 + 18x$
$= 3x(x^2) + 3x(5x) + 3x(6)$ • Factor out the GCF.
$= 3x(x^2 + 5x + 6)$ • Write the polynomial as a product of factors.

Factors of 6	Sum
1, 6	7
2, 3	5

  • Factor the trinomial $x^2 + 5x + 6$. Try only positive factors of 6.

$3x^3 + 15x^2 + 18x = 3x(x + 2)(x + 3)$

Check $3x(x + 2)(x + 3) = 3x(x^2 + 3x + 2x + 6)$
$$= 3x(x^2 + 5x + 6)$$
$$= 3x^3 + 15x^2 + 18x$$

Problem 3 Factor: $3a^2b - 18ab - 81b$

Solution See page S17.

Example 4 Factor: $x^2 + 9xy + 20y^2$

Solution   There is no common factor.

Factors of 20	Sum
1, 20	21
2, 10	12
4, 5	9

• Try only positive factors of 20.

Take Note

4y and 5y are placed in the binomials. This is necessary so that the middle term contains xy and the last term contains y^2.

$x^2 + 9xy + 20y^2 = (x + 4y)(x + 5y)$

Check $(x + 4y)(x + 5y) = x^2 + 5xy + 4xy + 20y^2$
$= x^2 + 9xy + 20y^2$

Problem 4 Factor: $4x^2 - 40xy + 84y^2$

Solution   See page S17.

8.2   Concept Review

Determine whether the statement is always true, sometimes true, or never true.

1. The value of b in the trinomial $x^2 - 8x + 7$ is 8.

2. The factored form of $x^2 - 5x - 14$ is $(x - 7)(x + 2)$.

3. To factor a trinomial of the form $x^2 + bx + c$ means to rewrite the polynomial as a product of two binomials.

4. In the factoring of a trinomial, if the constant term is positive, then the signs in both binomial factors will be the same.

5. In the factoring of a trinomial, if the constant term is negative, then the signs in both binomial factors will be negative.

6. The first step in factoring a trinomial is to determine whether the terms of the trinomial have a common factor.

8.2 | Exercises

1 Complete the table by listing the pairs of factors of the number and the sum of each pair.

1.

Factors of 18	Sum

2.

Factors of 12	Sum

3.

Factors of −21	Sum

4.

Factors of −10	Sum

5.

Factors of −28	Sum

6.

Factors of −32	Sum

Factor.

7. $x^2 + 3x + 2$

8. $x^2 + 5x + 6$

9. $x^2 - x - 2$

10. $x^2 + x - 6$

11. $a^2 + a - 12$

12. $a^2 - 2a - 35$

13. $a^2 - 3a + 2$

14. $a^2 - 5a + 4$

15. $a^2 + a - 2$

16. $a^2 - 2a - 3$

17. $b^2 - 6b + 9$

18. $b^2 + 8b + 16$

19. $b^2 + 7b - 8$

20. $y^2 - y - 6$

21. $y^2 + 6y - 55$

22. $z^2 - 4z - 45$

23. $y^2 - 5y + 6$

24. $y^2 - 8y + 15$

25. $z^2 - 14z + 45$

26. $z^2 - 14z + 49$

27. $z^2 - 12z - 160$

28. $p^2 + 2p - 35$

29. $p^2 + 12p + 27$

30. $p^2 - 6p + 8$

31. $x^2 + 20x + 100$

32. $x^2 + 18x + 81$

33. $b^2 + 9b + 20$

34. $b^2 + 13b + 40$

35. $x^2 - 11x - 42$

36. $x^2 + 9x - 70$

37. $b^2 - b - 20$

38. $b^2 + 3b - 40$

39. $y^2 - 14y - 51$

40. $y^2 - y - 72$

41. $p^2 - 4p - 21$

42. $p^2 + 16p + 39$

43. $y^2 - 8y + 32$

44. $y^2 - 9y + 81$

45. $x^2 - 20x + 75$

46. $p^2 + 24p + 63$

47. $x^2 - 15x + 56$

48. $x^2 + 21x + 38$

49. $x^2 + x - 56$

50. $x^2 + 5x - 36$

51. $a^2 - 21a - 72$

52. $a^2 - 7a - 44$

53. $a^2 - 15a + 36$

54. $a^2 - 21a + 54$

55. $z^2 - 9z - 136$

56. $z^2 + 14z - 147$

57. $c^2 - c - 90$

58. $c^2 - 3c - 180$

2 Factor.

59. $2x^2 + 6x + 4$

60. $3x^2 + 15x + 18$

61. $3a^2 + 3a - 18$

62. $4x^2 - 4x - 8$

63. $ab^2 + 2ab - 15a$

64. $ab^2 + 7ab - 8a$

65. $xy^2 - 5xy + 6x$

66. $xy^2 + 8xy + 15x$

67. $z^3 - 7z^2 + 12z$

68. $2a^3 + 6a^2 + 4a$

69. $3y^3 - 15y^2 + 18y$

70. $4y^3 + 12y^2 - 72y$

71. $3x^2 + 3x - 36$

72. $2x^3 - 2x^2 - 4x$

73. $5z^2 - 15z - 140$

74. $6z^2 + 12z - 90$

75. $2a^3 + 8a^2 - 64a$

76. $3a^3 - 9a^2 - 54a$

77. $x^2 - 5xy + 6y^2$

78. $x^2 + 4xy - 21y^2$

79. $a^2 - 9ab + 20b^2$

80. $a^2 - 15ab + 50b^2$

81. $x^2 - 3xy - 28y^2$

82. $s^2 + 2st - 48t^2$

83. $y^2 - 15yz - 41z^2$

84. $y^2 + 85yz + 36z^2$

85. $z^4 - 12z^3 + 35z^2$

86. $z^4 + 2z^3 - 80z^2$

87. $b^4 - 22b^3 + 120b^2$

88. $b^4 - 3b^3 - 10b^2$

89. $2y^4 - 26y^3 - 96y^2$

90. $3y^4 + 54y^3 + 135y^2$

91. $x^4 + 7x^3 - 8x^2$

92. $x^4 - 11x^3 - 12x^2$

93. $4x^2y + 20xy - 56y$

94. $3x^2y - 6xy - 45y$

95. $8y^2 - 32y + 24$

96. $10y^2 - 100y + 90$

97. $c^3 + 13c^2 + 30c$

98. $c^3 + 18c^2 - 40c$

99. $3x^3 - 36x^2 + 81x$

100. $4x^3 + 4x^2 - 24x$

101. $x^2 - 8xy + 15y^2$

102. $y^2 - 7xy - 8x^2$

103. $a^2 - 13ab + 42b^2$

104. $y^2 + 4yz - 21z^2$

105. $y^2 + 8yz + 7z^2$

106. $y^2 - 16yz + 15z^2$

107. $3x^2y + 60xy - 63y$

108. $4x^2y - 68xy - 72y$

109. $3x^3 + 3x^2 - 36x$

110. $4x^3 + 12x^2 - 160x$

111. $4z^3 + 32z^2 - 132z$

112. $5z^3 - 50z^2 - 120z$

8.2   Applying Concepts

Factor.

113. $20 + c^2 + 9c$

114. $x^2y - 54y - 3xy$

115. $45a^2 + a^2b^2 - 14a^2b$

116. $12p^2 - 96p + 3p^3$

Find all integers k such that the trinomial can be factored over the integers.

117. $x^2 + kx + 35$

118. $x^2 + kx + 18$

119. $x^2 - kx + 21$

120. $x^2 - kx + 14$

Determine the positive integer values of k for which the following polynomials are factorable over the integers.

121. $y^2 + 4y + k$

122. $z^2 + 7z + k$

123. $a^2 - 6a + k$

124. $c^2 - 7c + k$

125. $x^2 - 3x + k$

126. $y^2 + 5y + k$

127.  Exercises 121–126 included the requirement that $k > 0$. If k is allowed to be any integer, how many different values of k are possible for each polynomial? Explain your answer.

8.3 Factoring Polynomials of the Form $ax^2 + bx + c$

1 **Factor trinomials of the form $ax^2 + bx + c$ using trial factors**

Trinomials of the form $ax^2 + bx + c$, where a, b, and c are integers and $a \neq 0$, are shown at the right.

$$3x^2 - x + 4, \quad a = 3, \quad b = -1, \quad c = 4$$
$$4x^2 + 5x - 8, \quad a = 4, \quad b = 5, \quad c = -8$$

These trinomials differ from those in the previous section in that the coefficient of x^2 is not 1. There are various methods of factoring these trinomials. The method described in this objective is factoring trinomials using trial factors.

To **factor a trinomial of the form $ax^2 + bx + c$** means to express the polynomial as the product of two binomials. Factoring such polynomials by trial and error may require testing many trial factors. To reduce the number of trial factors, remember the following points.

> **Points to Remember in Factoring $ax^2 + bx + c$**
>
> **1.** If the terms of the trinomial do not have a common factor, then the terms of a binomial factor cannot have a common factor.
>
> **2.** When the constant term of the trinomial is positive, the constant terms of the binomials have the same sign as the coefficient of x in the trinomial.
>
> **3.** When the constant term of the trinomial is negative, the constant terms of the binomials have opposite signs.

Factor: $10x^2 - x - 3$

The terms of the trinomial do not have a common factor; therefore, a binomial factor will not have a common factor.

Because the constant term, c, of the trinomial is negative (-3), the constant terms of the binomial factors will have opposite signs [Point 3].

Find the factors of a (10) and the factors of c (-3).

Factors of 10	Factors of -3
1, 10	1, -3
2, 5	-1, 3

Using these factors, write trial factors, and use the Outer and Inner products of FOIL to check the middle term.

Trial Factors	Middle Term
$(x + 1)(10x - 3)$	$-3x + 10x = 7x$
$(x - 1)(10x + 3)$	$3x - 10x = -7x$
$(2x + 1)(5x - 3)$	$-6x + 5x = -x$
$(2x - 1)(5x + 3)$	$6x - 5x = x$
$(10x + 1)(x - 3)$	$-30x + x = -29x$
$(10x - 1)(x + 3)$	$30x - x = 29x$
$(5x + 1)(2x - 3)$	$-15x + 2x = -13x$
$(5x - 1)(2x + 3)$	$15x - 2x = 13x$

From the list of trial factors, $10x^2 - x - 3 = (2x + 1)(5x - 3)$.

Check: $(2x + 1)(5x - 3) = 10x^2 - 6x + 5x - 3$
$$= 10x^2 - x - 3$$

All the trial factors for this trinomial were listed in this example. However, once the correct binomial factors are found, it is not necessary to continue checking the remaining trial factors.

Factor: $4x^2 - 27x + 18$

The terms of the trinomial do not have a common factor; therefore, a binomial factor will not have a common factor.

Because the constant term, c, of the trinomial is positive (18), the constant terms of the binomial factors will have the same sign as the coefficient of x. Because the coefficient of x is -27, both signs will be negative [Point 2].

Find the factors of a (4) and the negative factors of c (18).

Factors of 4	Factors of 18
1, 4	-1, -18
2, 2	-2, -9
	-3, -6

Using these factors, write trial factors, and use the Outer and Inner products of FOIL to check the middle term.

Trial Factors	Middle Term
$(x - 1)(4x - 18)$	Common factor
$(x - 2)(4x - 9)$	$-9x - 8x = -17x$
$(x - 3)(4x - 6)$	Common factor
$(2x - 1)(2x - 18)$	Common factor
$(2x - 2)(2x - 9)$	Common factor
$(2x - 3)(2x - 6)$	Common factor
$(4x - 1)(x - 18)$	$-72x - x = -73x$
$(4x - 2)(x - 9)$	Common factor
$(4x - 3)(x - 6)$	$-24x - 3x = -27x$

The correct factors have been found.

$$4x^2 - 27x + 18 = (4x - 3)(x - 6)$$

This last example illustrates that many of the trial factors may have common factors and thus need not be tried. For the remainder of this chapter, the trial factors with a common factor will not be listed.

Example 1 Factor: $3x^2 + 20x + 12$

Solution

Factors of 3	Factors of 12
1, 3	1, 12
	2, 6
	3, 4

• Because 20 is positive, only the positive factors of 12 need be tried.

Trial Factors	Middle Term
$(x + 3)(3x + 4)$	$4x + 9x = 13x$
$(3x + 1)(x + 12)$	$36x + x = 37x$
$(3x + 2)(x + 6)$	$18x + 2x = 20x$

• Write the trial factors. Use FOIL to check the middle term.

$$3x^2 + 20x + 12 = (3x + 2)(x + 6)$$

Check $(3x + 2)(x + 6) = 3x^2 + 18x + 2x + 12$
$$= 3x^2 + 20x + 12$$

Problem 1 Factor: $6x^2 - 11x + 5$

Solution   See page S17.

Example 2 Factor: $6x^2 - 5x - 6$

Solution

Factors of 6	Factors of −6
1, 6	1, −6
2, 3	−1, 6
	2, −3
	−2, 3

• Find the factors of a (6) and the factors of c (−6).

Trial Factors	Middle Term
$(x - 6)(6x + 1)$	$x - 36x = -35x$
$(x + 6)(6x - 1)$	$-x + 36x = 35x$
$(2x - 3)(3x + 2)$	$4x - 9x = -5x$

• Write the trial factors. Use FOIL to check the middle term.

$$6x^2 - 5x - 6 = (2x - 3)(3x + 2)$$

Check $(2x - 3)(3x + 2) = 6x^2 + 4x - 9x - 6$
$$= 6x^2 - 5x - 6$$

Problem 2 Factor: $8x^2 + 14x - 15$

Solution   See page S18.

Example 3 Factor: $15 - 2x - x^2$

Solution

Factors of 15	Factors of -1
1, 15	1, -1
3, 5	

- The terms have no common factors. The coefficient of x^2 is -1.

Trial Factors	Middle Term
$(1 + x)(15 - x)$	$-x + 15x = 14x$
$(1 - x)(15 + x)$	$x - 15x = -14x$
$(3 + x)(5 - x)$	$-3x + 5x = 2x$
$(3 - x)(5 + x)$	$3x - 5x = -2x$

- Write the trial factors. Use FOIL to check the middle term.

$$15 - 2x - x^2 = (3 - x)(5 + x)$$

Check $(3 - x)(5 + x) = 15 + 3x - 5x - x^2$
$$= 15 - 2x - x^2$$

Problem 3 Factor: $24 - 2y - y^2$

Solution   See page S18.

The first step in factoring a trinomial is to determine whether there is a common factor. If there is a common factor, factor out the GCF of the terms.

Example 4 Factor: $3x^3 - 23x^2 + 14x$

Solution   The GCF of $3x^3$, $23x^2$, and $14x$ is x.

- Find the GCF of the terms of the polynomial.

$$3x^3 - 23x^2 + 14x = x(3x^2 - 23x + 14)$$

- Factor out the GCF.

Factors of 3	Factors of 14
1, 3	$-1, -14$
	$-2, -7$

- Factor the trinomial $3x^2 - 23x + 14$.

Trial Factors	Middle Term
$(x - 1)(3x - 14)$	$-14x - 3x = -17x$
$(x - 14)(3x - 1)$	$-x - 42x = -43x$
$(x - 2)(3x - 7)$	$-7x - 6x = -13x$
$(x - 7)(3x - 2)$	$-2x - 21x = -23x$

$$3x^3 - 23x^2 + 14x = x(x - 7)(3x - 2)$$

Check $x(x - 7)(3x - 2) = x(3x^2 - 2x - 21x + 14)$
$$= x(3x^2 - 23x + 14)$$
$$= 3x^3 - 23x^2 + 14x$$

Problem 4 Factor: $4a^2b^2 - 30a^2b + 14a^2$

Solution   See page S18.

2 **Factor trinomials of the form $ax^2 + bx + c$ by grouping**

Take Note

In this objective, we are using the skills taught in Objective 2 of Section 1 of this chapter. You may want to review that material before studying this objective.

In the previous objective, trinomials of the form $ax^2 + bx + c$ were factored using trial factors. In this objective, factoring by grouping is used.

To factor $ax^2 + bx + c$, first find two factors of $a \cdot c$ whose sum is b. Use the two factors to rewrite the middle term of the trinomial as the sum of two terms. Then use factoring by grouping to write the factorization of the trinomial.

Factor: $2x^2 + 13x + 15$

$a = 2, c = 15, a \cdot c = 2 \cdot 15 = 30$

Find two positive factors of 30 whose sum is 13.

Positive Factors of 30	Sum
1, 30	31
2, 15	17
3, 10	13
5, 6	11

The factors are 3 and 10.

Use the factors 3 and 10 to rewrite $13x$ as $3x + 10x$.

$$2x^2 + 13x + 15$$
$$= 2x^2 + 3x + 10x + 15$$

Factor by grouping.

$$= (2x^2 + 3x) + (10x + 15)$$
$$= x(2x + 3) + 5(2x + 3)$$
$$= (2x + 3)(x + 5)$$

Check: $(2x + 3)(x + 5) = 2x^2 + 10x + 3x + 15$
$$= 2x^2 + 13x + 15$$

Factor: $6x^2 - 11x - 10$

$a = 6, c = -10, a \cdot c = 6(-10) = -60$

Find two factors of -60 whose sum is -11.

Factors of -60	Sum
1, -60	-59
-1, 60	59
2, -30	-28
-2, 30	28
3, -20	-17
-3, 20	17
4, -15	-11

The required sum has been found. The remaining factors need not be checked. The factors are 4 and -15.

Use the factors 4 and -15 to rewrite $-11x$ as $4x - 15x$.

$$6x^2 - 11x - 10$$
$$= 6x^2 + 4x - 15x - 10$$

Factor by grouping.

Note: $-15x - 10 = -(15x + 10)$

$$= (6x^2 + 4x) - (15x + 10)$$
$$= 2x(3x + 2) - 5(3x + 2)$$
$$= (3x + 2)(2x - 5)$$

Check: $(3x + 2)(2x - 5) = 6x^2 - 15x + 4x - 10$
$$= 6x^2 - 11x - 10$$

Factor: $3x^2 - 2x - 4$

$a = 3, c = -4, a \cdot c = 3(-4) = -12$

Find two factors of -12 whose sum is -2.

Factors of -12	Sum
1, -12	-11
-1, 12	11
2, -6	-4
-2, 6	4
3, -4	-1
-3, 4	1

No integer factors of -12 have a sum of -2. Therefore, $3x^2 - 2x - 4$ is nonfactorable over the integers.

Example 5   Factor: $2x^2 + 19x - 10$

Solution $a \cdot c = 2(-10) = -20$ • Find $a \cdot c$.

$-1(20) = -20$ • Find two numbers whose product is
$-1 + 20 = 19$ -20 and whose sum is 19.

$2x^2 + 19x - 10$
$= 2x^2 - x + 20x - 10$ • Rewrite $19x$ as $-x + 20x$.
$= (2x^2 - x) + (20x - 10)$ • Factor by grouping.
$= x(2x - 1) + 10(2x - 1)$
$= (2x - 1)(x + 10)$

Problem 5   Factor: $2a^2 + 13a - 7$

Solution   See page S18.

Example 6   Factor: $8y^2 - 10y - 3$

Solution $a \cdot c = 8(-3) = -24$ • Find $a \cdot c$.

$2(-12) = -24$ • Find two numbers whose product is
$2 + (-12) = -10$ -24 and whose sum is -10.

$8y^2 - 10y - 3$
$= 8y^2 + 2y - 12y - 3$ • Rewrite $-10y$ as $2y - 12y$.
$= (8y^2 + 2y) - (12y + 3)$ • Factor by grouping.
$= 2y(4y + 1) - 3(4y + 1)$
$= (4y + 1)(2y - 3)$

Problem 6   Factor: $4a^2 - 11a - 3$

Solution   See page S18.

Remember that the first step in factoring a trinomial is to determine whether there is a common factor. If there is a common factor, factor out the GCF of the terms.

Example 7 Factor: $24x^2y - 76xy + 40y$

Solution

$24x^2y - 76xy + 40y$
$= 4y(6x^2 - 19x + 10)$

• The terms of the polynomial have a common factor, $4y$. Factor out the GCF.

$a \cdot c = 6(10) = 60$

• To factor $6x^2 - 19x + 10$, first find $a \cdot c$.

$-4(-15) = 60$
$-4 + (-15) = -19$

• Find two numbers whose product is 60 and whose sum is -19.

$6x^2 - 19x + 10$
$= 6x^2 - 4x - 15x + 10$
$= (6x^2 - 4x) - (15x - 10)$
$= 2x(3x - 2) - 5(3x - 2)$
$= (3x - 2)(2x - 5)$

• Rewrite $-19x$ as $-4x - 15x$.
• Factor by grouping.

$24x^2y - 76xy + 40y$
$= 4y(6x^2 - 19x + 10)$
$= 4y(3x - 2)(2x - 5)$

• Write the complete factorization of the given polynomial.

Problem 7 Factor: $15x^3 + 40x^2 - 80x$

Solution See page S18.

8.3 Concept Review

Determine whether the statement is always true, sometimes true, or never true.

1. The value of a in the trinomial $3x^2 + 8x - 7$ is -7.

2. The factored form of $2x^2 - 7x - 15$ is $(2x + 3)(x - 5)$.

3. To check the factorization of a trinomial of the form $ax^2 + bx + c$, multiply the two binomial factors.

4. The terms of the binomial $3x - 9$ have a common factor.

5. If the terms of a trinomial of the form $ax^2 + bx + c$ do not have a common factor, then one of its binomial factors cannot have a common factor.

6. To factor a trinomial of the form $ax^2 + bx + c$ by grouping, first find two numbers whose product is ac and whose sum is b.

8.3 Exercises

1 Factor by the method of using trial factors.

1. $2x^2 + 3x + 1$

2. $5x^2 + 6x + 1$

3. $2y^2 + 7y + 3$

4. $3y^2 + 7y + 2$

5. $2a^2 - 3a + 1$

6. $3a^2 - 4a + 1$

7. $2b^2 - 11b + 5$

8. $3b^2 - 13b + 4$

9. $2x^2 + x - 1$

10. $4x^2 - 3x - 1$

11. $2x^2 - 5x - 3$

12. $3x^2 + 5x - 2$

13. $6z^2 - 7z + 3$

14. $9z^2 + 3z + 2$

15. $6t^2 - 11t + 4$

16. $10t^2 + 11t + 3$

17. $8x^2 + 33x + 4$

18. $7x^2 + 50x + 7$

19. $3b^2 - 16b + 16$

20. $6b^2 - 19b + 15$

21. $2z^2 - 27z - 14$

22. $4z^2 + 5z - 6$

23. $3p^2 + 22p - 16$

24. $7p^2 + 19p + 10$

25. $6x^2 - 17x + 12$

26. $15x^2 - 19x + 6$

27. $5b^2 + 33b - 14$

28. $8x^2 - 30x + 25$

29. $6a^2 + 7a - 24$

30. $14a^2 + 15a - 9$

31. $18t^2 - 9t - 5$

32. $12t^2 + 28t - 5$

33. $15a^2 + 26a - 21$

34. $6a^2 + 23a + 21$

35. $8y^2 - 26y + 15$

36. $18y^2 - 27y + 4$

37. $8z^2 + 2z - 15$

38. $10z^2 + 3z - 4$

39. $3x^2 + 14x - 5$

40. $15a^2 - 22a + 8$

41. $12x^2 + 25x + 12$

42. $10b^2 + 43b - 9$

43. $3z^2 + 95z + 10$

44. $8z^2 - 36z + 1$

45. $3x^2 + xy - 2y^2$

46. $6x^2 + 10xy + 4y^2$

47. $28 + 3z - z^2$

48. $15 - 2z - z^2$

49. $8 - 7x - x^2$

50. $12 + 11x - x^2$

51. $9x^2 + 33x - 60$

52. $16x^2 - 16x - 12$

53. $24x^2 - 52x + 24$

54. $60x^2 + 95x + 20$

55. $35a^4 + 9a^3 - 2a^2$

56. $15a^4 + 26a^3 + 7a^2$

57. $15b^2 - 115b + 70$

58. $25b^2 + 35b - 30$

59. $10x^3 + 12x^2 + 2x$

60. $9x^3 - 39x^2 + 12x$

61. $10y^3 - 44y^2 + 16y$

62. $14y^3 + 94y^2 - 28y$

63. $4yz^3 + 5yz^2 - 6yz$

64. $2yz^3 - 17yz^2 + 8yz$

65. $20b^4 + 41b^3 + 20b^2$

66. $6b^4 - 13b^3 + 6b^2$

67. $9x^3y + 12x^2y + 4xy$

68. $9a^3b - 9a^2b^2 - 10ab^3$

2 Factor by grouping.

69. $2t^2 - t - 10$

70. $2t^2 + 5t - 12$

71. $3p^2 - 16p + 5$

72. $6p^2 + 5p + 1$

73. $12y^2 - 7y + 1$

74. $6y^2 - 5y + 1$

75. $5x^2 - 62x - 7$

76. $9x^2 - 13x - 4$

77. $12y^2 + 19y + 5$

78. $5y^2 - 22y + 8$

79. $7a^2 + 47a - 14$

80. $11a^2 - 54a - 5$

81. $4z^2 + 11z + 6$

82. $6z^2 - 25z + 14$

83. $22p^2 + 51p - 10$

84. $14p^2 - 41p + 15$

85. $8y^2 + 17y + 9$

86. $12y^2 - 145y + 12$

87. $6b^2 - 13b + 6$

88. $20b^2 + 37b + 15$

89. $33b^2 + 34b - 35$

90. $15b^2 - 43b + 22$

91. $18y^2 - 39y + 20$

92. $24y^2 + 41y + 12$

93. $15x^2 - 82x + 24$

94. $13z^2 + 49z - 8$

95. $10z^2 - 29z + 10$

96. $15z^2 - 44z + 32$

97. $36z^2 + 72z + 35$

98. $16z^2 + 8z - 35$

99. $14y^2 - 29y + 12$

100. $8y^2 + 30y + 25$

101. $6x^2 + 35x - 6$ **102.** $4x^2 + 6x + 2$ **103.** $12x^2 + 33x - 9$ **104.** $15y^2 - 50y + 35$

105. $30y^2 + 10y - 20$ **106.** $2x^3 - 11x^2 + 5x$ **107.** $2x^3 - 3x^2 - 5x$ **108.** $3a^2 + 5ab - 2b^2$

109. $2a^2 - 9ab + 9b^2$ **110.** $4y^2 - 11yz + 6z^2$ **111.** $2y^2 + 7yz + 5z^2$ **112.** $12 - x - x^2$

113. $18 + 17x - x^2$ **114.** $21 - 20x - x^2$ **115.** $360y^2 + 4y - 4$ **116.** $10t^2 - 5t - 50$

117. $16t^2 + 40t - 96$ **118.** $3p^3 - 16p^2 + 5p$ **119.** $6p^3 + 5p^2 + p$ **120.** $26z^2 + 98z - 24$

121. $30z^2 - 87z + 30$ **122.** $12a^3 + 14a^2 - 48a$ **123.** $42a^3 + 45a^2 - 27a$ **124.** $36p^2 - 9p^3 - p^4$

125. $9x^2y - 30xy^2 + 25y^3$ **126.** $8x^2y - 38xy^2 + 35y^3$

127. $9x^3y - 24x^2y^2 + 16xy^3$ **128.** $45a^3b - 78a^2b^2 + 24ab^3$

8.3   Applying Concepts

Factor.

129. $6y + 8y^3 - 26y^2$ **130.** $22p^2 - 3p^3 + 16p$ **131.** $a^3b - 24ab - 2a^2b$

132. $3xy^2 - 14xy + 2xy^3$ **133.** $25t^2 + 60t - 10t^3$ **134.** $3xy^3 + 2x^3y - 7x^2y^2$

Factor.

135. $2(y + 2)^2 - (y + 2) - 3$ **136.** $3(a + 2)^2 - (a + 2) - 4$

137. $10(x + 1)^2 - 11(x + 1) - 6$ **138.** $4(y - 1)^2 - 7(y - 1) - 2$

Find all integers k such that the trinomial can be factored over the integers.

139. $2x^2 + kx + 3$ **140.** $2x^2 + kx - 3$

141. $3x^2 + kx + 2$ **142.** $3x^2 + kx - 2$

143. $2x^2 + kx + 5$ **144.** $2x^2 + kx - 5$

145. Given that $x + 2$ is a factor of $x^3 - 2x^2 - 5x + 6$, factor $x^3 - 2x^2 - 5x + 6$ completely.

146. In your own words, explain how the signs of the last terms of the two binomial factors of a trinomial are determined.

147. **Geometry** The area of a rectangle is $(3x^2 + x - 2)$ ft^2. Find the dimensions of the rectangle in terms of the variable x. Given that $x > 0$, specify the dimension that is the length and the dimension that is the width. Can x be negative? Can $x = 0$? Explain your answers.

$A = 3x^2 + x - 2$

8.4 | **Special Factoring**

1 **Factor the difference of two squares and perfect-square trinomials**

Recall from Objective 4 in Section 3 of the chapter on Polynomials that the product of the sum and difference of the same two terms equals the square of the first term minus the square of the second term.

$$(a + b)(a - b) = a^2 - b^2$$

The expression $a^2 - b^2$ is the **difference of two squares.** The pattern above suggests the following rule for factoring the difference of two squares.

> **Rule for Factoring the Difference of Two Squares**
>
> | Difference of Two Squares | | Sum and Difference of Two Terms |
> |:---:|:---:|:---:|
> | $a^2 - b^2$ | $=$ | $(a + b)(a - b)$ |

$a^2 + b^2$ is the *sum* of two squares. It is nonfactorable over the integers.

Take Note

Convince yourself that the sum of two squares is nonfactorable over the integers by trying to factor $x^2 + 4$.

Example 1   Factor.   **A.** $x^2 - 16$ **B.** $x^2 - 10$ **C.** $z^6 - 25$

Solution   **A.** $x^2 - 16 = x^2 - 4^2$ • Write $x^2 - 16$ as the difference of two squares.

$= (x + 4)(x - 4)$ • The factors are the sum and difference of the terms x and 4.

B. $x^2 - 10$ is nonfactorable over the integers.   • Because 10 is not the square of an integer, $x^2 - 10$ cannot be written as the difference of two squares.

C. $z^6 - 25 = (z^3)^2 - 5^2$ • Write $z^6 - 25$ as the difference of two squares.

$= (z^3 + 5)(z^3 - 5)$ • The factors are the sum and difference of the terms z^3 and 5.

Problem 1   Factor.   **A.** $25a^2 - b^2$ **B.** $6x^2 - 1$ **C.** $n^8 - 36$

Solution   See page S18.

Example 2   Factor: $z^4 - 16$

Solution $z^4 - 16 = (z^2)^2 - (4)^2$ • This is the difference of two squares.

$= (z^2 + 4)(z^2 - 4)$ • The factors are the sum and difference of the terms z^2 and 4.

$= (z^2 + 4)(z + 2)(z - 2)$ • Factor $z^2 - 4$, which is the difference of two squares. $z^2 + 4$ is nonfactorable over the integers.

Problem 2   Factor: $n^4 - 81$

Solution   See page S18.

Recall from Objective 4 in Section 3 in the chapter on Polynomials the pattern for finding the square of a binomial.

$$(a + b)^2 = (a + b)(a + b) = a^2 + ab + ab + b^2$$
$$= a^2 + 2ab + b^2$$

Square the first term ─────────────┘
Twice the product of the two terms ─────┘
Square of last term ──────────────┘

The square of a binomial is a **perfect-square trinomial.** The pattern above suggests the following rule for factoring a perfect-square trinomial.

Rule for Factoring a Perfect-Square Trinomial		
Perfect-Square Trinomial		**Square of a Binomial**
$a^2 + 2ab + b^2$ =	$(a + b)(a + b)$ =	$(a + b)^2$
$a^2 - 2ab + b^2$ =	$(a - b)(a - b)$ =	$(a - b)^2$

Note in these patterns that the sign in the binomial is the sign of the middle term of the trinomial.

Factor: $4x^2 - 20x + 25$

Check that the first term and the last term are squares.

$4x^2 = (2x)^2, \ 25 = 5^2$

Use the squared terms to factor the trinomial as the square of a binomial. The sign of the binomial is the sign of the middle term of the trinomial.

$(2x - 5)^2$

Check the factorization.

$(2x - 5)^2$
$= (2x)^2 + 2(2x)(-5) + (-5)^2$
$= 4x^2 - 20x + 25$

The factorization is correct.

$4x^2 - 20x + 25 = (2x - 5)^2$

Factor: $9x^2 + 30x + 16$

Check that the first term and the last term are squares.

$9x^2 = (3x)^2, \ 16 = 4^2$

Use the squared terms to factor the trinomial as the square of a binomial. The sign of the binomial is the sign of the middle term of the trinomial.

$(3x + 4)^2$

Check the factorization.

$(3x + 4)^2$
$= (3x)^2 + 2(3x)(4) + 4^2$
$= 9x^2 + 24x + 16$

$9x^2 + 24x + 16 \neq 9x^2 + 30x + 16$

The proposed factorization is not correct.

In this case, the polynomial is not a perfect-square trinomial. It may, however, still factor. In fact, $9x^2 + 30x + 16 = (3x + 2)(3x + 8)$. If a trinomial does not check as a perfect-square trinomial, try to factor it by another method.

A perfect-square trinomial can always be factored using either of the methods presented in Section 3 of this chapter. However, noticing that a trinomial is a perfect-square trinomial can save you a considerable amount of time.

Example 3   Factor.   **A.** $9x^2 - 30x + 25$ **B.** $4x^2 + 37x + 9$

Solution   **A.** $9x^2 = (3x)^2$, $25 = 5^2$ • Check that the first and last terms are squares.

$(3x - 5)^2$ • Use the squared terms to factor the trinomial as the square of a binomial.

$(3x - 5)^2$
$= (3x)^2 + 2(3x)(-5) + (-5)^2$ • Check the factorization.
$= 9x^2 - 30x + 25$ • The factorization checks.
$9x^2 - 30x + 25 = (3x - 5)^2$

B. $4x^2 = (2x)^2$, $9 = 3^2$ • Check that the first and last terms are squares.

$(2x + 3)^2$ • Use the squared terms to factor the trinomial as the square of a binomial.

$(2x + 3)^2$
$= (2x)^2 + 2(2x)(3) + 3^2$ • Check the factorization.
$= 4x^2 + 12x + 9$ • The factorization does not check.

$4x^2 + 37x + 9$ • Use another method to factor the trinomial.
$= (4x + 1)(x + 9)$

Problem 3   Factor.   **A.** $16y^2 + 8y + 1$ **B.** $x^2 + 14x + 36$

Solution   See page S19.

2 Factor completely

When factoring a polynomial completely, ask yourself the following questions about the polynomial.

1. Is there a common factor? If so, factor out the common factor.
2. Is the polynomial the difference of two squares? If so, factor.
3. Is the polynomial a perfect-square trinomial? If so, factor.
4. Is the polynomial a trinomial that is the product of two binomials? If so, factor.
5. Does the polynomial contain four terms? If so, try factoring by grouping.
6. Is each binomial factor nonfactorable over the integers? If not, factor.

Take Note

Remember that you may have to factor more than once in order to write the polynomial as a product of factors, each of which is nonfactorable over the integers.

Example 4   Factor.   **A.** $3x^2 - 48$ **B.** $x^3 - 3x^2 - 4x + 12$
C. $4x^2y^2 + 12xy^2 + 9y^2$

Solution   **A.** $3x^2 - 48$
$= 3(x^2 - 16)$
$= 3(x + 4)(x - 4)$

• The GCF of the terms is 3. Factor out the common factor.
• Factor the difference of two squares.

B. $x^3 - 3x^2 - 4x + 12$
$= (x^3 - 3x^2) - (4x - 12)$
$= x^2(x - 3) - 4(x - 3)$
$= (x - 3)(x^2 - 4)$
$= (x - 3)(x + 2)(x - 2)$

• The polynomial contains four terms. Factor by grouping.

• Factor the difference of two squares.

C. $4x^2y^2 + 12xy^2 + 9y^2$
$= y^2(4x^2 + 12x + 9)$
$= y^2(2x + 3)^2$

• The GCF of the terms is y^2. Factor out the common factor.
• Factor the perfect-square trinomial.

Problem 4   Factor.   **A.** $12x^3 - 75x$ **B.** $a^2b - 7a^2 - b + 7$
C. $4x^3 + 28x^2 - 120x$

Solution   See page S19.

8.4   Concept Review

Determine whether the statement is always true, sometimes true, or never true.

1. The expression $x^2 - 12$ is an example of the difference of two squares.

2. The expression $(y + 8)(y - 8)$ is the product of the sum and difference of the same two terms. The two terms are y and 8.

3. A binomial is factorable.

4. A trinomial is factorable.

5. If a binomial is multiplied times itself, the result is a perfect-square trinomial.

6. In a perfect-square trinomial, the first and last terms are perfect squares.

7. If a polynomial contains four terms, try to factor it as a perfect-square trinomial.

8. The expression $x^2 + 9$ is the sum of two squares. It factors as $(x + 3)(x + 3)$.

8.4   **Exercises**

1 **1.** Provide an example of each of the following.
   a. the difference of two squares
   b. the product of the sum and difference of two terms
   c. a perfect-square trinomial
   d. the square of a binomial
   e. the sum of two squares

2. 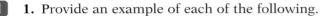   Explain the rule for factoring
   a. the difference of two squares.
   b. a perfect-square trinomial.

Which of the expressions are perfect squares?

3. 4; 8; $25x^6$; $12y^{10}$; $100x^4y^4$

4. 9; 18; $15a^8$; $49b^{12}$; $64a^{16}b^2$

Name the square root of the expression.

5. $16z^8$

6. $36d^{10}$

7. $81a^4y^6$

8. $25m^2n^{12}$

Factor.

9. $x^2 - 4$

10. $x^2 - 9$

11. $a^2 - 81$

12. $a^2 - 49$

13. $4x^2 - 1$

14. $9x^2 - 16$

15. $y^2 + 2y + 1$

16. $y^2 + 14y + 49$

17. $a^2 - 2a + 1$

18. $x^2 + 8x - 16$

19. $z^2 - 18z - 81$

20. $x^2 - 12x + 36$

21. $x^6 - 9$

22. $y^{12} - 121$

23. $25x^2 - 1$

24. $9x^2 - 1$

25. $1 - 49x^2$

26. $1 - 64x^2$

27. $x^2 + 2xy + y^2$

28. $x^2 + 6xy + 9y^2$

29. $4a^2 + 4a + 1$

30. $25x^2 + 10x + 1$

31. $64a^2 - 16a + 1$

32. $9a^2 + 6a + 1$

33. $t^2 + 36$

34. $x^2 + 64$

35. $x^4 - y^2$

36. $b^4 - 16a^2$

37. $9x^2 - 16y^2$

38. $25z^2 - y^2$

39. $16b^2 + 8b + 1$

40. $4a^2 - 20a + 25$

41. $4b^2 + 28b + 49$

42. $9a^2 - 42a + 49$

43. $25a^2 + 30ab + 9b^2$

44. $4a^2 - 12ab + 9b^2$

45. $x^2y^2 - 4$

46. $a^2b^2 - 25$

47. $16 - x^2y^2$

48. $49x^2 + 28xy + 4y^2$

49. $4y^2 - 36yz + 81z^2$

50. $64y^2 - 48yz + 9z^2$

51. $9a^2b^2 - 6ab + 1$

52. $16x^2y^2 - 24xy + 9$

53. $m^4 - 256$

54. $81 - t^4$

55. $9x^2 + 13x + 4$

56. $x^2 + 10x + 16$

57. $y^8 - 81$

58. $9 + 24a + 16a^2$

2 Factor.

59. $2x^2 - 18$

60. $y^3 - 10y^2 + 25y$

61. $x^4 + 2x^3 - 35x^2$

62. $a^4 - 11a^3 + 24a^2$

63. $5b^2 + 75b + 180$

64. $6y^2 - 48y + 72$

65. $3a^2 + 36a + 10$

66. $5a^2 - 30a + 4$

67. $2x^2y + 16xy - 66y$

68. $3a^2b + 21ab - 54b$

69. $x^3 - 6x^2 - 5x$

70. $b^3 - 8b^2 - 7b$

71. $3y^2 - 36$

72. $3y^2 - 147$

73. $20a^2 + 12a + 1$

74. $12a^2 - 36a + 27$

75. $x^2y^2 - 7xy^2 - 8y^2$

76. $a^2b^2 + 3a^2b - 88a^2$

77. $10a^2 - 5ab - 15b^2$

78. $16x^2 - 32xy + 12y^2$

79. $50 - 2x^2$

80. $72 - 2x^2$

81. $12a^3b - a^2b^2 - ab^3$

82. $2x^3y - 7x^2y^2 + 6xy^3$

83. $2ax - 2a + 2bx - 2b$

84. $4ax - 12a - 2bx + 6b$

85. $12a^3 - 12a^2 + 3a$

86. $18a^3 + 24a^2 + 8a$

87. $243 + 3a^2$

88. $75 + 27y^2$

89. $12a^3 - 46a^2 + 40a$

90. $24x^3 - 66x^2 + 15x$

91. $x^3 - 2x^2 - x + 2$

92. $ay^2 - by^2 - a + b$

93. $4a^3 + 20a^2 + 25a$

94. $2a^3 - 8a^2b + 8ab^2$

95. $27a^2b - 18ab + 3b$

96. $a^2b^2 - 6ab^2 + 9b^2$

97. $48 - 12x - 6x^2$

98. $21x^2 - 11x^3 - 2x^4$

99. $ax^2 - 4a + bx^2 - 4b$

100. $a^2x - b^2x - a^2y + b^2y$

101. $x^4 - x^2y^2$

102. $b^4 - a^2b^2$

103. $18a^3 + 24a^2 + 8a$

104. $32xy^2 - 48xy + 18x$

105. $2b + ab - 6a^2b$

106. $20x - 11xy - 3xy^2$

107. $4x - 20 - x^3 + 5x^2$

108. $ay^2 - by^2 - 9a + 9b$

109. $72xy^2 + 48xy + 8x$

110. $4x^2y + 8xy + 4y$

111. $15y^2 - 2xy^2 - x^2y^2$

112. $4x^4 - 38x^3 + 48x^2$

113. $y^3 - 9y$

114. $a^4 - 16$

115. $2x^4y^2 - 2x^2y^2$

116. $6x^5y - 6xy^5$

117. $x^9 - x^5$

118. $8b^5 - 2b^3$

119. $24x^3y + 14x^2y - 20xy$

120. $12x^3y - 60x^2y + 63xy$

121. $4x^4y^2 - 20x^3y^2 + 25x^2y^2$

122. $9x^4y^2 + 24x^3y^2 + 16x^2y^2$

123. $x^3 - 2x^2 - 4x + 8$

124. $24x^2y + 6x^3y - 45x^4y$

125. $8xy^2 - 20x^2y^2 + 12x^3y^2$

126. $45y^2 - 42y^3 - 24y^4$

127. $36a^3b - 62a^2b^2 + 12ab^3$

128. $18a^3b + 57a^2b^2 + 30ab^3$

129. $5x^2y^2 - 11x^3y^2 - 12x^4y^2$

130. $24x^2y^2 - 32x^3y^2 + 10x^4y^2$

131. $(4x - 3)^2 - y^2$

132. $(2a + 3)^2 - 25b^2$

133. $(x^2 - 4x + 4) - y^2$

134. $(4x^2 + 12x + 9) - 4y^2$

8.4   Applying Concepts

Find all integers k such that the trinomial is a perfect-square trinomial.

135. $4x^2 - kx + 9$

136. $25x^2 - kx + 1$

137. $36x^2 + kxy + y^2$

138. $64x^2 + kxy + y^2$

139. $x^2 + 6x + k$

140. $x^2 - 4x + k$

141. $x^2 - 2x + k$

142. $x^2 + 10x + k$

143. Number Problems The prime factorization of a number is $2^3 \cdot 3^2$. How many of its whole-number factors are perfect squares?

144. Number Problems The product of two numbers is 48. One of the two numbers is a perfect square. The other is a prime number. Find the sum of the two numbers.

145. Number Problems What is the smallest whole number by which 300 can be multiplied so that the product will be a perfect square?

146. Geometry The area of a square is $(16x^2 + 24x + 9)$ ft^2. Find the dimensions of the square in terms of the variable x. Can $x = 0$? What are the possible values of x?

$A = 16x^2 + 24x + 9$

The cube of an integer is a **perfect cube.** Because $2^3 = 8$, 8 is a perfect cube. Because $4^3 = 64$, 64 is a perfect cube. A variable expression can be a perfect cube; the exponents on variables of perfect cubes are multiples of 3. Therefore, x^3, x^6, and x^9 are perfect cubes. The sum and the difference of two perfect cubes are factorable. They can be written as the product of a binomial and a trinomal. Their factoring patterns are shown below.

$a^3 + b^3$ is the sum of two cubes.

$a^3 - b^3$ is the difference of two cubes.

$a^3 + b^3 = (a + b)(a^2 - ab + b^2)$
$a^3 - b^3 = (a - b)(a^2 + ab + b^2)$

To factor $x^3 - 8$, write the binomial as the difference of two perfect cubes. Use the factoring pattern shown above. Replace a with x and b with 2.

$x^3 - 8 = (x)^3 - (2)^3$
$\qquad = (x - 2)(x^2 + 2x + 4)$

Factor.

147. $x^3 + 8$

148. $y^3 + 27$

149. $y^3 - 27$

150. $x^3 - 1$

151. $y^3 + 64$

152. $x^3 - 125$

153. $8x^3 - 1$

154. $27y^3 + 1$

155. Select any odd integer greater than 1, square it, and then subtract 1. Is the result evenly divisible by 8? Prove that this procedure always produces a number divisible by 8. (*Suggestion:* Any odd integer greater than 1 can be expressed as $2n + 1$, where n is a natural number.)

8.5 Solving Equations

1 Solve equations by factoring

Recall that the Multiplication Property of Zero states that the product of a number and zero is zero.

If a is a real number, then $a \cdot 0 = 0$.

Consider the equation $a \cdot b = 0$. If this is a true equation, then either $a = 0$ or $b = 0$.

> **Principle of Zero Products**
>
> If the product of two factors is zero, then at least one of the factors must be zero.
>
> $$\text{If } a \cdot b = 0, \text{ then } a = 0 \text{ or } b = 0.$$

The Principle of Zero Products is used in solving equations.

Take Note

$x - 2$ is equal to a number. $x - 3$ is equal to a number. In $(x - 2)(x - 3)$, two numbers are being multiplied. Since their product is 0, one of the numbers must be equal to 0. The number $x - 2$ is equal to 0 or the number $x - 3$ is equal to 0.

Solve: $(x - 2)(x - 3) = 0$

If $(x - 2)(x - 3) = 0$, then $(x - 2) = 0$ or $(x - 3) = 0$.

$$(x - 2)(x - 3) = 0$$

Solve each equation for x.

$$x - 2 = 0 \qquad\qquad x - 3 = 0$$
$$x = 2 \qquad\qquad\quad x = 3$$

Check

$$(x - 2)(x - 3) = 0 \qquad (x - 2)(x - 3) = 0$$

$$\begin{array}{c|c} (2 - 2)(2 - 3) & 0 \\ 0(-1) & 0 \\ 0 & = 0 \end{array} \qquad \begin{array}{c|c} (3 - 2)(3 - 3) & 0 \\ 1(0) & 0 \\ 0 & = 0 \end{array}$$

A true equation   A true equation

Write the solutions.   The solutions are 2 and 3.

An equation of the form $ax^2 + bx + c = 0$, $a \neq 0$, is a **quadratic equation**. A quadratic equation is in **standard form** when the polynomial is in descending order and equal to zero.

$$3x^2 + 2x + 1 = 0$$

$$4x^2 - 3x + 2 = 0$$

A quadratic equation can be solved by using the Principle of Zero Products when the polynomial $ax^2 + bx + c$ is factorable.

Take Note

Note the steps involved in solving a quadratic equation by factoring:
(1) Write in standard form.
(2) Factor.
(3) Set each factor equal to 0.
(4) Solve each equation.
(5) Check.

Example 1 Solve: $2x^2 + x = 6$

Solution

$$2x^2 + x = 6$$
$$2x^2 + x - 6 = 0$$ • Write the equation in standard form.

$$(2x - 3)(x + 2) = 0$$ • Factor the trinomial.

$$2x - 3 = 0 \qquad x + 2 = 0$$ • Set each factor equal to zero (the Principle of Zero Products).

$$2x = 3 \qquad\qquad x = -2$$ • Solve each equation for x.
$$x = \frac{3}{2}$$

Check

$$\begin{array}{c|c} 2x^2 + x = 6 & 2x^2 + x = 6 \\ \hline 2\left(\dfrac{3}{2}\right)^2 + \dfrac{3}{2} \;\bigg|\; 6 & 2(-2)^2 + (-2) \;\bigg|\; 6 \\ 2\left(\dfrac{9}{4}\right) + \dfrac{3}{2} \;\bigg|\; 6 & 2\cdot 4 - 2 \;\bigg|\; 6 \\ \dfrac{9}{2} + \dfrac{3}{2} \;\bigg|\; 6 & 8 - 2 \;\bigg|\; 6 \\ 6 = 6 & 6 = 6 \end{array}$$

The solutions are $\dfrac{3}{2}$ and -2. • Write the solutions.

Problem 1 Solve: $2x^2 - 50 = 0$

Solution See page S19.

Take Note

The Principle of Zero Products cannot be used unless 0 is on one side of the equation.

Example 2 Solve: $(x - 3)(x - 10) = -10$

Solution

$$(x - 3)(x - 10) = -10$$
$$x^2 - 13x + 30 = -10$$ • Multiply $(x - 3)(x - 10)$.
$$x^2 - 13x + 40 = 0$$ • Write the equation in standard form.
$$(x - 8)(x - 5) = 0$$ • Factor.

$$x - 8 = 0 \qquad x - 5 = 0$$ • Set each factor equal to zero.
$$x = 8 \qquad\qquad x = 5$$ • Solve each equation for x.

The solutions are 8 and 5. • Write the solutions.

Problem 2 Solve: $(x + 2)(x - 7) = 52$

Solution See page S19.

2 ## Application problems

VIDEO & DVD   CD TUTOR   WEB   SSM

Example 3 The sum of the squares of two consecutive positive odd integers is equal to 130. Find the two integers.

Strategy ▶ First positive odd integer: n
Second positive odd integer: $n + 2$
Square of the first positive odd integer: n^2
Square of the second positive odd integer: $(n + 2)^2$

▶ The sum of the square of the first positive odd integer and the square of the second positive odd integer is 130.

Solution

$$n^2 + (n + 2)^2 = 130$$
$$n^2 + n^2 + 4n + 4 = 130$$
$$2n^2 + 4n - 126 = 0$$
$$2(n^2 + 2n - 63) = 0$$
$$n^2 + 2n - 63 = 0$$ • Divide each side of the equation by 2.
$$(n - 7)(n + 9) = 0$$

$$n - 7 = 0 \qquad n + 9 = 0$$ • Because −9 is not a positive odd
$$n = 7 \qquad\qquad n = -9$$ integer, it is not a solution.

$$n = 7$$ • The first positive odd integer is 7.

$$n + 2 = 7 + 2 = 9$$ • Substitute the value of *n* into the variable expression for the second positive odd integer and evaluate.

The two integers are 7 and 9.

Problem 3   The sum of the squares of two consecutive positive integers is 85. Find the two integers.

Solution   See page S19.

Example 4   A stone is thrown into a well with an initial velocity of 8 ft/s. The well is 440 ft deep. How many seconds later will the stone hit the bottom of the well? Use the equation $d = vt + 16t^2$, where d is the distance in feet, v is the initial velocity in feet per second, and t is the time in seconds.

Strategy   To find the time for the stone to drop to the bottom of the well, replace the variables d and v by their given values and solve for t.

Solution

$$d = vt + 16t^2$$
$$440 = 8t + 16t^2$$
$$0 = 16t^2 + 8t - 440$$ • Write the equation in standard
$$0 = 8(2t^2 + t - 55)$$ form.
$$0 = 2t^2 + t - 55$$ • Divide each side of the equation
$$0 = (2t + 11)(t - 5)$$ by 8.

$$2t + 11 = 0 \qquad t - 5 = 0$$
$$2t = -11 \qquad\quad t = 5$$
$$t = -\frac{11}{2}$$ • Because the time cannot be a negative number, $-\frac{11}{2}$ is not a solution.

The time is 5 s.

Problem 4   The length of a rectangle is 3 m more than twice the width. The area of the rectangle is 90 m². Find the length and width of the rectangle.

W

$2W + 3$

Solution   See page S19.

8.5 Concept Review

Determine whether the statement is always true, sometimes true, or never true.

1. If you multiply two numbers and the product is zero, then either one or both of the numbers must be zero.

2. The equation $2x^2 + 5x - 7$ is a quadratic equation.

3. The equation $4x^2 - 9 = 0$ is a quadratic equation.

4. The equation $3x + 1 = 0$ is a quadratic equation in standard form.

5. If $(x - 8)(x + 6) = 0$, then $x = -8$ or $x = 6$.

6. If a quadratic equation is not in standard form, the first step in solving the equation by factoring is to write it in standard form.

8.5 Exercises

1 **1.** What does the Principle of Zero Products state?

2. Why is it possible to solve some quadratic equations by using the Principle of Zero Products?

Solve.

3. $(y + 3)(y + 2) = 0$ **4.** $(y - 3)(y - 5) = 0$ **5.** $(z - 7)(z - 3) = 0$

6. $(z + 8)(z - 9) = 0$ **7.** $x(x - 5) = 0$ **8.** $x(x + 2) = 0$

9. $a(a - 9) = 0$ **10.** $a(a + 12) = 0$ **11.** $y(2y + 3) = 0$

12. $t(4t - 7) = 0$ **13.** $2a(3a - 2) = 0$ **14.** $4b(2b + 5) = 0$

15. $(b + 2)(b - 5) = 0$ **16.** $(b - 8)(b + 3) = 0$ **17.** $x^2 - 81 = 0$

18. $9x^2 - 1 = 0$ **19.** $16x^2 - 49 = 0$ **20.** $x^2 + 6x + 8 = 0$

21. $x^2 - 8x + 15 = 0$ **22.** $z^2 + 5z - 14 = 0$ **23.** $z^2 + z - 72 = 0$

24. $x^2 - 5x + 6 = 0$ **25.** $2y^2 - y - 1 = 0$ **26.** $2a^2 - 9a - 5 = 0$

27. $3a^2 + 14a + 8 = 0$ **28.** $2x^2 - 6x - 20 = 0$ **29.** $3y^2 + 12y - 63 = 0$

30. $a^2 - 5a = 0$ **31.** $x^2 - 7x = 0$ **32.** $2a^2 - 8a = 0$

33. $a^2 + 5a = -4$

34. $a^2 - 5a = 24$

35. $y^2 - 5y = -6$

36. $y^2 - 7y = 8$

37. $2t^2 + 7t = 4$

38. $3t^2 + t = 10$

39. $3t^2 - 13t = -4$

40. $5t^2 - 16t = -12$

41. $x(x - 12) = -27$

42. $x(x - 11) = 12$

43. $y(y - 7) = 18$

44. $y(y + 8) = -15$

45. $p(p + 3) = -2$

46. $p(p - 1) = 20$

47. $y(y + 4) = 45$

48. $y(y - 8) = -15$

49. $x(x + 3) = 28$

50. $p(p - 14) = 15$

51. $(x + 8)(x - 3) = -30$

52. $(x + 4)(x - 1) = 14$

53. $(y + 3)(y + 10) = -10$

54. $(z - 5)(z + 4) = 52$

55. $(z - 8)(z + 4) = -35$

56. $(z - 6)(z + 1) = -10$

57. $(a + 3)(a + 4) = 72$

58. $(a - 4)(a + 7) = -18$

59. $(z + 3)(z - 10) = -42$

60. $(2x + 5)(x + 1) = -1$

61. $(y + 3)(2y + 3) = 5$

62. $(y + 5)(3y - 2) = -14$

2 **63.** **Integer Problems** The square of a positive number is six more than five times the positive number. Find the number.

64. **Integer Problems** The square of a negative number is fifteen more than twice the negative number. Find the number.

65. **Integer Problems** The sum of two numbers is six. The sum of the squares of the two numbers is twenty. Find the two numbers.

66. **Integer Problems** The sum of two numbers is eight. The sum of the squares of the two numbers is thirty-four. Find the two numbers.

67. **Integer Problems** The sum of the squares of two consecutive positive integers is forty-one. Find the two integers.

68. **Integer Problems** The sum of the squares of two consecutive positive even integers is one hundred. Find the two integers.

69. **Integer Problems** The product of two consecutive positive integers is two hundred forty. Find the two integers.

70. **Integer Problems** The product of two consecutive positive even integers is one hundred sixty-eight. Find the two integers.

71. **Geometry** The length of the base of a triangle is three times the height. The area of the triangle is 54 ft². Find the base and height of the triangle.

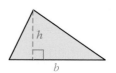

72. **Geometry** The height of a triangle is 4 m more than twice the length of the base. The area of the triangle is 35 m². Find the height of the triangle.

73. **Geometry** The length of a rectangle is two feet more than twice the width. The area is 144 ft². Find the length and width of the rectangle.

74. Geometry The width of a rectangle is 5 ft less than the length. The area of the rectangle is 176 ft². Find the length and width of the rectangle.

75. Geometry The length of each side of a square is extended 4 m. The area of the resulting square is 64 m². Find the length of a side of the original square.

76. Geometry The length of each side of a square is extended 2 cm. The area of the resulting square is 64 cm². Find the length of a side of the original square.

77. Geometry The radius of a circle is increased by 3 in., which increases the area by 100 in². Find the radius of the original circle. Round to the nearest hundredth.

78. Geometry A circle has a radius of 10 in. Find the increase in area that occurs when the radius is increased by 2 in. Round to the nearest hundredth.

79. Geometry The page of a book measures 6 in. by 9 in. A uniform border around the page leaves 28 in² for type. What are the dimensions of the type area?

80. Geometry A small garden measures 8 ft by 10 ft. A uniform border around the garden increases the total area to 143 ft². What is the width of the border?

Use the formula $d = vt + 16t^2$, where d is the distance in feet, v is the initial velocity in feet per second, and t is the time in seconds.

81. Physics An object is released from a plane at an altitude of 1600 ft. The initial velocity is 0 ft/s. How many seconds later will the object hit the ground?

82. Physics An object is released from the top of a building 320 ft high. The initial velocity is 16 ft/s. How many seconds later will the object hit the ground?

Use the formula $S = \dfrac{n^2 + n}{2}$, where S is the sum of the first n natural numbers.

83. Number Problems How many consecutive natural numbers beginning with 1 will give a sum of 78?

84. Number Problems How many consecutive natural numbers beginning with 1 will give a sum of 120?

Use the formula $N = \dfrac{t^2 - t}{2}$, where N is the number of football games that must be scheduled in a league with t teams if each team is to play every other team once.

85. **Sports** A league has 28 games scheduled. How many teams are in the league if each team plays every other team once?

86. **Sports** A league has 45 games scheduled. How many teams are in the league if each team plays every other team once?

Use the formula $h = vt - 16t^2$, where h is the height in feet an object will attain (neglecting air resistance) in t seconds and v is the initial velocity in feet per second.

87. **Sports** A baseball player hits a "Baltimore chop," meaning the ball bounces off home plate after the batter hits it. The ball leaves home plate with an initial upward velocity of 64 ft/s. How many seconds after the ball hits home plate will the ball be 64 ft above the ground?

88. **Sports** A golf ball is thrown onto a cement surface and rebounds straight up. The initial velocity of the rebound is 48 ft/s. How many seconds later will the golf ball return to the ground?

8.5   Applying Concepts

Solve.

89. $2y(y + 4) = -5(y + 3)$

90. $2y(y + 4) = 3(y + 4)$

91. $(a - 3)^2 = 36$

92. $(b + 5)^2 = 16$

93. $p^3 = 9p^2$

94. $p^3 = 7p^2$

95. $(2z - 3)(z + 5) = (z + 1)(z + 3)$

96. $(x + 3)(2x - 1) = (3 - x)(5 - 3x)$

97. Find $3n^2$ if $n(n + 5) = -4$.

98. Find $2n^3$ if $n(n + 3) = 4$.

99. **Geometry** The length of a rectangle is 7 cm, and the width is 4 cm. If both the length and the width are increased by equal amounts, the area of the rectangle is increased by 42 cm². Find the length and width of the larger rectangle.

7 cm

4 cm

100. **Geometry** A rectangular piece of cardboard is 10 in. longer than it is wide. Squares 2 in. on a side are to be cut from each corner, and then the sides will be folded up to make an open box with a volume of 192 in³. Find the length and width of the piece of cardboard.

101. ✎ Explain the error made in solving the equation at the right. Solve the equation correctly.

$$(x + 2)(x - 3) = 6$$
$$x + 2 = 6 \qquad x - 3 = 6$$
$$x = 4 \qquad\quad x = 9$$

102. ✎ Explain the error made in solving the equation at the right. Solve the equation correctly.

$$x^2 = x$$
$$\frac{x^2}{x} = \frac{x}{x}$$
$$x = 1$$

Focus on Problem Solving

Making a Table   Sometimes a table can be used to organize information so that it is in a useful form. In the chapter "Solving Equations and Inequalities," we used tables in the applications to organize the data. Tables are also useful in applications that require us to find all possible combinations in a given situation.

A basketball player scored 11 points in a game. The player can score 1 point for making a free throw, 2 points for making a field goal within the three-point line, and 3 points for making a field goal outside the three-point line. Find the possible combinations in which the player can score the 11 points.

The following table lists all the combinations of shots with which it is possible to score 11 points.

Free throws	0	2	1	3	5	0	2	4	6	8	1	3	5	7	9	11
2-point field goals	1	0	2	1	0	4	3	2	1	0	5	4	3	2	1	0
3-point field goals	3	3	2	2	2	1	1	1	1	1	0	0	0	0	0	0
Total points	11	11	11	11	11	11	11	11	11	11	11	11	11	11	11	11

There are 16 possible ways in which the basketball player could have scored 11 points.

1. A football team scores 17 points. A touchdown counts 6 points, an extra point scores 1 point, a field goal scores 3 points, and a safety scores 2 points. Find the possible combinations with which the team can score 17 points. Remember that the number of extra points cannot exceed the number of touchdowns scored.

2. Repeat Exercise 1. Assume that no safety was scored.

3. Repeat Exercise 1. Assume that no safety was scored and that the team scored two field goals.

4. Find the number of possible combinations of nickels, dimes, and quarters that one can get when receiving $.85 in change.

5. Repeat Exercise 4. Assume no combination contains coins that could be exchanged for a larger coin. That is, the combination of three quarters and two nickels would not be allowed, because the two nickels could be exchanged for a dime.

6. Find the number of possible combinations of $1, $5, $10, and $20 bills that one can get when receiving $33 in change.

The Trial-and-Error Method

A topic in Section 3 of this chapter is factoring trinomials using trial factors. This method involves writing trial factors and then using the FOIL method to determine which pair of factors is the correct one. The **trial-and-error method** of arriving at a solution to a problem involves performing repeated tests or experiments until a satisfactory conclusion is reached. However, not all problems solved using the trial-and-error method have a strategy by which to determine the answer. Here is an example:

Explain how you could cut through a cube so that the face of the resulting solid is **a.** a square, **b.** an equilateral triangle, **c.** a trapezoid, and **d.** a hexagon.

There is no formula to apply to this problem; there is no computation to perform. This problem requires picturing a cube and the results after cutting through it at different places on its surface and at different angles. For part (a), cutting perpendicular to the top and bottom of the cube and parallel to two of its sides will result in a square. The other shapes may prove more difficult.

When solving problems of this type, keep an open mind. Sometimes when using the trial-and-error method, we are hampered by narrowness of vision; we cannot expand our thinking to include other possibilities. Then when we see someone else's solution, it appears so obvious! For example, for the question above, it is necessary to conceive of cutting through the cube at places other than the top surface; we need to be open to the idea of beginning the cut at one of the corner points of the cube.

Look at the letter A printed at the left. If the letter were folded along line ℓ, the two sides of the letter would match exactly. This letter has **symmetry** with respect to line ℓ. Line ℓ is called the **axis of symmetry.** Now consider the letter H printed below at the left. Both lines ℓ_1 and ℓ_2 are axes of symmetry for this letter; the letter could be folded along either line and the two sides would match exactly. Does the letter A have more than one axis of symmetry? Find axes of symmetry for other capital letters of the alphabet. Which lower-case letters have one axis of symmetry? Do any of the lower-case letters have more than one axis of symmetry?

How many axes of symmetry does a square have? In determining lines of symmetry for a square, begin by drawing a square as shown at the right. The horizontal line of symmetry and the vertical line of symmetry may be immediately obvious to you. But there are two others. Do you see that a line drawn through opposite corners of the square is also a line of symmetry?

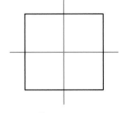

Find the number of axes of symmetry for each of the following figures: equilateral triangle, rectangle, regular pentagon, regular hexagon, and circle.

Many of the questions in this text that require an answer of "always true, sometimes true, or never true" are best solved by the trial-and-error method. For example, consider the following statement:

If two rectangles have the same area, then they have the same perimeter.

Try some numbers. Each of two rectangles, one measuring 6 units by 2 units and another measuring 4 units by 3 units, has an area of 12 square units, but the perimeter of the first is 16 units and the perimeter of the second is 14 units. Thus the answer "always true" has been eliminated. We still need to determine whether there is a case when the statement is true. After experimenting with a lot of numbers, you may come to realize that we are trying to determine whether it is possible for two different pairs of factors of a number to have the same sum. Is it?

Don't be afraid to make many experiments, and remember that *errors*, or tests that "don't work," are a part of the trial-and-*error* process.

Projects & Group Activities

Prime and Composite Numbers

A **prime number** is a natural number greater than 1 whose only natural number factors are itself and 1. The number 11 is a prime number because the only natural number factors of 11 are 11 and 1.

Eratosthenes, a Greek philosopher and astronomer who lived from 270 to 190 B.C., devised a method of identifying prime numbers. It is called the **Sieve of Eratosthenes.** The procedure is illustrated in the following paragraphs.

1	②	③	4	⑤	6	⑦	8	9	10
⑪	12	⑬	14	15	16	⑰	18	⑲	20
21	22	㉓	24	25	26	27	28	㉙	30
㉛	32	33	34	35	36	㊲	38	39	40
㊶	42	㊳	44	45	46	㊼	48	49	50
51	52	㊾	54	55	56	57	58	㊾	60
㊱	62	63	64	65	66	㊸	68	69	70
㊹	72	㊺	74	75	76	77	78	㊻	80
81	82	㊽	84	85	86	87	88	㊾	90
91	92	93	94	95	96	㊿	98	99	100

List all the natural numbers from 1 to 100. Cross out the number 1, because it is not a prime number. The number 2 is prime; circle it. Cross out all the other multiples of 2 (4, 6, 8, . . .), because they are not prime. The number 3 is prime; circle it. Cross out all the other multiples of 3 (6, 9, 12, . . .) that are not already crossed out. The number 4, the next consecutive number in the list, has already been crossed out. The number 5 is prime; circle it. Cross out all the other multiples of 5 that are not already crossed out. Continue in this manner until all the prime numbers less than 100 are circled.

A **composite number** is a natural number greater than 1 that has a natural-number factor other than itself and 1. The number 21 is a composite number because it has factors of 3 and 7. All the numbers crossed out in the table above, except the number 1, are composite numbers.

Solve.

1. Use the Sieve of Eratosthenes to find the prime numbers between 100 and 200.

2. How many prime numbers are even numbers?

3. Find the "twin primes" between 1 and 200. Twin primes are two prime numbers whose difference is 2. For instance, 3 and 5 are twin primes; 5 and 7 are also twin primes.

4. a. List two prime numbers that are consecutive natural numbers.

   b. Can there be any other pairs of prime numbers that are consecutive natural numbers?

5. Some primes are the sum of a square and 1. For example, $5 = 2^2 + 1$. Find another prime p such that $p = n^2 + 1$, where n is a natural number.

6. Find a prime p such that $p = n^2 - 1$, where n is a natural number.

7. a. $4!$ (which is read "4 factorial") is equal to $4 \cdot 3 \cdot 2 \cdot 1$. Show that $4! + 2$, $4! + 3$, and $4! + 4$ are all composite numbers.

   b. $5!$ (which is read "5 factorial") is equal to $5 \cdot 4 \cdot 3 \cdot 2 \cdot 1$. Will $5! + 2$, $5! + 3, 5! + 4$, and $5! + 5$ generate four consecutive composite numbers?

   c. Use the notation $6!$ to represent a list of five consecutive composite numbers.

Search the World Wide Web

At the address **http://www.utm.edu/research/primes/mersenne/**, you can find the history of, theorems on, and lists of Mersenne Primes. A prime number that can be written in the form $2^n - 1$, where n is also prime, is said to be a **Mersenne Prime.** In 2001, the largest known Mersenne Prime had more than 4 million digits. An interesting note is that the 25th and 26th Mersenne Primes were found by high school students Laura Nickel and Curt Noll.

Would you like to find the answers to questions such as these?

1. What are perfect numbers and how are they used?

2. Is zero a prime number, a composite number, or neither?

3. What is the difference between zero and nothing?

4. When something is divided by zero, why is the answer undefined?

5. What is infinity plus one?

6. Is infinity positive or negative?

These and many other questions are answered at the address

http://mathforum.org/dr.math/

The answers to these questions are presented in a clever and interesting fashion. The answer to the question "What is the purpose of the number zero?" is given below.

> The invention of zero was one of the most important breakthroughs in the history of civilization. More important, in my opinion, than the invention of the wheel. I think that it's a fairly deep concept.
>
> One crucial purpose that zero holds is as a placeholder in our system of notation. When we write the number 408, we're really using a shorthand notation. What we really mean by 408 is "4 times 100, plus 0 times 10, plus 8 times 1." Without the number zero, we wouldn't be able to tell the numbers 408, 48, 480, 408,000, and 4800 apart. So yes, zero is important.
>
> Another crucial role that zero plays in mathematics is that of an "additive identity element." What this means is that when you add zero to any number, you get the number that you started with. For instance, $5 + 0 = 5$. That may seem obvious and trivial, but it's actually quite important to have such a number. For instance, when you're manipulating some numerical quantity and you want to change its form but not its value, you might add some fancy version of zero to it, like this:
>
> $$x^2 + y^2 = x^2 + y^2 + 2xy - 2xy$$
> $$= x^2 + 2xy + y^2 - 2xy$$
> $$= (x + y)^2 - 2xy$$
>
> Now if we wanted to, we could use this as a proof that $(x + y)^2$ is always greater than $2xy$; the expression we started with was positive, so the one we ended up with must be positive too. Therefore, subtracting $2xy$ from $(x + y)^2$ must leave us with a positive number. Neat stuff.*

The Internet is a good source for the history of, and interesting facts about, mathematical concepts. You can use the address.

http://turnbull.dcs.st-and.ac.uk/history/HistTopics/Prime_numbers.html

to find the history of prime numbers. This Web address also contains a list of unsolved problems that involve prime numbers.

Problems Involving Falling Objects

When an object is dropped, the object accelerates due to Earth's gravitational pull. This means that the speed of the object continually increases. If we assume that there is little or no air resistance, then the height h, in feet, of an object as it drops is given by the equation

$$h = -16t^2 + h_0$$

where t is the number of seconds the object has been falling and h_0 is the initial height, in feet, from which the object was dropped.

Note that the equation above applies to an object that is *dropped*. If an object is *thrown* either up or down, then the equation must reflect the effect of the initial vertical velocity. Again assuming that there is little or no air resistance, the equation for the height h, in feet, of an object that is thrown is

$$h = -16t^2 + v_0 t + h_0$$

*Ask Dr. Math. Copyright © 1994–1997 *The Math Forum*. Used by permission.

where t is the number of seconds since the object was thrown, v_0 is the initial vertical velocity in feet per second, and h_0 is the initial height, in feet, from which the object was thrown. In this equation, **v_0 is positive for an object moving upward, negative for an object moving downward, and zero for an object that is stationary.**

Equations for Problems Involving Falling Objects

(neglecting air resistance)

For an object that is dropped: $h = -16t^2 + h_0$

For an object that is projected: $h = -16t^2 + v_0 t + h_0$

h = height, in feet
t = time in motion, in seconds
v_0 = initial vertical velocity, in feet per second
h_0 = initial height, in feet

Suppose you drop a basketball from the top of the gymnasium bleachers, a height of 16 ft, to the floor below. How long after you drop the basketball does it hit the ground?

$h = -16t^2 + h_0$	• Use the formula for an object that is dropped.
$0 = -16t^2 + 16$	• $h = 0$, the height of the basketball when it hits the ground. $h_0 = 16$, the initial height.
$0 = -16(t^2 - 1)$	• Factor -16 from $-16t^2 + 16$.
$0 = -16(t + 1)(t - 1)$	• Factor the difference of two squares.
$t + 1 = 0 \qquad t - 1 = 0$	• Use the Principle of Zero Products.
$\qquad t = -1 \qquad\quad t = 1$	• Solve each equation for t.

The time cannot be a negative number. The solution $t = -1$ is not possible.

The basketball hits the ground 1 s after you drop it.

An arrow is shot upward at an initial velocity of 144 ft/s. Find the time for the arrow to return to Earth.

$h = -16t^2 + v_0 t + h_0$	• Use the formula for an object that is projected.
$0 = -16t^2 + 144t + 0$	• $h = 0$, the height of the arrow when it returns to Earth. $v_0 = 144$, the initial velocity. (The initial velocity is positive because the object is moving upward.) $h_0 = 0$, the initial height.
$0 = -16t^2 + 144t$	
$0 = -16t(t - 9)$	• Factor $-16t$ from $-16t^2 + 144t$.
$-16t = 0 \qquad t - 9 = 0$	• Use the Principle of Zero Products.
$\qquad t = 0 \qquad\quad t = 9$	• Solve each equation for t.

The solution $t = 0$ is the time when the arrow is shot upward.

The arrow returns to Earth after 9 s.

Solve.

1. A bundle of leaflets is dropped from a plane at an altitude of 1600 ft. How many seconds later will the bundle hit the ground?

2. You drop your glasses from the top of a cliff that is 256 ft above the ocean. How many seconds later will your glasses hit the water?

3. A roofer tosses an old shingle from the roof of a building 320 ft high. The initial velocity is 16 ft/s. How many seconds later will the shingle hit the ground?

4. To test its bounce, a tennis ball is thrown onto a tennis court. It rebounds straight up with an initial velocity of 32 ft/s. How many seconds later will the tennis ball return to the ground?

5. In a slow-pitch softball game, the pitcher releases the ball at a height of 4 ft with an initial vertical velocity of 24 ft/s. The batter hits the ball when it is 4 ft off the ground. For how many seconds is the ball in the air before it is hit?

6. A basketball player shoots at a basket 25 ft away. The ball is released at a height of 8 ft with an initial vertical velocity of 33 ft/s. The height of the basket is 10 ft. How many seconds after the ball is released does it hit the basket? Explain what the solution you discarded represents.

7. An arrow is shot upward with an initial velocity of 128 ft/s. The equation $16t(t - a) = 0$, where t is the time in seconds and a is a constant, describes the times at which the arrow is on the ground. Suppose you know that the arrow is on the ground at 0 seconds and at 8 seconds. What is the value of a?

8. Suppose you double the height from which an object is dropped. Is the falling time of the object doubled? Why or why not?

9. Make up a problem involving an object that is dropped that can be solved by factoring a quadratic equation.

10. Make up a problem involving a projected object that can be solved by factoring a quadratic equation.

Chapter Summary

Key Words

The **greatest common factor** (GCF) of two or more integers is the greatest integer that is a factor of all the integers. [8.1.1, p. 386]

To **factor a polynomial** means to write the polynomial as a product of other polynomials. [8.1.1, p. 387]

To **factor a trinomial of the form** $ax^2 + bx + c$ means to express the trinomial as the product of two binomials. [8.3.1, p. 398]

A polynomial that does not factor using only integers is **nonfactorable over the integers.** [8.2.1, p. 394]

An equation of the form $ax^2 + bx + c = 0$, $a \neq 0$, is a **quadratic equation.** A quadratic equation is in **standard form** when the polynomial is in descending order and equal to zero. [8.5.1, p. 414]

Examples

The greatest common factor of 12 and 18 is 6 because 6 is the greatest integer that divides evenly into both 12 and 18.

To factor $8x + 12$ means to write it as the product $4(2x + 3)$. The expression $8x + 12$ is a sum. The expression $4(2x + 3)$ is a product (the polynomial $2x + 3$ is multiplied by 4).

To factor $3x^2 + 7x + 2$ means to write it as the product $(3x + 1)(x + 2)$.

The trinomial $x^2 + x + 2$ is nonfactorable over the integers. There are no two integers whose product is 2 and whose sum is 1.

The equation $2x^2 + 7x + 3 = 0$ is a quadratic equation in standard form.

Essential Rules and Procedures

Factoring the Difference of Two Squares
The difference of two squares is the product of the sum and difference of two terms.
$a^2 - b^2 = (a + b)(a - b)$ [8.4.1, p. 407]

$y^2 - 81 = (y + 9)(y - 9)$

Factoring a Perfect-Square Trinomial
A perfect-square trinomial is the square of a binomial.
$a^2 + 2ab + b^2 = (a + b)^2$
$a^2 - 2ab + b^2 = (a - b)^2$ [8.4.1, p. 408]

$x^2 + 10x + 25 = (x + 5)^2$
$x^2 - 10x + 25 = (x - 5)^2$

General Factoring Strategy   [8.4.2, p. 409]
1. Is there a common factor? If so, factor out the common factor.
2. Is the polynomial the difference of two squares? If so, factor.
3. Is the polynomial a perfect-square trinomial? If so, factor.
4. Is the polynomial a trinomial that is the product of two binomials? If so, factor.

$24x + 6 = 6(4x + 1)$
$2x^2y + 6xy + 8y = 2y(x^2 + 3x + 4)$
$4x^2 - 49 = (2x + 7)(2x - 7)$
$9x^2 + 6x + 1 = (3x + 1)^2$
$2x^2 + 7x + 5 = (2x + 5)(x + 1)$

5. Does the polynomial contain four terms? If so, try factoring by grouping.

$$\begin{aligned} x^3 - 3x^2 + 2x - 6 &= (x^3 - 3x^2) + (2x - 6) \\ &= x^2(x - 3) + 2(x - 3) \\ &= (x - 3)(x^2 + 2) \end{aligned}$$

6. Is each binomial factor nonfactorable over the integers? If not, factor.

$$\begin{aligned} x^4 - 16 &= (x^2 + 4)(x^2 - 4) \\ &= (x^2 + 4)(x + 2)(x - 2) \end{aligned}$$

The Principle of Zero Products

If the product of two factors is zero, then at least one of the factors must be zero.

If $a \cdot b = 0$, then $a = 0$ or $b = 0$.

The Principle of Zero Products is used to solve a quadratic equation by factoring.   [8.5.1, p. 414]

$$\begin{aligned} x^2 + x &= 12 \\ x^2 + x - 12 &= 0 \\ (x + 4)(x - 3) &= 0 \end{aligned}$$

$$\begin{array}{ll} x + 4 = 0 & x - 3 = 0 \\ x = -4 & x = 3 \end{array}$$

Chapter Review Exercises

1. Factor: $14y^9 - 49y^6 + 7y^3$

2. Factor: $3a^2 - 12a + ab - 4b$

3. Factor: $c^2 + 8c + 12$

4. Factor: $a^3 - 5a^2 + 6a$

5. Factor: $6x^2 - 29x + 28$

6. Factor: $3y^2 + 16y - 12$

7. Factor: $18a^2 - 3a - 10$

8. Factor: $a^2b^2 - 1$

9. Factor: $4y^2 - 16y + 16$

10. Solve: $a(5a + 1) = 0$

11. Factor: $12a^2b + 3ab^2$

12. Factor: $b^2 - 13b + 30$

13. Factor: $10x^2 + 25x + 4xy + 10y$

14. Factor: $3a^2 - 15a - 42$

15. Factor: $n^4 - 2n^3 - 3n^2$

16. Factor: $2x^2 - 5x + 6$

17. Factor: $6x^2 - 7x + 2$

18. Factor: $16x^2 + 49$

19. Solve: $(x - 2)(2x - 3) = 0$

20. Factor: $7x^2 - 7$

21. Factor: $3x^5 - 9x^4 - 4x^3$

22. Factor: $4x(x - 3) - 5(3 - x)$

23. Factor: $a^2 + 5a - 14$

24. Factor: $y^2 + 5y - 36$

25. Factor: $5x^2 - 50x - 120$

26. Solve: $(x + 1)(x - 5) = 16$

27. Factor: $7a^2 + 17a + 6$

28. Factor: $4x^2 + 83x + 60$

29. Factor: $9y^4 - 25z^2$

30. Factor: $5x^2 - 5x - 30$

31. Solve: $6 - 6y^2 = 5y$

32. Factor: $12b^3 - 58b^2 + 56b$

33. Factor: $5x^3 + 10x^2 + 35x$

34. Factor: $x^2 - 23x + 42$

35. Factor: $a(3a + 2) - 7(3a + 2)$

36. Factor: $8x^2 - 38x + 45$

37. Factor: $10a^2x - 130ax + 360x$

38. Factor: $2a^2 - 19a - 60$

39. Factor: $21ax - 35bx - 10by + 6ay$

40. Factor: $a^6 - 100$

41. Factor: $16a^2 + 8a + 1$

42. Solve: $4x^2 + 27x = 7$

43. Factor: $20a^2 + 10a - 280$

44. Factor: $6x - 18$

45. Factor: $3x^4y + 2x^3y + 6x^2y$

46. Factor: $d^2 + 3d - 40$

47. Factor: $24x^2 - 12xy + 10y - 20x$

48. Factor: $4x^3 - 20x^2 - 24x$

49. Solve: $x^2 - 8x - 20 = 0$

50. Factor: $3x^2 - 17x + 10$

51. Factor: $16x^2 - 94x + 33$

52. Factor: $9x^2 - 30x + 25$

53. Factor: $12y^2 + 16y - 3$

54. Factor: $3x^2 + 36x + 108$

55. The length of a playing field is twice the width. The area is 5000 yd². Find the length and width of the playing field.

56. The length of a hockey field is 20 yd less than twice the width. The area of the field is 6000 yd². Find the length and width of the hockey field.

57. The sum of the squares of two consecutive positive integers is forty-one. Find the two integers.

58. The size of a motion picture on the screen is given by the equation $S = d^2$, where d is the distance between the projector and the screen. Find the distance between the projector and the screen when the size of the picture is 400 ft².

59. A rectangular garden plot has dimensions 15 ft by 12 ft. A uniform path around the garden increases the total area to 270 ft². What is the width of the larger rectangle?

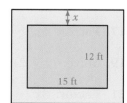

60. The length of each side of a square is extended 4 ft. The area of the resulting square is 576 ft². Find the length of a side of the original square.

Chapter Test

1. Factor: $6x^2y^2 + 9xy^2 + 12y^2$

2. Factor: $6x^3 - 8x^2 + 10x$

3. Factor: $p^2 + 5p + 6$

4. Factor: $a(x - 2) + b(2 - x)$

5. Solve: $(2a - 3)(a + 7) = 0$

6. Factor: $a^2 - 19a + 48$

7. Factor: $x^3 + 2x^2 - 15x$

8. Factor: $8x^2 + 20x - 48$

9. Factor: $ab + 6a - 3b - 18$

10. Solve: $4x^2 - 1 = 0$

11. Factor: $6x^2 + 19x + 8$

12. Factor: $x^2 - 9x - 36$

13. Factor: $2b^2 - 32$

14. Factor: $4a^2 - 12ab + 9b^2$

15. Factor: $px + x - p - 1$

16. Factor: $5x^2 - 45x - 15$

17. Factor: $2x^2 + 4x - 5$

18. Factor: $4x^2 - 49y^2$

19. Solve: $x(x - 8) = -15$

20. Factor: $p^2 + 12p + 36$

21. Factor: $18x^2 - 48xy + 32y^2$

22. Factor: $2y^4 - 14y^3 - 16y^2$

23. The length of a rectangle is 3 cm more than twice the width. The area of the rectangle is 90 cm². Find the length and width of the rectangle.

24. The length of the base of a triangle is three times the height. The area of the triangle is 24 in². Find the length of the base of the triangle.

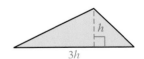

25. The product of two consecutive negative integers is one hundred fifty-six. Find the two integers.

Cumulative Review Exercises

1. Subtract: $4 - (-5) - 6 - 11$

2. Divide: $0.372 \div (-0.046)$
Round to the nearest tenth.

3. Simplify: $(3 - 7)^2 \div (-2) - 3 \cdot (-4)$

4. Evaluate $-2a^2 \div (2b) - c$ when $a = -4$, $b = 2$, and $c = -1$.

5. Identify the property that justifies the statement.
$(3 + 8) + 7 = 3 + (8 + 7)$

6. Multiply: $-\dfrac{3}{4}(-24x^2)$

7. Simplify: $-2[3x - 4(3 - 2x) - 8x]$

8. Solve: $-\dfrac{5}{7}x = -\dfrac{10}{21}$

9. Solve: $4 + 3(x - 2) = 13$

10. Solve: $3x - 2 = 12 - 5x$

11. Solve: $-2 + 4[3x - 2(4 - x) - 3] = 4x + 2$

12. 120% of what number is 42?

13. Solve: $-4x - 2 \geq 10$

14. Solve: $9 - 2(4x - 5) < 3(7 - 6x)$

15. Graph: $y = \dfrac{3}{4}x - 2$

16. Graph: $f(x) = -3x - 3$

17. Find the domain and range of the relation $\{(-5, -4), (-3, -2), (-1, 0), (1, 2), (3, 4)\}$. Is the relation a function?

18. Evaluate $f(x) = 6x - 5$ at $x = 11$.

19. Graph the solution set of $x + 3y > 2$.

20. Solve by substitution: $6x + y = 7$
$ x - 3y = 17$

21. Solve by the addition method: $2x - 3y = -4$
$ 5x + y = 7$

22. Add: $(3y^3 - 5y^2 - 6) + (2y^2 - 8y + 1)$

23. Simplify: $(-3a^4b^2)^3$

24. Multiply: $(x + 2)(x^2 - 5x + 4)$

25. Divide: $(8x^2 + 4x - 3) \div (2x - 3)$

26. Simplify: $(x^{-4}y^2)^3$

27. Factor: $3a - 3b - ax + bx$

28. Factor: $x^2 + 3xy - 10y^2$

29. Factor: $6a^4 + 22a^3 + 12a^2$

30. Factor: $25a^2 - 36b^2$

31. Factor: $12x^2 - 36xy + 27y^2$

32. Solve: $3x^2 + 11x - 20 = 0$

33. Find the range of the function given by the equation $f(x) = \dfrac{4}{5}x - 3$ if the domain is $\{-10, -5, 0, 5, 10\}$.

34. The daily high temperatures, in degrees Celsius, during one week were recorded as follows: $-4°, -7°, 2°, 0°, -1°, -6°, -5°$. Find the average daily high temperature for the week.

35. The width of a rectangle is 40% of the length. The perimeter of the rectangle is 42 cm. Find the length and width of the rectangle.

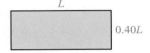

36. A board 10 ft long is cut into two pieces. Four times the length of the shorter piece is 2 ft less than three times the length of the longer piece. Find the length of each piece.

37. Company A rents cars for $6 a day and 25¢ for every mile driven. Company B rents cars for $15 a day and 10¢ per mile driven. You want to rent a car for 6 days. What is the maximum number of miles you can drive a Company A car if it is to cost you less than a Company B car?

38. An investment of $4000 is made at an annual simple interest rate of 8%. How much additional money must be invested at an annual simple interest rate of 11% so that the total interest earned is $1035?

39. A stereo that regularly sells for $165 is on sale for $99. Find the discount rate.

40. Find three consecutive even integers such that five times the middle integer is twelve more than twice the sum of the first and third.

*Rational equations are used to model light intensity. The farther light travels from its source, the lower its intensity. The standard unit of light intensity is the candela. When a baseball stadium is designed, the number of lights, their candela, and their placement must be properly determined so that no fan will miss a single pitch—at least not due to poor illumination. The topic of the **Project on pages 491–493** is intensity of illumination.*

Applications

Agriculture, *p. 473*
Art, *p. 472*
Business, *pp. 471, 472, 479, 480*
Cartography, *p. 472*
Catering, *p. 473*
Conservation, *p. 472*
Construction, *pp. 471, 472*
Cooking, *p. 471*
Demography, *p. 476*
Elections, *p. 471*
Electricity, *p. 480*
Energy, *p. 472*
Food industry, *p. 473*
Fuel efficiency, *p. 457*
Fundraising, *p. 476*
Gardening, *p. 498*
Geometry, *pp. 444, 468, 469, 473, 474, 475, 479, 497, 502*
History, *p. 476*
Intensity of illumination, *p. 493*
Interior decorating, *p. 466*
Investment problems, *p. 502*
Investments, *p. 466*
Loans, *pp. 466, 491*
Lotteries, *p. 475*
Mechanics, *p. 498*
Medicine, *p. 467*
Mixture problems, *p. 500*
Number problems, *pp. 475, 491, 502*
Paint mixtures, *p. 473*
Percent mixture problems, *p. 502*
Photography, *p. 476*
Redecorating, *p. 473*
Rocketry, *p. 472*
Solar system, *p. 493*
Sports, *p. 475*
Taxes, *p. 473*
Television, *p. 471*
Uniform motion problems, *pp. 436, 483, 484, 487, 488, 489, 498, 499, 500*
Work problems, *pp. 481, 482, 485, 486, 487, 489, 498, 499, 500, 502*

9

Rational Expressions

Objectives

Section 9.1
1. Simplify rational expressions
2. Multiply rational expressions
3. Divide rational expressions

Section 9.2
1. Find the least common multiple (LCM) of two or more polynomials
2. Express two fractions in terms of the LCM of their denominators

Section 9.3
1. Add and subtract rational expressions with the same denominator
2. Add and subtract rational expressions with different denominators

Section 9.4
1. Simplify complex fractions

Section 9.5
1. Solve equations containing fractions
2. Solve proportions
3. Applications of proportions
4. Problems involving similar triangles

Section 9.6
1. Solve a literal equation for one of the variables

Section 9.7
1. Work problems
2. Uniform motion problems

Need help? For online student resources, such as section quizzes, visit this web site: **math.college.hmco.com**

435

PrepTest

1. Find the least common multiple (LCM) of 12 and 18.

2. Simplify: $\dfrac{9x^3y^4}{3x^2y^7}$

3. Subtract: $\dfrac{3}{4} - \dfrac{8}{9}$

4. Divide: $\left(-\dfrac{8}{11}\right) \div \dfrac{4}{5}$

5. If a is a nonzero number, are the expressions $\dfrac{0}{a}$ and $\dfrac{a}{0}$ equal?

6. Solve: $\dfrac{2}{3}x - \dfrac{3}{4} = \dfrac{5}{6}$

Go Figure

A mouse proceeds clockwise along the sides of a 12 ft × 12 ft square traveling at a constant speed of 2 ft/s. Six seconds later a second mouse starts from the same point traveling clockwise along the same square traveling at a constant speed of 3 ft/s. How far apart are the two mice 18 s later?

7. Line ℓ_1 is parallel to line ℓ_2. Find the measure of angle a.

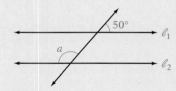

8. Factor: $x^2 - 4x - 12$

9. Factor: $2x^2 - x - 3$

10. At 9:00 A.M., Anthony begins jogging on a park trail at a rate of 9 ft/s. Ten minutes later his sister Jean begins jogging on the same trail in pursuit of her brother at a rate of 12 ft/s. At what time will Jean catch up to Anthony?

9.1 | Multiplication and Division of Rational Expressions

1 **Simplify rational expressions**

A fraction in which the numerator and denominator are polynomials is called a **rational expression.** Examples of rational expressions are shown at the right.

$\dfrac{5}{z} \qquad \dfrac{x^2+1}{2x-1} \qquad \dfrac{y^2-3}{3xy+1}$

Care must be exercised with rational expressions to ensure that when the variables are replaced with numbers, the resulting denominator is not zero.

Consider the rational expression at the right. The value of x cannot be 2 because the denominator would then be zero.

$$\dfrac{3x + 1}{2x - 4}$$

$$\dfrac{3 \cdot 2 + 1}{2 \cdot 2 - 4} = \dfrac{7}{0} \leftarrow \text{Not a real number}$$

A fraction is in simplest form when the numerator and denominator have no common factors other than 1. The Multiplication Property of One is used to write a rational expression in simplest form.

Simplify: $\dfrac{x^2 - 4}{x^2 - 2x - 8}$

Factor the numerator and denominator.

$$\dfrac{x^2 - 4}{x^2 - 2x - 8} = \dfrac{(x - 2)(x + 2)}{(x - 4)(x + 2)}$$

$$= \dfrac{x - 2}{x - 4} \cdot \dfrac{x + 2}{x + 2}$$

$$= \dfrac{x - 2}{x - 4} \cdot 1$$

The restrictions $x \neq -2$ and $x \neq 4$ are necessary to prevent division by zero.

$$= \dfrac{x - 2}{x - 4}, x \neq -2, 4$$

This simplification is usually shown with slashes through the common factors. The preceding simplification would be shown as follows.

$$\dfrac{x^2 - 4}{x^2 - 2x - 8} = \dfrac{(x - 2)\cancel{(x + 2)}^{1}}{(x - 4)\cancel{(x + 2)}_{1}} = \dfrac{x - 2}{x - 4}, x \neq -2, 4$$

In this problem, it is stated that $x \neq -2, 4$. Look at the factored form of the original rational expression.

$$\dfrac{(x - 2)(x + 2)}{(x - 4)(x + 2)}$$

If either of the factors in the denominator is zero, then the denominator is zero.

$$\begin{array}{cc} x + 2 = 0 & x - 4 = 0 \\ x = -2 & x = 4 \end{array}$$

When x is -2 or 4, the denominator is 0. Therefore, for this fraction, $x \neq -2, 4$.

For the remaining examples, we will omit the restrictions on the variables that prevent division by zero and assume the values of the variables are such that division by zero is not possible.

Simplify: $\dfrac{10 + 3x - x^2}{x^2 - 4x - 5}$

Factor the numerator and denominator.

$$\dfrac{10 + 3x - x^2}{x^2 - 4x - 5} = \dfrac{(5 - x)(2 + x)}{(x - 5)(x + 1)}$$

Divide by the common factors. Remember that $5 - x = -(x - 5)$. Therefore,

$$\frac{5 - x}{x - 5} = \frac{-(x - 5)}{x - 5} = \frac{-1}{1} = -1$$

Write the answer in simplest form.

$$= \frac{\overset{-1}{\cancel{(5 - x)}}(2 + x)}{\underset{1}{\cancel{(x - 5)}}(x + 1)}$$

$$= -\frac{x + 2}{x + 1}$$

Example 1 Simplify: **A.** $\dfrac{4x^3y^4}{6x^4y}$ **B.** $\dfrac{9 - x^2}{x^2 + x - 12}$

Solution **A.** $\dfrac{4x^3y^4}{6x^4y} = \dfrac{\overset{1}{\cancel{2}} \cdot 2x^3y^4}{\underset{1}{\cancel{2}} \cdot 3x^4y} = \dfrac{2y^3}{3x}$ • Simplify using the rules of exponents.

B. $\dfrac{9 - x^2}{x^2 + x - 12} = \dfrac{\overset{-1}{\cancel{(3 - x)}}(3 + x)}{\underset{1}{\cancel{(x - 3)}}(x + 4)} = -\dfrac{x + 3}{x + 4}$

Problem 1 Simplify. **A.** $\dfrac{6x^5y}{12x^2y^3}$ **B.** $\dfrac{x^2 + 2x - 24}{16 - x^2}$

Solution See page S19.

2 Multiply rational expressions

The product of two fractions is a fraction whose numerator is the product of the numerators of the two fractions and whose denominator is the product of the denominators of the two fractions.

$$\frac{a}{b} \cdot \frac{c}{d} = \frac{ac}{bd}$$

$$\frac{2}{3} \cdot \frac{4}{5} = \frac{8}{15}$$

$$\frac{3x}{y} \cdot \frac{2}{z} = \frac{6x}{yz}$$

$$\frac{x + 2}{x} \cdot \frac{3}{x - 2} = \frac{3(x + 2)}{x(x - 2)}$$

Multiply: $\dfrac{x^2 + 3x}{x^2 - 3x - 4} \cdot \dfrac{x^2 - 5x + 4}{x^2 + 2x - 3}$

$$\frac{x^2 + 3x}{x^2 - 3x - 4} \cdot \frac{x^2 - 5x + 4}{x^2 + 2x - 3}$$

Factor the numerator and denominator of each fraction.

$$= \frac{x(x + 3)}{(x - 4)(x + 1)} \cdot \frac{(x - 4)(x - 1)}{(x + 3)(x - 1)}$$

Multiply. Divide by the common factors.

$$= \frac{x\overset{1}{\cancel{(x + 3)}}\overset{1}{\cancel{(x - 4)}}\overset{1}{\cancel{(x - 1)}}}{\underset{1}{\cancel{(x - 4)}}(x + 1)\underset{1}{\cancel{(x + 3)}}\underset{1}{\cancel{(x - 1)}}}$$

Write the answer in simplest form.

$$= \frac{x}{x + 1}$$

Example 2 Multiply.

A. $\dfrac{10x^2 - 15x}{12x - 8} \cdot \dfrac{3x - 2}{20x - 25}$ **B.** $\dfrac{x^2 + x - 6}{x^2 + 7x + 12} \cdot \dfrac{x^2 + 3x - 4}{4 - x^2}$

Solution **A.** $\dfrac{10x^2 - 15x}{12x - 8} \cdot \dfrac{3x - 2}{20x - 25}$

$= \dfrac{5x(2x - 3)}{4(3x - 2)} \cdot \dfrac{(3x - 2)}{5(4x - 5)}$ • Factor the numerator and denominator of each fraction.

$= \dfrac{\overset{1}{\cancel{5}}x(2x - 3)\overset{1}{\cancel{(3x - 2)}}}{2 \cdot 2\underset{1}{\cancel{(3x - 2)}}\underset{1}{\cancel{5}}(4x - 5)}$ • Multiply. Divide by the common factors.

$= \dfrac{x(2x - 3)}{4(4x - 5)}$ • Write the answer in simplest form.

B. $\dfrac{x^2 + x - 6}{x^2 + 7x + 12} \cdot \dfrac{x^2 + 3x - 4}{4 - x^2}$

$= \dfrac{(x + 3)(x - 2)}{(x + 3)(x + 4)} \cdot \dfrac{(x + 4)(x - 1)}{(2 - x)(2 + x)}$

$= \dfrac{\overset{1}{\cancel{(x + 3)}}\overset{-1}{\cancel{(x - 2)}}\overset{1}{\cancel{(x + 4)}}(x - 1)}{\underset{1}{\cancel{(x + 3)}}\underset{1}{\cancel{(x + 4)}}\underset{1}{\cancel{(2 - x)}}(2 + x)}$

$= -\dfrac{x - 1}{x + 2}$

Problem 2 Multiply.

A. $\dfrac{12x^2 + 3x}{10x - 15} \cdot \dfrac{8x - 12}{9x + 18}$ **B.** $\dfrac{x^2 + 2x - 15}{9 - x^2} \cdot \dfrac{x^2 - 3x - 18}{x^2 - 7x + 6}$

Solution See page S20.

---

3 **Divide rational expressions**

The **reciprocal** of a fraction is a fraction with the numerator and denominator interchanged.

$$\text{Fraction} \left\{ \begin{array}{l} \dfrac{a}{b} \\[2mm] x^2 = \dfrac{x^2}{1} \\[2mm] \dfrac{x + 2}{x} \end{array} \right.\quad \left. \begin{array}{l} \dfrac{b}{a} \\[2mm] \dfrac{1}{x^2} \\[2mm] \dfrac{x}{x + 2} \end{array} \right\} \text{Reciprocal}$$

To divide two fractions, multiply by the reciprocal of the divisor.

$$\frac{a}{b} \div \frac{c}{d} = \frac{a}{b} \cdot \frac{d}{c} = \frac{ad}{bc}$$

$$\frac{4}{x} \div \frac{y}{5} = \frac{4}{x} \cdot \frac{5}{y} = \frac{20}{xy}$$

$$\frac{x+4}{x} \div \frac{x-2}{4} = \frac{x+4}{x} \cdot \frac{4}{x-2} = \frac{4(x+4)}{x(x-2)}$$

The basis for the division rule is shown below.

$$\underbrace{\frac{a}{b} \div \frac{c}{d}} = \frac{\dfrac{a}{b}}{\dfrac{c}{d}} = \frac{\dfrac{a}{b} \cdot \dfrac{d}{c}}{\dfrac{c}{d} \cdot \dfrac{d}{c}} = \frac{\dfrac{a}{b} \cdot \dfrac{d}{c}}{1} = \underbrace{\frac{a}{b} \cdot \frac{d}{c}}$$

Example 3   Divide.

A. $\dfrac{xy^2 - 3x^2y}{z^2} \div \dfrac{6x^2 - 2xy}{z^3}$

B. $\dfrac{2x^2 + 5x + 2}{2x^2 + 3x - 2} \div \dfrac{3x^2 + 13x + 4}{2x^2 + 7x - 4}$

Solution   **A.** $\dfrac{xy^2 - 3x^2y}{z^2} \div \dfrac{6x^2 - 2xy}{z^3} = \dfrac{xy^2 - 3x^2y}{z^2} \cdot \dfrac{z^3}{6x^2 - 2xy}$

$$= \frac{xy(y \overset{-1}{\cancel{- 3x}}) \cdot z^3}{z^2 \cdot 2x(3x \underset{1}{\cancel{- y}})}$$

$$= -\frac{yz}{2}$$

B. $\dfrac{2x^2 + 5x + 2}{2x^2 + 3x - 2} \div \dfrac{3x^2 + 13x + 4}{2x^2 + 7x - 4} = \dfrac{2x^2 + 5x + 2}{2x^2 + 3x - 2} \cdot \dfrac{2x^2 + 7x - 4}{3x^2 + 13x + 4}$

$$= \frac{(2x + 1)\overset{1}{\cancel{(x + 2)}}}{\underset{1}{\cancel{(2x - 1)}}\,\overset{1}{\cancel{(x + 2)}}} \cdot \frac{\overset{1}{\cancel{(2x - 1)}}\,\overset{1}{\cancel{(x + 4)}}}{(3x + 1)\underset{1}{\cancel{(x + 4)}}}$$

$$= \frac{2x + 1}{3x + 1}$$

Problem 3   Divide.

A. $\dfrac{a^2}{4bc^2 - 2b^2c} \div \dfrac{a}{6bc - 3b^2}$

B. $\dfrac{3x^2 + 26x + 16}{3x^2 - 7x - 6} \div \dfrac{2x^2 + 9x - 5}{x^2 + 2x - 15}$

Solution   See page S20.

9.1   Concept Review

Determine whether the statement is always true, sometimes true, or never true.

1. The numerator and denominator of a rational expression are polynomials.

2. A rational expression is in simplest form when the only factor common to both the numerator and the denominator is 1.

3. Before multiplying two rational expressions, we must write the expressions in terms of a common denominator.

4. The expression $\dfrac{x}{x + 2}$ is not a real number if $x = 0$.

5. The procedure for multiplying rational expressions is the same as that for multiplying arithmetic fractions.

6. When a rational expression is rewritten in simplest form, its value is less than it was before it was rewritten in simplest form.

7. To divide two rational expressions, multiply the reciprocal of the first expression by the second expression.

9.1   Exercises

1. 1. What is a rational expression? Provide an example.

2. When is a rational expression in simplest form?

3. For the rational expression $\dfrac{x + 7}{x - 4}$, explain why the value of x cannot be 4.

4. Explain why the following simplification is incorrect.

$$\frac{x + 3}{x} = \frac{\overset{1}{\cancel{x}} + 3}{\underset{1}{\cancel{x}}} = 4$$

Simplify.

5. $\dfrac{9x^3}{12x^4}$

6. $\dfrac{16x^2y}{24xy^3}$

7. $\dfrac{(x + 3)^2}{(x + 3)^3}$

8. $\dfrac{(2x - 1)^5}{(2x - 1)^4}$

9. $\dfrac{3n - 4}{4 - 3n}$

10. $\dfrac{5 - 2x}{2x - 5}$

11. $\dfrac{6y(y + 2)}{9y^2(y + 2)}$

12. $\dfrac{12x^2(3 - x)}{18x(3 - x)}$

13. $\dfrac{6x(x-5)}{8x^2(5-x)}$

14. $\dfrac{14x^3(7-3x)}{21x(3x-7)}$

15. $\dfrac{a^2+4a}{ab+4b}$

16. $\dfrac{x^2-3x}{2x-6}$

17. $\dfrac{4-6x}{3x^2-2x}$

18. $\dfrac{5xy-3y}{9-15x}$

19. $\dfrac{y^2-3y+2}{y^2-4y+3}$

20. $\dfrac{x^2+5x+6}{x^2+8x+15}$

21. $\dfrac{x^2+3x-10}{x^2+2x-8}$

22. $\dfrac{a^2+7a-8}{a^2+6a-7}$

23. $\dfrac{x^2+x-12}{x^2-6x+9}$

24. $\dfrac{x^2+8x+16}{x^2-2x-24}$

25. $\dfrac{x^2-3x-10}{25-x^2}$

26. $\dfrac{4-y^2}{y^2-3y-10}$

27. $\dfrac{2x^3+2x^2-4x}{x^3+2x^2-3x}$

28. $\dfrac{3x^3-12x}{6x^3-24x^2+24x}$

29. $\dfrac{6x^2-7x+2}{6x^2+5x-6}$

30. $\dfrac{2n^2-9n+4}{2n^2-5n-12}$

2 Multiply.

31. $\dfrac{8x^2}{9y^3}\cdot\dfrac{3y^2}{4x^3}$

32. $\dfrac{4a^2b^3}{15x^5y^2}\cdot\dfrac{25x^3y}{16ab}$

33. $\dfrac{12x^3y^4}{7a^2b^3}\cdot\dfrac{14a^3b^4}{9x^2y^2}$

34. $\dfrac{18a^4b^2}{25x^2y^3}\cdot\dfrac{50x^5y^6}{27a^6b^2}$

35. $\dfrac{3x-6}{5x-20}\cdot\dfrac{10x-40}{27x-54}$

36. $\dfrac{8x-12}{14x+7}\cdot\dfrac{42x+21}{32x-48}$

37. $\dfrac{3x^2+2x}{3xy-3y}\cdot\dfrac{3xy^3-3y^3}{3x^3+2x^2}$

38. $\dfrac{4a^2x-3a^2}{2by+5b}\cdot\dfrac{2b^3y+5b^3}{4ax-3a}$

39. $\dfrac{x^2+5x+4}{x^3y^2}\cdot\dfrac{x^2y^3}{x^2+2x+1}$

40. $\dfrac{x^2+x-2}{xy^2}\cdot\dfrac{x^3y}{x^2+5x+6}$

41. $\dfrac{x^4y^2}{x^2+3x-28}\cdot\dfrac{x^2-49}{xy^4}$

42. $\dfrac{x^5y^3}{x^2+13x+30}\cdot\dfrac{x^2+2x-3}{x^7y^2}$

43. $\dfrac{2x^2-5x}{2xy+y}\cdot\dfrac{2xy^2+y^2}{5x^2-2x^3}$

44. $\dfrac{3a^3+4a^2}{5ab-3b}\cdot\dfrac{3b^3-5ab^3}{3a^2+4a}$

45. $\dfrac{x^2-2x-24}{x^2-5x-6}\cdot\dfrac{x^2+5x+6}{x^2+6x+8}$

46. $\dfrac{x^2-8x+7}{x^2+3x-4}\cdot\dfrac{x^2+3x-10}{x^2-9x+14}$

47. $\dfrac{x^2+2x-35}{x^2+4x-21}\cdot\dfrac{x^2+3x-18}{x^2+9x+18}$

48. $\dfrac{y^2+y-20}{y^2+2y-15}\cdot\dfrac{y^2+4y-21}{y^2+3y-28}$

49. $\dfrac{x^2-3x-4}{x^2+6x+5}\cdot\dfrac{x^2+5x+6}{8+2x-x^2}$

50. $\dfrac{25-n^2}{n^2-2n-35}\cdot\dfrac{n^2-8n-20}{n^2-3n-10}$

51. $\dfrac{12x^2 - 6x}{x^2 + 6x + 5} \cdot \dfrac{2x^4 + 10x^3}{4x^2 - 1}$

52. $\dfrac{8x^3 + 4x^2}{x^2 - 3x + 2} \cdot \dfrac{x^2 - 4}{16x^2 + 8x}$

53. $\dfrac{16 + 6x - x^2}{x^2 - 10x - 24} \cdot \dfrac{x^2 - 6x - 27}{x^2 - 17x + 72}$

54. $\dfrac{x^2 - 11x + 28}{x^2 - 13x + 42} \cdot \dfrac{x^2 + 7x + 10}{20 - x - x^2}$

55. $\dfrac{2x^2 + 5x + 2}{2x^2 + 7x + 3} \cdot \dfrac{x^2 - 7x - 30}{x^2 - 6x - 40}$

56. $\dfrac{x^2 - 4x - 32}{x^2 - 8x - 48} \cdot \dfrac{3x^2 + 17x + 10}{3x^2 - 22x - 16}$

57. $\dfrac{2x^2 + x - 3}{2x^2 - x - 6} \cdot \dfrac{2x^2 - 9x + 10}{2x^2 - 3x + 1}$

58. $\dfrac{3y^2 + 14y + 8}{2y^2 + 7y - 4} \cdot \dfrac{2y^2 + 9y - 5}{3y^2 + 16y + 5}$

59. $\dfrac{6x^2 - 11x + 4}{6x^2 + x - 2} \cdot \dfrac{12x^2 + 11x + 2}{8x^2 + 14x + 3}$

60. $\dfrac{6 - x - 2x^2}{4x^2 + 3x - 10} \cdot \dfrac{3x^2 + 7x - 20}{2x^2 + 5x - 12}$

3 **61.** What is the reciprocal of a rational expression?

62. Explain how to divide two rational expressions.

Divide.

63. $\dfrac{4x^2y^3}{15a^2b^3} \div \dfrac{6xy}{5a^3b^5}$

64. $\dfrac{9x^3y^4}{16a^4b^2} \div \dfrac{45x^4y^2}{14a^7b}$

65. $\dfrac{6x - 12}{8x + 32} \div \dfrac{18x - 36}{10x + 40}$

66. $\dfrac{28x + 14}{45x - 30} \div \dfrac{14x + 7}{30x - 20}$

67. $\dfrac{6x^3 + 7x^2}{12x - 3} \div \dfrac{6x^2 + 7x}{36x - 9}$

68. $\dfrac{5a^2y + 3a^2}{2x^3 + 5x^2} \div \dfrac{10ay + 6a}{6x^3 + 15x^2}$

69. $\dfrac{x^2 + 4x + 3}{x^2y} \div \dfrac{x^2 + 2x + 1}{xy^2}$

70. $\dfrac{x^3y^2}{x^2 - 3x - 10} \div \dfrac{xy^4}{x^2 - x - 20}$

71. $\dfrac{x^2 - 49}{x^4y^3} \div \dfrac{x^2 - 14x + 49}{x^4y^3}$

72. $\dfrac{x^2y^5}{x^2 - 11x + 30} \div \dfrac{xy^6}{x^2 - 7x + 10}$

73. $\dfrac{4ax - 8a}{c^2} \div \dfrac{2y - xy}{c^3}$

74. $\dfrac{3x^2y - 9xy}{a^2b} \div \dfrac{3x^2 - x^3}{ab^2}$

75. $\dfrac{x^2 - 5x + 6}{x^2 - 9x + 18} \div \dfrac{x^2 - 6x + 8}{x^2 - 9x + 20}$

76. $\dfrac{x^2 + 3x - 40}{x^2 + 2x - 35} \div \dfrac{x^2 + 2x - 48}{x^2 + 3x - 18}$

77. $\dfrac{x^2 + 2x - 15}{x^2 - 4x - 45} \div \dfrac{x^2 + x - 12}{x^2 - 5x - 36}$

78. $\dfrac{y^2 - y - 56}{y^2 + 8y + 7} \div \dfrac{y^2 - 13y + 40}{y^2 - 4y - 5}$

79. $\dfrac{8 + 2x - x^2}{x^2 + 7x + 10} \div \dfrac{x^2 - 11x + 28}{x^2 - x - 42}$

80. $\dfrac{x^2 - x - 2}{x^2 - 7x + 10} \div \dfrac{x^2 - 3x - 4}{40 - 3x - x^2}$

81. $\dfrac{2x^2 - 3x - 20}{2x^2 - 7x - 30} \div \dfrac{2x^2 - 5x - 12}{4x^2 + 12x + 9}$

82. $\dfrac{6n^2 + 13n + 6}{4n^2 - 9} \div \dfrac{6n^2 + n - 2}{4n^2 - 1}$

83. $\dfrac{9x^2 - 16}{6x^2 - 11x + 4} \div \dfrac{6x^2 + 11x + 4}{8x^2 + 10x + 3}$

84. $\dfrac{15 - 14x - 8x^2}{4x^2 + 4x - 15} \div \dfrac{4x^2 + 13x - 12}{3x^2 + 13x + 4}$

85. $\dfrac{8x^2 + 18x - 5}{10x^2 - 9x + 2} \div \dfrac{8x^2 + 22x + 15}{10x^2 + 11x - 6}$

86. $\dfrac{10 + 7x - 12x^2}{8x^2 - 2x - 15} \div \dfrac{6x^2 - 13x + 5}{10x^2 - 13x + 4}$

9.1 Applying Concepts

For what values of x is the rational expression undefined? (*Hint:* Set the denominator equal to zero and solve for x.)

87. $\dfrac{x}{(x + 6)(x - 1)}$

88. $\dfrac{x}{(x - 2)(x + 5)}$

89. $\dfrac{8}{x^2 - 1}$

90. $\dfrac{7}{x^2 - 16}$

91. $\dfrac{x - 4}{x^2 - x - 6}$

92. $\dfrac{x + 5}{x^2 - 4x - 5}$

93. $\dfrac{3x}{x^2 + 6x + 9}$

94. $\dfrac{3x - 8}{3x^2 - 10x - 8}$

95. $\dfrac{4x + 7}{6x^2 - 5x - 4}$

Simplify.

96. $\dfrac{y^2}{x} \cdot \dfrac{x}{2} \div \dfrac{y}{x}$

97. $\dfrac{ab}{3} \cdot \dfrac{a}{b^2} \div \dfrac{a}{4}$

98. $\left(\dfrac{2x}{y}\right)^3 \div \left(\dfrac{x}{3y}\right)^2$

99. $\left(\dfrac{c}{3}\right)^2 \div \left(\dfrac{c}{2} \cdot \dfrac{c}{4}\right)$

100. $\left(\dfrac{a - 3}{b}\right)^2 \left(\dfrac{b}{3 - a}\right)^3$

101. $\left(\dfrac{x - 4}{y^2}\right)^3 \cdot \left(\dfrac{y}{4 - x}\right)^2$

102. $\dfrac{x^2 + 3x - 40}{x^2 + 2x - 35} \div \dfrac{x^2 + 2x - 48}{x^2 + 3x - 18} \cdot \dfrac{x^2 - 36}{x^2 - 9}$

103. $\dfrac{x^2 + x - 6}{x^2 + 7x + 12} \cdot \dfrac{x^2 + 3x - 4}{x^2 + x - 2} \div \dfrac{x^2 - 16}{x^2 - 4}$

Write in simplest form the ratio of the shaded area of the figure to the total area of the figure.

104.

105.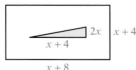

106. Given the expression $\dfrac{9}{x^2 + 1}$, choose some values of x and evaluate the expression for those values. Is it possible to choose a value of x for which the value of the expression is greater than 10? If so, give such a value. If not, explain why it is not possible.

107. Given the expression $\dfrac{1}{y - 3}$, choose some values of y and evaluate the expression for those values. Is it possible to choose a value of y for which the value of the expression is greater than 10,000,000? If so, give such a value. Explain your answer.

9.2 | Expressing Fractions in Terms of the Least Common Multiple of Their Denominators

1 **Find the least common multiple (LCM) of two or more polynomials**

The **least common multiple (LCM)** of two or more numbers is the smallest number that contains the prime factorization of each number.

The LCM of 12 and 18 is 36. 36 contains the prime factors of 12 and the prime factors of 18.

$12 = 2 \cdot 2 \cdot 3$
$18 = 2 \cdot 3 \cdot 3$

$$\text{LCM} = 36 = 2 \cdot 2 \cdot 3 \cdot 3$$

Factors of 12
Factors of 18

The least common multiple of two or more polynomials is the simplest polynomial that contains the factors of each polynomial.

To find the LCM of two or more polynomials, first factor each polynomial completely. The LCM is the product of each factor the greatest number of times it occurs in any one factorization.

Take Note

The LCM must contain the factors of each polynomial. As shown with braces at the right, the LCM contains the factors of $4x^2 + 4x$ and the factors of $x^2 + 2x + 1$.

Find the LCM of $4x^2 + 4x$ and $x^2 + 2x + 1$.

The LCM of $4x^2 + 4x$ and $x^2 + 2x + 1$ is the product of the LCM of the numerical coefficients and each variable factor the greatest number of times it occurs in any one factorization.

$4x^2 + 4x = 4x(x + 1) = 2 \cdot 2 \cdot x(x + 1)$
$x^2 + 2x + 1 = (x + 1)(x + 1)$

Factors of $4x^2 + 4x$

$$\text{LCM} = 2 \cdot 2 \cdot x(x + 1)(x + 1) = 4x(x + 1)(x + 1)$$

Factors of $x^2 + 2x + 1$

Example 1 Find the LCM of $4x^2y$ and $6xy^2$.

Solution
$$4x^2y = 2 \cdot 2 \cdot x \cdot x \cdot y$$
$$6xy^2 = 2 \cdot 3 \cdot x \cdot y \cdot y$$
• Factor each polynomial completely.

$$\text{LCM} = 2 \cdot 2 \cdot 3 \cdot x \cdot x \cdot y \cdot y$$
$$= 12x^2y^2$$
• Write the product of the LCM of the numerical coefficients and each variable factor the greatest number of times it occurs in any one factorization.

Problem 1 Find the LCM of $8uv^2$ and $12uw$.

Solution See page S20.

Example 2 Find the LCM of $x^2 - x - 6$ and $9 - x^2$.

Solution
$$x^2 - x - 6 = (x - 3)(x + 2)$$
$$9 - x^2 = -(x^2 - 9) = -(x + 3)(x - 3)$$

$$\text{LCM} = (x - 3)(x + 2)(x + 3)$$

Problem 2 Find the LCM of $m^2 - 6m + 9$ and $m^2 - 2m - 3$.

Solution See page S20.

---

2 ## Express two fractions in terms of the LCM of their denominators

VIDEO & DVD   CD TUTOR   WEB   SSM

When adding and subtracting fractions, it is often necessary to express two or more fractions in terms of a common denominator. This common denominator is the LCM of the denominators of the fractions.

Write the fractions $\dfrac{x + 1}{4x^2}$ and $\dfrac{x - 3}{6x^2 - 12x}$ in terms of the LCM of the denominators.

Find the LCM of the denominators.

The LCM is $12x^2(x - 2)$.

Take Note

$\dfrac{3(x - 2)}{3(x - 2)} = 1$ and $\dfrac{2x}{2x} = 1$. We are multiplying each fraction by 1, so we are not changing the value of either fraction.

For each fraction, multiple the numerator and denominator by the factors whose product with the denominator is the LCM.

$$\dfrac{x + 1}{4x^2} = \dfrac{x + 1}{4x^2} \cdot \dfrac{3(x - 2)}{3(x - 2)} = \dfrac{3x^2 - 3x - 6}{12x^2(x - 2)} \leftarrow$$

$$\dfrac{x - 3}{6x^2 - 12x} = \dfrac{x - 3}{6x(x - 2)} \cdot \dfrac{2x}{2x} = \dfrac{2x^2 - 6x}{12x^2(x - 2)} \leftarrow$$

$\left. \begin{array}{c} \\ \\ \end{array} \right\}$ LCM

Example 3 Write the fractions $\dfrac{x + 2}{3x^2}$ and $\dfrac{x - 1}{8xy}$ in terms of the LCM of the denominators.

Solution   The LCM is $24x^2y$.

$$\frac{x+2}{3x^2} = \frac{x+2}{3x^2} \cdot \frac{8y}{8y} = \frac{8xy+16y}{24x^2y}$$ • The product of $3x^2$ and $8y$ is the LCM.

$$\frac{x-1}{8xy} = \frac{x-1}{8xy} \cdot \frac{3x}{3x} = \frac{3x^2-3x}{24x^2y}$$ • The product of $8xy$ and $3x$ is the LCM.

Problem 3 Write the fractions $\dfrac{x-3}{4xy^2}$ and $\dfrac{2x+1}{9y^2z}$ in terms of the LCM of the denominators.

Solution   See page S20.

Example 4 Write the fractions $\dfrac{2x-1}{2x-x^2}$ and $\dfrac{x}{x^2+x-6}$ in terms of the LCM of the denominators.

Solution $\dfrac{2x-1}{2x-x^2} = \dfrac{2x-1}{-(x^2-2x)} = -\dfrac{2x-1}{x^2-2x}$ • Rewrite $\dfrac{2x-1}{2x-x^2}$ with a denominator of x^2-2x.

The LCM is $x(x-2)(x+3)$.

$$\frac{2x-1}{2x-x^2} = -\frac{2x-1}{x(x-2)} \cdot \frac{x+3}{x+3} = -\frac{2x^2+5x-3}{x(x-2)(x+3)}$$

$$\frac{x}{x^2+x-6} = \frac{x}{(x-2)(x+3)} \cdot \frac{x}{x} = \frac{x^2}{x(x-2)(x+3)}$$

Problem 4 Write the fractions $\dfrac{x+4}{x^2-3x-10}$ and $\dfrac{2x}{25-x^2}$ in terms of the LCM of the denominators.

Solution   See page S20.

9.2   Concept Review

Determine whether the statement is always true, sometimes true, or never true.

1. The least common multiple of two numbers is the smallest number that contains all the prime factors of both numbers.

2. The LCD is the least common multiple of the denominators of two or more fractions.

3. The LCM of x^2, x^5, and x^8 is x^2.

4. We can rewrite $\dfrac{x}{y}$ as $\dfrac{4x}{4y}$ by using the Multiplication Property of One.

5. Rewriting two fractions in terms of the LCM of their denominators is the reverse process of simplifying the fractions.

6. To rewrite a rational expression in terms of a common denominator, determine what factor you must multiply the denominator by so that the denominator will be the common denominator. Then multiply the numerator and denominator of the fraction by that factor.

9.2 | **Exercises**

1 Find the LCM of the expressions.

1. $8x^3y$
$12xy^2$

2. $6ab^2$
$18ab^3$

3. $10x^4y^2$
$15x^3y$

4. $12a^2b$
$18ab^3$

5. $8x^2$
$4x^2 + 8x$

6. $6y^2$
$4y + 12$

7. $2x^2y$
$3x^2 + 12x$

8. $4xy^2$
$6xy^2 + 12y^2$

9. $8x^2(x - 1)^2$
$10x^3(x - 1)$

10. $3x + 3$
$2x^2 + 4x + 2$

11. $4x - 12$
$2x^2 - 12x + 18$

12. $(x - 1)(x + 2)$
$(x - 1)(x + 3)$

13. $(2x - 1)(x + 4)$
$(2x + 1)(x + 4)$

14. $(2x + 3)^2$
$(2x + 3)(x - 5)$

15. $(x - 7)(x + 2)$
$(x - 7)^2$

16. $x - 1$
$x - 2$
$(x - 1)(x - 2)$

17. $(x + 4)(x - 3)$
$x + 4$
$x - 3$

18. $x^2 - x - 6$
$x^2 + x - 12$

19. $x^2 + 3x - 10$
$x^2 + 5x - 14$

20. $x^2 + 5x + 4$
$x^2 - 3x - 28$

21. $x^2 - 10x + 21$
$x^2 - 8x + 15$

22. $x^2 - 2x - 24$
$x^2 - 36$

23. $x^2 + 7x + 10$
$x^2 - 25$

24. $x^2 - 7x - 30$
$x^2 - 5x - 24$

25. $2x^2 - 7x + 3$
$2x^2 + x - 1$

26. $3x^2 - 11x + 6$
$3x^2 + 4x - 4$

27. $2x^2 - 9x + 10$
$2x^2 + x - 15$

28. $6 + x - x^2$
$x + 2$
$x - 3$

29. $15 + 2x - x^2$
$x - 5$
$x + 3$

30. $5 + 4x - x^2$
$x - 5$
$x + 1$

31. $x^2 + 3x - 18$
$3 - x$
$x + 6$

32. $x^2 - 5x + 6$
$1 - x$
$x - 6$

2 Write each fraction in terms of the LCM of the denominators.

33. $\dfrac{4}{x}; \dfrac{3}{x^2}$

34. $\dfrac{5}{ab^2}; \dfrac{6}{ab}$

35. $\dfrac{x}{3y^2}; \dfrac{z}{4y}$

36. $\dfrac{5y}{6x^2}; \dfrac{7}{9xy}$

37. $\dfrac{y}{x(x - 3)}; \dfrac{6}{x^2}$

38. $\dfrac{a}{y^2}; \dfrac{6}{y(y + 5)}$

39. $\dfrac{9}{(x - 1)^2}; \dfrac{6}{x(x - 1)}$

40. $\dfrac{a^2}{y(y + 7)}; \dfrac{a}{(y + 7)^2}$

41. $\dfrac{3}{x - 3}; -\dfrac{5}{x(3 - x)}$

42. $\dfrac{b}{y(y-4)}; \dfrac{b^2}{4-y}$

43. $\dfrac{3}{(x-5)^2}; \dfrac{2}{5-x}$

44. $\dfrac{3}{7-y}; \dfrac{2}{(y-7)^2}$

45. $\dfrac{3}{x^2+2x}; \dfrac{4}{x^2}$

46. $\dfrac{2}{y-3}; \dfrac{3}{y^3-3y^2}$

47. $\dfrac{x-2}{x+3}; \dfrac{x}{x-4}$

48. $\dfrac{x^2}{2x-1}; \dfrac{x+1}{x+4}$

49. $\dfrac{3}{x^2+x-2}; \dfrac{x}{x+2}$

50. $\dfrac{3x}{x-5}; \dfrac{4}{x^2-25}$

51. $\dfrac{5}{2x^2-9x+10}; \dfrac{x-1}{2x-5}$

52. $\dfrac{x-3}{3x^2+4x-4}; \dfrac{2}{x+2}$

53. $\dfrac{x}{x^2+x-6}; \dfrac{2x}{x^2-9}$

54. $\dfrac{x-1}{x^2+2x-15}; \dfrac{x}{x^2+6x+5}$

55. $\dfrac{x}{9-x^2}; \dfrac{x-1}{x^2-6x+9}$

56. $\dfrac{2x}{10+3x-x^2}; \dfrac{x+2}{x^2-8x+15}$

57. $\dfrac{3x}{x-5}; \dfrac{x}{x+4}; \dfrac{3}{20+x-x^2}$

58. $\dfrac{x+1}{x+5}; \dfrac{x+2}{x-7}; \dfrac{3}{35+2x-x^2}$

9.2 Applying Concepts

Write each expression in terms of the LCM of the denominators.

59. $\dfrac{3}{10^2}; \dfrac{5}{10^4}$

60. $\dfrac{8}{10^3}; \dfrac{9}{10^5}$

61. $b; \dfrac{5}{b}$

62. $3; \dfrac{2}{n}$

63. $1; \dfrac{y}{y-1}$

64. $x; \dfrac{x}{x^2-1}$

65. $\dfrac{x^2+1}{(x-1)^3}; \dfrac{x+1}{(x-1)^2}; \dfrac{1}{x-1}$

66. $\dfrac{a^2+a}{(a+1)^3}; \dfrac{a+1}{(a+1)^2}; \dfrac{1}{a+1}$

67. $\dfrac{b}{4a^2-4b^2}; \dfrac{a}{8a-8b}$

68. $\dfrac{c}{6c^2+7cd+d^2}; \dfrac{d}{3c^2-3d^2}$

69. $\dfrac{1}{x^2 + 2x + xy + 2y}, \dfrac{1}{x^2 + xy - 2x - 2y}$

70. $\dfrac{1}{ab + 3a - 3b - b^2}, \dfrac{1}{ab + 3a + 3b + b^2}$

71. When is the LCM of two expressions equal to their product?

9.3   Addition and Subtraction of Rational Expressions

1   **Add and subtract rational expressions with the same denominator**

When adding rational expressions in which the denominators are the same, add the numerators. The denominator of the sum is the common denominator.

$$\frac{a}{b} + \frac{c}{b} = \frac{a + c}{b}$$

$$\frac{5x}{18} + \frac{7x}{18} = \frac{12x}{18} = \frac{2x}{3}$$

$$\frac{x}{x^2 - 1} + \frac{1}{x^2 - 1} = \frac{x + 1}{x^2 - 1} = \frac{\overset{1}{\cancel{(x + 1)}}}{(x - 1)\underset{1}{\cancel{(x + 1)}}} = \frac{1}{x - 1}$$

Note that the sum is written in simplest form.

When subtracting rational expressions in which the denominators are the same, subtract the numerators. The denominator of the difference is the common denominator. Write the answer in simplest form.

> **Take Note**
>
> Be careful with signs when subtracting rational expressions. Note that we must subtract the *entire* numerator $2x + 3$.
>
> $(3x - 1) - (2x + 3) =$
> $3x - 1 - 2x - 3$

$$\frac{2x}{x - 2} - \frac{4}{x - 2} = \frac{2x - 4}{x - 2} = \frac{2\overset{1}{\cancel{(x - 2)}}}{\underset{1}{\cancel{x - 2}}} = 2$$

$$\frac{3x - 1}{x^2 - 5x + 4} - \frac{2x + 3}{x^2 - 5x + 4} = \frac{(3x - 1) - (2x + 3)}{x^2 - 5x + 4}$$

$$= \frac{x - 4}{x^2 - 5x + 4}$$

$$= \frac{\overset{1}{\cancel{(x - 4)}}}{\underset{1}{\cancel{(x - 4)}}(x - 1)}$$

$$= \frac{1}{x - 1}$$

Example 1   Add or subtract.   **A.** $\dfrac{7}{x^2} + \dfrac{9}{x^2}$ **B.** $\dfrac{3x^2}{x^2 - 1} - \dfrac{x + 4}{x^2 - 1}$

Solution   **A.** $\dfrac{7}{x^2} + \dfrac{9}{x^2} = \dfrac{7+9}{x^2}$ • The denominators are the same. Add the numerators.

$$= \dfrac{16}{x^2}$$

B. $\dfrac{3x^2}{x^2-1} - \dfrac{x+4}{x^2-1} = \dfrac{3x^2-(x+4)}{x^2-1}$ • The denominators are the same. Subtract the numerators.

$$= \dfrac{3x^2 - x - 4}{x^2 - 1}$$

$$= \dfrac{(3x-4)\overset{1}{\cancel{(x+1)}}}{(x-1)\underset{1}{\cancel{(x+1)}}}$$ • Write the answer in simplest form.

$$= \dfrac{3x-4}{x-1}$$

Problem 1   Add or subtract.   **A.** $\dfrac{3}{xy} + \dfrac{12}{xy}$ **B.** $\dfrac{2x^2}{x^2-x-12} - \dfrac{7x+4}{x^2-x-12}$

Solution   See page S20.

2   ## Add and subtract rational expressions with different denominators

VIDEO & DVD   CD TUTOR   WWW WEB   SSM

Before two fractions with different denominators can be added or subtracted, each fraction must be expressed in terms of a common denominator. This common denominator is the LCM of the denominators of the fractions.

Add: $\dfrac{x-3}{x^2-2x} + \dfrac{6}{x^2-4}$

Find the LCM of the denominators.

$x^2 - 2x = x(x-2)$
$x^2 - 4 = (x-2)(x+2)$

The LCM is $x(x-2)(x+2)$.

$$\dfrac{x-3}{x^2-2x} + \dfrac{6}{x^2-4}$$

Write each fraction in terms of the LCM of the denominators.

$$= \dfrac{x-3}{x(x-2)} \cdot \dfrac{x+2}{x+2} + \dfrac{6}{(x-2)(x+2)} \cdot \dfrac{x}{x}$$

Multiply the factors in the numerator.

$$= \dfrac{x^2 - x - 6}{x(x-2)(x+2)} + \dfrac{6x}{x(x-2)(x+2)}$$

Add the fractions.

$$= \dfrac{x^2 - x - 6 + 6x}{x(x-2)(x+2)}$$

$$= \dfrac{x^2 + 5x - 6}{x(x-2)(x+2)}$$

Factor the numerator to determine whether there are common factors in the numerator and denominator.

$$= \dfrac{(x+6)(x-1)}{x(x-2)(x+2)}$$

Example 2 Add or subtract.

A. $\dfrac{y}{x} - \dfrac{4y}{3x} + \dfrac{3y}{4x}$ B. $\dfrac{2x}{x-3} - \dfrac{5}{3-x}$ C. $x - \dfrac{3}{5x}$

Solution

A. The LCM of the denominators is $12x$.

• Find the LCM of the denominators.

$$\dfrac{y}{x} - \dfrac{4y}{3x} + \dfrac{3y}{4x}$$

$$= \dfrac{y}{x} \cdot \dfrac{12}{12} - \dfrac{4y}{3x} \cdot \dfrac{4}{4} + \dfrac{3y}{4x} \cdot \dfrac{3}{3}$$

• Write each fraction in terms of the LCM.

$$= \dfrac{12y}{12x} - \dfrac{16y}{12x} + \dfrac{9y}{12x}$$

$$= \dfrac{12y - 16y + 9y}{12x}$$

$$= \dfrac{5y}{12x}$$

B. The LCM of $x - 3$ and $3 - x$ is $x - 3$.

• $3 - x = -(x - 3)$

$$\dfrac{2x}{x-3} - \dfrac{5}{3-x}$$

$$= \dfrac{2x}{x-3} - \dfrac{5}{-(x-3)} \cdot \dfrac{-1}{-1}$$

• Multiply $\dfrac{5}{-(x-3)}$ by $\dfrac{-1}{-1}$ so that the denominator will be $x - 3$.

$$= \dfrac{2x}{x-3} - \dfrac{-5}{x-3}$$

$$= \dfrac{2x - (-5)}{x-3}$$

$$= \dfrac{2x + 5}{x-3}$$

C. The LCM of the denominators is $5x$.

$$x - \dfrac{3}{5x} = \dfrac{x}{1} - \dfrac{3}{5x}$$

$$= \dfrac{x}{1} \cdot \dfrac{5x}{5x} - \dfrac{3}{5x}$$

$$= \dfrac{5x^2}{5x} - \dfrac{3}{5x}$$

$$= \dfrac{5x^2 - 3}{5x}$$

Problem 2 Add or subtract.

A. $\dfrac{z}{8y} - \dfrac{4z}{3y} + \dfrac{5z}{4y}$ B. $\dfrac{5x}{x-2} - \dfrac{3}{2-x}$ C. $y + \dfrac{5}{y-7}$

Solution See page S20.

Example 3 Add or subtract.   **A.** $\dfrac{2x}{2x-3} - \dfrac{1}{x+1}$

B. $\dfrac{x+3}{x^2-2x-8} + \dfrac{3}{4-x}$

Solution   **A.** The LCM is $(2x-3)(x+1)$.

$$\dfrac{2x}{2x-3} - \dfrac{1}{x+1} = \dfrac{2x}{2x-3} \cdot \dfrac{x+1}{x+1} - \dfrac{1}{x+1} \cdot \dfrac{2x-3}{2x-3}$$

$$= \dfrac{2x^2+2x}{(2x-3)(x+1)} - \dfrac{2x-3}{(2x-3)(x+1)}$$

$$= \dfrac{(2x^2+2x)-(2x-3)}{(2x-3)(x+1)}$$

$$= \dfrac{2x^2+3}{(2x-3)(x+1)}$$

B. The LCM is $(x-4)(x+2)$.

$$\dfrac{x+3}{x^2-2x-8} + \dfrac{3}{4-x} = \dfrac{x+3}{(x-4)(x+2)} + \dfrac{3}{-(x-4)} \cdot \dfrac{-1 \cdot (x+2)}{-1 \cdot (x+2)}$$

$$= \dfrac{x+3}{(x-4)(x+2)} + \dfrac{-3(x+2)}{(x-4)(x+2)}$$

$$= \dfrac{(x+3)+(-3)(x+2)}{(x-4)(x+2)}$$

$$= \dfrac{x+3-3x-6}{(x-4)(x+2)}$$

$$= \dfrac{-2x-3}{(x-4)(x+2)}$$

Problem 3 Add or subtract.   **A.** $\dfrac{4x}{3x-1} - \dfrac{9}{x+4}$ **B.** $\dfrac{2x-1}{x^2-25} + \dfrac{2}{5-x}$

Solution   See page S21.

9.3 | Concept Review

Determine whether the statement is always true, sometimes true, or never true.

1. To add two fractions, add the numerators and the denominators.

2. The procedure for subtracting two rational expressions is the same as that for subtracting two arithmetic fractions.

3. To add two rational expressions, first multiply both expressions by the LCM of their denominators.

4. If $x \neq -2$ and $x \neq 0$, then $\dfrac{x}{x+2} + \dfrac{3}{x+2} = \dfrac{x+3}{x+2} = \dfrac{3}{2}$.

5. If $y \neq 8$, then $\dfrac{y}{y-8} - \dfrac{8}{y-8} = \dfrac{y-8}{y-8} = 1$.

6. If $x \neq 0$, then $\dfrac{4x-3}{x} - \dfrac{3x+1}{x} = \dfrac{4x-3-3x+1}{x} = \dfrac{x-2}{x}$.

9.3 Exercises

1 Add or subtract.

1. $\dfrac{3}{y^2} + \dfrac{8}{y^2}$

2. $\dfrac{6}{ab} - \dfrac{2}{ab}$

3. $\dfrac{3}{x+4} - \dfrac{10}{x+4}$

4. $\dfrac{x}{x+6} - \dfrac{2}{x+6}$

5. $\dfrac{3x}{2x+3} + \dfrac{5x}{2x+3}$

6. $\dfrac{6y}{4y+1} - \dfrac{11y}{4y+1}$

7. $\dfrac{2x+1}{x-3} + \dfrac{3x+6}{x-3}$

8. $\dfrac{4x+3}{2x-7} + \dfrac{3x-8}{2x-7}$

9. $\dfrac{5x-1}{x+9} - \dfrac{3x+4}{x+9}$

10. $\dfrac{6x-5}{x-10} - \dfrac{3x-4}{x-10}$

11. $\dfrac{x-7}{2x+7} - \dfrac{4x-3}{2x+7}$

12. $\dfrac{2n}{3n+4} - \dfrac{5n-3}{3n+4}$

13. $\dfrac{x}{x^2+2x-15} - \dfrac{3}{x^2+2x-15}$

14. $\dfrac{3x}{x^2+3x-10} - \dfrac{6}{x^2+3x-10}$

15. $\dfrac{2x+3}{x^2-x-30} - \dfrac{x-2}{x^2-x-30}$

16. $\dfrac{3x-1}{x^2+5x-6} - \dfrac{2x-7}{x^2+5x-6}$

17. $\dfrac{4y+7}{2y^2+7y-4} - \dfrac{y-5}{2y^2+7y-4}$

18. $\dfrac{x+1}{2x^2-5x-12} + \dfrac{x+2}{2x^2-5x-12}$

19. $\dfrac{2x^2+3x}{x^2-9x+20} + \dfrac{2x^2-3}{x^2-9x+20} - \dfrac{4x^2+2x+1}{x^2-9x+20}$

20. $\dfrac{2x^2+3x}{x^2-2x-63} - \dfrac{x^2-3x+21}{x^2-2x-63} - \dfrac{x-7}{x^2-2x-63}$

2 **21.** Why must two rational expressions have the same denominators before they can be added or subtracted?

22. Explain the procedure for adding rational expressions with different denominators.

Add or subtract.

23. $\dfrac{4}{x} + \dfrac{5}{y}$

24. $\dfrac{7}{a} + \dfrac{5}{b}$

25. $\dfrac{12}{x} - \dfrac{5}{2x}$

26. $\dfrac{5}{3a} - \dfrac{3}{4a}$

27. $\dfrac{1}{2x} - \dfrac{5}{4x} + \dfrac{7}{6x}$

28. $\dfrac{7}{4y} + \dfrac{11}{6y} - \dfrac{8}{3y}$

29. $\dfrac{5}{3x} - \dfrac{2}{x^2} + \dfrac{3}{2x}$

30. $\dfrac{6}{y^2} + \dfrac{3}{4y} - \dfrac{2}{5y}$

31. $\dfrac{2}{x} - \dfrac{3}{2y} + \dfrac{3}{5x} - \dfrac{1}{4y}$

32. $\dfrac{5}{2a} + \dfrac{7}{3b} - \dfrac{2}{b} - \dfrac{3}{4a}$

33. $\dfrac{2x + 1}{3x} + \dfrac{x - 1}{5x}$

34. $\dfrac{4x - 3}{6x} + \dfrac{2x + 3}{4x}$

35. $\dfrac{x - 3}{6x} + \dfrac{x + 4}{8x}$

36. $\dfrac{2x - 3}{2x} + \dfrac{x + 3}{3x}$

37. $\dfrac{2x + 9}{9x} - \dfrac{x - 5}{5x}$

38. $\dfrac{3y - 2}{12y} - \dfrac{y - 3}{18y}$

39. $\dfrac{x + 4}{2x} - \dfrac{x - 1}{x^2}$

40. $\dfrac{x - 2}{3x^2} - \dfrac{x + 4}{x}$

41. $\dfrac{x - 10}{4x^2} + \dfrac{x + 1}{2x}$

42. $\dfrac{x + 5}{3x^2} + \dfrac{2x + 1}{2x}$

43. $y + \dfrac{8}{3y}$

44. $\dfrac{7}{2n} - n$

45. $\dfrac{4}{x + 4} + x$

46. $x + \dfrac{3}{x + 2}$

47. $5 - \dfrac{x - 2}{x + 1}$

48. $3 + \dfrac{x - 1}{x + 1}$

49. $\dfrac{2x + 1}{6x^2} - \dfrac{x - 4}{4x}$

50. $\dfrac{x + 3}{6x} - \dfrac{x - 3}{8x^2}$

51. $\dfrac{x + 2}{xy} - \dfrac{3x - 2}{x^2y}$

52. $\dfrac{3x - 1}{xy^2} - \dfrac{2x + 3}{xy}$

53. $\dfrac{4x - 3}{3x^2y} + \dfrac{2x + 1}{4xy^2}$

54. $\dfrac{5x + 7}{6xy^2} - \dfrac{4x - 3}{8x^2y}$

55. $\dfrac{x-2}{8x^2} - \dfrac{x+7}{12xy}$

56. $\dfrac{3x-1}{6y^2} - \dfrac{x+5}{9xy}$

57. $\dfrac{4}{x-2} + \dfrac{5}{x+3}$

58. $\dfrac{2}{x-3} + \dfrac{5}{x-4}$

59. $\dfrac{6}{x-7} - \dfrac{4}{x+3}$

60. $\dfrac{3}{y+6} - \dfrac{4}{y-3}$

61. $\dfrac{2x}{x+1} + \dfrac{1}{x-3}$

62. $\dfrac{3x}{x-4} + \dfrac{2}{x+6}$

63. $\dfrac{4x}{2x-1} - \dfrac{5}{x-6}$

64. $\dfrac{6x}{x+5} - \dfrac{3}{2x+3}$

65. $\dfrac{2a}{a-7} + \dfrac{5}{7-a}$

66. $\dfrac{4x}{6-x} + \dfrac{5}{x-6}$

67. $\dfrac{x}{x^2-9} + \dfrac{3}{x-3}$

68. $\dfrac{y}{y^2-16} + \dfrac{1}{y-4}$

69. $\dfrac{2x}{x^2-x-6} - \dfrac{3}{x+2}$

70. $\dfrac{5x}{x^2+2x-8} - \dfrac{2}{x+4}$

71. $\dfrac{3x-1}{x^2-10x+25} - \dfrac{3}{x-5}$

72. $\dfrac{2a+3}{a^2-7a+12} - \dfrac{2}{a-3}$

73. $\dfrac{x+4}{x^2-x-42} + \dfrac{3}{7-x}$

74. $\dfrac{x+3}{x^2-3x-10} + \dfrac{2}{5-x}$

75. $\dfrac{x}{2x+4} - \dfrac{2}{x^2+2x}$

76. $\dfrac{x+2}{4x+16} - \dfrac{2}{x^2+4x}$

77. $\dfrac{x-1}{x^2-x-2} + \dfrac{3}{x^2-3x+2}$

78. $\dfrac{a+2}{a^2+a-2} + \dfrac{3}{a^2+2a-3}$

79. $\dfrac{1}{x+1} + \dfrac{x}{x-6} - \dfrac{5x-2}{x^2-5x-6}$

80. $\dfrac{x}{x-4} + \dfrac{5}{x+5} - \dfrac{11x-8}{x^2+x-20}$

81. $\dfrac{3x+1}{x-1} - \dfrac{x-1}{x-3} + \dfrac{x+1}{x^2-4x+3}$

82. $\dfrac{4x+1}{x-8} - \dfrac{3x+2}{x+4} - \dfrac{49x+4}{x^2-4x-32}$

9.3   Applying Concepts

Simplify.

83. $\dfrac{a}{a-b} + \dfrac{b}{b-a} + 1$

84. $\dfrac{y}{x-y} + 2 - \dfrac{x}{y-x}$

85. $b - 3 + \dfrac{5}{b+4}$

86. $2y - 1 + \dfrac{6}{y+5}$

87. $\dfrac{(n+1)^2}{(n-1)^2} - 1$

88. $1 - \dfrac{(y-2)^2}{(y+2)^2}$

89. $\dfrac{2x+9}{3-x} + \dfrac{x+5}{x+7} - \dfrac{2x^2+3x-3}{x^2+4x-21}$

90. $\dfrac{3x+5}{x+5} - \dfrac{x+1}{2-x} - \dfrac{4x^2-3x-1}{x^2+3x-10}$

91. $\dfrac{x^2+x-6}{x^2+2x-8} \cdot \dfrac{x^2+5x+4}{x^2+2x-3} - \dfrac{2}{x-1}$

92. $\dfrac{x^2+9x+20}{x^2+4x-5} \div \dfrac{x^2-49}{x^2+6x-7} - \dfrac{x}{x-7}$

93. $\dfrac{x^2-9}{x^2+6x+9} \div \dfrac{x^2+x-20}{x^2-x-12} + \dfrac{1}{x+1}$

94. $\dfrac{x^2-25}{x^2+10x+25} \cdot \dfrac{x^2-7x+10}{x^2-x-2} + \dfrac{1}{x+1}$

95. Find the sum of the following: $\dfrac{1}{1\cdot2} + \dfrac{1}{2\cdot3}$

$$\dfrac{1}{1\cdot2} + \dfrac{1}{2\cdot3} + \dfrac{1}{3\cdot4}$$

$$\dfrac{1}{1\cdot2} + \dfrac{1}{2\cdot3} + \dfrac{1}{3\cdot4} + \dfrac{1}{4\cdot5}$$

Note the pattern for these sums, and then find the sum of 50 terms, of 100 terms, and of 1000 terms.

Rewrite the expression as the sum of two fractions in simplest form.

96. $\dfrac{5b+4a}{ab}$

97. $\dfrac{6x+7y}{xy}$

98. $\dfrac{3x^2+4xy}{x^2y^2}$

99. $\dfrac{2mn^2+8m^2n}{m^3n^3}$

100. **Fuel Efficiency** Suppose that you drive about 12,000 miles per year and that the cost of gasoline averages $1.60 per gallon.
   a. Let x represent the number of miles per gallon your car gets. Write a variable expression for the amount you spend on gasoline in one year.
   b. Write and simplify a variable expression for the amount of money you will save each year if you can increase your gas mileage by 5 miles per gallon.
   c. If you currently get 25 miles per gallon and you increase your gas mileage by 5 miles per gallon, how much will you save in one year?

9.4 | Complex Fractions

1   Simplify complex fractions

A **complex fraction** is a fraction whose numerator or denominator contains one or more fractions. Examples of complex fractions are shown at the right.

$$\dfrac{3}{2 - \dfrac{1}{2}}, \qquad \dfrac{4 + \dfrac{1}{x}}{3 + \dfrac{2}{x}}, \qquad \dfrac{\dfrac{1}{x - 1} + x + 3}{x - 3 + \dfrac{1}{x + 4}}$$

Point of Interest

There are many instances of complex fractions in application problems. The fraction

$$\dfrac{1}{\dfrac{1}{r_1} + \dfrac{1}{r_2}}$$ is used to determine the

total resistance in certain electric circuits.

Simplify: $\dfrac{1 - \dfrac{4}{x^2}}{1 + \dfrac{2}{x}}$

Find the LCM of the denominators of the fractions in the numerator and denominator.

The LCM of x^2 and x is x^2.

Take Note

First, we are multiplying the complex fraction by $\dfrac{x^2}{x^2}$, which equals 1, so we are not changing the value of the fraction. Second, we are using the Distributive Property to multiply $\left(1 - \dfrac{4}{x^2}\right)x^2$ and $\left(1 + \dfrac{2}{x}\right)x^2$.

Multiply the numerator and denominator of the complex fraction by the LCM.

$$\dfrac{1 - \dfrac{4}{x^2}}{1 + \dfrac{2}{x}} = \dfrac{1 - \dfrac{4}{x^2}}{1 + \dfrac{2}{x}} \cdot \dfrac{x^2}{x^2}$$

$$= \dfrac{1 \cdot x^2 - \dfrac{4}{x^2} \cdot x^2}{1 \cdot x^2 + \dfrac{2}{x} \cdot x^2}$$

$$= \dfrac{x^2 - 4}{x^2 + 2x}$$

Simplify.

$$= \dfrac{(x - 2)\cancel{(x + 2)}}{x\cancel{(x + 2)}}$$

$$= \dfrac{x - 2}{x}$$

The method shown above of simplifying a complex fraction by multiplying the numerator and denominator by the LCM of the denominators is used in Example 1. However, a different approach is to rewrite the numerator and denominator of the complex fraction as single fractions and then divide the numerator by the denominator. The example shown above is simplified below using this alternate method.

Take Note

Recall that the fraction bar can be read "divided by."

Rewrite the numerator and denominator of the complex fraction as single fractions.

$$\dfrac{1 - \dfrac{4}{x^2}}{1 + \dfrac{2}{x}} = \dfrac{1 \cdot \dfrac{x^2}{x^2} - \dfrac{4}{x^2}}{1 \cdot \dfrac{x}{x} + \dfrac{2}{x}} = \dfrac{\dfrac{x^2}{x^2} - \dfrac{4}{x^2}}{\dfrac{x}{x} + \dfrac{2}{x}} = \dfrac{\dfrac{x^2 - 4}{x^2}}{\dfrac{x + 2}{x}}$$

Divide the numerator of the complex fraction by the denominator.

$$= \dfrac{x^2 - 4}{x^2} \div \dfrac{x + 2}{x} = \dfrac{x^2 - 4}{x^2} \cdot \dfrac{x}{x + 2}$$

Multiply the fractions.
Factor the numerator.

$$= \frac{(x^2 - 4)x}{x^2(x + 2)} = \frac{(x + 2)(x - 2)x}{x^2(x + 2)}$$

Simplify.

$$= \frac{x - 2}{x}$$

Note that this is the same result as shown above.

Example 1 Simplify.   **A.** $\dfrac{\dfrac{1}{x} + \dfrac{1}{2}}{\dfrac{1}{x^2} - \dfrac{1}{4}}$ **B.** $\dfrac{1 - \dfrac{2}{x} - \dfrac{15}{x^2}}{1 - \dfrac{11}{x} + \dfrac{30}{x^2}}$

Solution   **A.** The LCM of x, 2, x^2, and 4 is $4x^2$.

$$\frac{\dfrac{1}{x} + \dfrac{1}{2}}{\dfrac{1}{x^2} - \dfrac{1}{4}} = \frac{\dfrac{1}{x} + \dfrac{1}{2}}{\dfrac{1}{x^2} - \dfrac{1}{4}} \cdot \frac{4x^2}{4x^2}$$

$$= \frac{\dfrac{1}{x} \cdot 4x^2 + \dfrac{1}{2} \cdot 4x^2}{\dfrac{1}{x^2} \cdot 4x^2 - \dfrac{1}{4} \cdot 4x^2}$$

$$= \frac{4x + 2x^2}{4 - x^2}$$

$$= \frac{2x(2 \overset{1}{\cancel{+ x}})}{(2 - x)(2 \underset{1}{\cancel{+ x}})}$$

$$= \frac{2x}{2 - x}$$

B. The LCM of x and x^2 is x^2.

$$\frac{1 - \dfrac{2}{x} - \dfrac{15}{x^2}}{1 - \dfrac{11}{x} + \dfrac{30}{x^2}} = \frac{1 - \dfrac{2}{x} - \dfrac{15}{x^2}}{1 - \dfrac{11}{x} + \dfrac{30}{x^2}} \cdot \frac{x^2}{x^2}$$

$$= \frac{1 \cdot x^2 - \dfrac{2}{x} \cdot x^2 - \dfrac{15}{x^2} \cdot x^2}{1 \cdot x^2 - \dfrac{11}{x} \cdot x^2 + \dfrac{30}{x^2} \cdot x^2}$$

$$= \frac{x^2 - 2x - 15}{x^2 - 11x + 30}$$

$$= \frac{(\overset{1}{\cancel{x - 5}})(x + 3)}{(\underset{1}{\cancel{x - 5}})(x - 6)}$$

$$= \frac{x + 3}{x - 6}$$

Problem 1 Simplify. **A.** $\dfrac{\dfrac{1}{3} - \dfrac{1}{x}}{\dfrac{1}{9} - \dfrac{1}{x^2}}$ **B.** $\dfrac{1 + \dfrac{4}{x} + \dfrac{3}{x^2}}{1 + \dfrac{10}{x} + \dfrac{21}{x^2}}$

Solution See page S21.

9.4 Concept Review

Determine whether the statement is always true, sometimes true, or never true.

1. A complex fraction is a fraction that has fractions in the numerator and denominator.

2. We can simplify a complex fraction by first finding the least common multiple of the denominators of the fractions in the numerator and denominator of the complex fraction.

3. To simplify a complex fraction, multiply the complex fraction by the LCM of the denominators of the fractions in the numerator and denominator of the complex fraction.

4. When we multiply the numerator and denominator of a complex fraction by the same expression, we are using the Multiplication Property of One.

5. Our goal in simplifying a complex fraction is to rewrite it so that there are no fractions in the numerator or in the denominator. We then express the fraction in simplest form.

6. For the complex fraction $\dfrac{1 + \dfrac{4}{x} + \dfrac{3}{x^2}}{1 + \dfrac{10}{x} + \dfrac{21}{x^2}}$, x can be any real number except 0.

9.4 Exercises

1 Simplify.

1. $\dfrac{1 + \dfrac{3}{x}}{1 - \dfrac{9}{x^2}}$

2. $\dfrac{1 + \dfrac{4}{x}}{1 - \dfrac{16}{x^2}}$

3. $\dfrac{2 - \dfrac{8}{x + 4}}{3 - \dfrac{12}{x + 4}}$

4. $\dfrac{5 - \dfrac{25}{x + 5}}{1 - \dfrac{3}{x + 5}}$

5. $\dfrac{1 + \dfrac{5}{y - 2}}{1 - \dfrac{2}{y - 2}}$

6. $\dfrac{2 - \dfrac{11}{2x - 1}}{3 - \dfrac{17}{2x - 1}}$

7. $\dfrac{4 - \dfrac{2}{x + 7}}{5 + \dfrac{1}{x + 7}}$

8. $\dfrac{5 + \dfrac{3}{x - 8}}{2 - \dfrac{1}{x - 8}}$

9. $\dfrac{\dfrac{3}{x - 2} + 3}{\dfrac{4}{x - 2} + 4}$

10. $\dfrac{\dfrac{3}{2x + 1} - 3}{2 - \dfrac{4x}{2x + 1}}$

11. $\dfrac{2 - \dfrac{3}{x} - \dfrac{2}{x^2}}{2 + \dfrac{5}{x} + \dfrac{2}{x^2}}$

12. $\dfrac{2 + \dfrac{5}{x} - \dfrac{12}{x^2}}{4 - \dfrac{4}{x} - \dfrac{3}{x^2}}$

13. $\dfrac{1 - \dfrac{1}{x} - \dfrac{6}{x^2}}{1 - \dfrac{9}{x^2}}$

14. $\dfrac{1 + \dfrac{4}{x} + \dfrac{4}{x^2}}{1 - \dfrac{2}{x} - \dfrac{8}{x^2}}$

15. $\dfrac{1 - \dfrac{5}{x} - \dfrac{6}{x^2}}{1 + \dfrac{6}{x} + \dfrac{5}{x^2}}$

16. $\dfrac{1 - \dfrac{7}{a} + \dfrac{12}{a^2}}{1 + \dfrac{1}{a} - \dfrac{20}{a^2}}$

17. $\dfrac{1 - \dfrac{6}{x} + \dfrac{8}{x^2}}{\dfrac{4}{x^2} + \dfrac{3}{x} - 1}$

18. $\dfrac{1 + \dfrac{3}{x} - \dfrac{18}{x^2}}{\dfrac{21}{x^2} - \dfrac{4}{x} - 1}$

19. $\dfrac{x - \dfrac{4}{x + 3}}{1 + \dfrac{1}{x + 3}}$

20. $\dfrac{y + \dfrac{1}{y - 2}}{1 + \dfrac{1}{y - 2}}$

21. $\dfrac{1 - \dfrac{x}{2x + 1}}{x - \dfrac{1}{2x + 1}}$

22. $\dfrac{1 - \dfrac{2x - 2}{3x - 1}}{x - \dfrac{4}{3x - 1}}$

23. $\dfrac{x - 5 + \dfrac{14}{x + 4}}{x + 3 - \dfrac{2}{x + 4}}$

24. $\dfrac{a + 4 + \dfrac{5}{a - 2}}{a + 6 + \dfrac{15}{a - 2}}$

25. $\dfrac{x + 3 - \dfrac{10}{x - 6}}{x + 2 - \dfrac{20}{x - 6}}$

26. $\dfrac{x - 7 + \dfrac{5}{x - 1}}{x - 3 + \dfrac{1}{x - 1}}$

27. $\dfrac{y - 6 + \dfrac{22}{2y + 3}}{y - 5 + \dfrac{11}{2y + 3}}$

28. $\dfrac{x + 2 - \dfrac{12}{2x - 1}}{x + 1 - \dfrac{9}{2x - 1}}$

29. $\dfrac{x - \dfrac{2}{2x - 3}}{2x - 1 - \dfrac{8}{2x - 3}}$

30. $\dfrac{x + 3 - \dfrac{18}{2x + 1}}{x - \dfrac{6}{2x + 1}}$

31. $\dfrac{1 - \dfrac{2}{x + 1}}{1 + \dfrac{1}{x - 2}}$

32. $\dfrac{1 - \dfrac{1}{x + 2}}{1 + \dfrac{2}{x - 1}}$

33. $\dfrac{1 - \dfrac{2}{x + 4}}{1 + \dfrac{3}{x - 1}}$

34. $\dfrac{1 + \dfrac{1}{x - 2}}{1 - \dfrac{3}{x + 2}}$

35. $\dfrac{\dfrac{1}{x} - \dfrac{2}{x - 1}}{\dfrac{3}{x} + \dfrac{1}{x - 1}}$

36. $\dfrac{\dfrac{3}{n + 1} + \dfrac{1}{n}}{\dfrac{2}{n + 1} + \dfrac{3}{n}}$

37. $\dfrac{\dfrac{3}{2x-1}-\dfrac{1}{x}}{\dfrac{4}{x}+\dfrac{2}{2x-1}}$

38. $\dfrac{\dfrac{4}{3x+1}+\dfrac{3}{x}}{\dfrac{6}{x}-\dfrac{2}{3x+1}}$

39. $\dfrac{\dfrac{3}{b-4}-\dfrac{2}{b+1}}{\dfrac{5}{b+1}-\dfrac{1}{b-4}}$

40. $\dfrac{\dfrac{5}{x-5}-\dfrac{3}{x-1}}{\dfrac{6}{x-1}+\dfrac{2}{x-5}}$

9.4   Applying Concepts

Simplify.

41. $1+\dfrac{1}{1+\dfrac{1}{2}}$

42. $1+\dfrac{1}{1+\dfrac{1}{1+\dfrac{1}{2}}}$

43. $1-\dfrac{1}{1-\dfrac{1}{x}}$

44. $1-\dfrac{1}{1-\dfrac{1}{y+1}}$

45. $\dfrac{a^{-1}-b^{-1}}{a^{-2}-b^{-2}}$

46. $\dfrac{x^{-2}-y^{-2}}{x^{-2}y^{-2}}$

47. $\left(\dfrac{y}{4}-\dfrac{4}{y}\right)\div\left(\dfrac{4}{y}-3+\dfrac{y}{2}\right)$

48. $\left(\dfrac{b}{8}-\dfrac{8}{b}\right)\div\left(\dfrac{8}{b}-5+\dfrac{b}{2}\right)$

49.   How would you explain to a classmate why we multiply the numerator and denominator of a complex fraction by the LCM of the denominators of the fractions in the numerator and denominator?

9.5   Equations Containing Fractions

1   **Solve equations containing fractions**

In the chapter "Solving Equations and Inequalities," equations containing fractions were solved by the method of clearing denominators. Recall that to **clear denominators,** we multiply each side of an equation by the LCM of the denominators. The result is an equation that contains no fractions. In this section, we will again solve equations containing fractions by multiplying each side of the equation by the LCM of the denominators. The difference between this section and the chapter "Solving Equations and Inequalities" is that the fractions in these equations contain variables in the denominators.

Solve: $\dfrac{3x - 1}{4x} + \dfrac{2}{3x} = \dfrac{7}{6x}$

Find the LCM of the denominators.

The LCM of $4x$, $3x$, and $6x$ is $12x$.

$$\dfrac{3x - 1}{4x} + \dfrac{2}{3x} = \dfrac{7}{6x}$$

Multiply each side of the equation by the LCM of the denominators.

$$12x\left(\dfrac{3x - 1}{4x} + \dfrac{2}{3x}\right) = 12x\left(\dfrac{7}{6x}\right)$$

Simplify using the Distributive Property.

$$12x\left(\dfrac{3x - 1}{4x}\right) + 12x\left(\dfrac{2}{3x}\right) = 12x\left(\dfrac{7}{6x}\right)$$

$$\dfrac{12x}{1}\left(\dfrac{3x - 1}{4x}\right) + \dfrac{12x}{1}\left(\dfrac{2}{3x}\right) = \dfrac{12x}{1}\left(\dfrac{7}{6x}\right)$$

$$3(3x - 1) + 4(2) = 2(7)$$

Solve for x.

$$9x - 3 + 8 = 14$$
$$9x + 5 = 14$$
$$9x = 9$$
$$x = 1$$

1 checks as a solution.
The solution is 1.

Example 1   Solve: $\dfrac{4}{x} - \dfrac{x}{2} = \dfrac{7}{2}$

Solution

$$\dfrac{4}{x} - \dfrac{x}{2} = \dfrac{7}{2}$$

• The LCM of x and 2 is $2x$.

$$2x\left(\dfrac{4}{x} - \dfrac{x}{2}\right) = 2x\left(\dfrac{7}{2}\right)$$

$$\dfrac{2x}{1} \cdot \dfrac{4}{x} - \dfrac{2x}{1} \cdot \dfrac{x}{2} = \dfrac{2x}{1} \cdot \dfrac{7}{2}$$

$$8 - x^2 = 7x$$

• This is a quadratic equation.

$$0 = x^2 + 7x - 8$$
$$0 = (x + 8)(x - 1)$$

$$x + 8 = 0 \qquad x - 1 = 0$$
$$x = -8 \qquad x = 1$$

Both -8 and 1 check as solutions.
The solutions are -8 and 1.

Problem 1   Solve: $x + \dfrac{1}{3} = \dfrac{4}{3x}$

Solution   See page S21.

Occasionally, a value of a variable in a fractional equation makes one of the denominators zero. In this case, that value of the variable is not a solution of the equation.

Solve: $\dfrac{2x}{x-2} = 1 + \dfrac{4}{x-2}$

Find the LCM of the denominators.	The LCM is $x - 2$.

$$\dfrac{2x}{x-2} = 1 + \dfrac{4}{x-2}$$

Multiply each side of the equation by the LCM of the denominators.

$$(x-2)\dfrac{2x}{x-2} = (x-2)\left(1 + \dfrac{4}{x-2}\right)$$

Simplify using the Distributive Property.

$$(x-2)\left(\dfrac{2x}{x-2}\right) = (x-2) \cdot 1 + (x-2) \cdot \dfrac{4}{x-2}$$

Solve for x.

$$2x = x - 2 + 4$$
$$2x = x + 2$$
$$x = 2$$

When x is replaced by 2, the denominators of $\dfrac{2x}{x-2}$ and $\dfrac{4}{x-2}$ are zero. Therefore, 2 is not a solution of the equation.

The equation has no solution.

Example 2   Solve: $\dfrac{3x}{x-4} = 5 + \dfrac{12}{x-4}$

Solution

$$\dfrac{3x}{x-4} = 5 + \dfrac{12}{x-4}$$ • The LCM is $x - 4$.

$$(x-4) \cdot \dfrac{3x}{x-4} = (x-4)\left(5 + \dfrac{12}{x-4}\right)$$
$$3x = (x-4)5 + 12$$
$$3x = 5x - 20 + 12$$
$$3x = 5x - 8$$
$$-2x = -8$$
$$x = 4$$

4 does not check as a solution.
The equation has no solution.

Problem 2   Solve: $\dfrac{5x}{x+2} = 3 - \dfrac{10}{x+2}$

Solution   See page S21.

2   **Solve proportions**

Quantities such as 4 meters, 15 seconds, and 8 gallons are number quantities written with **units**. In these examples, the units are meters, seconds, and gallons.

A **ratio** is the quotient of two quantities that have the same unit.

The length of a living room is 16 ft, and the width is 12 ft. The ratio of the length to the width is written

$$\frac{16 \text{ ft}}{12 \text{ ft}} = \frac{16}{12} = \frac{4}{3}$$ A ratio is in simplest form when the two numbers do not have a common factor. Note that the units are not written.

A **rate** is the quotient of two quantities that have different units.

There are 2 lb of salt in 8 gal of water. The salt-to-water rate is

$$\frac{2 \text{ lb}}{8 \text{ gal}} = \frac{1 \text{ lb}}{4 \text{ gal}}$$ A rate is in simplest form when the two numbers do not have a common factor. The units are written as part of the rate.

A **proportion** is an equation that states the equality of two ratios or rates.

Examples of proportions are shown below.

$$\frac{30 \text{ mi}}{4 \text{ h}} = \frac{15 \text{ mi}}{2 \text{ h}} \qquad \frac{4}{6} = \frac{8}{12} \qquad \frac{3}{4} = \frac{x}{8}$$

Because a proportion is an equation containing fractions, the same method used to solve an equation containing fractions is used to solve a proportion. Multiply each side of the equation by the LCM of the denominators. Then solve for the variable.

Solve the proportion $\dfrac{4}{x} = \dfrac{2}{3}$.

$$\frac{4}{x} = \frac{2}{3}$$

Multiply each side of the proportion by the LCM of the denominators.

$$3x\left(\frac{4}{x}\right) = 3x\left(\frac{2}{3}\right)$$
$$12 = 2x$$

Solve the equation.

$$6 = x$$

The solution is 6.

Example 3   Solve.   **A.** $\dfrac{8}{x + 3} = \dfrac{4}{x}$ **B.** $\dfrac{6}{x + 4} = \dfrac{12}{5x - 13}$

Solution   **A.** $\dfrac{8}{x + 3} = \dfrac{4}{x}$

$$x(x + 3)\frac{8}{x + 3} = x(x + 3)\frac{4}{x}$$ • Multiply each side of the proportion by the LCM of the denominators.
$$8x = (x + 3)4$$
$$8x = 4x + 12$$
$$4x = 12$$
$$x = 3$$ • Remember to check the solution because it is a fractional equation.

The solution is 3.

B.

$$\frac{6}{x + 4} = \frac{12}{5x - 13}$$

$$(5x - 13)(x + 4)\frac{6}{x + 4} = (5x - 13)(x + 4)\frac{12}{5x - 13}$$

$$(5x - 13)6 = (x + 4)12$$

$$30x - 78 = 12x + 48$$

$$18x - 78 = 48$$

$$18x = 126$$

$$x = 7$$

The solution is 7.

Problem 3   Solve.   **A.** $\dfrac{2}{x + 3} = \dfrac{6}{5x + 5}$ **B.** $\dfrac{5}{2x - 3} = \dfrac{10}{x + 3}$

Solution   See pages S21–S22.

---

3   Applications of proportions

Example 4   The monthly loan payment for a car is $29.50 for each $1000 borrowed. At this rate, find the monthly payment for a $9000 car loan.

Strategy   To find the monthly payment, write and solve a proportion using P to represent the monthly car payment.

Solution

$$\frac{29.50}{1000} = \frac{P}{9000}$$

$$9000\left(\frac{29.50}{1000}\right) = 9000\left(\frac{P}{9000}\right)$$

$$265.50 = P$$

- The monthly payments are in the numerators. The loan amounts are in the denominators.

The monthly payment is $265.50.

> **Take Note**
>
> It is also correct to write the proportion with the loan amounts in the numerators and the monthly payments in the denominators. The solution will be the same.

Problem 4   Nine ceramic tiles are required to tile a 4 ft^2 area. At this rate, how many square feet can be tiled with 270 ceramic tiles?

Solution   See page S22.

Example 5   An investment of $1200 earns $96 each year. At the same rate, how much additional money must be invested to earn $128 each year?

Strategy   To find the additional amount of money that must be invested, write and solve a proportion using x to represent the additional money. Then $1200 + x$ is the total amount invested.

Solution

$$\frac{1200}{96} = \frac{1200 + x}{128}$$

$$\frac{25}{2} = \frac{1200 + x}{128}$$

$$128\left(\frac{25}{2}\right) = 128\left(\frac{1200 + x}{128}\right)$$

$$1600 = 1200 + x$$

$$400 = x$$

• The amounts invested are in the numerators. The amounts earned are in the denominators.

• Simplify $\frac{1200}{96}$.

An additional $400 must be invested.

Problem 5   Three ounces of a medication are required for a 120-pound adult. At the same rate, how many additional ounces of the medication are required for a 180-pound adult?

Solution   See page S22.

Take Note

It is also correct to write the proportion with the amounts earned in the numerators and the amounts invested in the denominators. The solution will be the same.

4 ## Problems involving similar triangles

VIDEO & DVD   CD TUTOR   WWW WEB   SSM

Similar objects have the same shape but not necessarily the same size. A tennis ball is similar to a basketball. A model ship is similar to an actual ship.

Similar objects have corresponding parts; for example, the rudder on the model ship corresponds to the rudder on the actual ship. The relationship between the sizes of each of the corresponding parts can be written as a ratio, and each ratio will be the same. If the rudder on the model ship is $\frac{1}{100}$ the size of the rudder on the actual ship, then the model wheelhouse is $\frac{1}{100}$ the size of the actual wheelhouse, the width of the model is $\frac{1}{100}$ the width of the actual ship, and so on.

The two triangles ABC and DEF shown at the right are similar. Side AB corresponds to DE, side BC corresponds to EF, and side AC corresponds to DF. The height CH corresponds to the height FK. The ratios of corresponding parts are equal.

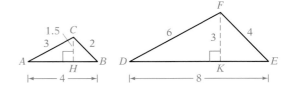

$$\frac{AB}{DE} = \frac{4}{8} = \frac{1}{2}, \qquad \frac{AC}{DF} = \frac{3}{6} = \frac{1}{2}, \qquad \frac{BC}{EF} = \frac{2}{4} = \frac{1}{2}, \qquad \text{and} \qquad \frac{CH}{FK} = \frac{1.5}{3} = \frac{1}{2}$$

Because the ratios of corresponding parts are equal, three proportions can be formed using the sides of the triangles.

$$\frac{AB}{DE} = \frac{AC}{DF}, \qquad \frac{AB}{DE} = \frac{BC}{EF}, \qquad \text{and} \qquad \frac{AC}{DF} = \frac{BC}{EF}$$

Three proportions can also be formed by using the sides and heights of the triangles.

$$\frac{AB}{DE} = \frac{CH}{FK}, \qquad \frac{AC}{DF} = \frac{CH}{FK}, \qquad \text{and} \qquad \frac{BC}{EF} = \frac{CH}{FK}$$

The corresponding angles in similar triangles are equal. Therefore,

$$\angle A = \angle D, \qquad \angle B = \angle E, \qquad \text{and} \qquad \angle C = \angle F$$

Triangles ABC and DEF at the right are similar. Find the area of triangle ABC.

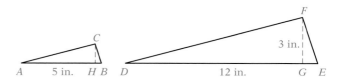

Solve a proportion to find the height of triangle ABC.

$$\frac{AB}{DE} = \frac{CH}{FG}$$

$$\frac{5}{12} = \frac{CH}{3}$$

$$12 \cdot \frac{5}{12} = 12 \cdot \frac{CH}{3}$$

$$5 = 4(CH)$$

The height is 1.25 in. The base is 5 in.

$$1.25 = CH$$

Use the formula for the area of a triangle.

$$A = \frac{1}{2}bh = \frac{1}{2}(5)(1.25) = 3.125$$

The area of triangle ABC is 3.125 in².

It is also true that if the three angles of one triangle are equal, respectively, to the three angles of another triangle, then the two triangles are similar.

A line DE is drawn parallel to the base AB in the triangle at the right. $\angle x = \angle m$ and $\angle y = \angle n$ because corresponding angles are equal. Because $\angle C = \angle C$, the three angles of triangle DEC are equal respectively to the three angles of triangle ABC. Triangle DEC is similar to triangle ABC.

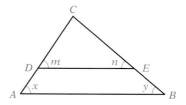

The sum of the three angles of a triangle is 180°. If two angles of one triangle are equal to two angles of another triangle, then the third angles must be equal. Thus we can say that if two angles of one triangle are equal to two angles of another triangle, then the two triangles are similar.

The lines AB and CD intersect at point O in the figure at the right. Angles C and D are right angles. Find the length of DO.

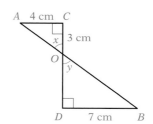

First determine whether triangle *AOC* is similar to triangle *BOD*.

Angle *C* = angle *D* because they are right angles. Angle *x* = angle *y* because they are vertical angles. Therefore, triangle *AOC* is similar to triangle *BOD* because two angles of one triangle are equal to two angles of the other triangle.

Use a proportion to find the length of the unknown side.

$$\frac{AC}{DB} = \frac{CO}{DO}$$

$$\frac{4}{7} = \frac{3}{DO}$$

$$7(DO)\frac{4}{7} = 7(DO)\frac{3}{DO}$$

$$4(DO) = 7(3)$$

$$4(DO) = 21$$

$$DO = 5.25$$

The length of *DO* is 5.25 cm.

Example 6   In the figure at the right, *AB* is parallel to *DC* and angles *B* and *D* are right angles. *AB* = 12 m, *DC* = 4 m, and *AC* = 18 m. Find the length of *CO*.

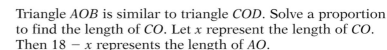

Strategy   Triangle *AOB* is similar to triangle *COD*. Solve a proportion to find the length of *CO*. Let *x* represent the length of *CO*. Then 18 − *x* represents the length of *AO*.

Solution

$$\frac{DC}{AB} = \frac{CO}{AO}$$

$$\frac{4}{12} = \frac{x}{18 - x}$$

$$12(18 - x) \cdot \frac{4}{12} = 12(18 - x) \cdot \frac{x}{18 - x}$$

$$(18 - x)4 = 12x$$

$$72 - 4x = 12x$$

$$72 = 16x$$

$$4.5 = x$$

The length of *CO* is 4.5 m.

Problem 6   In the figure at the right, *AB* is parallel to *DC* and angles *A* and *D* are right angles. *AB* = 10 cm, *CD* = 4 cm, and *DO* = 3 cm. Find the area of triangle *AOB*.

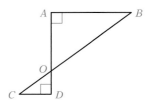

Solution   See page S22.

9.5 Concept Review

Determine whether the statement is always true, sometimes true, or never true.

1. The process of clearing denominators in an equation containing fractions is an application of the Multiplication Property of Equations.

2. If the denominator of a fraction is $x + 3$, then $x \neq 3$ for that fraction.

3. The first step in solving an equation containing fractions is to find the LCM of the denominators.

4. The solutions to an equation containing rational expressions must be checked because a solution might make one of the denominators equal to zero.

5. Suppose, after the clearing of denominators, a fractional equation results in a quadratic equation. Then the equation has two solutions.

6. Both sides of an equation containing fractions can be multiplied by the same number without changing the solution of the equation.

7. To solve a proportion, multiply each side of the proportion by the LCM of the denominators.

9.5 Exercises

1 1. In solving an equation containing fractions, why do we first multiply each side of the equation by the LCM of the denominators?

2. After solving an equation containing fractions, why must we check the solution?

Solve.

3. $2 + \dfrac{5}{x} = 7$

4. $3 + \dfrac{8}{n} = 5$

5. $1 - \dfrac{9}{x} = 4$

6. $3 - \dfrac{12}{x} = 7$

7. $\dfrac{2}{y} + 5 = 9$

8. $\dfrac{6}{x} + 3 = 11$

9. $\dfrac{4x}{x - 4} + 5 = \dfrac{5x}{x - 4}$

10. $\dfrac{2x}{x + 2} - 5 = \dfrac{7x}{x + 2}$

11. $2 + \dfrac{3}{a - 3} = \dfrac{a}{a - 3}$

12. $\dfrac{x}{x + 4} = 3 - \dfrac{4}{x + 4}$

13. $\dfrac{x}{x - 1} = \dfrac{8}{x + 2}$

14. $\dfrac{x}{x + 12} = \dfrac{1}{x + 5}$

15. $\dfrac{2x}{x + 4} = \dfrac{3}{x - 1}$

16. $\dfrac{5}{3n - 8} = \dfrac{n}{n + 2}$

17. $x + \dfrac{6}{x - 2} = \dfrac{3x}{x - 2}$

18. $x - \dfrac{6}{x - 3} = \dfrac{2x}{x - 3}$

19. $\dfrac{x}{x + 2} + \dfrac{2}{x - 2} = \dfrac{x + 6}{x^2 - 4}$

20. $\dfrac{x}{x + 4} = \dfrac{11}{x^2 - 16} + 2$

21. $\dfrac{8}{y} = \dfrac{2}{y - 2} + 1$

22. $\dfrac{8}{r} + \dfrac{3}{r - 1} = 3$

23. Can 2 be a solution of the equation $\dfrac{6x}{x + 1} - \dfrac{x}{x - 2} = 4$? Explain your answer.

2 Solve.

24. $\dfrac{x}{12} = \dfrac{3}{4}$

25. $\dfrac{6}{x} = \dfrac{2}{3}$

26. $\dfrac{4}{9} = \dfrac{x}{27}$

27. $\dfrac{16}{9} = \dfrac{64}{x}$

28. $\dfrac{x + 3}{12} = \dfrac{5}{6}$

29. $\dfrac{3}{5} = \dfrac{x - 4}{10}$

30. $\dfrac{18}{x + 4} = \dfrac{9}{5}$

31. $\dfrac{2}{11} = \dfrac{20}{x - 3}$

32. $\dfrac{2}{x} = \dfrac{4}{x + 1}$

33. $\dfrac{16}{x - 2} = \dfrac{8}{x}$

34. $\dfrac{x + 3}{4} = \dfrac{x}{8}$

35. $\dfrac{x - 6}{3} = \dfrac{x}{5}$

36. $\dfrac{2}{x - 1} = \dfrac{6}{2x + 1}$

37. $\dfrac{9}{x + 2} = \dfrac{3}{x - 2}$

38. $\dfrac{2x}{7} = \dfrac{x - 2}{14}$

3 **39.** **Elections** An exit poll showed that 4 out of every 7 voters cast a ballot in favor of an amendment to a city charter. At this rate, how many people voted in favor of the amendment if 35,000 people voted?

40. **Business** A quality control inspector found 3 defective transistors in a shipment of 500 transistors. At this rate, how many transistors would be defective in a shipment of 2000 transistors?

41. **Construction** An air conditioning specialist recommends 2 air vents for every 300 ft² of floor space. At this rate, how many air vents are required for an office building of 21,000 ft²?

42. **Television** In a city of 25,000 homes, a survey was taken to determine the number with cable television. Of the 300 homes surveyed, 210 had cable television. Estimate the number of homes in the city that have cable television.

43. **Cooking** A simple syrup is made by dissolving 2 c of sugar in $\frac{2}{3}$ c of boiling water. At this rate, how many cups of sugar are required for 2 c of boiling water?

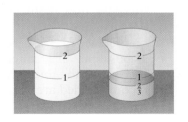

44. Energy The lighting for a billboard is provided by solar energy. If 3 energy panels generate 10 watts of power, how many panels are needed to provide 600 watts of power?

45. Conservation As part of a conservation effort for a lake, 40 fish were caught, tagged, and then released. Later, 80 fish were caught from the lake. Four of these 80 fish were found to have tags. Estimate the number of fish in the lake.

46. Business A company will accept a shipment of 10,000 computer disks if there are 2 or fewer defects in a sample of 100 randomly chosen disks. Assume that there are 300 defective disks in the shipment and that the rate of defective disks in the sample is the same as the rate in the shipment. Will the shipment be accepted?

47. Business A company will accept a shipment of 20,000 precision bearings if there are 3 or fewer defects in a sample of 100 randomly chosen bearings. Assume that there are 400 defective bearings in the shipment and that the rate of defective bearings in the sample is the same as the rate in the shipment. Will the shipment be accepted?

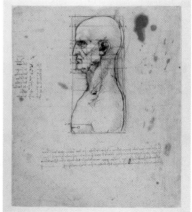

48. Art Leonardo da Vinci measured various distances on the human body in order to make accurate drawings. He determined that in general, the ratio of the kneeling height of a person to his or her standing height is $\frac{3}{4}$. Using this ratio, determine the standing height of a person who has a kneeling height of 48 in.

49. Art In one of Leonardo da Vinci's notebooks, he wrote that ". . . from the top to the bottom of the chin is the sixth part of a face, and it is the fifty-fourth part of the man." Suppose the distance from the top to the bottom of a person's chin is 1.25 in. Using da Vinci's measurements, find the height of the person.

50. Cartography On a map, two cities are $2\frac{5}{8}$ in. apart. If $\frac{3}{8}$ in. on the map represents 25 mi, find the number of miles between the two cities.

51. Cartography On a map, two cities are $5\frac{5}{8}$ in. apart. If $\frac{3}{4}$ in. on the map represents 100 mi, find the number of miles between the two cities.

52. Rocketry The engine of a small rocket burns 170,000 lb of fuel in 1 min. At this rate, how many pounds of fuel does the rocket burn in 45 s?

53. Construction To conserve energy and still allow for as much natural lighting as possible, an architect suggests that the ratio of the area of a window to the area of the total wall surface be 5 to 12. Using this ratio, determine the recommended area of a window to be installed in a wall that measures 8 ft by 12 ft.

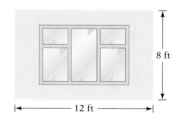

54. **Paint Mixtures** A green paint is created by mixing 3 parts of yellow with every 5 parts of blue. How many gallons of yellow paint are needed to make 60 gal of this green paint?

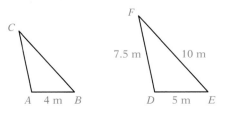

55. **Food Industry** A soft drink is made by mixing 4 parts of carbonated water with every 3 parts of syrup. How many milliliters of carbonated water are in 280 ml of soft drink?

56. **Redecorating** A painter estimates that 5 gal of paint will cover 1200 ft² of wall surface. How many additional gallons are required to cover 1680 ft²?

57. **Agriculture** A 50-acre field yields 1100 bushels of wheat annually. How many additional acres must be planted so that the annual yield will be 1320 bushels?

58. **Taxes** The sales tax on a car that sold for $12,000 is $780. At the same rate, how much higher is the sales tax on a car that sells for $13,500?

59. **Catering** A caterer estimates that 5 gal of coffee will serve 50 people. How much additional coffee is necessary to serve 70 people?

4 Solve. Triangles *ABC* and *DEF* in Exercises 60 to 67 are similar. Round answers to the nearest tenth.

60. Find side *AC*.

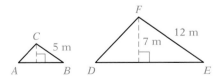

61. Find side *DE*.

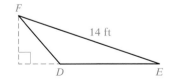

62. Find the height of triangle *ABC*.

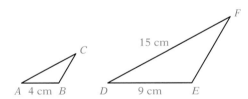

63. Find the height of triangle *DEF*.

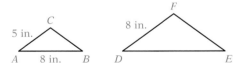

64. Find the perimeter of triangle *DEF*.

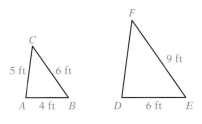

65. Find the perimeter of triangle *ABC*.

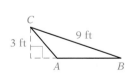

66. Find the area of triangle *ABC*.

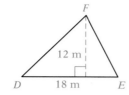

67. Find the area of triangle *ABC*.

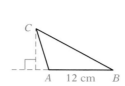
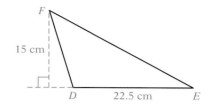

68. Given *BD*‖*AE*, *BD* measures 5 cm, *AE* measures 8 cm, and *AC* measures 10 cm, find the length of *BC*.

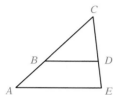

69. Given *AC*‖*DE*, *BD* measures 8 m, *AD* measures 12 m, and *BE* measures 6 m, find the length of *BC*.

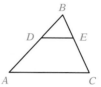

70. Given *DE*‖*AC*, *DE* measures 6 in., *AC* measures 10 in., and *AB* measures 15 in., find the length of *DA*.

71. Given *MP* and *NQ* intersect at *O*, *NO* measures 25 ft, *MO* measures 20 ft, and *PO* measures 8 ft, find the length of *QO*.

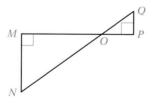

72. Given *MP* and *NQ* intersect at *O*, *NO* measures 24 cm, *MN* measures 10 cm, *MP* measures 39 cm, and *QO* measures 12 cm, find the length of *OP*.

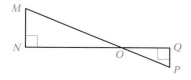

73. Given *MQ* and *NP* intersect at *O*, *NO* measures 12 m, *MN* measures 9 m, *PQ* measures 3 m, and *MQ* measures 20 m, find the perimeter of triangle *OPQ*.

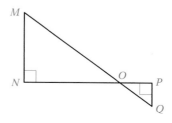

74. Similar triangles can be used as an indirect way to measure inaccessible distances. The diagram at the right represents a river of width *DC*. The triangles *AOB* and *DOC* are similar. The distances *AB*, *BO*, and *OC* can be measured. Find the width of the river.

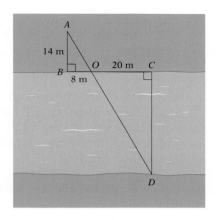

75. The sun's rays cast a shadow as shown in the diagram at the right. Find the height of the flagpole. Write the answer in terms of feet.

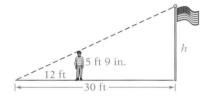

9.5   Applying Concepts

Solve.

76. $\frac{2}{3}(x + 2) - \frac{x + 1}{6} = \frac{1}{2}$

77. $\frac{3}{4}a = \frac{1}{2}(3 - a) + \frac{a - 2}{4}$

78. $\frac{x}{2x^2 - x - 1} = \frac{3}{x^2 - 1} + \frac{3}{2x + 1}$

79. $\frac{x + 1}{x^2 + x - 2} = \frac{x + 2}{x^2 - 1} + \frac{3}{x + 2}$

80. **Number Problems** The sum of a number and its reciprocal is $\frac{26}{5}$. Find the number.

81. **Number Problems** The sum of the multiplicative inverses of two consecutive integers is $-\frac{7}{12}$. Find the integers.

82. **Number Problems** The denominator of a fraction is 2 more than the numerator. If both the numerator and denominator of the fraction are increased by 3, the new fraction is $\frac{4}{5}$. Find the original fraction.

83. **Lotteries** Three people put their money together to buy lottery tickets. The first person put in $25, the second person put in $30, and the third person put in $35. One of their tickets was a winning ticket. If they won $4.5 million, what was the first person's share of the winnings?

84. **Sports** A basketball player has made 5 out of every 6 foul shots attempted. If 42 foul shots were missed in the player's career, how many foul shots were made in the player's career?

85. Photography The "sitting fee" for school pictures is $4. If 10 photos cost $10, including the sitting fee, what would 24 photos cost, including the sitting fee?

86. Fundraising No one belongs to both the Math Club and the Photography Club, but the two clubs join to hold a car wash. Ten members of the Math Club and 6 members of the Photography Club participate. The profits from the car wash are $120. If each club's profits are proportional to the number of members participating, what share of the profits does the Math Club receive?

87. History Eratosthenes, the fifth librarian of Alexandria (230 B.C.), was familiar with certain astronomical data that enabled him to calculate the circumference of Earth by using a proportion. He knew that on a midsummer day, the sun was directly overhead at Syene, as shown in the diagram. At the same time, at Alexandria, the sun was at a 7.5° angle from the zenith. The distance from Syene to Alexandria was 5000 stadia, or about 520 mi. Eratosthenes reasoned that the ratio of the 7.5° angle to one revolution was equal to the ratio of the arc length of 520 mi to the circumference of Earth. From this, he wrote and solved a proportion.
a. What did Eratosthenes calculate to be the circumference of Earth?

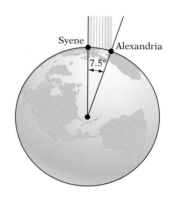

b. Find the difference between his calculation and the accepted value of 24,800 mi.

88. **Demography** The number of Americans aged 65 and over is increasing dramatically. This demographic fact is a concern for the medical profession because the ratio of the population who are caregivers (ages 50 to 64) to the population needing care (ages 85 and older) will be shrinking. The ratios are shown in the chart below.

Ratios of Caregivers to Those Needing Care	
1990	𝕚𝕚𝕚𝕚𝕚𝕚𝕚𝕚𝕚𝕚𝕚 to 𝕚
2010	𝕚𝕚𝕚𝕚𝕚𝕚𝕚𝕚𝕚𝕚 to 𝕚
2030	𝕚𝕚𝕚𝕚𝕚𝕚 to 𝕚
2050	𝕚𝕚𝕚𝕚 to 𝕚

Source: Copyright © 1996, *USA Today.* Reprinted with permission.

The "aging of America" is also a major concern for the Social Security Administration. Because the number of Americans aged 65 and over is increasing dramatically, the ratio of the number of workers contributing to Social Security to the number of beneficiaries will decrease dramatically during the next few decades.

Use a resource to find the present and predicted future ratios of the number of employees contributing to Social Security to the number of Social Security recipients. Write a paragraph describing a proposal for "fixing" the Social Security system.

9.6 Literal Equations

1 **Solve a literal equation for one of the variables**

A **literal equation** is an equation that contains more than one variable. Examples of literal equations are shown at the right.

$$2x + 3y = 6$$
$$4w - 2x + z = 0$$

Formulas are used to express a relationship among physical quantities. A **formula** is a literal equation that states rules about measurements. Examples of formulas are shown below.

$$\frac{1}{R_1} + \frac{1}{R_2} = \frac{1}{R} \qquad \text{(Physics)}$$
$$s = a + (n - 1)d \qquad \text{(Mathematics)}$$
$$A = P + Prt \qquad \text{(Business)}$$

The Addition and Multiplication Properties can be used to solve a literal equation for one of the variables. The goal is to rewrite the equation so that the letter being solved for is alone on one side of the equation and all numbers and other variables are on the other side.

Solve $A = P(1 + i)$ for i.

The goal is to rewrite the equation so that i is on one side of the equation and all other variables are on the other side.

$$A = P(1 + i)$$

Use the Distributive Property to remove parentheses.

$$A = P + Pi$$

Subtract P from each side of the equation.

$$A - P = P - P + Pi$$
$$A - P = Pi$$

Divide each side of the equation by P.

$$\frac{A - P}{P} = \frac{Pi}{P}$$

$$\frac{A - P}{P} = i$$

Example 1 **A.** Solve $I = \dfrac{E}{R + r}$ for R.   **B.** Solve $L = a(1 + ct)$ for c.

Solution **A.**
$$I = \frac{E}{R + r}$$
$$(R + r)I = (R + r)\frac{E}{R + r}$$
$$RI + rI = E$$
$$RI + rI - rI = E - rI$$
$$RI = E - rI$$
$$\frac{RI}{I} = \frac{E - rI}{I}$$
$$R = \frac{E - rI}{I}$$

B.
$$L = a(1 + ct)$$
$$L = a + act$$
$$L - a = a - a + act$$
$$L - a = act$$
$$\frac{L - a}{at} = \frac{act}{at}$$
$$\frac{L - a}{at} = c$$

Problem 1   **A.** Solve $s = \dfrac{A + L}{2}$ for L.   **B.** Solve $S = a + (n - 1)d$ for n.

Solution   See page S22.

Example 2   Solve $S = C - rC$ for C.

Solution
$$S = C - rC$$
$$S = C(1 - r) \qquad \text{• Factor } C \text{ from } C - rC.$$
$$\frac{S}{1 - r} = \frac{C(1 - r)}{1 - r} \qquad \text{• Divide each side of the equation by } 1 - r.$$
$$\frac{S}{1 - r} = C$$

Problem 2   Solve $S = rS + C$ for S.

Solution   See page S22.

9.6   Concept Review

Determine whether the statement is always true, sometimes true, or never true.

1. An equation that contains more than one variable is a literal equation.

2. The linear equation $y = 4x - 5$ is not a literal equation.

3. Literal equations are solved using the same properties of equations that are used to solve equations in one variable.

4. In solving a literal equation, the goal is to get the variable being solved for alone on one side of the equation and all numbers and other variables on the other side of the equation.

5. In solving $L = a(1 + ct)$ for c, you need to use the Distributive Property. In solving $S = C - rC$ for C, you need to factor.

6. We can divide both sides of a literal equation by the same nonzero expression.

9.6   Exercises

1 Solve the formula for the variable given.

1. $A = \dfrac{1}{2}bh$; h (Geometry)

2. $P = a + b + c$; b (Geometry)

3. $d = rt; t$ (Physics)

4. $E = IR; R$ (Physics)

5. $PV = nRT; T$ (Chemistry)

6. $A = bh; h$ (Geometry)

7. $P = 2L + 2W; L$ (Geometry)

8. $F = \dfrac{9}{5}C + 32; C$ (Temperature conversion)

9. $A = \dfrac{1}{2}h(b_1 + b_2); b_1$ (Geometry)

10. $C = \dfrac{5}{9}(F - 32); F$ (Temperature conversion)

11. $V = \dfrac{1}{3}Ah; h$ (Geometry)

12. $P = R - C; C$ (Business)

13. $R = \dfrac{C - S}{t}; S$ (Business)

14. $P = \dfrac{R - C}{n}; R$ (Business)

15. $A = P + Prt; P$ (Business)

16. $T = fm - gm; m$ (Engineering)

17. $A = Sw + w; w$ (Physics)

18. $a = S - Sr; S$ (Mathematics)

9.6 **Applying Concepts**

Geometry The surface area of a right circular cylinder is given by the formula $S = 2\pi rh + 2\pi r^2$, where r is the radius of the base and h is the height of the cylinder.

19. a. Solve the formula $S = 2\pi rh + 2\pi r^2$ for h.
  b. Use your answer to part (a) to find the height of a right circular cylinder when the surface area is 12π in^2 and the radius is 1 in.
  c. Use your answer to part (a) to find the height of a right circular cylinder when the surface area is 24π in^2 and the radius is 2 in.

1 in.

$S = 12\pi$ in^2

Business Break-even analysis is a method used to determine the sales volume required for a company to break even, or experience neither a profit nor a loss, on the sale of its product. The break-even point represents the number of units that must be made and sold for income from sales to equal the cost of producing the product. The break-even point can be calculated using the formula $B = \dfrac{F}{S - V}$, where F is the fixed costs, S is the selling price per unit, and V is the variable costs per unit.

20. a. Solve the formula $B = \dfrac{F}{S - V}$ for S.

    b. Use your answer to part (a) to find the selling price per desk required for a company to break even. The fixed costs are \$40,000, the variable costs per desk are \$160, and the company plans to make and sell 200 desks.

    c. Use your answer to part (a) to find the selling price per camera required for a company to break even. The fixed costs are \$30,000, the variable costs per camera are \$100, and the company plans to make and sell 600 cameras.

Business When markup is based on selling price, the selling price of a product is given by the formula $S = C + rC$, where C is the cost of the product and r is the markup rate.

21. a. Solve the formula $S = C + rC$ for r.

    b. Use your answer to part (a) to find the markup rate on a tennis racket when the cost is \$108 and the selling price is \$180.

    c. Use your answer to part (a) to find the markup rate on a car radio when the cost is \$120 and the selling price is \$172.

Electricity Resistors are used to control the flow of current. The total resistance of two resistors in a circuit can be given by the formula $R = \dfrac{1}{\dfrac{1}{R_1} + \dfrac{1}{R_2}}$, where

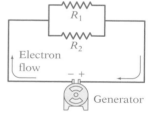

R_1 and R_2 are the resistances of the two resistors in the circuit. Resistance is measured in ohms.

22. a. Solve the formula $R = \dfrac{1}{\dfrac{1}{R_1} + \dfrac{1}{R_2}}$ for R_1.

    b. Use your answer to part (a) to find the resistance in R_1 if the resistance in R_2 is 30 ohms and the total resistance is 12 ohms.

    c. Use your answer to part (a) to find the resistance in R_1 if the resistance in R_2 is 15 ohms and the total resistance is 6 ohms.

9.7   Application Problems

1 **Work problems**

If a painter can paint a room in 4 h, then in 1 h the painter can paint $\dfrac{1}{4}$ of the room. The painter's rate of work is $\dfrac{1}{4}$ of the room each hour. The **rate of work** is that part of a task that is completed in 1 unit of time.

A pipe can fill a tank in 30 min. This pipe can fill $\dfrac{1}{30}$ of the tank in 1 min. The rate of work is $\dfrac{1}{30}$ of the tank each minute. If a second pipe can fill the tank in x min, the rate of work for the second pipe is $\dfrac{1}{x}$ of the tank each minute.

In solving a work problem, the goal is to determine the time it takes to complete a task. The basic equation that is used to solve work problems is

Rate of work · Time worked = Part of task completed

For example, if a faucet can fill a sink in 6 min, then in 5 min the faucet will fill $\frac{1}{6} \cdot 5 = \frac{5}{6}$ of the sink. In 5 min, the faucet completes $\frac{5}{6}$ of the task.

Solve: A painter can paint a ceiling in 60 min. The painter's apprentice can paint the same ceiling in 90 min. How long will it take them to paint the ceiling if they work together?

STRATEGY | *for solving a work problem*

▶ For each person or machine, write a numerical or variable expression for the rate of work, the time worked, and the part of the task completed. The results can be recorded in a table.

Unknown time to paint the ceiling working together: t

> ## Take Note
> Use the information given in the problem to fill in the rate and time columns of the table. Fill in the part of task completed column by multiplying the two expressions you wrote in each row.

	Rate of work	·	Time worked	=	Part of task completed
Painter	$\frac{1}{60}$	·	t	=	$\frac{t}{60}$
Apprentice	$\frac{1}{90}$	·	t	=	$\frac{t}{90}$

▶ Determine how the parts of the task completed are related. Use the fact that the sum of the parts of the task completed must equal 1, the complete task.

$$\frac{t}{60} + \frac{t}{90} = 1$$
• The part of the task completed by the painter plus the part of the task completed by the apprentice must equal 1.

$$180\left(\frac{t}{60} + \frac{t}{90}\right) = 180 \cdot 1$$
• Multiply each side of the equation by the LCM of 60 and 90.

$$3t + 2t = 180$$
$$5t = 180$$
$$t = 36$$

Working together, they will paint the ceiling in 36 min.

Example 1   A small water pipe takes four times longer to fill a tank than does a large water pipe. With both pipes open, it takes 3 h to fill the tank. Find the time it would take the small pipe, working alone, to fill the tank.

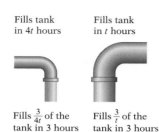

Fills tank
in 4t hours

Fills tank
in t hours

Fills $\frac{3}{4t}$ of the
tank in 3 hours

Fills $\frac{3}{t}$ of the
tank in 3 hours

Strategy   ▶ Time for large pipe to fill the tank: t

▶ Time for small pipe to fill the tank: $4t$

	Rate	Time	Part
Small pipe	$\dfrac{1}{4t}$	3	$\dfrac{3}{4t}$
Large pipe	$\dfrac{1}{t}$	3	$\dfrac{3}{t}$

▶ The sum of the parts of the task completed must equal 1.

Solution

$$\frac{3}{4t} + \frac{3}{t} = 1$$

• The part of the task completed by the small pipe plus the part of the task completed by the large pipe must equal 1.

$$4t\left(\frac{3}{4t} + \frac{3}{t}\right) = 4t \cdot 1$$

$$3 + 12 = 4t$$

$$15 = 4t$$

$$\frac{15}{4} = t$$

• t is the time for the large pipe to fill the tank.

$$4t = 4\left(\frac{15}{4}\right) = 15$$

• Substitute the value of t into the variable expression for the time for the small pipe to fill the tank.

The small pipe, working alone, takes 15 h to fill the tank.

Problem 1   Two computer printers that work at the same rate are working together to print the payroll checks for a corporation. After they work together for 3 h, one of the printers quits. The second requires 2 more hours to complete the checks. Find the time it would take one printer, working alone, to print the checks.

Solution   See page S22.

2 **Uniform motion problems**

A car that travels constantly in a straight line at 30 mph is in uniform motion. When an object is in **uniform motion**, its speed does not change.

The basic equation used to solve uniform motion problems is

$$\textbf{Distance} = \textbf{Rate} \cdot \textbf{Time}$$

An alternative form of this equation is written by solving the equation for time:

$$\frac{\textbf{Distance}}{\textbf{Rate}} = \textbf{Time}$$

This form of the equation is useful when the total time of travel for two objects is known or the times of travel for two objects are equal.

Solve: The speed of a boat in still water is 20 mph. The boat traveled 120 mi down a river in the same amount of time it took the boat to travel 80 mi up the river. Find the rate of the river's current.

> **STRATEGY** | *for solving a uniform motion problem*
>
> ▶ For each object, write a numerical or variable expression for the distance, rate, and time. The results can be recorded in a table.

The unknown rate of the river's current: r

Take Note

Use the information given in the problem to fill in the distance and rate columns of the table. Fill in the time column by dividing the two expressions you wrote in each row.

	Distance	÷	Rate	=	Time
Down river	120	÷	$20 + r$	=	$\dfrac{120}{20 + r}$
Up river	80	÷	$20 - r$	=	$\dfrac{80}{20 - r}$

> ▶ Determine how the times traveled by the two objects are related. For example, it may be known that the times are equal, or the total time may be known.

$$\frac{120}{20 + r} = \frac{80}{20 - r}$$

$$(20 + r)(20 - r)\frac{120}{20 + r} = (20 + r)(20 - r)\frac{80}{20 - r}$$

$$(20 - r)120 = (20 + r)80$$

$$2400 - 120r = 1600 + 80r$$

$$-200r = -800$$

$$r = 4$$

• The time spent traveling down the river is equal to the time spent traveling up the river.

The rate of the river's current is 4 mph.

┌─ **Example 2**   A cyclist rode the first 20 mi of a trip at a constant rate. For the next 16 mi, the cyclist reduced the speed by 2 mph. The total time for the 36 mi was 4 h. Find the rate of the cyclist for each leg of the trip.

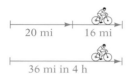

20 mi   16 mi

36 mi in 4 h

Strategy   ▶ Rate for the first 20 mi: r

▶ Rate for the next 16 mi: $r - 2$

	Distance	Rate	Time
First 20 mi	20	r	$\dfrac{20}{r}$
Next 16 mi	16	$r - 2$	$\dfrac{16}{r - 2}$

▶ The total time for the trip was 4 h.

Solution

$$\frac{20}{r} + \frac{16}{r - 2} = 4$$

$$r(r - 2)\left[\frac{20}{r} + \frac{16}{r - 2}\right] = r(r - 2) \cdot 4$$

$$(r - 2)20 + 16r = (r^2 - 2r)4$$

$$20r - 40 + 16r = 4r^2 - 8r$$

$$36r - 40 = 4r^2 - 8r$$

$$0 = 4r^2 - 44r + 40$$

$$0 = 4(r^2 - 11r + 10)$$

$$0 = r^2 - 11r + 10$$

$$0 = (r - 10)(r - 1)$$

$r - 10 = 0 \qquad r - 1 = 0$

$r = 10 \qquad\quad r = 1$

$r - 2 = 10 - 2 \qquad r - 2 = 1 - 2$

$\quad\quad = 8 \qquad\qquad\qquad = -1$

10 mph was the rate for the first 20 mi.
8 mph was the rate for the next 16 mi.

• The time spent riding the first 20 mi plus the time spent riding the next 16 mi is equal to 4 h.

• This is a quadratic equation.

• Divide each side by 4.
• Solve by factoring.

• Find the rate for the last 16 mi. The solution $r = 1$ mph is not possible, because the rate on the last 16 mi would be -1 mph.

Problem 2   The total time for a sailboat to sail back and forth across a lake 6 km wide was 3 h. The rate sailing back was twice the rate sailing across the lake. Find the rate of the sailboat going across the lake.

Solution   See page S23.

9.7   Concept Review

Determine whether the statement is always true, sometimes true, or never true.

1. The work problems in this section require use of the equation
Rate of work · Time worked = 1.

2. If it takes a janitorial crew 5 h to clean a company's offices, then in x hours the crew has completed $\frac{x}{5}$ of the job.

3. If it takes an automotive crew x min to service a car, then the rate of work is $\frac{1}{x}$ of the job each minute.

4. If only two people worked on a job and together they completed it, and one person completed $\frac{t}{30}$ of the job and the other person completed $\frac{t}{20}$ of the job, then $\frac{t}{30} + \frac{t}{20} = 1$.

5. Uniform motion problems are based on the equation
Distance · Rate = Time.

6. If a plane flies 300 mph in calm air and the rate of the wind is r mph, then the rate of the plane flying with the wind can be represented as $300 - r$ and the rate of the plane flying against the wind can be represented as $300 + r$.

9.7 | **Exercises**

1 Work Problems

1. ✎   Explain the meaning of the phrase "rate of work."

2. a. If $\frac{2}{5}$ of a room can be painted in 1 h, what is the rate of work?

  b. At the same rate, how long will it take to paint the entire room?

3. A park has two sprinklers that are used to fill a fountain. One sprinkler can fill the fountain in 3 h, whereas the second sprinkler can fill the fountain in 6 h. How long will it take to fill the fountain with both sprinklers operating?

4. One grocery clerk can stock a shelf in 20 min. A second clerk requires 30 min to stock the same shelf. How long would it take to stock the shelf if the two clerks worked together?

5. One person with a skiploader requires 12 h to remove a large quantity of earth. With a larger skiploader, the same amount of earth can be removed in 4 h. How long would it take to remove the earth if both skiploaders were operated together?

6. It takes Doug 6 days to reroof a house. If Doug's son helps him, the job can be completed in 4 days. How long would it take Doug's son, working alone, to do the job?

7. One computer can solve a complex prime factorization problem in 75 h. A second computer can solve the same problem in 50 h. How long would it take both computers, working together, to solve the problem?

8. A new machine makes 10,000 aluminum cans three times faster than an older machine. With both machines operating, it takes 9 h to make 10,000 cans. How long would it take the new machine, working alone, to make 10,000 cans?

9. A small air conditioner can cool a room 5°F in 60 min. A larger air conditioner can cool the room 5°F in 40 min. How long would it take to cool the room 5°F with both air conditioners working?

10. One printing press can print the first edition of a book in 55 min. A second printing press requires 66 min to print the same number of copies. How long would it take to print the first edition of the book with both presses operating?

11. Two welders working together can complete a job in 6 h. One of the welders, working alone, can complete the task in 10 h. How long would it take the second welder, working alone, to complete the task?

12. Working together, Pat and Chris can reseal a driveway in 6 h. Working alone, Pat can reseal the driveway in 15 h. How long would it take Chris, working alone, to reseal the driveway?

13. Two oil pipelines can fill a small tank in 30 min. Using one of the pipelines would require 45 min to fill the tank. How long would it take the second pipeline, working alone, to fill the tank?

14. A cement mason can construct a retaining wall in 8 h. A second mason requires 12 h to do the same job. After working alone for 4 h, the first mason quits. How long would it take the second mason to complete the wall?

15. With two reapers operating, a field can be harvested in 1 h. If only the newer reaper is used, the crop can be harvested in 1.5 h. How long would it take to harvest the field using only the older reaper?

16. A manufacturer of prefabricated homes has the company's employees work in teams. Team 1 can erect the Silvercrest model in 15 h. Team 2 can erect the same model in 10 h. How long would it take for Team 1 and Team 2, working together, to erect the Silvercrest model home?

17. One computer technician can wire a modem in 4 h. A second technician requires 6 h to do the same job. After working alone for 2 h, the first technician quits. How long will it take the second technician to complete the wiring?

18. A wallpaper hanger requires 2 h to hang the wallpaper on one wall of a room. A second wallpaper hanger requires 4 h to hang the same amount of wallpaper. The first wallpaper hanger works alone for 1 h and then quits. How long will it take the second hanger, working alone, to finish papering the wall?

19. A large heating unit and a small heating unit are being used to heat the water in a pool. The large unit, working alone, requires 8 h to heat the pool. After both units have been operating for 2 h, the large unit is turned off. The small unit requires 9 more hours to heat the pool. How long would it take the small unit, working alone, to heat the pool?

20. Two machines fill cereal boxes at the same rate. After the two machines work together for 7 h, one machine breaks down. The second machine requires 14 more hours to finish filling the boxes. How long would it have taken one of the machines, working alone, to fill the boxes?

21. Two welders who work at the same rate are riveting the girders of a building. After they work together for 10 h, one of the welders quits. The second welder requires 20 more hours to complete the welds. Find the time it would have taken one of the welders, working alone, to complete the welds.

22. A large drain and a small drain are opened to drain a pool. The large drain can empty the pool in 6 h. After both drains have been open 1 h, the large drain becomes clogged and is closed. The small drain remains open and requires 9 more hours to empty the pool. How long would it have taken the small drain, working alone, to empty the pool?

2 Uniform Motion Problems

23. A camper drove 90 mi to a recreational area and then hiked 5 mi into the woods. The rate of the camper while driving was nine times the rate while hiking. The time spent hiking and driving was 3 h. Find the rate at which the camper hiked.

24. The president of a company traveled 1800 mi by jet and 300 mi on a prop plane. The rate of the jet was four times the rate of the prop plane. The entire trip took 5 h. Find the rate of the jet plane.

25. To assess the damage done by a fire, a forest ranger traveled 1080 mi by jet and then an additional 180 mi by helicopter. The rate of the jet was four times the rate of the helicopter. The entire trip took 5 h. Find the rate of the jet.

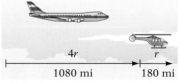

26. An engineer traveled 165 mi by car and then an additional 660 mi by plane. The rate of the plane was four times the rate of the car. The total trip took 6 h. Find the rate of the car.

27. After sailing 15 mi, a sailor changed direction and increased the boat's speed by 2 mph. An additional 19 mi was sailed at the increased speed. The total sailing time was 4 h. Find the rate of the boat for the first 15 mi.

28. On a recent trip, a trucker traveled 330 mi at a constant rate. Because of road conditions, the trucker then reduced the speed by 25 mph. An additional 30 mi was traveled at the reduced rate. The entire trip took 7 h. Find the rate of the trucker for the first 330 mi.

29. In calm water, the rate of a small rental motorboat is 15 mph. The rate of the current on the river is 3 mph. How far down the river can a family travel and still return the boat in 3 h?

30. The rate of a small aircraft in calm air is 125 mph. If the wind is currently blowing south at a rate of 15 mph, how far north can a pilot fly the plane and return it within 2 h?

31. Commuting from work to home, a lab technician traveled 10 mi at a constant rate through congested traffic. Upon reaching the expressway, the technician increased the speed by 20 mph. An additional 20 mi was traveled at the increased speed. The total time for the trip was 1 h. How fast did the technician travel through the congested traffic?

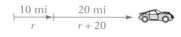

32. As part of a conditioning program, a jogger ran 8 mi in the same amount of time as it took a cyclist to ride 20 mi. The rate of the cyclist was 12 mph faster than the rate of the jogger. Find the rate of the jogger and the rate of the cyclist.

33. The speed of a boat in still water is 20 mph. The Jacksons traveled 75 mi down the Woodset River in the same amount of time as it took them to return 45 mi up the river. Find the rate of the river's current.

34. Crystal Lake is 6 mi wide. The total time it took Helena to kayak back and forth across the lake was 2 h. Her rate kayaking back was three times her rate going out across the lake. What was Helena's rate kayaking out across the lake?

35. An express train traveled 600 mi in the same amount of time as it took a freight train to travel 360 mi. The rate of the express train was 20 mph faster than the rate of the freight train. Find the rate of each train.

36. A twin-engine plane flies 800 mi in the same amount of time as it takes a single-engine plane to fly 600 mi. The rate of the twin-engine plane is 50 mph faster than the rate of the single-engine plane. Find the rate of the twin-engine plane.

37. A car is traveling at a rate that is 36 mph faster than the rate of a cyclist. The car travels 384 mi in the same time as it takes the cyclist to travel 96 mi. Find the rate of the car.

38. A small motor on a fishing boat can move the boat at a rate of 6 mph in calm water. Traveling with the current, the boat can travel 24 mi in the same amount of time as it takes to travel 12 mi against the current. Find the rate of the current.

39. A commercial jet can fly 550 mph in calm air. Traveling with the jet stream, the plane flew 2400 mi in the same amount of time as it takes to fly 2000 mi against the jet stream. Find the rate of the jet stream.

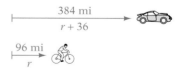

40. A cruise ship can sail 28 mph in calm water. Sailing with the Gulf Stream, the ship can sail 170 mi in the same amount of time as it takes to sail 110 mi against the Gulf Stream. Find the rate of the Gulf Stream.

41. Rowing with the current of a river, a rowing team can row 25 mi in the same amount of time as it takes to row 15 mi against the current. The rate of the rowing team in calm water is 20 mph. Find the rate of the current.

42. A plane can fly 180 mph in calm air. Flying with the wind, the plane can fly 600 mi in the same amount of time as it takes to fly 480 mi against the wind. Find the rate of the wind.

9.7   Applying Concepts

43. Work Problems One pipe can fill a tank in 2 h, a second pipe can fill the tank in 4 h, and a third pipe can fill the tank in 5 h. How long would it take to fill the tank with all three pipes operating?

44. Work Problems A mason can construct a retaining wall in 10 h. The mason's more experienced apprentice can do the same job in 15 h. How long would it take the mason's less experienced apprentice to do the job if, working together, all three can complete the wall in 5 h?

45. Uniform Motion Problems An Outing Club traveled 32 mi by canoe and then hiked 4 mi. The rate by boat was four times the rate on foot. The time spent walking was 1 h less than the time spent canoeing. Find the amount of time spent traveling by canoe.

46. Uniform Motion Problems A motorist drove 120 mi before running out of gas and walking 4 mi to a gas station. The rate of the motorist in the car was ten times the rate walking. The time spent walking was 2 h less than the time spent driving. How long did it take for the motorist to drive the 120 mi?

47. Uniform Motion Problems Because of bad weather, a bus driver reduced the usual speed along a 150-mile bus route by 10 mph. The bus arrived only 30 min later than its usual arrival time. How fast does the bus usually travel?

Focus on Problem Solving

Negations and *If . . . then . . .* Sentences

The sentence "George Washington was the first president of the United States" is a true sentence. The **negation** of that sentence is "George Washington was **not** the first president of the United States." That sentence is false. In general, the negation of a true sentence is a false sentence.

The negation of a false sentence is a true sentence. For instance, the sentence "The moon is made of green cheese" is a false sentence. The negation of that sentence, "The moon is **not** made of green cheese," is true.

The words *all, no* (or *none*) and *some* are called **quantifiers.** Writing the negation of a sentence that contains these words requires special attention. Consider the sentence "All pets are dogs." This sentence is not true because there are pets that are not dogs; cats, for example, are pets. Because the sentence is false, its negation must be true. You might be tempted to write "All pets are not dogs," but that sentence is not true because some pets are dogs. The correct negation of "All pets are dogs," is "Some pets are not dogs." Note the use of the word *some* in the negation.

Now consider the sentence "Some computers are portable." Because that sentence is true, its negation must be false. Writing "Some computers are not portable" as the negation is not correct, because that sentence is true. The negation of "Some computers are portable" is "No computers are portable."

The sentence "No flowers have red blooms" is false, because there is at least one flower (roses, for example) that has red blooms. Because the sentence is false, its negation must be true. The negation is "Some flowers have red blooms."

Statement	Negation
All A are B.	Some A are not B.
No A are B.	Some A are B.
Some A are B.	No A are B.
Some A are not B.	All A are B.

Write the negation of the sentence.

1. All cats like milk.

2. All computers need people.

3. Some trees are tall.

4. Some vegetables are good for you to eat.

5. No politicians are honest.

6. No houses have kitchens.

7. All police officers are tall.

8. All lakes are not polluted.

9. Some banks are not open on Sunday.

10. Some colleges do not offer night classes.

11. Some drivers are unsafe.

12. Some speeches are interesting.

13. All laws are good.

14. All Mark Twain books are funny.

15. All businesses are not profitable.

16. All motorcycles are not large.

A **premise** is a known or assumed fact. A premise can be stated using one of the quantifiers (*all, no, none,* or *some*) or using an *If . . . then . . .* sentence. For instance, the sentence "All triangles have three sides" can be written "If a figure is a triangle, *then* it has three sides."

We can write the sentence "No whole numbers are negative numbers" as an *If . . . then . . .* sentence: "If a number is a whole number, then it is not a negative number."

Write the sentence as an *If . . . then . . .* sentence.

17. All students at Barlock College must take a life science course.

18. All baseballs are round.

19. All computers need people.

20. All cats like milk.

21. No odd number is evenly divisible by 2.

22. No rectangles have five sides.

23. All roads lead to Rome.

24. All dogs have fleas.

25. No triangle has four angles.

26. No prime number greater than 2 is an even number.

Calculators

A calculator is an important tool for problem solving. It can be used as an aide to guessing or estimating a solution to a problem. Here are a few problems to solve with a calculator.

1. Choose any positive integer less than 9. Multiply the number by 1507. Now multiply the result by 7519. What is the answer? Choose another positive single-digit number and again multiply by 1507 and 7519. What is the answer? What pattern do you see? Why does this work?

2. Are there enough people in the United States so that if they held hands in a line, they would stretch around the world at the equator? To answer this question, begin by determining what information you need. What assumptions must you make?

3. Which of the reciprocals of the first 16 natural numbers have a terminating decimal representation and which have a repeating decimal representation?

4. What is the largest natural number n for which $4^n > 1 \cdot 2 \cdot \cdots \cdot n$?

5. Calculate 15^2, 35^2, 65^2, and 85^2. Study the results. Make a conjecture about a relationship between a number ending in 5 and its square. Use your conjecture to find 75^2 and 95^2. Does your conjecture work for 125^2?

6. Find the sum of the first 1000 natural numbers. (*Hint:* You could just start adding $1 + 2 + 3 + 4 + \cdots$, but even if you performed one operation each second, it would take over 15 minutes to find the sum. Instead, try pairing the numbers and then adding the numbers in each pair. Pair 1 and 1000, 2 and 999, 3 and 998, and so on. What is the sum of each pair? How many pairs are there? Use this information to answer the original question.)

7. For a borrower to qualify for a home loan, a bank requires that the monthly mortgage payment be less than 25% of the borrower's monthly take-home income. A laboratory technician has deductions for taxes, insurance, and retirement that amount to 25% of the technician's monthly gross income. What minimum gross monthly income must this technician earn to receive a bank loan that has a mortgage payment of $1200 per month?

Projects & Group Activities

Intensity of Illumination

You are already aware that the standard unit of length in the metric system is the meter (m) and that the standard unit of mass in the metric system is the kilogram (kg). You may not know that the standard unit of light intensity is the candela.

The rate at which light falls upon a one-square-unit area of surface is called the **intensity of illumination.** Intensity of illumination is measured in **lumens** (lm). A lumen is defined in the following illustration.

Picture a source of light equal to 1 candela positioned at the center of a hollow sphere that has a radius of 1 m. The rate at which light falls upon 1 m² of the inner surface of the sphere is equal to one lumen (1 lm). If a light source equal to 4 candelas is positioned at the center of the sphere, each square meter of the inner surface receives four times as much illumination, or 4 lm.

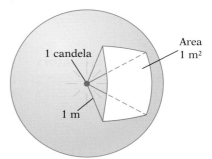

Light rays diverge as they leave a light source. The light that falls upon an area of 1 m² at a distance of 1 m from the source of light spreads out over an area of 4 m² when it is 2 m from the source. The same light spreads out over an area of 9 m² when it is 3 m from the light source and over an area of 16 m² when it is 4 m from the light source. Therefore, as a surface moves farther away from the source of light, the intensity of illumination on the surface decreases from its value at 1 m to $\left(\frac{1}{2}\right)^2$, or $\frac{1}{4}$, that value at 2 m; to $\left(\frac{1}{3}\right)^2$, or $\frac{1}{9}$, that value at 3 m; and to $\left(\frac{1}{4}\right)^2$, or $\frac{1}{16}$, that value at 4 m.

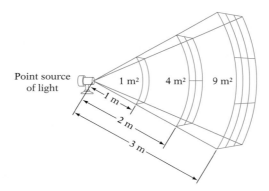

The formula for the intensity of illumination is

$$I = \frac{s}{r^2}$$

where I is the intensity of illumination in lumens, s is the strength of the light source in candelas, and r is the distance in meters between the light source and the illuminated surface.

A 30-candela lamp is 0.5 m above a desk. Find the illumination on the desk.

$$I = \frac{s}{r^2}$$

$$I = \frac{30}{(0.5)^2} = 120$$

The illumination on the desk is 120 lm.

1. A 100-candela light is hanging 5 m above a floor. What is the intensity of illumination on the floor beneath it?

2. A 25-candela source of light is 2 m above a desk. Find the intensity of illumination on the desk.

3. How strong a light source is needed to cast 20 lm of light on a surface 4 m from the source?

4. How strong a light source is needed to cast 80 lm of light on a surface 5 m from the source?

5. How far from the desk surface must a 40-candela light source be positioned if the desired intensity of illumination is 10 lm?

6. Find the distance between a 36-candela light source and a surface if the intensity of illumination on the surface is 0.01 lm.

7. Two lights cast the same intensity of illumination on a wall. One light is 6 m from the wall and has a rating of 36 candelas. The second light is 8 m from the wall. Find the candela rating of the second light.

8. A 40-candela light source and a 10-candela light source both throw the same intensity of illumination on a wall. The 10-candela light is 6 m from the wall. Find the distance from the 40-candela light to the wall.

Scale Model of the Solar System

The table below lists the diameters of the sun and each of its nine planets, as well as the average distance of each planet from the sun.

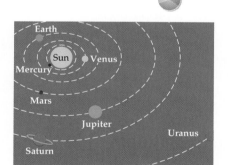

Body	Body Diameter (in kilometers)	Average Distance from the Sun (in millions of kilometers)
Sun	1,391,900	
Mercury	4,878	58.05
Venus	12,104	108.45
Earth	12,756	150.0
Mars	6,787	228.6
Jupiter	142,800	780.45
Saturn	120,660	1,430.85
Uranus	51,118	2,877.0
Neptune	49,528	4,509.0
Pluto	2,300	5,929.5

Source: **http://pds.jpl.nasa.gov/planets/welcome**

To make a scale model of our solar system, begin by choosing the diameter of the sun in the scale model. Use the chosen ratio to determine the diameters of the planets and the distances between the sun and each of the planets.

Chapter Summary

Key Words

A **rational expression** is a fraction in which the numerator or denominator is a variable expression. A rational expression is in simplest form when the numerator and denominator have no common factors other than 1.   [9.1.1, p. 436, 437]

The **reciprocal** of a fraction is that fraction with the numerator and denominator interchanged.   [9.1.3, p. 439]

The **least common multiple** (LCM) of two or more numbers is the smallest number that contains the prime factorization of each number.   [9.2.1, p. 445]

A **complex fraction** is a fraction whose numerator or denominator contains one or more fractions.   [9.4.1, p. 458]

A **ratio** is the quotient of two quantities that have the same unit. A **rate** is the quotient of two quantities that have different units. A **proportion** is an equation that states the equality of two ratios or rates.   [9.5.2, pp. 464–465]

A **literal equation** is an equation that contains more than one variable. A **formula** is a literal equation that states rules about measurements.   [9.6.1, p. 477]

Examples

$\dfrac{x+4}{x-3}$ is a rational expression in simplest form.

The reciprocal of $\dfrac{x-7}{y}$ is $\dfrac{y}{x-7}$.

The LCM of 8 and 12 is 24 because 24 is the smallest number that is a multiple of both 8 and 12.

$\dfrac{x+\dfrac{1}{x}}{2-\dfrac{3}{x}}$ is a complex fraction.

$\dfrac{3}{8}=\dfrac{15}{40}$ and $\dfrac{20\text{ m}}{5\text{ s}}=\dfrac{80\text{ m}}{20\text{ s}}$ are examples of proportions.

$2x+3y=6$ is an example of a literal equation. $A=\pi r^2$ is a literal equation that is also a formula.

Essential Rules and Procedures

Simplifying Rational Expressions
Factor the numerator and denominator. Divide by the common factors.   [9.1.1, p. 437]

$$\frac{x^2-3x-10}{x^2-25}=\frac{(x+2)(x-5)}{(x+5)(x-5)}$$
$$=\frac{x+2}{x+5}$$

Multiplying Rational Expressions
Multiply the numerators. Multiply the denominators. Write the answer in simplest form.
$\dfrac{a}{b}\cdot\dfrac{c}{d}=\dfrac{ac}{bd}$ [9.1.2, p. 438]

$$\frac{x^2-3x}{x^2+x}\cdot\frac{x^2+5x+4}{x^2-4x+3}$$
$$=\frac{x(x-3)}{x(x+1)}\cdot\frac{(x+4)(x+1)}{(x-3)(x-1)}$$
$$=\frac{x(x-3)(x+4)(x+1)}{x(x+1)(x-3)(x-1)}=\frac{x+4}{x-1}$$

Dividing Rational Expressions
To divide two fractions, multiply by the reciprocal of the divisor.
$$\frac{a}{b} \div \frac{c}{d} = \frac{a}{b} \cdot \frac{d}{c} = \frac{ad}{bc}$$ [9.1.3, p. 440]

$$\frac{3x^3y^2}{8a^4b^5} \div \frac{6x^4y}{4a^3b^2} = \frac{3x^3y^2}{8a^4b^5} \cdot \frac{4a^3b^2}{6x^4y}$$
$$= \frac{3x^3y^2 \cdot 4a^3b^2}{8a^4b^5 \cdot 6x^4y} = \frac{y}{4ab^3x}$$

Adding and Subtracting Rational Expressions

1. Find the LCM of the denominators.

2. Write each fraction with the LCM as the denominator.

3. Add or subtract the numerators. The denominator of the sum or difference is the common denominator.

4. Write the answer in simplest form.

[9.3.2, p. 451]

$$\frac{4x-2}{3x+12} - \frac{x-2}{x+4} = \frac{4x-2}{3(x+4)} - \frac{x-2}{x+4}$$
$$= \frac{4x-2}{3(x+4)} - \frac{x-2}{x+4} \cdot \frac{3}{3}$$
$$= \frac{4x-2}{3(x+4)} - \frac{3x-6}{3(x+4)}$$
$$= \frac{4x-2-(3x-6)}{3(x+4)}$$
$$= \frac{x+4}{3(x+4)} = \frac{1}{3}$$

Simplifying Complex Fractions
Multiply the numerator and denominator of the complex fraction by the LCM of the denominators of the fractions in the numerator and denominator.   [9.4.1, p. 458]

$$\frac{\frac{1}{x} + \frac{1}{y}}{\frac{1}{x} - \frac{1}{y}} = \frac{\frac{1}{x} + \frac{1}{y}}{\frac{1}{x} - \frac{1}{y}} \cdot \frac{xy}{xy} = \frac{\frac{1}{x} \cdot xy + \frac{1}{y} \cdot xy}{\frac{1}{x} \cdot xy - \frac{1}{y} \cdot xy}$$
$$= \frac{y+x}{y-x}$$

Solving Equations Containing Fractions
Clear denominators by multiplying each side of the equation by the LCM of the denominators. Then solve for the variable. [9.5.1, pp. 462–463]

$$\frac{1}{2a} = \frac{2}{a} - \frac{3}{8}$$
$$8a\left(\frac{1}{2a}\right) = 8a\left(\frac{2}{a} - \frac{3}{8}\right)$$
$$4 = 16 - 3a$$
$$-12 = -3a$$
$$4 = a$$

Similar Triangles
Similar triangles have the same shape but not necessarily the same size. The ratios of corresponding sides of similar triangles are equal. The ratio of corresponding heights is equal to the ratio of corresponding sides.   [9.5.4, pp. 467–468]

Triangles *ABC* and *DEF* are similar triangles. The ratios of corresponding sides and corresponding heights is $\frac{1}{3}$.

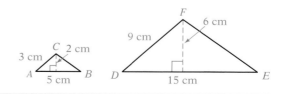

Solving Literal Equations
Rewrite the equation so that the letter being solved for is alone on one side of the equation and all numbers and other variables are on the other side.   [9.6.1, p. 477]

Solve $A = \frac{1}{2}bh$ for b.

$$2(A) = 2\left(\frac{1}{2}bh\right)$$
$$2A = bh$$
$$\frac{2A}{h} = \frac{bh}{h}$$
$$\frac{2A}{h} = b$$

Work Problems
Rate of work · Time worked = Part of task completed   [9.7.1, p. 481]

Pat can do a certain job in 3 h. Chris can do the same job in 5 h. How long would it take them, working together, to get the job done?

$$\frac{t}{3} + \frac{t}{5} = 1$$

Uniform Motion Problems
Distance ÷ Rate = Time   [9.7.2, p. 482]

Train A's speed is 15 mph faster than Train B's speed. Train A travels 150 mi in the same amount of time as it takes Train B to travel 120 mi. Find the rate of Train B.

$$\frac{120}{r} = \frac{150}{r + 15}$$

Chapter Review Exercises

1. Multiply: $\dfrac{8ab^2}{15x^3y} \cdot \dfrac{5xy^4}{16a^2b}$

2. Add: $\dfrac{5}{3x - 4} + \dfrac{4}{2x + 3}$

3. Solve $4x + 3y = 12$ for x.

4. Simplify: $\dfrac{16x^5y^3}{24xy^{10}}$

5. Divide: $\dfrac{20x^2 - 45x}{6x^3 + 4x^2} \div \dfrac{40x^3 - 90x^2}{12x^2 + 8x}$

6. Simplify: $\dfrac{x - \dfrac{16}{5x - 2}}{3x - 4 - \dfrac{88}{5x - 2}}$

7. Find the LCM of $24a^2b^5$ and $36a^3b$.

8. Subtract: $\dfrac{5x}{3x + 7} - \dfrac{x}{3x + 7}$

9. Write each fraction in terms of the LCM of the denominators.

$$\frac{3}{16x} ; \frac{5}{8x^2}$$

10. Simplify: $\dfrac{2x^2 - 13x - 45}{2x^2 - x - 15}$

11. Divide: $\dfrac{x^2 - 5x - 14}{x^2 - 3x - 10} \div \dfrac{x^2 - 4x - 21}{x^2 - 9x + 20}$

12. Add: $\dfrac{2y}{5y - 7} + \dfrac{3}{7 - 5y}$

13. Multiply: $\dfrac{3x^3 + 10x^2}{10x - 2} \cdot \dfrac{20x - 4}{6x^4 + 20x^3}$

14. Subtract: $\dfrac{5x + 3}{2x^2 + 5x - 3} - \dfrac{3x + 4}{2x^2 + 5x - 3}$

15. Find the LCM of $5x^4(x - 7)^2$ and $15x(x - 7)$.

16. Solve: $\dfrac{6}{x - 7} = \dfrac{8}{x - 6}$

17. Solve: $\dfrac{x + 8}{x + 4} = 1 + \dfrac{5}{x + 4}$

18. Simplify: $\dfrac{12a^2b(4x - 7)}{15ab^2(7 - 4x)}$

19. Simplify: $\dfrac{5x - 1}{x^2 - 9} + \dfrac{4x - 3}{x^2 - 9} - \dfrac{8x - 1}{x^2 - 9}$

20. Triangles ABC and DEF below are similar. Find the perimeter of triangle ABC.

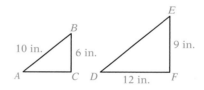

21. Solve: $\dfrac{20}{2x + 3} = \dfrac{17x}{2x + 3} - 5$

22. Simplify: $\dfrac{\dfrac{5}{x - 1} - \dfrac{3}{x + 3}}{\dfrac{6}{x + 3} + \dfrac{2}{x - 1}}$

23. Add: $\dfrac{x - 1}{x + 2} + \dfrac{3x - 2}{5 - x} + \dfrac{5x^2 + 15x - 11}{x^2 - 3x - 10}$

24. Divide: $\dfrac{18x^2 + 25x - 3}{9x^2 - 28x + 3} \div \dfrac{2x^2 + 9x + 9}{x^2 - 6x + 9}$

25. Simplify: $\dfrac{x^2 + x - 30}{15 + 2x - x^2}$

26. Solve: $\dfrac{5}{7} + \dfrac{x}{2} = 2 - \dfrac{x}{7}$

27. Simplify: $\dfrac{x + \dfrac{6}{x - 5}}{1 + \dfrac{2}{x - 5}}$

28. Multiply: $\dfrac{3x^2 + 4x - 15}{x^2 - 11x + 28} \cdot \dfrac{x^2 - 5x - 14}{3x^2 + x - 10}$

29. Solve: $\dfrac{x}{5} = \dfrac{x + 12}{9}$

30. Solve: $\dfrac{3}{20} = \dfrac{x}{80}$

31. Simplify: $\dfrac{1 - \dfrac{1}{x}}{1 - \dfrac{8x - 7}{x^2}}$

32. Solve $x - 2y = 15$ for x.

33. Add: $\dfrac{6}{a} + \dfrac{9}{b}$

34. Find the LCM of $10x^2 - 11x + 3$ and $20x^2 - 17x + 3$.

35. Solve $i = \dfrac{100m}{c}$ for c.

36. Solve: $\dfrac{15}{x} = \dfrac{3}{8}$

37. Solve: $\dfrac{22}{2x + 5} = 2$

38. Add: $\dfrac{x + 7}{15x} + \dfrac{x - 2}{20x}$

39. Multiply: $\dfrac{16a^2 - 9}{16a^2 - 24a + 9} \cdot \dfrac{8a^2 - 13a - 6}{4a^2 - 5a - 6}$

40. Write each fraction in terms of the LCM of the denominators.

$$\dfrac{x}{12x^2 + 16x - 3}, \dfrac{4x^2}{6x^2 + 7x - 3}$$

41. Add: $\dfrac{3}{4ab} + \dfrac{5}{4ab}$

42. Solve: $\dfrac{20}{x + 2} = \dfrac{5}{16}$

43. Solve: $\dfrac{5x}{3} - \dfrac{2}{5} = \dfrac{8x}{5}$

44. Divide: $\dfrac{6a^2b^7}{25x^3y} \div \dfrac{12a^3b^4}{5x^2y^2}$

45. A brick mason can construct a patio in 3 h. If the mason works with an apprentice, they can construct the patio in 2 h. How long would it take the apprentice, working alone, to construct the patio?

46. A weight of 21 lb stretches a spring 14 in. At the same rate, how far would a weight of 12 lb stretch the spring?

47. The rate of a jet is 400 mph in calm air. Traveling with the wind, the jet can fly 2100 mi in the same amount of time as it takes to fly 1900 mi against the wind. Find the rate of the wind.

48. A gardener uses 4 oz of insecticide to make 2 gal of garden spray. At this rate, how much additional insecticide is necessary to make 10 gal of the garden spray?

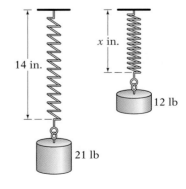

49. One hose can fill a pool in 15 h. A second hose can fill the pool in 10 h. How long would it take to fill the pool using both hoses?

50. A car travels 315 mi in the same amount of time as it takes a bus to travel 245 mi. The rate of the car is 10 mph faster than that of the bus. Find the rate of the car.

Chapter Test

1. Divide: $\dfrac{x^2 + 3x + 2}{x^2 + 5x + 4} \div \dfrac{x^2 - x - 6}{x^2 + 2x - 15}$

2. Subtract: $\dfrac{2x}{x^2 + 3x - 10} - \dfrac{4}{x^2 + 3x - 10}$

3. Find the LCM of $6x - 3$ and $2x^2 + x - 1$.

4. Solve: $\dfrac{3}{x + 4} = \dfrac{5}{x + 6}$

5. Multiply: $\dfrac{x^3 y^4}{x^2 - 4x + 4} \cdot \dfrac{x^2 - x - 2}{x^6 y^4}$

6. Simplify: $\dfrac{1 + \dfrac{1}{x} - \dfrac{12}{x^2}}{1 + \dfrac{2}{x} - \dfrac{8}{x^2}}$

7. Write each fraction in terms of the LCM of the denominators.

$$\dfrac{3}{x^2 - 2x}, \dfrac{x}{x^2 - 4}$$

8. Solve $3x + 5y + 15 = 0$ for x.

9. Solve: $\dfrac{6}{x} - 2 = 1$

10. Subtract: $\dfrac{2}{2x - 1} - \dfrac{3}{3x + 1}$

11. Divide: $\dfrac{x^2 - x - 56}{x^2 + 8x + 7} \div \dfrac{x^2 - 13x + 40}{x^2 - 4x - 5}$

12. Subtract: $\dfrac{3x}{x^2 + 5x - 24} - \dfrac{9}{x^2 + 5x - 24}$

13. Find the LCM of $3x^2 + 6x$ and $2x^2 + 8x + 8$.

14. Simplify: $\dfrac{x^2 - 7x + 10}{25 - x^2}$

15. Solve: $\dfrac{3x}{x - 3} - 2 = \dfrac{10}{x - 3}$

16. Solve $f = v + at$ for t.

17. Simplify: $\dfrac{12x^4 y^2}{18xy^7}$

18. Subtract: $\dfrac{2}{2x - 1} - \dfrac{1}{x + 1}$

19. Solve: $\dfrac{2}{x - 2} = \dfrac{12}{x + 3}$

20. Multiply: $\dfrac{x^5 y^3}{x^2 - x - 6} \cdot \dfrac{x^2 - 9}{x^2 y^4}$

21. Write each fraction in terms of the LCM of the denominators.

$$\frac{3y}{x(1-x)}, \ \frac{x}{(x+1)(x-1)}$$

22. Simplify: $\dfrac{1 - \dfrac{2}{x} - \dfrac{15}{x^2}}{1 - \dfrac{25}{x^2}}$

23. A salt solution is formed by mixing 4 lb of salt with 10 gal of water. At this rate, how many additional pounds of salt are required for 15 gal of water?

24. A small plane can fly at 110 mph in calm air. Flying with the wind, the plane can fly 260 mi in the same amount of time as it takes to fly 180 mi against the wind. Find the rate of the wind.

25. One pipe can fill a tank in 9 min. A second pipe requires 18 min to fill the tank. How long would it take both pipes, working together, to fill the tank?

Cumulative Review Exercises

1. Evaluate: $-|-17|$

2. Evaluate: $-\dfrac{3}{4} \cdot (2)^3$

3. Simplify: $\left(\dfrac{2}{3}\right)^2 \div \left(\dfrac{3}{2} - \dfrac{2}{3}\right) + \dfrac{1}{2}$

4. Evaluate $-a^2 + (a - b)^2$ when $a = -2$ and $b = 3$.

5. Simplify: $-2x - (-3y) + 7x - 5y$

6. Simplify: $2[3x - 7(x - 3) - 8]$

7. Solve: $3 - \dfrac{1}{4}x = 8$

8. Solve: $3[x - 2(x - 3)] = 2(3 - 2x)$

9. Find $16\dfrac{2}{3}\%$ of 60.

10. Solve: $\dfrac{5}{9}x < 1$

11. Solve: $x - 2 \geq 4x - 47$

12. Graph: $y = 2x - 1$

13. Graph: $f(x) = 3x + 2$

14. Graph: $x = 3$

15. Graph the solution set of $5x + 2y < 6$.

16. Find the range of the function given by the equation $f(x) = -2x + 11$ if the domain is $\{-8, -4, 0, 3, 7\}$.

17. Evaluate $f(x) = 4x - 3$ at $x = 10$.

18. Solve by substitution: $6x - y = 1$
$\qquad\qquad\qquad\qquad\qquad\quad y = 3x + 1$

19. Solve by the addition method: $2x - 3y = 4$
$\qquad\qquad\qquad\qquad\qquad\qquad\quad 4x + y = 1$

20. Multiply: $(3xy^4)(-2x^3y)$

21. Simplify: $(a^4b^3)^5$

22. Simplify: $\dfrac{a^2b^{-5}}{a^{-1}b^{-3}}$

23. Multiply: $(a - 3b)(a + 4b)$

24. Divide: $\dfrac{15b^4 - 5b^2 + 10b}{5b}$

25. Divide: $(x^3 - 8) \div (x - 2)$

26. Factor: $12x^2 - x - 1$

27. Factor: $y^2 - 7y + 6$

28. Factor: $2a^3 + 7a^2 - 15a$

29. Factor: $4b^2 - 100$

30. Solve: $(x + 3)(2x - 5) = 0$

31. Simplify: $\dfrac{x^2 + 3x - 28}{16 - x^2}$

32. Divide: $\dfrac{x^2 - 3x - 10}{x^2 - 4x - 12} \div \dfrac{x^2 - x - 20}{x^2 - 2x - 24}$

33. Subtract: $\dfrac{6}{3x - 1} - \dfrac{2}{x + 1}$

34. Solve: $\dfrac{4x}{x - 3} - 2 = \dfrac{8}{x - 3}$

35. Solve $f = v + at$ for a.

36. Translate "the difference between five times a number and eighteen is the opposite of three" into an equation and solve.

37. An investment of $5000 is made at an annual simple interest rate of 7%. How much additional money must be invested at an annual simple interest rate of 11% so that the total interest earned is 9% of the total investment?

38. A silversmith mixes 60 g of an alloy that is 40% silver with 120 g of another silver alloy. The resulting alloy is 60% silver. Find the percent of silver in the 120-gram alloy.

39. The length of the base of a triangle is 2 in. less than twice the height. The area of the triangle is 30 in². Find the base and height of the triangle.

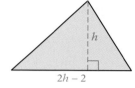

40. One water pipe can fill a tank in 12 min. A second pipe requires 24 min to fill the tank. How long would it take both pipes, working together, to fill the tank?

Civil engineers design expressway on-and off-ramps with both efficiency and safety in mind. A ramp cannot have too sharp a curve or the speed at which drivers take the curve will cause them to skid off the road. However, a ramp with too wide a curve requires more building material, takes longer to travel, and uses more land. A radical equation is used to determine the maximum speed at which a car can travel around a curve without skidding. This equation is used in the exercises in the **Project on pages 538–539**.

Applications

Aviation, pp. 513, 532

Civil engineering, pp. 538, 539

Communications, p. 531

Construction, pp. 528, 543

Deductive reasoning, pp. 534, 535

Earth science, p. 543

Education, p. 531

Geometry, pp. 513, 517, 524, 531, 533, 542

Home maintenance, pp. 529, 532, 544

Integer problems, pp. 530, 546

Markup, p. 546

Navigation, pp. 529, 531, 532, 538

Number problems, p. 546

Parks, p. 532

Percent mixture problems, p. 546

Physics, pp. 529, 531, 532, 542, 544

Space exploration, p. 542

Sports, p. 531

Statistics, pp. 536, 537

Television, pp. 532, 533

Traffic safety, p. 513

Work problems, p. 546

10
Radical Expressions

Objectives

Section 10.1
1. Simplify numerical radical expressions
2. Simplify variable radical expressions

Section 10.2
1. Add and subtract radical expressions

Section 10.3
1. Multiply radical expressions
2. Divide radical expressions

Section 10.4
1. Solve equations containing one or more radical expressions
2. Application problems

 Need help? For online student resources, such as section quizzes, visit this web site: **math.college.hmco.com**

PrepTest

1. Evaluate $-|-14|$.

2. Simplify: $3x^2y - 4xy^2 - 5x^2y$

3. Solve: $1.5h = 21$

4. Solve: $3x - 2 = 5 - 2x$

5. Simplify: $x^3 \cdot x^3$

6. Simplify: $(x + y)^2$

7. Simplify: $(2x - 3)^2$

8. Simplify: $(a - 5)(a + 5)$

9. Simplify: $(2 - 3v)(2 + 3v)$

10. Simplify: $\dfrac{2x^4y^3}{18x^2y}$

Go Figure

In a contest to guess how many jelly beans were in a jar, Hector guessed 223, Shannon guessed 215, Suki guessed 220, Deon guessed 221, Saul guessed 217, and Denise guessed 219. Two guesses were off by 4, two were off by 2, one was off by 1, and one was correct. What was the correct guess?

10.1 Introduction to Radical Expressions

1 ### Simplify numerical radical expressions

VIDEO & DVD   CD TUTOR   WEB   SSM

Point of Interest

The radical symbol was first used in 1525, when it was written as ✓. Some historians suggest that the radical symbol also developed into the symbols for "less than" and "greater than." Because typesetters of that time did not want to make additional symbols, the radical was rotated to the position ＼ and used as a "greater than" symbol and rotated to — and used for the "less than" symbol. Other evidence, however, suggests that the "less than" and "greater than" symbols were developed independently of the radical symbol.

A **square root** of a positive number x is a number whose square is x.

A square root of 16 is 4 because $4^2 = 16$.

A square root of 16 is -4 because $(-4)^2 = 16$.

Every positive number has two square roots, one a positive number and one a negative number. The symbol $\sqrt{}$, called a **radical sign**, is used to indicate the positive or **principal square root** of a number. For example, $\sqrt{16} = 4$ and $\sqrt{25} = 5$. The number or variable expression under the radical sign is called the **radicand**.

When the negative square root of a number is to be found, a negative sign is placed in front of the radical. For example, $-\sqrt{16} = -4$ and $-\sqrt{25} = -5$.

The square of an integer is a **perfect square**. 49, 81, and 144 are examples of perfect squares.

$7^2 = 49$
$9^2 = 81$
$12^2 = 144$

The principal square root of a perfect-square integer is a positive integer.

$$\sqrt{49} = 7$$
$$\sqrt{81} = 9$$
$$\sqrt{144} = 12$$

If a number is not a perfect square, its square root can only be approximated. For example, 2 and 7 are not perfect squares. Thus their square roots can only be approximated. These numbers are **irrational numbers.** Their decimal representations never terminate or repeat.

$$\sqrt{2} \approx 1.4142135\ldots$$

$$\sqrt{7} \approx 2.6457513\ldots$$

The approximate square roots of numbers that are not perfect squares can be found using a calculator. The square roots can be rounded to any given place value.

Take Note

Recall that a factor of a number divides the number evenly. For example, 6 is a factor of 18. 9 is also a factor of 18. Note that 9 is a *perfect-square factor* of 18, whereas 6 is not a perfect-square factor of 18.

Radical expressions that contain radicands that are not perfect squares are generally written in *simplest form*. A radical expression is in **simplest form** when the radicand contains no factor greater than 1 that is a perfect square. For example, $\sqrt{50}$ is not in simplest form because 25 is a perfect-square factor of 50. The radical expression $\sqrt{15}$ is in simplest form because there is no perfect-square factor of 15 that is greater than 1.

The Product Property of Square Roots and a knowledge of perfect squares is used to simplify radical expressions.

The Product Property of Square Roots

If a and b are positive real numbers, then $\sqrt{ab} = \sqrt{a} \cdot \sqrt{b}$.

The chart below shows square roots of some perfect squares.

Square Roots of Perfect Squares

$\sqrt{1} = 1$	$\sqrt{16} = 4$	$\sqrt{49} = 7$	$\sqrt{100} = 10$
$\sqrt{4} = 2$	$\sqrt{25} = 5$	$\sqrt{64} = 8$	$\sqrt{121} = 11$
$\sqrt{9} = 3$	$\sqrt{36} = 6$	$\sqrt{81} = 9$	$\sqrt{144} = 12$

Simplify: $\sqrt{72}$

$$\sqrt{72} = \sqrt{36 \cdot 2}$$

• Write the radicand as the product of a perfect square and a factor that does not contain a perfect square.

$$= \sqrt{36}\,\sqrt{2}$$

• Use the Product Property of Square Roots to write the expression as a product.

$$= 6\sqrt{2}$$

• Simplify $\sqrt{36}$.

Note in the example above that 72 must be written as the product of a perfect square and *a factor that does not contain a perfect square.* Therefore, it would not be correct to rewrite $\sqrt{72}$ as $\sqrt{9 \cdot 8}$ and simplify the expression as shown at the right. Although 9 is a perfect square factor of 72, 8 contains a perfect square ($8 = 4 \cdot 2$) and $\sqrt{4}$ can be simplified. Therefore, $\sqrt{8}$ is not in simplest form. Remember to find the *largest* perfect square that is a factor of the radicand.

$$\sqrt{72} = \sqrt{9 \cdot 8}$$
$$= \sqrt{9} \sqrt{8}$$
$$= 3\sqrt{8}$$
Not in simplest form

Simplify: $\sqrt{147}$

$$\sqrt{147} = \sqrt{49 \cdot 3}$$
- Write the radicand as the product of a perfect square and a factor that does not contain a perfect square.

$$= \sqrt{49} \sqrt{3}$$
- Use the Product Property of Square Roots to write the expression as a product.

$$= 7\sqrt{3}$$
- Simplify $\sqrt{49}$.

Simplify: $\sqrt{360}$

$$\sqrt{360} = \sqrt{36 \cdot 10}$$
- Write the radicand as the product of a perfect square and a factor that does not contain a perfect square.

$$= \sqrt{36} \sqrt{10}$$
- Use the Product Property of Square Roots to write the expression as a product.

$$= 6\sqrt{10}$$
- Simplify $\sqrt{36}$.

From the last example, note that $\sqrt{360} = 6\sqrt{10}$. The two expressions are different representations of the same number. Using a calculator, we find that $\sqrt{360} \approx 18.973666$ and $6\sqrt{10} \approx 6(3.1622777) = 18.973666$.

Simplify: $\sqrt{-16}$

Because the square of any real number is positive, there is no real number whose square is -16.

$\sqrt{-16}$ is not a real number.

Example 1 Simplify: $3\sqrt{90}$

Solution $3\sqrt{90} = 3\sqrt{9 \cdot 10}$
- Write the radicand as the product of a perfect square and a factor that does not contain a perfect square.

$$= 3\sqrt{9} \sqrt{10}$$
- Use the Product Property of Square Roots to write the expression as a product.

$$= 3 \cdot 3\sqrt{10}$$
- Simplify $\sqrt{9}$.

$$= 9\sqrt{10}$$
- Multiply $3 \cdot 3$.

Problem 1 Simplify: $-5\sqrt{32}$

Solution See page S23.

─Example 2 Simplify: $\sqrt{252}$

Solution $\sqrt{252} = \sqrt{36 \cdot 7}$ • Write the radicand as the product of a perfect square and a factor that does not contain a perfect square.

$= \sqrt{36} \sqrt{7}$ • Use the Product Property of Square Roots to write the expression as a product.

$= 6\sqrt{7}$ • Simplify $\sqrt{36}$.

Problem 2 Simplify: $\sqrt{216}$

Solution   See page S23.

2 Simplify variable radical expressions

Variable expressions that contain radicals do not always represent real numbers.

The variable expression at the right does not represent a real number when x is a negative number, such as -4.

$\sqrt{x^3}$
$\sqrt{(-4)^3} = \sqrt{-64} \leftarrow$ Not a real number

Now consider the expression $\sqrt{x^2}$ and evaluate this expression for $x = -2$ and $x = 2$.

$\sqrt{x^2}$
$\sqrt{(-2)^2} = \sqrt{4} = 2 = |-2|$
$\sqrt{2^2} \quad = \sqrt{4} = 2 = |2|$

This suggests the following.

The Square Root of a^2

For any real number a, $\sqrt{a^2} = |a|$.

If $a \geq 0$, then $\sqrt{a^2} = a$.

In order to avoid variable expressions that do not represent real numbers, and so that absolute value signs are not needed for certain expressions, the variables in this chapter will represent *positive* numbers unless otherwise stated.

A variable or a product of variables written in exponential form is a **perfect square** when each exponent is an even number.

To find the square root of a perfect square, remove the radical sign, and divide each exponent by 2.

⌐ Simplify: $\sqrt{a^6}$

Remove the radical sign, and divide the exponent by 2. $\sqrt{a^6} = a^3$

A variable radical expression is not in simplest form when the radicand contains a factor greater than 1 that is a perfect square.

Simplify: $\sqrt{x^7}$

Write x^7 as the product of x and a perfect square. $\sqrt{x^7} = \sqrt{x^6 \cdot x}$

Use the Product Property of Square Roots. $= \sqrt{x^6}\,\sqrt{x}$

Simplify the square root of the perfect square. $= x^3\sqrt{x}$

Example 3   Simplify: $\sqrt{b^{15}}$

Solution $\sqrt{b^{15}} = \sqrt{b^{14} \cdot b} = \sqrt{b^{14}} \cdot \sqrt{b} = b^7\sqrt{b}$

Problem 3   Simplify: $\sqrt{y^{19}}$

Solution   See page S23.

Simplify: $3x\sqrt{8x^3y^{13}}$

Write the radicand as the product of a perfect square and factors that do not contain a perfect square. $3x\sqrt{8x^3y^{13}} = 3x\sqrt{4x^2y^{12}(2xy)}$

Use the Product Property of Square Roots. $= 3x\sqrt{4x^2y^{12}}\,\sqrt{2xy}$

Simplify. $= 3x \cdot 2xy^6\sqrt{2xy}$

$= 6x^2y^6\sqrt{2xy}$

Example 4   Simplify.   **A.** $\sqrt{24x^5}$ **B.** $2a\sqrt{18a^3b^{10}}$

Solution   **A.** $\sqrt{24x^5} = \sqrt{4x^4 \cdot 6x}$ **B.** $2a\sqrt{18a^3b^{10}} = 2a\sqrt{9a^2b^{10} \cdot 2a}$

$= \sqrt{4x^4}\,\sqrt{6x}$ $= 2a\sqrt{9a^2b^{10}}\,\sqrt{2a}$

$= 2x^2\sqrt{6x}$ $= 2a \cdot 3ab^5\sqrt{2a}$

$= 6a^2b^5\sqrt{2a}$

Problem 4   Simplify.   **A.** $\sqrt{45b^7}$ **B.** $3a\sqrt{28a^9b^{18}}$

Solution   See page S23.

Simplify: $\sqrt{25(x + 2)^2}$

25 is a perfect square. $(x + 2)^2$ is a perfect square. $\sqrt{25(x + 2)^2} = 5(x + 2)$

$= 5x + 10$

Example 5   Simplify: $\sqrt{16(x + 5)^2}$

Solution $\sqrt{16(x + 5)^2} = 4(x + 5) = 4x + 20$

Problem 5   Simplify: $\sqrt{25(a + 3)^2}$

Solution   See page S23.

10.1 | Concept Review

Determine whether the statement is always true, sometimes true, or never true.

1. The square root of a positive number is a positive number, and the square root of a negative number is a negative number.

2. Every positive number has two square roots, one of which is the additive inverse of the other.

3. The square root of a number that is not a perfect square is an irrational number.

4. If the radicand of a radical expression is evenly divisible by a perfect square greater than 1, then the radical expression is not in simplest form.

5. Suppose a and b are real numbers. Then $\sqrt{ab} = \sqrt{a} \cdot \sqrt{b}$.

6. When a perfect square is written in exponential form, the exponents are multiples of 2.

10.1 | Exercises

1 **1.** Show by example that a positive number has two square roots.

2. Why is the square root of a negative number not a real number?

3. Which of the numbers 2, 9, 20, 25, 50, 81, and 100 are *not* perfect squares?

4. Write down a number that has a perfect-square factor that is greater than 1.

5. Write a sentence or two that you could email a friend to explain the concept of a perfect-square factor.

6. Name the perfect-square factors of 540. What is the largest perfect-square factor of 540?

Simplify.

7. $\sqrt{16}$ **8.** $\sqrt{64}$ **9.** $\sqrt{49}$ **10.** $\sqrt{144}$ **11.** $\sqrt{32}$ **12.** $\sqrt{50}$

13. $\sqrt{8}$ **14.** $\sqrt{12}$ **15.** $6\sqrt{18}$ **16.** $-3\sqrt{48}$ **17.** $5\sqrt{40}$ **18.** $2\sqrt{28}$

19. $\sqrt{15}$ **20.** $\sqrt{21}$ **21.** $\sqrt{29}$ **22.** $\sqrt{13}$ **23.** $-9\sqrt{72}$ **24.** $11\sqrt{80}$

25. $\sqrt{45}$ **26.** $\sqrt{225}$ **27.** $\sqrt{0}$ **28.** $\sqrt{210}$ **29.** $6\sqrt{128}$ **30.** $9\sqrt{288}$

31. $\sqrt{105}$ **32.** $\sqrt{55}$ **33.** $\sqrt{900}$ **34.** $\sqrt{300}$ **35.** $5\sqrt{180}$ **36.** $7\sqrt{98}$

37. $\sqrt{250}$ **38.** $\sqrt{120}$ **39.** $\sqrt{96}$ **40.** $\sqrt{160}$ **41.** $\sqrt{324}$ **42.** $\sqrt{444}$

Find the decimal approximation to the nearest thousandth.

43. $\sqrt{240}$ **44.** $\sqrt{300}$ **45.** $\sqrt{288}$ **46.** $\sqrt{600}$

47. $\sqrt{245}$ **48.** $\sqrt{525}$ **49.** $\sqrt{352}$ **50.** $\sqrt{363}$

2 **51.** ✎   How can you tell whether a variable exponential expression is a perfect square?

52. ✎   When is a radical expression in simplest form?

Simplify.

53. $\sqrt{x^6}$ **54.** $\sqrt{x^{12}}$ **55.** $\sqrt{y^{15}}$ **56.** $\sqrt{y^{11}}$

57. $\sqrt{a^{20}}$ **58.** $\sqrt{a^{16}}$ **59.** $\sqrt{x^4y^4}$ **60.** $\sqrt{x^{12}y^8}$

61. $\sqrt{4x^4}$ **62.** $\sqrt{25y^8}$ **63.** $\sqrt{24x^2}$ **64.** $\sqrt{x^3y^{15}}$

65. $\sqrt{x^3y^7}$ **66.** $\sqrt{a^{15}b^5}$ **67.** $\sqrt{a^3b^{11}}$ **68.** $\sqrt{24y^7}$

69. $\sqrt{60x^5}$ **70.** $\sqrt{72y^7}$ **71.** $\sqrt{49a^4b^8}$ **72.** $\sqrt{144x^2y^8}$

73. $\sqrt{18x^5y^7}$ **74.** $\sqrt{32a^5b^{15}}$ **75.** $\sqrt{40x^{11}y^7}$

76. $\sqrt{72x^9y^3}$ **77.** $\sqrt{80a^9b^{10}}$ **78.** $\sqrt{96a^5b^7}$

79. $2\sqrt{16a^2b^3}$ **80.** $5\sqrt{25a^4b^7}$ **81.** $x\sqrt{x^4y^2}$

82. $y\sqrt{x^3y^6}$ **83.** $4\sqrt{20a^4b^7}$ **84.** $5\sqrt{12a^3b^4}$

85. $3x\sqrt{12x^2y^7}$ **86.** $4y\sqrt{18x^5y^4}$ **87.** $2x^2\sqrt{8x^2y^3}$

88. $3y^2\sqrt{27x^4y^3}$ **89.** $\sqrt{25(a+4)^2}$ **90.** $\sqrt{81(x+y)^4}$

91. $\sqrt{4(x+2)^4}$ **92.** $\sqrt{9(x+2)^2}$ **93.** $\sqrt{x^2+4x+4}$

94. $\sqrt{b^2+8b+16}$ **95.** $\sqrt{y^2+2y+1}$ **96.** $\sqrt{a^2+6a+9}$

10.1 **Applying Concepts**

97. Traffic Safety Traffic accident investigators can estimate the speed S, in miles per hour, of a car from the length of its skid mark by using the formula $S = \sqrt{30\,fl}$, where f is the coefficient of friction (which depends on the type of road surface) and l is the length of the skid mark in feet. Suppose the coefficient of friction is 1.2 and the length of a skid mark is 60 ft. Determine the speed of the car **a.** as a radical in simplest form and **b.** rounded to the nearest integer.

98. Aviation The distance a pilot in an airplane can see to the horizon can be approximated by the equation $d = 1.2\sqrt{h}$, where d is the distance to the horizon in miles and h is the height of the plane in feet. For a pilot flying at an altitude of 5000 ft, what is the distance to the horizon? Round to the nearest tenth.

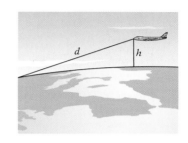

Simplify.

99. $\sqrt{0.0025a^3b^5}$

100. $-\dfrac{3y}{4}\sqrt{64x^4y^2}$

101. Given $f(x) = \sqrt{2x - 1}$, find each of the following. Write your answer in simplest form.
    a. $f(1)$ **b.** $f(5)$ **c.** $f(14)$

102. Use the roster method to list the whole numbers between $\sqrt{8}$ and $\sqrt{90}$.

103. Geometry The area of a square is 76 cm². Find the length of a side of the square. Round to the nearest tenth.

Assuming x can be any real number, for what values of x is the radical expression a real number? Write the answer as an inequality or write "all real numbers."

104. $\sqrt{x}$

105. $\sqrt{4x}$

106. $\sqrt{x - 2}$

107. $\sqrt{x + 5}$

108. $\sqrt{6 - 4x}$

109. $\sqrt{5 - 2x}$

110. $\sqrt{x^2 + 7}$

111. $\sqrt{x^2 + 1}$

112. Describe in your own words how to simplify a radical expression.

113. Explain why $2\sqrt{2}$ is in simplest form and $\sqrt{8}$ is not in simplest form.

10.2 Addition and Subtraction of Radical Expressions

1 **Add and subtract radical expressions**

The Distributive Property is used to simplify the sum or difference of radical expressions with the same radicand.

$$5\sqrt{2} + 3\sqrt{2} = (5 + 3)\sqrt{2} = 8\sqrt{2}$$
$$6\sqrt{2x} - 4\sqrt{2x} = (6 - 4)\sqrt{2x} = 2\sqrt{2x}$$

Radical expressions that are in simplest form and have different radicands cannot be simplified by the Distributive Property.

$2\sqrt{3} + 4\sqrt{2}$ cannot be simplified by the Distributive Property.

To simplify the sum or difference of radical expressions, simplify each term. Then use the Distributive Property.

Subtract: $4\sqrt{8} - 10\sqrt{2}$

Simplify each term.

$$
\begin{aligned}
4\sqrt{8} - 10\sqrt{2} &= 4\sqrt{4 \cdot 2} - 10\sqrt{2} \\
&= 4\sqrt{4}\sqrt{2} - 10\sqrt{2} \\
&= 4 \cdot 2\sqrt{2} - 10\sqrt{2} \\
&= 8\sqrt{2} - 10\sqrt{2}
\end{aligned}
$$

Subtract by using the Distributive Property.

$$
\begin{aligned}
&= (8 - 10)\sqrt{2} \\
&= -2\sqrt{2}
\end{aligned}
$$

Example 1 Simplify. **A.** $5\sqrt{2} - 3\sqrt{2} + 12\sqrt{2}$ **B.** $3\sqrt{12} - 5\sqrt{27}$

Solution **A.** $5\sqrt{2} - 3\sqrt{2} + 12\sqrt{2}$
$= (5 - 3 + 12)\sqrt{2}$ • Use the Distributive Property.
$= 14\sqrt{2}$

B. $3\sqrt{12} - 5\sqrt{27} = 3\sqrt{4 \cdot 3} - 5\sqrt{9 \cdot 3}$
$= 3\sqrt{4}\sqrt{3} - 5\sqrt{9}\sqrt{3}$
$= 3 \cdot 2\sqrt{3} - 5 \cdot 3\sqrt{3}$
$= 6\sqrt{3} - 15\sqrt{3}$
$= -9\sqrt{3}$

Problem 1 Simplify. **A.** $9\sqrt{3} + 3\sqrt{3} - 18\sqrt{3}$ **B.** $2\sqrt{50} - 5\sqrt{32}$

Solution See page S23.

Subtract: $8\sqrt{18x} - 2\sqrt{32x}$

Simplify each term.

$$
\begin{aligned}
8\sqrt{18x} - 2\sqrt{32x} &= 8\sqrt{9}\sqrt{2x} - 2\sqrt{16}\sqrt{2x} \\
&= 8 \cdot 3\sqrt{2x} - 2 \cdot 4\sqrt{2x} \\
&= 24\sqrt{2x} - 8\sqrt{2x}
\end{aligned}
$$

Subtract the radical expressions.

$$= 16\sqrt{2x}$$

Example 2 Simplify.

A. $3\sqrt{12x^3} - 2x\sqrt{3x}$ **B.** $2x\sqrt{8y} - 3\sqrt{2x^2y} + 2\sqrt{32x^2y}$

Solution **A.** $3\sqrt{12x^3} - 2x\sqrt{3x} = 3\sqrt{4x^2}\sqrt{3x} - 2x\sqrt{3x}$
$$= 3 \cdot 2x\sqrt{3x} - 2x\sqrt{3x}$$
$$= 6x\sqrt{3x} - 2x\sqrt{3x}$$
$$= 4x\sqrt{3x}$$

B. $2x\sqrt{8y} - 3\sqrt{2x^2y} + 2\sqrt{32x^2y}$
$$= 2x\sqrt{4}\sqrt{2y} - 3\sqrt{x^2}\sqrt{2y} + 2\sqrt{16x^2}\sqrt{2y}$$
$$= 2x \cdot 2\sqrt{2y} - 3 \cdot x\sqrt{2y} + 2 \cdot 4x\sqrt{2y}$$
$$= 4x\sqrt{2y} - 3x\sqrt{2y} + 8x\sqrt{2y}$$
$$= 9x\sqrt{2y}$$

Problem 2 Simplify.

A. $y\sqrt{28y} + 7\sqrt{63y^3}$ **B.** $2\sqrt{27a^5} - 4a\sqrt{12a^3} + a^2\sqrt{75a}$

Solution See page S23.

10.2 Concept Review

Determine whether the statement is always true, sometimes true, or never true.

1. $5\sqrt{2} + 6\sqrt{3} = 11\sqrt{5}$

2. The Distributive Property is used to add or subtract radical expressions with the same radicand.

3. $4\sqrt{7}$ and $\sqrt{7}$ are examples of like radical expressions.

4. The expressions $3\sqrt{12}$ and $5\sqrt{3}$ cannot be subtracted until $3\sqrt{12}$ is rewritten in simplest form.

5. $5\sqrt{32x} + 4\sqrt{18x} = 5(4\sqrt{2x}) + 4(3\sqrt{2x}) = 20\sqrt{2x} + 12\sqrt{2x} = 32\sqrt{2x}$

6. The expressions $9\sqrt{11}$ and $10\sqrt{13}$ can be added after each radical is written in simplest form.

10.2 Exercises

1 Simplify.

1. $2\sqrt{2} + \sqrt{2}$ **2.** $3\sqrt{5} + 8\sqrt{5}$ **3.** $-3\sqrt{7} + 2\sqrt{7}$

4. $4\sqrt{5} - 10\sqrt{5}$ **5.** $-3\sqrt{11} - 8\sqrt{11}$ **6.** $-3\sqrt{3} - 5\sqrt{3}$

7. $2\sqrt{x} + 8\sqrt{x}$ **8.** $3\sqrt{y} + 2\sqrt{y}$ **9.** $8\sqrt{y} - 10\sqrt{y}$

10. $-5\sqrt{2a} + 2\sqrt{2a}$

11. $-2\sqrt{3b} - 9\sqrt{3b}$

12. $-7\sqrt{5a} - 5\sqrt{5a}$

13. $3x\sqrt{2} - x\sqrt{2}$

14. $2y\sqrt{3} - 9y\sqrt{3}$

15. $2a\sqrt{3a} - 5a\sqrt{3a}$

16. $-5b\sqrt{3x} - 2b\sqrt{3x}$

17. $3\sqrt{xy} - 8\sqrt{xy}$

18. $-4\sqrt{xy} + 6\sqrt{xy}$

19. $\sqrt{45} + \sqrt{125}$

20. $\sqrt{32} - \sqrt{98}$

21. $2\sqrt{2} + 3\sqrt{8}$

22. $4\sqrt{128} - 3\sqrt{32}$

23. $5\sqrt{18} - 2\sqrt{75}$

24. $5\sqrt{75} - 2\sqrt{18}$

25. $5\sqrt{4x} - 3\sqrt{9x}$

26. $-3\sqrt{25y} + 8\sqrt{49y}$

27. $3\sqrt{3x^2} - 5\sqrt{27x^2}$

28. $-2\sqrt{8y^2} + 5\sqrt{32y^2}$

29. $2x\sqrt{xy^2} - 3y\sqrt{x^2y}$

30. $4a\sqrt{b^2a} - 3b\sqrt{a^2b}$

31. $3x\sqrt{12x} - 5\sqrt{27x^3}$

32. $2a\sqrt{50a} + 7\sqrt{32a^3}$

33. $4y\sqrt{8y^3} - 7\sqrt{18y^5}$

34. $2a\sqrt{8ab^2} - 2b\sqrt{2a^3}$

35. $b^2\sqrt{a^5b} + 3a^2\sqrt{ab^5}$

36. $y^2\sqrt{x^5y} + x\sqrt{x^3y^5}$

37. $4\sqrt{2} - 5\sqrt{2} + 8\sqrt{2}$

38. $3\sqrt{3} + 8\sqrt{3} - 16\sqrt{3}$

39. $5\sqrt{x} - 8\sqrt{x} + 9\sqrt{x}$

40. $\sqrt{x} - 7\sqrt{x} + 6\sqrt{x}$

41. $8\sqrt{2} - 3\sqrt{y} - 8\sqrt{2}$

42. $8\sqrt{3} - 5\sqrt{2} - 5\sqrt{3}$

43. $8\sqrt{8} - 4\sqrt{32} - 9\sqrt{50}$

44. $2\sqrt{12} - 4\sqrt{27} + \sqrt{75}$

45. $-2\sqrt{3} + 5\sqrt{27} - 4\sqrt{45}$

46. $-2\sqrt{8} - 3\sqrt{27} + 3\sqrt{50}$

47. $4\sqrt{75} + 3\sqrt{48} - \sqrt{99}$

48. $2\sqrt{75} - 5\sqrt{20} + 2\sqrt{45}$

49. $\sqrt{25x} - \sqrt{9x} + \sqrt{16x}$

50. $\sqrt{4x} - \sqrt{100x} - \sqrt{49x}$

51. $3\sqrt{3x} + \sqrt{27x} - 8\sqrt{75x}$

52. $5\sqrt{5x} + 2\sqrt{45x} - 3\sqrt{80x}$

53. $2a\sqrt{75b} - a\sqrt{20b} + 4a\sqrt{45b}$

54. $2b\sqrt{75a} - 5b\sqrt{27a} + 2b\sqrt{20a}$

55. $x\sqrt{3y^2} - 2y\sqrt{12x^2} + xy\sqrt{3}$

56. $a\sqrt{27b^2} + 3b\sqrt{147a^2} - ab\sqrt{3}$

57. $3\sqrt{ab^3} + 4a\sqrt{a^2b} - 5b\sqrt{4ab}$

58. $5\sqrt{a^3b} + a\sqrt{4ab} - 3\sqrt{49a^3b}$

59. $3a\sqrt{2ab^2} - \sqrt{a^2b^2} + 4b\sqrt{3a^2b}$

60. $2\sqrt{4a^2b^2} - 3a\sqrt{9ab^2} + 4b\sqrt{a^2b}$

10.2 **Applying Concepts**

Add or subtract.

61. $5\sqrt{x + 2} + 3\sqrt{x + 2}$

62. $8\sqrt{a + 5} - 4\sqrt{a + 5}$

63. $\dfrac{1}{2}\sqrt{8x^2y} + \dfrac{1}{3}\sqrt{18x^2y}$

64. $\dfrac{1}{4}\sqrt{48ab^2} + \dfrac{1}{5}\sqrt{75ab^2}$

65. $\dfrac{a}{3}\sqrt{54ab^3} + \dfrac{b}{4}\sqrt{96a^3b}$

66. $\dfrac{x}{6}\sqrt{72xy^5} + \dfrac{y}{7}\sqrt{98x^3y^3}$

67. $2\sqrt{8x + 4y} - 5\sqrt{18x + 9y}$

68. $3\sqrt{a^3 + a^2} + 5\sqrt{4a^3 + 4a^2}$

69. **Geometry** The lengths of the sides of a triangle are $4\sqrt{3}$ cm, $2\sqrt{3}$ cm, and $2\sqrt{15}$ cm. Find the perimeter of the triangle.

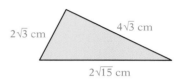

70. **Geometry** The length of a rectangle is $3\sqrt{2}$ cm. The width is $\sqrt{2}$ cm. Find the perimeter of the rectangle.

71. **Geometry** The length of a rectangle is $4\sqrt{5}$ cm. The width is $\sqrt{5}$ cm. Find the decimal approximation of the perimeter. Round to the nearest tenth.

72. Given $G(x) = \sqrt{x + 5} + \sqrt{5x + 3}$, write $G(3)$ in simplest form.

73. Use complete sentences to explain the steps in simplifying the radical expression $a\sqrt{32ab^2} + b\sqrt{50a^3}$.

10.3 | **Multiplication and Division of Radical Expressions**

1 **Multiply radical expressions**

VIDEO & DVD   CD TUTOR   WEB   SSM

The Product Property of Square Roots is used to multiply variable radical expressions.

$$\sqrt{2x}\,\sqrt{3y} = \sqrt{2x \cdot 3y}$$
$$= \sqrt{6xy}$$

> Multiply: $\sqrt{2x^2}\,\sqrt{32x^5}$
>
> Use the Product Property of Square Roots.
>
> $$\sqrt{2x^2}\,\sqrt{32x^5} = \sqrt{2x^2 \cdot 32x^5}$$
>
> Multiply the radicands.
>
> $$= \sqrt{64x^7}$$
>
> $$= \sqrt{64x^6}\,\sqrt{x}$$
>
> Simplify.
>
> $$= 8x^3\sqrt{x}$$

Example 1 Multiply: $\sqrt{3x^4}\,\sqrt{2x^2y}\,\sqrt{6xy^2}$

Solution $\sqrt{3x^4}\,\sqrt{2x^2y}\,\sqrt{6xy^2} = \sqrt{3x^4 \cdot 2x^2y \cdot 6xy^2}$
$\qquad\qquad\qquad\qquad\quad = \sqrt{36x^7y^3}$
$\qquad\qquad\qquad\qquad\quad = \sqrt{36x^6y^2}\,\sqrt{xy}$
$\qquad\qquad\qquad\qquad\quad = 6x^3y\sqrt{xy}$

Problem 1 Multiply: $\sqrt{5a}\,\sqrt{15a^3b^4}\,\sqrt{3b^5}$

Solution See page S23.

When the expression $\left(\sqrt{x}\right)^2$ is simplified by using the Product Property of Square Roots, the result is x.

$\begin{aligned} \left(\sqrt{x}\right)^2 &= \sqrt{x}\,\sqrt{x} \\ &= \sqrt{x \cdot x} \\ &= \sqrt{x^2} \\ &= x \end{aligned}$

For $a > 0$, $\left(\sqrt{a}\right)^2 = \sqrt{a^2} = a$.

Multiply: $\sqrt{2x}\left(x + \sqrt{2x}\right)$

Use the Distributive Property to remove parentheses.

$\begin{aligned} \sqrt{2x}\left(x + \sqrt{2x}\right) &= \sqrt{2x}\,(x) + \sqrt{2x}\,\sqrt{2x} \\ &= x\sqrt{2x} + \left(\sqrt{2x}\right)^2 \\ &= x\sqrt{2x} + 2x \end{aligned}$

Example 2 Multiply: $\sqrt{3ab}\left(\sqrt{3a} + \sqrt{9b}\right)$

Solution $\sqrt{3ab}\left(\sqrt{3a} + \sqrt{9b}\right) = \sqrt{3ab}\left(\sqrt{3a}\right) + \sqrt{3ab}\left(\sqrt{9b}\right)$
$\qquad\qquad\qquad\qquad\qquad\quad = \sqrt{9a^2b} + \sqrt{27ab^2}$
$\qquad\qquad\qquad\qquad\qquad\quad = \sqrt{9a^2}\,\sqrt{b} + \sqrt{9b^2}\,\sqrt{3a}$
$\qquad\qquad\qquad\qquad\qquad\quad = 3a\sqrt{b} + 3b\sqrt{3a}$

Problem 2 Multiply: $\sqrt{5x}\left(\sqrt{5x} - \sqrt{25y}\right)$

Solution See page S23.

To multiply $\left(\sqrt{2} - 3x\right)\left(\sqrt{2} + x\right)$, use the FOIL method.

$\begin{aligned} \left(\sqrt{2} - 3x\right)\left(\sqrt{2} + x\right) &= \left(\sqrt{2}\right)^2 + x\sqrt{2} - 3x\sqrt{2} - 3x^2 \\ &= 2 + (x - 3x)\sqrt{2} - 3x^2 \\ &= 2 - 2x\sqrt{2} - 3x^2 \end{aligned}$

Example 3 Multiply: $\left(2\sqrt{x} - \sqrt{y}\right)\left(5\sqrt{x} - 2\sqrt{y}\right)$

Solution $\left(2\sqrt{x} - \sqrt{y}\right)\left(5\sqrt{x} - 2\sqrt{y}\right)$
$\qquad = 10\left(\sqrt{x}\right)^2 - 4\sqrt{xy} - 5\sqrt{xy} + 2\left(\sqrt{y}\right)^2$
$\qquad = 10x - 9\sqrt{xy} + 2y$

Problem 3 Multiply: $\left(3\sqrt{x} - \sqrt{y}\right)\left(5\sqrt{x} - 2\sqrt{y}\right)$

Solution See page S23.

The expressions $a + b$ and $a - b$, which are the sum and difference of two terms, are called **conjugates** of each other.

The product of conjugates is the difference of two squares.

$$(a + b)(a - b) = a^2 - b^2$$
$$\left(2 + \sqrt{7}\right)\left(2 - \sqrt{7}\right) = 2^2 - \left(\sqrt{7}\right)^2 = 4 - 7 = -3$$
$$\left(3 + \sqrt{y}\right)\left(3 - \sqrt{y}\right) = 3^2 - \left(\sqrt{y}\right)^2 = 9 - y$$

Example 4 Multiply: $\left(\sqrt{a} - \sqrt{b}\right)\left(\sqrt{a} + \sqrt{b}\right)$

Solution $\left(\sqrt{a} - \sqrt{b}\right)\left(\sqrt{a} + \sqrt{b}\right) = \left(\sqrt{a}\right)^2 - \left(\sqrt{b}\right)^2 = a - b$

Problem 4 Multiply: $\left(2\sqrt{x} + 7\right)\left(2\sqrt{x} - 7\right)$

Solution See page S23.

2 ## Divide radical expressions

The square root of a quotient is equal to the quotient of the square roots.

> **The Quotient Property of Square Roots**
>
> If a and b are positive real numbers, then $\sqrt{\dfrac{a}{b}} = \dfrac{\sqrt{a}}{\sqrt{b}}$.

Point of Interest

A radical expression that occurs in Einstein's Theory of Relativity is

$$\dfrac{1}{\sqrt{1 - \dfrac{v^2}{c^2}}}$$

where v is the velocity of an object and c is the speed of light.

Simplify: $\sqrt{\dfrac{4x^2}{z^6}}$

Rewrite the radical expression as the quotient of the square roots.

Take the square roots of the perfect squares.

$$\sqrt{\dfrac{4x^2}{z^6}} = \dfrac{\sqrt{4x^2}}{\sqrt{z^6}}$$
$$= \dfrac{2x}{z^3}$$

Simplify: $\sqrt{\dfrac{24x^3y^7}{3x^7y^2}}$

Simplify the radicand.

Rewrite the radical expression as the quotient of the square roots.

Simplify.

$$\sqrt{\dfrac{24x^3y^7}{3x^7y^2}} = \sqrt{\dfrac{8y^5}{x^4}}$$
$$= \dfrac{\sqrt{8y^5}}{\sqrt{x^4}}$$
$$= \dfrac{\sqrt{4y^4}\,\sqrt{2y}}{\sqrt{x^4}}$$
$$= \dfrac{2y^2\sqrt{2y}}{x^2}$$

Simplify: $\dfrac{\sqrt{4x^2y}}{\sqrt{xy}}$

Use the Quotient Property of Square Roots.

Simplify the radicand.

Simplify the radical expression.

$$\dfrac{\sqrt{4x^2y}}{\sqrt{xy}} = \sqrt{\dfrac{4x^2y}{xy}}$$
$$= \sqrt{4x}$$
$$= \sqrt{4}\,\sqrt{x}$$
$$= 2\sqrt{x}$$

A radical expression is not in simplest form if a radical remains in the denominator. The procedure used to remove a radical from the denominator is called **rationalizing the denominator.**

⌐ Simplify: $\dfrac{2}{\sqrt{3}}$

The expression $\dfrac{2}{\sqrt{3}}$ has a radical expression in

the denominator. Multiply the expression by $\dfrac{\sqrt{3}}{\sqrt{3}}$,

which equals 1.

Simplify.

$$\dfrac{2}{\sqrt{3}} = \dfrac{2}{\sqrt{3}} \cdot \dfrac{\sqrt{3}}{\sqrt{3}}$$

$$= \dfrac{2\sqrt{3}}{(\sqrt{3})^2}$$

$$= \dfrac{2\sqrt{3}}{3}$$

Thus $\dfrac{2}{\sqrt{3}} = \dfrac{2\sqrt{3}}{3}$, but $\dfrac{2}{\sqrt{3}}$ is not in simplest form. Because no radical remains in the denominator and the radical in the numerator contains no perfect-square factors other than 1, $\dfrac{2\sqrt{3}}{3}$ is in simplest form.

When the denominator contains a radical expression with two terms, simplify the radical expression by multiplying the numerator and denominator by the conjugate of the denominator.

⌐ Simplify: $\dfrac{\sqrt{2y}}{\sqrt{y} + 3}$

Multiply the numerator and
denominator by $\sqrt{y} - 3$, the
conjugate of $\sqrt{y} + 3$.

Simplify.

$$\dfrac{\sqrt{2y}}{\sqrt{y} + 3} = \dfrac{\sqrt{2y}}{\sqrt{y} + 3} \cdot \dfrac{\sqrt{y} - 3}{\sqrt{y} - 3}$$

$$= \dfrac{\sqrt{2y^2} - 3\sqrt{2y}}{(\sqrt{y})^2 - 3^2}$$

$$= \dfrac{y\sqrt{2} - 3\sqrt{2y}}{y - 9}$$

The following list summarizes the discussions about radical expressions in simplest form.

Radical Expressions in Simplest Form

A radical expression is in simplest form if:

1. The radicand contains no factor greater than 1 that is a perfect square.

2. There is no fraction under the radical sign.

3. There is no radical in the denominator of a fraction.

Example 5   Simplify.   **A.** $\dfrac{\sqrt{4x^2y^5}}{\sqrt{3x^4y}}$ **B.** $\dfrac{\sqrt{2}}{\sqrt{2}-\sqrt{x}}$ **C.** $\dfrac{3-\sqrt{5}}{2+3\sqrt{5}}$

Solution   **A.** $\dfrac{\sqrt{4x^2y^5}}{\sqrt{3x^4y}} = \sqrt{\dfrac{4x^2y^5}{3x^4y}} = \sqrt{\dfrac{4y^4}{3x^2}} = \dfrac{\sqrt{4y^4}}{\sqrt{3x^2}} = \dfrac{2y^2}{x\sqrt{3}} \cdot \dfrac{\sqrt{3}}{\sqrt{3}} = \dfrac{2y^2\sqrt{3}}{3x}$

B. $\dfrac{\sqrt{2}}{\sqrt{2}-\sqrt{x}} = \dfrac{\sqrt{2}}{\sqrt{2}-\sqrt{x}} \cdot \dfrac{\sqrt{2}+\sqrt{x}}{\sqrt{2}+\sqrt{x}} = \dfrac{2+\sqrt{2x}}{2-x}$

C. $\dfrac{3-\sqrt{5}}{2+3\sqrt{5}} = \dfrac{3-\sqrt{5}}{2+3\sqrt{5}} \cdot \dfrac{2-3\sqrt{5}}{2-3\sqrt{5}}$

$= \dfrac{6 - 9\sqrt{5} - 2\sqrt{5} + 3 \cdot 5}{4 - 9 \cdot 5}$

$= \dfrac{21 - 11\sqrt{5}}{-41}$

$= -\dfrac{21 - 11\sqrt{5}}{41}$

Problem 5   Simplify.   **A.** $\sqrt{\dfrac{15x^6y^7}{3x^7y^9}}$ **B.** $\dfrac{\sqrt{y}}{\sqrt{y}+3}$ **C.** $\dfrac{5+\sqrt{y}}{1-2\sqrt{y}}$

Solution   See pages S23–S24.

10.3   Concept Review

Determine whether the statement is always true, sometimes true, or never true.

1. By the Product Property of Square Roots, if $a > 0$ and $b > 0$, then $\sqrt{a} \cdot \sqrt{b} = \sqrt{ab}$.

2. When we square a square root, the result is the radicand.

3. The procedure for rationalizing the denominator is used when a fraction has a radical expression in the denominator.

4. The square root of a fraction is equal to the square root of the numerator over the square root of the denominator.

5. A radical expression is in simplest form if the radicand contains no factor other than 1 that is a perfect square.

6. The conjugate of $5 - \sqrt{3}$ is $\sqrt{3} - 5$.

10.3 Exercises

1 Multiply.

1. $\sqrt{5}\,\sqrt{5}$ **2.** $\sqrt{11}\,\sqrt{11}$ **3.** $\sqrt{3}\,\sqrt{12}$ **4.** $\sqrt{2}\,\sqrt{8}$

5. $\sqrt{x}\,\sqrt{x}$ **6.** $\sqrt{y}\,\sqrt{y}$ **7.** $\sqrt{xy^3}\,\sqrt{x^5y}$ **8.** $\sqrt{a^3b^5}\,\sqrt{ab^5}$

9. $\sqrt{3a^2b^5}\,\sqrt{6ab^7}$ **10.** $\sqrt{5x^3y}\,\sqrt{10x^2y}$ **11.** $\sqrt{6a^3b^2}\,\sqrt{24a^5b}$ **12.** $\sqrt{8ab^5}\,\sqrt{12a^7b}$

13. $\sqrt{2}\left(\sqrt{2}-\sqrt{3}\right)$ **14.** $3\left(\sqrt{12}-\sqrt{3}\right)$ **15.** $\sqrt{x}\left(\sqrt{x}-\sqrt{y}\right)$ **16.** $\sqrt{b}\left(\sqrt{a}-\sqrt{b}\right)$

17. $\sqrt{5}\left(\sqrt{10}-\sqrt{x}\right)$ **18.** $\sqrt{6}\left(\sqrt{y}-\sqrt{18}\right)$ **19.** $\sqrt{8}\left(\sqrt{2}-\sqrt{5}\right)$ **20.** $\sqrt{10}\left(\sqrt{20}-\sqrt{a}\right)$

21. $\left(\sqrt{x}-3\right)^2$ **22.** $\left(2\sqrt{a}-y\right)^2$ **23.** $\sqrt{3a}\left(\sqrt{3a}-\sqrt{3b}\right)$

24. $\sqrt{5x}\left(\sqrt{10x}-\sqrt{x}\right)$ **25.** $\sqrt{2ac}\,\sqrt{5ab}\,\sqrt{10cb}$ **26.** $\sqrt{3xy}\,\sqrt{6x^3y}\,\sqrt{2y^2}$

27. $\left(3\sqrt{x}-2y\right)\left(5\sqrt{x}-4y\right)$ **28.** $\left(5\sqrt{x}+2\sqrt{y}\right)\left(3\sqrt{x}-\sqrt{y}\right)$ **29.** $\left(\sqrt{x}-\sqrt{y}\right)\left(\sqrt{x}+\sqrt{y}\right)$

30. $\left(\sqrt{3x}+y\right)\left(\sqrt{3x}-y\right)$ **31.** $\left(2\sqrt{x}+\sqrt{y}\right)\left(5\sqrt{x}+4\sqrt{y}\right)$ **32.** $\left(5\sqrt{x}-2\sqrt{y}\right)\left(3\sqrt{x}-4\sqrt{y}\right)$

2 **33.** Explain why $\dfrac{\sqrt{5}}{5}$ is in simplest form and $\dfrac{1}{\sqrt{5}}$ is not in simplest form.

34. Why can we multiply $\dfrac{1}{\sqrt{3}}$ by $\dfrac{\sqrt{3}}{\sqrt{3}}$ without changing the value of $\dfrac{1}{\sqrt{3}}$?

Simplify.

35. $\dfrac{\sqrt{32}}{\sqrt{2}}$ **36.** $\dfrac{\sqrt{45}}{\sqrt{5}}$ **37.** $\dfrac{\sqrt{98}}{\sqrt{2}}$ **38.** $\dfrac{\sqrt{48}}{\sqrt{3}}$

39. $\dfrac{\sqrt{27a}}{\sqrt{3a}}$ **40.** $\dfrac{\sqrt{72x^5}}{\sqrt{2x}}$ **41.** $\dfrac{\sqrt{15x^3y}}{\sqrt{3xy}}$ **42.** $\dfrac{\sqrt{40x^5y^2}}{\sqrt{5xy}}$

43. $\dfrac{\sqrt{2a^5b^4}}{\sqrt{98ab^4}}$ **44.** $\dfrac{\sqrt{48x^5y^2}}{\sqrt{3x^3y}}$ **45.** $\dfrac{1}{\sqrt{3}}$ **46.** $\dfrac{1}{\sqrt{8}}$

47. $\dfrac{15}{\sqrt{75}}$ **48.** $\dfrac{6}{\sqrt{72}}$ **49.** $\dfrac{6}{\sqrt{12x}}$ **50.** $\dfrac{14}{\sqrt{7y}}$

51. $\dfrac{8}{\sqrt{32x}}$ **52.** $\dfrac{15}{\sqrt{50x}}$ **53.** $\dfrac{3}{\sqrt{x}}$ **54.** $\dfrac{4}{\sqrt{2x}}$

55. $\dfrac{\sqrt{8x^2y}}{\sqrt{2x^4y^2}}$ **56.** $\dfrac{\sqrt{4x^2}}{\sqrt{9y}}$ **57.** $\dfrac{\sqrt{16a}}{\sqrt{49ab}}$ **58.** $\dfrac{5\sqrt{8}}{4\sqrt{50}}$

59. $\dfrac{5\sqrt{18}}{9\sqrt{27}}$

60. $\dfrac{\sqrt{12a^3b}}{\sqrt{24a^2b^2}}$

61. $\dfrac{\sqrt{3xy}}{\sqrt{27x^3y^2}}$

62. $\dfrac{\sqrt{9xy^2}}{\sqrt{27x}}$

63. $\dfrac{\sqrt{4x^2y}}{\sqrt{3xy^3}}$

64. $\dfrac{\sqrt{16x^3y^2}}{\sqrt{8x^3y}}$

65. $\dfrac{1}{\sqrt{2}-3}$

66. $\dfrac{5}{\sqrt{7}-3}$

67. $\dfrac{3}{5+\sqrt{5}}$

68. $\dfrac{7}{\sqrt{2}-7}$

69. $\dfrac{\sqrt{xy}}{\sqrt{x}-\sqrt{y}}$

70. $\dfrac{\sqrt{x}}{\sqrt{x}-\sqrt{y}}$

71. $\dfrac{5\sqrt{x^2y}}{\sqrt{75xy^2}}$

72. $\dfrac{3\sqrt{ab}}{a\sqrt{6b}}$

73. $\dfrac{\sqrt{2}}{\sqrt{2}-\sqrt{3}}$

74. $\dfrac{1+\sqrt{2}}{1-\sqrt{2}}$

75. $\dfrac{\sqrt{5}}{\sqrt{2}-\sqrt{5}}$

76. $\dfrac{\sqrt{6}}{\sqrt{3}-\sqrt{2}}$

77. $\dfrac{\sqrt{x}}{\sqrt{x}+3}$

78. $\dfrac{\sqrt{y}}{2-\sqrt{y}}$

79. $\dfrac{5\sqrt{3}-7\sqrt{3}}{4\sqrt{3}}$

80. $\dfrac{10\sqrt{7}-2\sqrt{7}}{2\sqrt{7}}$

81. $\dfrac{5\sqrt{8}-3\sqrt{2}}{\sqrt{2}}$

82. $\dfrac{5\sqrt{12}-\sqrt{3}}{\sqrt{27}}$

83. $\dfrac{3\sqrt{2}-8\sqrt{2}}{\sqrt{2}}$

84. $\dfrac{5\sqrt{3}-2\sqrt{3}}{2\sqrt{3}}$

85. $\dfrac{2\sqrt{8}+3\sqrt{2}}{\sqrt{32}}$

86. $\dfrac{3-\sqrt{6}}{5-2\sqrt{6}}$

87. $\dfrac{6-2\sqrt{3}}{4+3\sqrt{3}}$

88. $\dfrac{\sqrt{2}+2\sqrt{6}}{2\sqrt{2}-3\sqrt{6}}$

89. $\dfrac{2\sqrt{3}-\sqrt{6}}{5\sqrt{3}+2\sqrt{6}}$

90. $\dfrac{3+\sqrt{x}}{2-\sqrt{x}}$

91. $\dfrac{\sqrt{a}-4}{2\sqrt{a}+2}$

92. $\dfrac{3+2\sqrt{y}}{2-\sqrt{y}}$

93. $\dfrac{2+\sqrt{y}}{\sqrt{y}-3}$

94. $\dfrac{\sqrt{x}+\sqrt{y}}{\sqrt{x}-\sqrt{y}}$

10.3   Applying Concepts

Simplify.

95. $-\sqrt{1.3}\,\sqrt{1.3}$

96. $\sqrt{\dfrac{5}{8}}\,\sqrt{\dfrac{5}{8}}$

97. $-\sqrt{\dfrac{16}{81}}$

98. $\sqrt{1\dfrac{9}{16}}$

99. $\sqrt{2\dfrac{1}{4}}$

100. $-\sqrt{6\dfrac{1}{4}}$

Geometry Find the area of the geometric figure. All dimensions are given in meters.

101.

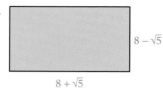

$8 - \sqrt{5}$

$8 + \sqrt{5}$

102.

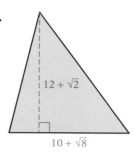

$12 + \sqrt{2}$

$10 + \sqrt{8}$

103. Answer true or false. If the equation is false, correct the right side.
 a. $\left(\sqrt{y}\right)^4 = y^2$ **b.** $\left(2\sqrt{x}\right)^3 = 8x\sqrt{x}$
 c. $\left(\sqrt{x} + 1\right)^2 = x + 1$ **d.** $\dfrac{1}{2 - \sqrt{3}} = 2 + \sqrt{3}$

104. Show that 2 is a solution of the equation $\sqrt{x + 2} + \sqrt{x - 1} = 3$.

105. Is 16 a solution of the equation $\sqrt{x} - \sqrt{x + 9} = 1$?

106. Show that $\left(1 + \sqrt{6}\right)$ and $\left(1 - \sqrt{6}\right)$ are solutions of the equation $x^2 - 2x - 5 = 0$.

107. In your own words, describe the process of rationalizing the denominator.

108. The number $\dfrac{\sqrt{5} + 1}{2}$ is called the golden ratio. Research the golden ratio and write a few paragraphs about this number and its applications.

10.4 Solving Equations Containing Radical Expressions

1 **Solve equations containing one or more radical expressions**

VIDEO & DVD   CD TUTOR   WEB   SSM

An equation that contains a variable expression in a radicand is a **radical equation**.

$\left.\begin{array}{l}\sqrt{x} = 4 \\ \sqrt{x + 2} = \sqrt{x - 7}\end{array}\right\}$ Radical equations

The following property of equality states that if two numbers are equal, then the squares of the numbers are equal. This property is used to solve radical equations.

> **_Property of Squaring Both Sides of an Equation_**
>
> If a and b are real numbers and $a = b$, then $a^2 = b^2$.

Solve: $\sqrt{x-2}-7=0$

Rewrite the equation with the radical on one side of the equation and the constant on the other side.

$$\sqrt{x-2}-7=0$$
$$\sqrt{x-2}=7$$

Square both sides of the equation.

$$\left(\sqrt{x-2}\right)^2=7^2$$

Solve the resulting equation.

$$x-2=49$$
$$x=51$$

Check the solution. When both sides of an equation are squared, the resulting equation may have a solution that is not a solution of the original equation.

Check
$$\sqrt{x-2}-7=0$$
$$\begin{array}{c|c}\sqrt{51-2}-7 & 0\\ \sqrt{49}-7 & 0\\ 7-7 & 0\\ 0 & = 0\end{array}$$

The solution is 51.

Take Note

Any time each side of an equation is squared, you must check the proposed solution of the equation.

Example 1 Solve: $\sqrt{3x}+2=5$

Solution
$$\sqrt{3x}+2=5$$
$$\sqrt{3x}=3$$

- Rewrite the equation so that the radical is alone on one side of the equation.

$$\left(\sqrt{3x}\right)^2=3^2$$
$$3x=9$$
$$x=3$$

- Square both sides of the equation.
- Solve for x.

Check
$$\sqrt{3x}+2=5$$
$$\begin{array}{c|c}\sqrt{3\cdot3}+2 & 5\\ \sqrt{9}+2 & 5\\ 3+2 & 5\\ 5 & = 5\end{array}$$

- Both sides of the equation were squared. The solution must be checked.

- This is a true equation. The solution checks.

The solution is 3.

Problem 1 Solve: $\sqrt{4x}+3=7$

Solution See page S24.

Example 2 Solve. **A.** $0=3-\sqrt{2x-3}$ **B.** $\sqrt{2x-5}+3=0$

Solution **A.**
$$0=3-\sqrt{2x-3}$$
$$\sqrt{2x-3}=3$$

- Rewrite the equation so that the radical is alone on one side of the equation.

$$\left(\sqrt{2x-3}\right)^2=3^2$$
$$2x-3=9$$
$$2x=12$$
$$x=6$$

- Square both sides of the equation.

Check

$$0 = 3 - \sqrt{2x - 3}$$

0	$3 - \sqrt{2 \cdot 6 - 3}$
0	$3 - \sqrt{12 - 3}$
0	$3 - \sqrt{9}$
0	$3 - 3$

$$0 = 0$$

• This is a true equation. The solution checks.

The solution is 6.

B. $\sqrt{2x - 5} + 3 = 0$

$$\sqrt{2x - 5} = -3$$

• Rewrite the equation so that the radical is alone on one side of the equation.

$$\left(\sqrt{2x - 5}\right)^2 = (-3)^2$$

• Square each side of the equation.

$$2x - 5 = 9$$

• Solve for x.

$$2x = 14$$
$$x = 7$$

Check

$$\sqrt{2x - 5} + 3 = 0$$

$\sqrt{2 \cdot 7 - 5} + 3$	0
$\sqrt{14 - 5} + 3$	0
$\sqrt{9} + 3$	0
$3 + 3$	0

$$6 \neq 0$$

• This is not a true equation. The solution does not check.

There is no solution.

Problem 2   Solve.   **A.** $\sqrt{3x - 2} - 5 = 0$ **B.** $\sqrt{4x - 7} + 5 = 0$

Solution   See page S24.

The following example illustrates the procedure for solving a radical equation containing two radical expressions. Note that the process of squaring both sides of the equation is perfomed twice.

Solve: $\sqrt{5 + x} + \sqrt{x} = 5$

Solve for one of the radical expressions. Square each side.

$$\sqrt{5 + x} + \sqrt{x} = 5$$
$$\sqrt{5 + x} = 5 - \sqrt{x}$$
$$\left(\sqrt{5 + x}\right)^2 = \left(5 - \sqrt{x}\right)^2$$

Recall that $(a - b)^2 = a^2 - 2ab + b^2$.

$$5 + x = 25 - 10\sqrt{x} + x$$

Simplify.

$$-20 = -10\sqrt{x}$$

This is still a radical equation.

$$2 = \sqrt{x}$$

Square each side.

$$2^2 = \left(\sqrt{x}\right)^2$$
$$4 = x$$

4 checks as the solution.   The solution is 4.

Example 3 Solve: $\sqrt{x} - \sqrt{x-5} = 1$

Solution

$$\sqrt{x} - \sqrt{x-5} = 1$$

$$\sqrt{x} = 1 + \sqrt{x-5}$$

$$(\sqrt{x})^2 = (1 + \sqrt{x-5})^2$$

$$x = 1 + 2\sqrt{x-5} + (x-5)$$

$$4 = 2\sqrt{x-5}$$

$$2 = \sqrt{x-5}$$

$$2^2 = (\sqrt{x-5})^2$$

$$4 = x-5$$

$$9 = x$$

- Solve for one of the radical expressions.
- Square each side.
- Simplify.

- This is a radical equation.
- Square each side.
- Simplify.

Check

$$\sqrt{x} - \sqrt{x-5} = 1$$

$\sqrt{9} - \sqrt{9-5}$	1
$3 - \sqrt{4}$	1
$3 - 2$	1
$1 = 1$	

The solution is 9.

Problem 3 Solve: $\sqrt{x} + \sqrt{x+9} = 9$

Solution See page S24.

2 ## Application problems

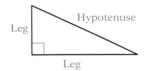

A right triangle contains one 90° angle. The side opposite the 90° angle is called the **hypotenuse.** The other two sides are called **legs.**

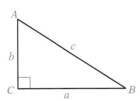

The angles in a right triangle are usually labeled with the capital letters *A*, *B*, and *C*, with *C* reserved for the right angle. The side opposite angle *A* is side *a*, the side opposite side *B* is side *b*, and *c* is the hypotenuse.

The Greek mathematician Pythagoras is generally credited with the discovery that the square of the hypotenuse of a right triangle is equal to the sum of the squares of the two legs. This is called the **Pythagorean Theorem.**

The figure at the left is a right triangle with legs measuring 3 units and 4 units and a hypotenuse measuring 5 units. Each side of the triangle is also the side of a square. The number of square units in the area of the largest square is equal to the sum of the numbers of square units in the areas of the smaller squares.

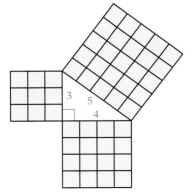

Square of the hypotenuse	=	Sum of the squares of the two legs

$$5^2 = 3^2 + 4^2$$

$$25 = 9 + 16$$

$$25 = 25$$

Pythagorean Theorem

If a and b are the lengths of the legs of a right triangle and c is the length of the hypotenuse, then $c^2 = a^2 + b^2$.

If the lengths of two sides of a right triangle are known, the Pythagorean Theorem can be used to find the length of the third side.

The Pythagorean Theorem is used to find the hypotenuse when the two legs are known.

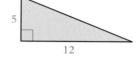

$$\text{Hypotenuse} = \sqrt{(\text{leg})^2 + (\text{leg})^2}$$
$$c = \sqrt{a^2 + b^2}$$
$$c = \sqrt{(5)^2 + (12)^2}$$
$$c = \sqrt{25 + 144}$$
$$c = \sqrt{169}$$
$$c = 13$$

Take Note

If we let $a = 12$ and $b = 5$, the result is the same.

The Pythagorean Theorem is used to find the length of a leg when one leg and the hypotenuse are known.

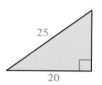

$$\text{Leg} = \sqrt{(\text{hypotenuse})^2 - (\text{leg})^2}$$
$$a = \sqrt{c^2 - b^2}$$
$$a = \sqrt{(25)^2 - (20)^2}$$
$$a = \sqrt{625 - 400}$$
$$a = \sqrt{225}$$
$$a = 15$$

Example 4 and Problem 4 illustrate the use of the Pythagorean Theorem. Example 5 and Problem 5 illustrate other applications of radical equations.

Example 4   A guy wire is attached to a point 22 m above the ground on a telephone pole that is perpendicular to the ground. The wire is anchored to the ground at a point 9 m from the base of the pole. Find the length of the guy wire. Round to the nearest hundredth.

Strategy   To find the length of the guy wire, use the Pythagorean Theorem. One leg is the distance from the bottom of the wire to the base of the telephone pole. The other leg is the distance from the top of the wire to the base of the telephone pole. The guy wire is the hypotenuse. Solve the Pythagorean Theorem for the hypotenuse.

22 m

9 m

Solution $c = \sqrt{a^2 + b^2}$
$c = \sqrt{(22)^2 + (9)^2}$
$c = \sqrt{484 + 81}$
$c = \sqrt{565}$
$c \approx 23.77$

The guy wire has a length of 23.77 m.

Problem 4   A ladder 12 ft long is resting against a building. How high on the building will the ladder reach when the bottom of the ladder is 5 ft from the building? Round to the nearest hundredth.

Solution   See page S24.

Example 5   How far would a submarine periscope have to be above the water for the lookout to locate a ship 5 mi away? The equation for the distance in miles that the lookout can see is $d = \sqrt{1.5h}$, where h is the height in feet above the surface of the water. Round to the nearest hundredth.

Strategy   To find the height above water, replace d in the equation with the given value. Then solve for h.

Solution
$$d = \sqrt{1.5h}$$
$$5 = \sqrt{1.5h}$$
$$5^2 = \left(\sqrt{1.5h}\right)^2$$
$$25 = 1.5h$$
$$\frac{25}{1.5} = h$$
$$16.67 \approx h$$

The periscope must be 16.67 ft above the water.

Problem 5   Find the length of a pendulum that makes one swing in 1.5 s. The equation for the time of one swing is $T = 2\pi\sqrt{\dfrac{L}{32}}$, where T is the time in seconds and L is the length in feet. Round to the nearest hundredth.

Solution   See page S24.

10.4   **Concept Review**

Determine whether the statement is always true, sometimes true, or never true.

1. A radical equation is an equation that contains a radical.

2. We can square both sides of an equation without changing the solutions of the equation.

3. We use the Property of Squaring Both Sides of an Equation in order to eliminate a radical expression from an equation.

4. The first step in solving a radical equation is to square both sides of the equation.

5. Suppose $a^2 = b^2$. Then $a = b$.

6. The Pythagorean Theorem is used to find the length of a side of a triangle.

10.4 Exercises

1 **1.** Which of the following equations are radical equations?
 a. $8 = \sqrt{5} + x$
 b. $\sqrt{x - 7} = 9$
 c. $\sqrt{x} + 4 = 6$
 d. $12 = \sqrt{3x}$

2. What does the Property of Squaring Both Sides of an Equation state?

Solve and check.

3. $\sqrt{x} = 5$ **4.** $\sqrt{y} = 7$ **5.** $\sqrt{a} = 12$ **6.** $\sqrt{a} = 9$

7. $\sqrt{5x} = 5$ **8.** $\sqrt{3x} = 4$ **9.** $\sqrt{4x} = 8$ **10.** $\sqrt{6x} = 3$

11. $\sqrt{2x} - 4 = 0$ **12.** $3 - \sqrt{5x} = 0$ **13.** $\sqrt{4x} + 5 = 2$ **14.** $\sqrt{3x} + 9 = 4$

15. $\sqrt{3x - 2} = 4$ **16.** $\sqrt{5x + 6} = 1$ **17.** $\sqrt{2x + 1} = 7$ **18.** $\sqrt{5x + 4} = 3$

19. $0 = 2 - \sqrt{3 - x}$ **20.** $0 = 5 - \sqrt{10 + x}$ **21.** $\sqrt{5x + 2} = 0$ **22.** $\sqrt{3x - 7} = 0$

23. $\sqrt{3x} - 6 = -4$ **24.** $\sqrt{5x} + 8 = 23$ **25.** $0 = \sqrt{3x - 9} - 6$ **26.** $0 = \sqrt{2x + 7} - 3$

27. $\sqrt{5x - 1} = \sqrt{3x + 9}$ **28.** $\sqrt{3x + 4} = \sqrt{12x - 14}$

29. $\sqrt{5x - 3} = \sqrt{4x - 2}$ **30.** $\sqrt{5x - 9} = \sqrt{2x - 3}$

31. $\sqrt{x^2 - 5x + 6} = \sqrt{x^2 - 8x + 9}$ **32.** $\sqrt{x^2 - 2x + 4} = \sqrt{x^2 + 5x - 12}$

33. $\sqrt{x} = \sqrt{x + 3} - 1$ **34.** $\sqrt{x + 5} = \sqrt{x} + 1$

35. $\sqrt{2x + 5} = 5 - \sqrt{2x}$ **36.** $\sqrt{2x} + \sqrt{2x + 9} = 9$

37. $\sqrt{3x} - \sqrt{3x + 7} = 1$ **38.** $\sqrt{x} - \sqrt{x + 9} = 1$

2 Solve. Round to the nearest hundredth.

39. **Integer Problem** Five added to the square root of the product of four and a number is equal to seven. Find the number.

40. **Integer Problem** The product of a number and the square root of three is equal to the square root of twenty-seven. Find the number.

41. **Integer Problem** Two added to the square root of the sum of a number and five is equal to six. Find the number.

42. **Integer Problem** The product of a number and the square root of seven is equal to the square root of twenty-eight. Find the number.

43. ✎ What does the Pythagorean Theorem state?

44. Label the right triangle shown at the right. Include the right angle symbol, the three angles, the two legs, and the hypotenuse.

45. Geometry The two legs of a right triangle measure 5 cm and 9 cm. Find the length of the hypotenuse.

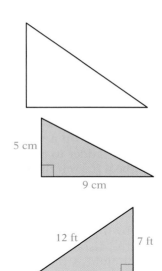

46. Geometry The two legs of a right triangle measure 8 in. and 4 in. Find the length of the hypotenuse.

47. Geometry The hypotenuse of a right triangle measures 12 ft. One leg of the triangle measures 7 ft. Find the length of the other leg of the triangle.

48. Geometry The hypotenuse of a right triangle measures 20 cm. One leg of the triangle measures 16 cm. Find the length of the other leg of the triangle.

49. Geometry The diagonal of a rectangle is a line drawn from one vertex to the opposite vertex. Find the length of the diagonal in the rectangle shown at the right. Round to the nearest tenth.

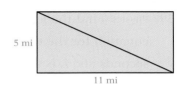

50. Education One method used to "curve" the grades on an exam is to use the formula $R = 10\sqrt{O}$, where R is the revised score and O is the original score. Use this formula to find the original score on an exam that has a revised score of 75. Round to the nearest whole number.

51. Physics A formula used in the study of shallow-water wave motion is $C = \sqrt{32H}$, where C is the wave velocity in feet per second and H is the depth in feet. Use this formula to find the depth of the water when the wave velocity is 20 ft/s.

52. ● Sports The infield of a baseball diamond is a square. The distance between successive bases is 90 ft. The pitcher's mound is on the diagonal between home plate and second base at a distance of 60.5 ft from home plate. Is the pitcher's mound more or less than halfway between home plate and second base?

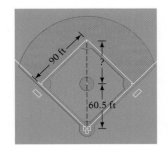

53. ● Sports The infield of a softball diamond is a square. The distance between successive bases is 60 ft. The pitcher's mound is on the diagonal between home plate and second base at a distance of 46 ft from home plate. Is the pitcher's mound more or less than halfway between home plate and second base?

54. Communications Marta Lightfoot leaves a dock in her sailboat and sails 2.5 mi due east. She then tacks and sails 4 mi due north. The walkie-talkie Marta has on board has a range of 5 mi. Will she be able to call a friend on the dock from her location using the walkie-talkie?

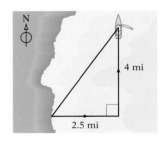

55. Navigation How far would a submarine periscope have to be above the water for the lookout to locate a ship 4 mi away? The equation for the distance in miles that the lookout can see is $d = \sqrt{1.5h}$, where h is the height in feet above the surface of the water.

56. Navigation How far would a submarine periscope have to be above the water for the lookout to locate a ship 7 mi away? The equation for the distance in miles that the lookout can see is $d = \sqrt{1.5h}$, where h is the height in feet above the surface of the water.

57. Home Maintenance Rick Wyman needs to clean the gutters of his home. The gutters are 24 ft above the ground. For safety, the distance a ladder reaches up a wall should be four times the distance from the bottom of the ladder to the base of the side of the house. Therefore, the ladder must be 6 ft from the base of the house. Will a 25-foot ladder be long enough to reach the gutters?

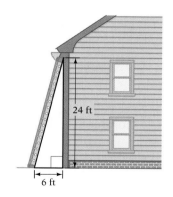

24 ft

6 ft

58. Physics The speed of a child riding a merry-go-round at a carnival is given by the equation $v = \sqrt{12r}$, where v is the speed in feet per second and r is the distance in feet from the center of the merry-go-round to the rider. If a child is moving at 15 ft/s, how far is the child from the center of the merry-go-round?

59. Physics Find the length of a pendulum that makes one swing in 3 s. The equation for the time of one swing is $T = 2\pi\sqrt{\dfrac{L}{32}}$, where T is the time in seconds and L is the length in feet.

60. Physics Find the length of a pendulum that makes one swing in 2 s. The equation for the time of one swing is $T = 2\pi\sqrt{\dfrac{L}{32}}$, where T is the time in seconds and L is the length in feet.

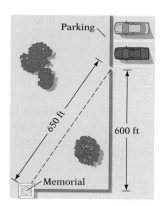

61. Parks An L-shaped sidewalk from the parking lot to a memorial is shown in the figure at the right. The distance directly across the grass to the memorial is 650 ft. The distance to the corner is 600 ft. Find the distance from the corner to the memorial.

62. Aviation A commuter plane leaves an airport traveling due south at 400 mph. Another plane leaving at the same time travels due east at 300 mph. Find the distance between the two planes after 2 h.

63. Physics A stone is dropped from a bridge and hits the water 2 s later. How high is the bridge? The equation for the distance an object falls in T seconds is $T = \sqrt{\dfrac{d}{16}}$, where d is the distance in feet.

Parking

650 ft

600 ft

Memorial

64. Physics A stone is dropped into a mine shaft and hits the bottom 3.5 s later. How deep is the mine shaft? The equation for the distance an object falls in T seconds is $T = \sqrt{\dfrac{d}{16}}$, where d is the distance in feet.

65. Television The measure of a television screen is given by the length of a diagonal across the screen. A 36-inch television has a width of 28.8 in. Find the height of the screen.

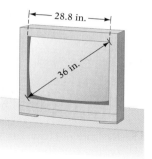

28.8 in.

36 in.

66. Television The measure of a television screen is given by the length of a diagonal across the screen. A 33-inch television has a width of 26.4 in. Find the height of the screen.

10.4 **Applying Concepts**

Solve.

67. $\sqrt{\dfrac{5y + 2}{3}} = 3$

68. $\sqrt{\dfrac{3y}{5}} - 1 = 2$

69. $\sqrt{9x^2 + 49} + 1 = 3x + 2$

70. Geometry In the coordinate plane, a triangle is formed by drawing lines between the points (0, 0) and (5, 0), (5, 0) and (5, 12), and (5, 12) and (0, 0). Find the number of units in the perimeter of the triangle.

71. Geometry The hypotenuse of a right triangle is $5\sqrt{2}$ cm, and one leg is $4\sqrt{2}$ cm.
  a. Find the perimeter of the triangle.   **b.** Find the area of the triangle.

72. Geometry Write an expression in factored form for the shaded region in the diagram at the right.

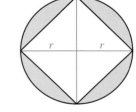

73. Geometry A circular fountain is being designed for a triangular plaza in a cultural center. The fountain is placed so that each side of the triangle touches the fountain as shown in the diagram at the right. Find the area of the fountain. The formula for the radius of the circle is given by

$$r = \sqrt{\dfrac{(s - a)(s - b)(s - c)}{s}}$$

where $s = \dfrac{1}{2}(a + b + c)$ and a, b, and c are the lengths of the sides of the triangle. Round to the nearest hundredth.

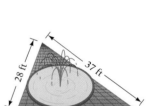

74. Can the Pythagorean Theorem be used to find the length of side c of the triangle at the right? If so, determine c. If not, explain why the theorem cannot be used.

75. What is a Pythagorean triple? Provide at least three examples of Pythagorean triples.

76. Complete the statement using the symbol $<$, $=$, or $>$. Explain how you determined which symbol to use.
  a. For an acute triangle with side c the longest side, $a^2 + b^2 \underline{\quad} c^2$.
  b. For a right triangle with side c the longest side, $a^2 + b^2 \underline{\quad} c^2$.
  c. For an obtuse triangle with side c the longest side, $a^2 + b^2 \underline{\quad} c^2$.

Focus on Problem Solving

Deductive Reasoning

Deductive reasoning uses a rule or statement of fact to reach a conclusion. In the section Equations Containing Fractions in the "Rational Expressions" chapter, we established the rule that if two angles of one triangle are equal to two angles of another triangle, then the two triangles are similar. Thus any time we establish this fact about two triangles, we know that the triangles are similar. Below are two examples of deductive reasoning.

Given that △△△ = ◊◊◊◊ and ◊◊◊◊ = ÓÓ, then △△△△△△ is equivalent to how many Ós?

Because 3 △s = 4 ◊s and 4 ◊s = 2 Ós, 3 △s = 2 Ós.

6 △s is twice 3 △s. We need to find twice 2 Ós, which is 4 Ós.

Therefore, △△△△△△ = ÓÓÓÓ.

Lomax, Parish, Thorpe, and Wong are neighbors. Each drives a different type of vehicle: a compact car, a sedan, a sports car, or a station wagon. From the following statements, determine which type of vehicle each of the neighbors drives.

a. Although the vehicle owned by Lomax has more mileage on it than does either the sedan or the sports car, it does not have the highest mileage of all four cars. (Use Xa in the chart below to eliminate whatever possibilities this statement enables you to rule out.)

b. Wong and the owner of the sports car live on one side of the street, and Thorpe and the owner of the compact car live on the other side of the street. (Use Xb to eliminate possibilities that this statement rules out.)

c. Thorpe owns the vehicle with the most mileage on it. (Use Xc to eliminate possibilities that this statement rules out.)

Take Note

To use the chart to solve this problem, write an X in a box to indicate that a possibility has been eliminated. Write a ✓ to show that a match has been found. When a row or column has 3 X's, a ✓ is written in the remaining open box in that row or column of the chart.

	Compact	Sedan	Sports Car	Wagon
Lomax	✓	Xa	Xa	Xb
Parish	Xb	Xb	✓	Xb
Thorpe	Xb	Xc	Xb	✓
Wong	Xb	✓	Xb	

Lomax drives the compact car, Parish drives the sports car, Thorpe drives the station wagon, and Wong drives the sedan.

Try the following exercises.

1. Given that ‡‡ = ••••• and ••••• = △△, then ‡‡‡‡‡ = how many △'s?

2. Given that ▽▽▽▽▽▽ = ◆◆◆◆ and ◆◆◆◆ = ⊕ ⊕, then ▽▽▽ = how many ⊕'s?

3. Given that ♣♣♣♣ = ⊗⊗⊗ and ⊗⊗⊗ = ♥♥, then ♥♥♥♥ = how many ♣'s?

4. Given that ◊◊◊◊◊ = ÓÓ and ÓÓ = ♠♠♠, then ♠♠♠♠♠♠ = how many ◊'s?

5. Anna, Kay, Megan, and Nicole decide to travel together during spring break, but they need to find a destination where each of them will be able to participate in her favorite sport (golf, horseback riding, sailing, or tennis). From the following statements, determine the favorite sport of each student.

   a. Anna and the student whose favorite sport is sailing both like to swim, whereas Nicole and the student whose favorite sport is tennis would prefer to scuba dive.

   b. Megan and the student whose favorite sport is sailing are roommates. Nicole and the student whose favorite sport is golf both live in a single.

6. Chang, Nick, Pablo, and Saul all take different forms of transportation (bus, car, subway, or taxi) from the office to the airport. From the following statements, determine which form of transportation each takes.

   a. Chang spent more on transportation than the fellow who took the bus but less than the fellow who took the taxi.

   b. Pablo, who did not travel by bus and who spent the least on transportation, arrived at the airport after Nick but before the fellow who took the subway.

   c. Saul spent less on transportation than either Chang or Nick.

Projects & Group Activities

Mean and Standard Deviation

An automotive engineer tests the miles-per-gallon ratings of 15 cars and records the results as follows:

25   22   21   27   25   35   29   31   25   26   21   39   34   32   28

The **mean** of the data is the sum of the measurements divided by the number of measurements. The symbol for the mean is $\bar{x}$.

$$\text{Mean} = \bar{x} = \frac{\text{sum of all data values}}{\text{number of data values}}$$

To find the mean for the engineer's data, add the numbers and then divide by 15.

$$\bar{x} = \frac{25 + 22 + 21 + 27 + 25 + 35 + 29 + 31 + 25 + 26 + 21 + 39 + 34 + 32 + 28}{15}$$

$$= \frac{420}{15} = 28$$

The mean number of miles per gallon for the 15 cars tested was 28 mi/gal.

The mean is one of the most frequently computed averages. It is the one that is commonly used to calculate a student's performance in a class.

The scores for a history student on 5 tests were 78, 82, 91, 87, and 93. What was the mean score for this student?

To find the mean, add the numbers. Then divide by 5.

$$\bar{x} = \frac{78 + 82 + 91 + 87 + 93}{5}$$

$$= \frac{431}{5} = 86.2$$

The mean score for the history student was 86.2.

Consider two students, each of whom has taken 5 exams.

Scores for Student A

84	86	83	85	87

$$\bar{x} = \frac{84 + 86 + 83 + 85 + 87}{5} = \frac{425}{5} = 85$$

The mean for Student A is 85.

Scores for Student B

90	75	94	68	98

$$\bar{x} = \frac{90 + 75 + 94 + 68 + 98}{5} = \frac{425}{5} = 85$$

The mean for Student B is 85.

For each of these students, the mean (average) for the 5 exams is 85. However, Student A has a more consistent record of scores than Student B. One way to measure the consistency, or "clustering" near the mean, of data is to use the **standard deviation.**

To calculate the standard deviation:

Step 1 Sum the squares of the differences between each value of the data and the mean.

Step 2 Divide the result in Step 1 by the number of items in the set of data.

Step 3 Take the square root of the result in Step 2.

The calculation for Student A is shown at the right.

Step 1

x	$x - \bar{x}$	$(x - \bar{x})^2$
84	84 − 85	$(-1)^2 = 1$
86	86 − 85	$1^2 = 1$
83	83 − 85	$(-2)^2 = 4$
85	85 − 85	$0^2 = 0$
87	87 − 85	$2^2 = 4$

Total = 10

The symbol for standard deviation is the lower case Greek letter *sigma, σ.*

Step 2 $\dfrac{10}{5} = 2$

Step 3 $\sigma = \sqrt{2} \approx 1.414$

The standard deviation for Student A's scores is approximately 1.414.

Following a similar procedure for Student B shows that the standard deviation for Student B's scores is approximately 11.524. Because the standard deviation of Student B's scores is greater than that of Student A's scores ($11.524 > 1.414$), Student B's scores are not as consistent as those of Student A.

1. The weights in ounces of 6 newborn infants were recorded by a hospital. The weights were 96, 105, 84, 90, 102, and 99. Find the standard deviation of the weights.

2. The numbers of rooms occupied in a hotel on 6 consecutive days were 234, 321, 222, 246, 312, and 396. Find the standard deviation of the numbers of rooms occupied.

3. Seven coins were tossed 100 times. The numbers of heads recorded for each coin were 56, 63, 49, 50, 48, 53, and 52. Find the standard deviation of the numbers of heads.

4. The temperatures for 11 consecutive days at a desert resort were 95°, 98°, 98°, 104°, 97°, 100°, 96°, 97°, 108°, 93°, and 104°. For the same days, temperatures in Antarctica were 27°, 28°, 28°, 30°, 28°, 27°, 30°, 25°, 24°, 26°, and 21°. Which location has the greater standard deviation of temperatures?

5. The scores for 5 college basketball games were 56, 68, 60, 72, and 64. The scores for 5 professional basketball games were 106, 118, 110, 122, and 114. Which scores have the greater standard deviation?

6. The weights in pounds of the 5-man front line of a college football team are 210, 245, 220, 230, and 225. Find the standard deviation of the weights.

7. One student received test scores of 85, 92, 86, and 89. A second student received scores of 90, 97, 91, and 94 (exactly 5 points more on each test). Are the means of the two students the same? If not, what is the relationship between the means of the two students? Are the standard deviations of the scores of the two students the same? If not, what is the relationship between the standard deviations of the scores of the two students?

8. A company is negotiating with its employees the terms of a raise in salary. One proposal would add $500 a year to each employee's salary. The second proposal would give each employee a 4% raise. Explain how each of these proposals would affect the current mean and standard deviation of salaries for the company.

Distance to the Horizon   In Section 4 of this chapter, we used the formula $d = \sqrt{1.5h}$ to calculate the approximate distance d (in miles) that a person could see who uses a periscope h feet above the water. This formula is derived by using the Pythagorean Theorem.

Consider the diagram (not to scale) at the right, which shows Earth as a sphere and the periscope extending h feet above the surface. From geometry, because AB is tangent to the circle and OA is a radius, triangle AOB is a right triangle. Therefore,

$$(OA)^2 + (AB)^2 = (OB)^2$$

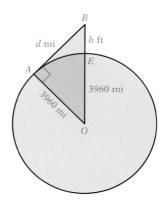

Substituting into this formula, we have

$$3960^2 + d^2 = \left(3960 + \frac{h}{5280}\right)^2$$

• Because h is in feet, $\dfrac{h}{5280}$ is in miles.

$$3960^2 + d^2 = 3960^2 + \frac{2 \cdot 3960}{5280}h + \left(\frac{h}{5280}\right)^2$$

$$d^2 = \frac{3}{2}h + \left(\frac{h}{5280}\right)^2$$

$$d = \sqrt{\frac{3}{2}h + \left(\frac{h}{5280}\right)^2}$$

At this point, an assumption is made that $\sqrt{\dfrac{3}{2}h + \left(\dfrac{h}{5280}\right)^2} \approx \sqrt{1.5h}$, where

we have written $\dfrac{3}{2}$ as 1.5. Thus $d \approx \sqrt{1.5h}$ is used to approximate the distance that can be seen using a periscope h feet above the water.

1. Write a paragraph that justifies the assumption that

$$\sqrt{\frac{3}{2}h + \left(\frac{h}{5280}\right)^2} \approx \sqrt{1.5h}$$

   (*Suggestion:* Evaluate each expression for various values of h. Because h is the height of a periscope above water, it is unlikely that $h > 25$ ft.)

2. The distance d is the distance from the top of the periscope to A. The distance along the surface of the water is given by arc AE. This distance, D, can be approximated by the equation

$$D \approx \sqrt{1.5h} + 0.306186\left(\sqrt{\frac{h}{5280}}\right)^3$$

   Using this formula, calculate D when $h = 10$.

**Expressway
On-Ramp Curves**

Highway engineers design expressway on-ramps with both efficiency and safety in mind. An on-ramp cannot be built with too sharp a curve, or the speed at which drivers take the curve will cause them to skid off the road. However, an on-ramp built with too wide a curve requires more building material, takes longer to travel, and uses more land. The following formula states the relationship between the maximum speed at which a car can travel around a curve without skidding: $v = \sqrt{2.5r}$. In this formula, v is speed in miles per hour and r is the radius, in feet, of an unbanked curve.

1. Use a graphing calculator to graph this equation. Use a window of Xmin $= 0$, Xmax $= 2000$, Ymin $= 0$, Ymax $= 100$. What do the values used for Xmin, Xmax, Ymin, and Ymax represent?

2. As the radius of the curve increases, does the maximum safe speed increase or decrease? Explain your answer on the basis of the graph of the equation.

3. Use the graph to approximate the maximum safe speed when the radius of the curve is 100 ft. Round to the nearest whole number.

4. Use the graph to approximate the radius of the curve for which the maximum safe speed is 40 mph. Round to the nearest whole number.

5. ⬤ Juan Montoya won the Indianapolis 500 in 2000 driving at an average speed of 167.607 mph. At this speed, should the radius of an unbanked curve be more or less than 1 mi?

Chapter Summary

Key Words

A **square root** of a positive number x is a number whose square is x. The square root of a negative number is not a real number.   [10.1.1, pp. 506–508]

Examples

A square root of 25 is 5 because $5^2 = 25$.

A square root of 25 is −5 because $(-5)^2 = 25$.

$\sqrt{-9}$ is not a real number.

The **principal square root** of a number is the positive square root. The symbol $\sqrt{}$ is called a **radical sign** and is used to indicate the principal square root of a number. The negative square root of a number is indicated by placing a negative sign in front of the radical. The **radicand** is the expression under the radical symbol.   [10.1.1, p. 506]

$\sqrt{25} = 5$

$-\sqrt{25} = -5$

In the expression $\sqrt{25}$, 25 is the radicand.

The square of an integer is a **perfect square**. If a number is not a perfect square, its square root can only be approximated. Such numbers are **irrational numbers**. Their decimal representations never terminate or repeat. [10.1.1, pp. 507, 508]

$2^2 = 4$, $3^2 = 9$, $4^2 = 16$, $5^2 = 25$, $6^2 = 36$, . . . 4, 9, 16, 25, 36, . . . are perfect squares.

5 is not a perfect square. $\sqrt{5}$ is an irrational number.

Conjugates are expressions with two terms that differ only in the sign of the second term. The expressions $a + b$ and $a - b$ are conjugates.   [10.3.1, p. 518]

$3 + \sqrt{7}$ and $3 - \sqrt{7}$ are conjugates.

$\sqrt{x} + 2$ and $\sqrt{x} - 2$ are conjugates.

A **radical equation** is an equation that contains a variable expression in a radicand.   [10.4.1, p. 524]

$\sqrt{2x} + 8 = 12$ is a radical equation. $2x + \sqrt{8} = 12$ is not a radical equation.

Essential Rules and Procedures

The Product Property of Square Roots
If a and b are positive real numbers, then $\sqrt{ab} = \sqrt{a} \cdot \sqrt{b}$. [10.1.1, p. 507]

$\sqrt{12} = \sqrt{4 \cdot 3} = \sqrt{4}\,\sqrt{3} = 2\sqrt{3}$

Adding and Subtracting Radical Expressions
The Distributive Property is used to simplify the sum or difference of like radical expressions.   [10.2.1, p. 514]

$16\sqrt{3x} - 4\sqrt{3x} = (16 - 4)\sqrt{3x}$
$= 12\sqrt{3x}$

Multiplying Radical Expressions

The Product Property of Square Roots is used to multiply radical expressions.

$$\sqrt{5x}\,\sqrt{7y} = \sqrt{5x \cdot 7y} = \sqrt{35xy}$$

Use the Distributive Property to remove parentheses.

$$\sqrt{y}\left(2 + \sqrt{3x}\right) = 2\sqrt{y} + \sqrt{3xy}$$

Use the FOIL method to multiply radical expressions with two terms.   [10.3.1, pp. 517–518]

$$\left(5 - \sqrt{x}\right)\left(11 + \sqrt{x}\right)$$
$$= 55 + 5\sqrt{x} - 11\sqrt{x} - \left(\sqrt{x}\right)^2$$
$$= 55 - 6\sqrt{x} - x$$

The Quotient Property of Square Roots

If a and b are positive real numbers, then $\sqrt{\dfrac{a}{b}} = \dfrac{\sqrt{a}}{\sqrt{b}}$.

[10.3.2, p. 519]

$$\sqrt{\frac{9x^2}{y^8}} = \frac{\sqrt{9x^2}}{\sqrt{y^8}} = \frac{3x}{y^4}$$

Dividing Radical Expressions

The Quotient Property of Square Roots is used to divide radical expressions.   [10.3.2, p. 519]

$$\frac{\sqrt{3x^5y}}{\sqrt{75xy^3}} = \sqrt{\frac{3x^5y}{75xy^3}} = \sqrt{\frac{x^4}{25y^2}}$$
$$= \frac{\sqrt{x^4}}{\sqrt{25y^2}} = \frac{x^2}{5y}$$

Rationalizing the Denominator

A radical is not in simplest form if there is a radical in the denominator. The procedure used to remove a radical from the denominator is called rationalizing the denominator. [10.3.2, p. 519]

$$\frac{5}{\sqrt{7}} = \frac{5}{\sqrt{7}} \cdot \frac{\sqrt{7}}{\sqrt{7}} = \frac{5\sqrt{7}}{7}$$

Radical Expressions in Simplest Form

A radical expression is in simplest form if:
1. The radicand contains no factor greater than 1 that is a perfect square.
2. There is no fraction under the radical sign.
3. There is no radical in the denominator of a fraction.
[10.3.2, p. 520]

$\sqrt{12}$, $\sqrt{\dfrac{3}{4}}$, and $\dfrac{1}{\sqrt{2}}$ are not in simplest form.

$2\sqrt{3}$, $\dfrac{\sqrt{3}}{2}$, and $\dfrac{\sqrt{2}}{2}$ are in simplest form.

Property of Squaring Both Sides of an Equation

If a and b are real numbers and $a = b$, then $a^2 = b^2$.
[10.4.1, p. 524]

$$\sqrt{x + 3} = 4$$
$$\left(\sqrt{x + 3}\right)^2 = 4^2$$
$$x + 3 = 16$$
$$x = 13$$

Solving Radical Equations

When an equation contains one radical expression, write the equation with the radical alone on one side of the equation. Square both sides of the equation. Solve for the variable. Whenever both sides of an equation are squared, the solutions must be checked.   [10.4.1, p. 525]

$$\sqrt{2x} - 1 = 5$$
$$\sqrt{2x} = 6$$
$$\left(\sqrt{2x}\right)^2 = 6^2$$
$$2x = 36$$
$$x = 18 \quad \text{The solution checks.}$$

Pythagorean Theorem
If a and b are the lengths of the legs of a right triangle and c is the length of the hypotenuse, then $c^2 = a^2 + b^2$. [10.4.2, p. 528]

Two legs of a right triangle measure 4 cm and 7 cm. To find the length of the hypotenuse, use the equation

$$c = \sqrt{a^2 + b^2}$$
$$c = \sqrt{4^2 + 7^2} = \sqrt{16 + 49} = \sqrt{65}$$

The hypotenuse measures $\sqrt{65}$ cm.

Chapter Review Exercises

1. Subtract: $5\sqrt{3} - 16\sqrt{3}$

2. Simplify: $\dfrac{2x}{\sqrt{3} - \sqrt{5}}$

3. Simplify: $\sqrt{x^2 + 16x + 64}$

4. Solve: $3 - \sqrt{7x} = 5$

5. Simplify: $\dfrac{5\sqrt{y} - 2\sqrt{y}}{3\sqrt{y}}$

6. Add: $6\sqrt{7} + \sqrt{7}$

7. Multiply: $\left(6\sqrt{a} + 5\sqrt{b}\right)\left(2\sqrt{a} + 3\sqrt{b}\right)$

8. Simplify: $\sqrt{49(x + 3)^4}$

9. Simplify: $2\sqrt{36}$

10. Solve: $\sqrt{b} = 4$

11. Subtract: $9x\sqrt{5} - 5x\sqrt{5}$

12. Multiply: $\left(\sqrt{5ab} - \sqrt{7}\right)\left(\sqrt{5ab} + \sqrt{7}\right)$

13. Solve: $\sqrt{2x + 9} = \sqrt{8x - 9}$

14. Simplify: $\sqrt{35}$

15. Add: $2x\sqrt{60x^3y^3} + 3x^2y\sqrt{15xy}$

16. Simplify: $\dfrac{\sqrt{3x^3y}}{\sqrt{27xy^5}}$

17. Simplify: $\left(3\sqrt{x} - \sqrt{y}\right)^2$

18. Simplify: $\sqrt{(a + 4)^2}$

19. Simplify: $5\sqrt{48}$

20. Add: $3\sqrt{12x} + 5\sqrt{48x}$

21. Simplify: $\dfrac{8}{\sqrt{x} - 3}$

22. Multiply: $\sqrt{6a}\left(\sqrt{3a} + \sqrt{2a}\right)$

23. Simplify: $3\sqrt{18a^5b}$

24. Simplify: $-3\sqrt{120}$

25. Subtract: $\sqrt{20a^5b^9} - 2ab^2\sqrt{45a^3b^5}$

26. Simplify: $\dfrac{\sqrt{98x^7y^9}}{\sqrt{2x^3y}}$

27. Solve: $\sqrt{5x + 1} = \sqrt{20x - 8}$

28. Simplify: $\sqrt{c^{18}}$

29. Simplify: $\sqrt{450}$

30. Simplify: $6a\sqrt{80b} - \sqrt{180a^2b} + 5a\sqrt{b}$

31. Simplify: $\dfrac{16}{\sqrt{a}}$

32. Solve: $6 - \sqrt{2y} = 2$

33. Multiply: $\sqrt{a^3b^4c}\ \sqrt{a^7b^2c^3}$

34. Simplify: $7\sqrt{630}$

35. Simplify: $y\sqrt{24y^6}$

36. Simplify: $\dfrac{\sqrt{250}}{\sqrt{10}}$

37. Solve: $\sqrt{x^2 + 5x + 4} = \sqrt{x^2 + 7x - 6}$

38. Multiply: $\left(4\sqrt{y} - \sqrt{5}\right)\left(2\sqrt{y} + 3\sqrt{5}\right)$

39. Find the decimal approximation of $\sqrt{9900}$ to the nearest thousandth.

40. Multiply: $\sqrt{7}\ \sqrt{7}$

41. Simplify: $5x\sqrt{150x^7}$

42. Simplify: $2x^2\sqrt{18x^2y^5} + 6y\sqrt{2x^6y^3} - 9xy^2\sqrt{8x^4y}$

43. Simplify: $\dfrac{\sqrt{54a^3}}{\sqrt{6a}}$

44. Simplify: $4\sqrt{250}$

45. Solve: $\sqrt{5x} = 10$

46. Simplify: $\dfrac{3a\sqrt{3} + 2\sqrt{12a^2}}{\sqrt{27}}$

47. Multiply: $\sqrt{2}\ \sqrt{50}$

48. Simplify: $\sqrt{36x^4y^5}$

49. Simplify: $4y\sqrt{243x^{17}y^9}$

50. Solve: $\sqrt{x + 1} - \sqrt{x - 2} = 1$

51. Simplify: $\sqrt{400}$

52. Simplify: $-4\sqrt{8x} + 7\sqrt{18x} - 3\sqrt{50x}$

53. Solve: $0 = \sqrt{10x + 4} - 8$

54. Multiply: $\sqrt{3}\left(\sqrt{12} - \sqrt{3}\right)$

55. The square root of the sum of two consecutive odd integers is equal to 10. Find the larger integer.

56.  The weight of an object is related to the object's distance from the surface of Earth. An equation for this relationship is $d = 4000\sqrt{\dfrac{W_0}{W_a}}$, where W_0 is the object's weight on the surface of Earth and W_a is the object's weight at a distance of d miles above Earth's surface. A space explorer weighs 36 lb when 8000 mi above Earth's surface. Find the explorer's weight on the surface of Earth.

57. The hypotenuse of a right triangle measures 18 cm. One leg of the triangle measures 11 cm. Find the length of the other leg of the triangle. Round to the nearest hundredth.

58. A bicycle will overturn if it rounds a corner too sharply or too quickly. The equation for the maximum velocity at which a cyclist can turn a corner without tipping over is given by the equation $v = 4\sqrt{r}$, where v is the velocity of the bicycle in miles per hour and r is the radius of the corner in feet. Find the radius of the sharpest corner that a cyclist can safely turn when riding at a speed of 20 mph.

59. A tsunami is a great sea wave produced by underwater earthquakes or volcanic eruption. The velocity of a tsunami as it approaches land is approximated by the equation $v = 3\sqrt{d}$, where v is the velocity in feet per second and d is the depth of the water in feet. Find the depth of the water when the velocity of a tsunami is 30 ft/s.

60. A guy wire is attached to a point 25 ft above the ground on a telephone pole that is perpendicular to the ground. The wire is anchored to the ground at a point 8 ft from the base of the pole. Find the length of the guy wire. Round to the nearest hundredth.

Chapter Test

1. Simplify: $\sqrt{121x^8y^2}$

2. Subtract: $5\sqrt{8} - 3\sqrt{50}$

3. Multiply: $\sqrt{3x^2y}\,\sqrt{6x^2}\,\sqrt{2x}$

4. Simplify: $\sqrt{45}$

5. Simplify: $\sqrt{72x^7y^2}$

6. Simplify: $3\sqrt{8y} - 2\sqrt{72x} + 5\sqrt{18x}$

7. Multiply: $\left(\sqrt{y} + 3\right)\left(\sqrt{y} + 5\right)$

8. Simplify: $\dfrac{4}{\sqrt{8}}$

9. Simplify: $\dfrac{\sqrt{162}}{\sqrt{2}}$

10. Solve: $\sqrt{5x - 6} = 7$

11. Find the decimal approximation of $\sqrt{500}$ to the nearest thousandth.

12. Simplify: $\sqrt{32a^5b^{11}}$

13. Multiply: $\sqrt{a}\left(\sqrt{a} - \sqrt{b}\right)$

14. Multiply: $\sqrt{8x^3y}\,\sqrt{10xy^4}$

15. Simplify: $\dfrac{\sqrt{98a^6b^4}}{\sqrt{2a^3b^2}}$

16. Solve: $\sqrt{9x} + 3 = 18$

17. Simplify: $\sqrt{192x^{13}y^5}$

18. Simplify: $2a\sqrt{2ab^3} + b\sqrt{8a^3b} - 5ab\sqrt{ab}$

19. Multiply: $\left(\sqrt{a} - 2\right)\left(\sqrt{a} + 2\right)$

20. Simplify: $\dfrac{3}{2 - \sqrt{5}}$

21. Solve: $3 = 8 - \sqrt{5x}$

22. Multiply: $\sqrt{3}\left(\sqrt{6} - \sqrt{x^2}\right)$

23. Subtract: $3\sqrt{a} - 9\sqrt{a}$

24. Simplify: $\sqrt{108}$

25. Find the decimal approximation of $\sqrt{63}$ to the nearest thousandth.

26. Simplify: $\dfrac{\sqrt{108a^7b^3}}{\sqrt{3a^4b}}$

27. Solve: $\sqrt{x} - \sqrt{x + 3} = 1$

28. The square root of the sum of two consecutive integers is equal to 9. Find the smaller integer.

29. Find the length of a pendulum that makes one swing in 2.5 s. The equation for the time of one swing of a pendulum is $T = 2\pi\sqrt{\dfrac{L}{32}}$, where T is the time in seconds and L is the length in feet. Round to the nearest hundredth.

30. A ladder 16 ft long is resting against a building. How high on the building will the ladder reach when the bottom of the ladder is 5 ft from the building? Round to the nearest tenth.

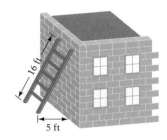

Cumulative Review Exercises

1. Simplify: $\left(\dfrac{2}{3}\right)^2\left(\dfrac{3}{4} - \dfrac{3}{2}\right) + \left(\dfrac{1}{2}\right)^2$

2. Simplify: $-3[x - 2(3 - 2x) - 5x] + 2x$

3. Solve: $2x - 4[3x - 2(1 - 3x)] = 2(3 - 4x)$

4. Solve: $3(x - 7) \geq 5x - 12$

5. Find the slope of the line that contains the points whose coordinates are $(2, -5)$ and $(-4, 3)$.

6. Find the equation of the line that contains the point whose coordinates are $(-2, -3)$ and has slope $\dfrac{1}{2}$.

7. Evaluate $f(x) = \dfrac{5}{2}x - 8$ at $x = -4$.

8. Graph: $f(x) = -4x + 2$

9. Graph the solution set of $2x + y < -2$.

10. Solve by graphing: $3x - 2y = 8$
$\quad\quad\quad\quad\quad\quad\quad\quad\quad 4x + 5y = 3$

11. Solve by substitution: $4x - 3y = 1$
$\quad\quad\quad\quad\quad\quad\quad\quad\quad\quad\quad 2x + y = 3$

12. Solve by the addition method: $5x + 4y = 7$
$\quad\quad\quad\quad\quad\quad\quad\quad\quad\quad\quad\quad\quad\quad 3x - 2y = 13$

13. Simplify: $(-3x^2y)(-2x^3y^{-4})$

14. Simplify: $\dfrac{12b^4 - 6b^2 + 2}{-6b^2}$

15. Factor: $12x^3y^2 - 9x^2y^3$

16. Factor: $9b^2 + 3b - 20$

17. Factor: $2a^3 - 16a^2 + 30a$

18. Multiply: $\dfrac{3x^3 - 6x^2}{4x^2 + 4x} \cdot \dfrac{3x - 9}{9x^3 - 45x^2 + 54x}$

19. Simplify: $\dfrac{1 - \dfrac{2}{x} - \dfrac{15}{x^2}}{1 - \dfrac{9}{x^2}}$

20. Subtract: $\dfrac{x + 2}{x - 4} - \dfrac{6}{(x - 4)(x - 3)}$

21. Solve: $\dfrac{x}{2x - 5} - 2 = \dfrac{3x}{2x - 5}$

22. Simplify: $2\sqrt{27a} - 5\sqrt{49a} + 8\sqrt{48a}$

23. Simplify: $\dfrac{\sqrt{320}}{\sqrt{5}}$

24. Solve: $\sqrt{2x - 3} - 5 = 0$

25. Three-eighths of a number is less than negative twelve. Find the largest integer that satisfies the inequality.

26. The selling price of a book is $29.40. The markup rate used by the bookstore is 20% of the cost. Find the cost of the book.

27. How many ounces of pure water must be added to 40 oz of a 12% salt solution to make a salt solution that is 5% salt?

28. The sum of two numbers is twenty-one. Three more than twice the smaller number is equal to the larger. Find the two numbers.

29. A small water pipe takes twice as long to fill a tank as does a larger water pipe. With both pipes open, it takes 16 h to fill the tank. Find the time it would take the small pipe, working alone, to fill the tank.

30. The square root of the sum of two consecutive integers is equal to 7. Find the smaller integer.

Businesses spend time and money doing research to determine the cost of producing and selling their products. The objective of any business is to earn a profit. Profit is the difference between a company's revenue (the total amount of money the company earns by selling its products or services) and its costs (the total amount of money the company spends in doing business). Quadratic equations are used in the **Focus on Problem Solving on pages 583–584** to determine the maximum profit companies can earn. You will be asked to determine the maximum monthly profit for a company that sells guitars such as the ones being crafted in this photo.

Applications

Automobiles, *p. 557*

Business, *p. 584*

Construction, *p. 595*

Education, *p. 595*

Energy, *p. 557*

Food industry, *p. 583*

Gardening, *p. 583*

Geometry, *pp. 580, 583, 592, 594*

Integer problems, *pp. 580, 583, 592, 594, 595*

Investments, *pp. 557, 595*

Metalwork, *p. 583*

Physics, *pp. 582, 592*

The Postal Service, *p. 557*

Recreation, *p. 581*

Safety, *p. 582*

Sports, *pp. 581, 582*

Taxes, *p. 595*

Travel, *p. 582*

Uniform motion problems, *pp. 578, 581, 582, 592, 594, 595*

Value mixture problems, *p. 595*

Work problems, *pp. 579, 581, 592*

11

Quadratic Equations

Objectives

Section 11.1
1. Solve quadratic equations by factoring
2. Solve quadratic equations by taking square roots

Section 11.2
1. Solve quadratic equations by completing the square

Section 11.3
1. Solve quadratic equations by using the quadratic formula

Section 11.4
1. Graph a quadratic equation of the form $y = ax^2 + bx + c$

Section 11.5
1. Application problems

 Need help? For online student resources, such as section quizzes, visit this web site: **math.college.hmco.com**

WEB

Prep Test

1. Evaluate $b^2 - 4ac$ when $a = 2$, $b = -3$, and $c = -4$.

2. Solve: $5x + 4 = 3$

3. Factor: $x^2 + x - 12$

4. Factor: $4x^2 - 12x + 9$

5. Is $x^2 - 10x + 25$ a perfect square trinomial?

6. Solve: $\dfrac{5}{x - 2} = \dfrac{15}{x}$

7. Graph: $y = -2x + 3$

8. Simplify: $\sqrt{28}$

Go Figure

Find the value of

$$\frac{1}{\sqrt{1} + \sqrt{2}} + \frac{1}{\sqrt{2} + \sqrt{3}} + \frac{1}{\sqrt{3} + \sqrt{4}} + \cdots + \frac{1}{\sqrt{8} + \sqrt{9}}.$$

9. If a is *any* real number, simplify $\sqrt{a^2}$.

10. Walking at a constant speed of 4.5 mph, Lucy and Sam walked from the beginning to the end of a hiking trail. When they reached the end, they immediately started back along the same path at a constant speed of 3 mph. If the round trip took 2 h, what is the length of the hiking trail?

11.1 Solving Quadratic Equations by Factoring or by Taking Square Roots

1 **Solve quadratic equations by factoring**

In Section 5 of the chapter on factoring, we solved quadratic equations by factoring. In this section, we will review that material and then solve quadratic equations by taking square roots.

An equation of the form $ax^2 + bx + c = 0$, $a \neq 0$, is a **quadratic equation.**

$$4x^2 - 3x + 1 = 0, \qquad a = 4, b = -3, c = 1$$
$$3x^2 - 4 = 0, \qquad a = 3, b = 0, \quad c = -4$$

A quadratic equation is also called a **second-degree equation.**

A quadratic equation is in **standard form** when the polynomial is in descending order and equal to zero.

Recall that the Principle of Zero Products states that if the product of two factors is zero, then at least one of the factors must be zero.

If $a \cdot b = 0$, then $a = 0$ or $b = 0$.

The Principle of Zero Products can be used in solving quadratic equations.

Solve by factoring: $2x^2 - x = 1$

$$2x^2 - x = 1$$

Write the equation in standard form.	$2x^2 - x - 1 = 0$
Factor the left side of the equation.	$(2x + 1)(x - 1) = 0$
If $(2x + 1)(x - 1) = 0$, then either $2x + 1 = 0$ or $x - 1 = 0$.	$2x + 1 = 0 \qquad x - 1 = 0$
Solve each equation for x.	$2x = -1 \qquad x = 1$
	$x = -\dfrac{1}{2}$

The solutions are $-\dfrac{1}{2}$ and 1.

A graphing calculator can be used to check the solutions of a quadratic equation. Consider the preceding example. The solutions appear to be $-\dfrac{1}{2}$ and 1. To check these solutions, store one value of x, $-\dfrac{1}{2}$, in the calculator.

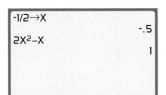

```
-1/2→X
                    -.5
2X²–X
                     1
```

Note that after you store $-\dfrac{1}{2}$ in the calculator's memory, your calculator rewrites the fraction as the decimal -0.5. Evaluate the expression on the left side of the original equation: $2x^2 - x$. The result is 1, which is the number on the right side of the original equation. The solution -0.5 checks. Now store the other value of x, 1, in the calculator. Evaluate the expression on the left side of the original equation: $2x^2 - x$. The result is 1, which is the number on the right side of the equation. The solution 1 checks.

Solve by factoring: $x^2 = 10x - 25$

$$x^2 = 10x - 25$$

Write the equation in standard form.	$x^2 - 10x + 25 = 0$
Factor the left side of the equation.	$(x - 5)(x - 5) = 0$
Let each factor equal 0.	$x - 5 = 0 \qquad x - 5 = 0$
Solve each equation for x.	$x = 5 \qquad x = 5$

The factorization in this example produced two identical factors. Because $x - 5$ occurs twice in the factored form of the equation, 5 is a **double root** of the equation.

Example 1   Solve by factoring: $\dfrac{z^2}{2} + \dfrac{z}{3} - \dfrac{1}{6} = 0$

Solution

$$\dfrac{z^2}{2} + \dfrac{z}{3} - \dfrac{1}{6} = 0$$

$$6\left(\dfrac{z^2}{2} + \dfrac{z}{3} - \dfrac{1}{6}\right) = 6(0)$$ • To eliminate the fractions, multiply each side of the equation by 6.

$$3z^2 + 2z - 1 = 0$$ • The quadratic equation is in standard form.

$$(3z - 1)(z + 1) = 0$$ • Factor the left side of the equation.

$$3z - 1 = 0 \qquad z + 1 = 0$$ • Let each factor equal zero.

$$3z = 1 \qquad\qquad z = -1$$ • Solve each equation for z.

$$z = \dfrac{1}{3}$$

The solutions are $\dfrac{1}{3}$ and -1.   • $\dfrac{1}{3}$ and -1 check as solutions.

Problem 1   Solve by factoring: $\dfrac{3y^2}{2} + y - \dfrac{1}{2} = 0$

Solution   See page S25.

2   Solve quadratic equations by taking square roots

Consider a quadratic equation of the form $x^2 = a$. This equation can be solved by factoring.

$$x^2 = 25$$
$$x^2 - 25 = 0$$
$$(x + 5)(x - 5) = 0$$

$$x + 5 = 0 \qquad x - 5 = 0$$
$$x = -5 \qquad\qquad x = 5$$

The solutions are -5 and 5.

Solutions that are plus or minus the same number are frequently written using $\pm$. For the example above, this would be written: "The solutions are ± 5." Because the solutions ± 5 can be written as $\pm\sqrt{25}$, an alternative method of solving this equation is suggested.

> ### **Take Note**
>
> Recall that the solution of the equation $|x| = 5$ is ± 5. This principle is used when solving an equation by taking square roots. Remember that $\sqrt{x^2} = |x|$. Therefore,
>
> $$x^2 = 25$$
> $$\sqrt{x^2} = \sqrt{25}$$
> $$|x| = 5 \qquad • \ \sqrt{x^2} = |x|$$
> $$x = \pm 5 \qquad • \text{ If } |x| = 5,$$
> $$\text{then } x = \pm 5.$$

> **Principle of Taking the Square Root of Each Side of an Equation**
>
> If $x^2 = a$, then $x = \pm\sqrt{a}$.

Solve by taking square roots: $x^2 = 25$

Take the square root of each side of the equation.

$$x^2 = 25$$
$$\sqrt{x^2} = \sqrt{25}$$

Simplify.

$$x = \pm\sqrt{25} = \pm 5$$

Write the solutions.

The solutions are 5 and -5.

Solve by taking square roots: $3x^2 = 36$

$$3x^2 = 36$$

Solve for x^2.

$$x^2 = 12$$

Take the square root of each side of the equation.

$$\sqrt{x^2} = \sqrt{12}$$

Simplify.

$$x = \pm\sqrt{12} = \pm2\sqrt{3}$$

Check

$$\begin{array}{c|c} 3x^2 = 36 \\ \hline 3\left(2\sqrt{3}\right)^2 & 36 \\ 3(12) & 36 \\ 36 = 36 \end{array} \qquad \begin{array}{c|c} 3x^2 = 36 \\ \hline 3\left(-2\sqrt{3}\right)^2 & 36 \\ 3(12) & 36 \\ 36 = 36 \end{array}$$

These are true equations. The solutions check.

Write the solutions.

The solutions are $2\sqrt{3}$ and $-2\sqrt{3}$.

> ## Take Note
> You should always check your solutions by substituting them back into the *original* equation.

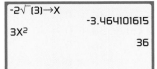
```
-2√(3)→X
              -3.464101615
3X²
                        36
```

A graphing calculator can be used to check irrational solutions of a quadratic equation. Consider the preceding example; the solutions $2\sqrt{3}$ and $-2\sqrt{3}$ checked. To check these solutions on a calculator, store one value of x, $2\sqrt{3}$, in the calculator. Note that after you store this number in the calculator's memory, the calculator rewrites the expression as the decimal 3.464101615, which is an approximation of $2\sqrt{3}$. Evaluate the expression on the left side of the original equation. The result is 36, which is the number on the right side of the original equation. The solution $2\sqrt{3}$ checks. Now store the other value of x, $-2\sqrt{3}$, in the calculator. Note that after you store this number in the calculator's memory, the calculator rewrites the expression as the decimal -3.464101615, which is an approximation of $-2\sqrt{3}$. Evaluate the expression on the left side of the original equation. The result is 36, which is the number on the right side of the equation. The solution $-2\sqrt{3}$ checks.

Example 2 Solve by taking square roots.

A. $2x^2 - 72 = 0$ **B.** $x^2 + 16 = 0$

Solution

A. $2x^2 - 72 = 0$

$$2x^2 = 72$$ • Solve for x^2.

$$x^2 = 36$$

$$\sqrt{x^2} = \sqrt{36}$$ • Take the square root of each

$$x = \pm\sqrt{36}$$ side of the equation.

$$x = \pm6$$

The solutions are 6 and -6.

B. $x^2 + 16 = 0$

$$x^2 = -16$$

$$\sqrt{x^2} = \sqrt{-16}$$ • $\sqrt{-16}$ is not a real number.

The equation has no real number solution.

Problem 2 Solve by taking square roots.

A. $4x^2 - 96 = 0$ **B.** $x^2 + 81 = 0$

Solution See page S25.

An equation containing the square of a binomial can be solved by taking square roots.

Solve by taking square roots: $2(x - 1)^2 - 36 = 0$

$$2(x - 1)^2 - 36 = 0$$

Solve for $(x - 1)^2$.

$$2(x - 1)^2 - 36 = 0$$
$$2(x - 1)^2 = 36$$
$$(x - 1)^2 = 18$$

Take the square root of each side of the equation.

$$\sqrt{(x - 1)^2} = \sqrt{18}$$

Simplify.

$$x - 1 = \pm\sqrt{18}$$
$$x - 1 = \pm3\sqrt{2}$$

Solve for x.

$$x - 1 = 3\sqrt{2} \qquad x - 1 = -3\sqrt{2}$$
$$x = 1 + 3\sqrt{2} \qquad x = 1 - 3\sqrt{2}$$

Check

$$\frac{2(x - 1)^2 - 36 = 0}{}$$

$$\begin{array}{r|l} 2(1 + 3\sqrt{2} - 1)^2 - 36 & 0 \\ 2(3\sqrt{2})^2 - 36 & 0 \\ 2(18) - 36 & 0 \\ 36 - 36 & 0 \\ 0 = 0 & \end{array}$$

$$\frac{2(x - 1)^2 - 36 = 0}{}$$

$$\begin{array}{r|l} 2(1 - 3\sqrt{2} - 1)^2 - 36 & 0 \\ 2(-3\sqrt{2})^2 - 36 & 0 \\ 2(18) - 36 & 0 \\ 36 - 36 & 0 \\ 0 = 0 & \end{array}$$

Write the solutions.   The solutions are $1 + 3\sqrt{2}$ and $1 - 3\sqrt{2}$.

Example 3 Solve by taking square roots: $(x - 6)^2 = 12$

Solution

$$(x - 6)^2 = 12$$
$$\sqrt{(x - 6)^2} = \sqrt{12}$$
$$x - 6 = \pm\sqrt{12}$$
$$x - 6 = \pm2\sqrt{3}$$

• Take the square root of each side of the equation.

$$x - 6 = 2\sqrt{3} \qquad x - 6 = -2\sqrt{3}$$
$$x = 6 + 2\sqrt{3} \qquad x = 6 - 2\sqrt{3}$$

• Solve for x.

The solutions are $6 + 2\sqrt{3}$ and $6 - 2\sqrt{3}$.

Problem 3 Solve by taking square roots: $(x + 5)^2 = 20$

Solution See page S25.

11.1   Concept Review

Determine whether the statement is always true, sometimes true, or never true.

1. $2x^2 - 3x + 9$ is a quadratic equation.

2. By the Principle of Zero Products, if $(3x + 4)(x - 7) = 0$, then $3x + 4 = 0$ or $x - 7 = 0$.

3. A quadratic equation has two distinct solutions.

4. To solve the equation $3x^2 - 26x = 9$ by factoring, first write the equation in standard form.

5. To say that the solutions are ± 6 means that the solutions are 6 and -6.

6. If we take the square root of the square of a binomial, the result is the binomial.

11.1   Exercises

1 For the given quadratic equation, find the values of a, b, and c.

1. $3x^2 - 4x + 1 = 0$

2. $x^2 + 2x - 5 = 0$

3. $2x^2 - 5 = 0$

4. $4x^2 + 1 = 0$

5. $6x^2 - 3x = 0$

6. $-x^2 + 7x = 0$

Write the quadratic equation in standard form.

7. $x^2 - 8 = 3x$

8. $2x^2 = 4x - 1$

9. $x^2 = 16$

10. $x + 5 = x(x - 3)$

11. $2(x + 3)^2 = 5$

12. $4(x - 1)^2 = 3$

13. How does a quadratic equation differ from a linear equation?

14. What does the Principle of Zero Products state?

15. Explain why the equation $x(x + 5) = 12$ cannot be solved by saying $x = 12$ or $x + 5 = 12$.

16. What is a double root of a quadratic equation?

Solve for x.

17. $(x + 3)(x - 5) = 0$ **18.** $(x - 9)(x + 11) = 0$ **19.** $x(x - 7) = 0$

20. $x(x + 10) = 0$ **21.** $(2x + 5)(3x - 1) = 0$ **22.** $(2x - 7)(3x + 4) = 0$

Solve by factoring.

23. $x^2 + 2x - 15 = 0$ **24.** $t^2 + 3t - 10 = 0$ **25.** $z^2 - 4z + 3 = 0$

26. $s^2 - 5s + 4 = 0$ **27.** $p^2 + 3p + 2 = 0$ **28.** $v^2 + 6v + 5 = 0$

29. $x^2 - 6x + 9 = 0$ **30.** $y^2 - 8y + 16 = 0$ **31.** $6x^2 - 9x = 0$

32. $12y^2 + 8y = 0$ **33.** $r^2 - 10 = 3r$ **34.** $t^2 - 12 = 4t$

35. $3v^2 - 5v + 2 = 0$ **36.** $2p^2 - 3p - 2 = 0$ **37.** $3s^2 + 8s = 3$

38. $3x^2 + 5x = 12$ **39.** $6r^2 = 12 - r$ **40.** $4t^2 = 4t + 3$

41. $5y^2 + 11y = 12$ **42.** $4v^2 - 4v + 1 = 0$ **43.** $9s^2 - 6s + 1 = 0$

44. $x^2 - 9 = 0$ **45.** $t^2 - 16 = 0$ **46.** $4y^2 - 1 = 0$

47. $9z^2 - 4 = 0$ **48.** $x + 15 = x(x - 1)$ **49.** $\dfrac{3x^2}{2} = 4x - 2$

50. $\dfrac{2x^2}{5} = 3x - 5$ **51.** $\dfrac{2x^2}{9} + x = 2$ **52.** $\dfrac{3x^2}{8} - x = 2$

53. $\dfrac{3}{4}z^2 - z = -\dfrac{1}{3}$ **54.** $\dfrac{r^2}{2} = 1 - \dfrac{r}{12}$

55. $p + 18 = p(p - 2)$ **56.** $r^2 - r - 2 = (2r - 1)(r - 3)$

57. $s^2 + 5s - 4 = (2s + 1)(s - 4)$ **58.** $x^2 + x + 5 = (3x + 2)(x - 4)$

2 Solve by taking square roots.

59. $x^2 = 36$ **60.** $y^2 = 49$ **61.** $v^2 - 1 = 0$ **62.** $z^2 - 64 = 0$

63. $4x^2 - 49 = 0$ **64.** $9w^2 - 64 = 0$ **65.** $9y^2 = 4$ **66.** $4z^2 = 25$

67. $16v^2 - 9 = 0$ **68.** $25x^2 - 64 = 0$ **69.** $y^2 + 81 = 0$ **70.** $z^2 + 49 = 0$

71. $w^2 - 24 = 0$ **72.** $v^2 - 48 = 0$ **73.** $(x - 1)^2 = 36$ **74.** $(y + 2)^2 = 49$

75. $2(x + 5)^2 = 8$ **76.** $4(z - 3)^2 = 100$ **77.** $2(x + 1)^2 = 50$

78. $3(x - 4)^2 = 27$ **79.** $4(x + 5)^2 = 64$ **80.** $9(x - 3)^2 = 81$

81. $2(x - 9)^2 = 98$

82. $5(x + 5)^2 = 125$

83. $12(x + 3)^2 = 27$

84. $8(x - 4)^2 = 50$

85. $9(x - 1)^2 - 16 = 0$

86. $4(y + 3)^2 - 81 = 0$

87. $49(v + 1)^2 - 25 = 0$

88. $81(y - 2)^2 - 64 = 0$

89. $(x - 4)^2 - 20 = 0$

90. $(y + 5)^2 - 50 = 0$

91. $(x + 1)^2 + 36 = 0$

92. $(y - 7)^2 + 49 = 0$

93. $2\left(z - \dfrac{1}{2}\right)^2 = 12$

94. $3\left(v + \dfrac{3}{4}\right)^2 = 36$

95. $4\left(x - \dfrac{2}{3}\right)^2 = 16$

11.1   Applying Concepts

96. Evaluate $2n^2 - 7n - 4$ given $n(n - 2) = 15$.

97. Evaluate $3y^2 + 5y - 2$ given $y(y + 3) = 28$.

Solve for x.

98. $(x^2 - 1)^2 = 9$

99. $(x^2 + 3)^2 = 25$

100. $(6x^2 - 5)^2 = 1$

101. $x^2 = x$

102. $ax^2 - bx = 0, a > 0$ and $b > 0$

103. $ax^2 - b = 0, a > 0$ and $b > 0$

104. **Investments** The value P of an initial investment of A dollars after 2 years is given by $P = A(1 + r)^2$, where r is the annual percentage rate earned by the investment. If an initial investment of \$1500 grew to a value of \$1782.15 in 2 years, what was the annual percentage rate?

105. **Energy** The kinetic energy of a moving body is given by $E = \dfrac{1}{2}mv^2$, where E is the kinetic energy in newton-meters, m is the mass in kilograms, and v is the speed in meters per second. What is the speed of a moving body whose mass is 5 kg and whose kinetic energy is 250 newton-meters?

106. **Automobiles** On a certain type of street surface, the equation $d = 0.055v^2$ can be used to approximate the distance d, in feet, a car traveling v miles per hour will slide when its brakes are applied. After applying the brakes, the owner of a car involved in an accident skidded 40 ft. Did the traffic officer investigating the accident issue the car owner a ticket for speeding if the speed limit was 30 mph?

11.2 Solving Quadratic Equations by Completing the Square

1 **Solve quadratic equations by completing the square**

VIDEO & DVD · CD TUTOR · WEB · SSM

Recall that a perfect-square trinomial is the square of a binomial.

Perfect-square Trinomial		Square of a Binomial
$x^2 + 6x + 9$	$=$	$(x + 3)^2$
$x^2 - 10x + 25$	$=$	$(x - 5)^2$
$x^2 + 8x + 16$	$=$	$(x + 4)^2$

For each perfect-square trinomial, the square of $\frac{1}{2}$ of the coefficient of x equals the constant term.

$$x^2 + 6x + 9, \qquad \left(\frac{1}{2} \cdot 6\right)^2 = 9$$

$$x^2 - 10x + 25, \qquad \left[\frac{1}{2}(-10)\right]^2 = 25 \qquad \left(\frac{1}{2} \text{ coefficient of } x\right)^2 = \textbf{Constant term}$$

$$x^2 + 8x + 16, \qquad \left(\frac{1}{2} \cdot 8\right)^2 = 16$$

This relationship can be used to write the constant term for a perfect-square trinomial. Adding to a binomial the constant term that makes it a perfect-square trinomial is called **completing the square.**

Complete the square of $x^2 - 8x$. Write the resulting perfect-square trinomial as the square of a binomial.

$$x^2 - 8x$$

Find the constant term. $\left[\frac{1}{2}(-8)\right]^2 = 16$

Complete the square of $x^2 - 8x$ by adding the constant term. $x^2 - 8x + 16$

Write the resulting perfect-square trinomial as the square of a binomial. $x^2 - 8x + 16 = (x - 4)^2$

Complete the square of $y^2 + 5y$. Write the resulting perfect-square trinomial as the square of a binomial.

$$y^2 + 5y$$

Find the constant term. $\left(\frac{1}{2} \cdot 5\right)^2 = \left(\frac{5}{2}\right)^2 = \frac{25}{4}$

Complete the square of $y^2 + 5y$ by adding the constant term.

$$y + 5y + \frac{25}{4}$$

Write the resulting perfect-square trinomial as the square of a binomial.

$$y^2 + 5y + \frac{25}{4} = \left(y + \frac{5}{2}\right)^2$$

A quadratic equation of the form $x^2 + bx + c = 0$, $x \neq 0$, that cannot be solved by factoring can be solved by completing the square. The procedure is:

1. Write the equation in the form $x^2 + bx = c$.

2. Add to each side of the equation the term that completes the square on $x^2 + bx$.

3. Factor the perfect-square trinomial. Write it as the square of a binomial.

4. Take the square root of each side of the equation.

5. Solve for x.

Point of Interest

Early mathematicians solved quadratic equations by literally completing the square. For these mathematicians, all equations had geometric interpretations. They found that a quadratic equation could be solved by making certain figures into squares. See the Projects and Group Activities at the end of this chapter for an idea of how this was done.

Solve by completing the square: $x^2 - 6x - 3 = 0$

Add the opposite of the constant term to each side of the equation.

$$x^2 - 6x - 3 = 0$$
$$x^2 - 6x = 3$$

Find the constant term that completes the square of $x^2 - 6x$.

$$\left[\frac{1}{2}(-6)\right]^2 = 9$$

Add this term to each side of the equation.

$$x^2 - 6x + 9 = 3 + 9$$

Factor the perfect-square trinomial.

$$(x - 3)^2 = 12$$

Take the square root of each side of the equation.

$$\sqrt{(x - 3)^2} = \sqrt{12}$$

Simplify.

$$x - 3 = \pm\sqrt{12}$$
$$x - 3 = \pm 2\sqrt{3}$$

Solve for x.

$$x - 3 = 2\sqrt{3} \qquad x - 3 = -2\sqrt{3}$$
$$x = 3 + 2\sqrt{3} \qquad x = 3 - 2\sqrt{3}$$

Check

$$x^2 - 6x - 3 = 0$$

$$\begin{array}{c|c} \left(3 + 2\sqrt{3}\right)^2 - 6\left(3 + 2\sqrt{3}\right) - 3 & 0 \\ 9 + 12\sqrt{3} + 12 - 18 - 12\sqrt{3} - 3 & 0 \\ & 0 = 0 \end{array}$$

$$x^2 - 6x - 3 = 0$$

$$\begin{array}{c|c} \left(3 - 2\sqrt{3}\right)^2 - 6\left(3 - 2\sqrt{3}\right) - 3 & 0 \\ 9 - 12\sqrt{3} + 12 - 18 + 12\sqrt{3} - 3 & 0 \\ & 0 = 0 \end{array}$$

Write the solutions.

The solutions are $3 + 2\sqrt{3}$ and $3 - 2\sqrt{3}$.

For the method of completing the square to be used, the coefficient of the x^2 term must be 1. If it is not 1, we must multiply each side of the equation by the reciprocal of the coefficient of x^2. This is illustrated in the example below.

Solve by completing the square: $2x^2 - x - 1 = 0$

Add the opposite of the constant term to each side of the equation.

$$2x^2 - x - 1 = 0$$
$$2x^2 - x = 1$$

The coefficient of the x^2 term must be 1. Multiply each side by the reciprocal of the coefficient of x^2.

$$\frac{1}{2}(2x^2 - x) = \frac{1}{2} \cdot 1$$
$$x^2 - \frac{1}{2}x = \frac{1}{2}$$

Find the constant term that completes the square of $x^2 - \frac{1}{2}x$.

$$\left[\frac{1}{2}\left(-\frac{1}{2}\right)\right]^2 = \left(-\frac{1}{4}\right)^2 = \frac{1}{16}$$

Add this term to each side of the equation.

$$x^2 - \frac{1}{2}x + \frac{1}{16} = \frac{1}{2} + \frac{1}{16}$$

Factor the perfect-square trinomial.

$$\left(x - \frac{1}{4}\right)^2 = \frac{9}{16}$$

Take the square root of each side of the equation.

$$\sqrt{\left(x - \frac{1}{4}\right)^2} = \sqrt{\frac{9}{16}}$$

Simplify.

$$x - \frac{1}{4} = \pm\sqrt{\frac{9}{16}}$$
$$x - \frac{1}{4} = \pm\frac{3}{4}$$

Solve for x.

$$x - \frac{1}{4} = \frac{3}{4} \qquad x - \frac{1}{4} = -\frac{3}{4}$$
$$x = 1 \qquad x = -\frac{1}{2}$$

Check

$$2x^2 - x - 1 = 0$$

$$\begin{array}{c|c} 2(1)^2 - 1 - 1 & 0 \\ 2(1) - 1 - 1 & 0 \\ 2 - 1 - 1 & 0 \\ 0 = 0 \end{array}$$

$$2x^2 - x - 1 = 0$$

$$\begin{array}{c|c} 2\left(-\frac{1}{2}\right)^2 - \left(-\frac{1}{2}\right) - 1 & 0 \\ 2\left(\frac{1}{4}\right) - \left(-\frac{1}{2}\right) - 1 & 0 \\ \frac{1}{2} + \frac{1}{2} - 1 & 0 \\ 0 = 0 \end{array}$$

Write the solutions.

The solutions are 1 and $-\frac{1}{2}$.

Example 1   Solve by completing the square.

     A. $2x^2 - 4x - 1 = 0$ **B.** $p^2 - 4p + 6 = 0$

Solution   **A.** $2x^2 - 4x - 1 = 0$

$$2x^2 - 4x = 1$$

    • Add the opposite of -1 to each side of the equation.

$$\frac{1}{2}(2x^2 - 4x) = \frac{1}{2} \cdot 1$$

    • The coefficient of the x^2 term must be 1.

$$x^2 - 2x = \frac{1}{2}$$

$$x^2 - 2x + 1 = \frac{1}{2} + 1$$

    • Complete the square of $x^2 - 2x$. Add 1 to each side of the equation.

$$(x - 1)^2 = \frac{3}{2}$$

    • Factor the perfect-square trinomial.

$$\sqrt{(x - 1)^2} = \sqrt{\frac{3}{2}}$$

    • Take the square root of each side of the equation.

$$x - 1 = \pm\sqrt{\frac{3}{2}}$$

$$x - 1 = \pm\frac{\sqrt{3}}{\sqrt{2}}$$

$$x - 1 = \pm\frac{\sqrt{6}}{2}$$

    • Rationalize the denominator.

$$\frac{\sqrt{3}}{\sqrt{2}} = \frac{\sqrt{3}}{\sqrt{2}} \cdot \frac{\sqrt{2}}{\sqrt{2}} = \frac{\sqrt{6}}{2}$$

$$x - 1 = \frac{\sqrt{6}}{2} \qquad\qquad x - 1 = -\frac{\sqrt{6}}{2}$$

$$x = 1 + \frac{\sqrt{6}}{2} \qquad\qquad x = 1 - \frac{\sqrt{6}}{2}$$

$$= \frac{2}{2} + \frac{\sqrt{6}}{2} \qquad\qquad = \frac{2}{2} - \frac{\sqrt{6}}{2}$$

$$= \frac{2 + \sqrt{6}}{2} \qquad\qquad = \frac{2 - \sqrt{6}}{2}$$

The solutions are $\dfrac{2 + \sqrt{6}}{2}$ and $\dfrac{2 - \sqrt{6}}{2}$.

    B. $p^2 - 4p + 6 = 0$

$$p^2 - 4p = -6$$

    • Add the opposite of the constant term to each side of the equation.

$$p^2 - 4p + 4 = -6 + 4$$

    • Complete the square of $p^2 - 4p$. Add 4 to each side of the equation.

$$(p - 2)^2 = -2$$

    • Factor the perfect-square trinomial.

$$\sqrt{(p - 2)^2} = \sqrt{-2}$$

    • Take the square root of each side of the equation.

$\sqrt{-2}$ is not a real number.

The equation has no real number solution.

Problem 1   Solve by completing the square.

    A. $3x^2 - 6x - 2 = 0$ **B.** $y^2 - 6y + 10 = 0$

Solution   See page S25.

Example 2   Solve by completing the square: $x^2 + 6x + 4 = 0$
Approximate the solutions to the nearest thousandth.

Solution

$$x^2 + 6x + 4 = 0$$
$$x^2 + 6x = -4$$
$$x^2 + 6x + 9 = -4 + 9 \qquad \bullet \text{ Complete the square of } x^2 + 6x. \text{ Add 9}$$
$$(x + 3)^2 = 5 \qquad\qquad\qquad \text{to each side of the equation.}$$
$$\sqrt{(x + 3)^2} = \sqrt{5}$$
$$x + 3 = \pm\sqrt{5}$$
$$x + 3 \approx \pm 2.2361$$

$$x + 3 \approx 2.2361 \qquad\qquad x + 3 \approx -2.2361$$
$$x \approx -3 + 2.2361 \qquad\qquad x \approx -3 - 2.2361$$
$$\approx -0.764 \qquad\qquad\qquad \approx -5.236$$

The solutions are approximately -0.764 and -5.236.

Problem 2   Solve by completing the square: $x^2 + 8x + 8 = 0$
Approximate the solutions to the nearest thousandth.

Solution   See page S25.

11.2   Concept Review

Determine whether the statement is always true, sometimes true, or never true.

1. When we square a binomial, the result is a perfect-square trinomial.

2. The constant term in a perfect-square trinomial is equal to the square of one-half the coefficient of x.

3. To "complete the square" means to add to $x^2 + bx$ the constant term that will make it a perfect-square trinomial.

4. To complete the square of $x^2 + 4x$, add 16 to it.

5. To complete the square of $3x^2 + 6x$, add 9 to it.

6. In solving a quadratic equation by completing the square, the next step after writing the equation in the form $(x + a)^2 = b$ is to take the square root of each side of the equation.

11.2   Exercises

1 Complete the square. Write the resulting perfect-square trinomial as the square of a binomial.

1. $x^2 + 12x$

2. $x^2 - 4x$

3. $x^2 + 10x$

4. $x^2 + 3x$ **5.** $x^2 - x$ **6.** $x^2 + 5x$

Solve by completing the square.

7. $x^2 + 2x - 3 = 0$ **8.** $y^2 + 4y - 5 = 0$ **9.** $v^2 + 4v + 1 = 0$ **10.** $y^2 - 2y - 5 = 0$

11. $v^2 - 6v + 13 = 0$ **12.** $x^2 + 4x + 13 = 0$ **13.** $x^2 + 6x = 5$ **14.** $w^2 - 8w = 3$

15. $x^2 = 4x - 4$ **16.** $z^2 = 8z - 16$ **17.** $z^2 = 2z + 1$ **18.** $y^2 = 10y - 20$

Solve. First try to solve the equation by factoring. If you are unable to solve the equation by factoring, solve the equation by completing the square.

19. $p^2 + 3p = 1$ **20.** $r^2 + 5r = 2$ **21.** $w^2 + 7w = 8$

22. $y^2 + 5y = -4$ **23.** $x^2 + 6x + 4 = 0$ **24.** $y^2 - 8y - 1 = 0$

25. $r^2 - 8r = -2$ **26.** $s^2 + 6s = 5$ **27.** $t^2 - 3t = -2$

28. $y^2 = 4y + 12$ **29.** $w^2 = 3w + 5$ **30.** $x^2 = 1 - 3x$

31. $x^2 - x - 1 = 0$ **32.** $x^2 - 7x = -3$ **33.** $y^2 - 5y + 3 = 0$

34. $z^2 - 5z = -2$ **35.** $v^2 + v - 3 = 0$ **36.** $x^2 - x = 1$

37. $y^2 = 7 - 10y$ **38.** $v^2 = 14 + 16v$ **39.** $s^2 + 3s = -1$

40. $r^2 - 3r = 5$ **41.** $t^2 - t = 4$ **42.** $y^2 + y - 4 = 0$

43. $x^2 - 3x + 5 = 0$ **44.** $z^2 + 5z + 7 = 0$ **45.** $2t^2 - 3t + 1 = 0$

46. $2x^2 - 7x + 3 = 0$ **47.** $2r^2 + 5r = 3$ **48.** $2y^2 - 3y = 9$

49. $4v^2 - 4v - 1 = 0$ **50.** $2s^2 - 4s - 1 = 0$ **51.** $4z^2 - 8z = 1$

52. $3r^2 - 2r = 2$ **53.** $3y - 5 = (y - 1)(y - 2)$ **54.** $4p + 2 = (p - 1)(p + 3)$

55. $\dfrac{x^2}{4} - \dfrac{x}{2} = 3$

56. $\dfrac{x^2}{6} - \dfrac{x}{3} = 1$

57. $\dfrac{2x^2}{3} = 2x + 3$

58. $\dfrac{3x^2}{2} = 3x + 2$

59. $\dfrac{x}{3} + \dfrac{3}{x} = \dfrac{8}{3}$

60. $\dfrac{x}{4} - \dfrac{2}{x} = \dfrac{3}{4}$

Solve by completing the square. Approximate the solutions to the nearest thousandth.

61. $y^2 + 3y = 5$

62. $w^2 + 5w = 2$

63. $2z^2 - 3z = 7$

64. $2x^2 + 3x = 11$

65. $4x^2 + 6x - 1 = 0$

66. $4x^2 + 2x - 3 = 0$

11.2   Applying Concepts

67. Evaluate $2b^2$ given $b^2 - 6b + 7 = 0$.

68. Evaluate $2y^2$ given $y^2 - 2y - 7 = 0$.

Solve.

69. $\sqrt{2x + 7} - 4 = x$

70. $\dfrac{x + 1}{2} + \dfrac{3}{x - 1} = 4$

71. $\dfrac{x - 2}{3} + \dfrac{2}{x + 2} = 4$

72. Explain why the equation $(x - 2)^2 = -4$ does not have a real number solution.

11.3   Solving Quadratic Equations by Using the Quadratic Formula

1   Solve quadratic equations by using the quadratic formula

Any quadratic equation can be solved by completing the square. Applying this method to the standard form of a quadratic equation produces a formula that can be used to solve any quadratic equation.

To solve $ax^2 + bx + c = 0$, $a \neq 0$, by completing the square, subtract the constant term from each side of the equation.

$$ax^2 + bx + c = 0$$
$$ax^2 + bx + c - c = 0 - c$$
$$ax^2 + bx = -c$$

Multiply each side of the equation by the reciprocal of a, the coefficient of x^2.

$$\frac{1}{a}(ax^2 + bx) = \frac{1}{a}(-c)$$
$$x^2 + \frac{b}{a}x = -\frac{c}{a}$$

Complete the square by adding $\left(\frac{1}{2} \cdot \frac{b}{a}\right)^2$ to each side of the equation.

$$x^2 + \frac{b}{a}x + \left(\frac{1}{2} \cdot \frac{b}{a}\right)^2 = \left(\frac{1}{2} \cdot \frac{b}{a}\right)^2 - \frac{c}{a}$$
$$x^2 + \frac{b}{a}x + \frac{b^2}{4a^2} = \frac{b^2}{4a^2} - \frac{c}{a}$$

Simplify the right side of the equation.

$$x^2 + \frac{b}{a}x + \frac{b^2}{4a^2} = \frac{b^2}{4a^2} - \left(\frac{c}{a} \cdot \frac{4a}{4a}\right)$$
$$x^2 + \frac{b}{a}x + \frac{b^2}{4a^2} = \frac{b^2}{4a^2} - \frac{4ac}{4a^2}$$
$$x^2 + \frac{b}{a}x + \frac{b^2}{4a^2} = \frac{b^2 - 4ac}{4a^2}$$

Factor the perfect-square trinomial on the left side of the equation.

$$\left(x + \frac{b}{2a}\right)^2 = \frac{b^2 - 4ac}{4a^2}$$

Take the square root of each side of the equation.

$$\sqrt{\left(x + \frac{b}{2a}\right)^2} = \sqrt{\frac{b^2 - 4ac}{4a^2}}$$
$$x + \frac{b}{2a} = \pm\frac{\sqrt{b^2 - 4ac}}{2a}$$

Solve for x.

$$x + \frac{b}{2a} = \frac{\sqrt{b^2 - 4ac}}{2a}$$
$$x = -\frac{b}{2a} + \frac{\sqrt{b^2 - 4ac}}{2a}$$
$$x = \frac{-b + \sqrt{b^2 - 4ac}}{2a}$$

$$x + \frac{b}{2a} = -\frac{\sqrt{b^2 - 4ac}}{2a}$$
$$x = -\frac{b}{2a} - \frac{\sqrt{b^2 - 4ac}}{2a}$$
$$x = \frac{-b - \sqrt{b^2 - 4ac}}{2a}$$

The Quadratic Formula

The solution of $ax^2 + bx + c = 0$, $a \neq 0$, is

$$x = \frac{-b + \sqrt{b^2 - 4ac}}{2a} \quad \text{or} \quad x = \frac{-b - \sqrt{b^2 - 4ac}}{2a}$$

The quadratic formula is frequently written in the form

$$x = \frac{-b \pm \sqrt{b^2 - 4ac}}{2a}$$

Solve by using the quadratic formula: $2x^2 = 4x - 1$

Write the equation in standard form. $a = 2$, $b = -4$, and $c = 1$.

$$2x^2 = 4x - 1$$
$$2x^2 - 4x + 1 = 0$$

Replace a, b, and c in the quadratic formula by their values.

Simplify.

$$x = \frac{-b \pm \sqrt{b^2 - 4ac}}{2a}$$

$$= \frac{-(-4) \pm \sqrt{(-4)^2 - 4 \cdot 2 \cdot 1}}{2 \cdot 2}$$

$$= \frac{4 \pm \sqrt{16 - 8}}{4}$$

$$= \frac{4 \pm \sqrt{8}}{4}$$

$$= \frac{4 \pm 2\sqrt{2}}{4} = \frac{2(2 \pm \sqrt{2})}{2 \cdot 2} = \frac{2 \pm \sqrt{2}}{2}$$

Check

$2x^2 = 4x - 1$	
$2\left(\dfrac{2 + \sqrt{2}}{2}\right)^2$	$4\left(\dfrac{2 + \sqrt{2}}{2}\right) - 1$
$2\left(\dfrac{4 + 4\sqrt{2} + 2}{4}\right)$	$2(2 + \sqrt{2}) - 1$
$2\left(\dfrac{6 + 4\sqrt{2}}{4}\right)$	$4 + 2\sqrt{2} - 1$
$\dfrac{6 + 4\sqrt{2}}{2}$	$3 + 2\sqrt{2}$
$3 + 2\sqrt{2} = 3 + 2\sqrt{2}$	

$2x^2 = 4x - 1$	
$2\left(\dfrac{2 - \sqrt{2}}{2}\right)^2$	$4\left(\dfrac{2 - \sqrt{2}}{2}\right) - 1$
$2\left(\dfrac{4 - 4\sqrt{2} + 2}{4}\right)$	$2(2 - \sqrt{2}) - 1$
$2\left(\dfrac{6 - 4\sqrt{2}}{4}\right)$	$4 - 2\sqrt{2} - 1$
$\dfrac{6 - 4\sqrt{2}}{2}$	$3 - 2\sqrt{2}$
$3 - 2\sqrt{2} = 3 - 2\sqrt{2}$	

Write the solutions.

The solutions are $\dfrac{2 + \sqrt{2}}{2}$ and $\dfrac{2 - \sqrt{2}}{2}$.

Example 1 Solve by using the quadratic formula.

A. $2x^2 - 3x + 1 = 0$ **B.** $2x^2 = 8x - 5$
C. $t^2 - 3t + 7 = 0$

Solution   **A.** $2x^2 - 3x + 1 = 0$ • $a = 2, b = -3, c = 1$

$$x = \frac{-(-3) \pm \sqrt{(-3)^2 - 4(2)(1)}}{2 \cdot 2}$$

$$= \frac{3 \pm \sqrt{9 - 8}}{4} = \frac{3 \pm \sqrt{1}}{4} = \frac{3 \pm 1}{4}$$

$$x = \frac{3 + 1}{4} \qquad x = \frac{3 - 1}{4}$$

$$= \frac{4}{4} = 1 \qquad = \frac{2}{4} = \frac{1}{2}$$

The solutions are 1 and $\dfrac{1}{2}$.

B.
$$2x^2 = 8x - 5$$
$$2x^2 - 8x + 5 = 0$$

• Write the equation in standard form. $a = 2$, $b = -8$, $c = 5$

$$x = \frac{-(-8) \pm \sqrt{(-8)^2 - 4(2)(5)}}{2 \cdot 2}$$

$$= \frac{8 \pm \sqrt{64 - 40}}{4}$$

$$= \frac{8 \pm \sqrt{24}}{4}$$

$$= \frac{8 \pm 2\sqrt{6}}{4}$$

$$= \frac{2(4 \pm \sqrt{6})}{2 \cdot 2} = \frac{4 \pm \sqrt{6}}{2}$$

The solutions are $\dfrac{4 + \sqrt{6}}{2}$ and $\dfrac{4 - \sqrt{6}}{2}$.

C. $t^2 - 3t + 7 = 0$

• $a = 1$, $b = -3$, $c = 7$

$$t = \frac{-(-3) \pm \sqrt{(-3)^2 - 4(1)(7)}}{2(1)}$$

$$= \frac{3 \pm \sqrt{9 - 28}}{2} = \frac{3 \pm \sqrt{-19}}{2}$$

$\sqrt{-19}$ is not a real number.

The equation has no real number solution.

Problem 1   Solve by using the quadratic formula.

A. $3x^2 + 4x - 4 = 0$ **B.** $x^2 + 2x = 1$

C. $z^2 + 2z + 6 = 0$

Solution   See pages S25–S26.

11.3   **Concept Review**

Determine whether the statement is always true, sometimes true, or never true.

1. Any quadratic equation can be solved by using the quadratic formula.

2. The solutions of a quadratic equation can be written as

$$x = \frac{-b + \sqrt{b^2 - 4ac}}{2a} \text{ and } x = \frac{-b - \sqrt{b^2 - 4ac}}{2a}.$$

3. The equation $4x^2 - 3x = 9$ is a quadratic equation in standard form.

4. In the quadratic formula $x = \dfrac{-b \pm \sqrt{b^2 - 4ac}}{2a}$, b is the coefficient of x^2.

5. $\dfrac{6 \pm 2\sqrt{3}}{2}$ simplifies to $3 \pm 2\sqrt{3}$.

11.3   Exercises

1 **1.** ✎ Write the quadratic formula. Explain what each variable in the formula represents.

2. ✎ Explain what the quadratic formula is used for.

Solve by using the quadratic formula.

3. $z^2 + 6z - 7 = 0$ **4.** $s^2 + 3s - 10 = 0$ **5.** $w^2 = 3w + 18$ **6.** $r^2 = 5 - 4r$

7. $t^2 - 2t = 5$ **8.** $y^2 - 4y = 6$ **9.** $t^2 + 6t - 1 = 0$ **10.** $z^2 + 4z + 1 = 0$

11. $w^2 + 3w + 5 = 0$ **12.** $x^2 - 2x + 6 = 0$ **13.** $w^2 = 4w + 9$ **14.** $y^2 = 8y + 3$

Solve. First try to solve the equation by factoring. If you are unable to solve the equation by factoring, solve the equation by using the quadratic formula.

15. $p^2 - p = 0$ **16.** $2v^2 + v = 0$ **17.** $4t^2 - 4t - 1 = 0$

18. $4x^2 - 8x - 1 = 0$ **19.** $4t^2 - 9 = 0$ **20.** $4s^2 - 25 = 0$

21. $3x^2 - 6x + 2 = 0$ **22.** $5x^2 - 6x = 3$ **23.** $3t^2 = 2t + 3$

24. $4n^2 = 7n - 2$ **25.** $2x^2 + x + 1 = 0$ **26.** $3r^2 - r + 2 = 0$

27. $2y^2 + 3 = 8y$ **28.** $5x^2 - 1 = x$ **29.** $3t^2 = 7t + 6$

30. $3x^2 = 10x + 8$ **31.** $3y^2 - 4 = 5y$ **32.** $6x^2 - 5 = 3x$

33. $3x^2 = x + 3$

34. $2n^2 = 7 - 3n$

35. $5d^2 - 2d - 8 = 0$

36. $x^2 - 7x - 10 = 0$

37. $5z^2 + 11z = 12$

38. $4v^2 = v + 3$

39. $v^2 + 6v + 1 = 0$

40. $s^2 + 4s - 8 = 0$

41. $4t^2 - 12t - 15 = 0$

42. $4w^2 - 20w + 5 = 0$

43. $2x^2 = 4x - 5$

44. $3r^2 = 5r - 6$

45. $9y^2 + 6y - 1 = 0$

46. $9s^2 - 6s - 2 = 0$

47. $6s^2 - s - 2 = 0$

48. $6y^2 + 5y - 4 = 0$

49. $4p^2 + 16p = -11$

50. $4y^2 - 12y = -1$

51. $4x^2 = 4x + 11$

52. $4s^2 + 12s = 3$

53. $4p^2 = -12p - 9$

54. $3y^2 + 6y = -3$

55. $9v^2 = -30v - 23$

56. $9t^2 = 30t + 17$

57. $\dfrac{x^2}{2} - \dfrac{x}{3} = 1$

58. $\dfrac{x^2}{4} - \dfrac{x}{2} = 5$

59. $\dfrac{2x^2}{5} = x + 1$

60. $\dfrac{3x^2}{2} + 2x = 1$

61. $\dfrac{x}{5} + \dfrac{5}{x} = \dfrac{12}{5}$

62. $\dfrac{x}{4} + \dfrac{3}{x} = \dfrac{5}{2}$

Solve by using the quadratic formula. Approximate the solutions to the nearest thousandth.

63. $x^2 - 2x - 21 = 0$

64. $y^2 + 4y - 11 = 0$

65. $s^2 - 6s - 13 = 0$

66. $w^2 + 8w - 15 = 0$

67. $2p^2 - 7p - 10 = 0$

68. $3t^2 - 8t - 1 = 0$

69. $4z^2 + 8z - 1 = 0$

70. $4x^2 + 7x + 1 = 0$

71. $5v^2 - v - 5 = 0$

11.3 **Applying Concepts**

Solve.

72. $\sqrt{x^2 + 2x + 1} = x - 1$

73. $\dfrac{x + 2}{3} - \dfrac{4}{x - 2} = 2$

74. $\dfrac{x + 1}{5} - \dfrac{3}{x - 1} = 2$

75. True or false?
    a. The equations $x = \sqrt{12 - x}$ and $x^2 = 12 - x$ have the same solution.
    b. If $\sqrt{a} + \sqrt{b} = c$, then $a + b = c^2$.
    c. $\sqrt{9} = \pm 3$
    d. $\sqrt{x^2} = |x|$

76. Factoring, completing the square, and using the quadratic formula are three methods of solving quadratic equations. Describe each method, and cite the advantages and disadvantages of using each.

77. Explain why the equation $0x^2 + 3x + 4 = 0$ cannot be solved by the quadratic formula.

11.4   Graphing Quadratic Equations in Two Variables

1   **Graph a quadratic equation of the form $y = ax^2 + bx + c$**

An equation of the form $y = ax^2 + bx + c$, $a \neq 0$, is a **quadratic equation in two variables.** Examples of quadratic equations in two variables are shown at the right.

$$y = 3x^2 - x + 1$$
$$y = -x^2 - 3$$
$$y = 2x^2 - 5x$$

For these equations, y is a function of x, and we can write $f(x) = ax^2 + bx + c$. This equation represents a **quadratic function.**

Evaluate $f(x) = 2x^2 - 3x + 4$ when $x = -2$.

$$f(x) = 2x^2 - 3x + 4$$
$$f(-2) = 2(-2)^2 - 3(-2) + 4 \qquad \bullet \text{ Replace } x \text{ with } -2.$$
$$= 2(4) - 3(-2) + 4 \qquad \bullet \text{ Simplify.}$$
$$= 8 + 6 + 4$$
$$= 18$$

The value of the function when $x = -2$ is 18.

The graph of $y = ax^2 + bx + c$ or $f(x) = ax^2 + bx + c$ is a **parabola.** The graph is ∪-shaped and opens up when a is positive and down when a is negative. The graphs of two parabolas are shown below.

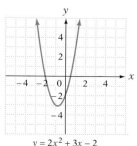

$y = 2x^2 + 3x - 2$
$a = 2$, a positive number
Parabola opens up

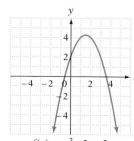

$f(x) = -x^2 + 3x + 2$
$a = -1$, a negative number
Parabola opens down

Graph $y = x^2 - 2x - 3$.

x	y
-2	5
-1	0
0	-3
1	-4
2	-3
3	0
4	5

• Find several solutions of the equation. Because the graph is not a straight line, several solutions must be found in order to determine the ∪-shape. Record the ordered pairs in a table.

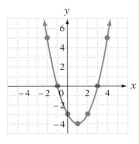

• Graph the ordered-pair solutions on a rectangular coordinate system. Draw a parabola through the points.

Note that the graph of $y = x^2 - 2x - 3$ crosses the x-axis at $(-1, 0)$ and at $(3, 0)$. These points are confirmed in the table of values. The x-intercepts of the graph are $(-1, 0)$ and $(3, 0)$.

The y-intercept is the point at which the graph crosses the y-axis. At this point, $x = 0$. From the graph or the table, we can see that the y-intercept is $(0, -3)$.

We can find the x-intercepts algebraically by letting $y = 0$ and solving for x.

$$y = x^2 - 2x - 3$$
$$0 = x^2 - 2x - 3$$
$$0 = (x + 1)(x - 3)$$

• Replace y with 0 and solve for x.
• This equation can be solved by factoring. However, it will be necessary to use the quadratic formula to solve some quadratic equations.

$$x + 1 = 0 \qquad x - 3 = 0$$
$$x = -1 \qquad x = 3$$

When $y = 0$, $x = -1$ or $x = 3$. The x-intercepts are $(-1, 0)$ and $(3, 0)$.

We can find the y-intercept algebraically by letting $x = 0$ and solving for y.

$$y = x^2 - 2x - 3$$
$$y = 0^2 - 2(0) - 3$$
$$y = 0 - 0 - 3$$
$$y = -3$$

• Replace x with 0 and simplify.

When $x = 0$, $y = -3$. The y-intercept is $(0, -3)$.

```
Plot1  Plot2  Plot3
\Y1 ■ X²–2X–3
\Y2 =
\Y3 =
\Y4 =
\Y5 =
\Y6 =
\Y7 =
```

Using a graphing utility, enter the equation $y = x^2 - 2x - 3$ and verify the graph shown above. (Refer to the Appendix for instructions on using a graphing calculator to graph a quadratic equation.) Verify that $(-1, 0)$, $(3, 0)$, and $(0, -3)$ are coordinates of points on the graph.

The graph of $y = -2x^2 + 1$ is shown below.

x	y
0	1
1	-1
-1	-1
2	-7
-2	-7

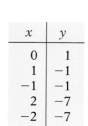

Plot1 Plot2 Plot3
\Y1 ■-2X²+1
\Y2 =
\Y3 =
\Y4 =
\Y5 =
\Y6 =
\Y7 =

> Use a graphing utility to verify the graph of $y = -2x^2 + 1$ shown above.

Note that the graph of $y = x^2 - 2x - 3$, shown on the previous page, opens up and that the coefficient of x^2 is positive. The graph of $y = -2x^2 + 1$ opens down, and the coefficient of x^2 is negative. As stated earlier, for any quadratic equation in two variables, the coefficient of x^2 determines whether the parabola opens up or down. **When a is positive, the parabola opens up. When a is negative, the parabola opens down.**

Point
of **Interest**

Mirrors in some telescopes are ground into the shape of a parabola. The mirror at the Mount Palomar Observatory is 2 ft thick at the ends and weighs 14.75 tons. The mirror has been ground to a true paraboloid (the three-dimensional version of a parabola) to within 0.0000015 in.

Every parabola has an **axis of symmetry** and a **vertex** that is on the axis of symmetry. If the parabola opens up, the vertex is the lowest point on the graph. If the parabola opens down, the vertex is the highest point on the graph.

To understand the axis of symmetry, think of folding the paper along that axis. The two halves of the graph will match up.

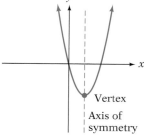

When graphing a quadratic equation in two variables, use the value of a to determine whether the parabola opens up or down. After graphing ordered-pair solutions of the equation, use symmetry to help you draw the parabola.

Example 1 Graph.   **A.** $y = x^2 - 2x$ **B.** $y = -x^2 + 4x - 4$

Solution   **A.** $a = 1$. a is positive.
The parabola opens up.

x	y
0	0
1	-1
-1	3
2	0
3	3

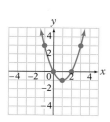

B. $a = -1$. a is negative.
The parabola opens down.

x	y
0	-4
1	-1
2	0
3	-1
4	-4

Problem 1 Graph. **A.** $y = x^2 + 2$ **B.** $y = -x^2 - 2x + 3$

Solution See page S26.

Example 2 Find the x- and y-intercepts of the graph of $y = x^2 - 2x - 5$.

Solution To find the x-intercepts, let $y = 0$ and solve for x.

$$y = x^2 - 2x - 5$$
$$0 = x^2 - 2x - 5$$

$$x = \frac{-b \pm \sqrt{b^2 - 4ac}}{2a}$$ • $x^2 - 2x - 5$ is nonfactorable over the integers. Use the quadratic formula.

$$= \frac{-(-2) \pm \sqrt{(-2)^2 - 4(1)(-5)}}{2(1)}$$

$$= \frac{2 \pm \sqrt{24}}{2} = \frac{2 \pm 2\sqrt{6}}{2} = 1 \pm \sqrt{6}$$

The x-intercepts are $\left(1 + \sqrt{6}, 0\right)$ and $\left(1 - \sqrt{6}, 0\right)$.

To find the y-intercept, let $x = 0$ and solve for y.

$$y = x^2 - 2x - 5$$
$$y = 0^2 - 2(0) - 5 = -5$$

The y-intercept is $(0, -5)$.

Problem 2 Find the x- and y-intercepts of the graph of $y = x^2 - 6x + 9$.

Solution See page S26.

11.4 Concept Review

Determine whether the statement is always true, sometimes true, or never true.

1. The equation $y = 3x^2 - 2x + 1$ is an example of a quadratic equation in two variables.

2. The graph of an equation of the form $y = ax^2 + bx + c$, $a \neq 0$, is a parabola.

3. The graph of an equation of the form $y = ax^2 + c$, $a \neq 0$, is a straight line.

4. For the equation $y = -x^2 + 2x - 1$, when $x = 1$, $y = 2$.

5. The graph of the equation $y = -\dfrac{1}{2}x^2 + 5x$ is a parabola opening up.

6. If a parabola opens down, then the vertex is the highest point on the graph.

11.4 | Exercises

1. What is a parabola?

2. Explain how you know whether the graph of the equation $y = -2x^2 + 3x + 4$ opens up or down.

3. Explain the axis of symmetry of a parabola.

4. What is the vertex of a parabola?

Determine whether the graph of the equation opens up or down.

5. $y = x^2 - 3$

6. $y = -x^2 + 4$

7. $y = \dfrac{1}{2}x^2 - 2$

8. $y = -\dfrac{1}{3}x^2 + 5$

9. $y = x^2 - 2x + 3$

10. $y = -x^2 + 4x - 1$

Evaluate the function for the given value of x.

11. $f(x) = x^2 - 2x + 1$; $x = 3$

12. $f(x) = 2x^2 + x - 1$; $x = -2$

13. $f(x) = 4 - x^2$; $x = -3$

14. $f(x) = x^2 + 6x + 9$; $x = -3$

15. $f(x) = -x^2 + 5x - 6$; $x = -4$

16. $f(x) = -2x^2 + 2x - 1$; $x = -1$

Graph.

17. $y = x^2$

18. $y = -x^2$

19. $y = -x^2 + 1$

20. $y = x^2 - 1$

21. $y = 2x^2$

22. $y = \frac{1}{2}x^2$

23. $y = -\frac{1}{2}x^2 + 1$

24. $y = 2x^2 - 1$

25. $y = x^2 - 4x$

26. $y = x^2 + 4x$

27. $y = x^2 - 2x + 3$

28. $y = x^2 - 4x + 2$

29. $y = -x^2 + 2x + 3$

30. $y = x^2 + 2x + 1$

31. $y = -x^2 + 3x - 4$

32. $y = -x^2 + 6x - 9$

33. $y = 2x^2 + x - 3$

34. $y = -2x^2 - 3x + 3$

Determine the x- and y-intercepts of the graph of the equation.

35. $y = x^2 - 5x + 6$

36. $y = x^2 + 5x - 6$

37. $f(x) = 9 - x^2$

38. $f(x) = x^2 + 12x + 36$

39. $y = x^2 + 2x - 6$

40. $y = x^2 + 4x - 2$

41. $f(x) = 2x^2 - x - 3$

42. $f(x) = 2x^2 - 13x + 15$

43. $y = 4 - x - x^2$

44. $y = 2 - 3x - 3x^2$

Graph using a graphing utility. Verify that the graph is a graph of a parabola opening up if a is positive or opening down if a is negative.

45. $y = x^2 - 2$

46. $y = -x^2 + 3$

47. $y = x^2 + 2x$

48. $y = -2x^2 + 4x$

49. $y = \frac{1}{2}x^2 - x$

50. $y = -\frac{1}{2}x^2 + 2$

51. $y = x^2 - x - 2$

52. $y = x^2 - 3x + 2$

53. $y = 2x^2 - x - 5$

54. $y = -2x^2 - 3x + 2$

55. $y = -x^2 - 2x - 1$

56. $y = \frac{1}{2}x^2 - 2x - 1$

11.4   Applying Concepts

Show that the equation is a quadratic equation in two variables by writing it in the form $y = ax^2 + bx + c$.

57. $y + 1 = (x - 4)^2$

58. $y - 2 = 3(x + 1)^2$

59. $y - 4 = 2(x - 3)^2$

60. $y + 3 = 3(x - 1)^2$

Find the x-intercepts of the graph of the equation.

61. $y = x^3 - x^2 - 6x$

62. $y = x^3 - 4x^2 - 5x$

63. $y = x^3 + x^2 - 4x - 4$ (*Hint:* Factor by grouping.)

64. $y = x^3 + 3x^2 - x - 3$ (*Hint:* Factor by grouping.)

State whether the graph is the graph of a linear function, a quadratic function, or neither.

65.

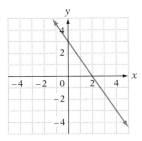

66.

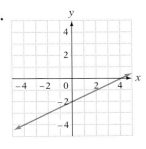

67.

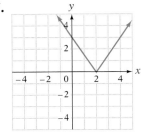

68.

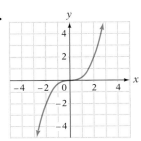

69.

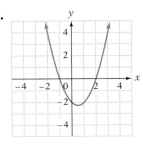

70.

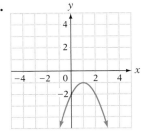

71. **The Postal Service** The graph below shows the cost of a first-class postage stamp from 1950 to 2002. A quadratic function that approximately models the data is $y = 0.0087x^2 - 0.6125x + 9.9096$, where $x \geq 50$ and $x = 50$ for the year 1950, and y is the cost in cents of a first-class stamp. What does the model equation predict will be the cost of a first-class stamp in 2020? Round to the nearest cent.

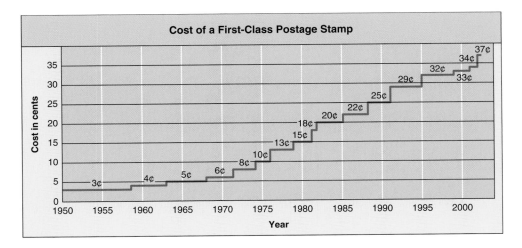

72. The point whose coordinates are (x_1, y_1) lies in quadrant I and is a point on the graph of the equation $y = 2x^2 - 2x + 1$. Given $y_1 = 13$, find x_1.

73. The point whose coordinates are (x_1, y_1) lies in quadrant II and is a point on the graph of the equation $y = 2x^2 - 3x - 2$. Given $y_1 = 12$, find x_1.

11.5   Application Problems

1   Application problems

The application problems in this section are varieties of those problems solved earlier in the text. Each of the strategies for the problems in this section results in a quadratic equation.

Solve: It took a motorboat a total of 7 h to travel 48 mi down a river and then 48 mi back again. The rate of the current was 2 mph. Find the rate of the motorboat in calm water.

> **STRATEGY**   *for solving an application problem*

> ▶ Determine the type of problem. For example, is it a uniform motion problem, a geometry problem, or a work problem?

The problem is a uniform motion problem.

> ▶ Choose a variable to represent the unknown quantity. Write numerical or variable expressions for all the remaining quantities. These results can be recorded in a table.

The unknown rate of the motorboat: r

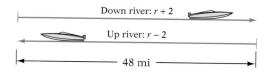

Down river: $r + 2$
Up river: $r - 2$
48 mi

	Distance	÷	Rate	=	Time
Down river	48	÷	$r + 2$	=	$\dfrac{48}{r + 2}$
Up river	48	÷	$r - 2$	=	$\dfrac{48}{r - 2}$

> ▶ Determine how the quantities are related. If necessary, review the strategies presented in the chapter "Solving Equations and Inequalities: Applications."

The total time of the trip was 7 h.

$$\frac{48}{r+2} + \frac{48}{r-2} = 7$$

$$(r+2)(r-2)\left(\frac{48}{r+2} + \frac{48}{r-2}\right) = (r+2)(r-2)7$$

$$(r-2)48 + (r+2)48 = (r^2-4)7$$

$$48r - 96 + 48r + 96 = 7r^2 - 28$$

$$96r = 7r^2 - 28$$

$$0 = 7r^2 - 96r - 28$$

$$0 = (7r+2)(r-14)$$

$$7r + 2 = 0 \qquad r - 14 = 0$$
$$7r = -2 \qquad r = 14$$
$$r = -\frac{2}{7}$$

The solution $-\frac{2}{7}$ is not possible, because the rate cannot be negative.

The rate of the motorboat in calm water is 14 mph.

Example 1 Working together, a painter and the painter's apprentice can paint a room in 4 h. Working alone, the apprentice requires 6 more hours to paint the room than the painter requires working alone. How long does it take the painter, working alone, to paint the room?

Strategy ▶ This is a work problem.

▶ Time for the painter to paint the room: t
Time for the apprentice to paint the room: $t + 6$

	Rate	Time	Part
Painter	$\frac{1}{t}$	4	$\frac{4}{t}$
Apprentice	$\frac{1}{t+6}$	4	$\frac{4}{t+6}$

▶ The sum of the parts of the task completed must equal 1.

Solution
$$\frac{4}{t} + \frac{4}{t+6} = 1$$

$$t(t+6)\left(\frac{4}{t} + \frac{4}{t+6}\right) = t(t+6) \cdot 1$$

$$(t+6)4 + t(4) = t(t+6)$$

$$4t + 24 + 4t = t^2 + 6t$$

$$0 = t^2 - 2t - 24$$

$$0 = (t-6)(t+4)$$

$$t - 6 = 0 \qquad t + 4 = 0$$
$$t = 6 \qquad t = -4 \qquad \text{• The solution } t = -4 \text{ is not possible.}$$

Working alone, the painter requires 6 h to paint the room.

> **Problem 1** The length of a rectangle is 3 m more than the width. The area is 40 m². Find the width.
>
> **Solution**   See page S26.

11.5 | Concept Review

Determine whether the statement is always true, sometimes true, or never true.

1. If the length of a rectangle is three more than twice the width and the width is represented by W, then the length is represented by $3W + 2$.

2. The sum of the squares of two consecutive odd integers can be represented as $n^2 + (n + 1)^2$.

3. If it takes one pipe 15 min longer to fill a tank than it does a second pipe, then the rate of work for the second pipe can be represented by $\frac{1}{t}$ and the rate of work for the first pipe can be represented by $\frac{1}{t + 15}$.

4. If a plane's rate of speed is r and the rate of the wind is 30 mph, then the plane's rate of speed flying with the wind is $30 + r$ and the plane's rate of speed flying against the wind is $30 - r$.

5. The graph of the equation $h = 48t - 16t^2$ is the graph of a parabola opening down.

6. A number that is a solution to a quadratic equation may not be a possible answer to the application problem it models. Such a solution is disregarded.

11.5 | Exercises

1

1. **Integer Problem** The sum of the squares of two positive consecutive odd integers is 130. Find the two integers.

2. **Integer Problem** The sum of two integers is 12. The product of the two integers is 35. Find the two integers.

3. **Integer Problem** The difference between two integers is 4. The product of the two integers is 60. Find the integers.

4. **Integer Problem** Twice an integer equals the square of the integer. Find the integer.

5.  Sports The area of the batter's box on a major-league baseball field is 24 ft². The length of the batter's box is 2 ft more than the width. Find the length and width of the rectangular batter's box.

6. Sports The length of the batter's box on a softball field is 1 ft more than twice the width. The area of the batter's box is 21 ft². Find the length and width of the rectangular batter's box.

7. Recreation The length of a children's playground is twice the width. The area is 5000 ft². Find the length and width of the rectangular playground.

8. Work Problem A tank has two drains. One drain takes 16 min longer to empty the tank than does the second drain. With both drains open, the tank is emptied in 6 min. How long would it take each drain, working alone, to empty the tank?

9. Work Problem One computer takes 21 min longer than a second computer to calculate the value of a complex equation. Working together, these computers can complete the calculation in 10 min. How long would it take each computer, working alone, to calculate the value?

10. 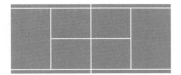 Sports The length of a singles tennis court is 24 ft more than twice the width. The area is 2106 ft². Find the length and width of the singles tennis court.

11. Uniform Motion Problem It takes 6 h longer to cross a channel in a ferryboat when one engine of the boat is used alone than when a second engine is used alone. Using both engines, the ferryboat can make the crossing in 4 h. How long would it take each engine, working alone, to power the ferryboat across the channel?

12. Work Problem An apprentice mason takes 8 h longer than an experienced mason to build a small fireplace. Working together, they can build the fireplace in 3 h. How long would it take the experienced mason, working alone, to build the fireplace?

13. Uniform Motion Problem It took a small plane 2 h longer to fly 375 mi against the wind than it took the plane to fly the same distance with the wind. The rate of the wind was 25 mph. Find the rate of the plane in calm air.

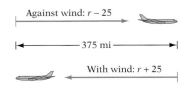

Against wind: $r - 25$

375 mi

With wind: $r + 25$

14. Uniform Motion Problem It took a motorboat 1 h longer to travel 36 mi against the current than it took the boat to travel 36 mi with the current. The rate of the current was 3 mph. Find the rate of the boat in calm water.

15. Uniform Motion Problem A motorcycle traveled 150 mi at a constant rate before its speed was decreased by 15 mph. Another 35 mi was driven at the decreased speed. The total time for the 185-mile trip was 4 h. Find the cyclist's rate during the first 150 mi.

16. Uniform Motion Problem A cruise ship sailed through a 20-mile inland passageway at a constant rate before its speed was increased by 15 mph. Another 75 mi was traveled at the increased rate. The total time for the 95-mile trip was 5 h. Find the rate of the ship during the last 75 mi.

17. Physics An arrow is projected into the air with an initial velocity of 48 ft/s. At what times will the arrow be 32 ft above the ground? Use the equation $h = 48t - 16t^2$, where h is the height, in feet, above the ground after t seconds.

32 ft

18. Physics A model rocket is launched with an initial velocity of 200 ft/s. The height, h, of the rocket t seconds after launch is given by $h = -16t^2 + 200t$. How many seconds after the launch will the rocket be 300 ft above the ground? Round to the nearest hundredth.

19. Travel In Germany there are no speed limits on some portions of the autobahn (highway). Other portions have a speed limit of 180 km/h (approximately 112 mph). The distance, d (in meters), required to stop a car traveling at v kilometers per hour is $d = 0.0056v^2 + 0.14v$. What is the maximum speed a driver can be traveling and still be able to stop within 150 m? Round to the nearest tenth.

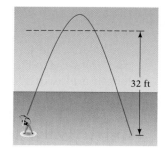

German Autobahn System

20. Sports In a slow-pitch softball game, the height of the ball thrown by a pitcher can be modeled by the equation $h = -16t^2 + 24t + 4$, where h is the height of the ball in feet, and t is the time, in seconds, since it was released by the pitcher. If the batter hits the ball when it is 2 ft off the ground, how many seconds has the ball been in the air? Round to the nearest hundredth.

21. Safety The path of water from a hose on a fire tugboat can be modeled by the equation $y = -0.005x^2 + 1.2x + 10$, where y is the height in feet of the water above the ocean when it is x feet from the tugboat. At what distance from the tugboat is the water from the hose when the water from the hose is 5 ft above the ocean? Round to the nearest tenth.

22. Sports A baseball player hits a ball. The height, in feet, of the ball above the ground at time t, in seconds, can be approximated by the equation $h = -16t^2 + 76t + 5$. When will the ball hit the ground? Round to the nearest hundredth. (*Hint:* The ball strikes the ground when $h = 0$ ft.)

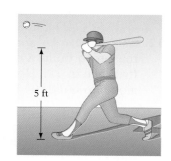

5 ft

23. Sports A basketball player shoots at a basket 25 ft away. The height h, in feet, of the ball above the ground at time t, in seconds, is given by $h = -16t^2 + 32t + 6.5$. How many seconds after the ball is released does it hit the basket? Round to the nearest hundredth. (*Hint:* When it hits the basket, $h = 10$ ft.)

24. Sports The hang time of a football that is kicked on the opening kickoff is given by $s = -16t^2 + 88t + 1$, where s is the height, in feet, of the football t seconds after leaving the kicker's foot. What is the hang time of a kickoff that hits the ground without being caught? Round to the nearest tenth.

11.5   Applying Concepts

25. **Geometry** A tin can has a volume of 1500 cm³ and a height of 15 cm. Find the radius of the base of the tin can. Round to the nearest hundredth.

15 cm

26. **Geometry** The hypotenuse of a right triangle measures √13 cm. One leg is 1 cm shorter than twice the length of the other leg. Find the lengths of the legs of the right triangle.

27. **Integer Problem** The sum of the squares of four consecutive integers is 86. Find the four integers.

28. **Geometry** Find the radius of a right circular cone that has a volume of 800 cm³ and a height of 12 cm. Round to the nearest hundredth.

12 cm

29. **Food Industry** The radius of a large pizza is 1 in. less than twice the radius of a small pizza. The difference between the areas of the two pizzas is 33π in². Find the radius of the large pizza.

30. **Gardening** The perimeter of a rectangular garden is 54 ft. The area of the garden is 180 ft². Find the length and width of the garden.

31. **Food Industry** A square piece of cardboard is to be formed into a box to transport pizzas. The box is formed by cutting 2-inch square corners from the cardboard and folding them up as shown in the figure at the right. If the volume of the box is 512 in³, what are the dimensions of the cardboard?

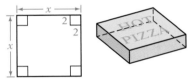

32. **Metalwork** A wire 8 ft long is cut into two pieces. A circle is formed from one piece and a square is formed from the other. The total area of both figures is given by $A = \frac{1}{16}(8 - x)^2 + \frac{x^2}{4\pi}$. What is the length of each piece of wire if the total area is 4.5 ft²? Round to the nearest thousandth.

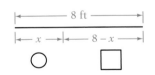

Focus on Problem Solving

Applying Algebraic Manipulation and Graphing Techniques to Business

Problem solving is often easier when we have both algebraic manipulation and graphing techniques at our disposal. Solving quadratic equations and graphing quadratic equations in two variables are used here to solve problems involving profit.

A company's **revenue**, R, is the total amount of money the company earned by selling its products. The **cost**, C, is the total amount of money the company spent to manufacture and sell its products. A company's **profit**, P, is the difference between the revenue and cost: $\boldsymbol{P = R - C}$. A company's revenue and cost may be represented by equations.

A company manufactures and sells wood stoves. The total monthly cost, in dollars, to produce n wood stoves is $C = 30n + 2000$. Write a variable expression for the company's monthly profit if the revenue, in dollars, obtained from selling all n wood stoves is $R = 150n - 0.4n^2$.

$$P = R - C$$
$$P = 150n - 0.4n^2 - (30n + 2000)$$ • Replace R by $150n - 0.4n^2$ and C by
$$P = -0.4n^2 + 120n - 2000$$ $30n + 2000$. Then simplify.

The company's monthly profit is $P = -0.4n^2 + 120n - 2000$.

How many wood stoves must the company manufacture and sell in order to make a profit of $6000 a month?

$$P = -0.4n^2 + 120n - 2000$$
$$6000 = -0.4n^2 + 120n - 2000$$ • Substitute 6000 for P.
$$0 = -0.4n^2 + 120n - 8000$$ • Write the equation in
                                                       standard form.
$$0 = n^2 - 300n + 20{,}000$$ • Divide each side of the
                                                       equation by -0.4.
$$0 = (n - 100)(n - 200)$$ • Factor.

$$n - 100 = 0 \qquad n - 200 = 0$$ • Solve for n.
$$n = 100 \qquad\quad n = 200$$

The company will make a monthly profit of $6000 if either 100 or 200 wood stoves are manufactured and sold.

The graph of $P = -0.4n^2 + 120n - 2000$ is shown at the left. Note that when $P = 6000$, the values of n are 100 and 200.

Also note that the coordinates of the highest point on the graph are $(150, 7000)$. This means that the company makes a *maximum* profit of $7000 per month when 150 wood stoves are manufactured and sold.

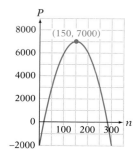

1. The total cost, in dollars, for a company to produce and sell n guitars per month is $C = 240n + 1200$. The company's revenue, in dollars, from selling all n guitars is $R = 400n - 2n^2$.

   a. How many guitars must the company produce and sell each month in order to make a monthly profit of $1200?
   b. Graph the profit equation. What is the maximum monthly profit the company can make?

2. A company's total monthly cost, in dollars, for manufacturing and selling n videotapes per month is $C = 35n + 2000$. The company's revenue, in dollars, from selling all n videotapes is $R = 175n - 0.2n^2$.

   a. How many videotapes must be produced and sold each month in order for the company to make a monthly profit of $18,000?
   b. Graph the profit equation. How many videotapes must the company produce and sell in order to make the maximum monthly profit?

Projects & Group Activities

Graphical Solutions of Quadratic Equations

A real number x is called a **zero of a function** if the value of the function when evaluated at x is 0. That is, if $f(x) = 0$, then x is called a zero of the function. For instance, evaluating $f(x) = x^2 + x - 6$ at $x = -3$, we have

$$f(x) = x^2 + x - 6$$
$$f(-3) = (-3)^2 + (-3) - 6 \qquad \bullet \text{ Replace } x \text{ with } -3.$$
$$f(-3) = 9 - 3 - 6$$
$$f(-3) = 0$$

For this function, $f(-3) = 0$, so -3 is a zero of the function.

1. Verify that 2 is also a zero of $f(x) = x^2 + x - 6$ by showing that $f(2) = 0$.

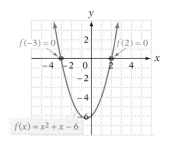

The graph of $y = x^2 + x - 6$ is shown at the left. Note that the graph crosses the x-axis at -3 and 2, the two zeros of the function. The points $(-3, 0)$ and $(2, 0)$ are x-intercepts of the graph.

Now consider the equation $0 = x^2 + x - 6$, which is $f(x) = x^2 + x - 6$ with $f(x)$ replaced with 0. Solving $0 = x^2 + x - 6$ for x, we have

$$0 = x^2 + x - 6$$
$$0 = (x + 3)(x - 2) \qquad \bullet \text{ Solve by factoring and using the}$$
$$\qquad\qquad\qquad\qquad\qquad \textbf{Principle of Zero Products.}$$
$$x + 3 = 0 \qquad x - 2 = 0$$
$$x = -3 \qquad x = 2$$

The solutions are -3 and 2.

Observe that the solutions of the equation are the zeros of the function. This important connection among the real zeros of a function, the x-intercepts of its graph, and the solutions of its equation is the basis for using a graphing calculator to solve an equation.

The following method of solving a quadratic equation using a graphing calculator is based on a TI-83/TI-83 Plus calculator. Other calculators will require a slightly different approach.

Approximate the solutions of $x^2 + 4x = 6$ by using a graphing calculator.

a. Write the equation in standard form.

$$x^2 + 4x = 6$$
$$x^2 + 4x - 6 = 0$$

b. Press $\boxed{\text{Y=}}$ and enter $x^2 + 4x - 6$ for Y1.

```
Plot1 Plot2 Plot3
\Y1 ◼ X²+4X–6
\Y2 =
\Y3 =
\Y4 =
\Y5 =
\Y6 =
\Y7 =
```

c. Press ⬚GRAPH⬚. If the graph does not appear on the screen, press ⬚ZOOM⬚ 6.

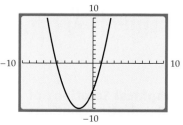

d. Press ⬚2nd⬚ CALC 2. Note that the selection for 2 says zero. This will begin the calculation of the zeros of the function, which are the solutions of the equation.

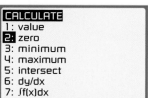

e. At the bottom of the screen you will see **Left Bound**? This is asking you to move the blinking cursor so that it is to the *left* of the first x-intercept. Use the left arrow key to move the cursor to the left of the first x-intercept. The values of x and y that appear on your calculator may be different from the ones shown here. Just be sure you are to the left of the x-intercept. When you have done that, press ⬚ENTER⬚.

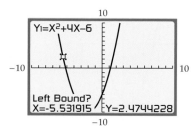

f. At the bottom of the screen you will see **Right Bound**? This is asking you to move the blinking cursor so that it is to the *right* of the x-intercept. Use the right arrow key to move the cursor to the right of the x-intercept. The values of x and y that appear on your calculator may be different from the ones shown here. Just be sure you are to the right of the x-intercept. When you have done that, press ⬚ENTER⬚.

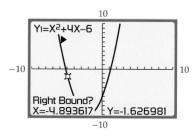

g. At the bottom of the screen you will see **Guess**? This is asking you to move the blinking cursor to the approximate x-intercept. Use the arrow keys to move the cursor to the approximate x-intercept. The values of x and y that appear on your calculator may be different from the ones shown here. When you have done that, press ⬚ENTER⬚.

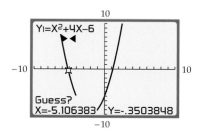

h. The zero of the function is approximately -5.162278. Thus one solution of $x^2 + 4x = 6$ is approximately -5.162278. Also note that the value of y is given as Y1 = -1E⁻12. This is the way the calculator writes a number in scientific notation. We would normally write Y1 $= -1.0 \times 10^{-12}$. This number is very close to zero.

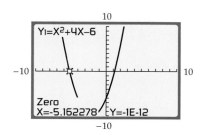

i. To find the other solution, repeat Steps **d** through **g**. The screens are shown below.

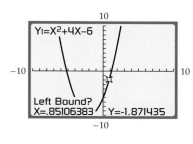

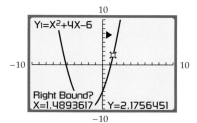

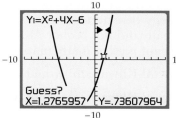

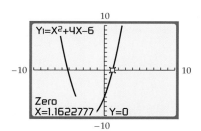

A second zero of the function is approximately 1.1622777.

Thus the two solutions of $x^2 + 4x = 6$ are approximately -5.162278 and 1.1622777.

Use a graphing calculator to approximate the solutions of the equation.

2. $x^2 + 3x - 4 = 0$ **3.** $2x^2 - 3x - 7 = 0$

4. $x^2 + 5x = 1$ **5.** $x^2 + 3.4x = 4.15$

6. $\pi x^2 - \sqrt{17}x - 2 = 0$ **7.** $\sqrt{2}x^2 + x = \sqrt{7}$

Geometric Construction of Completing the Square

Completing the square as a method of solving a quadratic equation has been known for centuries. The Persian mathematician Al-Khwarismi used this method in a textbook written around A.D. 825. The method was very geometric. That is, Al-Khwarismi literally completed a square. To understand how this method works, consider the following geometric shapes: a square whose area is x^2, a rectangle whose area is x, and another square whose area is 1.

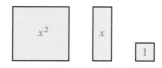

Now consider the expression $x^2 + 6x$. From our discussion in this chapter, to complete the square, we added $\left(\frac{1}{2} \cdot 6\right)^2 = 3^2 = 9$ to the expression. Here is the geometric construction that Al-Khwarismi used.

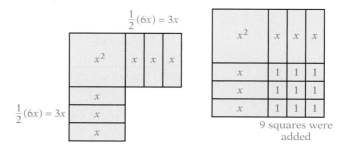

Note that it is necessary to add 9 squares to the figure to "complete the square." One of the difficulties of using a geometric method such as this is that it cannot easily be extended to $x^2 - 6x$. There is no way to draw an area of $-6x$! That really did not bother Al-Khwarismi much. Negative numbers were not a significant part of mathematics until well into the 13th century.

1. Show how Al-Khwarismi would have completed the square for $x^2 + 4x$.

2. Show how Al-Khwarismi would have completed the square for $x^2 + 10x$.

3. Do the geometric constructions for Exercises 1 and 2 correspond to the algebraic method shown in this chapter?

Chapter Summary

Key Words

A **quadratic equation** is an equation that can be written in the form $ax^2 + bx + c = 0$, $a \neq 0$. A quadratic equation is also called a **second-degree equation**. [11.1.1, p. 550]

A quadratic equation is in **standard form** when the polynomial is in descending order and equal to zero. [11.1.1, p. 550]

Examples

The equation $4x^2 - 3x + 6 = 0$ is a quadratic equation. In this equation, $a = 4$, $b = -3$, and $c = 6$.

The quadratic equation $5x^2 + 3x - 1 = 0$ is in standard form.

Adding to a binomial the constant term that makes it a perfect-square trinomial is called **completing the square**. [11.2.1, p. 558]

Adding to $x^2 - 6x$ the constant term 9 results in a perfect-square trinomial:

$$x^2 - 6x + 9 = (x - 3)(x - 3) = (x - 3)^2$$

The graph of an equation of the form $y = ax^2 + bx + c$ is a **parabola**. The graph is ∪-shaped and opens up when a is positive and down when a is negative. The **vertex** is the lowest point on a parabola that opens up, or the highest point on a parabola that opens down.   [11.4.1, pp. 570, 572]

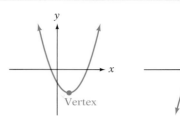

Parabola that opens up   Parabola that opens down

Essential Rules and Procedures

Solving a Quadratic Equation by Factoring
Write the equation in standard form, factor the left side of the equation, apply the Principle of Zero Products, and solve for the variable.   [11.1.1, p. 551]

$$x^2 + 2x = 15$$
$$x^2 + 2x - 15 = 0$$
$$(x - 3)(x + 5) = 0$$

$$x - 3 = 0 \qquad x + 5 = 0$$
$$x = 3 \qquad\quad x = -5$$

Principle of Taking the Square Root of Each Side of an Equation
If $x^2 = a$, then $x = \pm\sqrt{a}$.

This principle is used to solve quadratic equations by taking square roots.   [11.1.2, p. 552]

$$3x^2 - 48 = 0$$
$$3x^2 = 48$$
$$x^2 = 16$$
$$\sqrt{x^2} = \sqrt{16}$$
$$x = \pm\sqrt{16}$$
$$x = \pm 4$$

Solving a Quadratic Equation by Completing the Square
When the quadratic equation is in the form $x^2 + bx = c$, add to each side of the equation the term that completes the square on $x^2 + bx$. Factor the perfect-square trinomial, and write it as the square of a binomial. Take the square root of each side of the equation and solve for x.   [11.2.1, p. 559]

$$x^2 + 6x = 5$$
$$x^2 + 6x + 9 = 5 + 9$$
$$(x + 3)^2 = 14$$
$$\sqrt{(x + 3)^2} = \sqrt{14}$$
$$x + 3 = \pm\sqrt{14}$$
$$x = -3 \pm \sqrt{14}$$

The Quadratic Formula
The solutions of $ax^2 + bx + c = 0$, $a \neq 0$, are
$$x = \frac{-b \pm \sqrt{b^2 - 4ac}}{2a}.$$ [11.3.1, p. 565]

$$3x^2 = x + 5$$
$$3x^2 - x - 5 = 0$$
$$x = \frac{-b \pm \sqrt{b^2 - 4ac}}{2a}$$
$$= \frac{-(-1) \pm \sqrt{(-1)^2 - 4(3)(-5)}}{2(3)}$$
$$= \frac{1 \pm \sqrt{1 + 60}}{6} = \frac{1 \pm \sqrt{61}}{6}$$

Graphing a Quadratic Equation in Two Variables
Find several solutions of the equation. Graph the ordered-pair solutions on a rectangular coordinate system. Draw a parabola through the points.   [11.4.1, p. 571]

$$y = x^2 - x - 2$$

x	y
0	-2
1	-2
-1	0
2	0
-2	4
3	4

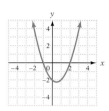

To find the x-intercepts of the graph of a parabola, let $y = 0$ and solve for x. The x-intercepts of the graph at the right are $(-1, 0)$ and $(2, 0)$. To find the y-intercept, let $x = 0$ and solve for y. The y-intercept of the graph at the right is $(0, -2)$. [11.4.1, p. 571]

Chapter Review Exercises

1. Solve: $b^2 - 16 = 0$

2. Solve: $x^2 - x - 3 = 0$

3. Solve: $x^2 - 3x - 5 = 0$

4. Solve: $49x^2 = 25$

5. Graph: $y = -\dfrac{1}{4}x^2$

6. Graph: $y = -3x^2$

7. Solve: $x^2 = 10x - 2$

8. Solve: $6x(x + 1) = x - 1$

9. Solve: $4y^2 + 9 = 0$

10. Solve: $5x^2 + 20x + 12 = 0$

11. Solve: $x^2 - 4x + 1 = 0$

12. Solve: $x^2 - x = 30$

13. Solve: $6x^2 + 13x - 28 = 0$

14. Solve: $x^2 = 40$

15. Solve: $3x^2 - 4x = 1$

16. Solve: $x^2 - 2x - 10 = 0$

17. Solve: $x^2 - 12x + 27 = 0$

18. Solve: $(x - 7)^2 = 81$

19. Graph: $y = 2x^2 + 1$

20. Graph: $y = \frac{1}{2}x^2 - 1$

21. Solve: $(y + 4)^2 - 25 = 0$

22. Solve: $4x^2 + 16x = 7$

23. Solve: $24x^2 + 34x + 5 = 0$

24. Solve: $x^2 = 4x - 8$

25. Solve: $x^2 - 5 = 8x$

26. Solve: $25(2x^2 - 2x + 1) = (x + 3)^2$

27. Solve: $\left(x - \frac{1}{2}\right)^2 = \frac{9}{4}$

28. Solve: $16x^2 = 30x - 9$

29. Solve: $x^2 + 7x = 3$

30. Solve: $12x^2 + 10 = 29x$

31. Solve: $4(x - 3)^2 = 20$

32. Solve: $x^2 + 8x - 3 = 0$

33. Graph: $y = x^2 - 3x$

34. Graph: $y = x^2 - 4x + 3$

35. Solve: $(x + 9)^2 = x + 11$

36. Solve: $(x - 2)^2 - 24 = 0$

37. Solve: $x^2 + 6x + 12 = 0$

38. Solve: $x^2 + 6x - 2 = 0$

39. Solve: $18x^2 - 52x = 6$

40. Solve: $2x^2 + 5x = 1$

41. Graph: $y = -x^2 + 4x - 5$

42. Solve: $x^2 + 3 = 9x$

43. Solve: $2x^2 + 5 = 7x$

44. Graph: $y = 4 - x^2$

45. It took an air balloon 1 h longer to fly 60 mi against the wind than it took to fly 60 mi with the wind. The rate of the wind was 5 mph. Find the rate of the air balloon in calm air.

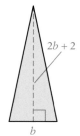

46. The height of a triangle is 2 m more than twice the length of the base. The area of the triangle is 20 m². Find the height of the triangle and the length of the base.

47. The sum of the squares of two consecutive positive odd integers is thirty-four. Find the two integers.

48. In 5 h, two campers rowed 12 mi down a stream and then rowed back to their campsite. The rate of the stream's current was 1 mph. Find the rate of the boat in still water.

49. An object is thrown into the air with an initial velocity of 32 ft/s. At what times will the object be 12 ft above the ground? Use the equation $h = 32t - 16t^2$, where h is the height, in feet, above the ground after t seconds.

50. A smaller drain takes 8 h longer to empty a tank than does a larger drain. Working together, the drains can empty the tank in 3 h. How long would it take each drain, working alone, to empty the tank?

Chapter Test

1. Solve: $3(x + 4)^2 - 60 = 0$

2. Solve: $2x^2 + 8x = 3$

3. Solve: $3x^2 + 7x = 20$

4. Solve: $x^2 - 3x = 6$

5. Solve: $x^2 + 4x - 16 = 0$

6. Graph: $y = x^2 + 2x - 4$

7. Solve: $x^2 + 4x + 2 = 0$

8. Solve: $x^2 + 3x = 8$

9. Solve: $2x^2 - 5x - 3 = 0$

10. Solve: $2x^2 - 6x + 1 = 0$

11. Solve: $3x^2 - x = 1$

12. Solve: $2(x - 5)^2 = 36$

13. Solve: $x^2 - 6x - 5 = 0$

14. Solve: $x^2 - 5x = 1$

15. Solve: $x^2 - 5x = 2$

16. Solve: $6x^2 - 17x = -5$

17. Solve: $x^2 + 3x - 7 = 0$

18. Solve: $2x^2 - 4x - 5 = 0$

19. Solve: $2x^2 - 3x - 2 = 0$

20. Graph: $y = x^2 - 2x - 3$

21. Solve: $3x^2 - 2x = 3$

22. The length of a rectangle is 2 ft less than twice the width. The area of the rectangle is 40 ft². Find the length and width of the rectangle.

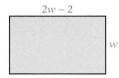

23. It took a motorboat 1 h more to travel 60 mi against the current than it took to go 60 mi with the current. The rate of the current was 1 mph. Find the rate of the boat in calm water.

24. The sum of the squares of three consecutive odd integers is 83. Find the middle odd integer.

25. A jogger ran 7 mi at a constant rate and then reduced the rate by 3 mph. An additional 8 mi was run at the reduced rate. The total time spent jogging the 15 mi was 3 h. Find the rate for the last 8 mi.

Cumulative Review Exercises

1. Simplify: $2x - 3[2x - 4(3 - 2x) + 2] - 3$

2. Solve: $-\dfrac{3}{5}x = -\dfrac{9}{10}$

3. Solve: $2x - 3(4x - 5) = -3x - 6$

4. Solve: $2x - 3(2 - 3x) > 2x - 5$

5. Find the x- and y-intercepts of the line $4x - 3y = 12$.

6. Find the equation of the line that contains the point whose coordinates are $(-3, 2)$ and has slope $-\dfrac{4}{3}$.

7. Find the domain and range of the relation $\{(-2, -8), (-1, -1), (0, 0), (1, 1), (2, 8)\}$. Is the relation a function?

8. Evaluate $f(x) = -3x + 10$ at $x = -9$.

9. Graph: $y = \dfrac{1}{4}x - 2$

10. Graph the solution set of $2x - 3y > 6$.

11. Solve by substitution: $3x - y = 5$
$\qquad\qquad\qquad\qquad\quad y = 2x - 3$

12. Solve by the addition method: $3x + 2y = 2$
$\qquad\qquad\qquad\qquad\qquad\qquad\quad 5x - 2y = 14$

13. Simplify: $\dfrac{(2a^{-2}b)^2}{-3a^{-5}b^4}$

14. Divide: $(x^2 - 8) \div (x - 2)$

15. Factor: $4y(x - 4) - 3(x - 4)$

16. Factor: $3x^3 + 2x^2 - 8x$

17. Divide: $\dfrac{3x^2 - 6x}{4x - 6} \div \dfrac{2x^2 + x - 6}{6x^2 - 24x}$

18. Subtract: $\dfrac{x}{2(x - 1)} - \dfrac{1}{(x - 1)(x + 1)}$

19. Simplify: $\dfrac{1 - \dfrac{7}{x} + \dfrac{12}{x^2}}{2 - \dfrac{1}{x} - \dfrac{15}{x^2}}$

20. Solve: $\dfrac{x}{x + 6} = \dfrac{3}{x}$

21. Multiply: $\left(\sqrt{a} - \sqrt{2}\right)\left(\sqrt{a} + \sqrt{2}\right)$

22. Solve: $3 = 8 - \sqrt{5x}$

23. Solve: $2x^2 - 7x = -3$

24. Solve: $3(x - 2)^2 = 36$

25. Solve: $3x^2 - 4x - 5 = 0$

26. Graph: $y = x^2 - 3x - 2$

27. In a certain state, the sales tax is $7\frac{1}{4}\%$. The sales tax on a chemistry textbook is $5.22. Find the cost of the textbook before the tax is added.

28. Find the cost per pound of a mixture made from 20 lb of cashews that cost $7 per pound and 50 lb of peanuts that cost $3.50 per pound.

29. A stock investment of 100 shares paid a dividend of $215. At this rate, how many additional shares are required to earn a dividend of $752.50?

30. A 720-mile trip from one city to another takes 3 h when a plane is flying with the wind. The return trip, against the wind, takes 4.5 h. Find the rate of the plane in calm air and the rate of the wind.

31. A student received a 70, a 91, an 85, and a 77 on four tests in a math class. What scores on the fifth test will enable the student to receive a minimum of 400 points?

32. A guy wire is attached to a point 30 m above the ground on a telephone pole that is perpendicular to the ground. The wire is anchored to the ground at a point 10 m from the base of the pole. Find the length of the guy wire. Round to the nearest hundredth.

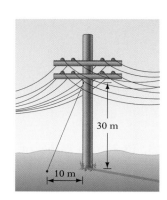

30 m

10 m

33. The sum of the squares of three consecutive odd integers is 155. Find the middle odd integer.

Final Exam

1. Evaluate: $-|-3|$

2. Subtract: $-15 - (-12) - 3$

3. Write $\dfrac{1}{8}$ as a percent.

4. Simplify: $-2^4 \cdot (-2)^4$

5. Simplify: $-7 - \dfrac{12 - 15}{2 - (-1)} \cdot (-4)$

6. Evaluate $\dfrac{a^2 - 3b}{2a - 2b^2}$ when $a = 3$ and $b = -2$.

7. Simplify: $6x - (-4y) - (-3x) + 2y$

8. Multiply: $(-15z)\left(-\dfrac{2}{5}\right)$

9. Simplify: $-2[5 - 3(2x - 7) - 2x]$

10. Solve: $20 = -\dfrac{2}{5}x$

11. Solve: $4 - 2(3x + 1) = 3(2 - x) + 5$

12. Find 19% of 80.

13. Solve: $4 - x \geq 7$

14. Solve: $2 - 2(y - 1) \leq 2y - 6$

15. Find the slope of the line that contains the points whose coordinates are $(-1, -3)$ and $(2, -1)$.

16. Find the equation of the line that contains the point whose coordinates are $(3, -4)$ and has slope $-\dfrac{2}{3}$.

17. Graph the line with slope $-\dfrac{1}{2}$ and y-intercept $(0, -3)$.

18. Graph: $f(x) = \dfrac{2}{3}x - 4$

19. Find the range of the function given by the equation $f(x) = -x + 5$ if the domain is $\{-6, -3, 0, 3, 6\}$.

20. Graph the solution set of $3x - 2y \geq 6$.

21. Solve by substitution: $y = 4x - 7$
$\qquad\qquad\qquad\quad y = 2x + 5$

22. Solve by the addition method: $4x - 3y = 11$
$\qquad\qquad\qquad\qquad\qquad\quad 2x + 5y = -1$

23. Subtract: $(2x^2 - 5x + 1) - (5x^2 - 2x - 7)$

24. Simplify: $(-3xy^3)^4$

25. Multiply: $(3x^2 - x - 2)(2x + 3)$

26. Simplify: $\dfrac{(-2x^2y^3)^3}{(-4x^{-1}y^4)^2}$

27. Simplify: $(4x^{-2}y)^3(2xy^{-2})^{-2}$

28. Divide: $\dfrac{12x^3y^2 - 16x^2y^2 - 20y^2}{4xy^2}$

29. Divide: $(5x^2 - 2x - 1) \div (x + 2)$

30. Write 0.000000039 in scientific notation.

31. Factor: $2a(4 - x) - 6(x - 4)$

32. Factor: $x^2 - 5x - 6$

33. Factor: $2x^2 - x - 3$

34. Factor: $6x^2 - 5x - 6$

35. Factor: $8x^3 - 28x^2 + 12x$

36. Factor: $25x^2 - 16$

37. Factor: $75y - 12x^2y$

38. Solve: $2x^2 = 7x - 3$

39. Multiply: $\dfrac{2x^2 - 3x + 1}{4x^2 - 2x} \cdot \dfrac{4x^2 + 4x}{x^2 - 2x + 1}$

40. Subtract: $\dfrac{5}{x + 3} - \dfrac{3x}{2x - 5}$

41. Simplify: $\dfrac{x - \dfrac{3}{2x - 1}}{1 - \dfrac{2}{2x - 1}}$

42. Solve: $\dfrac{5x}{3x - 5} - 3 = \dfrac{7}{3x - 5}$

43. Solve $a = 3a - 2b$ for a.

44. Simplify: $\sqrt{49x^6}$

45. Add: $2\sqrt{27a} + 8\sqrt{48a}$

46. Simplify: $\dfrac{\sqrt{3}}{\sqrt{5} - 2}$

47. Solve: $\sqrt{x + 4} - \sqrt{x - 1} = 1$

48. Solve: $(x - 3)^2 = 7$

49. Solve: $4x^2 - 2x - 1 = 0$

50. Graph: $y = x^2 - 4x + 3$

51. Translate and simplify "the sum of twice a number and three times the difference between the number and two."

52. Because of depreciation, the value of an office machine is now $2400. This is 80% of its original value. Find the original value of the machine.

53. A coin bank contains quarters and dimes. There are three times as many dimes as quarters. The total value of the coins in the bank is $11. Find the number of dimes in the bank.

54. One angle of a triangle is 10° more than the measure of the second angle. The third angle is 10° more than the measure of the first angle. Find the measure of each angle of the triangle.

55. The manufacturer's cost for a laser printer is $900. The manufacturer sells the printer for $1485. Find the markup rate.

56. An investment of $3000 is made at an annual simple interest rate of 8%. How much additional money must be invested at 11% so that the total interest earned is 10% of the total investment?

57. A grocer mixes 4 lb of peanuts that cost $2 per pound with 2 lb of walnuts that cost $5 per pound. What is the cost per pound of the resulting mixture?

58. A pharmacist mixes 20 L of a solution that is 60% acid with 30 L of a solution that is 20% acid. What is the percent concentration of acid in the resulting mixture?

59. A small plane flew at a constant rate for 1 h. The pilot then doubled the plane's speed. An additional 1.5 h were flown at the increased speed. If the entire flight was 860 km, how far did the plane travel during the first hour?

60. With the current, a motorboat travels 50 mi in 2.5 h. Against the current, it takes twice as long to travel 50 mi. Find the rate of the boat in calm water and the rate of the current.

61. The length of a rectangle is 5 m more than the width. The area of the rectangle is 50 m². Find the dimensions of the rectangle.

62. A paint formula requires 2 oz of dye for every 15 oz of base paint. How many ounces of dye are required for 120 oz of base paint?

63. It takes a chef 1 h to prepare a dinner. The chef's apprentice can prepare the same dinner in 1.5 h. How long would it take the chef and the apprentice, working together, to prepare the dinner?

64. The hypotenuse of a right triangle measures 14 cm. One leg of the triangle measures 8 cm. Find the length of the other leg of the triangle. Round to the nearest tenth.

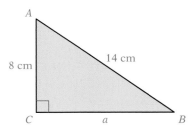

65. It took a plane $\frac{1}{2}$ h more to fly 500 mi against the wind than it took to fly the same distance with the wind. The rate of the plane in calm air is 225 mph. Find the rate of the wind.

Appendix

Table of Properties

Properties of Real Numbers

The Associative Property of Addition

If a, b, and c are real numbers, then
$(a + b) + c = a + (b + c)$.

The Commutative Property of Addition

If a and b are real numbers, then $a + b = b + a$.

The Addition Property of Zero

If a is a real number, then $a + 0 = 0 + a = a$.

The Multiplication Property of Zero

If a is a real number, then $a \cdot 0 = 0 \cdot a = 0$.

The Inverse Property of Addition

If a is a real number, then
$a + (-a) = (-a) + a = 0$.

The Associative Property of Multiplication

If a, b, and c are real numbers, then
$(a \cdot b) \cdot c = a \cdot (b \cdot c)$.

The Commutative Property of Multiplication

If a and b are real numbers, then $a \cdot b = b \cdot a$.

The Multiplication Property of One

If a is a real number, then $a \cdot 1 = 1 \cdot a = a$.

The Inverse Property of Multiplication

If a is a real number and $a \neq 0$, then
$a \cdot \dfrac{1}{a} = \dfrac{1}{a} \cdot a = 1$.

Distributive Property

If a, b, and c are real numbers, then
$a(b + c) = ab + ac$ or $(b + c)a = ba + ca$.

Properties of Equations

Addition Property of Equations

The same number or variable term can be added to each side of an equation without changing the solution of the equation.

Multiplication Property of Equations

Each side of an equation can be multiplied by the same nonzero number without changing the solution of the equation.

Properties of Inequalities

Addition Property of Inequalities

If $a > b$, then $a + c > b + c$.

If $a < b$, then $a + c < b + c$.

Multiplication Property of Inequalities

If $a > b$ and $c > 0$, then $ac > bc$.

If $a < b$ and $c > 0$, then $ac < bc$.

If $a > b$ and $c < 0$, then $ac < bc$.

If $a < b$ and $c < 0$, then $ac > bc$.

Table of Properties (*continued*)

Properties of Exponents

If m and n are integers, then $x^m \cdot x^n = x^{m+n}$.

If m and n are integers, then $(x^m)^n = x^{mn}$.

If $x \neq 0$, then $x^0 = 1$.

If m and n are integers and $x \neq 0$, then
$$\frac{x^m}{x^n} = x^{m-n}.$$

If m, n, and p are integers, then
$(x^m \cdot y^n)^p = x^{mp} y^{np}$.

If n is a positive integer and $x \neq 0$, then
$$x^{-n} = \frac{1}{x^n} \text{ and } \frac{1}{x^{-n}} = x^n.$$

If m, n, and p are integers and $y \neq 0$, then
$$\left(\frac{x^m}{y^n}\right)^p = \frac{x^{mp}}{y^{np}}.$$

Principle of Zero Products

If $a \cdot b = 0$, then $a = 0$ or $b = 0$.

Properties of Radical Expressions

If a and b are positive real numbers, then
$\sqrt[n]{ab} = \sqrt[n]{a}\,\sqrt[n]{b}$.

If a and b are positive real numbers, then
$$\sqrt[n]{\frac{a}{b}} = \frac{\sqrt[n]{a}}{\sqrt[n]{b}}.$$

Property of Squaring Both Sides of an Equation

If a and b are real numbers and $a = b$, then $a^2 = b^2$.

Calculator Guide for the TI-83 and TI-83 Plus

To evaluate an expression

a. Press the $\boxed{\text{Y=}}$ key. A menu showing \Y₁ = through \Y₇ = will be displayed vertically with a blinking cursor to the right of \Y₁ =. Press $\boxed{\text{CLEAR}}$, if necessary, to delete an unwanted expression.

b. Input the expression to be evaluated. For example, to input the expression $-3a^2b - 4c$, use the following keystrokes:

$\boxed{\text{(−)}}$ 3 $\boxed{\text{ALPHA}}$ A $\boxed{\text{X}^2}$ $\boxed{\text{ALPHA}}$ B $\boxed{-}$ 4 $\boxed{\text{ALPHA}}$ C $\boxed{\text{2nd}}$ QUIT

Note the difference between the keys for a negative sign $\boxed{\text{(−)}}$ and a minus sign $\boxed{-}$.

c. Store the value of each variable that will be used in the expression. For example, to evaluate the expression above when $a = 3$, $b = -2$, and $c = -4$, use the following keystrokes:

3 $\boxed{\text{STO▷}}$ $\boxed{\text{ALPHA}}$ A $\boxed{\text{ENTER}}$ $\boxed{\text{(−)}}$ 2 $\boxed{\text{STO▷}}$ $\boxed{\text{ALPHA}}$ B $\boxed{\text{ENTER}}$ $\boxed{\text{(−)}}$ 4 $\boxed{\text{STO▷}}$ $\boxed{\text{ALPHA}}$ C $\boxed{\text{ENTER}}$

These steps store the value of each variable.

d. Press $\boxed{\text{VARS}}$ $\boxed{▷}$ $\boxed{1}$ $\boxed{1}$ $\boxed{\text{ENTER}}$. The value for the expression, Y₁, for the given values is displayed; in this case, Y₁ = 70.

To graph a function

a. Press the $\boxed{\text{Y=}}$ key. A menu showing \Y₁ = through \Y₇ = will be displayed vertically with a blinking cursor to the right of \Y₁ =. Press $\boxed{\text{CLEAR}}$, if necessary, to delete an unwanted expression.

b. Input the expression for each function that is to be graphed. Press $\boxed{\text{X,T,θ,}n}$ to input x. For example, to input $y = x^3 + 2x^2 - 5x - 6$, use the following keystrokes:

$\boxed{\text{X,T,θ,}n}$ $\boxed{\wedge}$ 3 $\boxed{+}$ 2 $\boxed{\text{X,T,θ,}n}$ $\boxed{\text{X}^2}$ $\boxed{-}$ 5 $\boxed{\text{X,T,θ,}n}$ $\boxed{-}$ 6

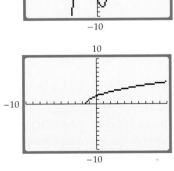

c. Set the viewing window by pressing $\boxed{\text{WINDOW}}$. Enter the values for the minimum x-value (Xmin), the maximum x-value (Xmax), the distance between tick marks on the x-axis (Xscl), the minimum y-value (Ymin), the maximum y-value (Ymax), and the distance between tick marks on the y-axis (Yscl). Now press $\boxed{\text{GRAPH}}$. For the graph shown at the left, Xmin = −10, Xmax = 10, Xscl = 1, Ymin = −10, Ymax = 10, and Yscl = 1. This is called the standard viewing window. Pressing $\boxed{\text{ZOOM}}$ $\boxed{6}$ is a quick way to set the calculator to the standard viewing window. *Note:* This will also immediately graph the function in that window.

d. Press the $\boxed{\text{Y=}}$ key. The equal sign has a black rectangle around it. This indicates that the function is active and will be graphed when the $\boxed{\text{GRAPH}}$ key is pressed. A function is deactivated by using the arrow keys. Move the cursor over the equal sign and press $\boxed{\text{ENTER}}$. When the cursor is moved to the right, the black rectangle will not be present and that equation will not be active.

e. Graphing some radical equations requires special care. To graph the equation $y = \sqrt{2x + 3}$ shown at the left, enter the following keystrokes:

$\boxed{\text{Y=}}$ $\boxed{\text{CLEAR}}$ $\boxed{\text{2nd}}$ $\sqrt{}$ 2 $\boxed{\text{X,T,θ,}n}$ $\boxed{+}$ 3 $\boxed{)}$ $\boxed{\text{GRAPH}}$

To display the *x*-coordinates of rectangular coordinates as integers

a. Set the viewing window as follows: Xmin = −47, Xmax = 47, Xscl = 10, Ymin = −31, Ymax = 31, Yscl = 10.

b. Graph the function and use the TRACE feature. Press TRACE and then move the cursor with the ◁ and ▷ keys. The values of x and $y = f(x)$ displayed on the bottom of the screen are the coordinates of a point on the graph.

To display the *x*-coordinates of rectangular coordinates in tenths

a. Set the viewing window as follows: ZOOM 4

b. Graph the function. Press TRACE and then move the cursor with the ◁ and ▷ keys. The values of x and $y = f(x)$ displayed on the bottom of the screen are the coordinates of a point on the graph.

To evaluate a function for a given value of *x*, or to produce ordered pairs of a function

a. Input the equation; for example, input $Y_1 = 2x^3 − 3x + 2$.

b. Press 2nd QUIT.

c. To evaluate the function when $x = 3$, press VARS ▷ 1 1 (3) ENTER. The value for the function for the given x-value is displayed, in this case, $Y_1(3) = 47$. An ordered pair of the function is (3, 47).

d. Repeat step **c.** to produce as many pairs as desired.

The TABLE feature can also be used to determine ordered pairs.

To use a table

a. Press 2nd TBLSET to activate the table setup menu.

b. TblStart is the beginning number for the table; ΔTbl is the difference between any two successive x-values in the table.

c. The portion of the table that appears as Indpnt: [Auto] Ask / Depend: [Auto] Ask allows you to choose between having the calculator automatically produce the results (Auto) or having the calculator ask you for values of x. You can choose Ask by using the arrow keys.

d. Once a table has been set up, enter an expression for Y_1. Now select TABLE by pressing 2nd TABLE. A table showing ordered pair solutions of the equation will be displayed on the screen.

Zoom Features

To zoom in or out on a graph

Here are two methods of using ZOOM.

a. The first method uses the built-in features of the calculator. Press TRACE and then move the cursor to a point on the graph that is of interest. Press ZOOM. The ZOOM menu will appear. Press 2 ENTER to zoom in on the graph by the amount shown under the SET FACTORS menu. The center of the new graph is the location at which you placed the cursor. Press ZOOM 3 ENTER to zoom out on the graph by the amount under the SET FACTORS menu. (The SET FACTORS menu is accessed by pressing ZOOM ▷ 4.)

b. The second method uses the ZBOX option under the ZOOM menu. To use this method, press ZOOM 1. A cursor will appear on the graph. Use the arrow keys to move the cursor to a portion of the graph that is of interest. Press ENTER. Now use the arrow keys to draw a box around the portion of the graph you wish to zoom in on. Press ENTER. The portion of the graph defined by the box will be drawn.

c. Pressing ZOOM 6 resets the window to the standard viewing window.

Solving Equations

This discussion is based on the fact that the real number solutions of an equation are related to the x-intercepts of a graph. For instance, the real solutions of the equation $x^2 = x + 1$ are the x-intercepts of the graph of $f(x) = x^2 - x - 1$, which are the zeros of f.

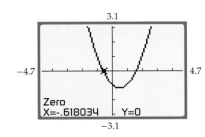

To solve $x^2 = x + 1$, rewrite the equation with all terms on one side: $x^2 - x - 1 = 0$. Think of this equation as $Y_1 = x^2 - x - 1$. The x-intercepts of the graph of Y_1 are the solutions of the equation $x^2 = x + 1$.

a. Enter $x^2 - x - 1$ into Y_1.

b. Graph the equation. You may need to adjust the viewing window so that the x-intercepts are visible.

c. Press 2nd CALC 2.

d. Move the cursor to a point on the curve that is to the left of an x-intercept. Press ENTER.

e. Move the cursor to a point on the curve that is to the right of the same x-intercept. Press ENTER ENTER.

f. The root is shown as the x-coordinate on the bottom of the screen; in this case, the root is approximately -0.618034. To find the next intercept, repeat steps **c.** through **e.** The SOLVER feature under the MATH menu can also be used to find solutions of equations.

Solving Systems of Equations in Two Variables

To solve a system of equations

To solve
$$y = x^2 - 1$$
$$\frac{1}{2}x + y = 1$$

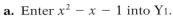

a. If necessary, solve one or both of the equations for y.

b. Enter the first equation as Y_1: $Y_1 = x^2 - 1$.

c. Enter the second equation as Y_2: $Y_2 = 1 - .5x$.

d. Graph both equations. (*Note:* The point of intersection must appear on the screen. It may be necessary to adjust the viewing window so that the points of intersection are displayed.)

e. Press 2nd CALC 5.

f. Move the cursor close to the first point of intersection. Press ENTER ENTER ENTER.

g. The first point of intersection is $(-1.686141, 1.8430703)$.

h. Repeat steps **e.** and **f.** for each point of intersection.

Finding Minimum or Maximum Values of a Function

a. Enter the function into Y₁. The equation $y = x^2 - x - 1$ is used here.

b. Graph the equation. You may need to adjust the viewing window so that the maximum or minimum points are visible.

c. Press $\boxed{\text{2nd}}$ CALC $\boxed{3}$ to determine a minimum value or press $\boxed{\text{2nd}}$ CALC $\boxed{4}$ to determine a maximum value.

d. Move the cursor to a point on the curve that is to the left of the minimum (maximum). Press $\boxed{\text{ENTER}}$.

e. Move the cursor to a point on the curve that is to the right of the minimum (maximum). Press $\boxed{\text{ENTER}}$ $\boxed{\text{ENTER}}$.

f. The minimum (maximum) is shown as the y-coordinate on the bottom of the screen; in this case the minimum value is -1.25.

Statistics

To calculate a linear regression equation

a. Press $\boxed{\text{STAT}}$ to access the statistics menu. Press 1 to Edit or enter a new list of data. To delete data already in a list, press the up arrow to highlight the list name. For instance, to delete data in L₁, highlight L₁. Then press $\boxed{\text{CLEAR}}$ and $\boxed{\text{ENTER}}$. Now enter each value of the independent variable in L₁. Enter each value of the dependent variable in L₂.

b. When all the data has been entered, press $\boxed{\text{STAT}}$ $\boxed{\triangleright}$ 4 $\boxed{\text{ENTER}}$. The values of the slope and y-intercept of the linear regression equation will be displayed on the screen.

c. To graph the linear regression equation, modify step **b.** with these keystrokes: $\boxed{\text{STAT}}$ $\boxed{\triangleright}$ 4 $\boxed{\text{VARS}}$ $\boxed{\triangleright}$ 1 1 $\boxed{\text{ENTER}}$. This will store the regression equation in Y₁. Now press $\boxed{\text{GRAPH}}$. It may be necessary to adjust the viewing window.

d. To evaluate the regression equation for a value of x, complete step **c.** but do not graph the equation. Now press $\boxed{\text{VARS}}$ $\boxed{\triangleright}$ 1 1 $\boxed{(}$ A $\boxed{)}$ $\boxed{\text{ENTER}}$, where A is replaced by the number at which you want to evaluate the expression.

Solutions to Chapter Problems

Solutions to Chapter 1 Problems

SECTION 1.1

Problem 1

$A = \{1, 2, 3, 4\}$

Problem 2

$-5 < -1$

$-1 = -1$

$5 > -1$

The element 5 is greater than -1.

Problem 3

A. 9   **B.** -62

Problem 4

A. $|-5| = 5$ **B.** $-|-9| = -9$

SECTION 1.2

Problem 1

A. $-162 + 98 = -64$

B. $-154 + (-37) = -191$

C. $-36 + 17 + (-21) = -19 + (-21)$
$= -40$

Problem 2

$-8 - 14 = -8 + (-14)$
$= -22$

Problem 3

$4 - (-3) - 12 - (-7) - 20$
$= 4 + 3 + (-12) + 7 + (-20)$
$= 7 + (-12) + 7 + (-20)$
$= -5 + 7 + (-20)$
$= 2 + (-20)$
$= -18$

Problem 4

A. $-38 \cdot 51 = -1938$

B. $-7(-8)(9)(-2) = 56(9)(-2)$
$= 504(-2)$
$= -1008$

Problem 5

A. $(-135) \div (-9) = 15$

B. $\dfrac{84}{-6} = -14$

C. $-\dfrac{36}{-12} = -(-3) = 3$

Problem 6

Strategy   To find the average daily high temperature:
▶ Add the seven temperature readings.
▶ Divide the sum by 7.

Solution $-5 + (-6) + 3 + 0 + (-4) + (-7) + (-2)$
$= -11 + 3 + 0 + (-4) + (-7) + (-2)$
$= -8 + 0 + (-4) + (-7) + (-2)$
$= -8 + (-4) + (-7) + (-2)$
$= -12 + (-7) + (-2)$
$= -19 + (-2)$
$= -21$

$-21 \div 7 = -3$

The average daily high temperature was $-3°C$.

SECTION 1.3

Problem 1

$$
\begin{array}{r}
0.16 \\
25\overline{)4.00} \\
-2\,5 \\
\hline
1\,50 \\
-1\,50 \\
\hline
0
\end{array}
$$

$\dfrac{4}{25} = 0.16$

Problem 2

$$
\begin{array}{r}
0.444 \\
9\overline{)4.000} \\
-3\,6 \\
\hline
40 \\
-36 \\
\hline
40 \\
-36 \\
\hline
4
\end{array}
$$

$\dfrac{4}{9} = 0.\overline{4}$

Problem 3

Prime factorization of 9 and 12:

$9 = 3 \cdot 3 \qquad 12 = 2 \cdot 2 \cdot 3$

$\text{LCM} = 2 \cdot 2 \cdot 3 \cdot 3 = 36$

$\dfrac{5}{9} - \dfrac{11}{12} = \dfrac{20}{36} - \dfrac{33}{36} = \dfrac{20}{36} + \dfrac{-33}{36}$
$= \dfrac{20 + (-33)}{36} = \dfrac{-13}{36} = -\dfrac{13}{36}$

Problem 4

$$-\frac{7}{8} - \frac{5}{6} + \frac{1}{2} = -\frac{21}{24} - \frac{20}{24} + \frac{12}{24}$$

$$= \frac{-21}{24} + \frac{-20}{24} + \frac{12}{24}$$

$$= \frac{-21 + (-20) + 12}{24} = \frac{-29}{24} = -\frac{29}{24}$$

Problem 5

$$\begin{array}{r} 3.097 \\ 4.9 \\ +\ 3.09 \\ \hline 11.087 \end{array}$$

Problem 6

$$\begin{array}{r} 67.910 \\ -16.127 \\ \hline 51.783 \end{array}$$

$$16.127 - 67.91 = -51.783$$

Problem 7

$$-\frac{7}{12} \cdot \frac{9}{14} = -\frac{7 \cdot 9}{12 \cdot 14} = -\frac{\overset{1}{\cancel{7}} \cdot \overset{1}{\cancel{3}} \cdot 3}{2 \cdot 2 \cdot \cancel{3} \cdot 2 \cdot \cancel{7}}$$

$$= -\frac{3}{8}$$

Problem 8

$$-\frac{3}{8} \div \left(-\frac{5}{12}\right) = \frac{3}{8} \cdot \frac{12}{5}$$

$$= \frac{3 \cdot 12}{8 \cdot 5} = \frac{3 \cdot \overset{1}{\cancel{2}} \cdot \overset{1}{\cancel{2}} \cdot 3}{\cancel{2} \cdot \cancel{2} \cdot 2 \cdot 5} = \frac{9}{10}$$

Problem 9

$$\begin{array}{r} 5.44 \\ \times\ \ \ 3.8 \\ \hline 4352 \\ 1632 \\ \hline 20.672 \end{array}$$

$$(-5.44)(3.8) = -20.672$$

Problem 10

$$\begin{array}{r} 0.231 \\ 1.7.\overline{)0.3.940} \\ -3\ 4 \\ \hline 54 \\ -51 \\ \hline 30 \\ -17 \\ \hline 13 \end{array}$$

$$-0.394 \div 1.7 \approx -0.23$$

Problem 11

$$125\% = 125\left(\frac{1}{100}\right) = \frac{125}{100} = 1\frac{1}{4}$$

$$125\% = 125(0.01) = 1.25$$

Problem 12

$$16\frac{2}{3}\% = 16\frac{2}{3}\left(\frac{1}{100}\right) = \frac{50}{3}\left(\frac{1}{100}\right) = \frac{1}{6}$$

Problem 13

$$6.08\% = 6.08(0.01) = 0.0608$$

Problem 14

A. $0.043 = 0.043(100\%) = 4.3\%$

B. $2.57 = 2.57(100\%) = 257\%$

Problem 15

$$\frac{5}{9} = \frac{5}{9}(100\%) = \frac{500}{9}\% \approx 55.6\%$$

Problem 16

$$\frac{9}{16} = \frac{9}{16}(100\%) = \frac{900}{16}\% = 56\frac{1}{4}\%$$

SECTION 1.4

Problem 1

$$(-5)^3 = (-5)(-5)(-5) = 25(-5) = -125$$
$$-5^3 = -(5 \cdot 5 \cdot 5) = -(25 \cdot 5) = -125$$

Problem 2

$$(-3)^3 = (-3)(-3)(-3) = 9(-3) = -27$$
$$(-3)^4 = (-3)(-3)(-3)(-3)$$
$$= 9(-3)(-3) = -27(-3) = 81$$

Problem 3

$$(3^3)(-2)^3 = (3)(3)(3) \cdot (-2)(-2)(-2)$$
$$= 27 \cdot (-8) = -216$$

$$\left(-\frac{2}{5}\right)^2 = \left(-\frac{2}{5}\right)\left(-\frac{2}{5}\right) = \frac{2 \cdot 2}{5 \cdot 5} = \frac{4}{25}$$

Problem 4

$$36 \div (8 - 5)^2 - (-3)^2 \cdot 2$$
$$= 36 \div (3)^2 - (-3)^2 \cdot 2$$
$$= 36 \div 9 - 9 \cdot 2$$
$$= 4 - 9 \cdot 2$$
$$= 4 - 18$$
$$= -14$$

Problem 5

$$27 \div 3^2 + (-3)^2 \cdot 4$$
$$= 27 \div 9 + 9 \cdot 4$$
$$= 3 + 9 \cdot 4$$
$$= 3 + 36$$
$$= 39$$

Problem 6

$$4 - 3[4 - 2(6 - 3)] \div 2$$
$$= 4 - 3[4 - 2(3)] \div 2$$
$$= 4 - 3[4 - 6] \div 2$$
$$= 4 - 3(-2) \div 2$$
$$= 4 - (-6) \div 2$$
$$= 4 - (-3)$$
$$= 7$$

Solutions to Chapter 2 Problems

SECTION 2.1

Problem 1

-4

Problem 2

$2xy + y^2$

$2(-4)(2) + (2)^2 = 2(-4)(2) + 4$

$\qquad = -8(2) + 4$

$\qquad = -16 + 4$

$\qquad = -12$

Problem 3

$\dfrac{a^2 + b^2}{a + b}$

$\dfrac{(5)^2 + (-3)^2}{5 + (-3)} = \dfrac{25 + 9}{5 + (-3)}$

$\qquad = \dfrac{34}{2}$

$\qquad = 17$

Problem 4

$x^3 - 2(x + y) + z^2$

$(2)^3 - 2[2 + (-4)] + (-3)^2$

$\quad = (2)^3 - 2(-2) + (-3)^2$

$\quad = 8 - 2(-2) + 9$

$\quad = 8 + 4 + 9$

$\quad = 12 + 9$

$\quad = 21$

Problem 5

$V = \dfrac{1}{3}\pi r^2 h$

$V = \dfrac{1}{3}\pi(4.5)^2(9.5)$ $\qquad \bullet\, r = \dfrac{1}{2}d = \dfrac{1}{2}(9) = 4.5$

$V = \dfrac{1}{3}\pi(20.25)(9.5)$

$V \approx 201.5$

The volume is approximately 201.5 cm³.

SECTION 2.2

Problem 1

$7 + (-7) = 0$

Problem 2

The Associative Property of Addition

Problem 3

A. $9x + 6x = (9 + 6)x = 15x$

B. $-4y - 7y = [-4 + (-7)]y = -11y$

Problem 4

A. $3a - 2b + 5a = 3a + 5a - 2b$

$\qquad = (3a + 5a) - 2b$

$\qquad = 8a - 2b$

B. $x^2 - 7 + 9x^2 - 14$

$\quad = x^2 + 9x^2 - 7 - 14$

$\quad = (x^2 + 9x^2) + (-7 - 14)$

$\quad = 10x^2 - 21$

Problem 5

A. $-7(-2a) = [-7(-2)]a$

$\qquad\qquad = 14a$

B. $-\dfrac{5}{6}(-30y^2) = \left[-\dfrac{5}{6}(-30)\right]y^2$

$\qquad\qquad\quad = 25y^2$

C. $(-5x)(-2) = (-2)(-5x)$

$\qquad\qquad\quad = [-2(-5)]x$

$\qquad\qquad\quad = 10x$

Problem 6

A. $7(4 + 2y) = 7(4) + 7(2y)$

$\qquad\qquad = 28 + 14y$

B. $-(5x - 12) = -1(5x - 12)$

$\qquad\qquad = -1(5x) - (-1)(12)$

$\qquad\qquad = -5x + 12$

C. $(3a - 1)5 = (3a)(5) - (1)(5)$

$\qquad\qquad = 15a - 5$

D. $-3(6a^2 - 8a + 9)$

$\quad = -3(6a^2) - (-3)(8a) + (-3)(9)$

$\quad = -18a^2 + 24a - 27$

Problem 7

$7(x - 2y) - 3(-x - 2y)$

$\quad = 7x - 14y + 3x + 6y$

$\quad = 10x - 8y$

Problem 8

$3y - 2[x - 4(2 - 3y)]$

$\quad = 3y - 2[x - 8 + 12y]$

$\quad = 3y - 2x + 16 - 24y$

$\quad = -2x - 21y + 16$

SECTION 2.3

Problem 1

A. eighteen <u>less than</u> the <u>cube</u> of x

$x^3 - 18$

B. y <u>decreased by</u> the <u>sum</u> of z and nine

$y - (z + 9)$

C. the <u>difference between</u> the <u>square</u> of q

and the <u>sum</u> of r and t

$q^2 - (r + t)$

Problem 2

the unknown number: n

the square of the number: n^2

the product of five and the square of the number: $5n^2$

$5n^2 + n$

Problem 3

the unknown number: n

twice the number: $2n$

the sum of seven and twice the number: $7 + 2n$

$3(7 + 2n)$

Problem 4

the unknown number: n

twice the number: $2n$

the difference between twice the number and 17:

$2n - 17$

$n - (2n - 17) = n - 2n + 17$
$$= -n + 17$$

Problem 5

the unknown number: n

three-fourths of the number: $\dfrac{3}{4}n$

one-fifth of the number: $\dfrac{1}{5}n$

$\dfrac{3}{4}n + \dfrac{1}{5}n = \dfrac{15}{20}n + \dfrac{4}{20}n$
$$= \dfrac{19}{20}n$$

Problem 6

the time required by the newer model: t

the time required by the older model is twice the time required for the newer model: $2t$

Problem 7

the length of one piece: L

the length of the second piece: $6 - L$

Solutions to Chapter 3 Problems

SECTION 3.1

Problem 1

$$5 - 4x = 8x + 2$$

$5 - 4\left(\dfrac{1}{4}\right)$	$8\left(\dfrac{1}{4}\right) + 2$
$5 - 1$	$2 + 2$
$4 = 4$	

Yes, $\dfrac{1}{4}$ is a solution.

Problem 2

$$10x - x^2 = 3x - 10$$

$10(5) - (5)^2$	$3(5) - 10$
$50 - 25$	$15 - 10$
$25 \neq 5$	

No, 5 is not a solution.

Problem 3

$$x - \frac{1}{3} = -\frac{3}{4}$$
$$x - \frac{1}{3} + \frac{1}{3} = -\frac{3}{4} + \frac{1}{3}$$
$$x + 0 = -\frac{9}{12} + \frac{4}{12}$$
$$x = -\frac{5}{12}$$

Check $\quad x - \dfrac{1}{3} = -\dfrac{3}{4}$

$$-\frac{5}{12} - \frac{1}{3} \;\Big|\; -\frac{3}{4}$$
$$-\frac{3}{4} = -\frac{3}{4}$$

The solution is $-\dfrac{5}{12}$.

Problem 4

$$-8 = 5 + x$$
$$-8 - 5 = 5 - 5 + x$$
$$-13 = x$$

The solution is -13.

Problem 5

$$-\frac{2x}{5} = 6$$
$$\left(-\frac{5}{2}\right)\left(-\frac{2}{5}x\right) = \left(-\frac{5}{2}\right)(6)$$
$$1x = -15$$
$$x = -15$$

The solution is -15.

Problem 6

$6x = 10$	**Check**	$6x = 10$

$$\frac{6x}{6} = \frac{10}{6}$$
$$x = \frac{5}{3}$$

$$6\left(\frac{5}{3}\right) \;\Big|\; 10$$
$$10 = 10$$

The solution is $\dfrac{5}{3}$.

Problem 7

$$4x - 8x = 16$$
$$-4x = 16$$
$$\frac{-4x}{-4} = \frac{16}{-4}$$
$$x = -4$$

The solution is -4.

Problem 8

Strategy The distance is 80 mi. Therefore, $d = 80$. The cyclists are moving in opposite directions, so the rate at which the distance between them is changing is the sum of the rates of the two cyclists. The rate is 18 mph + 14 mph = 32 mph. Therefore, $r = 32$. To find the time, solve the equation $d = rt$ for t.

Solution
$$d = rt$$
$$80 = 32t$$
$$\frac{80}{32} = \frac{32t}{32}$$
$$2.5 = t$$

They will meet 2.5 h after they begin.

Problem 9

Strategy Because the plane is flying against the wind, the rate of the plane is its speed in calm air (250 mph) minus the speed of the headwind (25 mph): 250 mph − 25 mph = 225 mph. Therefore, $r = 225$. The time is 3 h, so $t = 3$. To find the distance, solve the equation $d = rt$ for d.

Solution
$$d = rt$$
$$d = 225(3)$$
$$d = 675$$

The plane can fly 675 mi in 3 h.

Problem 10

$$PB = A$$
$$P(60) = 27$$
$$60P = 27$$
$$\frac{60P}{60} = \frac{27}{60}$$
$$P = 0.45$$

27 is 45% of 60.

Problem 11

Strategy To find the percent of the questions answered correctly, solve the basic percent equation using $B = 80$ and $A = 72$. The percent is unknown.

Solution
$$PB = A$$
$$P(80) = 72$$
$$80P = 72$$
$$\frac{80P}{80} = \frac{72}{80}$$
$$P = 0.9$$

90% of the questions were answered correctly.

Problem 12

Strategy To determine the sales tax, solve the basic percent equation using $B = 895$ and $P = 6\% = 0.06$. The amount is unknown.

Solution
$$PB = A$$
$$895(0.06) = A$$
$$53.7 = A$$

The sales tax is $53.70.

Problem 13

Strategy To determine how much she must deposit into the bank account:

▶ Find the amount of interest earned on the municipal bond by solving the equation $I = Prt$ for I using $P = 1000$, $r = 6.4\% = 0.064$, and $t = 1$.

▶ Solve $I = Prt$ for P using the amount of interest earned on the municipal bond for I, $r = 8\% = 0.08$, and $t = 1$.

Solution
$$I = Prt$$
$$I = 1000(0.064)(1)$$
$$I = 64$$

The interest earned on the municipal bond is $64.

$$I = Prt$$
$$64 = P(0.08)(1)$$
$$64 = 0.08P$$
$$\frac{64}{0.08} = \frac{0.08P}{0.08}$$
$$800 = P$$

Clarissa must deposit $800 into the bank account.

Problem 14

Strategy To find the number of ounces of cereal in the bowl, solve the equation $Q = Ar$ for A using $Q = 2$ and $r = 25\% = 0.25$.

Solution
$$Q = Ar$$
$$2 = A(0.25)$$
$$\frac{2}{0.25} = \frac{A(0.25)}{0.25}$$
$$8 = A$$

The cereal bowl contains 8 oz of cereal.

SECTION 3.2

Problem 1

$$5x + 7 = 10$$
$$5x + 7 - 7 = 10 - 7$$
$$5x = 3$$
$$\frac{5x}{5} = \frac{3}{5}$$
$$x = \frac{3}{5}$$

The solution is $\frac{3}{5}$.

Problem 2

$$11 = 11 + 3x$$
$$11 - 11 = 11 - 11 + 3x$$
$$0 = 3x$$
$$\frac{0}{3} = \frac{3x}{3}$$
$$0 = x$$

The solution is 0.

Problem 3

$$\frac{5}{2} - \frac{2}{3}x = \frac{1}{2}$$
$$6\left(\frac{5}{2} - \frac{2}{3}x\right) = 6\left(\frac{1}{2}\right)$$
$$6\left(\frac{5}{2}\right) - 6\left(\frac{2}{3}x\right) = 3$$
$$15 - 4x = 3$$
$$15 - 15 - 4x = 3 - 15$$
$$-4x = -12$$
$$\frac{-4x}{-4} = \frac{-12}{-4}$$
$$x = 3$$

The solution is 3.

Problem 4

$$5x + 4 = 6 + 10x$$
$$5x - 10x + 4 = 6 + 10x - 10x$$
$$-5x + 4 = 6$$
$$-5x + 4 - 4 = 6 - 4$$
$$-5x = 2$$
$$\frac{-5x}{-5} = \frac{2}{-5}$$
$$x = -\frac{2}{5}$$

The solution is $-\frac{2}{5}$.

Problem 5

$$5x - 4(3 - 2x) = 2(3x - 2) + 6$$
$$5x - 12 + 8x = 6x - 4 + 6$$
$$13x - 12 = 6x + 2$$
$$13x - 6x - 12 = 6x - 6x + 2$$
$$7x - 12 = 2$$
$$7x - 12 + 12 = 2 + 12$$
$$7x = 14$$
$$\frac{7x}{7} = \frac{14}{7}$$
$$x = 2$$

The solution is 2.

Problem 6

$$-2[3x - 5(2x - 3)] = 3x - 8$$
$$-2[3x - 10x + 15] = 3x - 8$$
$$-2[-7x + 15] = 3x - 8$$
$$14x - 30 = 3x - 8$$
$$14x - 3x - 30 = 3x - 3x - 8$$
$$11x - 30 = -8$$
$$11x - 30 + 30 = -8 + 30$$
$$11x = 22$$
$$\frac{11x}{11} = \frac{22}{11}$$
$$x = 2$$

The solution is 2.

Problem 7

Strategy To find the number of years, replace V with 10,200 in the given equation and solve for t.

Solution
$$V = 450t + 7500$$
$$10,200 = 450t + 7500$$
$$10,200 - 7500 = 450t + 7500 - 7500$$
$$2700 = 450t$$
$$\frac{2700}{450} = \frac{450t}{450}$$
$$6 = t$$

The value of the investment will be $10,200 in 6 years.

Problem 8

Strategy The lever is 14 ft long, so $d = 14$. One force is 6 ft from the fulcrum, so $x = 6$. The one force is 40 lb, so $F_1 = 40$. To find the other force when the system balances, replace the variables F_1, x, and d in the lever system equation with the given values, and solve for F_2.

Solution
$$F_1 x = F_2(d - x)$$
$$40(6) = F_2(14 - 6)$$
$$240 = F_2(8)$$
$$240 = 8F_2$$
$$\frac{240}{8} = \frac{8F_2}{8}$$
$$30 = F_2$$

A 30-pound force must be applied to the other end.

SECTION 3.3

Problem 1

The solution set is the numbers greater than -2.

Problem 2

$$x + 2 < -2$$
$$x + 2 - 2 < -2 - 2$$
$$x < -4$$

Problem 3

$$5x + 3 > 4x + 5$$
$$5x - 4x + 3 > 4x - 4x + 5$$
$$x + 3 > 5$$
$$x + 3 - 3 > 5 - 3$$
$$x > 2$$

Problem 4

$$3x < 9$$
$$\frac{3x}{3} < \frac{9}{3}$$
$$x < 3$$

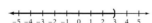

Problem 5

$$-\frac{3}{4}x \geq 18$$
$$-\frac{4}{3}\left(-\frac{3}{4}x\right) \leq -\frac{4}{3}(18)$$
$$x \leq -24$$

Problem 6

$$5 - 4x > 9 - 8x$$
$$5 - 4x + 8x > 9 - 8x + 8x$$
$$5 + 4x > 9$$
$$5 - 5 + 4x > 9 - 5$$
$$4x > 4$$
$$\frac{4x}{4} > \frac{4}{4}$$
$$x > 1$$

Problem 7

$$8 - 4(3x + 5) \leq 6(x - 8)$$
$$8 - 12x - 20 \leq 6x - 48$$
$$-12 - 12x \leq 6x - 48$$
$$-12 - 12x - 6x \leq 6x - 6x - 48$$
$$-12 - 18x \leq -48$$
$$-12 + 12 - 18x \leq -48 + 12$$
$$-18x \leq -36$$
$$\frac{-18x}{-18} \geq \frac{-36}{-18}$$
$$x \geq 2$$

Soltuions to Chapter 4 Problems

SECTION 4.1

Problem 1

the unknown number: n

nine less than twice a number	is	five times the sum of the number and twelve

$$2n - 9 = 5(n + 12)$$
$$2n - 9 = 5n + 60$$
$$2n - 5n - 9 = 5n - 5n + 60$$
$$-3n - 9 = 60$$
$$-3n - 9 + 9 = 60 + 9$$
$$-3n = 69$$
$$\frac{-3n}{-3} = \frac{69}{-3}$$
$$n = -23$$

The number is -23.

Problem 2

Strategy   To find the number of carbon atoms, write and solve an equation using n to represent the number of carbon atoms.

Solution

eight	represents	twice the number of carbon atoms

$$8 = 2n$$
$$\frac{8}{2} = \frac{2n}{2}$$
$$4 = n$$

There are 4 carbon atoms in a butane molecule.

Problem 3

Strategy   To find the number of 10-speed bicycles made each day, write and solve an equation using n to represent the number of 10-speed bicycles and $160 - n$ to represent the number of 3-speed bicycles.

Solution

four times the number of 3-speed bicycles made	equals	30 less than the number of 10-speed bicycles made

$$4(160 - n) = n - 30$$
$$640 - 4n = n - 30$$
$$640 - 4n - n = n - n - 30$$
$$640 - 5n = -30$$
$$640 - 640 - 5n = -30 - 640$$
$$-5n = -670$$
$$\frac{-5n}{-5} = \frac{-670}{-5}$$
$$n = 134$$

There are 134 10-speed bicycles made each day.

SECTION 4.2

Problem 1

Strategy ▶ First consecutive integer: n
Second consecutive integer: $n + 1$
Third consecutive integer: $n + 2$

▶ The sum of the three integers is -12.

Solution $n + (n + 1) + (n + 2) = -12$
$$3n + 3 = -12$$
$$3n = -15$$
$$n = -5$$

$n + 1 = -5 + 1 = -4$
$n + 2 = -5 + 2 = -3$

The three consecutive integers are -5, -4, and -3.

Problem 2

Strategy ▶ Number of dimes: x
Number of nickels: $5x$
Number of quarters: $x + 6$

Coin	Number	Value	Total value
Dime	x	10	$10x$
Nickel	$5x$	5	$5(5x)$
Quarter	$x + 6$	25	$25(x + 6)$

▶ The sum of the total values of the individual denominations of coins equals the total value of all the coins (630 cents).

Solution $10x + 5(5x) + 25(x + 6) = 630$
$$10x + 25x + 25x + 150 = 630$$
$$60x + 150 = 630$$
$$60x = 480$$
$$x = 8$$

$5x = 5(8) = 40$
$x + 6 = 8 + 6 = 14$

The bank contains 8 dimes, 40 nickels, and 14 quarters.

SECTION 4.3

Problem 1

Strategy ▶ Each equal side: s

▶ Use the equation for the perimeter of a square.

Solution $P = 4s$
$$52 = 4s$$
$$13 = s$$

The length of each side is 13 ft.

Problem 2

Strategy To find the supplement, let x represent the supplement of a 107° angle. Use the fact that supplementary angles are two angles whose sum is 180° to write an equation. Solve for x.

Solution $107° + x = 180°$
$$x = 73°$$

The supplement of a 107° angle is a 73° angle.

Problem 3

Strategy The angles labeled are adjacent angles of intersecting lines and are therefore supplementary angles. To find x, write an equation and solve for x.

Solution $x + (3x + 20°) = 180°$
$$4x + 20° = 180°$$
$$4x = 160°$$
$$x = 40°$$

Problem 4

Strategy $2x = y$ because alternate exterior angles have the same measure.

$y + (x + 15°) = 180°$ because adjacent angles of intersecting lines are supplementary angles. Substitute $2x$ for y and solve for x.

Solution $y + (x + 15°) = 180°$
$$2x + (x + 15°) = 180°$$
$$3x + 15° = 180°$$
$$3x = 165°$$
$$x = 55°$$

Problem 5

Strategy ▶ To find the measure of $\angle a$, use the fact that $\angle a$ and $\angle y$ are vertical angles.

▶ To find the measure of $\angle b$, use the fact that the sum of the interior angles of a triangle is 180°.

▶ To find the measure of $\angle d$, use the fact that $\angle b$ and $\angle d$ are supplementary angles.

Solution $\angle a = \angle y = 55°$

$\angle a + \angle b + 90° = 180°$
$55° + \angle b + 90° = 180°$
$145° + \angle b = 180°$
$\angle b = 35°$

$\angle b + \angle d = 180°$
$35° + \angle d = 180°$
$\angle d = 145°$

Problem 6

Strategy To find the measure of the third angle, use the fact that the sum of the measures of the interior angles of a triangle is 180°. Write an equation using x to represent the measure of the third angle. Solve the equation for x.

Solution $x + 90° + 27° = 180°$
$x + 117° = 180°$
$x = 63°$

The measure of the third angle is 63°.

SECTION 4.4

Problem 1

Strategy   Given: $C = \$120$
$S = \$180$
Unknown markup rate: r
Use the equation $S = C + rC$.

Solution $S = C + rC$
$180 = 120 + 120r$
$60 = 120r$
$0.5 = r$

The markup rate is 50%.

Problem 2

Strategy   Given: $r = 40\% = 0.40$
$S = \$266$
Unknown cost: C
Use the equation $S = C + rC$.

Solution $S = C + rC$
$266 = C + 0.4C$
$266 = 1.40C$
$190 = C$

The cost is \$190.

Problem 3

Strategy   Given: $R = \$39.80$
$S = \$29.85$
Unknown discount rate: r
Use the equation $S = R - rR$.

Solution $S = R - rR$
$29.85 = 39.80 - 39.80r$
$-9.95 = -39.80r$
$0.25 = r$

The discount rate is 25%.

Problem 4

Strategy   Given: $S = \$43.50$
$r = 25\% = 0.25$
Unknown regular price: R
Use the equation $S = R - rR$.

Solution $S = R - rR$
$43.50 = R - 0.25R$
$43.50 = 0.75R$
$58 = R$

The regular price is \$58.

SECTION 4.5

Problem 1

Strategy   ▶ Additional amount: x

	Principal	·	Rate	=	Interest
Amount at 7%	2500	·	0.07	=	0.07(2500)
Amount at 10%	x	·	0.10	=	0.10x
Amount at 9%	$2500 + x$	·	0.09	=	0.09(2500 + x)

▶ The sum of the interest earned by the two investments equals the interest earned on the total investment.

Solution $0.07(2500) + 0.10x = 0.09(2500 + x)$
$175 + 0.10x = 225 + 0.09x$
$175 + 0.01x = 225$
$0.01x = 50$
$x = 5000$

\$5000 more must be invested at 10%.

SECTION 4.6

Problem 1

Strategy   ▶ Pounds of \$.75 fertilizer: x

	Amount	Cost	Value
\$.90 fertilizer	20	0.90	0.90(20)
\$.75 fertilizer	x	0.75	0.75x
\$.85 fertilizer	$20 + x$	0.85	0.85(20 + x)

▶ The sum of the values before mixing equals the value after mixing.

Solution $0.90(20) + 0.75x = 0.85(20 + x)$
$18 + 0.75x = 17 + 0.85x$
$18 - 0.10x = 17$
$-0.10x = -1$
$x = 10$

10 lb of the \$.75 fertilizer must be added.

Problem 2

Strategy   ▶ Pure orange juice is 100% orange juice. $100\% = 1.00$

Amount of pure orange juice: x

	Amount	Percent	Quantity
Pure orange juice	x	1.00	$1.00x$
Fruit drink	5	0.10	$0.10(5)$
Orange drink	$x + 5$	0.25	$0.25(x + 5)$

▶ The sum of the quantities before mixing is equal to the quantity after mixing.

Solution
$$1.00x + 0.10(5) = 0.25(x + 5)$$
$$1.00x + 0.5 = 0.25x + 1.25$$
$$0.75x + 0.5 = 1.25$$
$$0.75x = 0.75$$
$$x = 1$$

To make the orange drink, 1 qt of pure orange juice is added to the fruit drink.

SECTION 4.7

Problem 1

Strategy ▶ Rate of the first train: r
Rate of the second train: $2r$

	Rate	Time	Distance
First train	r	3	$3r$
Second train	$2r$	3	$3(2r)$

▶ The sum of the distances traveled by each train equals 306 mi.

Solution
$$3r + 3(2r) = 306$$
$$3r + 6r = 306$$
$$9r = 306$$
$$r = 34$$

$2r = 2(34) = 68$

The first train is traveling at 34 mph. The second train is traveling at 68 mph.

Problem 2

Strategy ▶ Time spent flying out: t
Time spent flying back: $7 - t$

	Rate	Time	Distance
Out	120	t	$120t$
Back	90	$7 - t$	$90(7 - t)$

▶ The distance out equals the distance back.

Solution
$$120t = 90(7 - t)$$
$$120t = 630 - 90t$$
$$210t = 630$$
$$t = 3 \quad \text{(The time out was 3 h.)}$$

Distance $= 120t = 120(3) = 360$ mi

The parcel of land was 360 mi away.

SECTION 4.8

Problem 1

Strategy To find the minimum selling price, write and solve an inequality using S to represent the selling price.

Solution
$$340 < 0.70S$$
$$\frac{340}{0.70} < \frac{0.70S}{0.70}$$
$$485.71429 < S$$

The minimum selling price is \$485.72.

Problem 2

Strategy To find the maximum number of miles:

▶ Write an expression for the cost of each car, using x to represent the number of miles driven during the week.

▶ Write and solve an inequality.

Solution

cost of a Company A car	is less than	cost of a Company B car

$$9(7) + 0.10x < 12(7) + 0.08x$$
$$63 + 0.10x < 84 + 0.08x$$
$$63 + 0.10x - 0.08x < 84 + 0.08x - 0.08x$$
$$63 + 0.02x < 84$$
$$63 - 63 + 0.02x < 84 - 63$$
$$0.02x < 21$$
$$\frac{0.02x}{0.02} < \frac{21}{0.02}$$
$$x < 1050$$

The maximum number of miles is 1049.

Solutions to Chapter 5 Problems

SECTION 5.1

Problem 1

Problem 2

$A(4, 2)$, $B(-3, 4)$, $C(-3, 0)$, $D(0, 0)$

Problem 3

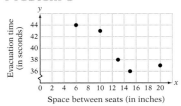

Problem 4

In 1900, the population was 76 million: (1900, 76)
In 2000, the population was 281 million: (2000, 281)

Average rate of change

$$= \frac{\text{change in } y}{\text{change in } x}$$

$$= \frac{\text{population in 2000} - \text{population in 1900}}{2000 - 1900}$$

$$= \frac{281 - 76}{2000 - 1900}$$

$$= \frac{205}{100} = 2.05$$

The average rate of change in the population was 2.05 million people per year.

Problem 5

28 weeks after conception, the weight is 3 lb: (28, 3)
38 weeks after conception, the weight is 7.5 lb: (38, 7.5)

Average rate of change

$$= \frac{\text{change in } y}{\text{change in } x}$$

$$= \frac{\text{weight after 38 weeks} - \text{weight after 28 weeks}}{38 - 28}$$

$$= \frac{7.5 - 3}{38 - 28}$$

$$= \frac{4.5}{10} = 0.45$$

The average rate of change in weight is 0.45 lb per week.

SECTION 5.2

Problem 1

$$y = -\frac{1}{2}x - 3$$

-4	$-\frac{1}{2}(2) - 3$
	$-1 - 3$
	-4
$-4 = -4$	

Yes, $(2, -4)$ is a solution of $y = -\frac{1}{2}x - 3$.

Problem 2

$$y = -\frac{1}{4}x + 1$$

$$= -\frac{1}{4}(4) + 1$$

$$= -1 + 1$$

$$= 0$$

The ordered-pair solution is (4, 0).

Problem 3

Problem 4

Problem 5

$$5x - 2y = 10$$

$$5x - 5x - 2y = -5x + 10$$

$$-2y = -5x + 10$$

$$\frac{-2y}{-2} = \frac{-5x + 10}{-2}$$

$$y = \frac{-5x}{-2} + \frac{10}{-2}$$

$$y = \frac{5}{2}x - 5$$

Problem 6

A. $5x - 2y = 10$

$-2y = -5x + 10$

$y = \dfrac{5}{2}x - 5$

B. $x - 3y = 9$

$-3y = -x + 9$

$y = \dfrac{1}{3}x - 3$

Problem 7

x-intercept: $4x - y = 4$

$4x - 0 = 4$

$4x = 4$

$x = 1$

The x-intercept is $(1, 0)$.

y-intercept: $4x - y = 4$

$4(0) - y = 4$

$-y = 4$

$y = -4$

The y-intercept is $(0, -4)$.

Problem 8

A.   **B.**

SECTION 5.3

Problem 1

A. $m = \dfrac{y_2 - y_1}{x_2 - x_1} = \dfrac{3 - 2}{1 - (-1)} = \dfrac{1}{2}$

The slope is $\dfrac{1}{2}$.

B. $m = \dfrac{y_2 - y_1}{x_2 - x_1} = \dfrac{-5 - 2}{4 - 1} = \dfrac{-7}{3}$

The slope is $-\dfrac{7}{3}$.

C. $m = \dfrac{y_2 - y_1}{x_2 - x_1} = \dfrac{7 - 3}{2 - 2} = \dfrac{4}{0}$

The slope is undefined.

D. $m = \dfrac{y_2 - y_1}{x_2 - x_1} = \dfrac{-3 - (-3)}{-5 - 1} = \dfrac{0}{-6} = 0$

The slope is 0.

Problem 2

Find the slope of the line through $(-2, -3)$ and $(7, 1)$.

$\dfrac{y_2 - y_1}{x_2 - x_1} = \dfrac{1 - (-3)}{7 - (-2)} = \dfrac{4}{9}$

Find the slope of the line through $(1, 4)$ and $(-5, 6)$.

$\dfrac{y_2 - y_1}{x_2 - x_1} = \dfrac{6 - 4}{-5 - 1} = \dfrac{2}{-6} = -\dfrac{1}{3}$

$\dfrac{4}{9} \neq -\dfrac{1}{3}$

The slopes are not equal.
The lines are not parallel.

Problem 3

$m = \dfrac{8650 - 6100}{1 - 4} = \dfrac{2550}{-3} = -850$

A slope of -850 means that the value of the car is decreasing at a rate of \$850 per year.

Problem 4

y-intercept $= (0, b) = (0, -1)$

$m = -\dfrac{1}{4}$

Problem 5

Solve the equation for y.

$x - 2y = 4$

$-2y = -x + 4$

$y = \dfrac{1}{2}x - 2$

y-intercept $= (0, b) = (0, -2)$

$m = \dfrac{1}{2}$

SECTION 5.4

Problem 1

$$y = \frac{3}{2}x + b$$

$$-2 = \frac{3}{2}(4) + b$$

$$-2 = 6 + b$$

$$-8 = b$$

$$y = \frac{3}{2}x - 8$$

Problem 2

$$m = \frac{2}{5} \quad (x_1, y_1) = (5, 4)$$

$$y - y_1 = m(x - x_1)$$

$$y - 4 = \frac{2}{5}(x - 5)$$

$$y - 4 = \frac{2}{5}x - 2$$

$$y = \frac{2}{5}x + 2$$

SECTION 5.5

Problem 1

The domain is $\{1\}$.
The range is $\{0, 1, 2, 3, 4\}$.
There are ordered pairs with the same first coordinate and different second coordinates. The relation is not a function.

Problem 2

$$f(x) = -5x + 1$$
$$f(2) = -5(2) + 1$$
$$f(2) = -10 + 1$$
$$f(2) = -9$$

The ordered pair $(2, -9)$ is an element of the function.

Problem 3

$$f(x) = 4x - 3$$
$$f(-5) = 4(-5) - 3 = -20 - 3 = -23$$
$$f(-3) = 4(-3) - 3 = -12 - 3 = -15$$
$$f(-1) = 4(-1) - 3 = -4 - 3 = -7$$
$$f(1) = 4(1) - 3 = 4 - 3 = 1$$

The range is $\{-23, -15, -7, 1\}$.
The ordered pairs $(-5, -23)$, $(-3, -15)$, $(-1, -7)$, and $(1, 1)$ belong to the function.

Problem 4

$$f(x) = -\frac{1}{2}x - 3$$

$$y = -\frac{1}{2}x - 3$$

Problem 5

A. $d = 40t$ **B.**
$f(t) = 40t$

C. The ordered pair $(3, 120)$ means that in 3 h, the car travels a distance of 120 mi.

SECTION 5.6

Problem 1

$$x - 3y < 2$$
$$x - x - 3y < -x + 2$$
$$-3y < -x + 2$$
$$\frac{-3y}{-3} > \frac{-x + 2}{-3}$$
$$y > \frac{1}{3}x - \frac{2}{3}$$

Problem 2

$$x < 3$$

Solutions to Chapter 6 Problems

SECTION 6.1

Problem 1

$$\frac{2x - 5y = 8}{\begin{array}{c|c} 2(-1) - 5(-2) & 8 \\ -2 + 10 & 8 \\ 8 = 8 \end{array}} \qquad \frac{-x + 3y = -5}{\begin{array}{c|c} -(-1) + 3(-2) & -5 \\ 1 + (-6) & -5 \\ -5 = -5 \end{array}}$$

Yes, $(-1, -2)$ is a solution of the system of equations.

Problem 2

A.

The solution is $(-3, 2)$.

B.

No solution

The lines are parallel and therefore do not intersect. The system of equations is inconsistent and has no solution.

SECTION 6.2

Problem 1

(1) $\qquad 7x - y = 4$
(2) $\qquad 3x + 2y = 9$

$\qquad\qquad 7x - y = 4$ • **Solve equation (1) for**
$\qquad\qquad -y = -7x + 4$ **y.**
$\qquad\qquad y = 7x - 4$

$\qquad\qquad 3x + 2y = 9$ • **Substitute $7x - 4$ for**
$\qquad 3x + 2(7x - 4) = 9$ **y in equation (2).**
$\qquad 3x + 14x - 8 = 9$
$\qquad\qquad 17x - 8 = 9$
$\qquad\qquad\qquad 17x = 17$
$\qquad\qquad\qquad x = 1$

$\qquad\qquad 7x - y = 4$ • **Substitute the value of**
$\qquad\qquad 7(1) - y = 4$ **x in equation (1).**
$\qquad\qquad 7 - y = 4$
$\qquad\qquad -y = -3$
$\qquad\qquad y = 3$

The solution is $(1, 3)$.

Problem 2

$$3x - y = 4$$
$$y = 3x + 2$$

$$3x - y = 4$$
$$3x - (3x + 2) = 4$$
$$3x - 3x - 2 = 4$$
$$-2 = 4$$

The system of equations is inconsistent and has no solution.

Problem 3

$$y = -2x + 1$$
$$6x + 3y = 3$$

$$6x + 3y = 3$$
$$6x + 3(-2x + 1) = 3$$
$$6x - 6x + 3 = 3$$
$$3 = 3$$

The system of equations is dependent. The solutions are the ordered pairs that satisfy the equation $y = -2x + 1$.

SECTION 6.3

Problem 1

(1) $\qquad x - 2y = 1$
(2) $\qquad 2x + 4y = 0$

$\qquad 2(x - 2y) = 2 \cdot 1$ • **Eliminate y.**
$\qquad\qquad 2x + 4y = 0$

$\qquad\qquad 2x - 4y = 2$
$\qquad\qquad 2x + 4y = 0$
$\qquad\qquad\qquad 4x = 2$ • **Add the equations.**
$\qquad\qquad\qquad x = \dfrac{2}{4} = \dfrac{1}{2}$

$\qquad\qquad 2x + 4y = 0$ • **Replace x in equation (2).**
$\qquad\qquad 2\left(\dfrac{1}{2}\right) + 4y = 0$
$\qquad\qquad 1 + 4y = 0$
$\qquad\qquad 4y = -1$
$\qquad\qquad y = -\dfrac{1}{4}$

The solution is $\left(\dfrac{1}{2}, -\dfrac{1}{4}\right)$.

Problem 2

(1) $\qquad 2x - 3y = 4$
(2) $\qquad -4x + 6y = -8$

$\qquad 2(2x - 3y) = 2 \cdot 4$ • **Eliminate y.**
$\qquad\qquad -4x + 6y = -8$

$\qquad\qquad 4x - 6y = 8$
$\qquad\qquad -4x + 6y = -8$
$\qquad\qquad\qquad 0 + 0 = 0$ • **Add the equations.**
$\qquad\qquad\qquad 0 = 0$

The system of equations is dependent. The solutions are the ordered pairs that satisfy the equation $2x - 3y = 4$.

SECTION 6.4

Problem 1

Strategy ▶ Rate of the current: c
Rate of the canoeist in calm water: r

	Rate	Time	Distance
With current	$r + c$	3	$3(r + c)$
Against current	$r - c$	4	$4(r - c)$

▶ The distance traveled with the current is 24 mi.
The distance traveled against the current is 24 mi.

Solution

$$3(r + c) = 24 \qquad \frac{3(r + c)}{3} = \frac{24}{3} \qquad r + c = 8$$

$$4(r - c) = 24 \qquad \frac{4(r - c)}{4} = \frac{24}{4} \qquad r - c = 6$$

$$2r = 14$$
$$r = 7$$

$$r + c = 8$$
$$7 + c = 8$$
$$c = 1$$

The rate of the current is 1 mph.
The rate of the canoeist in calm water is 7 mph.

Problem 2

Strategy ▶ The number of dimes in the first bank: d
The number of quarters in the first bank: q

First bank

	Number	Value	Total value
Dimes	d	10	$10d$
Quarters	q	25	$25q$

Second bank

	Number	Value	Total value
Dimes	$2d$	10	$20d$
Quarters	$\frac{1}{2}q$	25	$\frac{25}{2}q$

▶ The total value of the coins in the first bank is \$3.90.
The total value of the coins in the second bank is \$3.30.

Solution

$$10d + 25q = 390 \qquad 10d + 25q = 390 \qquad 10d + 25q = 390$$

$$20d + \frac{25}{2}q = 330 \qquad -2\left(20d + \frac{25}{2}q\right) = -2(330) \qquad -40d - 25q = -660$$

$$-30d = -270$$
$$d = 9$$

$$10d + 25q = 390$$
$$10(9) + 25q = 390$$
$$90 + 25q = 390$$
$$25q = 300$$
$$q = 12$$

There are 9 dimes and 12 quarters in the first bank.

Solutions to Chapter 7 Problems

SECTION 7.1

Problem 1

$$2x^2 + 4x - 3$$
$$\underline{5x^2 - 6x}$$
$$7x^2 - 2x - 3$$

Problem 2

$$(-4x^2 - 3xy + 2y^2) + (3x^2 - 4y^2)$$
$$= (-4x^2 + 3x^2) - 3xy + (2y^2 - 4y^2)$$
$$= -x^2 - 3xy - 2y^2$$

Problem 3

The opposite of $2y^2 - xy + 5x^2$ is $-2y^2 + xy - 5x^2$.

$$8y^2 - 4xy + x^2$$
$$\underline{-2y^2 + xy - 5x^2}$$
$$6y^2 - 3xy - 4x^2$$

Problem 4

$$(-3a^2 - 4a + 2) - (5a^3 + 2a - 6)$$
$$= (-3a^2 - 4a + 2) + (-5a^3 - 2a + 6)$$
$$= -5a^3 - 3a^2 + (-4a - 2a) + (2 + 6)$$
$$= -5a^3 - 3a^2 - 6a + 8$$

SECTION 7.2

Problem 1

$$(3x^2)(6x^3) = (3 \cdot 6)(x^2 \cdot x^3) = 18x^5$$

Problem 2

$$(-3xy^2)(-4x^2y^3)$$
$$= [(-3)(-4)](x \cdot x^2)(y^2 \cdot y^3)$$
$$= 12x^3y^5$$

Problem 3

$$(3x)(2x^2y)^3 = (3x)(2^3x^6y^3) = (3x)(8x^6y^3)$$
$$= (3 \cdot 8)(x \cdot x^6)y^3 = 24x^7y^3$$

SECTION 7.3

Problem 1

A. $(-2y + 3)(-4y) = 8y^2 - 12y$

B. $-a^2(3a^2 + 2a - 7) = -3a^4 - 2a^3 + 7a^2$

Problem 2

$$
\begin{array}{r}
2y^3 + 2y^2 - 3 \\
3y - 1 \\
\hline
-2y^3 - 2y^2 + 3 \\
6y^4 + 6y^3 - 9y \\
\hline
6y^4 + 4y^3 - 2y^2 - 9y + 3
\end{array}
$$

Problem 3

$$
\begin{array}{r}
3x^3 - 2x^2 + x - 3 \\
2x + 5 \\
\hline
15x^3 - 10x^2 + 5x - 15 \\
6x^4 - 4x^3 + 2x^2 - 6x \\
\hline
6x^4 + 11x^3 - 8x^2 - x - 15
\end{array}
$$

Problem 4

$(4y - 5)(3y - 3)$
$= 12y^2 - 12y - 15y + 15$
$= 12y^2 - 27y + 15$

Problem 5

$(3a + 2b)(3a - 5b)$
$= 9a^2 - 15ab + 6ab - 10b^2$
$= 9a^2 - 9ab - 10b^2$

Problem 6

$(2a + 5c)(2a - 5c) = (2a)^2 - (5c)^2$
$= 4a^2 - 25c^2$

Problem 7

$(3x + 2y)^2 = (3x)^2 + 2(3x)(2y) + (2y)^2$
$= 9x^2 + 12xy + 4y^2$

Problem 8

$(6x - y)^2 = (6x)^2 + 2(6x)(-y) + (-y)^2$
$= 36x^2 - 12xy + y^2$

Problem 9

Strategy To find the area in terms of x, replace the variables L and W in the equation $A = LW$ with the given values, and solve for A.

Solution $A = LW$
$A = (x + 7)(x - 4)$
$A = x^2 - 4x + 7x - 28$
$A = x^2 + 3x - 28$
The area is $(x^2 + 3x - 28)$ m².

Problem 10

Strategy To find the area of the triangle in terms of x, replace the variables b and h in the equation $A = \frac{1}{2}bh$ with the given values, and solve for A.

Solution $A = \frac{1}{2}bh$

$A = \frac{1}{2}(x + 3)(4x - 6)$

$A = \frac{1}{2}(4x^2 + 6x - 18)$

$A = 2x^2 + 3x - 9$

The area is $(2x^2 + 3x - 9)$ cm².

SECTION 7.4

Problem 1

$$\frac{2^{-2}}{2^3} = 2^{-2-3} = 2^{-5} = \frac{1}{2^5} = \frac{1}{32}$$

Problem 2

$$\frac{b^8}{a^{-5}b^6} = a^5b^{8-6} = a^5b^2$$

Problem 3

A. $\dfrac{12x^{-8}y^4}{-16xy^{-3}} = -\dfrac{3x^{-9}y^7}{4} = -\dfrac{3y^7}{4x^9}$

B. $(-3ab)(2a^3b^{-2})^{-3} = (-3ab)(2^{-3}a^{-9}b^6)$
$= -\dfrac{3a^{-8}b^7}{2^3} = -\dfrac{3b^7}{8a^8}$

Problem 4

A. $57,000,000,000 = 5.7 \cdot 10^{10}$

B. $0.000000017 = 1.7 \cdot 10^{-8}$

Problem 5

A. $5 \cdot 10^{12} = 5,000,000,000,000$

B. $4.0162 \cdot 10^{-9} = 0.0000000040162$

Problem 6

A. $(2.4 \cdot 10^{-9})(1.6 \cdot 10^3) = 3.84 \cdot 10^{-6}$

B. $\dfrac{5.4 \cdot 10^{-2}}{1.8 \cdot 10^{-4}} = 3 \cdot 10^2$

SECTION 7.5

Problem 1

$\dfrac{4x^3y + 8x^2y^2 - 4xy^3}{2xy} = \dfrac{4x^3y}{2xy} + \dfrac{8x^2y^2}{2xy} - \dfrac{4xy^3}{2xy}$
$= 2x^2 + 4xy - 2y^2$

Problem 2

$\dfrac{24x^2y^2 - 18xy + 6y}{6xy} = \dfrac{24x^2y^2}{6xy} - \dfrac{18xy}{6xy} + \dfrac{6y}{6xy}$

$= 4xy - 3 + \dfrac{1}{x}$

Problem 3

$$\begin{array}{r} x^2 + 2x + 2 \\ x - 2\overline{)x^3 + 0x^2 - 2x - 4} \\ \underline{x^3 - 2x^2} \\ 2x^2 - 2x \\ \underline{2x^2 - 4x} \\ 2x - 4 \\ \underline{2x - 4} \\ 0 \end{array}$$

$(x^3 - 2x - 4) \div (x - 2) = x^2 + 2x + 2$

Solutions to Chapter 8 Problems

SECTION 8.1

Problem 1

$4x^6y = 2 \cdot 2 \cdot x^6 \cdot y$
$18x^2y^6 = 2 \cdot 3 \cdot 3 \cdot x^2 \cdot y^6$
$\text{GCF} = 2 \cdot x^2 \cdot y = 2x^2y$

Problem 2

A. $14a^2 = 2 \cdot 7 \cdot a^2$
$21a^4b = 3 \cdot 7 \cdot a^4 \cdot b$
The GCF is $7a^2$.

$\begin{aligned} 14a^2 - 21a^4b &= 7a^2(2) + 7a^2(-3a^2b) \\ &= 7a^2(2 - 3a^2b) \end{aligned}$

B. $6x^4y^2 = 2 \cdot 3 \cdot x^4 \cdot y^2$
$9x^3y^2 = 3 \cdot 3 \cdot x^3 \cdot y^2$
$12x^2y^4 = 2 \cdot 2 \cdot 3 \cdot x^2 \cdot y^4$
The GCF is $3x^2y^2$.

$\begin{aligned} &6x^4y^2 - 9x^3y^2 + 12x^2y^4 \\ &= 3x^2y^2(2x^2) + 3x^2y^2(-3x) + 3x^2y^2(4y^2) \\ &= 3x^2y^2(2x^2 - 3x + 4y^2) \end{aligned}$

Problem 3

$a(b - 7) + b(b - 7) = (b - 7)(a + b)$

Problem 4

$\begin{aligned} &3y(5x - 2) - 4(2 - 5x) \\ &= 3y(5x - 2) + 4(5x - 2) \\ &= (5x - 2)(3y + 4) \end{aligned}$

Problem 5

$\begin{aligned} &y^5 - 5y^3 + 4y^2 - 20 \\ &= (y^5 - 5y^3) + (4y^2 - 20) \\ &= y^3(y^2 - 5) + 4(y^2 - 5) \\ &= (y^2 - 5)(y^3 + 4) \end{aligned}$

SECTION 8.2

Problem 1

Factors of 15	Sum
$-1, -15$	-16
$-3, -5$	-8

$x^2 - 8x + 15 = (x - 3)(x - 5)$

Problem 2

Factors of -18	Sum
$1, -18$	-17
$-1, 18$	17
$2, -9$	-7
$-2, 9$	7
$3, -6$	-3
$-3, 6$	3

$x^2 + 3x - 18 = (x + 6)(x - 3)$

Problem 3

The GCF is $3b$.

$3a^2b - 18ab - 81b = 3b(a^2 - 6a - 27)$

Factors of -27	Sum
$1, -27$	-26
$-1, 27$	26
$3, -9$	-6
$-3, 9$	6

$3a^2b - 18ab - 81b = 3b(a + 3)(a - 9)$

Problem 4

The GCF is 4.

$\begin{aligned} &4x^2 - 40xy + 84y^2 \\ &= 4(x^2 - 10xy + 21y^2) \end{aligned}$

Factors of 21	Sum
$-1, -21$	-22
$-3, -7$	-10

$4x^2 - 40xy + 84y^2 = 4(x - 3y)(x - 7y)$

SECTION 8.3

Problem 1

Factors of 6	Factors of 5
$1, 6$	$-1, -5$
$2, 3$	

Trial Factors	Middle Term
$(x - 1)(6x - 5)$	$-5x - 6x = -11x$

$6x^2 - 11x + 5 = (x - 1)(6x - 5)$

Problem 2

Factors of 8	Factors of -15
1, 8	1, -15
2, 4	-1, 15
	3, -5
	-3, 5

Trial Factors	Middle Term
$(x + 1)(8x - 15)$	$-15x + 8x = -7x$
$(x - 1)(8x + 15)$	$15x - 8x = 7x$
$(x + 3)(8x - 5)$	$-5x + 24x = 19x$
$(x - 3)(8x + 5)$	$5x - 24x = -19x$
$(2x + 1)(4x - 15)$	$-30x + 4x = -26x$
$(2x - 1)(4x + 15)$	$30x - 4x = 26x$
$(2x + 3)(4x - 5)$	$-10x + 12x = 2x$
$(2x - 3)(4x + 5)$	$10x - 12x = -2x$
$(8x + 1)(x - 15)$	$-120x + x = -119x$
$(8x - 1)(x + 15)$	$120x - x = 119x$
$(8x + 3)(x - 5)$	$-40x + 3x = -37x$
$(8x - 3)(x + 5)$	$40x - 3x = 37x$
$(4x + 1)(2x - 15)$	$-60x + 2x = -58x$
$(4x - 1)(2x + 15)$	$60x - 2x = 58x$
$(4x + 3)(2x - 5)$	$-20x + 6x = -14x$
$(4x - 3)(2x + 5)$	$20x - 6x = 14x$

$8x^2 + 14x - 15 = (4x - 3)(2x + 5)$

Problem 3

Factors of 24	Factors of -1
1, 24	1, -1
2, 12	
3, 8	
4, 6	

Trial Factors	Middle Term
$(1 - y)(24 + y)$	$y - 24y = -23y$
$(2 - y)(12 + y)$	$2y - 12y = -10y$
$(3 - y)(8 + y)$	$3y - 8y = -5y$
$(4 - y)(6 + y)$	$4y - 6y = -2y$

$24 - 2y - y^2 = (4 - y)(6 + y)$

Problem 4

The GCF is $2a^2$.

$4a^2b^2 - 30a^2b + 14a^2$
$= 2a^2(2b^2 - 15b + 7)$

Factors of 2	Factors of 7
1, 2	$-1, -7$

Trial Factors	Middle Term
$(b - 1)(2b - 7)$	$-7b - 2b = -9b$
$(b - 7)(2b - 1)$	$-b - 14b = -15b$

$4a^2b^2 - 30a^2b + 14a^2$
$= 2a^2(b - 7)(2b - 1)$

Problem 5

$a \cdot c = -14$
$-1(14) = -14, -1 + 14 = 13$

$2a^2 + 13a - 7$
$= 2a^2 - a + 14a - 7$
$= (2a^2 - a) + (14a - 7)$
$= a(2a - 1) + 7(2a - 1)$
$= (2a - 1)(a + 7)$

Problem 6

$a \cdot c = -12$
$1(-12) = -12, 1 - 12 = -11$

$4a^2 - 11a - 3$
$= 4a^2 + a - 12a - 3$
$= (4a^2 + a) - (12a + 3)$
$= a(4a + 1) - 3(4a + 1)$
$= (4a + 1)(a - 3)$

Problem 7

The GCF is $5x$.

$15x^3 + 40x^2 - 80x = 5x(3x^2 + 8x - 16)$
$3(-16) = -48$
$-4(12) = -48, -4 + 12 = 8$

$3x^2 + 8x - 16$
$= 3x^2 - 4x + 12x - 16$
$= (3x^2 - 4x) + (12x - 16)$
$= x(3x - 4) + 4(3x - 4)$
$= (3x - 4)(x + 4)$

$15x^3 + 40x^2 - 80x = 5x(3x^2 + 8x - 16)$
$\qquad\qquad\qquad\quad = 5x(3x - 4)(x + 4)$

SECTION 8.4

Problem 1

A. $25a^2 - b^2 = (5a)^2 - b^2$
$\qquad\qquad\quad = (5a + b)(5a - b)$

B. $6x^2 - 1$ is nonfactorable over the integers.

C. $n^8 - 36 = (n^4)^2 - 6^2$
$\qquad\qquad = (n^4 + 6)(n^4 - 6)$

Problem 2

$n^4 - 81 = (n^2 + 9)(n^2 - 9)$
$\qquad\quad = (n^2 + 9)(n + 3)(n - 3)$

Problem 3

A. $16y^2 = (4y)^2$, $1 = 1^2$

$(4y + 1)^2 = (4y)^2 + 2(4y)(1) + (1)^2$
$\qquad = 16y^2 + 8y + 1$

$16y^2 + 8y + 1 = (4y + 1)^2$

B. $x^2 = (x)^2$, $36 = 6^2$

$(x + 6)^2 = x^2 + 2(x)(6) + 6^2$
$\qquad = x^2 + 12x + 36$

The polynomial is not a perfect square.

$x^2 + 14x + 36$ is nonfactorable over the integers.

Problem 4

A. $12x^3 - 75x = 3x(4x^2 - 25)$
$\qquad = 3x(2x + 5)(2x - 5)$

B. $a^2b - 7a^2 - b + 7$
$\qquad = (a^2b - 7a^2) - (b - 7)$
$\qquad = a^2(b - 7) - (b - 7)$
$\qquad = (b - 7)(a^2 - 1)$
$\qquad = (b - 7)(a + 1)(a - 1)$

C. $4x^3 + 28x^2 - 120x = 4x(x^2 + 7x - 30)$
$\qquad = 4x(x + 10)(x - 3)$

SECTION 8.5

Problem 1

$2x^2 - 50 = 0$
$2(x^2 - 25) = 0$
$x^2 - 25 = 0$
$(x + 5)(x - 5) = 0$

$x + 5 = 0 \qquad x - 5 = 0$
$\quad x = -5 \qquad \quad x = 5$

The solutions are -5 and 5.

Problem 2

$(x + 2)(x - 7) = 52$
$x^2 - 5x - 14 = 52$
$x^2 - 5x - 66 = 0$
$(x - 11)(x + 6) = 0$

$x - 11 = 0 \qquad x + 6 = 0$
$\quad x = 11 \qquad \quad x = -6$

The solutions are -6 and 11.

Problem 3

Strategy First positive integer: n
Second positive integer: $n + 1$
Square of the first positive integer: n^2
Square of the second positive integer: $(n + 1)^2$

The sum of the squares of the two integers is 85.

Solution
$n^2 + (n + 1)^2 = 85$
$n^2 + n^2 + 2n + 1 = 85$
$2n^2 + 2n - 84 = 0$
$2(n^2 + n - 42) = 0$
$n^2 + n - 42 = 0$
$(n + 7)(n - 6) = 0$

$n + 7 = 0 \qquad n - 6 = 0$
$\quad n = -7 \qquad \quad n = 6$

-7 is not a positive integer.

$n + 1 = 6 + 1 = 7$

The two integers are 6 and 7.

Problem 4

Strategy Width $= W$
Length $= 2W + 3$

The area of the rectangle is 90 m².
Use the equation $A = LW$.

Solution
$A = LW$
$90 = (2W + 3)W$
$90 = 2W^2 + 3W$
$0 = 2W^2 + 3W - 90$
$0 = (2W + 15)(W - 6)$

$2W + 15 = 0 \qquad W - 6 = 0$
$\quad 2W = -15 \qquad \quad W = 6$
$\quad W = -\dfrac{15}{2}$

The width cannot be a negative number.

$2W + 3 = 2(6) + 3 = 12 + 3 = 15$

The width is 6 m. The length is 15 m.

Solutions to Chapter 9 Problems

SECTION 9.1

Problem 1

A. $\dfrac{6x^5y}{12x^2y^3} = \dfrac{\overset{1}{\cancel{2}} \cdot \overset{1}{\cancel{3}} \cdot x^5y}{\underset{1}{\cancel{2}} \cdot 2 \cdot \underset{1}{\cancel{3}} \cdot x^2y^3} = \dfrac{x^3}{2y^2}$

B. $\dfrac{x^2 + 2x - 24}{16 - x^2} = \dfrac{\overset{-1}{\cancel{(x - 4)}}(x + 6)}{\underset{1}{\cancel{(4 - x)}}(4 + x)} = -\dfrac{x + 6}{x + 4}$

Problem 2

A. $\dfrac{12x^2 + 3x}{10x - 15} \cdot \dfrac{8x - 12}{9x + 18}$

$= \dfrac{3x(4x + 1)}{5(2x - 3)} \cdot \dfrac{4(2x - 3)}{9(x + 2)}$

$= \dfrac{\overset{1}{3x}(4x + 1) \cdot 2 \cdot 2(\cancel{2x - 3})}{5(\cancel{2x - 3}) \cdot \underset{1}{\cancel{3}} \cdot 3(x + 2)}$

$= \dfrac{4x(4x + 1)}{15(x + 2)}$

B. $\dfrac{x^2 + 2x - 15}{9 - x^2} \cdot \dfrac{x^2 - 3x - 18}{x^2 - 7x + 6}$

$= \dfrac{(x - 3)(x + 5)}{(3 - x)(3 + x)} \cdot \dfrac{(x + 3)(x - 6)}{(x - 1)(x - 6)}$

$= \dfrac{\overset{-1}{\cancel{(x - 3)}}(x + 5) \cdot \cancel{(x + 3)}\cancel{(x - 6)}}{\underset{1}{\cancel{(3 - x)}}\underset{1}{\cancel{(3 + x)}} \cdot (x - 1)\underset{1}{\cancel{(x - 6)}}}$

$= -\dfrac{x + 5}{x - 1}$

Problem 3

A. $\dfrac{a^2}{4bc^2 - 2b^2c} \div \dfrac{a}{6bc - 3b^2}$

$= \dfrac{a^2}{4bc^2 - 2b^2c} \cdot \dfrac{6bc - 3b^2}{a}$

$= \dfrac{a^2 \cdot 3b\overset{1}{\cancel{(2c - b)}}}{2bc\underset{1}{\cancel{(2c - b)}} \cdot a} = \dfrac{3a}{2c}$

B. $\dfrac{3x^2 + 26x + 16}{3x^2 - 7x - 6} \div \dfrac{2x^2 + 9x - 5}{x^2 + 2x - 15}$

$= \dfrac{3x^2 + 26x + 16}{3x^2 - 7x - 6} \cdot \dfrac{x^2 + 2x - 15}{2x^2 + 9x - 5}$

$= \dfrac{\cancel{(3x + 2)}(x + 8)}{\cancel{(3x + 2)}\underset{1}{\cancel{(x - 3)}}} \cdot \dfrac{\underset{1}{\cancel{(x + 5)}}\cancel{(x - 3)}}{(2x - 1)\underset{1}{\cancel{(x + 5)}}}$

$= \dfrac{x + 8}{2x - 1}$

SECTION 9.2

Problem 1

$8uv^2 = 2 \cdot 2 \cdot 2 \cdot u \cdot v \cdot v$
$12uw = 2 \cdot 2 \cdot 3 \cdot u \cdot w$

$\text{LCM} = 2 \cdot 2 \cdot 2 \cdot 3 \cdot u \cdot v \cdot v \cdot w$
$\qquad = 24uv^2w$

Problem 2

$m^2 - 6m + 9 = (m - 3)(m - 3)$
$m^2 - 2m - 3 = (m + 1)(m - 3)$

$\text{LCM} = (m - 3)(m - 3)(m + 1)$

Problem 3

The LCM is $36xy^2z$.

$\dfrac{x - 3}{4xy^2} = \dfrac{x - 3}{4xy^2} \cdot \dfrac{9z}{9z} = \dfrac{9xz - 27z}{36xy^2z}$

$\dfrac{2x + 1}{9y^2z} = \dfrac{2x + 1}{9y^2z} \cdot \dfrac{4x}{4x} = \dfrac{8x^2 + 4x}{36xy^2z}$

Problem 4

$\dfrac{2x}{25 - x^2} = \dfrac{2x}{-(x^2 - 25)} = -\dfrac{2x}{x^2 - 25}$

The LCM is $(x + 2)(x - 5)(x + 5)$.

$\dfrac{x + 4}{x^2 - 3x - 10} = \dfrac{x + 4}{(x + 2)(x - 5)} \cdot \dfrac{x + 5}{x + 5}$

$\qquad\qquad = \dfrac{x^2 + 9x + 20}{(x + 2)(x - 5)(x + 5)}$

$\dfrac{2x}{25 - x^2} = -\dfrac{2x}{(x - 5)(x + 5)} \cdot \dfrac{x + 2}{x + 2}$

$\qquad\qquad = -\dfrac{2x^2 + 4x}{(x + 2)(x - 5)(x + 5)}$

SECTION 9.3

Problem 1

A. $\dfrac{3}{xy} + \dfrac{12}{xy} = \dfrac{3 + 12}{xy} = \dfrac{15}{xy}$

B. $\dfrac{2x^2}{x^2 - x - 12} - \dfrac{7x + 4}{x^2 - x - 12} = \dfrac{2x^2 - (7x + 4)}{x^2 - x - 12}$

$= \dfrac{2x^2 - 7x - 4}{x^2 - x - 12} = \dfrac{(2x + 1)\overset{1}{\cancel{(x - 4)}}}{(x + 3)\underset{1}{\cancel{(x - 4)}}} = \dfrac{2x + 1}{x + 3}$

Problem 2

A. The LCM of the denominators is $24y$.

$\dfrac{z}{8y} - \dfrac{4z}{3y} + \dfrac{5z}{4y} = \dfrac{z}{8y} \cdot \dfrac{3}{3} - \dfrac{4z}{3y} \cdot \dfrac{8}{8} + \dfrac{5z}{4y} \cdot \dfrac{6}{6}$

$= \dfrac{3z}{24y} - \dfrac{32z}{24y} + \dfrac{30z}{24y} = \dfrac{3z - 32z + 30z}{24y} = \dfrac{z}{24y}$

B. The LCM is $x - 2$.

$\dfrac{5x}{x - 2} - \dfrac{3}{2 - x} = \dfrac{5x}{x - 2} - \dfrac{3}{-(x - 2)} \cdot \dfrac{-1}{-1}$

$= \dfrac{5x}{x - 2} - \dfrac{-3}{x - 2} = \dfrac{5x - (-3)}{x - 2} = \dfrac{5x + 3}{x - 2}$

C. The LCM is $y - 7$.

$y + \dfrac{5}{y - 7} = \dfrac{y}{1} + \dfrac{5}{y - 7} = \dfrac{y}{1} \cdot \dfrac{y - 7}{y - 7} + \dfrac{5}{y - 7}$

$= \dfrac{y^2 - 7y}{y - 7} + \dfrac{5}{y - 7} = \dfrac{y^2 - 7y + 5}{y - 7}$

Problem 3

A. The LCM is $(3x - 1)(x + 4)$.

$$\frac{4x}{3x - 1} - \frac{9}{x + 4}$$

$$= \frac{4x}{3x - 1} \cdot \frac{x + 4}{x + 4} - \frac{9}{x + 4} \cdot \frac{3x - 1}{3x - 1}$$

$$= \frac{4x^2 + 16x}{(3x - 1)(x + 4)} - \frac{27x - 9}{(3x - 1)(x + 4)}$$

$$= \frac{(4x^2 + 16x) - (27x - 9)}{(3x - 1)(x + 4)}$$

$$= \frac{4x^2 + 16x - 27x + 9}{(3x - 1)(x + 4)} = \frac{4x^2 - 11x + 9}{(3x - 1)(x + 4)}$$

B. The LCM is $(x + 5)(x - 5)$.

$$\frac{2x - 1}{x^2 - 25} + \frac{2}{5 - x}$$

$$= \frac{2x - 1}{(x + 5)(x - 5)} + \frac{2}{-(x - 5)} \cdot \frac{-1(x + 5)}{-1(x + 5)}$$

$$= \frac{2x - 1}{(x + 5)(x - 5)} + \frac{-2(x + 5)}{(x + 5)(x - 5)}$$

$$= \frac{(2x - 1) + (-2)(x + 5)}{(x + 5)(x - 5)}$$

$$= \frac{2x - 1 - 2x - 10}{(x + 5)(x - 5)}$$

$$= \frac{-11}{(x + 5)(x - 5)} = -\frac{11}{(x + 5)(x - 5)}$$

SECTION 9.4

Problem 1

A. The LCM of 3, x, 9, and x^2 is $9x^2$.

$$\frac{\dfrac{1}{3} - \dfrac{1}{x}}{\dfrac{1}{9} - \dfrac{1}{x^2}} = \frac{\dfrac{1}{3} - \dfrac{1}{x}}{\dfrac{1}{9} - \dfrac{1}{x^2}} \cdot \frac{9x^2}{9x^2} = \frac{\dfrac{1}{3} \cdot 9x^2 - \dfrac{1}{x} \cdot 9x^2}{\dfrac{1}{9} \cdot 9x^2 - \dfrac{1}{x^2} \cdot 9x^2}$$

$$= \frac{3x^2 - 9x}{x^2 - 9} = \frac{3x(x - 3)}{(x - 3)(x + 3)} = \frac{3x}{x + 3}$$

B. The LCM of x and x^2 is x^2.

$$\frac{1 + \dfrac{4}{x} + \dfrac{3}{x^2}}{1 + \dfrac{10}{x} + \dfrac{21}{x^2}} = \frac{1 + \dfrac{4}{x} + \dfrac{3}{x^2}}{1 + \dfrac{10}{x} + \dfrac{21}{x^2}} \cdot \frac{x^2}{x^2}$$

$$= \frac{1 \cdot x^2 + \dfrac{4}{x} \cdot x^2 + \dfrac{3}{x^2} \cdot x^2}{1 \cdot x^2 + \dfrac{10}{x} \cdot x^2 + \dfrac{21}{x^2} \cdot x^2}$$

$$= \frac{x^2 + 4x + 3}{x^2 + 10x + 21}$$

$$= \frac{(x + 1)(x + 3)}{(x + 3)(x + 7)} = \frac{x + 1}{x + 7}$$

SECTION 9.5

Problem 1

The LCM of 3 and $3x$ is $3x$.

$$x + \frac{1}{3} = \frac{4}{3x}$$

$$3x\left(x + \frac{1}{3}\right) = 3x\left(\frac{4}{3x}\right)$$

$$3x \cdot x + \frac{3x}{1} \cdot \frac{1}{3} = \frac{3x}{1} \cdot \frac{4}{3x}$$

$$3x^2 + x = 4$$

$$3x^2 + x - 4 = 0$$

$$(3x + 4)(x - 1) = 0$$

$$3x + 4 = 0 \qquad x - 1 = 0$$

$$3x = -4 \qquad x = 1$$

$$x = -\frac{4}{3}$$

Both $-\dfrac{4}{3}$ and 1 check as solutions. The solutions are $-\dfrac{4}{3}$ and 1.

Problem 2

The LCM is $x + 2$.

$$\frac{5x}{x + 2} = 3 - \frac{10}{x + 2}$$

$$\frac{(x + 2)}{1} \cdot \frac{5x}{x + 2} = \frac{(x + 2)}{1}\left(3 - \frac{10}{x + 2}\right)$$

$$5x = (x + 2)3 - 10$$

$$5x = 3x + 6 - 10$$

$$5x = 3x - 4$$

$$2x = -4$$

$$x = -2$$

-2 does not check as a solution. The equation has no solution.

Problem 3

A.

$$\frac{2}{x + 3} = \frac{6}{5x + 5}$$

$$(x + 3)(5x + 5)\frac{2}{x + 3} = (x + 3)(5x + 5)\frac{6}{5x + 5}$$

$$(5x + 5)2 = (x + 3)6$$

$$10x + 10 = 6x + 18$$

$$4x + 10 = 18$$

$$4x = 8$$

$$x = 2$$

The solution is 2.

B.
$$\frac{5}{2x - 3} = \frac{10}{x + 3}$$

$$(x + 3)(2x - 3)\frac{5}{2x - 3} = (x + 3)(2x - 3)\frac{10}{x + 3}$$

$$(x + 3)5 = (2x - 3)10$$
$$5x + 15 = 20x - 30$$
$$-15x + 15 = -30$$
$$-15x = -45$$
$$x = 3$$

The solution is 3.

Problem 4

Strategy To find the total area that 270 ceramic tiles will cover, write and solve a proportion using x to represent the number of square feet that 270 tiles will cover.

Solution
$$\frac{4}{9} = \frac{x}{270}$$

$$270\left(\frac{4}{9}\right) = 270\left(\frac{x}{270}\right)$$
$$120 = x$$

A 120 ft² area can be tiled using 270 ceramic tiles.

Problem 5

Strategy To find the additional amount of medication required for a 180-pound adult, write and solve a proportion using x to represent the additional medication. Then $3 + x$ is the total amount required for a 180-pound adult.

Solution
$$\frac{120}{3} = \frac{180}{3 + x}$$

$$\frac{40}{1} = \frac{180}{3 + x}$$

$$(3 + x) \cdot 40 = (3 + x) \cdot \frac{180}{3 + x}$$
$$120 + 40x = 180$$
$$40x = 60$$
$$x = 1.5$$

1.5 additional ounces are required for a 180-pound adult.

Problem 6

Strategy Triangle AOB is similar to triangle DOC. Solve a proportion to find the length of AO. Then use the formula for the area of a triangle to find the area of triangle AOB.

Solution
$$\frac{CD}{AB} = \frac{DO}{AO}$$

$$\frac{4}{10} = \frac{3}{x}$$

$$10x\left(\frac{4}{10}\right) = 10x\left(\frac{3}{x}\right)$$
$$4x = 30$$
$$x = 7.5$$

$$A = \frac{1}{2}bh$$

$$A = \frac{1}{2}(10)(7.5)$$

$$A = 37.5$$

The area of triangle AOB is 37.5 cm².

SECTION 9.6

Problem 1

A.
$$s = \frac{A + L}{2}$$

$$2 \cdot s = 2\left(\frac{A + L}{2}\right)$$
$$2s = A + L$$
$$2s - A = A - A + L$$
$$2s - A = L$$

B.
$$S = a + (n - 1)d$$
$$S = a + nd - d$$
$$S - a = a - a + nd - d$$
$$S - a = nd - d$$
$$S - a + d = nd - d + d$$
$$S - a + d = nd$$
$$\frac{S - a + d}{d} = \frac{nd}{d}$$
$$\frac{S - a + d}{d} = n$$

Problem 2

$$S = rS + C$$
$$S - rS = C$$
$$S(1 - r) = C$$
$$S = \frac{C}{1 - r}$$

SECTION 9.7

Problem 1

Strategy ▶ Time for one printer to complete the job: t

	Rate	Time	Part
1st printer	$\frac{1}{t}$	3	$\frac{3}{t}$
2nd printer	$\frac{1}{t}$	5	$\frac{5}{t}$

▶ The sum of the parts of the task completed must equal 1.

Solution
$$\frac{3}{t} + \frac{5}{t} = 1$$
$$t\left(\frac{3}{t} + \frac{5}{t}\right) = t \cdot 1$$
$$3 + 5 = t$$
$$8 = t$$

Working alone, one printer takes 8 h to print the payroll.

Problem 2

Strategy ▶ Rate sailing across the lake: r
Rate sailing back: $2r$

	Distance	Rate	Time
Across	6	r	$\dfrac{6}{r}$
Back	6	$2r$	$\dfrac{6}{2r}$

▶ The total time for the trip was 3 h.

Solution
$$\frac{6}{r} + \frac{6}{2r} = 3$$
$$2r\left(\frac{6}{r} + \frac{6}{2r}\right) = 2r(3)$$
$$2r \cdot \frac{6}{r} + 2r \cdot \frac{6}{2r} = 6r$$
$$12 + 6 = 6r$$
$$18 = 6r$$
$$3 = r$$

The rate across the lake was 3 km/h.

Solutions to Chapter 10 Problems

SECTION 10.1

Problem 1
$$-5\sqrt{32} = -5\sqrt{16 \cdot 2}$$
$$= -5\sqrt{16}\sqrt{2}$$
$$= -5 \cdot 4\sqrt{2}$$
$$= -20\sqrt{2}$$

Problem 2
$$\sqrt{216} = \sqrt{36 \cdot 6}$$
$$= \sqrt{36}\sqrt{6}$$
$$= 6\sqrt{6}$$

Problem 3
$$\sqrt{y^{19}} = \sqrt{y^{18} \cdot y} = \sqrt{y^{18}}\sqrt{y} = y^9\sqrt{y}$$

Problem 4
A. $\sqrt{45b^7} = \sqrt{9b^6 \cdot 5b} = \sqrt{9b^6}\sqrt{5b} = 3b^3\sqrt{5b}$

B. $3a\sqrt{28a^9b^{18}} = 3a\sqrt{4a^8b^{18} \cdot 7a}$
$$= 3a\sqrt{4a^8b^{18}}\sqrt{7a}$$
$$= 3a \cdot 2a^4b^9\sqrt{7a}$$
$$= 6a^5b^9\sqrt{7a}$$

Problem 5
$$\sqrt{25(a + 3)^2} = 5(a + 3) = 5a + 15$$

SECTION 10.2

Problem 1
A. $9\sqrt{3} + 3\sqrt{3} - 18\sqrt{3}$
$$= (9 + 3 - 18)\sqrt{3} = -6\sqrt{3}$$

B. $2\sqrt{50} - 5\sqrt{32} = 2\sqrt{25 \cdot 2} - 5\sqrt{16 \cdot 2}$
$$= 2\sqrt{25}\sqrt{2} - 5\sqrt{16}\sqrt{2}$$
$$= 2 \cdot 5\sqrt{2} - 5 \cdot 4\sqrt{2}$$
$$= 10\sqrt{2} - 20\sqrt{2} = -10\sqrt{2}$$

Problem 2
A. $y\sqrt{28y} + 7\sqrt{63y^3}$
$$= y\sqrt{4}\sqrt{7y} + 7\sqrt{9y^2}\sqrt{7y}$$
$$= y \cdot 2\sqrt{7y} + 7 \cdot 3y\sqrt{7y}$$
$$= 2y\sqrt{7y} + 21y\sqrt{7y} = 23y\sqrt{7y}$$

B. $2\sqrt{27a^5} - 4a\sqrt{12a^3} + a^2\sqrt{75a}$
$$= 2\sqrt{9a^4}\sqrt{3a} - 4a\sqrt{4a^2}\sqrt{3a} + a^2\sqrt{25}\sqrt{3a}$$
$$= 2 \cdot 3a^2\sqrt{3a} - 4a \cdot 2a\sqrt{3a} + a^2 \cdot 5\sqrt{3a}$$
$$= 6a^2\sqrt{3a} - 8a^2\sqrt{3a} + 5a^2\sqrt{3a} = 3a^2\sqrt{3a}$$

SECTION 10.3

Problem 1
$$\sqrt{5a}\sqrt{15a^3b^4}\sqrt{3b^5} = \sqrt{225a^4b^9}$$
$$= \sqrt{225a^4b^8}\sqrt{b}$$
$$= 15a^2b^4\sqrt{b}$$

Problem 2
$$\sqrt{5x}\left(\sqrt{5x} - \sqrt{25y}\right) = \left(\sqrt{5x}\right)^2 - \sqrt{125xy}$$
$$= \sqrt{25x^2} - \sqrt{25}\sqrt{5xy} = 5x - 5\sqrt{5xy}$$

Problem 3
$$\left(3\sqrt{x} - \sqrt{y}\right)\left(5\sqrt{x} - 2\sqrt{y}\right)$$
$$= 15\left(\sqrt{x}\right)^2 - 6\sqrt{xy} - 5\sqrt{xy} + 2\left(\sqrt{y}\right)^2$$
$$= 15x - 11\sqrt{xy} + 2y$$

Problem 4
$$\left(2\sqrt{x} + 7\right)\left(2\sqrt{x} - 7\right)$$
$$= \left(2\sqrt{x}\right)^2 - 7^2 = 4x - 49$$

Problem 5
A. $\dfrac{\sqrt{15x^6y^7}}{\sqrt{3x^7y^9}} = \sqrt{\dfrac{15x^6y^7}{3x^7y^9}} = \sqrt{\dfrac{5}{xy^2}} = \dfrac{\sqrt{5}}{\sqrt{xy^2}}$
$$= \frac{\sqrt{5}}{y\sqrt{x}} = \frac{\sqrt{5}}{y\sqrt{x}} \cdot \frac{\sqrt{x}}{\sqrt{x}} = \frac{\sqrt{5x}}{xy}$$

B. $\dfrac{\sqrt{y}}{\sqrt{y}+3} = \dfrac{\sqrt{y}}{\sqrt{y}+3} \cdot \dfrac{\sqrt{y}-3}{\sqrt{y}-3} = \dfrac{y-3\sqrt{y}}{y-9}$

C. $\dfrac{5+\sqrt{y}}{1-2\sqrt{y}} = \dfrac{5+\sqrt{y}}{1-2\sqrt{y}} \cdot \dfrac{1+2\sqrt{y}}{1+2\sqrt{y}}$

$\qquad = \dfrac{5+10\sqrt{y}+\sqrt{y}+2y}{1-4y}$

$\qquad = \dfrac{5+11\sqrt{y}+2y}{1-4y}$

SECTION 10.4

Problem 1

$\sqrt{4x}+3 = 7$
$\sqrt{4x} = 4$
$(\sqrt{4x})^2 = 4^2$
$4x = 16$
$x = 4$

Check $\quad \sqrt{4x}+3 = 7$

$$\begin{array}{c|c} \sqrt{4\cdot 4}+3 & 7 \\ \sqrt{16}+3 & 7 \\ 4+3 & 7 \\ & 7 = 7 \end{array}$$

The solution is 4.

Problem 2

A. $\sqrt{3x-2}-5 = 0$
$\sqrt{3x-2} = 5$
$(\sqrt{3x-2})^2 = 5^2$
$3x-2 = 25$
$3x = 27$
$x = 9$

Check $\quad \sqrt{3x-2}-5 = 0$

$$\begin{array}{c|c} \sqrt{3\cdot 9-2}-5 & 0 \\ \sqrt{27-2}-5 & 0 \\ \sqrt{25}-5 & 0 \\ 5-5 & 0 \\ & 0 = 0 \end{array}$$

The solution is 9.

B. $\sqrt{4x-7}+5 = 0$
$\sqrt{4x-7} = -5$
$(\sqrt{4x-7})^2 = (-5)^2$
$4x-7 = 25$
$4x = 32$
$x = 8$

Check $\quad \sqrt{4x-7}+5 = 0$

$$\begin{array}{c|c} \sqrt{4\cdot 8-7}+5 & 0 \\ \sqrt{32-7}+5 & 0 \\ \sqrt{25}+5 & 0 \\ 5+5 & 0 \\ & 10 \neq 0 \end{array}$$

There is no solution.

Problem 3

$\sqrt{x}+\sqrt{x+9} = 9$
$\sqrt{x} = 9-\sqrt{x+9}$
$(\sqrt{x})^2 = (9-\sqrt{x+9})^2$
$x = 81-18\sqrt{x+9}+(x+9)$
$18\sqrt{x+9} = 90$
$\sqrt{x+9} = 5$
$(\sqrt{x+9})^2 = 5^2$
$x+9 = 25$
$x = 16$

Check $\quad \sqrt{x}+\sqrt{x+9} = 9$

$$\begin{array}{c|c} \sqrt{16}+\sqrt{16+9} & 9 \\ 4+\sqrt{25} & 9 \\ 4+5 & 9 \\ & 9 = 9 \end{array}$$

The solution is 16.

Problem 4

Strategy To find the distance, use the Pythagorean Theorem. The hypotenuse is the length of the ladder. One leg is the distance from the bottom of the ladder to the base of the building. The distance along the building from the ground to the top of the ladder is the unknown leg.

Solution $a = \sqrt{c^2 - b^2}$
$a = \sqrt{(12)^2 - (5)^2}$
$a = \sqrt{144 - 25}$
$a = \sqrt{119}$
$a \approx 10.91$

The distance is 10.91 ft.

Problem 5

Strategy To find the length of the pendulum, replace T in the equation with the given value and solve for L.

Solution

$$T = 2\pi\sqrt{\dfrac{L}{32}}$$

$$1.5 = 2\pi\sqrt{\dfrac{L}{32}}$$

$$\dfrac{1.5}{2\pi} = \sqrt{\dfrac{L}{32}}$$

$$\left(\dfrac{1.5}{2\pi}\right)^2 = \left(\sqrt{\dfrac{L}{32}}\right)^2$$

$$\left(\dfrac{1.5}{2\pi}\right)^2 = \dfrac{L}{32}$$

$$32\left(\dfrac{1.5}{2\pi}\right)^2 = L \qquad \text{• Use the } \pi \text{ key on your calculator.}$$

$$1.82 \approx L$$

The length of the pendulum is 1.82 ft.

Solutions to Chapter 11 Problems

SECTION 11.1

Problem 1

$$\frac{3y^2}{2} + y - \frac{1}{2} = 0$$

$$2\left(\frac{3y^2}{2} + y - \frac{1}{2}\right) = 2(0)$$

$$3y^2 + 2y - 1 = 0$$

$$(3y - 1)(y + 1) = 0$$

$$3y - 1 = 0 \qquad y + 1 = 0$$
$$3y = 1 \qquad\qquad y = -1$$
$$y = \frac{1}{3}$$

The solutions are $\frac{1}{3}$ and -1.

Problem 2

A. $4x^2 - 96 = 0$

$$4x^2 = 96$$
$$x^2 = 24$$
$$\sqrt{x^2} = \sqrt{24}$$
$$x = \pm\sqrt{24}$$
$$x = \pm 2\sqrt{6}$$

The solutions are $2\sqrt{6}$ and $-2\sqrt{6}$.

B. $x^2 + 81 = 0$

$$x^2 = -81$$
$$\sqrt{x^2} = \sqrt{-81} \quad \text{(not a real number)}$$

The equation has no real number solution.

Problem 3

$$(x + 5)^2 = 20$$
$$\sqrt{(x + 5)^2} = \sqrt{20}$$
$$x + 5 = \pm\sqrt{20}$$
$$x + 5 = \pm 2\sqrt{5}$$

$$x + 5 = 2\sqrt{5} \qquad x + 5 = -2\sqrt{5}$$
$$x = -5 + 2\sqrt{5} \qquad x = -5 - 2\sqrt{5}$$

The solutions are $-5 + 2\sqrt{5}$ and $-5 - 2\sqrt{5}$.

SECTION 11.2

Problem 1

A. $3x^2 - 6x - 2 = 0$

$$3x^2 - 6x = 2$$
$$\frac{1}{3}(3x^2 - 6x) = \frac{1}{3} \cdot 2$$
$$x^2 - 2x = \frac{2}{3}$$

Complete the square.

$$x^2 - 2x + 1 = \frac{2}{3} + 1$$
$$(x - 1)^2 = \frac{5}{3}$$
$$\sqrt{(x - 1)^2} = \sqrt{\frac{5}{3}}$$
$$x - 1 = \pm\sqrt{\frac{5}{3}}$$
$$x - 1 = \pm\frac{\sqrt{15}}{3}$$

$$x - 1 = \frac{\sqrt{15}}{3} \qquad x - 1 = -\frac{\sqrt{15}}{3}$$
$$x = 1 + \frac{\sqrt{15}}{3} \qquad x = 1 - \frac{\sqrt{15}}{3}$$
$$= \frac{3 + \sqrt{15}}{3} \qquad = \frac{3 - \sqrt{15}}{3}$$

The solutions are $\frac{3 + \sqrt{15}}{3}$ and $\frac{3 - \sqrt{15}}{3}$.

B. $y^2 - 6y + 10 = 0$

$$y^2 - 6y = -10$$

Complete the square.

$$y^2 - 6y + 9 = -10 + 9$$
$$(y - 3)^2 = -1$$
$$\sqrt{(y - 3)^2} = \sqrt{-1}$$

$\sqrt{-1}$ is not a real number.

The equation has no real number solution.

Problem 2

$$x^2 + 8x + 8 = 0$$
$$x^2 + 8x = -8$$
$$x^2 + 8x + 16 = -8 + 16$$
$$(x + 4)^2 = 8$$
$$\sqrt{(x + 4)^2} = \sqrt{8}$$
$$x + 4 = \pm\sqrt{8}$$
$$x + 4 \approx \pm 2.828$$

$$x + 4 \approx 2.828 \qquad x + 4 \approx -2.828$$
$$x \approx -4 + 2.828 \qquad x \approx -4 - 2.828$$
$$\approx -1.172 \qquad\qquad \approx -6.828$$

The solutions are approximately -1.172 and -6.828.

SECTION 11.3

Problem 1

A. $3x^2 + 4x - 4 = 0$

$$a = 3, b = 4, c = -4$$
$$x = \frac{-(4) \pm \sqrt{(4)^2 - 4(3)(-4)}}{2 \cdot 3}$$
$$= \frac{-4 \pm \sqrt{16 + 48}}{6}$$

$$= \frac{-4 \pm \sqrt{64}}{6} = \frac{-4 \pm 8}{6}$$

$$x = \frac{-4 + 8}{6} \qquad x = \frac{-4 - 8}{6}$$

$$= \frac{4}{6} = \frac{2}{3} \qquad\qquad = \frac{-12}{6} = -2$$

The solutions are $\frac{2}{3}$ and -2.

B. $\quad x^2 + 2x = 1$
$\quad\ \ x^2 + 2x - 1 = 0$

$\quad a = 1, b = 2, c = -1$

$$x = \frac{-(2) \pm \sqrt{(2)^2 - 4(1)(-1)}}{2 \cdot 1}$$

$$= \frac{-2 \pm \sqrt{4 + 4}}{2}$$

$$= \frac{-2 \pm \sqrt{8}}{2}$$

$$= \frac{-2 \pm 2\sqrt{2}}{2}$$

$$= \frac{2(-1 \pm \sqrt{2})}{2} = -1 \pm \sqrt{2}$$

The solutions are $-1 + \sqrt{2}$ and $-1 - \sqrt{2}$.

C. $\ z^2 + 2z + 6 = 0$

$\quad a = 1, b = 2, c = 6$

$$z = \frac{-(2) \pm \sqrt{2^2 - 4(1)(6)}}{2(1)}$$

$$= \frac{-2 \pm \sqrt{4 - 24}}{2} = \frac{-2 \pm \sqrt{-20}}{2}$$

$\sqrt{-20}$ is not a real number.

The equation has no real number solution.

SECTION 11.4

Problem 1

A. $y = x^2 + 2$
$\quad a = 1.$ a is positive.
The parabola opens up.

x	y
0	2
1	3
−1	3
2	6
−2	6

B. $y = -x^2 - 2x + 3$

$\quad a = -1.$ a is negative.

The parabola opens down.

x	y
0	3
1	0
−1	4
2	−5
−2	3
−3	0
−4	−5

Problem 2

To find the x-intercepts, let $y = 0$ and solve for x.

$y = x^2 - 6x + 9$
$0 = x^2 - 6x + 9$
$0 = (x - 3)(x - 3)$

$$x - 3 = 0 \qquad x - 3 = 0$$
$$x = 3 \qquad\quad x = 3$$

The x-intercept is $(3, 0)$.

To find the y-intercept, let $x = 0$ and solve for y.

$y = x^2 - 6x + 9$
$y = 0^2 - 6(0) + 9 = 9$

The y-intercept is $(0, 9)$.

SECTION 11.5

Problem 1

Strategy   ▶ This is a geometry problem.

         ▶ Width of the rectangle: W
            Length of the rectangle: $W + 3$

         ▶ Use the equation $A = LW$.

Solution $A = LW$
          $40 = (W + 3)W$
          $40 = W^2 + 3W$
           $0 = W^2 + 3W - 40$
           $0 = (W + 8)(W - 5)$

$$W + 8 = 0 \qquad W - 5 = 0$$
$$W = -8 \qquad\quad W = 5$$

The solution -8 is not possible.
The width is 5 m.

Answers to Selected Exercises

Answers to Chapter 1 Selected Exercises

PREP TEST

1. 127.16   **2.** 49,743   **3.** 4517   **4.** 11,396   **5.** 24   **6.** 24   **7.** 4   **8.** $3 \cdot 7$
9. $\dfrac{2}{5}$

CONCEPT REVIEW 1.1

1. Sometimes true   **3.** Never true   **5.** Sometimes true   **7.** Always true   **9.** Always true

SECTION 1.1

3. $>$ **5.** $<$ **7.** $>$ **9.** $>$ **11.** $<$ **13.** $>$ **15.** $>$ **17.** $<$
19. {1, 2, 3, 4, 5, 6, 7, 8}   **21.** {1, 2, 3, 4, 5, 6, 7, 8}   **23.** {−6, −5, −4, −3, −2, −1}   **25.** 5
27. −23, −18   **29.** 21, 37   **31.** −52, −46, 0   **33.** −17, 0, 4, 29   **35.** 5, 6, 7, 8, 9
37. −10, −9, −8, −7, −6, −5   **41.** −22   **43.** 31   **45.** 168   **47.** −630   **49.** 18
51. −49   **53.** 16   **55.** 12   **57.** −29   **59.** −14   **61.** 0   **63.** −34
65. **a.** 8, 5, 2, −1, −3   **b.** 8, 5, 2, 1, 3   **67.** $>$ **69.** $<$ **71.** $>$ **73.** $<$
75. −19, −|−8|, |−5|, 6   **77.** −22, −(−3), |−14|, |−25| **79.** **a.** 5°F with a 20 mph wind feels colder.
b. −15°F with a 20 mph wind feels colder.   **81.** −4, 4   **83.** −3, 11   **85.** Negative   **87.** 0
89. True

CONCEPT REVIEW 1.2

1. Sometimes true   **3.** Always true   **5.** Always true   **7.** Never true   **9.** Never true

SECTION 1.2

3. −11   **5.** −9   **7.** −3   **9.** 1   **11.** −5   **13.** −30   **15.** 9   **17.** 1   **19.** −10
21. −28   **23.** −392   **27.** 8   **29.** −7   **31.** 7   **33.** −2   **35.** −28   **37.** −13
39. 6   **41.** −9   **43.** 2   **45.** −138   **47.** −8   **53.** 42   **55.** −20   **57.** −16
59. 25   **61.** 0   **63.** −72   **65.** −102   **67.** 140   **69.** −70   **71.** 162   **73.** 120
75. 36   **77.** 192   **79.** −108   **81.** 0   **83.** $3(-12) = -36$ **85.** $-5(11) = -55$ **87.** −2
89. 8   **91.** 0   **93.** −9   **95.** −9   **97.** 9   **99.** −24   **101.** −12   **103.** −13
105. −18   **107.** 19   **109.** 26   **111.** 11   **113.** −13   **115.** 13   **117.** The temperature
is 3°C.   **119.** The difference is 14°C.   **121.** The difference is 399°C.   **123.** The difference in elevation
is 5662 m.   **125.** The difference in elevation is 6028 m.   **127.** The difference in elevation is 9248 m.
129. The average daily high temperature is −3°C.   **131.** The temperature dropped 100°F.
133. The average temperature at 30,000 ft is 36°F colder than at 20,000 ft.   **135.** The student's score is 93.
137. The 5-day moving average for the Disney stock is −24, −14, −12, 4, 8, 20.   **139.** 17   **141.** 3
143. −4, −9, −14   **145.** −16, 4, −1   **147.** 5436   **149.** b   **151.** a
153. No. $10 - (-8) = 18$

CONCEPT REVIEW 1.3

1. Never true   **3.** Sometimes true   **5.** Always true   **7.** Never true   **9.** Always true

SECTION 1.3

1. $0.\overline{3}$ **3.** 0.25 **5.** 0.4 **7.** $0.1\overline{6}$ **9.** 0.125 **11.** $0.\overline{2}$ **13.** $0.\overline{45}$ **15.** $0.58\overline{3}$

17. $0.2\overline{6}$ **19.** 0.4375 **21.** 0.24 **23.** 0.225 **25.** $0.6\overline{81}$ **27.** $0.458\overline{3}$ **29.** $0.\overline{15}$

31. $0.0\overline{81}$ **33.** $\dfrac{13}{12}$ **35.** $-\dfrac{5}{24}$ **37.** $-\dfrac{19}{24}$ **39.** $\dfrac{5}{26}$ **41.** $\dfrac{7}{24}$ **43.** 0 **45.** $-\dfrac{7}{16}$

47. $\dfrac{11}{24}$ **49.** 1 **51.** $\dfrac{11}{8}$ **53.** $\dfrac{169}{315}$ **55.** -38.8 **57.** -6.192 **59.** 13.355

61. 4.676 **63.** -10.03 **65.** -37.19 **67.** -17.5 **69.** 853.2594 **71.** $-\dfrac{3}{8}$ **73.** $\dfrac{1}{10}$

75. $\dfrac{15}{64}$ **77.** $\dfrac{3}{2}$ **79.** $-\dfrac{8}{9}$ **81.** $\dfrac{2}{3}$ **83.** 4.164 **85.** 4.347 **87.** -4.028

89. -2.22 **91.** 12.26448 **93.** 0.75 **95.** -2060.55 **97.** 0.09 **101.** $\dfrac{3}{4}$, 0.75

103. $\dfrac{1}{2}$, 0.5 **105.** $\dfrac{16}{25}$, 0.64 **107.** $1\dfrac{3}{4}$, 1.75 **109.** $\dfrac{19}{100}$, 0.19 **111.** $\dfrac{1}{20}$, 0.05

113. $4\dfrac{1}{2}$, 4.5 **115.** $\dfrac{2}{25}$, 0.08 **117.** $\dfrac{1}{9}$ **119.** $\dfrac{5}{16}$ **121.** $\dfrac{1}{200}$ **123.** $\dfrac{1}{16}$ **125.** 0.073

127. 0.158 **129.** 0.0915 **131.** 0.1823 **133.** 15% **135.** 5% **137.** 17.5%

139. 115% **141.** 0.8% **143.** 6.5% **145.** 54% **147.** 33.3% **149.** 44.4%

151. 250% **153.** $37\dfrac{1}{2}\%$ **155.** $35\dfrac{5}{7}\%$ **157.** 125% **159.** $155\dfrac{5}{9}\%$ **161.** $\dfrac{1}{25}$ of the

respondents found their most recent jobs on the Internet.   **163.** Fewer than one-quarter of the respondents found their most recent jobs through a newspaper ad.   **165.** Natural number, integer, positive integer, rational number, real number   **167.** Rational number, real number   **169.** Irrational number, real number

171. The cost is $1.29.   **173.** The difference is $138.499 billion.   **175.** The deficit in 1985 was 4 times greater than in 1975.   **177.** The average deficit for the years 1995 through 2000 was $-$4,579 million.

179. The new fraction is greater than $\dfrac{2}{5}$.   **181.** $1.06x$ **183.** $53¢$ **185.** Yes. For example, $c = \dfrac{a+b}{2}$.

CONCEPT REVIEW 1.4

1. Never true   **3.** Never true   **5.** Always true   **7.** Never true   **9.** Never true

SECTION 1.4

1. 36 **3.** -49 **5.** 9 **7.** 81 **9.** $\dfrac{1}{4}$ **11.** 0.09 **13.** 12 **15.** 0.216

17. -12 **19.** 16 **21.** -864 **23.** -1008 **25.** 3 **27.** $-77{,}760$ **31.** 9 **33.** 12

35. 1 **37.** 7 **39.** -36 **41.** 13 **43.** 4 **45.** 15 **47.** -1 **49.** 4

51. 0.51 **53.** 1.7 **55.** $\dfrac{17}{48}$ **57.** $<$ **59.** $>$ **61.** 100 **63.** 120 **65.** It will take the computer 17 s.   **67.** 6 **69.** 9

CHAPTER REVIEW EXERCISES*

1. {1, 2, 3, 4, 5, 6} [1.1.1]   **2.** 62.5% [1.3.4]   **3.** −4 [1.1.2]   **4.** 4 [1.2.2]   **5.** 17 [1.2.4]
6. $0.\overline{7}$ [1.3.1]   **7.** −5.3578 [1.3.3]   **8.** 8 [1.4.2]   **9.** 4 [1.1.2]   **10.** −14 [1.2.2]

11. 67.2% [1.3.4]   **12.** $\dfrac{159}{200}$ [1.3.4]   **13.** −9 [1.2.4]   **14.** 0.85 [1.3.1]   **15.** $-\dfrac{1}{2}$ [1.3.3]

16. 9 [1.4.2]   **17.** < [1.1.1]   **18.** −16 [1.2.1]   **19.** 90 [1.2.3]   **20.** −6.881 [1.3.2]
21. −5, −3 [1.1.1]   **22.** 0.07 [1.3.4]   **23.** 12 [1.4.1]   **24.** > [1.1.1]   **25.** −3 [1.2.1]

26. −48 [1.2.3]   **27.** $\dfrac{1}{15}$ [1.3.2]   **28.** −108 [1.4.1]   **29.** 277.8% [1.3.4]   **30.** 2.4 [1.3.4]

31. 2 [1.1.2]   **32.** −14 [1.2.2]   **33.** −8 [1.2.4]   **34.** 0.35 [1.3.1]   **35.** $-\dfrac{7}{6}$ [1.3.3]

36. 12 [1.4.2]   **37.** 3 [1.1.2]   **38.** 18 [1.2.2]   **39.** $288\dfrac{8}{9}\%$ [1.3.4]   **40.** **a.** 12, 8, 1, −7
b. 12, 8, 1, 7 [1.1.2]   **41.** 12 [1.2.4]   **42.** $0.\overline{63}$ [1.3.1]   **43.** −11.5 [1.3.3]   **44.** 8 [1.4.2]

45. > [1.1.1]   **46.** −8 [1.2.1]   **47.** 72 [1.2.3]   **48.** $-\dfrac{11}{24}$ [1.3.2]   **49.** −17, −9, 0, 4 [1.1.1]

50. 0.2% [1.3.4]   **51.** −4 [1.4.1]   **52.** 3.561 [1.3.2]   **53.** −17 [1.1.2]   **54.** −27 [1.2.2]

55. $-\dfrac{1}{10}$ [1.3.3]   **56.** < [1.1.1]   **57.** 0.72 [1.3.1]   **58.** −128 [1.4.1]   **59.** 7.5% [1.3.4]

60. $54\dfrac{2}{7}\%$ [1.3.4]   **61.** −18 [1.2.1]   **62.** 0 [1.2.3]   **63.** 9 [1.4.2]   **64.** $\dfrac{7}{8}$ [1.3.2]

65. 16 [1.2.4]   **66.** < [1.1.1]   **67.** −3 [1.4.1]   **68.** −6 [1.2.1]   **69.** 300 [1.2.3]
70. {−3, −2, −1} [1.1.1]   **71.** The temperature is 8°C. [1.2.5]   **72.** The average low temperature was
−2°C. [1.2.5]   **73.** The difference is 106°F. [1.2.5]   **74.** The temperature is −6°C. [1.2.5]
75. The difference is 714°C. [1.2.5]

CHAPTER TEST

1. $\dfrac{11}{20}$ [1.3.4]   **2.** −6, −8 [1.1.1]   **3.** −3 [1.2.2]   **4.** 0.15 [1.3.1]   **5.** $-\dfrac{1}{14}$ [1.3.3]

6. −15 [1.2.4]   **7.** $-\dfrac{8}{3}$ [1.4.1]   **8.** 2 [1.2.1]   **9.** {1, 2, 3, 4, 5, 6} [1.1.1]   **10.** 159% [1.3.4]

11. 29 [1.1.2]   **12.** > [1.1.1]   **13.** $-\dfrac{23}{18}$ [1.3.2]   **14.** 258 [1.2.3]   **15.** 3 [1.4.2]

16. $\dfrac{5}{6}$ [1.3.3]   **17.** 23.1% [1.3.4]   **18.** 0.062 [1.3.4]   **19.** 14 [1.2.2]   **20.** $0.4\overline{3}$ [1.3.1]

21. −2.43 [1.3.3]   **22.** 15 [1.2.4]   **23.** 640 [1.4.1]   **24.** −12 [1.2.1]   **25.** −34 [1.1.2]

26. > [1.1.1]   **27.** **a.** 17, 6, −5, −9   **b.** 17, 6, 5, 9 [1.1.2]   **28.** $69\dfrac{13}{23}\%$ [1.3.4]

29. −11.384 [1.3.2]   **30.** 160 [1.2.3]   **31.** −2 [1.4.2]   **32.** The temperature is 4°C. [1.2.5]
33. The average is −48°F. [1.2.5]

*The numbers in brackets following the answers in the Chapter Review Exercises refer to the objective that corresponds to that problem. For example, the reference [1.2.1] stands for Section 1.2, Objective 1. This notation will be used for all Pretests, Chapter Review Exercises, Chapter Tests, and Cumulative Review Exercises throughout the text.

Answers to Chapter 2 Selected Exercises

PREP TEST

1. 3 [1.2.2] **2.** 4 [1.2.4] **3.** $\frac{1}{12}$ [1.3.2] **4.** $-\frac{4}{9}$ [1.3.3] **5.** $\frac{3}{10}$ [1.3.3]

6. −16 [1.4.1] **7.** $\frac{8}{27}$ [1.4.1] **8.** 48 [1.4.1] **9.** 1 [1.4.2] **10.** 12 [1.4.2]

CONCEPT REVIEW 2.1

1. Always true **3.** Sometimes true **5.** Always true

SECTION 2.1

1. $2x^2, 5x, \underline{-8}$ **3.** $-a^4, \underline{6}$ **5.** $7x^2y, 6xy^2$ **7.** $1, -9$ **9.** $1, -4, -1$ **13.** 10 **15.** 32
17. 21 **19.** 16 **21.** −9 **23.** 3 **25.** −7 **27.** 13 **29.** −15 **31.** 41
33. 1 **35.** 5 **37.** 1 **39.** 57 **41.** 5 **43.** 12 **45.** 6 **47.** 10 **49.** 8
51. −3 **53.** −2 **55.** −22 **57.** 4 **59.** 20 **61.** 24 **63.** 4.96 **65.** −5.68
67. The volume is 25.8 in³. **69.** The area is 93.7 cm². **71.** The volume is 484.9 m³. **73.** 41

75. 21 **77.** 24 **79.** $\frac{1}{2}$ **81.** 13 **83.** −1 **85.** 4 **87.** 81

CONCEPT REVIEW 2.2

1. Never true **3.** Always true **5.** Always true **7.** Never true

SECTION 2.2

1. 2 **3.** 5 **5.** 6 **7.** −8 **9.** −4 **11.** The Inverse Property of Addition
13. The Commutative Property of Addition **15.** The Associative Property of Addition
17. The Commutative Property of Multiplication **19.** The Associative Property of Multiplication
23. $3x, 5x$ **25.** $14x$ **27.** $5a$ **29.** $-6y$ **31.** $-3b - 7$ **33.** $5a$ **35.** $-2ab$

37. $5xy$ **39.** 0 **41.** $-\frac{5}{6}x$ **43.** $-\frac{1}{24}x^2$ **45.** $11x$ **47.** $7a$ **49.** $-14x^2$

51. $-x + 3y$ **53.** $17x - 3y$ **55.** $-2a - 6b$ **57.** $-3x - 8y$ **59.** $-4x^2 - 2x$ **61.** $12x$
63. $-21a$ **65.** $6y$ **67.** $8x$ **69.** $-6a$ **71.** $12b$ **73.** $-15x^2$ **75.** x^2 **77.** x
79. n **81.** x **83.** n **85.** $2x$ **87.** $-2x$ **89.** $-15a^2$ **91.** $6y$ **93.** $3y$
95. $-2x$ **97.** $-x - 2$ **99.** $8x - 6$ **101.** $-2a - 14$ **103.** $-6y + 24$ **105.** $35 - 21b$
107. $-9 + 15x$ **109.** $15x^2 + 6x$ **111.** $2y - 18$ **113.** $-15x - 30$ **115.** $-6x^2 - 28$
117. $-6y^2 + 21$ **119.** $-6a^2 + 7b^2$ **121.** $4x^2 - 12x + 20$ **123.** $-3y^2 + 9y + 21$
125. $-12a^2 - 20a + 28$ **127.** $12x^2 - 9x + 12$ **129.** $10x^2 - 20xy - 5y^2$ **131.** $-8b^2 + 6b - 9$
133. $a - 7$ **135.** $-11x + 13$ **137.** $-4y - 4$ **139.** $-2x - 16$ **141.** $18y - 51$
143. $a + 7b$ **145.** $6x + 28$ **147.** $5x - 75$ **149.** $4x - 4$ **151.** $38x - 63$ **153.** b
155. $20x + y$ **157.** $-3x + y$ **159.** 0 **161.** $-a + b$ **163.** **a.** Yes **b.** No

CONCEPT REVIEW 2.3

1. Never true **3.** Never true **5.** Always true **7.** Never true

SECTION 2.3

1. $19 - d$ **3.** $r - 12$ **5.** $28a$ **7.** $5(n - 7)$ **9.** $y - 3y$ **11.** $-6b$ **13.** $\dfrac{4}{p - 6}$

15. $\dfrac{x - 9}{2x}$ **17.** $-4s - 21$ **19.** $\dfrac{d + 8}{d}$ **21.** $\dfrac{3}{8}(t + 15)$ **23.** $w + \dfrac{7}{w}$ **25.** $d + (16d - 3)$

27. $\dfrac{n}{19}$ **29.** $n + 40$ **31.** $(n - 90)^2$ **33.** $\dfrac{4}{9}n + 20$ **35.** $n(n + 10)$ **37.** $7n + 14$

39. $\dfrac{12}{n + 2}$ **41.** $\dfrac{2}{n + 1}$ **43.** $60 - \dfrac{n}{50}$ **45.** $n^2 + 3n$ **47.** $(n + 3) + n^3$ **49.** $n^2 - \dfrac{1}{4}n$

51. $2n - \dfrac{7}{n}$ **53.** $n^3 - 12n$ **55.** $n + (n + 10)$; $2n + 10$ **57.** $n - (9 - n)$; $2n - 9$

59. $\dfrac{1}{5}n - \dfrac{3}{8}n$; $-\dfrac{7}{40}n$ **61.** $(n + 9) + 4$; $n + 13$ **63.** $2(3n + 40)$; $6n + 80$ **65.** $7(5n)$; $35n$

67. $17n + 2n$; $19n$ **69.** $n + 12n$; $13n$ **71.** $3(n^2 + 4)$; $3n^2 + 12$ **73.** $\dfrac{3}{4}(16n + 4)$; $12n + 3$

75. $16 - (n + 9)$; $-n + 7$ **77.** $\dfrac{4n}{2} + 5$; $2n + 5$ **79.** $6(n + 8)$; $6n + 48$ **81.** $7 - (n + 2)$; $-n + 5$

83. $\dfrac{1}{3}(n + 6n)$; $\dfrac{7n}{3}$ **85.** $(8 + n^3) + 2n^3$; $3n^3 + 8$ **87.** $(n - 6) + (n + 12)$; $2n + 6$

89. $(n - 20) + (n + 9)$; $2n - 11$ **91.** $14 + (n - 3)10$; $10n - 16$ **93.** Let x be one number; x and $18 - x$
95. Let G be the number of genes in the roundworm genome; $G + 11{,}000$ **97.** Let N be the total number of Americans; $\dfrac{3}{4}N$ **99.** Let s be the number of points awarded for a safety; $3s$ **101.** Let m be the number of moons of Jupiter; $m + 9$ **103.** Let p be the pounds of pecans produced in Texas; $\dfrac{1}{2}p$ **105.** Let p be the world population in 1980; $2p$ **107.** Let L be the measure of the largest angle; $\dfrac{1}{2}L - 10$ **109.** Let n be either the number of nickels or the number of dimes; n and $35 - n$ **111.** Let h be the number of hours of overtime worked; $720 + 27h$ **113.** Let m be the number of minutes of phone calls; $19.95 + 0.08m$
115. $\dfrac{1}{4}x$

CHAPTER REVIEW EXERCISES

1. y^2 [2.2.2]   **2.** $3x$ [2.2.3]   **3.** $-10a$ [2.2.3]   **4.** $-4x + 8$ [2.2.4]   **5.** $7x + 38$ [2.2.5]
6. 16 [2.1.1]   **7.** 9 [2.2.1]   **8.** $36y$ [2.2.3]   **9.** $6y - 18$ [2.2.4]   **10.** $-3x + 21y$ [2.2.5]
11. $-8x^2 + 12y^2$ [2.2.4]   **12.** $5x$ [2.2.2]   **13.** 22 [2.1.1]   **14.** $2x$ [2.2.3]
15. $15 - 35b$ [2.2.4]   **16.** $-7x + 33$ [2.2.5]   **17.** The Commutative Property of Multiplication [2.2.1]
18. $24 - 6x$ [2.2.4]   **19.** $5x^2$ [2.2.2]   **20.** $-7x + 14$ [2.2.5]   **21.** $-9y^2 + 9y + 21$ [2.2.4]
22. $2x + y$ [2.2.5]   **23.** 3 [2.1.1]   **24.** $36y$ [2.2.3]   **25.** $5x - 43$ [2.2.5]   **26.** $2x$ [2.2.3]
27. $-6x^2 + 21y^2$ [2.2.4]   **28.** 6 [2.1.1]   **29.** $-x + 6$ [2.2.5]   **30.** $-5a - 2b$ [2.2.2]
31. $-10x^2 + 15x - 30$ [2.2.4]   **32.** $-9x - 7y$ [2.2.2]   **33.** $6a$ [2.2.3]   **34.** $17x - 24$ [2.2.5]
35. $-2x - 5y$ [2.2.2]   **36.** $30b$ [2.2.3]   **37.** 21 [2.2.1]   **38.** $-2x^2 + 4x$ [2.2.2]
39. $-6x^2$ [2.2.3]   **40.** $15x - 27$ [2.2.5]   **41.** $-8a^2 + 3b^2$ [2.2.4]
42. The Multiplication Property of Zero [2.2.1]   **43.** $b - 7b$ [2.3.1]   **44.** $n + 2n^2$ [2.3.1]
45. $\dfrac{6}{n} - 3$ [2.3.1]   **46.** $\dfrac{10}{y - 2}$ [2.3.1]   **47.** $8\left(\dfrac{2n}{16}\right)$; n [2.3.2]   **48.** $4(2 + 5n)$; $8 + 20n$ [2.3.2]

49. Let h be the height of the triangle; $h + 15$ [2.3.3] **50.** Let b be the amount of either espresso beans or mocha java beans; b and $20 - b$ [2.3.3]

CHAPTER TEST

1. $36y$ [2.2.3] **2.** $4x - 3y$ [2.2.2] **3.** $10n - 6$ [2.2.5] **4.** 2 [2.1.1] **5.** The Multiplication Property of One [2.2.1] **6.** $4x - 40$ [2.2.4] **7.** $\frac{1}{12}x^2$ [2.2.2] **8.** $4x$ [2.2.3]

9. $-24y^2 + 48$ [2.2.4] **10.** 19 [2.2.1] **11.** 6 [2.1.1] **12.** $-3x + 13y$ [2.2.5]
13. b [2.2.2] **14.** $78a$ [2.2.3] **15.** $3x^2 - 15x + 12$ [2.2.4] **16.** -32 [2.1.1]

17. $37x - 5y$ [2.2.5] **18.** $\frac{n + 8}{17}$ [2.3.1] **19.** $(a + b) - b^2$ [2.3.1] **20.** $n^2 + 11n$ [2.3.1]

21. $20(n + 9); 20n + 180$ [2.3.2] **22.** $(n - 3) + (n + 2); 2n - 1$ [2.3.2] **23.** $n - \frac{1}{4}(2n); \frac{1}{2}n$ [2.3.2]

24. Let d be the distance from Earth to the sun; $30d$ [2.3.3]

25. Let L be the length of one piece; L and $9 - L$ [2.3.3]

CUMULATIVE REVIEW EXERCISES

1. -7 [1.2.1] **2.** 5 [1.2.2] **3.** 24 [1.2.3] **4.** -5 [1.2.4] **5.** 1.25 [1.3.1]

6. $\frac{3}{5}, 0.60$ [1.3.4] **7.** $\{-4, -3, -2, -1\}$ [1.1.1] **8.** 8% [1.3.4] **9.** $\frac{11}{48}$ [1.3.2]

10. $\frac{5}{18}$ [1.3.3] **11.** $\frac{1}{4}$ [1.3.3] **12.** $\frac{8}{3}$ [1.4.1] **13.** -5 [1.4.2] **14.** $\frac{53}{48}$ [1.4.2]

15. -8 [2.1.1] **16.** $5x^2$ [2.2.2] **17.** $-a - 12b$ [2.2.2] **18.** $3a$ [2.2.3] **19.** $20b$ [2.2.3]
20. $20 - 10x$ [2.2.4] **21.** $6y - 21$ [2.2.4] **22.** $-6x^2 + 8y^2$ [2.2.4] **23.** $-8y^2 + 20y + 32$ [2.2.4]
24. $-10x + 15$ [2.2.5] **25.** $5x - 17$ [2.2.5] **26.** $13x - 16$ [2.2.5] **27.** $6x + 29y$ [2.2.5]
28. $6 - 12n$ [2.3.1] **29.** $5 + (n - 7); n - 2$ [2.3.2]
30. Let s be the speed of the dial-up connection; $10s$ [2.3.3]

Answers to Chapter 3 Selected Exercises

PREP TEST

1. -4 [1.2.2] **2.** 1 [1.3.3] **3.** -10 [1.3.3] **4.** 0.9 [1.3.4] **5.** 75% [1.3.4]
6. 63 [2.1.1] **7.** $10x - 5$ [2.2.2] **8.** -9 [2.2.2] **9.** $9x - 18$ [2.2.5]

CONCEPT REVIEW 3.1

1. Sometimes true **3.** Always true **5.** Always true **7.** Always true

SECTION 3.1

3. Yes **5.** No **7.** No **9.** Yes **11.** Yes **13.** Yes **15.** No **17.** No
19. No **21.** Yes **23.** No **25.** No **31.** 2 **33.** 15 **35.** 6 **37.** -6
39. 3 **41.** 0 **43.** -2 **45.** -7 **47.** -7 **49.** -12 **51.** 2 **53.** -5

55. 15 **57.** 9 **59.** 14 **61.** -1 **63.** 1 **65.** $-\frac{1}{2}$ **67.** $-\frac{3}{4}$ **69.** $\frac{1}{12}$

71. $-\frac{7}{12}$ **73.** $-\frac{5}{7}$ **75.** $-\frac{1}{3}$ **77.** 1.869 **79.** 0.884 **85.** 7 **87.** -7 **89.** -4

91. 5 **93.** 9 **95.** −8 **97.** 0 **99.** −7 **101.** 8 **103.** 12 **105.** 12
107. −18 **109.** −18 **111.** 6 **113.** −12 **115.** 3 **117.** −24 **119.** 9
121. $\dfrac{1}{3}$ **123.** $\dfrac{15}{7}$ **125.** 4 **127.** 3 **129.** 4.48 **131.** 2.06 **133.** −2.1
135. The train travels 245 mi in the 5-hour period. **137.** The dietician's average rate of speed is 30 mph.
139. It will take 4 h to complete the trip. **141.** It would take 31.25 s to walk from one end of the moving
sidewalk to the other. **143.** The two joggers will meet 40 min after they start. **145.** It will take her 0.5 h
to travel 4 mi upstream. **149.** 24% **151.** 7.2 **153.** 400 **155.** 9 **157.** 25%
159. 5 **161.** 200% **163.** 400 **165.** 7.7 **167.** 200 **169.** 400 **171.** 20
173. 80% **175.** 40% of 80 is equal to 80% of 40. **177.** 250 seats are reserved for wheelchair accessibility.
179. In the average single-family home, 74 gal of water are used per person per day.
181. 51% of a typical traveler's vacation bill is paid in cash. **183.** There is insufficient information.
185. 12% of the deaths were not attributed to motor vehicle accidents.
187. In this country 96.1 billion kilowatt-hours of electricity are used for home lighting per year.
189. **a.** 22% of the people watching the commercial would have to buy one can of the soft drink.
 b. 3.7% of the people watching the commercial would have to purchase one six-pack of the soft drink.
191. The price of the less expensive model is $1423.41. **193.** The annual simple interest rate is 9%.
195. Sal earned $240 in interest from the two accounts. **197.** Makana earned $63 in one year.
199. The interest rate on the combined investment is between 6% and 9%.
201. The percent concentration of the hydrogen perioxide is 2%. **203.** Apple Dan's has the greater
concentration of apple juice. **205.** 12.5 g of the cream are not glycerine. **207.** The percent concentration
of salt in the remaining solution is 12.5%. **209.** −21 **211.** −27 **213.** 21 **215.** 22°
217. The cost of the dinner was $80. **219.** The new value is two times the original value.
221. One possible answer is $5x = -10$.

CONCEPT REVIEW 3.2

1. Always true **3.** Never true **5.** Never true **7.** Sometimes true

SECTION 3.2

1. 3 **3.** 6 **5.** −1 **7.** 1 **9.** −9 **11.** 3 **13.** 3 **15.** 3 **17.** 2
19. 2 **21.** 4 **23.** $\dfrac{6}{7}$ **25.** $\dfrac{2}{3}$ **27.** −1 **29.** $\dfrac{3}{4}$ **31.** $\dfrac{1}{3}$ **33.** $-\dfrac{3}{4}$ **35.** $\dfrac{1}{3}$
37. $-\dfrac{1}{6}$ **39.** 1 **41.** 0 **43.** $\dfrac{2}{5}$ **45.** $-\dfrac{3}{2}$ **47.** 18 **49.** 8 **51.** −16
53. 25 **55.** 21 **57.** 15 **59.** −16 **61.** $\dfrac{7}{6}$ **63.** $\dfrac{1}{8}$ **65.** $\dfrac{15}{2}$ **67.** $-\dfrac{18}{5}$
69. 2 **71.** 3 **73.** 1 **75.** 2 **77.** 3.95 **79.** −0.8 **81.** −11 **83.** 0
85. 3 **87.** 8 **89.** 2 **91.** −2 **93.** −3 **95.** 2 **97.** −2 **99.** −2
101. −7 **103.** 0 **105.** −2 **107.** −2 **109.** 4 **111.** 10 **113.** 3 **115.** $\dfrac{3}{4}$
117. $\dfrac{2}{7}$ **119.** $-\dfrac{3}{4}$ **121.** 3 **123.** 10 **125.** $\dfrac{4}{3}$ **127.** 3 **129.** −14 **131.** 7
133. 3 **135.** 4 **137.** 2 **139.** 2 **141.** 2 **143.** −7 **145.** $\dfrac{4}{7}$ **147.** $\dfrac{1}{2}$
149. $-\dfrac{1}{3}$ **151.** $\dfrac{10}{3}$ **153.** $-\dfrac{1}{4}$ **155.** 0.5 **157.** 0 **159.** −1

161. The car will slide 168 ft.   **163.** The depth of the diver is 40 ft.   **165.** The initial velocity is 8 ft/s.
167. The length of the humerus is approximately 31.8 in.   **169.** The passenger was driven 6 mi.
171. He made 8 errors.   **173.** The population is approximately 51,000 people.   **175.** The break-even point is 350 television sets.   **179.** No, the seesaw is not balanced.   **181.** A force of 25 lb must be applied to the other end of the lever.   **183.** The fulcrum must be placed 10 ft from the 128-pound acrobat.

185. No solution   **187.** $-\dfrac{11}{4}$ **189.** 62   **191.** One possible answer is $3x - 6 = 2x - 2$.

CONCEPT REVIEW 3.3

1. Always true   **3.** Sometimes true   **5.** Never true   **7.** Always true   **9.** Sometimes true

SECTION 3.3

1. 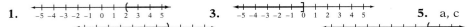   **3.** ⟨number line⟩   **5.** a, c
7. $x < 2$ ⟨number line⟩   **9.** $x > 3$ ⟨number line⟩
11. $n \geq 3$ ⟨number line⟩   **13.** $x \leq -4$ ⟨number line⟩
15. $x \geq -1$ ⟨number line⟩   **17.** $y \geq -9$ **19.** $x < 12$ **21.** $x \geq 5$ **23.** $x < -11$
25. $x \leq 10$ **27.** $x \geq -6$ **29.** $x > 2$ **31.** $d < -\dfrac{1}{6}$ **33.** $x \geq -\dfrac{31}{24}$ **35.** $x < \dfrac{5}{4}$
37. $x > \dfrac{5}{24}$ **39.** $x \leq -1.2$ **41.** $x \leq 0.70$ **43.** $x < -7.3$ **45.** a, b, c
47. $x \leq -3$ ⟨number line⟩   **49.** $x > -2$ ⟨number line⟩
51. $x > 0$ ⟨number line⟩   **53.** $n \geq 4$ ⟨number line⟩
55. $x > -2$ ⟨number line⟩   **57.** $x < \dfrac{5}{3}$ **59.** $x \geq 5$ **61.** $x > -\dfrac{5}{2}$ **63.** $x \leq -\dfrac{2}{3}$
65. $x < -18$ **67.** $x > -16$ **69.** $y \geq 6$ **71.** $x \geq -6$ **73.** $b \leq 33$ **75.** $n < \dfrac{3}{4}$
77. $x \leq -\dfrac{6}{7}$ **79.** $y \leq \dfrac{5}{6}$ **81.** $x \leq \dfrac{27}{28}$ **83.** $y > -\dfrac{3}{2}$ **85.** $x > -0.5$
87. $y \geq -0.8$ **89.** $x \leq 4.2$ **91.** $x \leq 5$ **93.** $x < -5.4$ **97.** $x < 4$ **99.** $x < -4$
101. $x \geq 1$ **103.** $x \leq 5$ **105.** $x < 0$ **107.** $x < 20$ **109.** $x > 500$ **111.** $x > 2$
113. $x \leq -5$ **115.** $y \leq \dfrac{5}{2}$ **117.** $x < \dfrac{25}{11}$ **119.** $x > 11$ **121.** $n \leq \dfrac{11}{18}$ **123.** $x \geq 6$
125. $x \leq \dfrac{2}{5}$ **127.** $t < 1$ **129.** $n > \dfrac{7}{10}$ **131.** 3   **133.** $\{1, 2\}$ **135.** $\{1, 2, 3\}$
137. $\{3, 4, 5\}$ **139.** $\{10, 11, 12, 13\}$ **141.** ⟨number line⟩   **143.** ⟨number line⟩

CHAPTER REVIEW EXERCISES

1. No [3.1.1]   **2.** 20 [3.1.2]   **3.** -7 [3.1.3]   **4.** 7 [3.2.1]   **5.** 4 [3.2.2]
6. $-\dfrac{1}{5}$ [3.2.3]   **7.** 405 [3.1.5]   **8.** 25 [3.1.5]   **9.** 67.5% [3.1.5]
10. 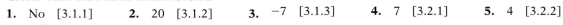 [3.3.1]   **11.** $x > 2$ ⟨number line⟩ [3.3.1]
12. $x > -4$ ⟨number line⟩ [3.3.2]   **13.** $x \geq -4$ [3.3.3]   **14.** $x \geq 4$ [3.3.3]
15. Yes [3.1.1]   **16.** 2.5 [3.1.2]   **17.** -49 [3.1.3]   **18.** $\dfrac{1}{2}$ [3.2.1]

19. $\frac{1}{3}$ [3.2.2] **20.** 10 [3.2.3] **21.** 16 [3.1.5] **22.** 125 [3.1.5] **23.** $16\frac{2}{3}\%$ [3.1.5]

24. $x < -4$ [3.3.1] **25.** $x \le -2$ [3.3.2]

26. -2 [3.2.3] **27.** $\frac{5}{6}$ [3.1.2] **28.** 20 [3.1.3] **29.** 6 [3.2.1] **30.** 0 [3.2.3]

31. $x < 12$ [3.3.3] **32.** 15 [3.1.5] **33.** 5 [3.2.1] **34.** $x > 5$ [3.3.3] **35.** 4% [3.1.5]

36. $x > -18$ [3.3.3] **37.** $x < \frac{1}{2}$ [3.3.3] **38.** The measure of the third angle is 110°. [3.2.4]

39. A force of 24 lb must be applied to the other end of the lever. [3.2.4] **40.** The width is 6 ft. [3.2.4]
41. The discount is $31.99. [3.2.4] **42.** Approximately 11,065 plants and animals are at risk of extinction on Earth. [3.1.5] **43.** The depth is 80 ft. [3.2.4] **44.** The fulcrum is 3 ft from the 25-lb force. [3.2.4]
45. The length of the rectangle is 24 ft. [3.2.4] **46.** It will take the motorboat 1.5 h to travel 30 mi. [3.1.4]
47. She must invest $625 in an account that earns an annual simple interest rate of 8%. [3.1.5]
48. The percent concentration of hydrochloric acid is 6%. [3.1.5]

CHAPTER TEST

1. -12 [3.1.3] **2.** $-\frac{1}{2}$ [3.2.2] **3.** -3 [3.2.1] **4.** No [3.1.1] **5.** $\frac{1}{8}$ [3.1.2]

6. $-\frac{1}{3}$ [3.2.3] **7.** 5 [3.2.1] **8.** $\frac{1}{2}$ [3.2.2] **9.** -5 [3.1.2] **10.** -5 [3.2.2]

11. $-\frac{40}{3}$ [3.1.3] **12.** $-\frac{22}{7}$ [3.2.3] **13.** 2 [3.2.3] **14.** $\frac{12}{11}$ [3.2.3] **15.** -3 [3.2.3]

16. 125% [3.1.5] **17.** 40 [3.1.5] **18.** [3.3.1]

19. $x \le -1$ [3.3.1] **20.** $x > -2$ [3.3.2]

21. $x > \frac{1}{2}$ [3.3.1] **22.** $x \le -\frac{9}{2}$ [3.3.3] **23.** $x \ge -16$ [3.3.2] **24.** $x \le 2$ [3.3.3]

25. $x \le -3$ [3.3.3] **26.** $x > 2$ [3.3.3] **27.** 4 [3.2.2] **28.** 24 [3.1.5]

29. $x \ge 3$ [3.3.2] **30.** Yes [3.1.1] **31.** The astronaut would weigh 30 lb on the moon. [3.1.5] **32.** The final temperature of the water after mixing is 60°C. [3.2.4]
33. The number of calculators produced was 200. [3.2.4] **34.** They will meet 2 h after they begin. [3.1.4]
35. He must invest $930 in the second account. [3.1.5] **36.** The percent concentration of chocolate syrup in the chocolate milk is 25%. [3.1.5]

CUMULATIVE REVIEW EXERCISES

1. 6 [1.2.2] **2.** -48 [1.2.3] **3.** $-\frac{19}{48}$ [1.3.2] **4.** -2 [1.3.3] **5.** 54 [1.4.1]

6. 24 [1.4.2] **7.** 6 [2.1.1] **8.** $-17x$ [2.2.2] **9.** $-5a - 2b$ [2.2.2] **10.** $2x$ [2.2.3]
11. $36y$ [2.2.3] **12.** $2x^2 + 6x - 4$ [2.2.4] **13.** $-4x + 14$ [2.2.5] **14.** $6x - 34$ [2.2.5]

15. $\{-7, -6, -5, -4, -3, -2, -1\}$ [1.1.1] **16.** $87\frac{1}{2}\%$ [1.3.4] **17.** 3.42 [1.3.4] **18.** $\frac{5}{8}$ [1.3.4]

19. Yes [3.1.1] **20.** -5 [3.1.2] **21.** -25 [3.1.3] **22.** 3 [3.2.1] **23.** -3 [3.2.2]

24. 13 [3.2.3] **25.** $x < -\frac{8}{9}$ [3.3.2] **26.** $x \ge 12$ [3.3.3] **27.** $x > 9$ [3.3.3] **28.** $8 - \frac{n}{12}$

[2.3.1] **29.** $n + (n + 2)$; $2n + 2$ [2.3.2] **30.** b and $35 - b$ [2.3.3] **31.** Let L be the length of the longer piece; $3 - L$ [2.3.3] **32.** 17% of the computer programmer's salary is deducted for income tax. [3.1.5]

33. The equation predicts that the first 4-minute mile was run in 1952.   [3.2.4]
34. The final temperature of the water after mixing is 60°C.   [3.2.4]   **35.** A force of 24 lb must be applied to the other end of the lever.   [3.2.4]

Answers to Chapter 4 Selected Exercises

PREP TEST
1. $0.65R$ [2.2.2]   **2.** $0.03x + 20$ [2.2.5]   **3.** $3n + 6$ [2.2.2]   **4.** $5 - 2x$ [2.3.1]
5. 40%   [1.3.4]   **6.** 2   [3.2.3]   **7.** 0.25   [3.2.1]   **8.** $x < 4$ [3.3.3]   **9.** $20 - n$ [2.3.3]

CONCEPT REVIEW 4.1
1. Always true   **3.** Always true   **5.** Never true

SECTION 4.1
1. $n - 15 = 7; n = 22$ **3.** $7n = -21; n = -3$ **5.** $3n - 4 = 5; n = 3$ **7.** $4(2n + 3) = 12; n = 0$
9. $12 = 6(n - 3); n = 5$ **11.** $22 = 6n - 2; n = 4$ **13.** $4n + 7 = 2n + 3; n = -2$
15. $5n - 8 = 8n + 4; n = -4$ **17.** $2(n - 25) = 3n; n = -50$ **19.** $3n = 2(20 - n)$; 8 and 12
21. $3n + 2(18 - n) = 44$; 8 and 10   **23.** The original value was $16,000.   **25.** There are 58 calories in a medium-size orange.   **27.** The amount of mulch is 15 lb.   **29.** The intensity of the sound is 140 decibels.
31. The monthly payment is $117.75.   **33.** The length is 80 ft. The width is 50 ft.   **35.** To replace the water pump, 5 h of labor were required.   **37.** You are purchasing 9 tickets.   **39.** The length of the patio is 15 ft.   **41.** The union member worked 168 h during the month of March.   **43.** The customer used the service for 11 min.   **45.** One-third of the container is filled at 3:39 P.M.   **47.** The package weighs more than 12 oz but not over 13 oz.   **49.** There are 60 coins in the bank.

CONCEPT REVIEW 4.2
1. Always true   **3.** Always true   **5.** Never true

SECTION 4.2
3. The integers are 17, 18, and 19.   **5.** The integers are 26, 28, and 30.   **7.** The integers are 17, 19, and 21.
9. The integers are 8 and 10.   **11.** The integers are 7 and 9.   **13.** The integers are $-9, -8$, and -7.
15. The integers are 10, 12, and 14.   **17.** No solution   **21.** There are 12 dimes and 15 quarters in the bank.   **23.** The executive bought eight 23¢ stamps and thirty-two 37¢ stamps.   **25.** There are five 29¢ stamps in the drawer.   **27.** There are 28 quarters in the bank.   **29.** There are 20 one-dollar bills and 6 five-dollar bills in the cash box.   **31.** There are 11 pennies in the bank.   **33.** There are twenty-seven 22¢ stamps in the collection.   **35.** There are thirteen 3¢ stamps, eighteen 7¢ stamps, and nine 12¢ stamps in the collection.   **37.** There are eighteen 6¢ stamps, six 8¢ stamps, and twenty-four 15¢ stamps.
39. The integers are $-12, -10, -8$, and -6.   **41.** There is $6.70 in the bank.   **43.** For any three consecutive odd integers, the sum of the first and third integers is twice the second integer.

CONCEPT REVIEW 4.3
1. Always true   **3.** Always true   **5.** Never true

SECTION 4.3
3. The sides measure 50 ft, 50 ft, and 25 ft.   **5.** The length is 13 m. The width is 8 m.   **7.** The length is 40 ft. The width is 20 ft.   **9.** The sides measure 40 cm, 20 cm, and 50 cm.   **11.** The length is 130 ft. The

width is 39 ft.   **13.**   The width is 12 ft.   **15.**   Each side measures 12 in.   **17. a.**   90°   **b.**   180°
c.   360°   **d.**   between 0° and 90°   **e.**   between 90° and 180°   **f.**   90°   **g.**   180°   **19.**   The complement of a 28°
angle is 62°.   **21.**   The supplement of a 73° angle is 107°.   **23.**   35°   **25.**   20°   **27.**   53°
29.   121°   **31.**   15°   **33.**   18°   **35.**   45°   **37.**   49°   **39.**   12°   **41.** $\angle a = 122°$, $\angle b = 58°$
43. $\angle a = 44°$; $\angle b = 136°$ **45.**   20°   **47.**   40°   **49.**   128°   **51.** $\angle x = 160°$, $\angle y = 145°$
53. $\angle a = 40°$, $\angle b = 140°$ **55.** $75° - x$ **57.**   The measure of the third angle is 45°.
59.   The measure of the third angle is 73°.   **61.**   The measure of the third angle is 43°.
63.   The angles measure 38°, 38°, and 104°.   **65.**   The angles measure 60°, 30°, and 90°.
67.   The length is 9 cm. The width is 4 cm.

CONCEPT REVIEW 4.4

1.   Always true   **3.**   Never true   **5.**   Always true

SECTION 4.4

3.   The selling price is $56.   **5.**   The selling price is $565.64.   **7.**   The markup rate is 75%.
9.   The markup rate is 44.4%.   **11.**   The markup rate is 60%.   **13.**   The cost of the compact disc player
is $120.   **15.**   The cost of the basketball is $59.   **17.**   The cost of the computer is $1750.
21.   The sale price is $101.25.   **23.**   The discount price is $218.50.   **25.**   The discount rate is 25%.
27.   The discount rate is 23.2%.   **29.**   The markdown rate is 38%.   **31.**   The regular price is $510.
33.   The regular price is $300.   **35.**   The regular price is $275.   **37.**   The markup is $18.
39.   The markup rate is 40%.   **41.**   The cost of the camera is $230.   **43.**   The regular price is $80.
45.   No. The single discount that would give the same sale price is 28%.

CONCEPT REVIEW 4.5

1.   Always true   **3.**   Always true   **5.**   Never true

SECTION 4.5

1. $9000 was invested at 7%, and $6000 was invested at 6.5%.   **3.** $1500 was invested in the mutual fund.
5. $200,000 was deposited at 10%, and $100,000 was deposited at 8.5%.   **7.**   Teresa has $3000 invested in bonds
that earn 8% annual simple interest.   **9.**   She has $2500 invested at 11%.   **11.** $40,500 was invested at 8%,
and $13,500 was invested at 12%.   **13.**   The total amount invested was $650,000.   **15.**   The total amount
invested was $500,000.   **17.**   The amount of the research consultant's investment is $45,000.
19.   The total annual interest received was $3040.   **21.**   The value of the investment in three years is $3831.87.
23. a.   By age 55, the couple should have $170,000 saved for retirement.

CONCEPT REVIEW 4.6

1.   Always true   **3.**   Always true   **5.**   Never true   **7.**   Always true

SECTION 4.6

3.   The clinic should use 2 lb of dog food and 3 lb of vitamin supplement.   **5.**   The mixture costs $6.98 per
pound.   **7.**   56 oz of the $4.30 alloy and 144 oz of the $1.80 alloy were used.   **9.**   You can add 1 lb of
blueberries.   **11.**   10 kg of hard candy must be mixed with the jelly beans.   **13.**   To make the mixture, 3 lb
of caramel are needed.   **15.**   To make the mixture, 300 lb of the $.80 meal are needed.   **17.**   37 lb of
almonds and 63 lb of walnuts were used.   **19.**   The cost per pound is $1.40.   **21.**   The cost per ounce is $3.
25.   The percent concentration is 24%.   **27.**   20 gal of the 15% solution must be mixed with 5 gal of the 20%
solution.   **29.**   30 lb of 25% wool yarn is used.   **31.**   6.25 gal of the plant food that is 9% nitrogen are

combined with the plant food that is 25% nitrogen.   **33.**   The percent concentration is 19%.   **35.**   30 oz of the potpourri that is 60% lavender are used.   **37.**   To make the solution, 100 ml of the 7% solution and 200 ml of the 4% solution are used.   **39.**   150 oz of pure chocolate are added.   **41.**   The percent concentration is 12%.   **43.**   1.2 oz of dried apricots must be added.   **45.**   The cost is $3.65 per ounce.   **47.**   10 oz of water evaporated from the solution.   **49.**   75 g of pure water must be added.   **51.**   85 adults and 35 children attended the performance.

CONCEPT REVIEW 4.7

1. Never true   **3.** Never true   **5.** Always true

SECTION 4.7

1.   The speed of the first plane is 105 mph. The speed of the second plane is 130 mph.   **3.**   The second skater overtakes the first skater after 40 s.   **5.**   Michael's boat will be alongside the tour boat in 2 h.
7.   The airport is 120 mi from the corporate offices.   **9.**   The sailboat traveled 36 mi in the first 3 h.
11.   The passenger train is traveling at 50 mph. The freight train is traveling at 30 mph.   **13.**   It takes the second ship 1 h to catch up to the first ship.   **15.**   The two trains will pass each other in 4 h.   **17.**   The average speed on the winding road was 34 mph.   **19.**   The car overtakes the cyclist 48 mi from the starting point.
21.   The rate of the cyclist is 13 mph.   **23.**   The rate of the car is 60 mph.   **25.**   The joggers will meet at 7:48 A.M.   **27.**   The cyclist's average speed is $13\frac{1}{3}$ mph.

CONCEPT REVIEW 4.8

1. Never true   **3.** Always true   **5.** Never true

SECTION 4.8

1.   The smallest integer is 2.   **3.**   The amount of income tax to be paid is $3150 or more.   **5.**   The organization must collect more than 440 lb.   **7.**   The student must score 78 or better.   **9.**   The dollar amount in sales must be more than $5714.   **11.**   The dollar amount the agent expects to sell is $20,000 or less.
13.   A person must use this service for more than 60 min.   **15.**   8 ounces or less must be added.
17.   The ski area is more than 38 mi away.   **19.**   The maximum number of miles you can drive is 166.
21.   The integers are 1, 3, and 5 or 3, 5, or 7.   **23.**   The crew can prepare 8 to 12 aircraft in this period of time.
25.   The minimum number is 467 calendars.

CHAPTER REVIEW EXERCISES

1. $2x - 7 = x$; 7   [4.1.1]   **2.**   The length of the shorter piece is 14 in.   [4.1.2]   **3.**   The two numbers are 8 and 13.   [4.1.2]   **4.** $2x + 6 = 4(x - 2)$; 7   [4.1.1]   **5.**   The discount rate is $33\frac{1}{3}$%.   [4.4.2]
6. $8000 is invested at 6%, and $7000 is invested at 7%.   [4.5.1]   **7.**   The angles measure 65°, 65°, and 50°.
[4.3.3]   **8.**   The rate of the motorcyclist is 45 mph.   [4.7.1]   **9.**   The measures of the three angles are 16°, 82°, and 82°.   [4.3.3]   **10.**   The markup rate is 80%.   [4.4.1]   **11.**   The concentration of butterfat is 14%.
[4.6.2]   **12.**   The length is 80 ft. The width is 20 ft.   [4.3.1]   **13.**   A minimum of 1151 copies can be ordered.
[4.8.1]   **14.**   The regular price is $32.   [4.4.2]   **15.**   7 qt of cranberry juice and 3 qt of apple juice were used.
[4.6.1]   **16.**   The cost is $25.95.   [4.4.1]   **17.**   The two numbers are 6 and 30.   [4.1.2]   **18.** $5600 is deposited in the 12% account.   [4.5.1]   **19.**   The cards are the 7 and 8 of hearts.   [4.2.1]   **20.**   One liter of pure water should be added.   [4.6.2]   **21.**   The lowest score the student can receive is 72.   [4.8.1]
22. $-7 = \frac{1}{2}x - 10$; 6   [4.1.1]   **23.**   The height is 1063 ft.   [4.1.2]   **24.**   The measures of the three angles are 75°, 60°, and 45°.   [4.3.3]   **25.**   There are 128 one-dollar bills.   [4.2.2]   **26.**   The length of the longer

piece is 7 ft.   [4.1.2]   **27.** The number of hours of consultation was 8.   [4.1.2]   **28.** There are 77 12¢ stamps and 37 17¢ stamps.   [4.2.2]   **29.** The cost is $671.25.   [4.4.1]   **30.** They will meet after 4 min. [4.7.1]   **31.** The measures of the three sides are 8 in., 12 in., and 15 in.   [4.3.1]   **32.** The integers are −17, −15, and −13.   [4.2.1]   **33.** The maximum width is 11 ft.   [4.8.1]   **34.** $\angle a = 138°$, $\angle b = 42°$ [4.3.2]

CHAPTER TEST

1. $6n + 13 = 3n − 5; n = −6$ [4.1.1]   **2.** $3n − 15 = 27; n = 14$ [4.1.1]   **3.** The numbers are 8 and 10. [4.1.2]   **4.** The lengths are 6 ft and 12 ft.   [4.1.2]   **5.** The cost is $200.   [4.4.1]   **6.** The discount rate is 20%.   [4.4.2]   **7.** 20 gal of the 15% solution must be added.   [4.6.2]   **8.** The length is 14 m. The width is 5 m.   [4.3.1]   **9.** The integers are 5, 7, and 9.   [4.2.1]   **10.** Five or more residents are in the nursing home.   [4.8.1]   **11.** $5000 is invested at 10%, and $2000 is invested at 15%.   [4.5.1]   **12.** 8 lb of the $7 coffee and 4 lb of the $4 coffee should be used.   [4.6.1]   **13.** The rate of the first plane is 225 mph. The rate of the second plane is 125 mph.   [4.7.1]   **14.** The measures of the three angles are 48°, 33°, and 99°.   [4.3.3] **15.** There are 15 nickels and 35 quarters in the bank.   [4.2.2]   **16.** The amounts to be deposited are $1400 at 6.75% and $1000 at 9.45%.   [4.5.1]   **17.** The minimum length is 24 ft.   [4.8.1]   **18.** The sale price is $79.20.   [4.4.2]

CUMULATIVE REVIEW EXERCISES

1. $−12, −6$ [1.1.1]   **2.** 6   [1.2.2]   **3.** $−\dfrac{1}{6}$ [1.4.1]   **4.** $−\dfrac{11}{6}$ [1.4.2]   **5.** $−18$ [1.1.2]

6. $−24$ [2.1.1]   **7.** $9x + 4y$ [2.2.2]   **8.** $−12 + 8x + 20x^3$ [2.2.4]   **9.** $4x + 4$ [2.2.5] **10.** $6x^2$ [2.2.2]   **11.** No   [3.1.1]   **12.** $−3$ [3.2.1]   **13.** $−15$ [3.1.3]   **14.** 3   [3.2.1]

15. $−10$ [3.2.3]   **16.** $\dfrac{2}{5}$ [1.3.4]   **17.** $x \ge 4$    [3.3.2]   **18.** $x \ge −3$

[3.3.2]   **19.** $x > 18$ [3.3.3]   **20.** $x \ge 4$ [3.3.3]   **21.** 2.5%   [1.3.4]   **22.** 12%   [1.3.4] **23.** 3   [3.1.5]   **24.** 45   [3.1.5]   **25.** $8n + 12 = 4n; n = −3$ [4.1.1]   **26.** The area of the garage is 600 ft².   [4.1.2]   **27.** The number of hours of labor is 5.   [4.1.2]   **28.** 20% of the libraries had the reference book.   [3.1.5]   **29.** The amount of money deposited is $2000.   [4.5.1]   **30.** The markup rate is 75%.   [4.4.1]   **31.** 60 g of the gold alloy must be used.   [4.6.1]   **32.** 30 oz of pure water must be added. [4.6.2]   **33.** The measure of one of the equal angles is 47°.   [4.3.3]   **34.** The middle even integer is 14. [4.2.1]   **35.** There are 7 dimes in the bank.   [4.2.2]

Answers to Chapter 5 Selected Exercises

PREP TEST

1. $−3$ [1.4.2]   **2.** $−1$ [2.1.1]   **3.** $−3x + 12$ [2.2.4]   **4.** $−2$ [3.2.1]   **5.** 5   [3.2.1]

6. $−2$ [3.2.1]   **7.** 13   [2.1.1]   **8.** $\dfrac{3}{4}x − 4$ [2.2.4]   **9.** a, b, c   [3.3.1]

CONCEPT REVIEW 5.1

1. Always true   **3.** Never true   **5.** Never true   **7.** Always true

SECTION 5.1

3.   **5.**   **7.**   **9.** $A(2, 3)$, $B(4, 0)$, $C(−4, 1)$, $D(−2, −2)$

11. $A(-2, 5)$, $B(3, 4)$, $C(0, 0)$, $D(-3, -2)$ **13. a.** 2; −4 **b.** 1; −3   **15.**

17.   **19.**   **21.** The average rate of change is 4° per hour.

23. a. The average rate of change was $.10 per year. **b.** The average rate of change was $.07 per year. This was less than the average annual rate of change from 1978 to 1990.   **25. a.** The average rate of change is $-2.1\overline{3}$ million people per year. **b.** The average rate of change is −1.3 million people per year.   **27. a.** The average rate of change was $66 thousand per year. **b.** The average annual rate of change from 1981 to 2001 was greater by $64.85 thousand.   **29.** 1 unit   **31.** 0 units   **33.** 1 unit   **35.** (0, 0)

CONCEPT REVIEW 5.2

1. Never true   **3.** Always true   **5.** Never true   **7.** Never true   **9.** Always true

SECTION 5.2

1. $m = 4, b = 1$ **3.** $m = \frac{5}{6}, b = \frac{1}{6}$ **5.** Yes   **7.** No   **9.** No   **11.** Yes   **13.** No

15. (3, 7)   **17.** (6, 3)   **19.** (0, 1)   **21.** (−5, 0)   **23.** 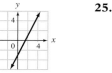   **25.**

27.   **29.**   **31.**   **33.**   **35.**

37.   **39.**   **41.**   **43.**   **45.**

47.   **49.**   **51.**   **53.** $y = -3x + 10$ **55.** $y = 4x - 3$

57. $y = -\dfrac{3}{2}x + 3$　　**59.** $y = \dfrac{2}{5}x - 2$　　**61.** $y = -\dfrac{2}{7}x + 2$　　**63.** $y = -\dfrac{1}{3}x + 2$　　**65.**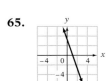

67. 　　**69.** 　　**71.** 　　**73.** 　　**75.**

77. (3, 0), (0, −3)　　**79.** (3, 0), (0, −6)　　**81.** (10, 0), (0, −2)　　**83.** (−4, 0), (0, 12)　　**85.** (0, 0), (0, 0)

87. (−6, 0), (0, 3)　　**89.** 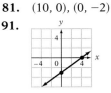　　**91.** 　　**93.** 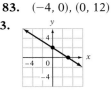　　**95.**

97. 　　**99.** 　　**101.** 　　**103.** 　　**105.**

107. 　　**109.** 　　**111.** **a.** $y = -\dfrac{1}{2}x - 5$　**b.** (−2, −4)　　**113.** 1 unit

CONCEPT REVIEW 5.3

1. Always true　　**3.** Never true　　**5.** Never true

SECTION 5.3

3. −2　　**5.** $\dfrac{1}{3}$　　**7.** $-\dfrac{5}{2}$　　**9.** $-\dfrac{1}{2}$　　**11.** −1　　**13.** Undefined　　**15.** 0　　**17.** $-\dfrac{1}{3}$

19. 0　　**21.** −5　　**23.** Undefined　　**25.** $-\dfrac{2}{3}$　　**27.** Yes　　**29.** No　　**31.** $m = 50$. The electronic mail sorter sorts 50 pieces of mail per minute.　　**33.** $m = -0.02$. The car uses 0.02 gal of gasoline per mile driven.　　**37.** 　　**39.** 　　**41.** 　　**43.**

45. 　　**47.** 　　**49.** 　　**51.** 　　**53.**

55. Increasing the coefficient of x increases the slope.　　**57.** Increasing the constant term increases the y-intercept.　　**59.** i and D; ii and C; iii and B; iv and F; v and E; vi and A　　**61.** No, not all graphs of straight lines have a y-intercept. In general, vertical lines do not have a y-intercept. For example, $x = 2$ does not have a y-intercept.

CONCEPT REVIEW 5.4

1. Always true   **3.** Sometimes true   **5.** Always true

SECTION 5.4

1. $y = 2x + 2$ **3.** $y = -3x - 1$ **5.** $y = \dfrac{1}{3}x$ **7.** $y = \dfrac{3}{4}x - 5$ **9.** $y = -\dfrac{3}{5}x$

11. $y = \dfrac{1}{4}x + \dfrac{5}{2}$ **13.** $y = -\dfrac{2}{3}x - 7$ **17.** $y = 2x - 3$ **19.** $y = -2x - 3$ **21.** $y = \dfrac{2}{3}x$

23. $y = \dfrac{1}{2}x + 2$ **25.** $y = -\dfrac{3}{4}x - 2$ **27.** $y = \dfrac{3}{4}x + \dfrac{5}{2}$ **29.** $y = -\dfrac{4}{3}x - 9$ **31.** Yes; $y = -x + 6$

33. No   **35.** 7   **37.** −1   **39.** $F = \dfrac{9}{5}C + 32$ **41.** $y = -\dfrac{2}{3}x + \dfrac{5}{3}$

CONCEPT REVIEW 5.5

1. Always true   **3.** Never true   **5.** Always true   **7.** Always true

SECTION 5.5

3. {(44, 40), (3, 3), (16, 28), (13, 13), (15, 13), (20, 17)}; yes
5. {(197, 125), (205, 122), (257, 498), (226, 108), (205, 150)}; no   **7.** {(55, 273), (65, 355), (75, 447), (85, 549)}; yes
9. Domain: {0, 2, 4, 6}; range: {0}; yes   **11.** Domain: {2}; range: {2, 4, 6, 8}; no   **13.** Domain: {0, 1, 2, 3};
range {0, 1, 2, 3}; yes   **15.** Domain: {−3, −2, −1, 1, 2, 3}; range: {−3, 2, 3, 4}; yes   **17.** 40; (10, 40)

19. −11; (−6, −11)   **21.** 12; (−2, 12)   **23.** $\dfrac{7}{2}; \left(\dfrac{1}{2}, \dfrac{7}{2}\right)$ **25.** 2; (−5, 2)   **27.** 32; (−4, 32)

29. Range: {−19, −13, −7, −1, 5}; the ordered pairs (−5, −19), (−3, −13), (−1, −7), (1, −1), and (3, 5) belong to the
function.   **31.** Range: {1, 2, 3, 4, 5}; the ordered pairs (−4, 1), (−2, 2), (0, 3), (2, 4), and (4, 5) belong to the
function.   **33.** Range: {6, 7, 15}; the ordered pairs (−3, 15), (−1, 7), (0, 6), (1, 7), and (3, 15) belong to the function.

35.   **37.**   **39.**   **41.**   **43.**

45.   **47.**   **49.**   **51.**   **53.**

55.   **57.** **a.** $f(x) = 30{,}000 - 5000x$ **b.**

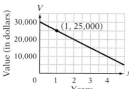

c. The value of the computer after 1 year is $25,000.   **59.** **a.** $f(m) = 0.18m + 50$
b.   **c.** The cost to drive this car 500 mi is $140.

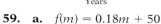

CONCEPT REVIEW 5.6

1. Never true **3.** Always true **5.** Always true

SECTION 5.6

3. **5.** **7.** **9.** **11.**

13. **15.** **17.** **19.** $y \geq 2x + 2$ **21.** $y > 2$

23. **25.**

CHAPTER REVIEW EXERCISES

1. $(3, 0)$ [5.2.1] **2.** $y = 3x - 1$ [5.4.1/5.4.2] **3.** [5.2.3] **4.** [5.1.1]

5. 79 [5.5.1] **6.** 0 [5.3.1] **7.** [5.2.2] **8.** [5.2.3]

9. $y = -\dfrac{2}{3}x + \dfrac{4}{3}$ [5.4.1/5.4.2] **10.** $(2, 0), (0, -3)$ [5.2.3] **11.** [5.3.2]

12. [5.5.2] **13.** $y = \dfrac{2}{3}x + 3$ [5.4.1/5.4.2] **14.** 2 [5.3.1] **15.** -4 [5.5.1]

16. Domain: $\{-20, -10, 0, 10\}$; range: $\{-10, -5, 0, 5\}$; yes [5.5.1] **17.** 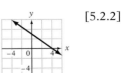 [5.2.2]

18. [5.3.2] **19.** [5.1.1] **20.** [5.5.2]

21. $(6, 0)$, $(0, -4)$ [5.2.3]   **22.** $y = 2x + 2$ [5.4.1/5.4.2]   **23.** 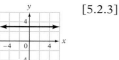 [5.2.3]

24. [5.6.1]   **25.** Yes [5.3.1]   **26.** $y = \dfrac{1}{2}x + 2$ [5.4.1/5.4.2]

27. [5.3.2]   **28.** [5.2.3]   **29.** $(-2, -5)$ [5.2.1]

30. $y = -3x + 2$ [5.4.1/5.4.2]   **31.** [5.5.2]   **32.** 0 [5.5.1]   **33.** Undefined [5.3.1]

34. $y = \dfrac{1}{2}x - 2$ [5.4.1/5.4.2]   **35.** Yes [5.2.1]   **36.** $(2, -1)$ [5.2.1]   **37.** $(0, 0)$; $(0, 0)$ [5.2.3]

38. $y = 3$ [5.4.1/5.4.2]   **39.** [5.6.1]   **40.** [5.3.2]

41. $y = 3x - 4$ [5.4.1/5.4.2]   **42.** [5.2.3]

43. Domain: $\{-10, -5, 5\}$; range $\{-5, 0\}$; no [5.5.1]   **44.** Range: $\{-53, -23, 7, 37, 67\}$ [5.5.1]

45. Range: $\{2, 3, 4, 5, 6\}$ [5.5.1]   **46.** 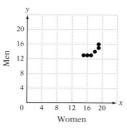 [5.1.2]

47. 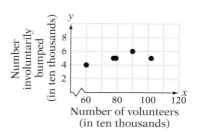 [5.1.2]   **48.** $\{(25, 4.8), (35, 3.5), (40, 2.1), (20, 5.5), (45, 1.0)\}$; yes [5.5.1]

49. **a.** $f(s) = 70s + 40,000$ **b.**

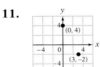

c. The contractor's cost to build a house of 1500 ft² is $145,000.   [5.5.2]

50. The average annual rate of change was $1277.94.   [5.1.3]

CHAPTER TEST

1. $y = -\dfrac{1}{3}x$ [5.4.1/5.4.2]   **2.** $\dfrac{7}{11}$ [5.3.1]   **3.** $(8, 0); (0, -12)$ [5.2.3]   **4.** $(9, -13)$ [5.2.1]

5.

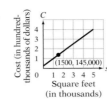

[5.2.3]   **6.**

[5.2.2]   **7.** 10 [5.5.1]   **8.** No [5.2.1]

9.

[5.3.2]   **10.**

[5.6.1]   **11.**

[5.1.1]

(0, 4)
(3, −2)

12.

[5.3.2]   **13.** 0 [5.3.1]   **14.** $y = -\dfrac{2}{5}x + 7$ [5.4.1/5.4.2]

15. $y = 4x - 7$ [5.4.1/5.4.2]   **16.** 13 [5.5.1]   **17.**

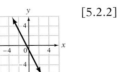

[5.2.2]

18.

[5.6.1]   **19.**

[5.2.3]   **20.**

[5.3.2]

21.

[5.5.2]   **22.**

[5.5.2]   **23.** 25 [5.5.1]

24. $\{-22, -7, 5, 11, 26\}$ [5.5.1]   **25.** No [5.3.1]   **26.** The average annual rate of change was 0.5 million drivers per year. [5.1.3]   **27.**

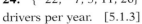

[5.1.2]   **28.** **a.** $f(t) = 8t + 1000$

b.

c.   The cost to manufacture 340 toasters is \$3720.   [5.5.2]

29.   {(8.5, 64), (9.4, 68), (10.1, 76), (11.4, 87), (12.0, 92)}; yes   [5.5.1]   **30.**   The slope is 36. The cost of lumber has been increasing by \$36 per month.   [5.3.1]

CUMULATIVE REVIEW EXERCISES

1. -12 [1.4.2]   **2.** $-\dfrac{5}{8}$ [2.1.1]   **3.** $-17x + 28$ [2.2.5]   **4.** $\dfrac{3}{2}$ [3.2.1]   **5.**   1   [3.2.3]

6. $\dfrac{1}{15}$ [1.3.4]   **7.**   {1, 2, 3, 4, 5, 6, 7, 8}   [1.1.1]   **8.** $-4, 0, 5$ [1.1.1]   **9.** $a \geq -1$ [3.3.3]

10. $y = \dfrac{4}{5}x - 3$ [5.2.3]   **11.** $(-2, -7)$ [5.2.1]   **12.**   0   [5.3.1]   **13.**   (4, 0); (0, 10)   [5.2.3]

14. $y = -x + 5$ [5.4.1/5.4.2]   **15.**   [5.2.2]   **16.**   [5.2.3]

17.   [5.5.2]   **18.**   [5.6.1]

19.   Domain: {0, 1, 2, 3, 4}; range: {0, 1, 2, 3, 4}; yes   [5.5.1]   **20.** -11 [5.5.1]
21.   {$-7, -2, 3, 8, 13, 18$}   [5.5.1]   **22.**   7 and 17   [4.1.2]   **23.**   The fulcrum is 7 ft from the 80-pound force.
[3.2.4]   **24.**   The length of the first side is 22 ft.   [4.3.1]   **25.**   The sale price is \$62.30.   [4.4.2]

Answers to Chapter 6 Selected Exercises

PREP TEST

1. $y = \dfrac{3}{4}x - 6$ [5.2.3]   **2.**   1000   [3.2.3]   **3.**   33y   [2.2.5]   **4.** $10x - 10$ [2.2.5]   **5.**   Yes   [3.1.1]

6.   (4, 0) and (0, -3)   [5.2.3]   **7.**   Yes   [5.3.1]   **8.**   [5.2.2]   **9.**   1.5 h   [4.7.1]

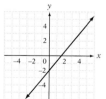

CONCEPT REVIEW 6.1

1.   Always true   **3.**   Sometimes true   **5.**   Never true

SECTION 6.1

3.   Yes   **5.**   Yes   **7.**   No   **9.**   No   **11.**   No   **13.**   Yes   **15.**   Yes   **19.**   Independent
21.   Dependent   **23.**   Inconsistent   **25.**   (2, -1)   **27.**   No solution   **29.**   (-2, 4)   **31.**   (4, 1)
33.   (4, 1)   **35.**   (4, 3)   **37.**   (3, -2)   **39.**   (2, -2)   **41.**   The system of equations is inconsistent and

has no solution. **43.** The system of equations is dependent. The solutions are the ordered pairs that satisfy the equation $y = 2x - 2$. **45.** $(1, -4)$ **47.** $(0, 0)$ **49.** The system of equations is inconsistent and has no solution. **51.** $(0, -2)$ **53.** $(1, -1)$ **55.** $(2, 0)$ **57.** $(-3, -2)$
59. The system of equations is inconsistent and has no solution.
61. $y = 2$ **63.** $y = x$ **65. a.** Always true **b.** Never true **c.** Always true
$\quad\ y = x + 4$ $\qquad y = -x + 2$

CONCEPT REVIEW 6.2

1. Always true **3.** Never true **5.** Never true

SECTION 6.2

1. $(2, 1)$ **3.** $(4, 1)$ **5.** $(-1, 1)$ **7.** $(3, 1)$ **9.** $(1, 1)$ **11.** $(-1, 1)$ **13.** The system of equations is inconsistent and has no solution. **15.** The system of equations is inconsistent and has no solution.
17. $\left(-\dfrac{3}{4}, -\dfrac{3}{4}\right)$ **19.** $\left(\dfrac{9}{5}, \dfrac{6}{5}\right)$ **21.** $(-7, -23)$ **23.** $(1, 1)$ **25.** $(2, 0)$ **27.** $(2, 1)$
29. $\left(\dfrac{9}{19}, -\dfrac{13}{19}\right)$ **31.** $(5, 7)$ **33.** $(1, 7)$ **35.** $\left(\dfrac{17}{5}, -\dfrac{7}{5}\right)$ **37.** $\left(-\dfrac{6}{11}, \dfrac{31}{11}\right)$ **39.** $(2, 3)$
41. $(0, 0)$ **43.** The system of equations is dependent. The solutions are the ordered pairs that satisfy the equation $3x + y = 4$. **45.** $\left(\dfrac{20}{17}, -\dfrac{15}{17}\right)$ **47.** $(5, 2)$ **49.** $(-17, -8)$ **51.** $\left(-\dfrac{5}{7}, \dfrac{13}{7}\right)$
53. $(3, -2)$ **55.** $\left(-\dfrac{11}{20}, \dfrac{23}{40}\right)$ **57.** $(3, -2)$ **59.** $(1, 7)$ **61.** 2 **63.** 2

CONCEPT REVIEW 6.3

1. Always true **3.** Never true **5.** Never true

SECTION 6.3

1. $(5, -1)$ **3.** $(1, 3)$ **5.** $(1, 1)$ **7.** $(3, -2)$ **9.** The system of equations is dependent. The solutions are the ordered pairs that satisfy the equation $2x - y = 1$. **11.** $(3, 1)$ **13.** The system of equations is inconsistent and has no solution. **15.** $(1, 3)$ **17.** $(2, 0)$ **19.** $(0, 0)$ **21.** $(5, -2)$
23. $(-17, 7)$ **25.** $(3, 4)$ **27.** $(1, -1)$ **29.** The system of equations is dependent. The solutions are the ordered pairs that satisfy the equation $x + 3y = 4$. **31.** $(3, 1)$ **33.** $\left(1, \dfrac{1}{3}\right)$ **35.** $(1, 1)$
37. $\left(\dfrac{1}{2}, -\dfrac{1}{2}\right)$ **39.** $\left(\dfrac{2}{3}, \dfrac{1}{9}\right)$ **41.** $\left(\dfrac{7}{25}, -\dfrac{1}{25}\right)$ **43.** $\left(\dfrac{11}{12}, -\dfrac{7}{6}\right)$ **45.** $(1, 4)$ **47.** $(1, 2)$
49. $A = 3, B = -1$ **51.** $A = 2$ **53. a.** $k \neq 1$ **b.** $k \neq \dfrac{3}{2}$ **c.** $k \neq 4$

CONCEPT REVIEW 6.4

1. Always true **3.** Never true **5.** Never true

SECTION 6.4

1. The speed of the whale in calm water is 35 mph. The rate of the current is 5 mph. **3.** The team's rowing rate in calm water was 14 km/h. The rate of the current was 6 km/h. **5.** The rate of the boat in calm water was 7 mph. The rate of the current was 3 mph. **7.** The speed of the jet in calm air was 525 mph. The rate of the wind was 35 mph. **9.** The rate of the plane in calm air was 100 mph. The rate of the wind was 20 mph. **11.** The speed of the helicopter in calm air was 225 mph. The rate of the wind was 45 mph. **13.** The cost is $245.

15. The cost of the wheat is $.65 per pound. The cost of the rye is $.70 per pound. **17.** The daytime rate is $7 per hour. The nighttime rate is $9 per hour. **19.** The team scored 21 two-point baskets and 15 three-point baskets. **21.** There are 8 nickels and 10 quarters in the first bank. **23.** There are 8 dimes and 10 quarters in the bank. **25.** The two angles measure 55° and 125°. **27.** $6000 was invested in the 9% account, and $4000 was invested in the 8% account. **29.** There are 0 dimes and 5 nickels, or 1 dime and 3 nickels, or 2 dimes and 1 nickel in the bank.

CHAPTER REVIEW EXERCISES

1. $(-1, 1)$ [6.2.1] **2.** $(2, -3)$ [6.1.1] **3.** $(-3, 1)$ [6.3.1] **4.** $(1, 6)$ [6.2.1] **5.** $(1, 1)$ [6.1.1] **6.** $(1, -5)$ [6.3.1] **7.** $(3, 2)$ [6.2.1] **8.** The system of equations is dependent. The solutions are the ordered pairs that satisfy the equation $8x - y = 25$. [6.3.1] **9.** $(3, -1)$ [6.3.1] **10.** $(1, -3)$ [6.1.1] **11.** $(-2, -7)$ [6.3.1] **12.** $(4, 0)$ [6.2.1] **13.** Yes [6.1.1] **14.** $\left(-\dfrac{5}{6}, \dfrac{1}{2}\right)$ [6.3.1]

15. $(-1, -2)$ [6.3.1] **16.** $(3, -3)$ [6.1.1] **17.** The system of equations is inconsistent and has no solution. [6.3.1] **18.** $\left(-\dfrac{1}{3}, \dfrac{1}{6}\right)$ [6.2.1] **19.** The system of equations is dependent. The solutions are the ordered pairs that satisfy the equation $y = 2x - 4$. [6.1.1] **20.** The system of equations is dependent. The solutions are the ordered pairs that satisfy the equation $3x + y = -2$. [6.3.1] **21.** $(-2, -13)$ [6.3.1] **22.** The system of equations is dependent. The solutions are the ordered pairs that satisfy the equation $4x + 3y = 12$. [6.2.1] **23.** $(4, 2)$ [6.1.1] **24.** The system of equations is inconsistent and has no solution. [6.3.1] **25.** $\left(\dfrac{1}{2}, -1\right)$ [6.2.1] **26.** No [6.1.1] **27.** $\left(\dfrac{2}{3}, -\dfrac{1}{6}\right)$ [6.3.1] **28.** The system of equations is inconsistent and has no solution. [6.2.1] **29.** $\left(\dfrac{1}{3}, 6\right)$ [6.2.1] **30.** The system of equations is inconsistent and has no solution. [6.1.1] **31.** $(0, -1)$ [6.3.1] **32.** $(-1, -3)$ [6.2.1]

33. $(0, -2)$ [6.2.1] **34.** $(2, 0)$ [6.3.1] **35.** $(-4, 2)$ [6.3.1] **36.** $(1, 6)$ [6.2.1] **37.** The rate of the plane in calm air is 180 mph. The rate of the wind is 20 mph. [6.4.1] **38.** There were 60 adult tickets and 140 children's tickets sold. [6.4.2] **39.** The rate of the canoeist in calm water was 8 mph. The rate of the current was 2 mph. [6.4.1] **40.** There were 130 mailings requiring 33¢ in postage. [6.4.2] **41.** The rate of the boat in calm water was 14 km/h. The rate of the current was 2 km/h. [6.4.1] **42.** The customer purchased 4 compact discs at $15 each and 6 at $10 each. [6.4.2] **43.** The rate of the plane in calm air was 125 km/h. The rate of the wind was 15 km/h. [6.4.1] **44.** The rate of the paddle boat in calm water was 3 mph. The rate of the current was 1 mph. [6.4.1] **45.** There are 350 bushels of lentils and 200 bushels of corn in the silo. [6.4.2] **46.** The rate of the plane in calm air was 105 mph. The rate of the wind was 15 mph. [6.4.1] **47.** There are 4 dimes in the coin purse. [6.4.2] **48.** The rate of the plane in calm air was 325 mph. The rate of the wind was 25 mph. [6.4.1] **49.** The investor bought 1300 of the $6 shares and 200 of the $25 shares. [6.4.2] **50.** The rate of the sculling team in calm water was 9 mph. The rate of the current was 3 mph. [6.4.1]

CHAPTER TEST

1. $(3, 1)$ [6.2.1] **2.** $(2, 1)$ [6.3.1] **3.** Yes [6.1.1] **4.** $(1, -1)$ [6.2.1] **5.** $\left(\dfrac{1}{2}, -1\right)$ [6.3.1] **6.** $(-2, 6)$ [6.1.1] **7.** The system of equations is inconsistent and has no solution. [6.2.1]

8. $(2, -1)$ [6.2.1] **9.** $(2, -1)$ [6.3.1] **10.** $\left(\dfrac{22}{7}, -\dfrac{5}{7}\right)$ [6.2.1] **11.** $(1, -2)$ [6.3.1]

12. No [6.1.1] **13.** $(2, 1)$ [6.2.1] **14.** $(2, -2)$ [6.3.1] **15.** $(2, 0)$ [6.1.1] **16.** $(4, 1)$ [6.2.1]

17. $(3, -2)$ [6.3.1] **18.** The system of equations is dependent. The solutions are the ordered pairs that satisfy the equation $3x + 6y = 2$. [6.1.1] **19.** $(1, 1)$ [6.2.1] **20.** $(4, -3)$ [6.3.1] **21.** $(-6, 1)$ [6.2.1]

22. $(1, -4)$ [6.3.1] **23.** The rate of the plane in calm air is 100 mph. The rate of the wind is 20 mph. [6.4.1]

24. The rate of the boat in calm water is 14 mph. The rate of the current is 2 mph. [6.4.1] **25.** There are 40 dimes and 30 nickels in the first bank. [6.4.2]

CUMULATIVE REVIEW EXERCISES

1. $-8, -4$ [1.1.1] **2.** $\{1, 2, 3, 4, 5, 6, 7, 8, 9, 10\}$ [1.1.1] **3.** 10 [1.4.2] **4.** $40a - 28$ [2.2.5]

5. $\dfrac{3}{2}$ [2.1.1] **6.** $-\dfrac{3}{2}$ [3.1.3] **7.** $-\dfrac{7}{2}$ [3.2.3] **8.** $-\dfrac{2}{9}$ [3.2.3] **9.** $x < -5$ [3.3.3]

10. $x \le 3$ [3.3.3] **11.** 24% [3.1.5] **12.** $(4, 0), (0, -2)$ [5.2.3] **13.** $-\dfrac{7}{5}$ [5.3.1]

14. $y = -\dfrac{3}{2}x$ [5.4.1/5.4.2] **15.** [5.2.3] **16.** [5.2.2]

17. [5.6.1] **18.** [5.5.2] **19.** Domain: $\{-5, 0, 1, 5\}$; range: $\{5\}$; yes [5.5.1]

20. 3 [5.5.1] **21.** Yes [6.1.1] **22.** $(-2, 1)$ [6.2.1] **23.** $(0, 2)$ [6.1.1] **24.** $(2, 1)$ [6.3.1]

25. $\{-22, -12, -2, 8, 18\}$ [5.5.1] **26.** 200 cameras were produced during the month. [3.2.4]

27. $3750 should be invested at 9.6\%$, and $5000 should be invested at 7.2\%. [4.5.1] **28.** The rate of the plane is 160 mph. The rate of the wind is 30 mph. [6.4.1] **29.** The rate of the boat in calm water is 10 mph. [6.4.1]

30. There are 40 dimes in the first bank. [6.4.2]

Answers to Chapter 7 Selected Exercises

PREP TEST

1. 1 [1.2.2] **2.** -18 [1.2.3] **3.** $\dfrac{2}{3}$ [1.2.4] **4.** 48 [2.1.1] **5.** 0 [1.2.4]

6. No [2.2.2] **7.** $5x^2 - 9x - 6$ [2.2.2] **8.** 0 [2.2.2] **9.** $-6x + 24$ [2.2.4]

10. $-7xy + 10y$ [2.2.5]

CONCEPT REVIEW 7.1

1. Always true **3.** Never true **5.** Always true

SECTION 7.1

1. Binomial **3.** Monomial **5.** No **7.** Yes **9.** Yes **11.** $-2x^2 + 3x$ **13.** $y^2 - 8$

15. $5x^2 + 7x + 20$ **17.** $x^3 + 2x^2 - 6x - 6$ **19.** $2a^3 - 3a^2 - 11a + 2$ **21.** $5x^2 + 8x$

23. $7x^2 + xy - 4y^2$ **25.** $3a^2 - 3a + 17$ **27.** $5x^3 + 10x^2 - x - 4$ **29.** $3r^3 + 2r^2 - 11r + 7$

31. $-2x^3 + 3x^2 + 10x + 11$ **33.** $4x$ **35.** $3y^2 - 4y - 2$ **37.** $-7x - 7$ **39.** $4x^3 + 3x^2 + 3x + 1$
41. $y^3 - y^2 + 6y - 6$ **43.** $-y^2 - 13xy$ **45.** $2x^2 - 3x - 1$ **47.** $-2x^3 + x^2 + 2$ **49.** $3a^3 - 2$
51. $4y^3 - 2y^2 + 2y - 4$ **53.** $-a^2 - 1$ **55.** $-4x^2 + 6x + 3$ **57.** $4x^2 - 6x + 3$
59. Yes. For example, $(3x^3 - 2x^2 + 3x - 4) - (3x^3 + 4x^2 - 6x + 5) = -6x^2 + 9x - 9$

CONCEPT REVIEW 7.2

1. Never true   **3.** Always true   **5.** Never true

SECTION 7.2

3. **a.** Product of two exponential expressions **b.** Power of an exponential expression
5. **a.** Power of an exponential expression **b.** Product of two exponential expressions
7. $2x^2$ **9.** $12x^2$ **11.** $6a^7$ **13.** x^3y^5 **15.** $-10x^9y$ **17.** x^7y^8 **19.** $-6x^3y^5$
21. x^4y^5z **23.** $a^3b^5c^4$ **25.** $-a^5b^8$ **27.** $-6a^5b$ **29.** $40y^{10}z^6$ **31.** $-20a^2b^3$
33. $x^3y^5z^3$ **35.** $-12a^{10}b^7$ **37.** 64 **39.** 4 **41.** -64 **43.** x^9 **45.** x^{14} **47.** x^4
49. $4x^2$ **51.** $-8x^6$ **53.** x^4y^6 **55.** $9x^4y^2$ **57.** $27a^8$ **59.** $-8x^7$ **61.** x^8y^4
63. a^4b^6 **65.** $16x^{10}y^3$ **67.** $-18x^3y^4$ **69.** $-8a^7b^5$ **71.** $-54a^9b^3$ **73.** $-72a^5b^5$
75. $32x^3$ **77.** $-3a^3b^3$ **79.** $13x^5y^5$ **81.** $27a^5b^3$ **83.** $-5x^7y^4$ **85.** a^{2n} **87.** a^{2n}
89. True   **91.** False; $x^{2 \cdot 5} = x^{10}$ **93.** No; $2^{(3^2)}$ is larger. $(2^3)^2 = 8^2 = 64$, whereas $2^{(3^2)} = 2^9 = 512$.
95. $x = y$ when n is odd.

CONCEPT REVIEW 7.3

1. Always true   **3.** Sometimes true   **5.** Always true   **7.** Sometimes true

SECTION 7.3

1. $x^2 - 2x$ **3.** $-x^2 - 7x$ **5.** $3a^3 - 6a^2$ **7.** $-5x^4 + 5x^3$ **9.** $-3x^5 + 7x^3$ **11.** $12x^3 - 6x^2$
13. $6x^2 - 12x$ **15.** $-x^3y + xy^3$ **17.** $2x^4 - 3x^2 + 2x$ **19.** $2a^3 + 3a^2 + 2a$ **21.** $3x^6 - 3x^4 - 2x^2$
23. $-6y^4 - 12y^3 + 14y^2$ **25.** $-2a^3 - 6a^2 + 8a$ **27.** $6y^4 - 3y^3 + 6y^2$ **29.** $x^3y - 3x^2y^2 + xy^3$
31. $x^3 + 4x^2 + 5x + 2$ **33.** $a^3 - 6a^2 + 13a - 12$ **35.** $-2b^3 + 7b^2 + 19b - 20$
37. $-6x^3 + 31x^2 - 41x + 10$ **39.** $x^4 - 4x^3 - 3x^2 + 14x - 8$ **41.** $15y^3 - 16y^2 - 70y + 16$
43. $5a^4 - 20a^3 - 15a^2 + 62a - 8$ **45.** $y^4 + 4y^3 + y^2 - 5y + 2$ **47.** **a.** $y, 2y$ **b.** $y, 3$ **c.** $-8, 2y$
d. $-8, 3$ **49.** $x^2 + 4x + 3$ **51.** $a^2 + a - 12$ **53.** $y^2 - 5y - 24$ **55.** $y^2 - 10y + 21$
57. $2x^2 + 15x + 7$ **59.** $3x^2 + 11x - 4$ **61.** $4x^2 - 31x + 21$ **63.** $3y^2 - 2y - 16$
65. $9x^2 + 54x + 77$ **67.** $21a^2 - 83a + 80$ **69.** $15b^2 + 47b - 78$ **71.** $2a^2 + 7ab + 3b^2$
73. $6a^2 + ab - 2b^2$ **75.** $2x^2 - 3xy - 2y^2$ **77.** $10x^2 + 29xy + 21y^2$ **79.** $6a^2 - 25ab + 14b^2$
81. $2a^2 - 11ab - 63b^2$ **83.** $100a^2 - 100ab + 21b^2$ **85.** $15x^2 + 56xy + 48y^2$ **87.** $14x^2 - 97xy - 60y^2$
89. $56x^2 - 61xy + 15y^2$ **91.** $y^2 - 25$ **93.** $4x^2 - 9$ **95.** $x^2 + 2x + 1$ **97.** $9a^2 - 30a + 25$
99. $9x^2 - 49$ **101.** $4a^2 + 4ab + b^2$ **103.** $x^2 - 4xy + 4y^2$ **105.** $16 - 9y^2$
107. $25x^2 + 20xy + 4y^2$ **109.** The area is $(10x^2 - 35x)$ ft².   **111.** The area is $(18x^2 + 12x + 2)$ in².
113. The area $(4x^2 + 4x + 1)$ km².   **115.** The area is $(4x^2 + 10x)$ m².   **117.** The area is
$(\pi x^2 + 8\pi x + 16\pi)$ cm².   **119.** The area is $(90x + 2025)$ ft².   **121.** $4ab$
123. $9a^4 - 24a^3 + 28a^2 - 16a + 4$ **125.** $24x^3 - 3x^2$ **127.** $x^{2n} + x^n$ **129.** $x^{2n} + 2x^n + 1$
131. $x^2 - 1$ **133.** $x^4 - 1$ **135.** $x^6 - 1$ **137.** 1024 **139.** $12x^2 - x - 20$
141. $7x^2 - 11x - 8$

CONCEPT REVIEW 7.4

1. Never true   **3.** Always true   **5.** Never true   **7.** Never true

SECTION 7.4

1. $\dfrac{1}{25}$ **3.** 64 **5.** $\dfrac{1}{27}$ **7.** 1 **11.** y^4 **13.** a^3 **15.** p^4 **17.** $2x^3$ **19.** $2k$

21. $m^5 n^2$ **23.** $\dfrac{3r^2}{2}$ **25.** $-\dfrac{2a}{3}$ **27.** $\dfrac{1}{x^2}$ **29.** a^6 **31.** $\dfrac{4}{x^7}$ **33.** $5b^8$ **35.** $\dfrac{x^2}{3}$

37. 1 **39.** $\dfrac{1}{y^5}$ **41.** $\dfrac{1}{a^6}$ **43.** $\dfrac{1}{3x^3}$ **45.** $\dfrac{2}{3x^5}$ **47.** $\dfrac{y^4}{x^2}$ **49.** $\dfrac{2}{5m^3 n^8}$ **51.** $\dfrac{1}{p^3 q}$

53. $\dfrac{1}{2y^3}$ **55.** $\dfrac{7xz}{8y^3}$ **57.** $\dfrac{p^2}{2m^3}$ **59.** $-\dfrac{8x^3}{y^6}$ **61.** $\dfrac{9}{x^2 y^4}$ **63.** $\dfrac{2}{x^4}$ **65.** $-\dfrac{5}{a^8}$ **67.** $-\dfrac{a^5}{8b^4}$

69. $\dfrac{10y^3}{x^4}$ **71.** $\dfrac{1}{2x^3}$ **73.** $\dfrac{3}{x^3}$ **75.** $\dfrac{1}{2x^2 y^6}$ **77.** $-\dfrac{1}{6x^3}$ **81.** No. 39.4 is not between 1 and 10.

83. No. 2.4 is not an integer. **85.** $2.37 \cdot 10^6$ **87.** $4.5 \cdot 10^{-4}$ **89.** $3.09 \cdot 10^5$ **91.** $6.01 \cdot 10^{-7}$

93. 710,000 **95.** 0.000043 **97.** 671,000,000 **99.** 0.00000713 **101.** $1.6 \cdot 10^{10}$

103. $5.98 \cdot 10^{24}$ **105.** $1.6 \cdot 10^{-19}$ **107.** $1 \cdot 10^{-12}$ **109.** $6.65 \cdot 10^{19}$ **111.** $3.22 \cdot 10^{-14}$

113. $3.6 \cdot 10^5$ **115.** $1.8 \cdot 10^{-18}$ **117.** $\dfrac{3}{64}$ **119.** For 2^x: $\dfrac{1}{4}, \dfrac{1}{2}, 1, 2, 4$. For 2^{-x}: $4, 2, 1, \dfrac{1}{2}, \dfrac{1}{4}$.

121. 0.0625 **123.** xy^2 **125.** $\dfrac{x^6 z^6}{4y^5}$ **127.** 1 **129.** 0 **131.** False; $(2a)^{-3} = \dfrac{1}{8a^3}$

133. False; $(2 + 3)^{-1} = 5^{-1} = \dfrac{1}{5}$

CONCEPT REVIEW 7.5

1. Always true **3.** Never true

SECTION 7.5

1. $x + 1$ **3.** $2a - 5$ **5.** $3a + 2$ **7.** $4b^2 - 3$ **9.** $x - 2$ **11.** $-x + 2$

13. $x^2 + 3x - 5$ **15.** $x^4 - 3x^2 - 1$ **17.** $xy + 2$ **19.** $-3y^3 + 5$ **21.** $3x - 2 + \dfrac{1}{x}$

23. $-3x + 7 - \dfrac{6}{x}$ **25.** $4a - 5 + 6b$ **27.** $9x + 6 - 3y$ **29.** $x + 2$ **31.** $2x + 1$ **33.** $2x - 4$

35. $x + 1 + \dfrac{2}{x - 1}$ **37.** $2x - 1 - \dfrac{2}{3x - 2}$ **39.** $5x + 12 + \dfrac{12}{x - 1}$ **41.** $a + 3 + \dfrac{4}{a + 2}$

43. $y - 6 + \dfrac{26}{2y + 3}$ **45.** $2x + 5 + \dfrac{8}{2x - 1}$ **47.** $5a - 6 + \dfrac{4}{3a + 2}$ **49.** $3x - 5$ **51.** $x^2 + 2x + 3$

53. $x^2 - 3$ **55.** $(x + 2), (x - 3)$ **57.** $3a^2$ **59.** $4x^2 - 6x + 3$

CHAPTER REVIEW EXERCISES

1. $21y^2 + 4y - 1$ [7.1.1] **2.** $-20x^3 y^5$ [7.2.1] **3.** $-8x^3 - 14x^2 + 18x$ [7.3.1]

4. $10a^2 + 31a - 63$ [7.3.3] **5.** $6x^2 - 7x + 10$ [7.5.1] **6.** $2x^2 + 3x - 8$ [7.1.2] **7.** -729 [7.2.2]

8. $x^3 - 6x^2 + 7x - 2$ [7.3.2] **9.** $a^2 - 49$ [7.3.4] **10.** 1296 [7.4.1] **11.** $x - 6$ [7.5.2]

12. $2x^3 + 9x^2 - 3x - 12$ [7.1.1] **13.** $18a^8 b^6$ [7.2.1] **14.** $3x^4 y - 2x^3 y + 12x^2 y$ [7.3.1]

15. $8b^2 - 2b - 15$ [7.3.3] **16.** $-4y + 8$ [7.5.1] **17.** $13y^3 - 12y^2 - 5y - 1$ [7.1.2]

18. 64 [7.2.2] **19.** $6y^3 + 17y^2 - 2y - 21$ [7.3.2] **20.** $4b^2 - 81$ [7.3.4] **21.** $\dfrac{b^6 c^2}{a^4}$ [7.4.1]

22. $2y - 9$ [7.5.2] **23.** $3.97 \cdot 10^{-6}$ [7.4.2] **24.** $x^4 y^8 z^4$ [7.2.1] **25.** $-12y^5 + 4y^4 - 18y^3$ [7.3.1]

26. $18x^2 - 48x + 24$ [7.3.3] **27.** 0.0000623 [7.4.2] **28.** $-7a^2 - a + 4$ [7.1.2] **29.** $9x^4 y^6$ [7.2.2]

30. $12a^3 - 8a^2 - 9a + 6$ [7.3.3]   **31.** $25y^2 - 70y + 49$ [7.3.4]   **32.** $\dfrac{x^4y^6}{9}$ [7.4.1]

33. $x + 5 + \dfrac{4}{x + 12}$ [7.5.2]   **34.** 240,000 [7.4.2]   **35.** $a^6b^{11}c^9$ [7.2.1]

36. $8a^3b^3 - 4a^2b^4 + 6ab^6$ [7.3.1]   **37.** $6x^2 - 7xy - 20y^2$ [7.3.3]   **38.** $4b^4 + 12b^2 - 1$ [7.5.1]

39. $-4b^2 - 13b + 28$ [7.1.2]   **40.** $100a^{15}b^{13}$ [7.2.2]   **41.** $12b^5 - 4b^4 - 6b^3 - 8b^2 + 5$ [7.3.2]

42. $36 - 25x^2$ [7.3.4]   **43.** $\dfrac{2y^3}{x^3}$ [7.4.1]   **44.** $a^2 - 2a + 6$ [7.5.2]   **45.** $4b^3 - 5b^2 - 9b + 7$ [7.1.1]

46. $-54a^{13}b^5c^7$ [7.2.1]   **47.** $-18x^4 - 27x^3 + 63x^2$ [7.3.1]   **48.** $30y^2 - 109y + 30$ [7.3.3]

49. $9.176 \cdot 10^{12}$ [7.4.2]   **50.** $-2y^2 + y - 5$ [7.1.2]   **51.** $144x^{14}y^{18}z^{16}$ [7.2.2]

52. $-12x^4 - 17x^3 - 2x^2 - 33x - 27$ [7.3.2]   **53.** $64a^2 + 16a + 1$ [7.3.4]   **54.** $\dfrac{2}{ab^6}$ [7.4.1]

55. $b^2 + 5b + 2 + \dfrac{7}{b - 7}$ [7.5.2]   **56.** The area is $(20x^2 - 35x)$ m^2. [7.3.5]

57. The area is $(25x^2 + 40x + 16)$ in^2. [7.3.5]   **58.** The area is $(9x^2 - 4)$ ft^2 [7.3.5]   **59.** The area is $(\pi x^2 - 12\pi x + 36\pi)$ cm^2. [7.3.5]   **60.** The area is $(15x^2 - 28x - 32)$ mi^2. [7.3.5]

CHAPTER TEST

1. $3x^3 + 6x^2 - 8x + 3$ [7.1.1]   **2.** $-4x^4 + 8x^3 - 3x^2 - 14x + 21$ [7.3.2]   **3.** $4x^3 - 6x^2$ [7.3.1]

4. $-8a^6b^3$ [7.2.2]   **5.** $-4x^6$ [7.4.1]   **6.** $\dfrac{6b}{a}$ [7.4.1]   **7.** $-5a^3 + 3a^2 - 4a + 3$ [7.1.2]

8. $a^2 + 3ab - 10b^2$ [7.3.3]   **9.** $4x^4 - 2x^2 + 5$ [7.5.1]   **10.** $2x + 3 + \dfrac{2}{2x - 3}$ [7.5.2]

11. $-6x^3y^6$ [7.2.1]   **12.** $6y^4 - 9y^3 + 18y^2$ [7.3.1]   **13.** $\dfrac{9}{x^3}$ [7.4.1]   **14.** $4x^2 - 20x + 25$ [7.3.4]

15. $10x^2 - 43xy + 28y^2$ [7.3.3]   **16.** $x^3 - 7x^2 + 17x - 15$ [7.3.2]   **17.** $\dfrac{a^4}{b^6}$ [7.4.1]

18. $2.9 \cdot 10^{-5}$ [7.4.2]   **19.** $16y^2 - 9$ [7.3.4]   **20.** $3y^3 + 2y^2 - 10y$ [7.1.2]   **21.** $9a^6b^4$ [7.2.2]

22. $10a^3 - 39a^2 + 20a - 21$ [7.3.2]   **23.** $9b^2 + 12b + 4$ [7.3.4]   **24.** $-\dfrac{1}{4a^2b^5}$ [7.4.1]

25. $4x + 8 + \dfrac{21}{2x - 3}$ [7.5.2]   **26.** a^3b^7 [7.2.1]   **27.** $a^2 + ab - 12b^2$ [7.3.3]

28. 0.000000035 [7.4.2]   **29.** The area is $(4x^2 + 12x + 9)$ m^2. [7.3.5]   **30.** The area is $(\pi x^2 - 10\pi x + 25\pi)$ in^2. [7.3.5]

CUMULATIVE REVIEW EXERCISES

1. $\dfrac{1}{144}$ [1.3.2]   **2.** $\dfrac{25}{9}$ [1.4.1]   **3.** $\dfrac{4}{5}$ [1.4.2]   **4.** 87 [1.1.2]   **5.** 0.775 [1.3.1]

6. $-\dfrac{27}{4}$ [2.1.1]   **7.** $-x - 4xy$ [2.2.2]   **8.** $-12x$ [2.2.3]   **9.** $-22x + 20$ [2.2.5]

10. 8 [2.2.1]   **11.** -18 [3.1.3]   **12.** 16 [3.2.2]   **13.** 12 [3.2.3]

14. 80 [3.2.2]   **15.** 24% [3.1.5]   **16.** $x \geq -3$ [3.3.3]   **17.** $-\dfrac{9}{5}$ [5.3.1]

18. $y = -\dfrac{3}{2}x - \dfrac{3}{2}$ [5.4.1/5.4.2]   **19.**  [5.2.3]   **20.** 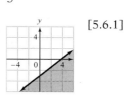 [5.6.1]

21. Domain: {−8, −6, −4, −2}; range: {−7, −5, −2, 0}; yes [5.5.1] **22.** −2 [5.5.1] **23.** (4, 1) [6.2.1]
24. (1, −4) [6.3.1] **25.** $5b^3 − 7b^2 + 8b − 10$ [7.1.2] **26.** $15x^3 − 26x^2 + 11x − 4$ [7.3.2]

27. $20b^2 − 47b + 24$ [7.3.3] **28.** $25b^2 + 30b + 9$ [7.3.4] **29.** $-\dfrac{b^4}{4a}$ [7.4.1]

30. $5y − 4 + \dfrac{1}{y}$ [7.5.1] **31.** $a − 7$ [7.5.2] **32.** $\dfrac{9y^2}{x^6}$ [7.4.1] **33.** {1, 9, 17, 21, 25} [5.5.1]

34. $5(n − 12); 5n − 60$ [2.3.2] **35.** $8n − 2n = 18; n = 3$ [4.1.1] **36.** The length is 15 m. The width is
6 m. [4.3.1] **37.** The selling price is $43.20. [4.4.1] **38.** The concentration of orange juice is 28%.
[4.6.2] **39.** The car overtakes the cyclist 25 mi from the starting point. [4.7.1] **40.** The area is
$(9x^2 + 12x + 4)$ ft². [7.3.5]

Answers to Chapter 8 Selected Exercises

PREP TEST
1. $2 \cdot 3 \cdot 5$ [1.3.2] **2.** $−12y + 15$ [2.2.4] **3.** $−a + b$ [2.2.4] **4.** $−3a + 3b$ [2.2.5]

5. 0 [3.1.3] **6.** $−\dfrac{1}{2}$ [3.2.1] **7.** $x^2 − 2x − 24$ [7.3.3] **8.** $6x^2 − 11x − 10$ [7.3.3]

9. x^3 [7.4.1] **10.** $3x^3y$ [7.4.1]

CONCEPT REVIEW 8.1
1. Always true **3.** Always true **5.** Never true **7.** Never true

SECTION 8.1
3. x^3 **5.** xy^4 **7.** xy^4z^2 **9.** $7a^3$ **11.** 1 **13.** $3a^2b^2$ **15.** ab **17.** $2x$
19. $3x$ **21.** $5(a + 1)$ **23.** $8(2 − a^2)$ **25.** $4(2x + 3)$ **27.** $6(5a − 1)$ **29.** $x(7x − 3)$
31. $a^2(3 + 5a^3)$ **33.** $y(14y + 11)$ **35.** $2x(x^3 − 2)$ **37.** $2x^2(5x^2 − 6)$ **39.** $xy(x − y^2)$
41. $2a^5b + 3xy^3$ **43.** $6b^2(a^2b − 2)$ **45.** $2abc(3a + 2b)$ **47.** $9(2x^2y^2 − a^2b^2)$
49. $6x^3y^3(1 − 2x^3y^3)$ **51.** $x(x^2 − 3x − 1)$ **53.** $2(x^2 + 4x − 6)$ **55.** $b(b^2 − 5b − 7)$
57. $4(2y^2 − 3y + 8)$ **59.** $5y(y^2 − 4y + 2)$ **61.** $3y^2(y^2 − 3y − 2)$ **63.** $3y(y^2 − 3y + 8)$
65. $a^2(6a^3 − 3a − 2)$ **67.** $x^2(8y^2 − 4y + 1)$ **69.** $x^3y^3(4x^2y^2 − 8xy + 1)$ **71.** $(a + b)(x + 2)$
73. $(b + 2)(x − y)$ **75.** $(y − 4)(a − b)$ **77.** $a(x − 2) + b(x − 2)$ **79.** $b(a − 7) − 3(a − 7)$
81. $x(a − 2b) + y(a − 2b)$ **83.** $(x − 2)(a − 5)$ **85.** $(y − 3)(b − 3)$ **87.** $(x − y)(a + 2)$
89. $(c + 5)(z − 8)$ **91.** $(3x − 4)(w + 1)$ **93.** $(x + 4)(x^2 + 3)$ **95.** $(y + 2)(2y^2 + 3)$
97. $(a + 3)(b − 2)$ **99.** $(a − 2)(x^2 − 3)$ **101.** $(a − b)(3x − 2y)$ **103.** $(x − 3)(x + 4a)$
105. $(x − 5)(y − 2)$ **107.** $(7x + 2y)(3x − 7)$ **109.** $(2r + a)(a − 1)$ **111.** $(4x + 3y)(x − 4)$
113. $(2y − 3)(5xy + 3)$ **115. a.** $(x + 3)(2x + 5)$ **b.** $(2x + 5)(x + 3)$ **117. a.** $(a − b)(2a − 3b)$
b. $(2a − 3b)(a − b)$ **119.** 28 **121.** P doubles

CONCEPT REVIEW 8.2
1. Never true **3.** Always true **5.** Never true

SECTION 8.2

1.

Factors of 18	Sum
1, 18	19
2, 9	11
3, 6	9

3.

Factors of −21	Sum
1, −21	−20
−1, 21	20
3, −7	−4
−3, 7	4

5.

Factors of −28	Sum
1, −28	−27
−1, 28	27
2, −14	−12
−2, 14	12
4, −7	−3
−4, 7	3

7. $(x + 1)(x + 2)$ **9.** $(x + 1)(x - 2)$ **11.** $(a + 4)(a - 3)$ **13.** $(a - 1)(a - 2)$ **15.** $(a + 2)(a - 1)$

17. $(b - 3)(b - 3)$ **19.** $(b + 8)(b - 1)$ **21.** $(y + 11)(y - 5)$ **23.** $(y - 2)(y - 3)$

25. $(z - 5)(z - 9)$ **27.** $(z + 8)(z - 20)$ **29.** $(p + 3)(p + 9)$ **31.** $(x + 10)(x + 10)$

33. $(b + 4)(b + 5)$ **35.** $(x + 3)(x - 14)$ **37.** $(b + 4)(b - 5)$ **39.** $(y + 3)(y - 17)$

41. $(p + 3)(p - 7)$ **43.** Nonfactorable over the integers   **45.** $(x - 5)(x - 15)$ **47.** $(x - 7)(x - 8)$

49. $(x + 8)(x - 7)$ **51.** $(a + 3)(a - 24)$ **53.** $(a - 3)(a - 12)$ **55.** $(z + 8)(z - 17)$

57. $(c + 9)(c - 10)$ **59.** $2(x + 1)(x + 2)$ **61.** $3(a + 3)(a - 2)$ **63.** $a(b + 5)(b - 3)$

65. $x(y - 2)(y - 3)$ **67.** $z(z - 3)(z - 4)$ **69.** $3y(y - 2)(y - 3)$ **71.** $3(x + 4)(x - 3)$

73. $5(z + 4)(z - 7)$ **75.** $2a(a + 8)(a - 4)$ **77.** $(x - 2y)(x - 3y)$ **79.** $(a - 4b)(a - 5b)$

81. $(x + 4y)(x - 7y)$ **83.** Nonfactorable over the integers   **85.** $z^2(z - 5)(z - 7)$

87. $b^2(b - 10)(b - 12)$ **89.** $2y^2(y + 3)(y - 16)$ **91.** $x^2(x + 8)(x - 1)$ **93.** $4y(x + 7)(x - 2)$

95. $8(y - 1)(y - 3)$ **97.** $c(c + 3)(c + 10)$ **99.** $3x(x - 3)(x - 9)$ **101.** $(x - 3y)(x - 5y)$

103. $(a - 6b)(a - 7b)$ **105.** $(y + z)(y + 7z)$ **107.** $3y(x + 21)(x - 1)$ **109.** $3x(x + 4)(x - 3)$

111. $4z(z + 11)(z - 3)$ **113.** $(c + 4)(c + 5)$ **115.** $a^2(b - 5)(b - 9)$ **117.** $36, 12, -12, -36$

119. $22, 10, -10, -22$ **121.** $3, 4$ **123.** $5, 8, 9$ **125.** 2

CONCEPT REVIEW 8.3

1. Never true   **3.** Sometimes true   **5.** Always true

SECTION 8.3

1. $(x + 1)(2x + 1)$ **3.** $(y + 3)(2y + 1)$ **5.** $(a - 1)(2a - 1)$ **7.** $(b - 5)(2b - 1)$

9. $(x + 1)(2x - 1)$ **11.** $(x - 3)(2x + 1)$ **13.** Nonfactorable over the integers   **15.** $(2t - 1)(3t - 4)$

17. $(x + 4)(8x + 1)$ **19.** $(b - 4)(3b - 4)$ **21.** $(z - 14)(2z + 1)$ **23.** $(p + 8)(3p - 2)$

25. $(2x - 3)(3x - 4)$ **27.** $(b + 7)(5b - 2)$ **29.** $(2a - 3)(3a + 8)$ **31.** $(3t + 1)(6t - 5)$

33. $(3a + 7)(5a - 3)$ **35.** $(2y - 5)(4y - 3)$ **37.** $(2z + 3)(4z - 5)$ **39.** $(x + 5)(3x - 1)$

41. $(3x + 4)(4x + 3)$ **43.** Nonfactorable over the integers   **45.** $(x + y)(3x - 2y)$ **47.** $(4 + z)(7 - z)$

49. $(1 - x)(8 + x)$ **51.** $3(x + 5)(3x - 4)$ **53.** $4(2x - 3)(3x - 2)$ **55.** $a^2(5a + 2)(7a - 1)$

57. $5(b - 7)(3b - 2)$ **59.** $2x(x + 1)(5x + 1)$ **61.** $2y(y - 4)(5y - 2)$ **63.** $yz(z + 2)(4z - 3)$

65. $b^2(4b + 5)(5b + 4)$ **67.** $xy(3x + 2)(3x + 2)$ **69.** $(t + 2)(2t - 5)$ **71.** $(p - 5)(3p - 1)$

73. $(3y - 1)(4y - 1)$ **75.** Nonfactorable over the integers   **77.** $(3y + 1)(4y + 5)$ **79.** $(a + 7)(7a - 2)$

81. $(z + 2)(4z + 3)$ **83.** $(2p + 5)(11p - 2)$ **85.** $(y + 1)(8y + 9)$ **87.** $(2b - 3)(3b - 2)$

89. $(3b + 5)(11b - 7)$ **91.** $(3y - 4)(6y - 5)$ **93.** Nonfactorable over the integers

95. $(2z - 5)(5z - 2)$ **97.** $(6z + 5)(6z + 7)$ **99.** $(2y - 3)(7y - 4)$ **101.** $(x + 6)(6x - 1)$

103. $3(x + 3)(4x - 1)$ **105.** $10(y + 1)(3y - 2)$ **107.** $x(x + 1)(2x - 5)$ **109.** $(a - 3b)(2a - 3b)$

111. $(y + z)(2y + 5z)$ **113.** $(1 + x)(18 - x)$ **115.** $4(9y + 1)(10y - 1)$ **117.** $8(t + 4)(2t - 3)$

119. $p(2p + 1)(3p + 1)$ **121.** $3(2z - 5)(5z - 2)$ **123.** $3a(2a + 3)(7a - 3)$ **125.** $y(3x - 5y)(3x - 5y)$

127. $xy(3x - 4y)(3x - 4y)$ **129.** $2y(y - 3)(4y - 1)$ **131.** $ab(a + 4)(a - 6)$ **133.** $5t(3 + 2t)(4 - t)$

135. $(y + 3)(2y + 1)$ **137.** $(2x - 1)(5x + 7)$ **139.** $7, -7, 5, -5$ **141.** $7, -7, 5, -5$
143. $11, -11, 7, -7$ **145.** $(x + 2)(x - 3)(x - 1)$

CONCEPT REVIEW 8.4

1. Never true   **3.** Sometimes true   **5.** Always true   **7.** Never true

SECTION 8.4

1. Answers will vary. **a.** $x^2 - 16$ **b.** $(x + 7)(x - 7)$ **c.** $x^2 + 10x + 25$ **d.** $(x - 3)^2$ **e.** $x^2 + 36$
3. $4; 25x^6; 100x^4y^4$ **5.** $4z^4$ **7.** $9a^2y^3$ **9.** $(x + 2)(x - 2)$ **11.** $(a + 9)(a - 9)$
13. $(2x + 1)(2x - 1)$ **15.** $(y + 1)^2$ **17.** $(a - 1)^2$ **19.** Nonfactorable over the integers
21. $(x^3 + 3)(x^3 - 3)$ **23.** $(5x + 1)(5x - 1)$ **25.** $(1 + 7x)(1 - 7x)$ **27.** $(x + y)^2$ **29.** $(2a + 1)^2$
31. $(8a - 1)^2$ **33.** Nonfactorable over the integers   **35.** $(x^2 + y)(x^2 - y)$ **37.** $(3x + 4y)(3x - 4y)$
39. $(4b + 1)^2$ **41.** $(2b + 7)^2$ **43.** $(5a + 3b)^2$ **45.** $(xy + 2)(xy - 2)$ **47.** $(4 + xy)(4 - xy)$
49. $(2y - 9z)^2$ **51.** $(3ab - 1)^2$ **53.** $(m^2 + 16)(m + 4)(m - 4)$ **55.** $(x + 1)(9x + 4)$
57. $(y^4 + 9)(y^2 + 3)(y^2 - 3)$ **59.** $2(x + 3)(x - 3)$ **61.** $x^2(x + 7)(x - 5)$ **63.** $5(b + 3)(b + 12)$
65. Nonfactorable over the integers   **67.** $2y(x + 11)(x - 3)$ **69.** $x(x^2 - 6x - 5)$ **71.** $3(y^2 - 12)$
73. $(2a + 1)(10a + 1)$ **75.** $y^2(x + 1)(x - 8)$ **77.** $5(a + b)(2a - 3b)$ **79.** $2(5 + x)(5 - x)$
81. $ab(3a - b)(4a + b)$ **83.** $2(x - 1)(a + b)$ **85.** $3a(2a - 1)^2$ **87.** $3(81 + a^2)$
89. $2a(2a - 5)(3a - 4)$ **91.** $(x - 2)(x + 1)(x - 1)$ **93.** $a(2a + 5)^2$ **95.** $3b(3a - 1)^2$
97. $6(4 + x)(2 - x)$ **99.** $(x + 2)(x - 2)(a + b)$ **101.** $x^2(x + y)(x - y)$ **103.** $2a(3a + 2)^2$
105. $b(2 - 3a)(1 + 2a)$ **107.** $(x - 5)(2 + x)(2 - x)$ **109.** $8x(3y + 1)^2$ **111.** $y^2(5 + x)(3 - x)$
113. $y(y + 3)(y - 3)$ **115.** $2x^2y^2(x + 1)(x - 1)$ **117.** $x^5(x^2 + 1)(x + 1)(x - 1)$
119. $2xy(3x - 2)(4x + 5)$ **121.** $x^2y^2(2x - 5)^2$ **123.** $(x - 2)^2(x + 2)$ **125.** $4xy^2(1 - x)(2 - 3x)$
127. $2ab(2a - 3b)(9a - 2b)$ **129.** $x^2y^2(1 - 3x)(5 + 4x)$ **131.** $(4x - 3 + y)(4x - 3 - y)$
133. $(x - 2 + y)(x - 2 - y)$ **135.** $12, -12$ **137.** $12, -12$ **139.** 9 **141.** 1 **143.** 4
145. 3 **147.** $(x + 2)(x^2 - 2x + 4)$ **149.** $(y - 3)(y^2 + 3y + 9)$ **151.** $(y + 4)(y^2 - 4y + 16)$
153. $(2x - 1)(4x^2 + 2x + 1)$

CONCEPT REVIEW 8.5

1. Always true   **3.** Always true   **5.** Never true

SECTION 8.5

3. $-3, -2$ **5.** $7, 3$ **7.** $0, 5$ **9.** $0, 9$ **11.** $0, -\dfrac{3}{2}$ **13.** $0, \dfrac{2}{3}$ **15.** $-2, 5$

17. $9, -9$ **19.** $\dfrac{7}{4}, -\dfrac{7}{4}$ **21.** $3, 5$ **23.** $8, -9$ **25.** $-\dfrac{1}{2}, 1$ **27.** $-\dfrac{2}{3}, -4$ **29.** $-7, 3$

31. $0, 7$ **33.** $-1, -4$ **35.** $2, 3$ **37.** $\dfrac{1}{2}, -4$ **39.** $\dfrac{1}{3}, 4$ **41.** $3, 9$ **43.** $9, -2$

45. $-1, -2$ **47.** $5, -9$ **49.** $4, -7$ **51.** $-2, -3$ **53.** $-8, -5$ **55.** $1, 3$ **57.** $-12, 5$

59. $3, 4$ **61.** $-\dfrac{1}{2}, -4$ **63.** The number is 6.   **65.** The numbers are 2 and 4.

67. The integers are 4 and 5.   **69.** The integers are 15 and 16.   **71.** The base is 18 ft. The height is 6 ft.
73. The length is 18 ft. The width is 8 ft.   **75.** The length of a side of the original square is 4 m.
77. The radius of the original circle is 3.81 in.   **79.** The dimensions are 4 in. by 7 in.   **81.** The object will
hit the ground 10 s later.   **83.** 12 consecutive natural numbers beginning with 1 will give a sum of 78.
85. There are 8 teams in the league.   **87.** The ball will be 64 ft above the ground 2 s after it hits home plate.

89. $-\dfrac{3}{2}, -5$ **91.** $9, -3$ **93.** $0, 9$ **95.** $-6, 3$ **97.** 48 or 3   **99.** The length is 10 cm. The width is 7 cm.

CHAPTER REVIEW EXERCISES

1. $7y^3(2y^6 - 7y^3 + 1)$ [8.1.1]   **2.** $(a - 4)(3a + b)$ [8.1.2]   **3.** $(c + 2)(c + 6)$ [8.2.1]
4. $a(a - 2)(a - 3)$ [8.2.2]   **5.** $(2x - 7)(3x - 4)$ [8.3.1/8.3.2]   **6.** $(y + 6)(3y - 2)$ [8.3.1/8.3.2]
7. $(3a + 2)(6a - 5)$ [8.3.1/8.3.2]   **8.** $(ab + 1)(ab - 1)$ [8.4.1]   **9.** $4(y - 2)^2$ [8.4.2]
10. $0, -\dfrac{1}{5}$ [8.5.1]   **11.** $3ab(4a + b)$ [8.1.1]   **12.** $(b - 3)(b - 10)$ [8.2.1]
13. $(2x + 5)(5x + 2y)$ [8.1.2]   **14.** $3(a + 2)(a - 7)$ [8.2.2]   **15.** $n^2(n + 1)(n - 3)$ [8.2.2]
16. Nonfactorable over the integers [8.3.1/8.3.2]   **17.** $(2x - 1)(3x - 2)$ [8.3.1/8.3.2]
18. Nonfactorable over the integers [8.4.1]   **19.** $2, \dfrac{3}{2}$ [8.5.1]   **20.** $7(x + 1)(x - 1)$ [8.4.2]
21. $x^3(3x^2 - 9x - 4)$ [8.1.1]   **22.** $(x - 3)(4x + 5)$ [8.1.2]   **23.** $(a + 7)(a - 2)$ [8.2.1]
24. $(y + 9)(y - 4)$ [8.2.1]   **25.** $5(x - 12)(x + 2)$ [8.2.2]   **26.** $-3, 7$ [8.5.1]
27. $(a + 2)(7a + 3)$ [8.3.1/8.3.2]   **28.** $(x + 20)(4x + 3)$ [8.3.1/8.3.2]   **29.** $(3y^2 + 5z)(3y^2 - 5z)$ [8.4.1]
30. $5(x + 2)(x - 3)$ [8.2.2]   **31.** $-\dfrac{3}{2}, \dfrac{2}{3}$ [8.5.1]   **32.** $2b(2b - 7)(3b - 4)$ [8.3.1/8.3.2]
33. $5x(x^2 + 2x + 7)$ [8.1.1]   **34.** $(x - 2)(x - 21)$ [8.2.1]   **35.** $(3a + 2)(a - 7)$ [8.1.2]
36. $(2x - 5)(4x - 9)$ [8.3.1/8.3.2]   **37.** $10x(a - 4)(a - 9)$ [8.2.2]   **38.** $(a - 12)(2a + 5)$ [8.3.1/8.3.2]
39. $(3a - 5b)(7x + 2y)$ [8.1.2]   **40.** $(a^3 + 10)(a^3 - 10)$ [8.4.1]   **41.** $(4a + 1)^2$ [8.4.1]
42. $\dfrac{1}{4}, -7$ [8.5.1]   **43.** $10(a + 4)(2a - 7)$ [8.3.1/8.3.2]   **44.** $6(x - 3)$ [8.1.1]
45. $x^2y(3x^2 + 2x + 6)$ [8.1.1]   **46.** $(d - 5)(d + 8)$ [8.2.1]   **47.** $2(2x - y)(6x - 5)$ [8.1.2]
48. $4x(x + 1)(x - 6)$ [8.2.2]   **49.** $-2, 10$ [8.5.1]   **50.** $(x - 5)(3x - 2)$ [8.3.1/8.3.2]
51. $(2x - 11)(8x - 3)$ [8.3.1/8.3.2]   **52.** $(3x - 5)^2$ [8.4.1]   **53.** $(2y + 3)(6y - 1)$ [8.3.1/8.3.2]
54. $3(x + 6)^2$ [8.4.2]   **55.** The length is 100 yd. The width is 50 yd. [8.5.2]   **56.** The length is 100 yd. The width is 60 yd. [8.5.2]   **57.** The two integers are 4 and 5. [8.5.2]   **58.** The distance is 20 ft. [8.5.2]
59. The width of the larger rectangle is 15 ft. [8.5.2]   **60.** The length of a side of the original square was 20 ft. [8.5.2]

CHAPTER TEST

1. $3y^2(2x^2 + 3x + 4)$ [8.1.1]   **2.** $2x(3x^2 - 4x + 5)$ [8.1.1]   **3.** $(p + 2)(p + 3)$ [8.2.1]
4. $(x - 2)(a - b)$ [8.1.2]   **5.** $\dfrac{3}{2}, -7$ [8.5.1]   **6.** $(a - 3)(a - 16)$ [8.2.1]   **7.** $x(x + 5)(x - 3)$ [8.2.2]
8. $4(x + 4)(2x - 3)$ [8.3.1/8.3.2]   **9.** $(b + 6)(a - 3)$ [8.1.2]   **10.** $\dfrac{1}{2}, -\dfrac{1}{2}$ [8.5.1]
11. $(2x + 1)(3x + 8)$ [8.3.1/8.3.2]   **12.** $(x + 3)(x - 12)$ [8.2.1]   **13.** $2(b + 4)(b - 4)$ [8.4.2]
14. $(2a - 3b)^2$ [8.4.1]   **15.** $(p + 1)(x - 1)$ [8.1.2]   **16.** $5(x^2 - 9x - 3)$ [8.1.1]
17. Nonfactorable over the integers [8.3.1/8.3.2]   **18.** $(2x + 7y)(2x - 7y)$ [8.4.1]   **19.** $3, 5$ [8.5.1]
20. $(p + 6)^2$ [8.4.1]   **21.** $2(3x - 4y)^2$ [8.4.2]   **22.** $2y^2(y + 1)(y - 8)$ [8.2.2]
23. The length is 15 cm. The width is 6 cm. [8.5.2]   **24.** The length of the base is 12 in. [8.5.2]
25. The two integers are -13 and -12. [8.5.2]

CUMULATIVE REVIEW EXERCISES

1. -8 [1.2.2]   **2.** -8.1 [1.3.3]   **3.** 4 [1.4.2]   **4.** -7 [2.1.1]   **5.** The Associative Property of

Addition [2.2.1]   **6.** $18x^2$ [2.2.3]   **7.** $-6x + 24$ [2.2.5]   **8.** $\dfrac{2}{3}$ [3.1.3]   **9.** 5 [3.2.3]

10. $\dfrac{7}{4}$ [3.2.2]   **11.** 3 [3.2.3]   **12.** 35 [3.1.5]   **13.** $x \le -3$ [3.3.3]   **14.** $x < \dfrac{1}{5}$ [3.3.3]

15.   [5.2.2]   **16.**   [5.5.2]

17. Domain: $\{-5, -3, -1, 1, 3\}$; range: $\{-4, -2, 0, 2, 4\}$; yes [5.5.1]   **18.** 61 [5.5.1]

19.   [5.6.1]   **20.** $(2, -5)$ [6.2.1]   **21.** $(1, 2)$ [6.3.1]   **22.** $3y^3 - 3y^2 - 8y - 5$ [7.1.1]

23. $-27a^{12}b^6$ [7.2.2]   **24.** $x^3 - 3x^2 - 6x + 8$ [7.3.2]   **25.** $4x + 8 + \dfrac{21}{2x - 3}$ [7.5.2]

26. $\dfrac{y^6}{x^{12}}$ [7.4.1]   **27.** $(a - b)(3 - x)$ [8.1.2]   **28.** $(x + 5y)(x - 2y)$ [8.2.2]

29. $2a^2(3a + 2)(a + 3)$ [8.3.1/8.3.2]   **30.** $(5a + 6b)(5a - 6b)$ [8.4.1]   **31.** $3(2x - 3y)^2$ [8.4.2]

32. $\dfrac{4}{3}, -5$ [8.5.1]   **33.** $\{-11, -7, -3, 1, 5\}$ [5.5.1]   **34.** The average daily high temperature

was -3°C. [1.2.5]   **35.** The length is 15 cm. The width is 6 cm. [4.3.1]   **36.** The pieces measure 4 ft and

6 ft. [4.1.2]   **37.** The maximum distance you can drive a Company A car is 359 mi. [4.8.1]

38. $6500 must be invested at an annual simple interest rate of 11%. [4.5.1]

39. The discount rate is 40%. [4.4.2]   **40.** The three integers are 10, 12, and 14. [4.2.1]

Answers to Chapter 9 Selected Exercises

PREP TEST

1. 36 [1.3.2]   **2.** $\dfrac{3x}{y^3}$ [7.4.1]   **3.** $-\dfrac{5}{36}$ [1.3.2]   **4.** $-\dfrac{10}{11}$ [1.3.3]   **5.** No. $\dfrac{0}{a} = 0$ but $\dfrac{a}{0}$ is

undefined. [1.2.4]   **6.** $\dfrac{19}{8}$ [3.2.1]   **7.** 130° [4.3.2]   **8.** $(x - 6)(x + 2)$ [8.2.1]

9. $(2x - 3)(x + 1)$ [8.3.1/8.3.2]   **10.** Jean will catch up to Anthony at 9:40 A.M. [4.7.1]

CONCEPT REVIEW 9.1

1. Always true   **3.** Never true   **5.** Always true   **7.** Never true

SECTION 9.1

5. $\dfrac{3}{4x}$ **7.** $\dfrac{1}{x + 3}$ **9.** -1 **11.** $\dfrac{2}{3y}$ **13.** $-\dfrac{3}{4x}$ **15.** $\dfrac{a}{b}$ **17.** $-\dfrac{2}{x}$ **19.** $\dfrac{y - 2}{y - 3}$

21. $\dfrac{x + 5}{x + 4}$ **23.** $\dfrac{x + 4}{x - 3}$ **25.** $-\dfrac{x + 2}{x + 5}$ **27.** $\dfrac{2(x + 2)}{x + 3}$ **29.** $\dfrac{2x - 1}{2x + 3}$ **31.** $\dfrac{2}{3xy}$

33. $\dfrac{8xy^2ab}{3}$ **35.** $\dfrac{2}{9}$ **37.** $\dfrac{y^2}{x}$ **39.** $\dfrac{y(x+4)}{x(x+1)}$ **41.** $\dfrac{x^3(x-7)}{y^2(x-4)}$ **43.** $-\dfrac{y}{x}$ **45.** $\dfrac{x+3}{x+1}$

47. $\dfrac{x-5}{x+3}$ **49.** $-\dfrac{x+3}{x+5}$ **51.** $\dfrac{12x^4}{(x+1)(2x+1)}$ **53.** $-\dfrac{x+3}{x-12}$ **55.** $\dfrac{x+2}{x+4}$ **57.** $\dfrac{2x-5}{2x-1}$

59. $\dfrac{3x-4}{2x+3}$ **63.** $\dfrac{2xy^2ab^2}{9}$ **65.** $\dfrac{5}{12}$ **67.** $3x$ **69.** $\dfrac{y(x+3)}{x(x+1)}$ **71.** $\dfrac{x+7}{x-7}$ **73.** $-\dfrac{4ac}{y}$

75. $\dfrac{x-5}{x-6}$ **77.** 1 **79.** $-\dfrac{x+6}{x+5}$ **81.** $\dfrac{2x+3}{x-6}$ **83.** $\dfrac{4x+3}{2x-1}$ **85.** $\dfrac{(2x+5)(4x-1)}{(2x-1)(4x+5)}$

87. $-6,\,1$ **89.** $-1,\,1$ **91.** $3,\,-2$ **93.** -3 **95.** $\dfrac{4}{3},\,-\dfrac{1}{2}$ **97.** $\dfrac{4a}{3b}$ **99.** $\dfrac{8}{9}$

101. $\dfrac{x-4}{y^4}$ **103.** $\dfrac{(x-2)(x-2)}{(x-4)(x+4)}$ **105.** $\dfrac{x}{x+8}$

CONCEPT REVIEW 9.2

1. Always true   **3.** Never true   **5.** Always true

SECTION 9.2

1. $24x^3y^2$ **3.** $30x^4y^2$ **5.** $8x^2(x+2)$ **7.** $6x^2y(x+4)$ **9.** $40x^3(x-1)^2$ **11.** $4(x-3)^2$

13. $(2x-1)(2x+1)(x+4)$ **15.** $(x-7)^2(x+2)$ **17.** $(x+4)(x-3)$ **19.** $(x+5)(x-2)(x+7)$

21. $(x-7)(x-3)(x-5)$ **23.** $(x+2)(x+5)(x-5)$ **25.** $(2x-1)(x-3)(x+1)$

27. $(2x-5)(x-2)(x+3)$ **29.** $(x+3)(x-5)$ **31.** $(x+6)(x-3)$ **33.** $\dfrac{4x}{x^2};\dfrac{3}{x^2}$ **35.** $\dfrac{4x}{12y^2};\dfrac{3yz}{12y^2}$

37. $\dfrac{xy}{x^2(x-3)};\dfrac{6x-18}{x^2(x-3)}$ **39.** $\dfrac{9x}{x(x-1)^2};\dfrac{6x-6}{x(x-1)^2}$ **41.** $\dfrac{3x}{x(x-3)};\dfrac{5}{x(x-3)}$ **43.** $\dfrac{3}{(x-5)^2};-\dfrac{2x-10}{(x-5)^2}$

45. $\dfrac{3x}{x^2(x+2)};\dfrac{4x+8}{x^2(x+2)}$ **47.** $\dfrac{x^2-6x+8}{(x+3)(x-4)};\dfrac{x^2+3x}{(x+3)(x-4)}$ **49.** $\dfrac{3}{(x+2)(x-1)};\dfrac{x^2-x}{(x+2)(x-1)}$

51. $\dfrac{5}{(2x-5)(x-2)};\dfrac{x^2-3x+2}{(2x-5)(x-2)}$ **53.** $\dfrac{x^2-3x}{(x+3)(x-3)(x-2)};\dfrac{2x^2-4x}{(x+3)(x-3)(x-2)}$

55. $-\dfrac{x^2-3x}{(x-3)^2(x+3)};\dfrac{x^2+2x-3}{(x-3)^2(x+3)}$ **57.** $\dfrac{3x^2+12x}{(x-5)(x+4)};\dfrac{x^2-5x}{(x-5)(x+4)};-\dfrac{3}{(x-5)(x+4)}$ **59.** $\dfrac{300}{10^4};\dfrac{5}{10^4}$

61. $\dfrac{b^2}{b};\dfrac{5}{b}$ **63.** $\dfrac{y-1}{y-1};\dfrac{y}{y-1}$ **65.** $\dfrac{x^2+1}{(x-1)^3};\dfrac{x^2-1}{(x-1)^3};\dfrac{x^2-2x+1}{(x-1)^3}$ **67.** $\dfrac{2b}{8(a+b)(a-b)};$

$\dfrac{a^2+ab}{8(a+b)(a-b)}$ **69.** $\dfrac{x-2}{(x+y)(x+2)(x-2)};\dfrac{x+2}{(x+y)(x+2)(x-2)}$

CONCEPT REVIEW 9.3

1. Never true   **3.** Never true   **5.** Always true

SECTION 9.3

1. $\dfrac{11}{y^2}$ **3.** $-\dfrac{7}{x+4}$ **5.** $\dfrac{8x}{2x+3}$ **7.** $\dfrac{5x+7}{x-3}$ **9.** $\dfrac{2x-5}{x+9}$ **11.** $\dfrac{-3x-4}{2x+7}$ **13.** $\dfrac{1}{x+5}$

15. $\dfrac{1}{x-6}$ **17.** $\dfrac{3}{2y-1}$ **19.** $\dfrac{1}{x-5}$ **23.** $\dfrac{4y+5x}{xy}$ **25.** $\dfrac{19}{2x}$ **27.** $\dfrac{5}{12x}$ **29.** $\dfrac{19x-12}{6x^2}$

31. $\dfrac{52y-35x}{20xy}$ **33.** $\dfrac{13x+2}{15x}$ **35.** $\dfrac{7}{24}$ **37.** $\dfrac{x+90}{45x}$ **39.** $\dfrac{x^2+2x+2}{2x^2}$ **41.** $\dfrac{2x^2+3x-10}{4x^2}$

43. $\dfrac{3y^2+8}{3y}$ **45.** $\dfrac{x^2+4x+4}{x+4}$ **47.** $\dfrac{4x+7}{x+1}$ **49.** $\dfrac{-3x^2+16x+2}{12x^2}$ **51.** $\dfrac{x^2-x+2}{x^2y}$

53. $\dfrac{16xy - 12y + 6x^2 + 3x}{12x^2y^2}$ **55.** $\dfrac{3xy - 6y - 2x^2 - 14x}{24x^2y}$ **57.** $\dfrac{9x + 2}{(x - 2)(x + 3)}$ **59.** $\dfrac{2(x + 23)}{(x - 7)(x + 3)}$

61. $\dfrac{2x^2 - 5x + 1}{(x + 1)(x - 3)}$ **63.** $\dfrac{4x^2 - 34x + 5}{(2x - 1)(x - 6)}$ **65.** $\dfrac{2a - 5}{a - 7}$ **67.** $\dfrac{4x + 9}{(x + 3)(x - 3)}$ **69.** $\dfrac{-x + 9}{(x + 2)(x - 3)}$

71. $\dfrac{14}{(x - 5)^2}$ **73.** $\dfrac{-2(x + 7)}{(x + 6)(x - 7)}$ **75.** $\dfrac{x - 2}{2x}$ **77.** $\dfrac{x^2 + x + 4}{(x - 2)(x + 1)(x - 1)}$ **79.** $\dfrac{x - 4}{x - 6}$

81. $\dfrac{2x + 1}{x - 1}$ **83.** 2   **85.** $\dfrac{b^2 + b - 7}{b + 4}$ **87.** $\dfrac{4n}{(n - 1)^2}$ **89.** $-\dfrac{3(x^2 + 8x + 25)}{(x - 3)(x + 7)}$ **91.** 1

93. $\dfrac{x^2 - x + 2}{(x + 5)(x + 1)}$ **95.** $\dfrac{2}{3}, \dfrac{3}{4}, \dfrac{4}{5}, \dfrac{50}{51}, \dfrac{100}{101}, \dfrac{1000}{1001}$ **97.** $\dfrac{6}{y} + \dfrac{7}{x}$ **99.** $\dfrac{2}{m^2n} + \dfrac{8}{mn^2}$

CONCEPT REVIEW 9.4

1. Sometimes true   **3.** Never true   **5.** Always true

SECTION 9.4

1. $\dfrac{x}{x - 3}$ **3.** $\dfrac{2}{3}$ **5.** $\dfrac{y + 3}{y - 4}$ **7.** $\dfrac{2(2x + 13)}{5x + 36}$ **9.** $\dfrac{3}{4}$ **11.** $\dfrac{x - 2}{x + 2}$ **13.** $\dfrac{x + 2}{x + 3}$

15. $\dfrac{x - 6}{x + 5}$ **17.** $-\dfrac{x - 2}{x + 1}$ **19.** $x - 1$ **21.** $\dfrac{1}{2x - 1}$ **23.** $\dfrac{x - 3}{x + 5}$ **25.** $\dfrac{x - 7}{x - 8}$

27. $\dfrac{2y - 1}{2y + 1}$ **29.** $\dfrac{x - 2}{2x - 5}$ **31.** $\dfrac{x - 2}{x + 1}$ **33.** $\dfrac{x - 1}{x + 4}$ **35.** $\dfrac{-x - 1}{4x - 3}$ **37.** $\dfrac{x + 1}{2(5x - 2)}$

39. $\dfrac{b + 11}{4b - 21}$ **41.** $\dfrac{5}{3}$ **43.** $-\dfrac{1}{x - 1}$ **45.** $\dfrac{ab}{b + a}$ **47.** $\dfrac{y + 4}{2(y - 2)}$

CONCEPT REVIEW 9.5

1. Always true   **3.** Always true   **5.** Sometimes true   **7.** Always true

SECTION 9.5

3. 1   **5.** -3 **7.** $\dfrac{1}{2}$ **9.** 5   **11.** No solution   **13.** 2, 4   **15.** $-\dfrac{3}{2}, 4$ **17.** 3

19. -1 **21.** 4   **25.** 9   **27.** 36   **29.** 10   **31.** 113   **33.** -2 **35.** 15

37. 4   **39.** 20,000 people voted in favor of the amendment.   **41.** The office building requires 140 air vents.

43. For 2 c of boiling water, 6 c of sugar are required.   **45.** There are about 800 fish in the lake.

47. Yes, the shipment will be accepted.   **49.** The person's height is 67.5 in.   **51.** The distance between the two cities is 750 mi.   **53.** The recommended area of a window is 40 ft².   **55.** There are 160 ml of carbonated water in 280 ml of soft drink.   **57.** For an annual yield of 1320 bushels, 10 additional acres must be planted.   **59.** To serve 70 people, 2 additional gallons are necessary.   **61.** The measure of side DE is 12.8 in.   **63.** The height of triangle DEF is 4.7 ft.   **65.** The perimeter of triangle ABC is 18 m.   **67.** The area of triangle ABC is 48 cm².   **69.** The length of BC is 15 m.   **71.** The length of QO is 10 ft.   **73.** The perimeter of triangle OPQ is 12 m.   **75.** The height of the flagpole is 14.375 ft.   **77.** 1

79. $0, -\dfrac{2}{3}$ **81.** -3 and -4 **83.** The first person's share of the winnings was $1.25 million.

85. The cost would be $18.40.   **87. a.** He calculated the circumference to be 24,960 mi.
b. The difference is 160 mi.

CONCEPT REVIEW 9.6

1. Always true   **3.** Always true   **5.** Always true

SECTION 9.6

1. $h = \dfrac{2A}{b}$ **3.** $t = \dfrac{d}{r}$ **5.** $T = \dfrac{PV}{nR}$ **7.** $L = \dfrac{P - 2W}{2}$ **9.** $b_1 = \dfrac{2A - hb_2}{h}$ **11.** $h = \dfrac{3V}{A}$

13. $S = C - Rt$ **15.** $P = \dfrac{A}{1 + rt}$ **17.** $w = \dfrac{A}{S + 1}$ **19. a.** $h = \dfrac{S - 2\pi r^2}{2\pi r}$ **b.** 5 in. **c.** 4 in.

21. a. $r = \dfrac{S - C}{C}$ **b.** $66.\overline{6}\%$ **c.** $43.\overline{3}\%$

CONCEPT REVIEW 9.7

1. Never true **3.** Always true **5.** Never true

SECTION 9.7

3. It will take 2 h to fill the fountain with both sprinklers operating. **5.** It would take 3 h to remove the earth.
7. It would take the computers 30 h. **9.** It would take 24 min to cool the room 5°F. **11.** It would take
the second welder 15 h. **13.** It would take the second pipeline 90 min to fill the tank. **15.** It would take
3 h to harvest the field using only the older reaper. **17.** It will take the second technician 3 h.

19. It would take the small unit $14\frac{2}{3}$ h. **21.** It would have taken one of the welders 40 h.

23. The camper hiked at a rate of 5 mph. **25.** The rate of the jet was 360 mph. **27.** The rate of the boat
for the first 15 mi was 7.5 mph. **29.** The family can travel 21.6 mi down the river. **31.** The technician
traveled at a rate of 20 mph through the congested traffic. **33.** The rate of the river's current was 5 mph.
35. The rate of the freight train was 30 mph. The rate of the express train was 50 mph. **37.** The rate of the car
is 48 mph. **39.** The rate of the jet stream was 50 mph. **41.** The rate of the current is 5 mph.

43. It would take $1\frac{1}{19}$ h to fill the tank. **45.** The amount of time spent traveling by canoe was 2 h.

47. The bus usually travels at a rate of 60 mph.

CHAPTER REVIEW EXERCISES

1. $\dfrac{by^3}{6ax^2}$ [9.1.2] **2.** $\dfrac{22x - 1}{(3x - 4)(2x + 3)}$ [9.3.2] **3.** $x = -\dfrac{3}{4}y + 3$ [9.6.1] **4.** $\dfrac{2x^4}{3y^7}$ [9.1.1]

5. $\dfrac{1}{x^2}$ [9.1.3] **6.** $\dfrac{x - 2}{3x - 10}$ [9.4.1] **7.** $72a^3b^5$ [9.2.1] **8.** $\dfrac{4x}{3x + 7}$ [9.3.1]

9. $\dfrac{3x}{16x^2} ; \dfrac{10}{16x^2}$ [9.2.2] **10.** $\dfrac{x - 9}{x - 3}$ [9.1.1] **11.** $\dfrac{x - 4}{x + 3}$ [9.1.3] **12.** $\dfrac{2y - 3}{5y - 7}$ [9.3.2]

13. $\dfrac{1}{x}$ [9.1.2] **14.** $\dfrac{1}{x + 3}$ [9.3.1] **15.** $15x^4(x - 7)^2$ [9.2.1] **16.** 10 [9.5.2]

17. No solution [9.5.1] **18.** $-\dfrac{4a}{5b}$ [9.1.1] **19.** $\dfrac{1}{x + 3}$ [9.3.1] **20.** 24 in. [9.5.4]

21. 5 [9.5.1] **22.** $\dfrac{x + 9}{4x}$ [9.4.1] **23.** $\dfrac{3x - 1}{x - 5}$ [9.3.2] **24.** $\dfrac{x - 3}{x + 3}$ [9.1.3]

25. $-\dfrac{x + 6}{x + 3}$ [9.1.1] **26.** 2 [9.5.1] **27.** $x - 2$ [9.4.1] **28.** $\dfrac{x + 3}{x - 4}$ [9.1.2] **29.** 15 [9.5.2]

30. 12 [9.5.2] **31.** $\dfrac{x}{x - 7}$ [9.4.1] **32.** $x = 2y + 15$ [9.6.1] **33.** $\dfrac{3(3a + 2b)}{ab}$ [9.3.2]

34. $(5x - 3)(2x - 1)(4x - 1)$ [9.2.1] **35.** $c = \dfrac{100m}{i}$ [9.6.1] **36.** 40 [9.5.2] **37.** 3 [9.5.2]

38. $\dfrac{7x + 22}{60x}$ [9.3.2] **39.** $\dfrac{8a + 3}{4a - 3}$ [9.1.2] **40.** $\dfrac{3x^2 - x}{(6x - 1)(2x + 3)(3x - 1)}$; $\dfrac{24x^3 - 4x^2}{(6x - 1)(2x + 3)(3x - 1)}$ [9.2.2]

41. $\dfrac{2}{ab}$ [9.3.1] **42.** 62 [9.5.2] **43.** 6 [9.5.1] **44.** $\dfrac{b^3 y}{10ax}$ [9.1.3] **45.** It would take the

apprentice 6 h. [9.7.1] **46.** The spring would stretch 8 in. [9.5.3] **47.** The rate of the wind is 20 mph.
[9.7.2] **48.** To make 10 gal of the garden spray will require 16 additional ounces. [9.5.3] **49.** Using both
hoses, it would take 6 h to fill the pool. [9.7.1] **50.** The rate of the car is 45 mph. [9.7.2]

CHAPTER TEST

1. $\dfrac{x + 5}{x + 4}$ [9.1.3] **2.** $\dfrac{2}{x + 5}$ [9.3.1] **3.** $3(2x - 1)(x + 1)$ [9.2.1] **4.** -1 [9.5.2]

5. $\dfrac{x + 1}{x^3(x - 2)}$ [9.1.2] **6.** $\dfrac{x - 3}{x - 2}$ [9.4.1] **7.** $\dfrac{3x + 6}{x(x - 2)(x + 2)}$; $\dfrac{x^2}{x(x - 2)(x + 2)}$ [9.2.2]

8. $x = -\dfrac{5}{3}y - 5$ [9.6.1] **9.** 2 [9.5.1] **10.** $\dfrac{5}{(2x - 1)(3x + 1)}$ [9.3.2] **11.** 1 [9.1.3]

12. $\dfrac{3}{x + 8}$ [9.3.1] **13.** $6x(x + 2)^2$ [9.2.1] **14.** $-\dfrac{x - 2}{x + 5}$ [9.1.1] **15.** 4 [9.5.1]

16. $t = \dfrac{f - v}{a}$ [9.6.1] **17.** $\dfrac{2x^3}{3y^5}$ [9.1.1] **18.** $\dfrac{3}{(2x - 1)(x + 1)}$ [9.3.2] **19.** 3 [9.5.2]

20. $\dfrac{x^3(x + 3)}{y(x + 2)}$ [9.1.2] **21.** $-\dfrac{3xy + 3y}{x(x + 1)(x - 1)}$; $\dfrac{x^2}{x(x + 1)(x - 1)}$ [9.2.2] **22.** $\dfrac{x + 3}{x + 5}$ [9.4.1]

23. For 15 gal of water, 2 lb of additional salt will be required. [9.5.3] **24.** The rate of the wind is 20 mph.
[9.7.2] **25.** It would take both pipes 6 min to fill the tank. [9.7.1]

CUMULATIVE REVIEW EXERCISES

1. -17 [1.1.2] **2.** -6 [1.4.1] **3.** $\dfrac{31}{30}$ [1.4.2] **4.** 21 [2.1.1] **5.** $5x - 2y$ [2.2.2]

6. $-8x + 26$ [2.2.5] **7.** -20 [3.2.1] **8.** -12 [3.2.3] **9.** 10 [3.1.5] **10.** $x < \dfrac{9}{5}$ [3.3.2]

11. $x \le 15$ [3.3.3] **12.** [5.2.2] **13.** [5.5.2] **14.** [5.2.3]

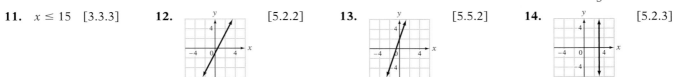

15. [5.6.1] **16.** $\{-3, 5, 11, 19, 27\}$ [5.5.1] **17.** 37 [5.5.1] **18.** $\left(\dfrac{2}{3}, 3\right)$ [6.2.1]

19. $\left(\dfrac{1}{2}, -1\right)$ [6.3.1] **20.** $-6x^4 y^5$ [7.2.1] **21.** $a^{20} b^{15}$ [7.2.2] **22.** $\dfrac{a^3}{b^2}$ [7.4.1]

23. $a^2 + ab - 12b^2$ [7.3.3] **24.** $3b^3 - b + 2$ [7.5.1] **25.** $x^2 + 2x + 4$ [7.5.2]
26. $(4x + 1)(3x - 1)$ [8.3.1/8.3.2] **27.** $(y - 6)(y - 1)$ [8.2.1] **28.** $a(2a - 3)(a + 5)$ [8.3.1/8.3.2]

29. $4(b + 5)(b - 5)$ [8.4.2] **30.** -3 and $\dfrac{5}{2}$ [8.5.1] **31.** $-\dfrac{x + 7}{x + 4}$ [9.1.1] **32.** 1 [9.1.3]

33. $\dfrac{8}{(3x - 1)(x + 1)}$ [9.3.2] **34.** 1 [9.5.1] **35.** $a = \dfrac{f - v}{t}$ [9.6.1]

36. $5x - 18 = -3; x = 3$ [4.1.1]   **37.** \$5000 must be invested at an annual simple interest rate of 11%.   [4.5.1]
38. There is 70% silver in the 120-gram alloy.   [4.6.2]   **39.** The base is 10 in. The height is 6 in.   [8.5.2]
40. It would take both pipes 8 min to fill the tank.   [9.7.1]

Answers to Chapter 10 Selected Exercises

PREP TEST

1. -14 [1.1.2]   **2.** $-2x^2y - 4xy^2$ [2.2.2]   **3.** 14   [3.1.3]   **4.** $\dfrac{7}{5}$ [3.2.2]   **5.** x^6 [7.2.1]

6. $x^2 + 2xy + y^2$ [7.3.4]   **7.** $4x^2 - 12x + 9$ [7.3.4]   **8.** $a^2 - 25$ [7.3.4]   **9.** $4 - 9v^2$ [7.3.4]

10. $\dfrac{x^2y^2}{9}$ [7.4.1]

CONCEPT REVIEW 10.1

1. Never true   **3.** Always true   **5.** Sometimes true

SECTION 10.1

3. 2, 20, 50   **7.** 4   **9.** 7   **11.** $4\sqrt{2}$ **13.** $2\sqrt{2}$ **15.** $18\sqrt{2}$ **17.** $10\sqrt{10}$
19. $\sqrt{15}$ **21.** $\sqrt{29}$ **23.** $-54\sqrt{2}$ **25.** $3\sqrt{5}$ **27.** 0   **29.** $48\sqrt{2}$ **31.** $\sqrt{105}$
33. 30   **35.** $30\sqrt{5}$ **37.** $5\sqrt{10}$ **39.** $4\sqrt{6}$ **41.** 18   **43.** 15.492   **45.** 16.971
47. 15.652   **49.** 18.762   **53.** x^3 **55.** $y^7\sqrt{y}$ **57.** a^{10} **59.** x^2y^2 **61.** $2x^2$
63. $2x\sqrt{6}$ **65.** $xy^3\sqrt{xy}$ **67.** $ab^5\sqrt{ab}$ **69.** $2x^2\sqrt{15x}$ **71.** $7a^2b^4$ **73.** $3x^2y^3\sqrt{2xy}$
75. $2x^5y^3\sqrt{10xy}$ **77.** $4a^4b^5\sqrt{5a}$ **79.** $8ab\sqrt{b}$ **81.** x^3y **83.** $8a^2b^3\sqrt{5b}$ **85.** $6x^2y^3\sqrt{3y}$
87. $4x^3y\sqrt{2y}$ **89.** $5a + 20$ **91.** $2x^2 + 8x + 8$ **93.** $x + 2$ **95.** $y + 1$
97. a. The speed of the car is $12\sqrt{15}$ mph.   **b.** The speed of the car is approximately 46 mph.
99. $0.05ab^2\sqrt{ab}$ **101. a.** 1   **b.** 3   **c.** $3\sqrt{3}$ **103.** The length of a side of the square is 8.7 cm.

105. $x \geq 0$ **107.** $x \geq -5$ **109.** $x \leq \dfrac{5}{2}$ **111.** All real numbers

CONCEPT REVIEW 10.2

1. Never true   **3.** Always true   **5.** Always true

SECTION 10.2

1. $3\sqrt{2}$ **3.** $-\sqrt{7}$ **5.** $-11\sqrt{11}$ **7.** $10\sqrt{x}$ **9.** $-2\sqrt{y}$ **11.** $-11\sqrt{3b}$ **13.** $2x\sqrt{2}$
15. $-3a\sqrt{3a}$ **17.** $-5\sqrt{xy}$ **19.** $8\sqrt{5}$ **21.** $8\sqrt{2}$ **23.** $15\sqrt{2} - 10\sqrt{3}$ **25.** $\sqrt{x}$
27. $-12x\sqrt{3}$ **29.** $2xy\sqrt{x} - 3xy\sqrt{y}$ **31.** $-9x\sqrt{3x}$ **33.** $-13y^2\sqrt{2y}$ **35.** $4a^2b^2\sqrt{ab}$
37. $7\sqrt{2}$ **39.** $6\sqrt{x}$ **41.** $-3\sqrt{y}$ **43.** $-45\sqrt{2}$ **45.** $13\sqrt{3} - 12\sqrt{5}$ **47.** $32\sqrt{3} - 3\sqrt{11}$
49. $6\sqrt{x}$ **51.** $-34\sqrt{3x}$ **53.** $10a\sqrt{3b} + 10a\sqrt{5b}$ **55.** $-2xy\sqrt{3}$ **57.** $-7b\sqrt{ab} + 4a^2\sqrt{b}$
59. $3ab\sqrt{2a} - ab + 4ab\sqrt{3b}$ **61.** $8\sqrt{x + 2}$ **63.** $2x\sqrt{2y}$ **65.** $2ab\sqrt{6ab}$
67. $-11\sqrt{2x + y}$ **69.** The perimeter is $\left(6\sqrt{3} + 2\sqrt{15}\right)$ cm.   **71.** The perimeter is 22.4 cm.

CONCEPT REVIEW 10.3

1. Always true   **3.** Always true   **5.** Sometimes true

SECTION 10.3

1. 5   **3.** 6   **5.** x **7.** x^3y^2 **9.** $3ab^6\sqrt{2a}$ **11.** $12a^4b\sqrt{b}$ **13.** $2 - \sqrt{6}$
15. $x - \sqrt{xy}$ **17.** $5\sqrt{2} - \sqrt{5x}$ **19.** $4 - 2\sqrt{10}$ **21.** $x - 6\sqrt{x} + 9$ **23.** $3a - 3\sqrt{ab}$

25. $10abc$ **27.** $15x - 22y\sqrt{x} + 8y^2$ **29.** $x - y$ **31.** $10x + 13\sqrt{xy} + 4y$ **35.** 4

37. 7   **39.** 3   **41.** $x\sqrt{5}$ **43.** $\dfrac{a^2}{7}$ **45.** $\dfrac{\sqrt{3}}{3}$ **47.** $\sqrt{3}$ **49.** $\dfrac{\sqrt{3x}}{x}$ **51.** $\dfrac{\sqrt{2x}}{x}$

53. $\dfrac{3\sqrt{x}}{x}$ **55.** $\dfrac{2\sqrt{y}}{xy}$ **57.** $\dfrac{4\sqrt{b}}{7b}$ **59.** $\dfrac{5\sqrt{6}}{27}$ **61.** $\dfrac{\sqrt{y}}{3xy}$ **63.** $\dfrac{2\sqrt{3x}}{3y}$ **65.** $-\dfrac{\sqrt{2}+3}{7}$

67. $\dfrac{15 - 3\sqrt{5}}{20}$ **69.** $\dfrac{x\sqrt{y} + y\sqrt{x}}{x - y}$ **71.** $\dfrac{\sqrt{3xy}}{3y}$ **73.** $-2 - \sqrt{6}$ **75.** $-\dfrac{\sqrt{10} + 5}{3}$

77. $\dfrac{x - 3\sqrt{x}}{x - 9}$ **79.** $-\dfrac{1}{2}$ **81.** 7   **83.** -5 **85.** $\dfrac{7}{4}$ **87.** $-\dfrac{42 - 26\sqrt{3}}{11}$

89. $\dfrac{14 - 9\sqrt{2}}{17}$ **91.** $\dfrac{a - 5\sqrt{a} + 4}{2a - 2}$ **93.** $\dfrac{y + 5\sqrt{y} + 6}{y - 9}$ **95.** -1.3 **97.** $-\dfrac{4}{9}$

99. $\dfrac{3}{2}$ **101.** The area is 59 m². **103. a.** True   **b.** True   **c.** False; $x + 2\sqrt{x} + 1$

d. True   **105.** No

CONCEPT REVIEW 10.4

1. Sometimes true   **3.** Always true   **5.** Sometimes true

SECTION 10.4

1. b and c   **3.** 25   **5.** 144   **7.** 5   **9.** 16   **11.** 8   **13.** No solution   **15.** 6

17. 24   **19.** -1 **21.** $-\dfrac{2}{5}$ **23.** $\dfrac{4}{3}$ **25.** 15   **27.** 5   **29.** 1   **31.** 1   **33.** 1

35. 2   **37.** No solution   **39.** The number is 1.   **41.** The number is 11.   **45.** The length of the hypotenuse is 10.30 cm.   **47.** The length of the other leg of the triangle is 9.75 ft.   **49.** The length of the diagonal is 12.1 mi.   **51.** The depth of the water is 12.5 ft.   **53.** The pitcher's mound is more than halfway between home plate and second base.   **55.** The periscope has to be 10.67 ft above the water.   **57.** Yes, the ladder will be long enough to reach the gutters.   **59.** The length of the pendulum is 7.30 ft.

61. The distance to the memorial is 250 ft.   **63.** The bridge is 64 ft high.   **65.** The height of the screen is 21.6 in.   **67.** 5   **69.** 8   **71. a.** The perimeter is $12\sqrt{2}$ cm.   **b.** The area is 12 cm².

73. The area is 244.78 ft².

CHAPTER REVIEW EXERCISES

1. $-11\sqrt{3}$ [10.2.1]   **2.** $-x\sqrt{3} - x\sqrt{5}$ [10.3.2]   **3.** $x + 8$ [10.1.2]   **4.** No solution [10.4.1]

5. 1 [10.3.2]   **6.** $7\sqrt{7}$ [10.2.1]   **7.** $12a + 28\sqrt{ab} + 15b$ [10.3.1]   **8.** $7x^2 + 42x + 63$ [10.1.2]

9. 12 [10.1.1]   **10.** 16 [10.4.1]   **11.** $4x\sqrt{5}$ [10.2.1]   **12.** $5ab - 7$ [10.3.1]

13. 3 [10.4.1]   **14.** $\sqrt{35}$ [10.1.1]   **15.** $7x^2 y\sqrt{15xy}$ [10.2.1]   **16.** $\dfrac{x}{3y^2}$ [10.3.2]

17. $9x - 6\sqrt{xy} + y$ [10.3.1]   **18.** $a + 4$ [10.1.2]   **19.** $20\sqrt{3}$ [10.1.1]   **20.** $26\sqrt{3x}$ [10.2.1]

21. $\dfrac{8\sqrt{x} + 24}{x - 9}$ [10.3.2]   **22.** $3a\sqrt{2} + 2a\sqrt{3}$ [10.3.1]   **23.** $9a^2\sqrt{2ab}$ [10.1.2]

24. $-6\sqrt{30}$ [10.1.1]   **25.** $-4a^2 b^4\sqrt{5ab}$ [10.2.1]   **26.** $7x^2 y^4$ [10.3.2]   **27.** $\dfrac{3}{5}$ [10.4.1]

28. c^9 [10.1.2]   **29.** $15\sqrt{2}$ [10.1.1]   **30.** $18a\sqrt{5b} + 5a\sqrt{b}$ [10.2.1]   **31.** $\dfrac{16\sqrt{a}}{a}$ [10.3.2]

32. 8 [10.4.1]   **33.** $a^5 b^3 c^2$ [10.3.1]   **34.** $21\sqrt{70}$ [10.1.1]   **35.** $2y^4\sqrt{6}$ [10.1.2]

36. 5 [10.3.2]   **37.** 5 [10.4.1]   **38.** $8y + 10\sqrt{5y} - 15$ [10.3.1]   **39.** 99.499 [10.1.1]

40. 7 [10.3.1]   **41.** $25x^4\sqrt{6x}$ [10.1.2]   **42.** $-6x^3y^2\sqrt{2y}$ [10.2.1]   **43.** $3a$ [10.3.2]

44. $20\sqrt{10}$ [10.1.1]   **45.** 20 [10.4.1]   **46.** $\dfrac{7a}{3}$ [10.3.2]   **47.** 10 [10.3.1]

48. $6x^2y^2\sqrt{y}$ [10.1.2]   **49.** $36x^8y^5\sqrt{3xy}$ [10.1.2]   **50.** 3 [10.4.1]   **51.** 20 [10.1.1]
52. $-2\sqrt{2x}$ [10.2.1]   **53.** 6 [10.4.1]   **54.** 3 [10.3.1]   **55.** The larger integer is 51. [10.4.2]
56. The explorer's weight on the surface of Earth is 144 lb. [10.4.2]   **57.** The length of the other leg is
14.25 cm. [10.4.2]   **58.** The radius of the sharpest corner is 25 ft. [10.4.2]
59. The depth is 100 ft. [10.4.2]   **60.** The length of the guy wire is 26.25 ft. [10.4.2]

CHAPTER TEST

1. $11x^4y$ [10.1.2]   **2.** $-5\sqrt{2}$ [10.2.1]   **3.** $6x^2\sqrt{xy}$ [10.3.1]   **4.** $3\sqrt{5}$ [10.1.1]
5. $6x^3y\sqrt{2x}$ [10.1.2]   **6.** $6\sqrt{2y} + 3\sqrt{2x}$ [10.2.1]   **7.** $y + 8\sqrt{y} + 15$ [10.3.1]   **8.** $\sqrt{2}$ [10.3.2]
9. 9 [10.3.2]   **10.** 11 [10.4.1]   **11.** 22.361 [10.1.1]   **12.** $4a^2b^5\sqrt{2ab}$ [10.1.2]
13. $a - \sqrt{ab}$ [10.3.1]   **14.** $4x^2y^2\sqrt{5y}$ [10.3.1]   **15.** $7ab\sqrt{a}$ [10.3.2]   **16.** 25 [10.4.1]
17. $8x^6y^2\sqrt{3xy}$ [10.1.2]   **18.** $4ab\sqrt{2ab} - 5ab\sqrt{ab}$ [10.2.1]   **19.** $a - 4$ [10.3.1]
20. $-6 - 3\sqrt{5}$ [10.3.2]   **21.** 5 [10.4.1]   **22.** $3\sqrt{2} - x\sqrt{3}$ [10.3.1]   **23.** $-6\sqrt{a}$ [10.2.1]
24. $6\sqrt{3}$ [10.1.1]   **25.** 7.937 [10.1.1]   **26.** $6ab\sqrt{a}$ [10.3.2]   **27.** No solution [10.4.1]
28. The smaller integer is 40. [10.4.2]   **29.** The length of the pendulum is 5.07 ft. [10.4.2]
30. The ladder is 15.2 ft high on the building. [10.4.2]

CUMULATIVE REVIEW EXERCISES

1. $-\dfrac{1}{12}$ [1.4.2]   **2.** $2x + 18$ [2.2.5]   **3.** $\dfrac{1}{13}$ [3.2.3]   **4.** $x \le -\dfrac{9}{2}$ [3.3.3]   **5.** $-\dfrac{4}{3}$ [5.3.1]

6. $y = \dfrac{1}{2}x - 2$ [5.4.1/5.4.2]   **7.** -18 [5.5.1]   **8.** [5.5.2]   **9.** [5.6.1]

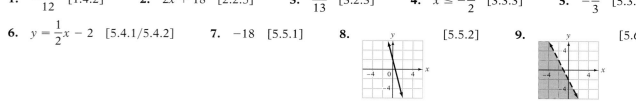

10. $(2, -1)$ [6.1.1]   **11.** $(1, 1)$ [6.2.1]   **12.** $(3, -2)$ [6.3.1]   **13.** $\dfrac{6x^5}{y^3}$ [7.4.1]

14. $-2b^2 + 1 - \dfrac{1}{3b^2}$ [7.5.1]   **15.** $3x^2y^2(4x - 3y)$ [8.1.1]   **16.** $(3b + 5)(3b - 4)$ [8.3.1/8.3.2]

17. $2a(a - 3)(a - 5)$ [8.2.2]   **18.** $\dfrac{1}{4(x + 1)}$ [9.1.2]   **19.** $\dfrac{x - 5}{x - 3}$ [9.4.1]   **20.** $\dfrac{x + 3}{x - 3}$ [9.3.2]

21. $\dfrac{5}{3}$ [9.5.1]   **22.** $38\sqrt{3a} - 35\sqrt{a}$ [10.2.1]   **23.** 8 [10.3.2]   **24.** 14 [10.4.1]

25. The largest integer that satisfies the inequality is -33. [4.8.1]   **26.** The cost is $24.50. [4.4.1]
27. 56 oz of pure water must be added. [4.6.2]   **28.** The two numbers are 6 and 15. [4.1.2]
29. It would take the small pipe 48 h to fill the tank. [9.7.1]   **30.** The smaller integer is 24. [10.4.2]

Answers to Chapter 11 Selected Exercises

PREP TEST

1. 41 [2.1.1]   **2.** $-\dfrac{1}{5}$ [3.2.1]   **3.** $(x + 4)(x - 3)$ [8.2.1]   **4.** $(2x - 3)^2$ [8.4.1]

5. Yes [8.4.1]   **6.** 3 [9.5.2]   **7.** 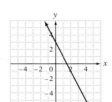 [5.2.2]   **8.** $2\sqrt{7}$ [10.1.1]

9. $|a|$ [10.1.2]   **10.** The hiking trail is 3.6 mi long. [4.7.1]

CONCEPT REVIEW 11.1

1. Never true   **3.** Sometimes true   **5.** Always true

SECTION 11.1

1. $a = 3, b = -4, c = 1$ **3.** $a = 2, b = 0, c = -5$ **5.** $a = 6, b = -3, c = 0$ **7.** $x^2 - 3x - 8 = 0$

9. $x^2 - 16 = 0$ **11.** $2x^2 + 12x + 13 = 0$ **17.** -3 and 5 **19.** 0 and 7 **21.** $-\frac{5}{2}$ and $\frac{1}{3}$

23. -5 and 3 **25.** 1 and 3 **27.** -1 and -2 **29.** 3 **31.** 0 and $\frac{3}{2}$ **33.** -2 and 5

35. $\frac{2}{3}$ and 1 **37.** -3 and $\frac{1}{3}$ **39.** $-\frac{3}{2}$ and $\frac{4}{3}$ **41.** -3 and $\frac{4}{5}$ **43.** $\frac{1}{3}$ **45.** -4 and 4

47. $-\frac{2}{3}$ and $\frac{2}{3}$ **49.** $\frac{2}{3}$ and 2 **51.** -6 and $\frac{3}{2}$ **53.** $\frac{2}{3}$ **55.** -3 and 6 **57.** 0 and 12

59. -6 and 6 **61.** -1 and 1 **63.** $-\frac{7}{2}$ and $\frac{7}{2}$ **65.** $-\frac{2}{3}$ and $\frac{2}{3}$ **67.** $-\frac{3}{4}$ and $\frac{3}{4}$

69. There is no real number solution.   **71.** $2\sqrt{6}$ and $-2\sqrt{6}$ **73.** -5 and 7 **75.** -7 and -3

77. -6 and 4 **79.** -9 and -1 **81.** 2 and 16 **83.** $-\frac{9}{2}$ and $-\frac{3}{2}$ **85.** $-\frac{1}{3}$ and $\frac{7}{3}$

87. $-\frac{12}{7}$ and $-\frac{2}{7}$ **89.** $4 + 2\sqrt{5}$ and $4 - 2\sqrt{5}$ **91.** There is no real number solution.

93. $\frac{1}{2} + \sqrt{6}$ and $\frac{1}{2} - \sqrt{6}$ **95.** $-\frac{4}{3}$ and $\frac{8}{3}$ **97.** 66 and 110 **99.** $\sqrt{2}$ and $-\sqrt{2}$

101. 0 and 1 **103.** $\frac{\sqrt{ab}}{a}$ and $-\frac{\sqrt{ab}}{a}$ **105.** The speed is 10 m/s.

CONCEPT REVIEW 11.2

1. Always true   **3.** Always true   **5.** Never true

SECTION 11.2

1. $x^2 + 12x + 36; (x + 6)^2$ **3.** $x^2 + 10x + 25; (x + 5)^2$ **5.** $x^2 - x + \frac{1}{4}; \left(x - \frac{1}{2}\right)^2$

7. -3 and 1 **9.** $-2 + \sqrt{3}$ and $-2 - \sqrt{3}$ **11.** There is no real number solution.   **13.** $-3 + \sqrt{14}$ and

$-3 - \sqrt{14}$ **15.** 2 **17.** $1 + \sqrt{2}$ and $1 - \sqrt{2}$ **19.** $\frac{-3 + \sqrt{13}}{2}$ and $\frac{-3 - \sqrt{13}}{2}$ **21.** -8 and 1

23. $-3 + \sqrt{5}$ and $-3 - \sqrt{5}$ **25.** $4 + \sqrt{14}$ and $4 - \sqrt{14}$ **27.** 1 and 2 **29.** $\frac{3 + \sqrt{29}}{2}$ and $\frac{3 - \sqrt{29}}{2}$

31. $\frac{1 + \sqrt{5}}{2}$ and $\frac{1 - \sqrt{5}}{2}$ **33.** $\frac{5 + \sqrt{13}}{2}$ and $\frac{5 - \sqrt{13}}{2}$ **35.** $\frac{-1 + \sqrt{13}}{2}$ and $\frac{-1 - \sqrt{13}}{2}$

37. $-5 + 4\sqrt{2}$ and $-5 - 4\sqrt{2}$ **39.** $\dfrac{-3 + \sqrt{5}}{2}$ and $\dfrac{-3 - \sqrt{5}}{2}$ **41.** $\dfrac{1 + \sqrt{17}}{2}$ and $\dfrac{1 - \sqrt{17}}{2}$

43. There is no real number solution. **45.** $\dfrac{1}{2}$ and 1 **47.** -3 and $\dfrac{1}{2}$ **49.** $\dfrac{1 + \sqrt{2}}{2}$ and $\dfrac{1 - \sqrt{2}}{2}$

51. $\dfrac{2 + \sqrt{5}}{2}$ and $\dfrac{2 - \sqrt{5}}{2}$ **53.** $3 + \sqrt{2}$ and $3 - \sqrt{2}$ **55.** $1 + \sqrt{13}$ and $1 - \sqrt{13}$ **57.** $\dfrac{3 + 3\sqrt{3}}{2}$ and

$\dfrac{3 - 3\sqrt{3}}{2}$ **59.** $4 + \sqrt{7}$ and $4 - \sqrt{7}$ **61.** 1.193 and -4.193 **63.** 2.766 and -1.266

65. 0.151 and -1.651 **67.** $22 + 12\sqrt{2}$ and $22 - 12\sqrt{2}$ **69.** -3 **71.** $6 + \sqrt{58}$ and $6 - \sqrt{58}$

CONCEPT REVIEW 11.3

1. Always true **3.** Never true **5.** Never true

SECTION 11.3

3. -7 and 1 **5.** -3 and 6 **7.** $1 + \sqrt{6}$ and $1 - \sqrt{6}$ **9.** $-3 + \sqrt{10}$ and $-3 - \sqrt{10}$

11. There is no real number solution. **13.** $2 + \sqrt{13}$ and $2 - \sqrt{13}$ **15.** 0 and 1 **17.** $\dfrac{1 + \sqrt{2}}{2}$ and

$\dfrac{1 - \sqrt{2}}{2}$ **19.** $-\dfrac{3}{2}$ and $\dfrac{3}{2}$ **21.** $\dfrac{3 + \sqrt{3}}{3}$ and $\dfrac{3 - \sqrt{3}}{3}$ **23.** $\dfrac{1 + \sqrt{10}}{3}$ and $\dfrac{1 - \sqrt{10}}{3}$ **25.** There is

no real number solution. **27.** $\dfrac{4 + \sqrt{10}}{2}$ and $\dfrac{4 - \sqrt{10}}{2}$ **29.** $-\dfrac{2}{3}$ and 3 **31.** $\dfrac{5 + \sqrt{73}}{6}$ and $\dfrac{5 - \sqrt{73}}{6}$

33. $\dfrac{1 + \sqrt{37}}{6}$ and $\dfrac{1 - \sqrt{37}}{6}$ **35.** $\dfrac{1 + \sqrt{41}}{5}$ and $\dfrac{1 - \sqrt{41}}{5}$ **37.** -3 and $\dfrac{4}{5}$ **39.** $-3 + 2\sqrt{2}$ and

$-3 - 2\sqrt{2}$ **41.** $\dfrac{3 + 2\sqrt{6}}{2}$ and $\dfrac{3 - 2\sqrt{6}}{2}$ **43.** There is no real number solution. **45.** $\dfrac{-1 + \sqrt{2}}{3}$ and

$\dfrac{-1 - \sqrt{2}}{3}$ **47.** $-\dfrac{1}{2}$ and $\dfrac{2}{3}$ **49.** $\dfrac{-4 + \sqrt{5}}{2}$ and $\dfrac{-4 - \sqrt{5}}{2}$ **51.** $\dfrac{1 + 2\sqrt{3}}{2}$ and $\dfrac{1 - 2\sqrt{3}}{2}$

53. $-\dfrac{3}{2}$ **55.** $\dfrac{-5 + \sqrt{2}}{3}$ and $\dfrac{-5 - \sqrt{2}}{3}$ **57.** $\dfrac{1 + \sqrt{19}}{3}$ and $\dfrac{1 - \sqrt{19}}{3}$ **59.** $\dfrac{5 + \sqrt{65}}{4}$ and $\dfrac{5 - \sqrt{65}}{4}$

61. $6 + \sqrt{11}$ and $6 - \sqrt{11}$ **63.** 5.690 and -3.690 **65.** 7.690 and -1.690 **67.** 4.589 and -1.089

69. 0.118 and -2.118 **71.** 1.105 and -0.905 **73.** $3 + \sqrt{13}$ and $3 - \sqrt{13}$ **75. a.** False

b. False **c.** False **d.** True

CONCEPT REVIEW 11.4

1. Always true **3.** Never true **5.** Never true

SECTION 11.4

5. Up **7.** Up **9.** Up **11.** 4 **13.** -5 **15.** -42 **17.**

19. **21.** **23.** **25.** **27.**

29. **31.** **33.** **35.** The *x*-intercepts are (2, 0) and (3, 0).

The *y*-intercept is (0, 6). **37.** The *x*-intercepts are (−3, 0) and (3, 0). The *y*-intercept is (0, 9).
39. The *x*-intercepts are $\left(-1 + \sqrt{7}, 0\right)$ and $\left(-1 - \sqrt{7}, 0\right)$. The *y*-intercept is (0, −6). **41.** The *x*-intercepts are $\left(\frac{3}{2}, 0\right)$ and (−1, 0). The *y*-intercept is (0, −3). **43.** The *x*-intercepts are $\left(\frac{1 + \sqrt{17}}{-2}, 0\right)$ and $\left(\frac{1 - \sqrt{17}}{-2}, 0\right)$.

The *y*-intercept is (0, 4). **45.** **47.** **49.**

51. **53.** **55.** **57.** $y = x^2 - 8x + 15$

59. $y = 2x^2 - 12x + 22$ **61.** (−2, 0), (0, 0), and (3, 0) **63.** (−2, 0), (−1, 0), and (2, 0)
65. Linear function **67.** Neither **69.** Quadratic function
71. The model predicts the cost will be 62¢. **73.** −2

CONCEPT REVIEW 11.5

1. Never true **3.** Always true **5.** Always true

SECTION 11.5

1. The two integers are 7 and 9. **3.** The integers are 6 and 10, or −10 and −6. **5.** The length is 6 ft. The width is 4 ft. **7.** The length is 100 ft. The width is 50 ft. **9.** It would take the first computer 35 min and the second computer 14 min. **11.** It would take the first engine 12 h and the second engine 6 h.
13. The rate of the plane in calm air was 100 mph. **15.** The cyclist's rate during the first 150 mi was 50 mph.
17. The arrow will be 32 ft above the ground at 1 s and at 2 s. **19.** The maximum speed is 151.6 km/h.
21. The water from the hose is 244.1 ft from the tugboat. **23.** The ball hits the basket 1.88 s after it is released.
25. The radius of the base is 5.64 cm. **27.** The integers are −6, −5, −4, and −3, or 3, 4, 5, and 6.
29. The radius of the large pizza is 7 in. **31.** The dimensions of the cardboard are 20 in. by 20 in.

CHAPTER REVIEW EXERCISES

1. 4 and −4 [11.1.1/11.1.2] **2.** $\frac{1 + \sqrt{13}}{2}$ and $\frac{1 - \sqrt{13}}{2}$ [11.2.1/11.3.1] **3.** $\frac{3 + \sqrt{29}}{2}$ and $\frac{3 - \sqrt{29}}{2}$

[11.2.1/11.3.1] **4.** $\frac{5}{7}$ and $-\frac{5}{7}$ [11.1.1/11.1.2] **5.** [11.4.1] **6.** [11.4.1]

7. $5 + \sqrt{23}$ and $5 - \sqrt{23}$ [11.2.1/11.3.1] **8.** $-\frac{1}{2}$ and $-\frac{1}{3}$ [11.1.1] **9.** There is no real number solution.

[11.1.2] **10.** $\frac{-10 + 2\sqrt{10}}{5}$ and $\frac{-10 - 2\sqrt{10}}{5}$ [11.2.1/11.3.1] **11.** $2 + \sqrt{3}$ and $2 - \sqrt{3}$ [11.2.1/11.3.1]

12. 6 and −5 [11.1.1]   **13.** $\dfrac{4}{3}$ and $-\dfrac{7}{2}$ [11.1.1]   **14.** $2\sqrt{10}$ and $-2\sqrt{10}$ [11.1.2]

15. $\dfrac{2+\sqrt{7}}{3}$ and $\dfrac{2-\sqrt{7}}{3}$ [11.2.1/11.3.1]   **16.** $1+\sqrt{11}$ and $1-\sqrt{11}$ [11.2.1/11.3.1]   **17.** 3 and 9

[11.1.1]   **18.** 16 and −2 [11.1.2]   **19.**  [11.4.1]   **20.** [11.4.1]

21. 1 and −9 [11.1.2]   **22.** $\dfrac{-4+\sqrt{23}}{2}$ and $\dfrac{-4-\sqrt{23}}{2}$ [11.2.1/11.3.1]   **23.** $-\dfrac{1}{6}$ and $-\dfrac{5}{4}$ [11.1.1]

24. There is no real number solution. [11.2.1/11.3.1]   **25.** $4+\sqrt{21}$ and $4-\sqrt{21}$ [11.2.1/11.3.1]

26. $\dfrac{4}{7}$ [11.1.1]   **27.** 2 and −1 [11.1.2]   **28.** $\dfrac{3}{8}$ and $\dfrac{3}{2}$ [11.1.1]   **29.** $\dfrac{-7+\sqrt{61}}{2}$ and

$\dfrac{-7-\sqrt{61}}{2}$ [11.2.1/11.3.1]   **30.** 2 and $\dfrac{5}{12}$ [11.1.1]   **31.** $3+\sqrt{5}$ and $3-\sqrt{5}$ [11.1.2]

32. $-4+\sqrt{19}$ and $-4-\sqrt{19}$ [11.2.1/11.3.1]   **33.** [11.4.1]   **34.** 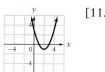 [11.4.1]

35. −7 and −10 [11.1.1]   **36.** $2+2\sqrt{6}$ and $2-2\sqrt{6}$ [11.1.2]   **37.** There is no real number solution.

[11.2.1/11.3.1]   **38.** $-3+\sqrt{11}$ and $-3-\sqrt{11}$ [11.2.1/11.3.1]   **39.** 3 and $-\dfrac{1}{9}$ [11.1.1]

40. $\dfrac{-5+\sqrt{33}}{4}$ and $\dfrac{-5-\sqrt{33}}{4}$ [11.2.1/11.3.1]   **41.** 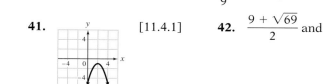 [11.4.1]   **42.** $\dfrac{9+\sqrt{69}}{2}$ and

$\dfrac{9-\sqrt{69}}{2}$ [11.2.1/11.3.1]   **43.** 1 and $\dfrac{5}{2}$ [11.1.1]   **44.** [11.4.1]   **45.** The rate of the air

balloon is 25 mph. [11.5.1]   **46.** The height is 10 m. The base is 4 m. [11.5.1]   **47.** The two integers

are 3 and 5. [11.5.1]   **48.** The rate of the boat in still water is 5 mph. [11.5.1]   **49.** The object will be

12 ft above the ground at $\dfrac{1}{2}$ s and $1\dfrac{1}{2}$ s. [11.5.1]   **50.** It would take the smaller drain 12 h and the larger drain

4 h. [11.5.1]

CHAPTER TEST

1. $-4+2\sqrt{5}$ and $-4-2\sqrt{5}$ [11.1.2]   **2.** $\dfrac{-4+\sqrt{22}}{2}$ and $\dfrac{-4-\sqrt{22}}{2}$ [11.2.1/11.3.1]

3. −4 and $\dfrac{5}{3}$ [11.1.1]   **4.** $\dfrac{3+\sqrt{33}}{2}$ and $\dfrac{3-\sqrt{33}}{2}$ [11.2.1/11.3.1]   **5.** $-2+2\sqrt{5}$ and $-2-2\sqrt{5}$

[11.2.1/11.3.1]   **6.**   [11.4.1]   **7.** $-2 + \sqrt{2}$ and $-2 - \sqrt{2}$ [11.2.1/11.3.1]

8. $\dfrac{-3 + \sqrt{41}}{2}$ and $\dfrac{-3 - \sqrt{41}}{2}$ [11.2.1/11.3.1]   **9.** $-\dfrac{1}{2}$ and 3   [11.1.1]   **10.** $\dfrac{3 + \sqrt{7}}{2}$ and $\dfrac{3 - \sqrt{7}}{2}$

[11.2.1/11.3.1]   **11.** $\dfrac{1 + \sqrt{13}}{6}$ and $\dfrac{1 - \sqrt{13}}{6}$ [11.2.1/11.3.1]   **12.** $5 + 3\sqrt{2}$ and $5 - 3\sqrt{2}$ [11.1.2]

13. $3 + \sqrt{14}$ and $3 - \sqrt{14}$ [11.2.1/11.3.1]   **14.** $\dfrac{5 + \sqrt{29}}{2}$ and $\dfrac{5 - \sqrt{29}}{2}$ [11.2.1/11.3.1]

15. $\dfrac{5 + \sqrt{33}}{2}$ and $\dfrac{5 - \sqrt{33}}{2}$ [11.2.1/11.3.1]   **16.** $\dfrac{5}{2}$ and $\dfrac{1}{3}$ [11.1.1]   **17.** $\dfrac{-3 + \sqrt{37}}{2}$ and $\dfrac{-3 - \sqrt{37}}{2}$

[11.2.1/11.3.1]   **18.** $\dfrac{2 + \sqrt{14}}{2}$ and $\dfrac{2 - \sqrt{14}}{2}$ [11.2.1/11.3.1]   **19.** $-\dfrac{1}{2}$ and 2   [11.1.1]

20.   [11.4.1]   **21.** $\dfrac{1 + \sqrt{10}}{3}$ and $\dfrac{1 - \sqrt{10}}{3}$ [11.2.1/11.3.1]   **22.**   The length is 8 ft.

The width is 5 ft.   [11.5.1]   **23.**   The rate of the boat in calm water is 11 mph.   [11.5.1]

24.   The middle odd integer is 5 or −5.   [11.5.1]   **25.**   The rate for the last 8 mi was 4 mph.   [11.5.1]

CUMULATIVE REVIEW EXERCISES

1. $-28x + 27$ [2.2.5]   **2.** $\dfrac{3}{2}$ [3.1.3]   **3.** 3   [3.2.3]   **4.** $x > \dfrac{1}{9}$ [3.3.3]   **5.** (3, 0) and (0, −4)

[5.2.3]   **6.** $y = -\dfrac{4}{3}x - 2$ [5.4.1/5.4.2]   **7.**   Domain: {−2, −1, 0, 1, 2}; range: {−8, −1, 0, 1, 8}; yes   [5.5.1]

8. 37   [5.5.1]   **9.** 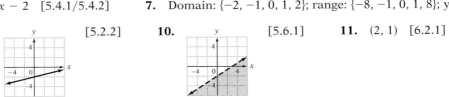   [5.2.2]   **10.**   [5.6.1]   **11.** (2, 1)   [6.2.1]

12. (2, −2)   [6.3.1]   **13.** $-\dfrac{4a}{3b^2}$ [7.4.1]   **14.** $x + 2 - \dfrac{4}{x - 2}$ [7.5.2]   **15.** $(x - 4)(4y - 3)$ [8.1.2]

16. $x(3x - 4)(x + 2)$ [8.3.1/8.3.2]   **17.** $\dfrac{9x^2(x - 2)(x - 4)}{(2x - 3)^2(x + 2)}$ [9.1.3]   **18.** $\dfrac{x + 2}{2(x + 1)}$ [9.3.2]

19. $\dfrac{x - 4}{2x + 5}$ [9.4.1]   **20.** −3 and 6   [9.5.2]   **21.** $a - 2$ [10.3.1]   **22.** 5   [10.4.1]

23. $\dfrac{1}{2}$ and 3   [11.1.1]   **24.** $2 + 2\sqrt{3}$ and $2 - 2\sqrt{3}$ [11.1.2]   **25.** $\dfrac{2 + \sqrt{19}}{3}$ and $\dfrac{2 - \sqrt{19}}{3}$ [11.2.1/11.3.1]

26.   [11.4.1]   **27.**   The cost is $72.   [3.1.5]   **28.**   The cost is $4.50 per pound.   [4.6.1]

29. To earn a dividend of \$752.50, 250 additional shares are required. [9.5.2]   **30.** The rate of the plane in calm air is 200 mph. The rate of the wind is 40 mph. [6.4.1]   **31.** The student must receive a score of 77 or better. [4.8.1]   **32.** The length of the guy wire is 31.62 m. [10.4.2]   **33.** The middle odd integer is -7 or 7. [11.5.1]

Answers to Final Exam

1. -3 [1.1.2]   **2.** -6 [1.2.2]   **3.** 12.5% [1.3.4]   **4.** -256 [1.4.1]   **5.** -11 [1.4.2]

6. $-\dfrac{15}{2}$ [2.1.1]   **7.** $9x + 6y$ [2.2.2]   **8.** $6z$ [2.2.3]   **9.** $16x - 52$ [2.2.5]   **10.** -50 [3.1.3]

11. -3 [3.2.3]   **12.** 15.2 [3.1.5]   **13.** $x \le -3$ [3.3.3]   **14.** $y \ge \dfrac{5}{2}$ [3.3.3]   **15.** $\dfrac{2}{3}$ [5.3.1]

16. $y = -\dfrac{2}{3}x - 2$ [5.4.1/5.4.2]   **17.** [5.3.2]   **18.** [5.5.2]

19. $\{-1, 2, 5, 8, 11\}$ [5.5.1]   **20.** [5.6.1]   **21.** $(6, 17)$ [6.2.1]   **22.** $(2, -1)$ [6.3.1]

23. $-3x^2 - 3x + 8$ [7.1.2]   **24.** $81x^4y^{12}$ [7.2.2]   **25.** $6x^3 + 7x^2 - 7x - 6$ [7.3.2]

26. $-\dfrac{x^8y}{2}$ [7.4.1]   **27.** $\dfrac{16y^7}{x^8}$ [7.4.1]   **28.** $3x^2 - 4x - \dfrac{5}{x}$ [7.5.1]   **29.** $5x - 12 + \dfrac{23}{x + 2}$ [7.5.2]

30. $3.9 \cdot 10^{-8}$ [7.4.2]   **31.** $2(4 - x)(a + 3)$ [8.1.2]   **32.** $(x - 6)(x + 1)$ [8.2.1]

33. $(2x - 3)(x + 1)$ [8.3.1/8.3.2]   **34.** $(3x + 2)(2x - 3)$ [8.3.1/8.3.2]   **35.** $4x(2x - 1)(x - 3)$ [8.3.1/8.3.2]   **36.** $(5x + 4)(5x - 4)$ [8.4.1]   **37.** $3y(5 + 2x)(5 - 2x)$ [8.4.2]   **38.** $\dfrac{1}{2}$ and 3 [8.5.1/11.1.1]   **39.** $\dfrac{2(x + 1)}{x - 1}$ [9.1.2]   **40.** $\dfrac{-3x^2 + x - 25}{(2x - 5)(x + 3)}$ [9.3.2]   **41.** $x + 1$ [9.4.1]

42. 2 [9.5.1]   **43.** $a = b$ [9.6.1]   **44.** $7x^3$ [10.1.2]   **45.** $38\sqrt{3a}$ [10.2.1]

46. $\sqrt{15} + 2\sqrt{3}$ [10.3.2]   **47.** 5 [10.4.1]   **48.** $3 + \sqrt{7}$ and $3 - \sqrt{7}$ [11.1.2]

49. $\dfrac{1 + \sqrt{5}}{4}$ and $\dfrac{1 - \sqrt{5}}{4}$ [11.2.1/11.3.1]   **50.** [11.4.1]   **51.** $2x + 3(x - 2); 5x - 6$ [2.3.2]

52. The original value of the machine was \$3000. [3.1.5]   **53.** There are 60 dimes in the bank. [4.2.2]   **54.** The angles measure 60°, 50°, and 70°. [4.3.3]   **55.** The markup rate is 65%. [4.4.1]   **56.** \$6000 must be invested at 11%. [4.5.1]   **57.** The cost is \$3 per pound. [4.6.1]   **58.** The percent concentration is 36%. [4.6.2]   **59.** The plane traveled 215 km during the first hour. [4.7.1]   **60.** The rate of the boat is 15 mph. The rate of the current is 5 mph. [6.4.1]   **61.** The length is 10 m. The width is 5 m. [8.5.2]   **62.** 16 oz of dye are required. [9.5.3]   **63.** It would take them 0.6 h to prepare the dinner. [9.7.1]   **64.** The length of the other leg is 11.5 cm. [10.4.2]   **65.** The rate of the wind is 25 mph. [11.5.1]

Glossary

abscissa The first number in an ordered pair. It measures a horizontal distance and is also called the first coordinate. (Sec. 5.1)

absolute value of a number The distance of the number from zero on the number line. (Sec. 1.1)

acute angle An angle whose measure is between 0° and 90°. (Sec. 4.3)

addend In addition, a number being added. (Sec. 1.2)

addition The process of finding the total of two numbers. (Sec. 1.2)

addition method An algebraic method of finding an exact solution of a system of linear equations. (Sec. 6.3)

additive inverses Numbers that are the same distance from zero on the number line, but on opposite sides; also called opposites. (Sec. 1.1, 2.2)

adjacent angles Two angles that share a common side. (Sec. 4.3)

alternate exterior angles Two nonadjacent angles that are on opposite sides of the transversal and outside the parallel lines. (Sec. 4.3)

alternate interior angles Two nonadjacent angles that are on opposite sides of the transversal and between the parallel lines. (Sec. 4.3)

analytic geometry Geometry in which a coordinate system is used to study the relationships between variables. (Sec. 5.1)

arithmetic mean of values Average determined by calculating the sum of the values and then dividing that result by the number of values. (Sec. 1.2)

axes The two number lines that form a rectangular coordinate system; also called coordinate axes. (Sec. 5.1)

base In exponential notation, the factor that is multiplied the number of times shown by the exponent. (Sec. 1.4)

basic percent equation Percent times base equals amount. (Sec. 3.1)

binomial A polynomial of two terms. (Sec. 7.1)

clearing denominators Removing denominators from an equation that contains fractions by multiplying each side of the equation by the LCM of the denominators. (Sec. 3.2, 9.5)

coefficient The number part of a variable term. (Sec. 2.1)

combining like terms Using the Distributive Property to add the coefficients of like variable terms; adding like terms of a variable expression. (Sec. 2.2)

complementary angles Two angles whose sum is 90°. (Sec. 4.3)

completing the square Adding to a binomial the constant term that makes it a perfect-square trinomial. (Sec. 11.2)

complex fraction A fraction whose numerator or denominator contains one or more fractions. (Sec. 9.4)

conjugates Binomial expressions that differ only in the sign of a term. The expressions $a + b$ and $a - b$ are conjugates. (Sec. 10.3)

consecutive even integers Even integers that follow one another in order. (Sec. 4.2)

consecutive integers Integers that follow one another in order. (Sec. 4.2)

consecutive odd integers Odd integers that follow one another in order. (Sec. 4.2)

constant term A term that includes no variable part; also called a constant. (Sec. 2.1)

coordinate axes The two number lines that form a rectangular coordinate system; also simply called axes. (Sec. 5.1)

coordinates of a point The numbers in an ordered pair that is associated with a point. (Sec. 5.1)

corresponding angles Two angles that are on the same side of the transversal and are both acute angles or are both obtuse angles. (Sec. 4.3)

cost The price that a business pays for a product. (Sec. 4.4)

decimal notation Notation in which a number consists of a whole-number part, a decimal point, and a decimal part. (Sec. 1.3)

degree A unit used to measure angles. (Sec. 4.3)

degree of a polynomial in one variable The largest exponent that appears on the variable. (Sec. 7.1)

dependent system of equations A system of equations that has an infinite number of solutions. (Sec. 6.1)

dependent variable In a function, the variable whose value depends on the value of another variable known as the independent variable. (Sec. 5.5)

descending order The terms of a polynomial in one variable arranged so that the exponents on the variable decrease from left to right. The polynomial

$9x^5 - 2x^4 + 7x^3 + x^2 - 8x + 1$ is in descending order. (Sec. 7.1)

discount The amount by which a retailer reduces the regular price of a product for a promotional sale. (Sec. 4.4)

discount rate The percent of the regular price that the discount represents. (Sec. 4.4)

domain The set of first coordinates of the ordered pairs in a relation. (Sec. 5.5)

element of a set One of the objects in a set. (Sec. 1.1)

equation A statement of the equality of two mathematical expressions. (Sec. 3.1)

equilateral triangle A triangle in which all three sides are of equal length. (Sec. 4.3)

evaluating a function Replacing x in $f(x)$ with some value and then simplifying the numerical expression that results. (Sec. 5.5)

evaluating a variable expression Replacing each variable by its value and then simplifying the resulting numerical expression. (Sec. 2.1)

even integer An integer that is divisible by 2. (Sec. 4.2)

exponent In exponential notation, the elevated number that indicates how many times the base occurs in the multiplication. (Sec. 1.4)

exponential form The expression 2^5 is in exponential form. Compare *factored form*. (Sec. 1.4)

factor In multiplication, a number being multiplied. (Sec. 1.2)

factor a polynomial To write the polynomial as a product of other polynomials. (Sec. 8.1)

factor a trinomial of the form $ax^2 + bx + c$ To express the trinomial as the product of two binomials. (Sec. 8.3)

factored form The expression $2 \cdot 2 \cdot 2 \cdot 2 \cdot 2$ is in factored form. Compare *exponential form*. (Sec. 1.4)

FOIL A method of finding the product of two binomials; the letters stand for First, Outer, Inner, and Last. (Sec. 7.3)

formula A literal equation that states rules about measurements. (Sec. 9.6)

function A relation in which no two ordered pairs that have the same first coordinate have different second coordinates. (Sec. 5.5)

functional notation A function designated by $f(x)$, which is the value of the function at x. (Sec. 5.5)

graph of a function A graph of the ordered pairs (x, y) of the function. (Sec. 5.5)

graph of a relation The graph of the ordered pairs that belong to the relation. (Sec. 5.5)

graph of an equation in two variables A graph of the ordered-pair solutions of the equation. (Sec. 5.2)

graph of an integer A heavy dot directly above that number on the number line. (Sec. 1.1)

graph of an ordered pair The dot drawn at the coordinates of the point in the plane. (Sec. 5.1)

graphing a point in the plane Placing a dot at the location given by the ordered pair; also called plotting a point in the plane. (Sec. 5.1)

greater than A number a is greater than another number b, written $a > b$, if a is to the right of b on the number line. (Sec. 1.1)

greater than or equal to The symbol $\geq$ means "is greater than or equal to." (Sec. 1.1)

greatest common factor The greatest common factor (GCF) of two or more integers is the greatest integer that is a factor of all the integers. The greatest common factor of two or more monomials is the product of the GCF of the coefficients and the common variable factors. (Sec. 8.1)

half-plane The solution set of an inequality in two variables. (Sec. 5.6)

hypotenuse In a right triangle, the side opposite the 90° angle. (Sec. 10.4)

inconsistent system of equations A system of equations that has no solution. (Sec. 6.1)

independent system of equations A system of equations that has one solution. (Sec. 6.1)

independent variable In a function, the variable that varies independently and whose value determines the value of the dependent variable. (Sec. 5.5)

inequality An expression that contains the symbol $>$, $<$, $\geq$ (is greater than or equal to), or $\leq$ (is less than or equal to). (Sec. 3.3)

integers The numbers $\ldots, -3, -2, -1, 0, 1, 2, 3, \ldots$ (Sec. 1.1)

irrational number The decimal representation of an irrational number never repeats or terminates and can only be approximated. (Sec. 1.3, 10.1)

isosceles triangle A triangle that has two equal angles and two equal sides. (Sec. 4.3)

least common denominator The smallest number that is a multiple of each denominator in question. (Sec. 1.3)

least common multiple (LCM) The LCM of two or more numbers is the smallest number that contains the prime factorization of each number. The LCM of two or more polynomials is the polynomial of least degree that contains the factors of each polynomial. (Sec. 1.3, 9.2)

legs In a right triangle, the sides opposite the acute angles, or the two shortest sides of the triangle. (Sec. 10.4)

less than A number a is less than another number b, written $a < b$, if a is to the left of b on the number line. (Sec. 1.1)

less than or equal to The symbol $\leq$ means "is less than or equal to." (Sec. 1.1)

like terms Terms of a variable expression that have the same variable part. (Sec. 2.2)

linear equation in two variables An equation of the form $y = mx + b$, where m is the coefficient of x and b is a constant; also called a linear function. (Sec. 5.2)

linear function An equation of the form $f(x) = mx + b$ or $y = mx + b$, where m is the coefficient of x and b is a constant; also called a linear equation in two variables. (Sec. 5.5)

literal equation An equation that contains more than one variable. (Sec. 9.6)

markup The difference between selling price and cost. (Sec. 4.4)

markup rate The percent of a retailer's cost that the markup represents. (Sec. 4.4)

monomial A number, a variable, or a product of numbers and variables; a polynomial of one term. (Sec. 7.1)

multiplication The process of finding the product of two numbers. (Sec. 1.2)

multiplicative inverse of a number The reciprocal of a number. (Sec. 2.2)

natural numbers The numbers 1, 2, 3, ... (Sec. 1.1)

negative integers The numbers . . . , $-4, -3, -2, -1$. (Sec. 1.1)

negative slope A property of a line that slants downward to the right. (Sec. 5.3)

nonfactorable over the integers A polynomial that does not factor when only integers are used. (Sec. 8.1)

numerical coefficient The number part of a variable term. When the numerical coefficient is 1 or -1, the 1 is usually not written. (Sec. 2.1)

obtuse angle An angle whose measure is between $90°$ and $180°$. (Sec. 4.3)

odd integer An integer that is not divisible by 2. (Sec. 4.2)

opposite of a polynomial The polynomial created when the sign of each term of the original polynomial is changed. (Sec. 7.1)

opposites Numbers that are the same distance from zero on the number line, but on opposite sides; also called additive inverses. (Sec. 1.1)

ordered pair A pair of numbers, such as (a, b) that can be used to identify a point in the plane determined by the axes of a rectangular coordinate system. (Sec. 5.1)

Order of Operations Agreement A set of rules that tells us in what order to perform the operations that occur in a numerical expression. (Sec. 1.4)

ordinate The second number in an ordered pair. It measures a vertical distance and is also called the second coordinate. (Sec. 5.1)

origin The point of intersection of the two coordinate axes that form a rectangular coordinate system. (Sec. 5.1)

parabola The graph of a quadratic equation in two variables. (Sec. 11.4)

parallel lines Lines that never meet; the distance between them is always the same. (Sec. 4.3)

percent Parts of 100. (Sec. 1.3)

perfect cube The cube of an integer. (Sec. 8.4)

perfect square The square of an integer. (Sec. 10.1)

perfect-square trinomial A trinomial that is a product of a binomial and itself. (Sec. 8.4)

perimeter The distance around a plane geometric figure. (Sec. 4.3)

perpendicular lines Intersecting lines that form right angles. (Sec. 4.3)

plane Flat surface determined by the intersection of two lines. (Sec. 5.1)

plotting a point in the plane Placing a dot at the location given by the ordered pair; also called graphing a point in the plane. (Sec. 5.1)

point-slope formula If (x_1, y_1) is a point on a line with slope m, then $y - y_1 = m(x - x_1)$. (Sec. 5.4)

polynomial A variable expression in which the terms are monomials. (Sec. 7.1)

positive integers The integers, 1, 2, 3, 4, ... (Sec. 1.1)

positive slope A property of a line that slants upward to the right. (Sec. 5.3)

principal square root The positive square root of a number. (Sec. 10.1)

product In multiplication, the result of multiplying two numbers. (Sec. 1.2)

proportion An equation that states the equality of two ratios or rates. (Sec. 9.5)

Pythagorean Theorem If a and b are the lengths of the legs of a right triangle and c is the length of the hypotenuse, then $c^2 = a^2 + b^2$. (Sec. 10.4)

quadrant One of the four regions into which the two axes of a rectangular coordinate system divide the plane. (Sec. 5.1)

quadratic equation An equation of the form $ax^2 + bx + c, a \neq 0$; also called a second-degree equation. (Sec. 8.5, 11.1)

quadratic equation in two variables An equation of the form $y = ax^2 + bx + c$, where a is not equal to zero. (Sec. 11.4)

quadratic function A quadratic function is given by $f(x) = ax^2 + bx + c$, where a is not equal to zero. (Sec. 11.4)

radical equation An equation that contains a variable expression in a radicand. (Sec. 10.4)

radical sign The symbol $\sqrt{\ }$, which is used to indicate the positive, or principal, square root of a number. (Sec. 10.1)

radicand In a radical expression, the expression under the radical sign. (Sec. 10.1)

range The set of second coordinates of the ordered pairs in a relation. (Sec. 5.5)

rate The quotient of two quantities that have different units. (Sec. 9.5)

rate of work That part of a task that is completed in one unit of time. (Sec. 9.7)

ratio The quotient of two quantities that have the same unit. (Sec. 9.5)

rational expression A fraction in which the numerator and denominator are polynomials. (Sec. 9.1)

rational number A number that can be written in the form a/b, where a and b are integers and b is not equal to zero. (Sec. 1.3)

rationalizing the denominator The procedure used to remove a radical from the denominator of a fraction. (Sec. 10.3)

real numbers The rational numbers and the irrational numbers. (Sec. 1.3)

reciprocal Interchanging the numerator and denominator of a rational number yields that number's reciprocal. (Sec. 2.2, 9.1)

rectangular coordinate system System formed by two number lines, one horizontal and one vertical, that intersect at the zero point of each line. (Sec. 5.1)

relation Any set of ordered pairs. (Sec. 5.5)

repeating decimal Decimal that is formed when dividing the numerator of its fractional counterpart by the denominator results in a decimal part wherein one or more digits repeat infinitely. (Sec. 1.3)

right angle An angle whose measure is 90°. (Sec. 4.3)

roster method Method of writing a set by enclosing a list of the elements in braces. (Sec. 1.1)

scatter diagram A graph of collected data as points in a coordinate system. (Sec. 5.1)

scientific notation Notation in which each number is expressed as the product of two factors, one a number between 1 and 10 and the other a power of ten. (Sec. 7.4)

second-degree equation An equation of the form $ax^2 + bx + c = 0, a \neq 0$; also called a quadratic equation. (Sec. 11.1)

selling price The price for which a business sells a product to a customer. (Sec. 4.4)

set A collection of objects. (Sec. 1.1)

similar objects Similar objects have the same shape but not necessarily the same size. (Sec. 9.5)

simplest form of a fraction A fraction is in simplest form when there are no common factors in the numerator and the denominator. (Sec. 9.1)

slope The measure of the slant of a line. The symbol for slope is m. (Sec. 5.3)

slope-intercept form The slope-intercept form of an equation of a straight line is $y = mx + b$. (Sec. 5.3)

solution of a system of equations in two variables An ordered pair that is a solution of each equation of the system. (Sec. 6.1)

solution of an equation A number that, when substituted for the variable, results in a true equation. (Sec. 3.1)

solution of an equation in two variables An ordered pair whose coordinates make the equation a true statement. (Sec. 5.2)

solution set of an inequality A set of numbers, each element of which, when substituted for the variable, results in a true inequality. (Sec. 3.3)

solving an equation Finding a solution of the equation. (Sec. 3.1)

square root A square root of a positive number x is a number a for which $a^2 = x$. (Sec. 10.1)

standard form A quadratic equation is in standard form when the polynomial is in descending order and equal to zero. $ax^2 + bx + c = 0$ is in standard form. (Sec. 8.5, 11.1)

straight angle An angle whose measure is 180°. (Sec. 4.3)

substitution method An algebraic method of finding an exact solution of a system of equations. (Sec. 6.2)

subtraction The process of finding the difference between two numbers. (Sec. 1.2)

sum In addition, the total of two or more numbers. (Sec. 1.2)

supplementary angles Two angles whose sum is 180°. (Sec. 4.3)

system of equations Equations that are considered together. (Sec. 6.1)

terminating decimal A decimal that is formed when dividing the numerator of its fractional counterpart by the denominator results in a remainder of zero. (Sec. 1.3)

terms of a variable expression The addends of the expression. (Sec. 2.1)

transversal A line intersecting two other lines at two different points. (Sec. 4.3)

triangle A three-sided closed figure. (Sec. 4.3)

trinomial A polynomial of three terms. (Sec. 7.1)

undefined slope A property of a vertical line. (Sec. 5.3)

uniform motion The motion of a moving object whose speed and direction do not change. (Sec. 4.7, 9.7)

variable A letter of the alphabet used to stand for a number that is unknown or that can change. (Sec. 2.1)

variable expression An expression that contains one or more variables. (Sec. 2.1)

variable part In a variable term, the variable or variables and their exponents. (Sec. 2.1)

variable term A term composed of a numerical coefficient and a variable part. (Sec. 2.1)

vertex of a parabola The lowest point on the graph of a parabola if the parabola opens up. The highest point on the graph of a parabola if the parabola opens down. (Sec. 11.4)

vertical angles Two angles that are on opposite sides of the intersection of two lines. (Sec. 4.3)

whole numbers The whole numbers are 0, 1, 2, 3, ... (Sec. 1.1)

x-coordinate The abscissa in an xy-coordinate system. (Sec. 5.1)

x-intercept The point at which a graph crosses the x-axis. (Sec. 5.2)

y-coordinate The ordinate in an xy-coordinate system. (Sec. 5.1)

y-intercept The point at which a graph crosses the y-axis. (Sec. 5.2)

zero slope A property of a horizontal line. (Sec. 5.3)

Index

Abscissa, 225
Absolute value, 5
Abundant number, 328
Acute angle, 164
Addend, 9
Addition, 9
  of additive inverses, 60
  Associative Property of, 59
  Commutative Property of, 58
  of decimals, 23
  Distributive Property, 60
  of integers, 9–10
  Inverse Property of, 59–60
  of polynomials, 343
  of radical expressions, 514
  of rational expressions, 450–451
  of rational numbers, 22–23
  verbal phrases for, 68
Addition method of solving systems of
      equations, 314–316
Addition Property of Equations, 91
Addition Property of Inequalities, 126
Addition Property of Zero, 59
Additive inverse, 4, 60
  of polynomials, 343
Adjacent angles, 165
Algebraic fraction(s), see Rational
      expression(s)
al Hassar, 21
Alternate exterior angles, 167
Alternate interior angles, 166
Analytic geometry, 224
Angles
  acute, 164
  adjacent, 165
  alternate exterior, 167
  alternate interior, 166
  complementary, 164
  corresponding, 167
  exterior, 168
  interior, 168
  obtuse, 164
  problems involving, 164–167
  right, 164
  straight, 164
  supplementary, 164
  of a triangle, 168
  vertical, 165
Application problems, see the Index of
      Applications on the inside front
      cover.
  translating, 68–71, 148
Arithmetic mean, 14
Associative Property of Addition, 59
Associative Property of Multiplication, 59

Average
  arithmetic mean, 14
  moving, 14, 41
  rate of change, 228–229
Axes, 224
Axis of symmetry, 422
  of parabola, 572

Base
  in exponential expression, 33, 347
  in percent equation, 98
Basic percent equation, 98
Binomial factors, 387
Binomial(s), 342
  expanding, 376
  factors, 387
  multiplication of, 352–353
  square of a, 354, 558
Brahmagupta, 21
Break-even analysis, 330
Break-even point, 330

Calculators
  checking solutions of an equation with,
      116, 551, 553
  evaluating variable expressions with,
      55
  graphing functions with, 271
  graphing linear equations with, 239,
      240, 256, 284
  graphing quadratic equations with, 571
  and the Order of Operations
      Agreement, 40
  plus/minus key, 40
  and problem solving, 491
  simplifying numerical expressions
      with, 40
  solving quadratic equations with,
      585–587
  solving systems of equations with, 303,
      330–332
  viewing window, 284
Chapter Review Exercises, 44, 83, 140,
      215, 289, 334, 379, 429, 496, 541,
      590
Chapter Summary, 42, 82, 138, 213, 286,
      333, 378, 428, 494, 539, 588
Chapter Test, 47, 85, 142, 218, 292, 337,
      381, 431, 499, 543, 593
Clearing denominators, 113, 462
Coefficient, 53
Coin problems, 156–157
Combining like terms, 61
Common denominator, 23
  least common multiple, 22, 23

Common factor, 387
Commutative Property of Addition, 58
Commutative Property of Multiplication,
      58
Complementary angles, 164
Completing the square, 558
  geometric method, 587–588
  quadratic formula derivation, 565
  solving quadratic equations by,
      559–560
Complex fraction, 458
Composite number, 424
Conjugate, 518
Consecutive integers, 155
Constant, 52
Constant term, 52
  of a perfect square trinomial, 558
Consumer Price Index, 137
Coordinate axes, 224
Coordinates, 225
Coordinate system, rectangular, 224
Corresponding angles, 167
Cost, 176, 583
Counterexamples, 283
Cube
  of a number, 33
  perfect, 413
Cubes, sum and difference of, factoring,
      413
Cumulative Review Exercises, 86, 143,
      220, 295, 338, 382, 431, 500, 544,
      594
Current or rate-of-wind problems,
      319–320

da Vinci, Leonardo, 162
Decimal(s)
  and fractions, 21
  operations on, 23, 25
  and percent, 26, 27
  repeating, 22
  representing irrational numbers, 22
  representing rational numbers, 21–22
  and scientific notation, 363–364
  terminating, 21
Deductive reasoning, 534
Deficient number, 328
Degree
  angle measure, 164
  of a polynomial, 343
Demography, 135
Denominator(s)
  clearing, 113, 462
  least common multiple of, 23
  of rational expressions, 436–437

rationalization of, 519–520
Dependent system, 302, 303
Dependent variable, 270
Descartes, René, 33
Descending order, 343
Diagramming problems, 210
Difference of two squares, 407
Dimensional analysis, 374–375
Diophantus, 348
Discount, 178
Discount rate, 178
Distance-rate problems, 95–97, 197–198,
     482–483, 578–579
Distributive Property, 60
   FOIL method, 352–353
   use in simplifying expressions, 63–64
   use in solving equations, 115–116
   use in solving inequalities, 130
Division
   of decimals, 25
   exponential expressions and, 360–362
   of integers, 13
   of monomials, 360
   one and, 13
   of polynomials, 370–371
   of radical expressions, 519–520
   of rational expressions, 440
   of rational numbers, 24–25
   and reciprocals, 24–25
   as related to multiplication, 13
   verbal phrases for, 69
   zero and, 13
Domain, 267

Einstein, Albert, 90
Element of a set, 2
Equals sign, verbal phrases for, 148
Equation(s), 90
   Addition Property of, 91
   basic percent, 98
   first-degree, 91–95, 112–116
      applications of, 95–97, 117, 149,
         155–157, 162–169, 176–178,
         183–184, 189–191, 197–198
   formulas, see Formula(s)
   fractional, 113
   graphs of, see Graph(s)
   linear, see Linear equations
   literal, 477
   Multiplication Property of, 93
   of parabola, 570
   containing parentheses, 115–117
   quadratic, see Quadratic equation(s)
   radical, 524–526
   rational, 462–464
   solution of, 90
   solving, see Solving equations
   squaring both sides of, 524
   systems, see Systems of equations
   translating into, from sentences, 148
   in two variables, 237
Equilateral triangle, 162

Eratosthenes, 423
Evaluating expressions
   absolute value, 5
   exponential, 34
   numerical, 35–36
   variable, 53
Evaluating functions, 268
Even integers, 80
   consecutive, 155
Expanding a binomial, 376
Exponent(s), 33, 347
   geometric interpretation of, 34
   integers as, 360, 361
   negative, 361
   of perfect cubes, 413
   of perfect squares, 509
   raising a power to a power, 348
   rules of, 362
   zero as, 360
Exponential expression(s), 33, 347
   base of, 33
   division of, 360–362
   evaluating, 34
   factored form of, 33
   multiplication of, 347
   as repeated multiplication, 33, 347
   simplifying, 348
Exponential form, 33
Expression
   translating from verbal to variable,
      68–71
   variable, 52
Exterior angle, 168

Factor
   binomial, 387
   common binomial, 387
   common monomial, 387
   greatest common, 386
   in multiplication, 11
   proper, 329
Factored form of an exponential
   expression, 33
Factoring
   applications of, 415–416
   common binomial factor, 387
   common factors, 386–387
   completely, 394, 409
   difference of two squares, 407
   by grouping, 387–388
   perfect-square trinomial, 408–409
   solving equations by, 414–415
   sum or difference of two cubes, 413
   trinomials, 392–394, 398–403
Fibonacci, 3
Final exam, 596
First-degree equations, solving, 91–95,
   112–116
First-degree inequalities, 125–130
Focus on Problem Solving, 39, 79, 135,
   209, 283, 328, 374, 421, 489, 534,
   583

FOIL method, 352–353
Formula(s), 477
   for distance (rate, time), 95, 197
   for perimeter, 162
   point-slope, 262
   quadratic, 565
   simple interest, 100
   slope, 251
Fraction(s), 21
   addition and subtraction of, 2
   algebraic, see Rational expression(s)
   complex, 458
   and decimals, 21–22
   division of, 24
   multiplication of, 24
   and percent, 26, 27
Fractional equations, 113, 462–464
Function(s), 267, 570
   applications of, 271

Geometric contruction of completing the
   square, 587–588
Geometry problems, 162–169
Go Figure, 2, 52, 90, 148, 224, 300, 342,
   386, 436, 506, 550
Golden rectangle, 162
Graph(s)
   of functions, 269–270, 570–572
   of horizontal line, 244
   of inequalities, 126
   of integers, 3
   of linear equation in two variables,
      238–244
   of linear functions, 269–271
   of linear inequalities, 278–280
   of ordered pair, 225
   of parabolas, 570–572, 584
   of quadratic equations, 570–572, 584
      intercepts of, 571
   of relations, 267
   scatter diagrams, 227
   of systems of linear equations,
      300–304
   of vertical line, 244
   using x- and y-intercepts, 243–244
Greater than, 3
Greater than or equal to, 4
Greatest common factor (GCF), 23, 386
Grouping, factoring by, 387–388
Grouping symbols, 35

Half plane, 278
Harriot, Thomas, 4
Horizontal axis, 224
Horizontal line, 244
Hypotenuse, 527

If...then...statements, 490
Illumination intensity, 492
Inconsistent system, 302, 303
Independent system, 302, 303
Independent variable, 270

Inductive reasoning, 39
Inequality(ies), 125
    Addition Property of, 126
    applications of, 203
    graphing, 126, 278–280
    Multiplication Property of, 128
    parentheses in,130
    solution set of, 128, 280
    solving, *see* Solving inequalitites
Inequality symbols, 3, 4, 203
Integer problems, 155–156
Integer(s), 3
    addition of, 9–10
    applications of, 14–15
    consecutive, 155
    division of, 13
    even and odd, 80
    as exponents, 360, 361
    as fractions, 21
    multiplication of, 11–12
    negative, 3
    positive, 3
    subtraction of, 10–11
Intercepts
    graphing using, 243–244
    x- and *y*-, 243
Interest, simple, 100, 183
Interest rate, 100, 183
Interior angle, 168
Intersection, point of, for a system of
        equations, 301
Inverse
    additive, 4, 60
        of a polynomial, 343
    multiplicative, 60
Inverse Property of Addition, 59–60
Inverse Property of Multiplication, 60
Investment problems, 100, 183–184, 332
Irrational number, 22, 507
Isosceles triangle, 162

Least common denominator, 23
Least common multiple (LCM), 22, 23,
        445
    of polynomials, 445
Legs of a right triangle, 527
Less than, 3
Less than or equal to, 3
Lever problems, 118
Like terms, 61
Linear equation(s)
    graphing, 238–244
    slope, 250–252
    solution of, 237
    systems of, *see* Systems of Equations
    in two variables, 237
    writing, given the slope and a point on
        the line, 261–262
    x- and *y*-intercepts of, 243
Linear function, 270
    graphs of, 269–271
Linear inequalities, 278–280

Line(s)
    equations of, *see* Linear equations
    horizontal, 244
    parallel, 165, 253
    perpendicular, 165
    slope of, 250
    vertical, 244
Literal equation, 477
Lumens, 492

Markup, 176
Markup rate, 176
Mean, 14, 535
Mersenne prime, 424
Mixture problems
    percent mixture, 101, 190–191, 212
    value mixture,189–190, 322
Monomial(s), 342
    products of, 347
    quotients of, 360
    simplifying powers of, 348
Motion problems, 95–97, 197–198,
        482–483, 578–579
Moving average, 14, 41
Multiple, least common, 22, 23
Multiplication, 11
    Associative Property of, 59
    of binoimials, 352–354
    Commutative Property of, 58
    of decimals, 25
    Distributive Property, 60
    of exponential expressions, 347
    FOIL method, 352–353
    of integers, 11–12
    Inverse Property of, 60
    of monomials, 347
    by one, 59
    of polynomials, 351–354
        applications of, 355
    of radical expressions, 517–518
    of rational expressions, 438
    of rational numbers, 24
    verbal phrases for, 69
    by zero, 59
Multiplication Property of Equations, 93
Multiplication Property of Inequalities,
        128
Multiplication Property of One, 59
Multiplication Property of Zero, 59, 414
Multiplicative inverse, 60

Natural number, 2
Negative exponent, 361
Negative integer, 3
Neilsen ratings, 211
Negation, 489
Newton, Isaac, 361
Nonfactorable over the integers, 394
Number line, 3
Number(s)
    absolute value of, 5
    abundant, 328

composite, 424
deficient, 328
integer, 3
irrational, 22, 507
natural, 2
perfect, 328
prime, 328, 423
rational, 21
real, 22
triangular, 81
whole, 3
Numerical coefficient, 53

Obtuse angle, 164
Odd integers, 80
    consecutive, 155
One
    in division, 13
    Multiplication Property of, 59
Opposite, 4
    of polynomial, 343
Ordered pair, 255
Order of Operations Agreement, 35
    and calculators, 40
Ordinate, 225
Origin, 224

Parabola, 570
Parallel lines, 165, 253
Pascal, Blaise, 376
Pascal's triangle, 376
Patterns in mathematics, 41, 81, 328, 329
Percent, 26–27
Percent equation, 98
Percent mixture problems, 101, 190–191,
        212
Perfect cube, 413
Perfect number, 328
Perfect square, 506, 509
Perfect-square trinomial, 408, 558
Perimeter, 162
    formulas for, 162
Perpendicular lines, 165
Plane, 224
Plot a point in the plane, 225
Point-slope formula, 262
Polynomial(s), 342
    addition and subtraction of, 343
    degree of, 343
    division of, 370–371
    factoring, *see* Factoring
    greatest common factor of the terms
        of, 386
    least common multiple of, 445
    multiplication of, 351–354
        applications of, 355
    nonfactorable over the integers, 394
    properties of, 377
Positive integer, 3
Power(s), 33
    of exponential expressions, 348
    of products, 348

simplifying, of binomials, 376
verbal phrases for, 69
Premise, 490
Prep Test, 2, 52, 90, 148, 224, 300, 342,
    386, 436, 506, 550
Prime factorization, 328
Prime number, 328, 423
    Mersenne, 424
Principal, 100, 183
Principal square root, 506
Principle of Taking the Square Root of
    Each Side of an Equation, 552
Principle of Zero Products, 414, 551
Product Property of Square Roots, 507
Product(s)
    in multiplication, 11
    of multiplicative inverses, 60
    of radical expressions, 517–518
    simplifying powers of, 348
    of the square of a binomial, 354
    of the sum and difference of two terms,
        352
    of two binomials, 352–353
    of two rational expressions, 438
Profit, 583
Projects and Group Activities, 40, 80, 136,
    211, 284, 329, 376, 423, 491, 535,
    585
Proper factors, 329
Property(ies)
    Associative
        of Addition, 59
        of Multiplication, 59
    Commutative
        of Addition, 58
        of Multiplication, 58
    Distributive, 60
    of Equations
        Addition, 91
        Multiplication, 93
    of Inequalities
        of Addition, 126
        of Multiplication, 128
    Inverse, of Addition, 59–60
    Inverse, of Multiplication, 60
    of One, Multiplication, 59
    Product, of Square Roots, 507
    Quotient, of Square Roots, 519
    of Squaring Both Sides of an Equation,
        524
    of Zero, 59, 114
Proportion, 465
    applications of, 466–469
Ptolemy, 3
Pythagorean Theorem, 528

Quadrant, 225
Quadratic equation(s), 414, 550
    applications of, 415–416, 425–427,
        578–579, 584
    graphing, 570–572
    solving

by completing the square, 559–560
by factoring, 414–415, 551
by quadratic formula, 564–566
by taking square roots, 552–554
standard form, 414, 550
Quadratic formula, 565
Quadratic function, 570
Quantifier, 489
Quotient Property of Square Roots, 519

Radical equation, 524
    applications of, 527–528
Radical expression(s)
    addition and subtraction of, 514
    division of, 519–520
    multiplication of, 517–518
    radicand of, 506
    rationalizing denominators of, 519–520
    simplest form of, 509, 520
    simplifying, 506–510
Radical sign, 506
Radicand, 506
Range, 267
Rate
    of change, 229
    interest, 100, 183
    motion problems, 95
    in proportions, 465
    in work problems, 480
Rate-of-wind or current problems,
    319–320
Ratio, 464–465
Rational equations, 462–464
Rational expression(s), 436
    addition and subtraction of, 450–451
    division of, 439–440
    expressing in terms of the LCD, 446
    multiplication of, 448
    simplifying, 436–438
    solving equations containing, 462–464
Rationalizing a denominator, 519–520
Rational number(s), 21
    addition and subtraction of, 22–23
    decimal notation for, 21–22
    multiplication and division of, 24–25
Real number(s), 22
    Properties of, 58–60
Reasoning
    deductive, 534
    inductive, 39
Reciprocal(s), 24, 60
    and division, 24, 440
    of rational expressions, 439
Rectangle, 162
    golden, 162
    perimeter of, 162
Rectangular coordinate system, 224
Relation, 267
Repeating decimal, 22
Revenue, 583
Right angle, 164
Right triangle, 527

Roster method, 2
Rule for Dividing Exponential
    Expressions, 361
Rule for Multiplying Exponential
    Expressions, 347
Rule of Negative Exponents, 361
Rule for Simplifying Powers of
    Exponential Expressions, 348
Rule for Simplifying Powers of Products,
    348
Rules of exponents, 362

Scatter diagrams, 226–227
Scientific notation, 363–365
Second-degree equation, 550
Selling price, 176
Sequence, 39
Set(s), 2
    element of, 2
    solution, 125
    writing, using roster method, 2
Sieve of Eratosthenes, 423
Similar triangles, 467–469
Simple interest, 100, 183
Simplest form
    of an exponential expression, 362
    of a radical expression, 507, 520
    of a rate, 465
    of a ratio, 465
    of a rational expression, 437
Simplifying
    exponential expressions, 34, 348
    numerical expressions, 35–36
    radical expressions, 506–510
    rational expressions, 436–438
    variable expressions, 61–64
Slope of a line, 250–252
    applications of, 254
Slope-intercept form of a straight line,
    255
Solution(s)
    of equations, 90
    of inequalities, *see* Solution set of an
        inequality
    of linear equations in two variables,
        237
    of quadratic equation, *see* Quadratic
        equation(s)
    of system of equations, 301
Solution set of an inequality, 125, 280
Solving equations
    using the Addition Property, 91–93
    by factoring, 414–415
    first-degree, 91–95, 112–116
        applications of, 95–97, 117, 149,
            155–157, 162–169, 176–178,
            183–184, 189–191, 197–198
    fractional, 113
    literal, 477
    using the Multiplication Property,
        93–95
    containing parentheses, 115–116

proportions, 465
    applications of, 466–469
quadratic, *see* Quadratic equation(s)
containing radical expressions,
    524–525
rational, 462–464
systems of, *see* Systems of equations
Solving inequalities
    using the Addition Property, 126–127,
        129–130
    applications of, 203
    using the Multiplication Property,
        128–130
    containing parentheses, 130
Solving proportions, 465
    applications of, 466–469
Square
    of a binomial, 354, 558
    of an integer, 506
    perfect, trinomial, 408, 558
    perimeter of, 162
    of a whole number, 33
Square root(s), 506
    of a^2, 509
    of negative numbers, 508
    of a perfect square, 506, 509
    principal, 506
    Product Property of, 507
    Quotient Property of, 519
    taking, in solving equations, 552–554
Stamp problems, 156–157
Standard deviation, 536
Standard form of a quadratic equation,
    414, 550
Straight angle, 164
Substitution method for solving systems
    of equations, 309–311
Subtraction, 10
    of decimals, 23
    of integers, 10–11
    of polynomials, 343
    of radical expressions, 514
    of rational expressions, 450–451
    of rational numbers, 22–23
    verbal phrases for, 68
Sum, 9
    of additive inverses, 60
Sum and difference of two cubes, 413
Sum and difference of two terms, 353
Supplementary angles, 164

Symbols
    approximately equal to, 22
    degree, 164
    grouping, 35
    inequality, 3, 4
    radical sign, 506
    slope, 250
Symmetry, 422
Systems of equations, 300
    applications of, 319–322
    dependent, 302, 303
    inconsistent, 302, 303
    independent, 302, 303
    solution of, 301
    solving
        by addition method, 314–316
        with a calculator, 303
        by graphing, 300–304
        by substitution method, 309–311

Target heartbeat rate, 136
Terminating decimal, 21
Term(s), 52
    coefficient of, 53
    combining like, 61
    constant, 52
    like, 61
    of a sequence, 39
    variable, 52
Transversal, 166
Translating sentences into equations, 148
Translating verbal expressions into
    variable expressions, 68–71
Trial-and-error approach to problem
    solving, 209, 422
Triangle(s)
    equilateral, 162
    isosceles, 162
    Pascal's, 376
    perimeter of, 162
    problems involving the angles of,
        168–169
    right, 527
    similar, 467–469
Triangular numbers, 81
Trinomial, 342
    factoring, 392–394, 398–403
    perfect-square, 408, 558
Turning points, 377
Twin primes, 425

Uniform motion problems, 95–97,
    197–198, 482–483, 578–579
Units, 464

Value mixture problems, 189–190, 322
Value of a function, 268
Variable(s), 3, 52
    in abstract problems, 79
    dependent, 270
    independent, 270
Variable expression(s), 52
    evaluating, 53
    like terms of, 61
    simplifying, 61–64
    translating into, from verbal
        expressions, 68–71
Variable part of a variable term, 53
Variable term, 52
Verbal expressions
    for inequalities, 203
    translating into variable expressions,
        68–71
Vertex of a parabola, 572
Vertical angles, 165
Vertical axis, 224
Vertical line, 244

Whole numbers, 3
Work problems, 480–481

x-axis, 224
x-coordinate, 225
x-intercept, 243
    of a parabola, 571

y-axis, 224
y-coordinate, 225
y-intercept, 243
    of a parabola, 571

Zero(s), 3
    absolute value of, 5
    Addition Property of, 59
    in the denominator of a rational
        expression, 436–437
    division and, 13
    as an exponent, 360
    Multiplication Property of, 59, 414

Table of Equations and Formulas

Slope of a Straight Line

$$\text{Slope} = m = \frac{y_2 - y_1}{x_2 - x_1}, \; x_1 \neq x_2$$

Slope-intercept Form of a Straight Line

$$y = mx + b$$

Point-slope Formula for a Line

$$y - y_1 = m(x - x_1)$$

Quadratic Formula

$$x = \frac{-b \pm \sqrt{b^2 - 4ac}}{2a}$$

Perimeter and Area of a Triangle, and Sum of the Measures of the Angles

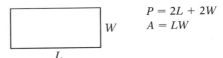

$$P = a + b + c$$
$$A = \frac{1}{2}bh$$
$$\angle A + \angle B + \angle C = 180°$$

Pythagorean Theorem

$$a^2 + b^2 = c^2$$

Perimeter and Area of a Rectangle

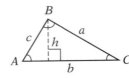

$$P = 2L + 2W$$
$$A = LW$$

Perimeter and Area of a Square

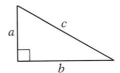

$$P = 4s$$
$$A = s^2$$

Circumference and Area of a Circle

$$C = 2\pi r \quad \text{or} \quad C = \pi d$$
$$A = \pi r^2$$

Area of a Trapezoid

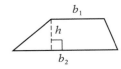

$$A = \frac{1}{2}h(b_1 + b_2)$$

Volume and Surface Area of a Rectangular Solid

$$V = LWH$$
$$SA = 2LW + 2LH + 2WH$$

Volume and Surface Area of a Cube

$$V = s^3$$
$$SA = 6s^2$$

Volume and Surface Area of a Right Circular Cylinder

$$V = \pi r^2 h$$
$$SA = 2\pi r^2 + 2\pi rh$$

Volume and Surface Area of a Right Circular Cone

$$V = \frac{1}{3}\pi r^2 h$$
$$SA = \pi r^2 + \pi rl$$

Volume and Surface Area of a Sphere

$$V = \frac{4}{3}\pi r^3$$
$$SA = 4\pi r^2$$

Camels Don't Ski

Written by
Francesca Simon

Illustrated by
Ailie Busby

LEVINSON BOOKS

For my favourite skiers
Joel, Donna, Alicia and David Simon
FS

For Nick
AB

Calamity **hated** heat. Calamity **hated** sun. Calamity **hated** sand.
But most of all Calamity **hated** carrying.

Every day Calamity travelled across the desert, carrying
her heavy load.
Up to the Malabar sand-dunes, down to the Blue Oasis.

Up to the Malabar sand-dunes, down to the Blue Oasis.
Back and forth, back and forth Calamity trudged through
the heat and the dust.

Every day it got **hotter**,
and **hotter**, and **hotter**,
and her load felt **heavier**,
and **heavier**, and **heavier**.
And every day Calamity
complained **more**,
and **more**, and **more**.

"**It's too hilly**," she grumbled.

"**It's too heavy**," she whined.

"**I'm too hot**," she moaned.

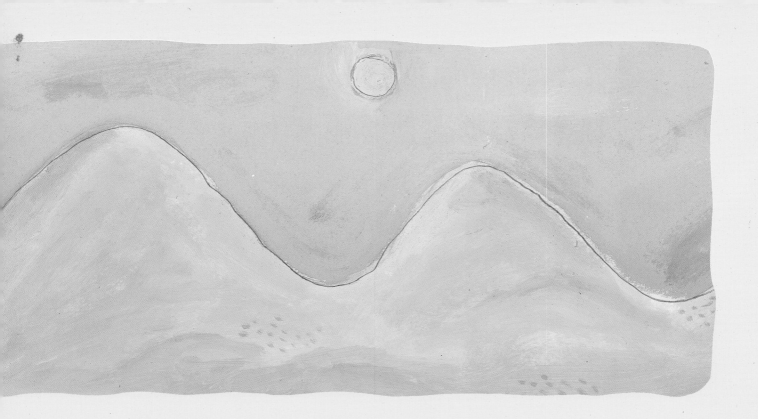

But Calamity
wasn't happy
even when she
wasn't carrying.

"I'm bored,"
she would wail.
"There's nothing
to do."

One morning Calamity woke up panting.
The sun seemed more sweltering than ever.
She was hot, sticky, sweaty and dusty.

She stretched out a dainty hoof and touched
the sizzling sand. "EEEEEEK!" shrieked Calamity.
"It's h–h–h–h–h–hot!" "Of course it's hot,"
said her sister Madge. **"It's the desert."**

"Caravan, get ready!" said the Leader.

"I can't possibly work today," said Calamity. "It's much too hot."

"Nonsense," said Madge.

"I can't carry all this," said Calamity. "It's too heavy."

"Stop making such a fuss," said Madge. "I'm carrying twice as much."

"**Caravan, start moving!**" said the Leader.
The camels plodded on their way.

"**Are we there yet?**" said Calamity.

"**No!**" said Madge.

"**I'm thirsty,**" said Calamity.

"**You can have a drink at the Blue Oasis,**" said Gus.

"**My back hurts,**" Calamity whined.

"**My feet ache,**" she moaned.

"**My legs are wobbly,**" she grumbled.

"**Stop whining!**" said Gloria.

Then things got **worse**.

And **worse**.

And **worse.**

"I've had it!" shrieked Calamity. "I need a holiday! A cold weather holiday! I'm going skiing."

"Camels don't ski," said Madge.

"I'm going to learn," said Calamity.

"You'll freeze," said Gloria.

"No I won't," said Calamity.
"I am a cold weather camel."

Calamity packed her bags and set off.

Finally she arrived.

"Yippee, I'm free!" shouted Calamity.
"No more heat! No more sandstorms!
No more sweating! And no more carrying!"

"There are just a few things you'll need,"
said the ski instructor.

She stretched out a dainty hoof
and touched the icy snow.
"EEEEEK!" shrieked Calamity.
"It's c-c-c-c-c-c-cold!"

Shivering, Calamity trudged up the slippery slope.
Up to the top of the mountain, down to the bottom,
up to the top of the mountain, down to the bottom.

Up and up she trudged through the snow
and the sleet.
Then down down down she skidded.

Every day it got **colder**,
and **colder**, and **colder**,
and her load felt **heavier**,
and **heavier**, and **heavier**.

Then things got **worse**.

And **worse**.

And **worse**.

Calamity turned and looked at her load.
"**That's it?**" she said.

Everyone stared.

"**What did you say?**" said Madge.

"**These bags feel light as a feather
after what I've been carrying,**"
said Calamity.

"**Caravan, start moving!**" said the Leader.

Calamity bounced along the sand.

"Yippee, I'm free!" she shouted.
"No more cold! No more snowstorms!
No more shivering! And no more…
well, a lot less carrying!"

First published in Great Britain in 1998 by Levinson Children's Books, a division of David & Charles Ltd.,
Winchester House; 259-269 Old Marylebone Road, London NW1 5XJ

2 4 6 8 10 9 7 5 3 1

Text copyright © Francesca Simon 1998
Illustrations copyright © Ailie Busby 1988

The right of Francesca Simon and Ailie Busby to be identified as the author and illustrator of this
work has been asserted by them in accordance with the Copyright Design and Patents Act 1988

Hardback ISBN 1 899607 59 5
Paperback ISBN 1 86233 052 2

A CIP record for this book is available from the British Library

Printed and bound in Italy